10th International Conference on Thermal Equipment, Renewable Energy and Rural Development (TE-RE-RD 2021)

E3S Web of Conferences Volume 286 (2021)

Online
10 - 12 June 2021

Editors:

Gabriel Paul Negreanu
Nicolae-Valentin Vladut
Mohand Tazerout
Can Ertekin

ISBN: 978-1-7138-3478-6

Printed from e-media with permission by:

Curran Associates, Inc.
57 Morehouse Lane
Red Hook, NY 12571

Some format issues inherent in the e-media version may also appear in this print version.

For additional information, please contact EDP Sciences – Web of Conferences
at the address below.

EDP Sciences – Web of Conferences
17, Avenue du Hoggar
Parc d'Activité de Courtabœuf
BP 112
F-91944 Les Ulis Cedex A
France

Phone: +33 (0) 1 69 18 75 75
Fax: +33 (0) 1 69 28 84 91

contact-edps@webofconferences.org

Additional copies of this publication are available from:

Curran Associates, Inc.
57 Morehouse Lane
Red Hook, NY 12571 USA
Phone: 845-758-0400
Fax: 845-758-2633
Email: curran@proceedings.com
Web: www.proceedings.com

TABLE OF CONTENTS

PREFACE

The 10-th (Jubilee) International Conference on Thermal Equipment, Renewable Energy and Rural Development (TE-RE-RD 2019 - http://www.tererd.pub.ro/) was launched in December 2020 in order to be organized in the period 26-27 June 2020 at University Politehnica of Bucharest, but, due to the restrictions caused by the COVID-19 pandemics, it was shifted Online (powered by Microsoft Teams)

TE-RE-RD is an International pluridisciplinary Conference aimed to be a forum for information exchange between scientists from academic, research and development communities around the world on topics related to thermal machines, equipments and technologies, equipments, technologies and policies for rural development, renewable energy sources conversion and other technical domains.

Another target of the Conference is to offer to the young people involved in research activities (PhD, Master, and regular students) the opportunity to disseminate their work results and learn from experimented scientists.

We also welcome people and entities from industry, agriculture, administration, trade, politics and media who want to cooperate with academic and R&D organizations in the fields of the Conference.

The main Organizers, University POLITEHNICA of Bucharest (Faculty of Mechanical Engineering and Mechatronics and Faculty of Biotechnical Systems Engineering), National Institute of Research – Development for Machines and Installations Designed to Agriculture and Food Industry and the Romanian Society of Agricultural Mechanical Engineers produce every year since 2012 an itinerant Conference in academic cities or touristic resorts. Unfortunately, the social distance restrictions in last two editions lead the Organizers to move the sessions online.

The main topics of the Conference were:

- Processes in thermal equipment
- Steam and hot water generators
- Steam and gas turbines
- Internal combustion engines
- Environmental protection
- Solar, wind, geothermal, biomass
- Building efficiency
- Equipment for rural development
- Politics of rural development
- Agricultural engineering
- Energy of rural space

For this year the Organizers continued to use the Conference Management Toolkit provided by Microsoft in order to manage the paper submissions and reviewing process. We received 82 papers distributed to 90 reviewers. 68 papers were accepted to be presented and 14 papers were rejected or retired by the authors after revision process (double peer review).

The Conference was developed on 4 tracks:

- 12 papers in the section Thermal Equipments
- 17 papers in the section Renewable Energy
- 25 papers in the section Rural Development
- 14 papers in the section Miscellaneous

Besides the representatives of the Organizers, the Conference also welcomed researchers from University Angel Kanchev of Ruse (Bulgaria), Technical University of Varna (Bulgaria), Technical University of Sofia (Bulgaria), University of Chemical Technology and Metallurgy Sofia (Bulgaria), University of Food Technologies of Plovdiv (Bulgaria), The Kielce University of Technology (Poland), Al - Furat Al- Awsat Technical University (Irak), University of Bonab (Iran), IHU University Tehran, (Iran), Urmia University (Iran), Institute of Market Problems and Economic-Ecological Research of the National Academy of Science of Ukraine (Ukraine), University of Craiova (Romania), Naval Academy Mircea cel Batran Constanța (România), Maritime University of Constanta (Romania), INCD Comoti (Romania), INCD ICEMENERG (Romania), INOE2000-IHP Bucharest (Romania), Agricultural Research and Development Secuieni (Romania), ELSACO Esco (Romania), etc.
We hope that next edition the number and the value of papers will increase, the current participants will come back and new authors will attend. This hope is supported by the indexing of TE-RE-RD 2019 in Web of Science.

See you again in 2022,

Prof. Gabriel Paul Negreanu

Dr. Nicolae-Valentin Vlăduț

Prof. Mohand Tazerout

Prof. Can Ertekin

Editors of the Conference Proceedings TE-RE-RD 2021

OPENING ADDRESS

Dear participants to the 10^{-th} International Conference on Thermal Equipment, Renewable Energy and Rural Development (TE-RE-RD 2021).

On behalf of the Organizing Committee, allow me to welcome you to this event, organized each year by a consortium of University, Research Institute and NGO.

Ten years ago we join our forces and resources in order to transform a National Conference into an International one. Each year we changed the location attracting new co-organizers, we welcomed new participants who remained in our community as authors or reviewers, and we improved the scientific level of the papers and provided conditions for networking and cooperation.

This year, the host of the event should be University Politehnica of Bucharest. But we had to consider sanitary restrictions and special conditions associated to the COVID-19 pandemic, to eliminate the possibility of infecting any participant. For the second year consecutively, we took the hard decision to move the event online.

Thus, we use today four parallel channels of Microsoft Teams, corresponding to the four Conference Tracks, and I hope that the connections will keep stable and will allow the voice and image streaming during all the presentations.

Ladies and gentlemen, dear colleagues and friends

In this edition of our Conference, we will maintain the agreement signed with the prestigious Publishing House - E3S Web of Conferences, stipulating that accepted and presented papers will be published and indexed in Clarivate Analytics Web of Sciences, Scopus and other important international scientific databases.

Finally, I wish to express my gratitude to all participants to the Conference for their contribution to the event's success, and especially to the members of the Scientific and Organizing Committees, who worked hard to make it possible.

I also hope that we will be able to organize the 2022 edition in the "face-to-face" format.

Prof. Lucian Mihăescu

Chairman of TE-RE-RD 2021

Statement of Peer review

In submitting conference proceedings to *Web of Conferences*, the editors of the proceedings certify to the Publisher that

1. They adhere to its **Policy on Publishing Integrity** in order to safeguard good scientific practice in publishing.
2. All articles have been subjected to peer review administered by the proceedings editors.
3. Reviews have been conducted by expert referees, who have been requested to provide unbiased and constructive comments aimed, whenever possible, at improving the work.
4. Proceedings editors have taken all reasonable steps to ensure the quality of the materials they publish and their decision to accept or reject a paper for publication has been based only on the merits of the work and the relevance to the journal.

Title, date and place of the conference

10-th International Conference on Thermal Equipment, Renewable Energy and Rural Development (TE-RE-RD 2021), 10-12 June,
Online powered by Microsoft Teams

Proceedings editor(s):

Prof. Gabriel Paul Negreanu, Dr. Nicolae-Valentin Vlăduţ, Prof. Mohand Tazerout, Prof. Can Ertekin

Date and editor's signature 22.06.2021

17, avenue du Hoggar - PA de Courtabœuf – BP 112 - 91944 Les Ulis Cedex A (France)
Tél. : 33 (0)1 69 18 75 75 - Fax : 33(0)1 69 07 45 17 – www.edpsciences.org

Researches on increasing the dissolved oxygen concentration in stationary waters

Marilena Monica Boltinescu (Roza)[1], *Nicolae Vlad* Sima[1], *Dorina Nicoleta* Albu[1], *Mihaela* Petroşel (Bănică)[1] and *Mihaela* Constantin[1,*]

[1]University Politehnica of Bucharest, Faculty of Mechanical and Mechatronics Engineering, Splaiul Independenţei, no. 313, Bucharest, Romania

Abstract. This paper presents a way to solve the transfer rate equation of oxygen in water. After running a calculation program, the graph of the function that indicates the increase of the dissolved oxygen concentration in the water as a function of time is plotted. Constructed in an original manner, the experimental installation allows the measurement of the increase of the dissolved oxygen concentration in water. The experimental obtained results are compared with the theoretically obtained data.

1Introduction

The oxygen quantity that is dissolved in wateris a function depending on salinity, the temperature and the pressure [1].

Hot water has less O_2than cold water, salt watercontains less O_2than fresh water, and lower-pressure waterhas less oxygen than high-pressure water. Recommended dissolvedO_2degreefor fishes is from 5.0 mg / $dm^3$to 9.0 mg / dm^3. In the case that thedissolved O_2degree is under 3.0 mg / dm^3, due to lack of oxygen, the fishes can die. If the dissolved oxygen level is over 9.0 mg/dm^3, it can also be deadly to fishes.

Figure 1 shows thatO_2 can be:

- O_2 bound to H_2
- free oxygen, namely: oxygen dissolved in water.

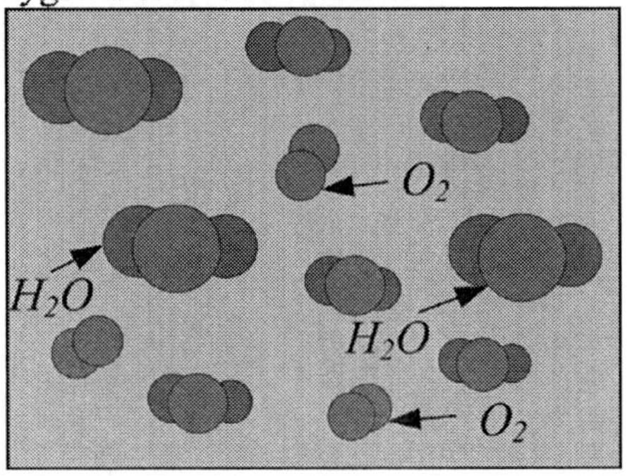

Fig. 1. The O_2 molecules dissolved in water.

*Corresponding author: i.mihaelaconstantin@gmail.com

 https://doi.org/10.1051/e3sconf/202128601001

The oxygen solubility in wateris a function depending on the atmospheric pressure, the temperature, water turbulenceandthe magnitude of the air-water area.

Each water type source has its own biological and physicochemical specific featuresand differ from one area to anotheraccording to the composition of the mineralsin the covered region, the contact duration, the temperatureand the climatic regime.

The method of oxygenation of water is made on the basis of the transfer between air and water;from the air, oxygen is transferred through various methods. The air bubbles emitted by a system that forms it are placed in a water tank. Water oxygenation systems that generate very fine air bubblesare the most effective. From the literature [2] [3], it is known that the rate of O_2transfer to water is higher as the diameter of the air bubbles decreases; the diameter of the air bubble depends on the diameter of the orifice in the perforated plate of a FBG.

The air bubbles can be arranged in as follows (Figure 2):

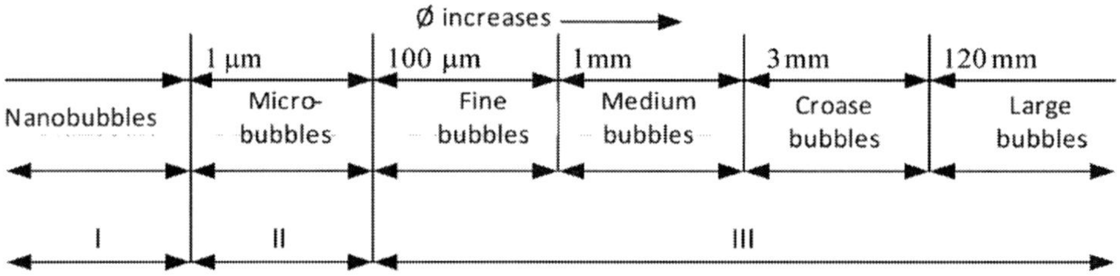

Fig. 2. Gas bubbles classification infunction of Ø:
I - the region where the air bubblescan be seen under a microscope;
II - the region where the air bubblesare difficult to see;
III - the region where the air bubblescan be seen with the naked eye.

Theperformance of awater oxygenation system is specified by the following two parameters[4][5]:

- the water oxygenation efficiency;
- the water oxygenation efficiency.

Themethodof aeration (oxygenation) of water is a fundamental field of Technical Thermodynamics, a discipline materialized by teaching the following courses: Mass and heat transfer, Gas Dynamics, Technical Thermodynamics.

2 Solving the oxygen transfer rate equation

In water oxygenationmethod, the temperature influences both the oxygen conditions, the dissolved O_2saturation concentration and the mass transfer coefficient. The saturation concentration values are directly influenced by the water temperature [6].

The dissolved O_2transfer rate in waterdepending on the temperature is given by the relation [7]:

$$\frac{dC}{d\tau} = \left(a \cdot k_L\right)_{20^0} \cdot \left(C_s - C\right) \; [kg/m^3 s] \quad (1)$$

where:

- ak_L–the transfer coefficient of the oxygen $[s^{-1}]$;
- Cs –the mass oxygen concentration at saturation$[kg/m^3]$;
- C –the current mass concentrationof the $O_2$$[kg/m^3]$.

After integrating relation 1 applying the limit condition $C = C_0$ for $\tau = 0$, it results [6], [7]:

$$\frac{dC}{C_s - C} = a \cdot k_L \, d\tau \quad (2)$$

In the hypothesis that C <C_s, after preforming the integration, it results:

$$-ln(C_s - C) = a \cdot k_L \cdot \tau + ct \qquad (3)$$

From the limit condition, the constant C is obtained:

$$C = C_0 \text{when} \tau = 0 \ (4)$$

resulting:

$$ct = -ln(C_s - C_0) \ (5)$$

Introducing5 in 3 one can obtain:

$$-ln(C_s - C_0) = a \cdot k_L \cdot \tau - ln(C_s - C_0) \ (6)$$

From relation (6) one can obtain:

$$C = C_s - (C_s - C_0) \cdot e^{-a \cdot k_L \cdot \tau} \ (7)$$

In equation (7), the following must be known: C_0, C_S, ak_L.

For the determination ofthevalues of $C = f(\tau)$, a computation program [8], [9] was elaborated, presented in figure 3.

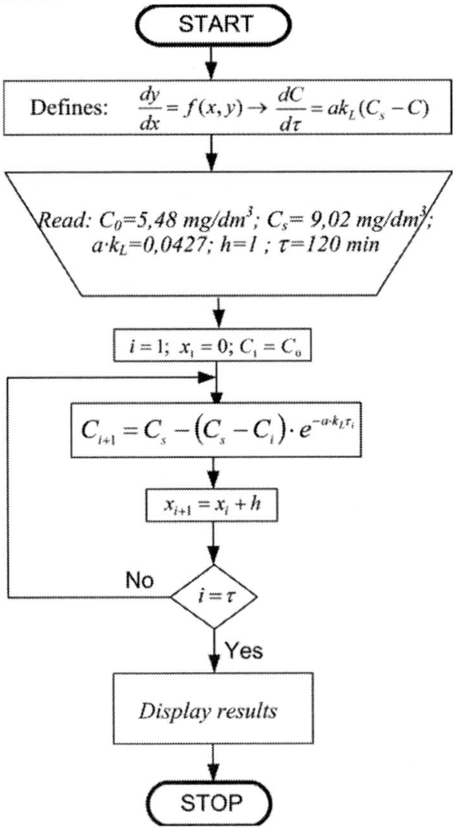

Fig. 3. Logical computation scheme for determining the change of the dissolvedO_2concentration.

After executing a calculation program(fig. 3), numerical values were obtained based on which, the curve in figure 4 was drawn.

 https://doi.org/10.1051/e3sconf/202128601001

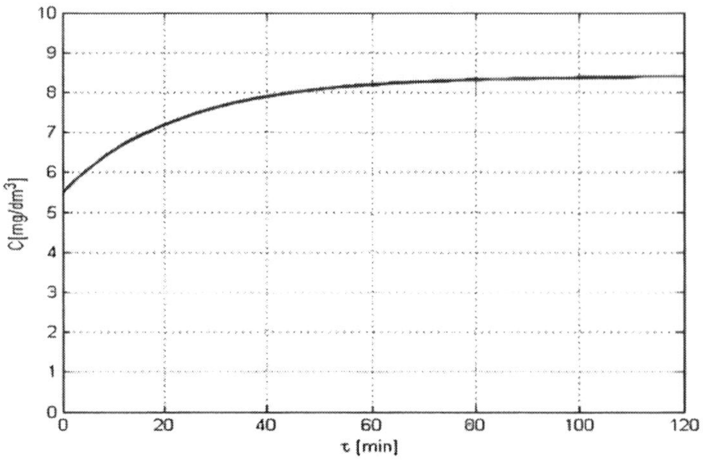

Fig. 4. The variationof thedissolved O_2concentration.

The main data that are maintained in the case of experimental researches are:

- the air flowrate: $\dot{V}$ = 600 dm^3/h; the inlet air pressure in the fine bubble generator: p = 573 mmH_2O; hydrostatic load: H = 500 mmH_2O; duration of the experience: τ = 120 min; C_0 = 5.84 mg / dm^3; t_{H2O} = 23.7°C → C_s = 9.02 mg / dm^3.

3 The presentation of the experimental installation

In figure 5 one can see the sketch of the experimental installationfor measuring theconcentration of the dissolved O_2in water, in time.

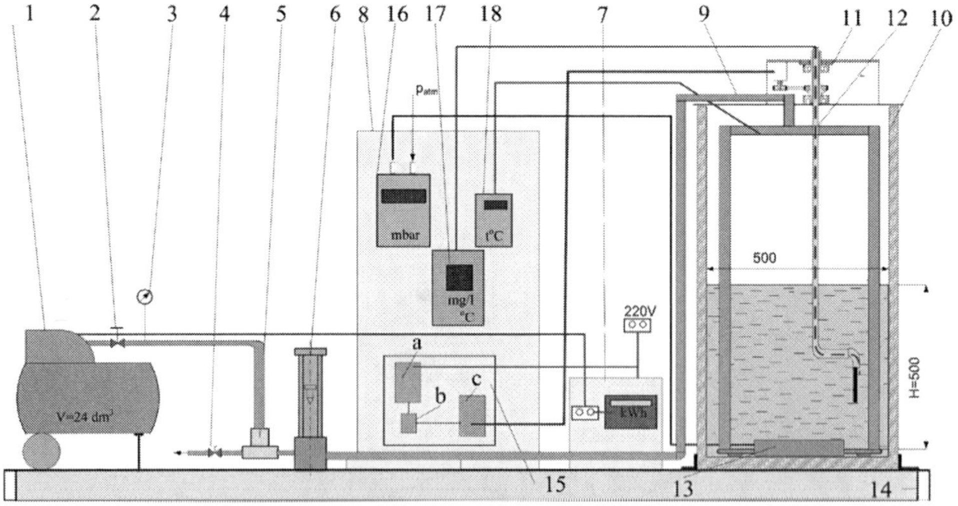

Fig. 5. Scheme of the experimental stand for researches related to water oxygenation

1 - electrocompressor with air reservoir; 2 - pressure reducer; 3 - pressure gauge; 4 - connection for evacuating air outside; 5 –T-bend; 6 - flowmeter; 7 - electrical instrument panel; 8 - gauge board; 9 - piping for transporting compressed air to the fine bubble generator; 10 - water tank;11 - machineryof actuation of the probe; 12 - oxygenometer probe;13 - fine bubble generator with orifices Ø 0.1mm; 14 - installation support; 15 - indicator: a - main inlet,b-interruptor, c - pilot cell, 16 - digital pressure gauge; 17 - oxygenometer 18 - digital thermometer.

The air compressed by the electrocompressor (1), traversethe flowmeter (6) and subsequently goes inthe FBG (13).

During the experiment, the following were measured: air flow rate, air pressure and the concentration of the dissolved oxygen in water.

The experimental installation (figure 5)comprises a series of devices mounted on the panel (8):

- digital thermometer withhigh accuracy, with thermistor sensor;
- differential pressure gauge, supplied with a piezoresistive transducer;
-oxygenometer with polarographic probe and microprocessor with digital board.

The electromechanical machineryfor actuating the oxygenometer probe (11), (12), is supplied with a pecker motor; this machineryadmitsthe cycle of the oxygenometer probe in the water mass, with a speed of 0.38 m /s [10][11].

Figure 6 shows an overview of the experimental stand, designed and constructed in theLaboratory of the Department of Thermotechnics, Engines, Thermal and Refrigeration Equipment's.

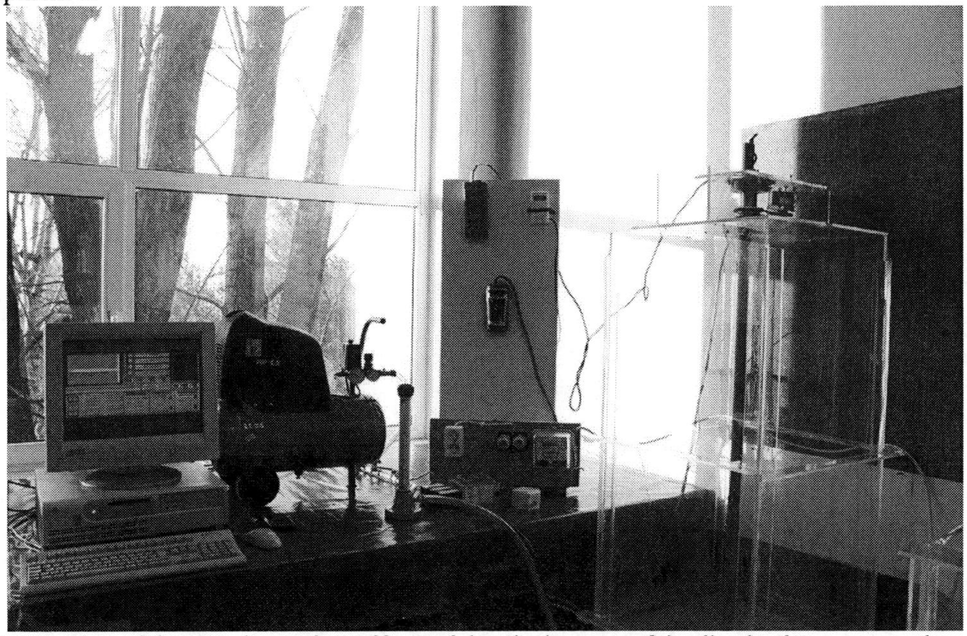

Fig.6. Overview of the experimental standfor studying the increase of the dissolved O_2concentration.

One can observe that on the leftwards of the image (figure 6), there is a computer that controls the actuation machineryof the oxygenometer probe; next to the computer is an electro-compressor with a compressed air tank and the devices panel. To the right of the image, is the water tank, made of clear plexiglass.

4 Experimental researches, obtained results

The steps to be followed in experimental researches are the following:

1. Introduce compressed air into the fine bubble generator and observe if the 152 orifices with Ø 0.1mm are not blocked;
2. The tank is filled with water (H = 500 mmH_2O) and measure t_{H2O}, C_0, t_{air};
3. The FBG (fine bubble generator)is introduced in water and the time at which the experienceis measured;

4. Introduce compressed air into the FBG and after 15 minutes pull out the FBG from the tank and insert the oxygenometer probe. This operation is performed every 15 minutes, until $\tau = 120$ minutes is reached;

5. From previous researches [12][13][14], it was found that after a time of $\tau = 120$ minutes, the value of C_0 tends to C_s;

6. Finally, clean the oxygenometer probe and drain the water.

Following the measurements, the data in table 1 were obtained.

Table 1. The values of dissolved O_2concentration in water,in time.

$\tau\ [min]$	0	15	30	45	60	75	90	105	120
$\dot{V}_{air}\ [dm^3/h]$	600	600	600	600	600	600	600	600	600
$\dot{V}_{IO_2} = 126\ [dm^3/h]$	126	126	126	126	126	126	126	126	126
$\dot{V}_{O_2}$ from other sources	0	0	0	0	0	0	0	0	0
$t_{H_2O}\ [^\circ C]$	23.7	23.7	23.7	23.7	23.7	23.7	23.7	23.7	23.7
$t_{air}\ [^\circ C]$	24.1	24.1	24.1	24.1	24.1	24.1	24.1	24.1	24.1
$C_0\ [mg/dm^3]$	5.84	5.84	5.84	5.84	5.84	5.84	5.84	5.84	5.84
$C_s\ [mg/dm^3]$	8.4	8.4	8.4	8.4	8.4	8.4	8.4	8.4	8.4
$C\ [mg/dm^3]$	5.84	6.89	7.65	8.01	8.10	8.26	8.31	8.35	8.39

In conformity with the data in table 1, the curveform figure 7 was drawn.

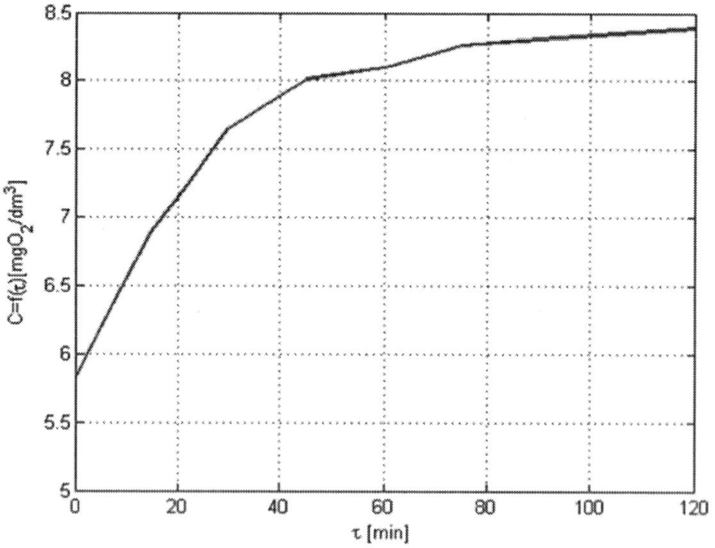

Fig. 7. The C = f (τ)function.

From figure 7, one can observe that, with the passage of time the value of C_0 tends to C_s.

Figure 8 compares the modification of the dissolved O_2 concentration in water in the two cases:

1- graph plottedaccording to the theoretical data;
2- graph plottedaccording to theexperimental data.

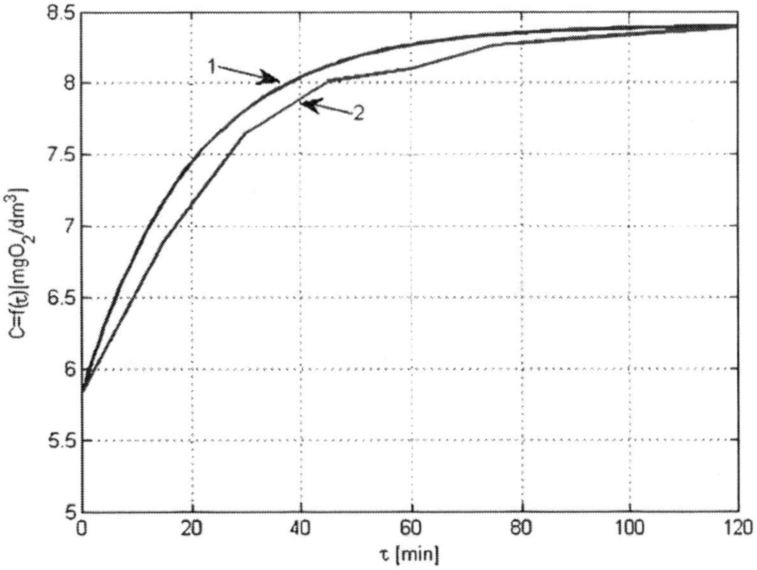

Fig. 8. The modification of the dissolved O_2concentration.

Figure 8 shows a good coincidence of the two graphs, which demonstrates the accuracy and rigor of theoretical and experimental researches.

5Conclusions

1. In the paper, the differential equation of the oxygen transfer rate in waterwas solved, by numerical integration.
2. A computation program was performed and the curve $C = f(\tau)$ was drawn; the theoretical results are similar to those obtained in other papers on the field of study.
3. The paper presents an original installation, designed and built-intosee the change in the dissolved O_2concentration in water.
4. Experimental researches has demonstrated the accuracy of theoretical calculations; for the graphs $C = f(\tau)$, the difference between theoretical and experimental values is small.

References

1. I.M.Călușaru, The influence of the physical properties of the liquid on the efficiency of the oxygenation processes, PhD thesis, University Politehnica of Bucharest, Faculty of Mechanical and Mechatronic Engineering (2014).
2. T.Miyahara, Y.Matsuha, T. Tekahashi, The size of bubbles generated from perforated plates, International Chemical Engineering, **23**,517-523 (1983).
3. R. Mlisan (Cusma), Influence of the bubble generators architecture on the concentration growth of dissolved oxygen into the water, PhD thesis, University Politehnica of Bucharest, Faculty of Mechanical and Mechatronic Engineering (2017).

4. Al. S. Pătulea, The influence of the function al parameters and of the architecture of the fine bubble generators on the efficiency of the aeration installations, PhD thesis, University Politehnica of Bucharest, Faculty of Mechanical and Mechatronic Engineering (2012).
5. Gh.Băran, I.Pinacovschi, F.Bunea, Performance of fine bubble generators, (in Romanian), Hidrotehnica Magazine,**53**, no.3-4, 27-32 (2008).
6. A.Halanay, R.Gologan, D.Timotiu, *Elements of mathematical analysis*, (in Romanian), MATRIX ROM Publishing House, Bucharest (2008).
7. G.Reckeenwald, *Numerical Integration of Ordinary Differential Equation for Initial Value Problems*, Portland State University Department of Mechanical Engineering, (2006).
8. H.M.Antia, *Numerical Methods for Scientist andEngineers*,Birkhauser Publishing,Limited Basel, Switzerland (2002).
9. D.Houcque, R.McCormick, Ap*plication of MATLAB: Ordinary Differential Equation (ODE)*, School of Engineering and Applied Science Northwestern University, Illinois(2007).
10. I.M.Călușaru, N.Băran, Al. Pătulea, Researches regarding the transfer of oxygen in water, the 3rd International Conference on Mechanic Automation and Control Engineering (MACE 2012), IEEE Computer Society CPS, indexed by Ei Compendex, Baotou, China, 2617 – 2620 (2012).
11. Al. Pătulea, N.Băran and I.M.Călușaru, Measurements of DissolvedOxygen Concentration in Stationary Water, World Environment **2**, no 4,106-109 (2012).
12. M.Călușaru (Constantin), E.B. Tănase, N.Băran,R. Mlisan (Cusma), Researches Regarding the Modification of Dissolved Oxygen Concentration in Water, IJISET-International Journal of Innovative Science, Engineering & Technology B, no. 6,228 – 231 (2014).
13. N.Băran, Gh. Băran, G. Mateescu, D. Besnea, Experimental Research Regarding a New Type of Fine Bubble Generator, Proceedings of the International Conference, MECHAHITECH'09 Bucharest (2009).
14. Al. Pătulea, I.M.Călușaru, N.Băran, Researches Regarding the Measurements of the Dissolved Concentration in Water, Advanced Materials Research,**550-553**,3388-3394 (2012).

Numerical investigation on combustion and NO_x emissions of a CH_4 / kerosene afterburner

Andreea Cristina Mangra[1,*] and *Ene* Barbu[1]

[1]Romanian Research & Development Institute for Gas Turbine COMOTI Bucharest, Romania

Abstract. Cogeneration groups equipped with gas turbines usually operate with natural gas. The new requirements regarding the flexibility of functioning both with conventional and alternative fuels have led to the development of new solutions. An afterburning installation facilities, on the one hand, the combustion of fuels which can not normally be used in gas turbines, and on the other hand, allows a rapid adaptation of the cogeneration group to the user's variable thermal energy requirements. In some cases, an afterburning installation may contribute to the reduction of NOx emissions. It also increases the group's adaptability to changing environmental conditions, especially during the summer. Having in mind these tendencies and starting from the data collected from the 2xST 18 – Suplacu de Barcau cogeneration plant, this paper presents a numerical analysis of an afterburner in order to switch from natural gas to natural gas / kerosene functioning, with minimal modifications to the existing solution. A 3D steady RANS numerical integration of the Navier-Stokes equations has been carried out using the commercial software ANSYS CFX. For future research work, it will be taken into consideration the use of more complex reaction mechanisms and the variation of the spray angle will be considered. Also, in order to validate the CFD results a combustion experiments campaign will be conducted.

1 Introduction

Cogeneration groups equipped with modern gas turbines have to meet the requirements regarding the flexibility of functioning both with conventional and alternative fuels. Also, the pollutant emissions, have to be kept at a low level, even in partial load operation conditions.

The high oxygen content in the exhaust gases coming from the gas turbine, around 11-16 % vol. [1], can be used in the combustion of an additional amount of fuel (afterburning).

Afterburning is generally used on supersonic military aircrafts at take-off or during combat [2].

The utilization of an afterburning installation on cogeneration groups equipped with gas turbines increases the flexibility regarding of the used fuel and allows a rapid adaptation to the user's variable thermal agent requirements [3].

* Corresponding author: andreea.petcu@comoti.ro

The control of the combustion process and of the afferent pollutants emissions, while complying with increasingly restrictive legislative requirements, remains one of the main challenges of designing a combustion system. In the field of cogeneration groups with gas turbines, combustion process modelling inside the afterburner becomes an integral part of the entire group design.

In the scientific literature, afterburning numerical modelling addresses, besides pollutant emissions reduction issues, aspects related to efficiency gains, compensation of environmental parameters variations, impact on the recovery boiler, optimization of flow conditions etc.

Sullivan-Lewis [4] analyzes several afterburner geometrical configurations regarding the functioning with low NO_x and CO emissions, at a given fuel/air ratio and the geometry's influence on the combustion stability. Ganjehkaviri [5] has conducted a study on the impact of introducing an afterburner into a combined cycle power plant. Elmegaard et al. [6] studied the possibility of using a polycarbonate afterburning installation - functioning on natural gas / biogas or pyrolysis gas - to increase the efficiency of a power plant with indirect combustion gas turbine. Mehrabani et al. [7] proposed using an afterburning installation to compensate for the environmental parameters variation and to increase the efficiency of combined cycle power groups. Backlund [8] shows that parallel flow burners lead to: larger steam flows, a high efficiency of the gas turbine power plant, an efficient use of the oxygen from the exhaust gases coming from the gas turbine, a larger range of operating regimes, low NO_x and CO emissions. Ahamed [9] analyzes various methods to improve the blending of fuel with exhaust gases from a turbo-engine, as well as the impact on the recovery boiler. Shi [10] optimizes the velocity and temperature fields upstream and downstream of the afterburner using numerical modelling. To determine the performance of an afterburner, Lezsovits et al. [11] used the commercial software ANSYS CFX 5.6.

The construction of the 2xST 18 - Suplacu de Barcau cogeneration plant, which has an afterburner installation functioning on natural gas with multiple parallel flows, led to researches regarding performance improvement. The afterburner was designed to function using as oxidant exhaust gases from a turbo-engine. However, in emergency cases (when there is a problem with the turbo-engine), it can also function with pure air as oxidant, air delivered by a fan.

Similar to the afterburner in the 2xST 18 cogeneration plant, a test bench for a 3-modules afterburner with a single gas supply ramp has been built in INCDT COMOTI [12] (Figure 1 a)).

Each afterburner module is composed of a deflector and 3 fuel nozzles. The central fuel nozzle, with a diameter of 6 mm, is the primary one. The other two are secondary ones with a role in flame propagation from one module to another. The deflector is fixed to the fuel ramp by the means of the 2 secondary fuel nozzles. The deflector (Figure 1 b)) is composed of a horizontal swirler plate and two vertical plates with the role of limiting the flame.

This configuration of the afterburner module insures a good mixing of the fuel (natural gas) with the oxidizer (exhaust gases from the turbo-engine / pure air). It insures the cooling of the area surrounding the afterburner, the stabilization of the flame front and leads to obtaining higher temperatures downstream, with low pollutants levels.

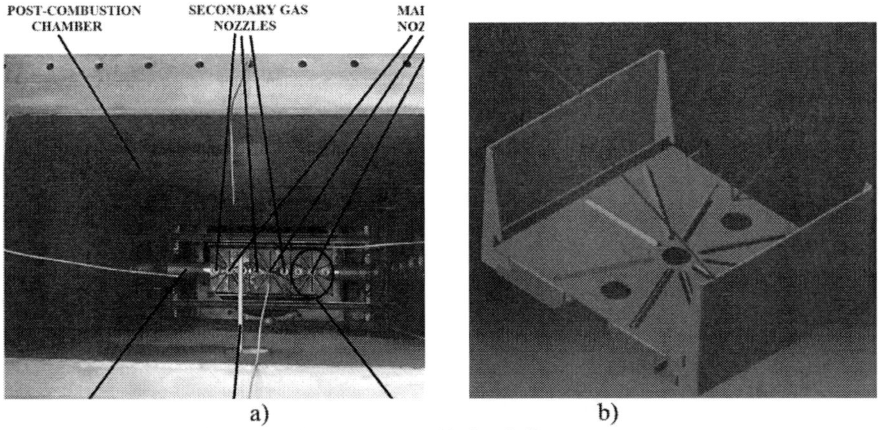

Fig. 1: a) The 3-modules afterburner in COMOTI; b) the deflector

The test bench has the possibility of testing the afterburner using as oxidant either exhaust gases coming from a turbo-engine or pure air (Figure 2).

The test bench is mainly composed of: a TA2 turbo-engine running on natural gas (this is a TV2-117A turbo-engine converted by researchers from COMOTI to function with natural gas instead of kerosene), positioned horizontally, the afterburner and afterburning chamber positioned vertically in the turbo-engines exhaust chimney. The exhaust gases temperature and composition are monitored at a height of 1895 mm from the afterburner.

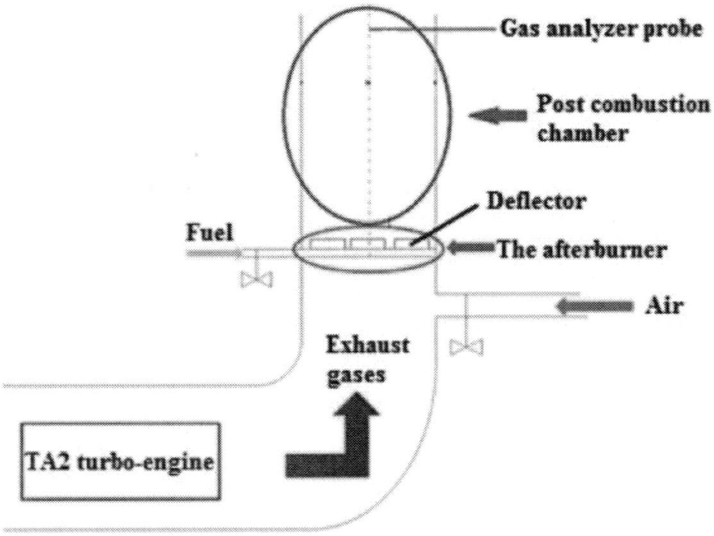

Fig. 2 Median longitudinal plan section through the afterburner test bench in COMOTI

In the context of the shown tendencies, the article presents a numerical analysis regarding changing the afterburner functioning on natural gas to functioning on natural gas/ kerosene, with minimal modifications to the existing solution.

2 Numerical simulations

2.1 Geometry

The proposed solution consists of modifying the central module of the afterburner presented in Figure 1 in order to function on kerosene, the adjacent natural gas modules remaining unchanged.

To better understand the implications of the proposed solution, there have been conducted first numerical simulations with natural gas as fuel and exhaust gases/ pure air as oxidant, for the configuration presented in Figures 1-3.

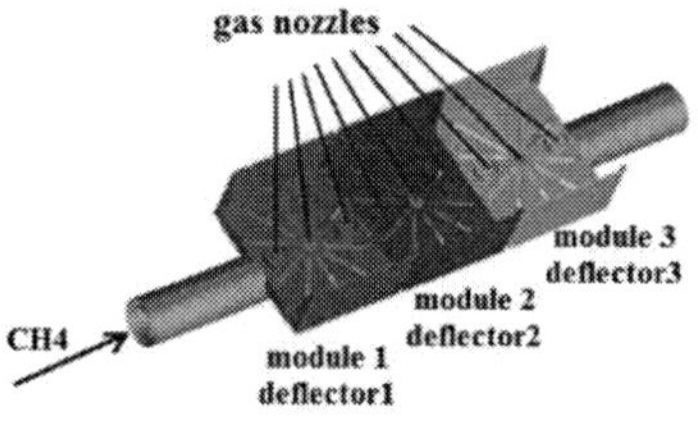

Fig. 3 Afterburner modules configuration

Based on the results obtained from the numerical simulations on natural gas, there have been proposed an afterburner configuration functioning on kerosene. The central natural gas nozzle corresponding to the central module has been replaced with a liquid fuel atomizer. The atomizer exit nozzle diameter is of 0.85 mm. The other natural gas nozzles are blocked.

2.2 Computational domain and mesh

In Figure 4 a)-b) is presented the computational domain. For this domain an unstructured computational grid of 7.678.803 tetrahedral elements and 1.341.681 nodes has been created using ICEM CFD.

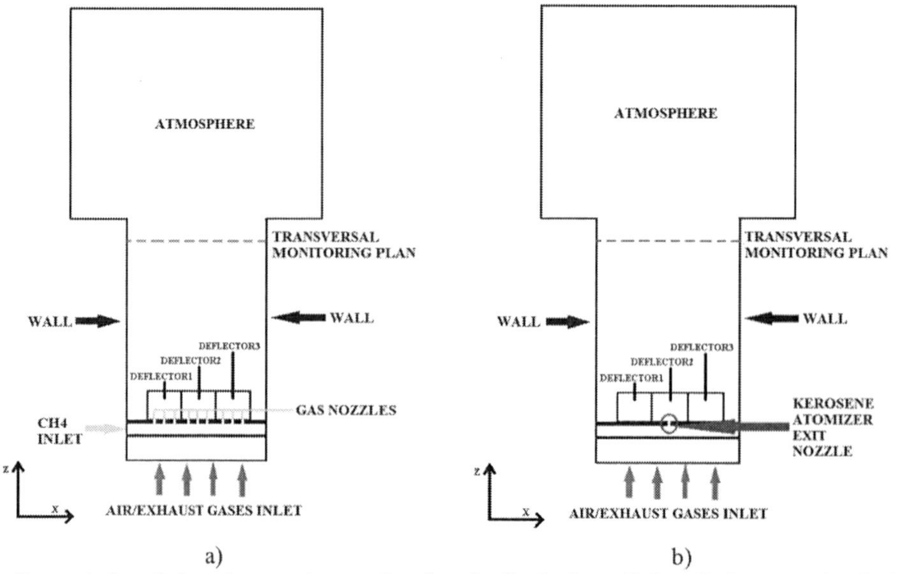

Fig. 4 Computational domain, on the median longitudinal plan of the afterburner: a) afterburner functioning on methane gas; b) afterburner functioning on kerosene

2.3 CFD settings

A three-dimensional steady RANS numerical integration of the Navier-Stokes equations has been carried out using the commercial software ANSYS CFX.

In the numerical simulations it has been considered that the natural gas is composed only of methane gas (CH_4).

The methane gas mass flow corresponding to one module has been chosen in accordance with the design parameters of the afterburner from the 2xST 18 cogeneration plant, functioning on natural gas [12]. It has been considered that the total fuel mass flow is equally distributed to each module.

The composition of the exhaust gases, coming from the TA2 turbo-engine, used as oxidant is presented in Table 1.

Table 1. Exhaust gas composition

	Mass fraction
CO_2	0.0465
H_2O	0.0418
O_2	0.1623
N_2	0.7494

The kerosene mass flow, for the afterburner functioning with exhaust gases coming from the turbo-engine, was calculated according to relation (1) [13]:

$$\dot{m}_{kerosene} = \dot{m}_{methane\_gas} \cdot \frac{q_{methane\_gas}}{q_{kerosene}} \quad (1)$$

where: $\dot{m}_{kerosene}$ - kerosene mass flow (kg/s), $\dot{m}_{methane\_gas}$ - methane gas mass flow (kg/s), $q_{methane\_gas}$ - natural gas lower heat value [14], $q_{kerosene}$ - kerosene gas lower heat value [15].

The kerosene mass flow, when the afterburner functions on fresh air (emergency regime), was calculated taking into account the maximum allowed thermal power for an afterburner module (350 kW) and the lower calorific value of kerosene [15].

In practice, the measurement of the flue gas temperature is performed in the centre of the afterburner chamber, on the monitoring plane positioned at 1895 mm above the burner, using the gas analyzer's integrated temperature probe. The flue gas reference temperature, resulting from the cogeneration plant design data, is 1043 K [12]. Preliminary experiments on natural gas, performed in COMOTI (Fig. 5) [16], showed a strongly turbulent flame. Thus in the numerical simulations the average temperature of the flue gases on the monitoring plane was compared with the reference temperature of 1043 K.

The fuel mass flow being thus set, the oxidant mass flow has been determined by varying it until an average temperature on the monitoring plan of 1043K, with an error of ± 1%, has been obtained.

Fig. 5 Flame structure during the preliminary experiments on the 3 modules afterburner functioning on natural gas

In Table 2 are presented the boundary conditions for the considered cases.

Table 2. Boundary conditions

Fuel	Fuel mass flow [kg/s]	Fuel temp. [K]	Oxidant	Oxidant mass flow [kg/s]	Oxidant temp. [K]
CH4	0.006	293	Exhaust gases	0.813	797
CH4	0.014	293	Air	0.65	293
Kerosene	0.007	293	Exhaust gases	0.9	797
Kerosene	0.008	293	Air	0.23	293

The outlet has been considered of opening type, setting the pressure to be the atmospheric one.

In these numerical simulations an Eddy Dissipation combustion Model (EDM) and a k-ε turbulence model have been used. The combustion model was based on a two steps methane-air reaction mechanism, respectively a two steps kerosene-air reaction mechanism, both imported from the ANSYS library [17]. Also, both used reaction mechanisms take into consideration the formation of thermal NO and prompt NO.

In the case of functioning with kerosene, the fuel was uniformly injected into the atmosphere through the atomizer exit nozzle, under a total spray cone angle of 45°. The initial fuel droplets diameter was set at 0.1 mm. The break-up of the fuel droplets into smaller droplets was simulated using the CAB second break-up model from the ANSYS library. To simulate the transformation of the Jet A (kerosene) droplets into Jet A vapours a Liquid Evaporation Model from the ANSYS library was used.

3 Results and discussions

In Fig. 6 a)-b) and 7 a)-b) are presented the temperature field, respectively the NO mass fraction field on the afterburning chamber's median longitudinal plan.

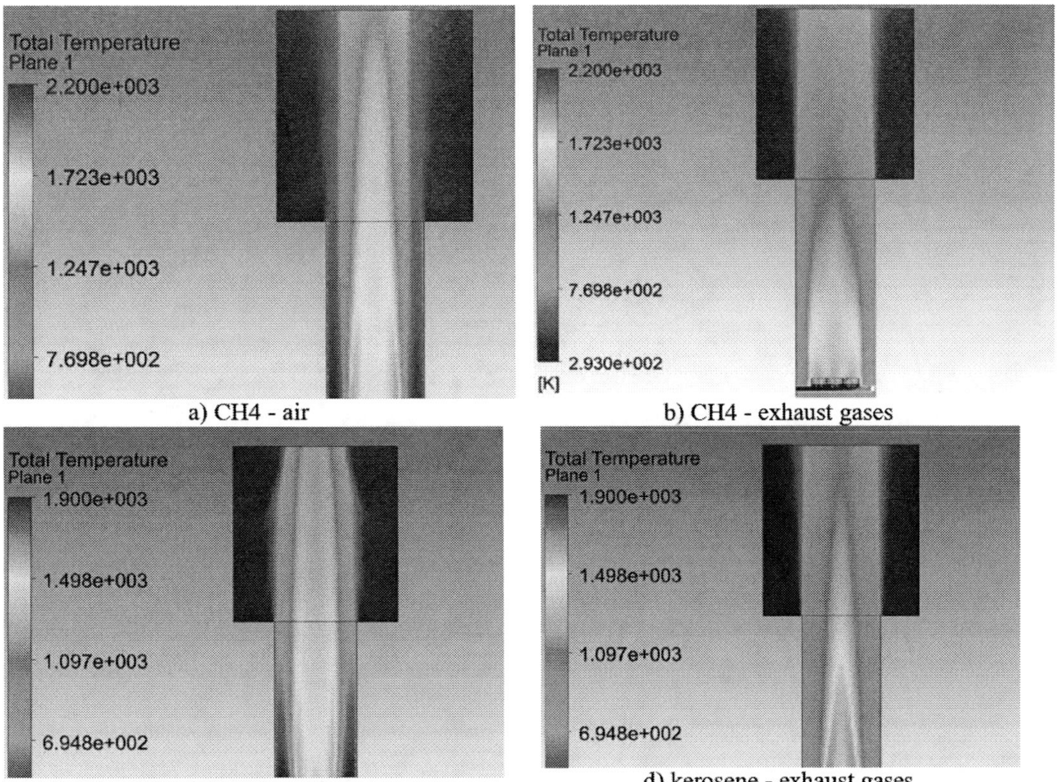

a) CH4 - air b) CH4 - exhaust gases

c) kerosene - air d) kerosene - exhaust gases

Fig. 6 The temperature field on the median longitudinal plan

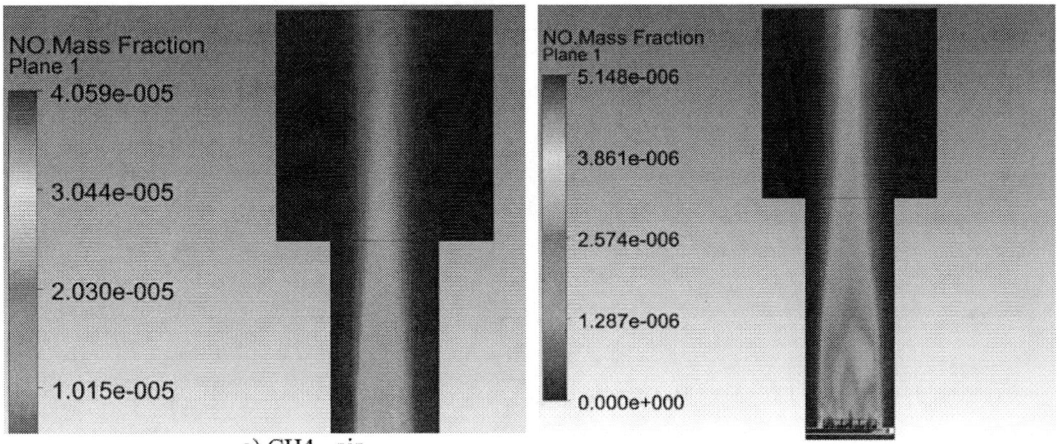

a) CH4 - air b) CH4 – exhaust gases

https://doi.org/10.1051/e3sconf/202128601002

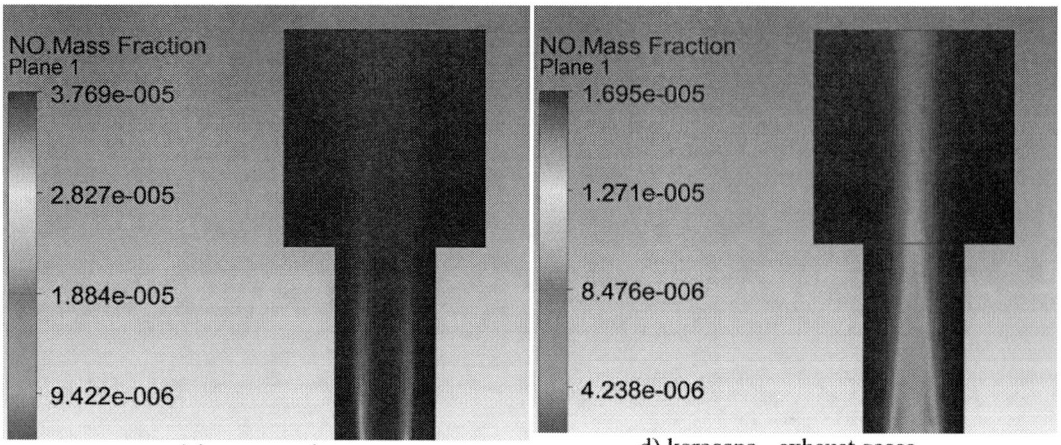

c) kerosene -air
d) kerosene - exhaust gases

Fig. 7 The NO mass fraction field on the median longitudinal plan

The trend followed by the NO mass fraction in the exhaust gases is consistent with the behaviour of the exhaust gases temperature. The NO concentration increasing with the rise of the temperature. The NO concentration is higher when fresh air is used as oxidant, both when CH_4 was used as fuel, as well as when kerosene was used as fuel. The high exhaust gases temperature favours the dissociation of the nitrogen molecules, which leads to the formation of nitrogen oxides [18].

In Fig. 8 and 9 are presented the temperature and NO mass fraction variations, along the central axis of the afterburner combustion chamber.

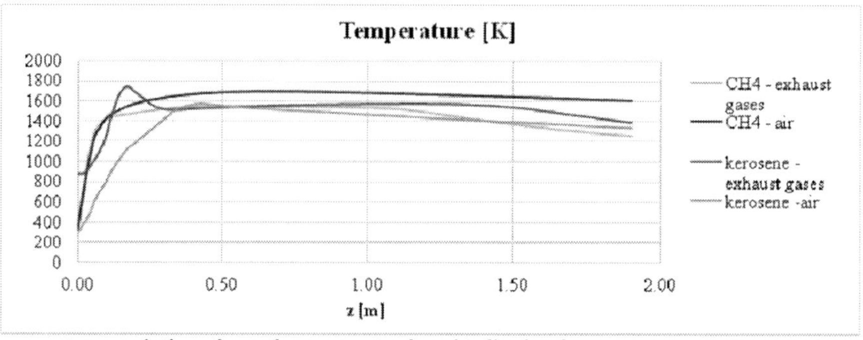

Fig. 8 Temperature variation along the symmetry longitudinal axis

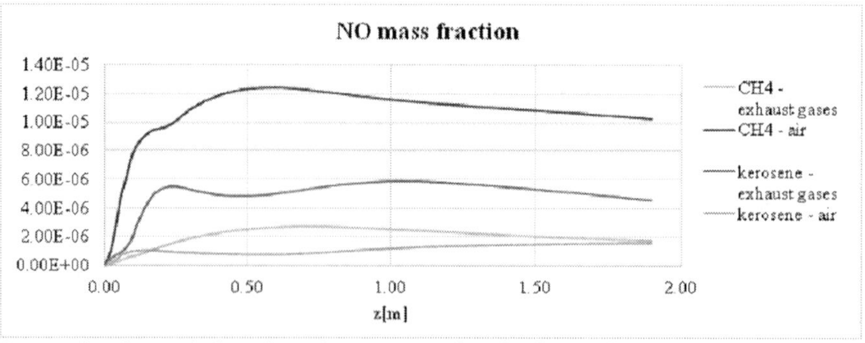

Fig. 9 NO mass fraction variation along the symmetry longitudinal axis

As was expected the NO mass fraction increases with the temperature rise. The NOx emissions are significantly higher in the functioning regime CH_4 - air compared with the functioning regime CH_4 - exhaust gases. This confirms that the afterburner has been optimized to function in CH_4 - exhaust gases regime.

In Fig. 10 and 11 present the temperature variation on the transversal axis on the monitoring plan (positioned at 1895 mm from the afterburner).

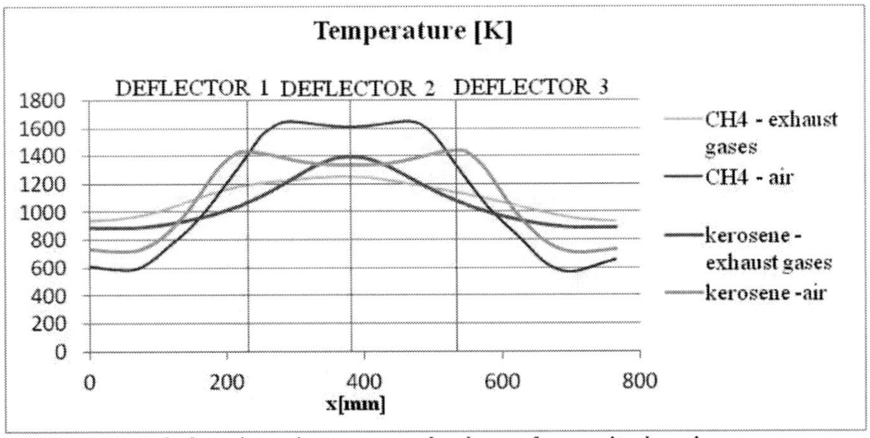

Fig. 10 Temperature variation along the transversal axis, on the monitoring plan

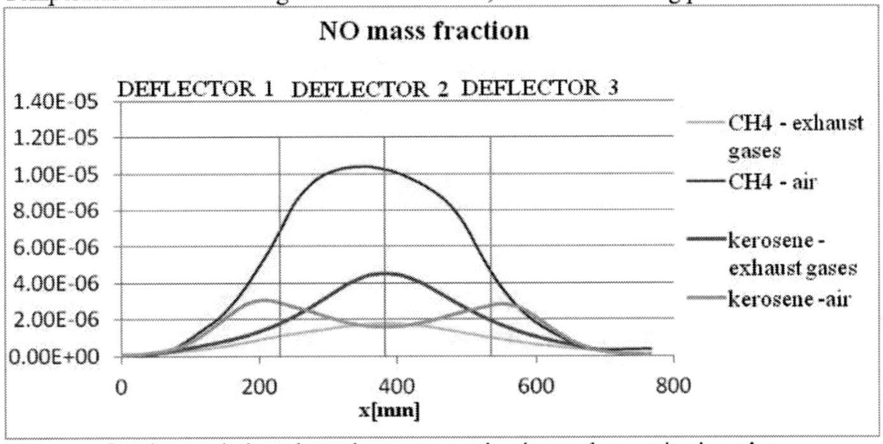

Fig. 11 NO mass fraction variation along the transversal axis, on the monitoring plan

Fig. 10 and 11 reconfirm the normal dependence between the NO emissions and the gases temperature. The lowest NO emissions are obtained for the functioning regime CH_4-exhaust gases. This is consistent with the fact that the afterburner was designed to function optimally in this regime. The highest NO_x emissions are obtained when operating on CH_4 - air, as emergency regime. This is observed also in the results presented in Fig. 9.

Regarding the functioning on kerosene, further numerical investigations are necessary in order to optimize the afterburner. Interestingly, NOx emissions values, when functioning on kerosene - air, are comparable to the values obtained for functioning on CH_4-exhaust gases from the turbo-engine.

In Table 3 are presented the averaged temperature values obtained through numerical simulations, on the monitoring plan where the gas analyser probe is positioned (see Fig. 2).

Table 3. Averaged temperature on the monitoring plan

Fuel	Fuel mass flow [kg/s]	Oxidant	Oxidant mass flow [kg/s]	Temp. [K]	Temp. error [%]
CH4	0.006	Exhaust gases	0.813	1042	0.095
CH4	0.014	Air	0.65	1033	0.958
kerosene	0.007	Exhaust gases	0.9	1039	0.383
kerosene	0.008	Air	0.23	1034	0.862

From the results in Table 3, it can be seen that, for the considered numerical simulation conditions, an error under 1 % for the average temperature on the monitoring plan has been obtained.

4 Conclusions

In order to study the possibility of switching the afterburner from functioning on gaseous fuel to liquid fuel (kerosene) functioning, a CFD analysis has been carried out. During all numerical simulations, it has been intended that the average temperature on the monitoring plan to be 1043 K ± 1%.

The numerical modelling, confirmed that the constructive afterburner solution has been optimized for functioning on gaseous fuel with exhaust gases from the turbo-engine, the lowest NO_x emissions been obtained for this functioning regime.

Theoretical data and preliminary experiments have highlighted the turbulent structure of the flame, imprinted by the afterburner's modules' swirls, and the flame stability in the range of the analyzed CH_4 mass flow (0.006 - 0.014 kg/s). In the case of functioning on kerosene, the flame stability and the low NOx emissions will have to be confirmed through experiments. The transition to kerosene functioning will be based on an original technical solution, which will be the subject of a patent application.

For future research work it will be taken into consideration the possibility of the afterburner functioning simultaneous with CH_4 and kerosene. Also, the use of more complex reaction mechanisms and the variation of the spray angle will be considered. In order to validate the CFD results, in parallel with the numerical simulations the combustion experiments campaign will be continued.

References

[1] Baukal E. Ch. Baukal, *The John Zink combustion handbook* (CRC Press, 2001)
[2] R. Anand, International Journal of Advanced Research in Engineering & Management (IJAREM), vol. 3, issue 5, pp. 20-37 (2017)
[3] E. I. Caceres, M. R. Montanes, O. L. Nord, Journal of Power Technologies, vol. 98, issue 9, pp. 188–197 (2018)
[4] E. Sullivan-Lewis , R. Hack, V. McDonell, Assessment of a rich-burn, quick-mix, lean-burn–based supplemental burner system in a vitiated air stream, Combustion Science and Technology, ISSN: 0010-2202 (2015)
[5] A. Ganjehkaviri, M. Yusof, M. N. M. Jaafar, Jurnal Teknologi (Sciences & Engineering), vol. 79, pp. 27-32 (2017)
[6] B. Elmegaard, U. Henriksen, B. Qvale, Int. J. Thermodynamics, vol.6, No.2, pp. 85-92 (2003)
[7] M. K. Mehrabani et al. Thermal Science, vol. 21, No. 6B, pp. 3011-3023 (2017)
[8] C. J. Backlund, Fiorenza E. Fiorenza, Gas Turbine and Aeroengine Congress, Amsterdam, The Netherlands, June 6-9 (1988)

[9] T. M. Ahamed, S. A. Nathan, International Journal of Multidisciplinary Research and Modern Education (IJMRME), vol. II, issue I, pp. 632-639 (2016)
[10] X. Shi, W. Chriswindarto, D. Boyce, Proceedings of ASME 2009 Power Conference ASME Power 2009, July 21-23, Albuquerque, New Mexico, USA, pp. 1-8 (2009)
[11] F. Lezsovits, S. Könczöl, K. Sztankó, Thermal Science, vol. 14, No. 3, pp. 845-854 (2010)
[12] E. Barbu, et al., WSEAS Transactions on Environment and Development, vol. 6, issue 6, pp. 405-416 (2010)
[13] I. Staffell, The energy and fuel data sheet, University of Birmingham, UK (2011)
[14] V. Rathod, P. V. Bhale, 4th International Conference on Advances in Energy Research 2013, ICAER 2013
[15] T. Giampaolo, *Gas Turbine Handbook: Principles and Practice* (CTC Press)
[16] E. Barbu et al., The influence of inlet air cooling and afterburning on gas turbine cogeneration groups performance, Gas Turbines - Materials, Modelling and Performance, Dr. Gurrappa Injeti (Ed.), InTech (2015)
[17] ***.2016. ANSYS CFX-Solver Modelling Guide, release 17, ANSYS INC
[18] H. A. Lefebvre, R. D. Ballal, *Gas Turbine Combustion. Alternative Fuels and Emissions*, 3rd edition, (CRC Press, 2010)

Numerical modelling of heat convection processes in the ducts of two-stage heat exchanger

Zhivko Kolev[1,*] and *Seher* Kadirova[2]

[1]Department of Heat, Hydraulics and Environmental Engineering, University of Ruse, Ruse 7017, Bulgaria
[2]Department of Electronics, University of Ruse, Ruse 7017, Bulgaria

Abstract. The paper presents results from numerical modelling in ABAQUS of heat convection processes in the ducts of two-stage "water to air" heat exchanger, used as a heat consumer in reversible heat pump installation. A comparison between the values of heat convection coefficient has been done, as they have been determined on the basis of simulation results, and criterion equations. The heat convection processes have been modelled on the base of boundary conditions, determined by measuring of main parameters. The computing mesh has been verified on the basis of certain temperature values, by successively increasing the number of computing cells. The heat convection processes in the ducts of the heat exchanger have been simulated in "heating" and "cooling" modes of the installation. The study has been performed in laboratory conditions.

1 Introduction

In heat exchangers, the processes of heat transfer at forced convection are very common and are the subject of many studies and calculations. An experimental study of a cross-current heat exchanger is presented in [1]. The heat flows, heat convection coefficients and overall heat transfer coefficients, have been calculated. The research [2] focuses on the experimental determination of dependencies between certain criterion parameters, used for calculations of a heat exchanger with spirally finned tubes. In [3] is presented the methodology of experimental determination of heat convection coefficients and heat flows in a two-pipe heat exchanger. A tube-in-tube heat exchanger is studied in [4]. The main relations between different criterion parameters have been determined. Investigation [5] presents calculation of thermal and hydraulic parameters of a heat exchanger with wavy finned surfaces. In [6] a methodology for investigation of the hydrodynamics of a floating head heat exchanger is presented. Article [7] focuses on an experimental study of a heat exchanger used in an air-conditioning system. The heat flows exchanged under operation of the installation in "heating" mode, have been calculated.

One of the most common methods for studying various heat transfer and hydrodynamic processes is the numerical modelling. For this purpose, various computer programs and

* Corresponding author: zkolev@uni-ruse.bg

approaches are used. In investigation [8] is presented the numerical modelling in ANSYS of heat transfer processes in pipes with different finning. In another research a computer simulation in ANSYS of processes of convective heat transfer in pipes with different possibilities for turbulisation of the fluid flowing in them, is considered [9]. Paper [10] focuses on simulation of heat transfer processes in tubes with internal finning in the simulation environment ANSYS. Article [11] presents computer modelling in COMSOL Multiphysics of thermal conductivity processes in a finned tube heat exchanger. In [12] a mathematical modelling of thermal conductivity processes in the fins of a heat exchange apparatus operating according to the "cross-flow" scheme is considered. Paper [13] focuses on computer modelling in COMSOL Multiphysics of processes in a finned-tube heat exchanger. Investigation [14] presents a mathematical modelling of a heat exchanger with finned tubes, applied in drying processes. Another approach of computer modelling in ANSYS of heat transfer processes in the fins of a heat exchanger used in road vehicles is presented in [15]. Article [16] focuses on computer simulation of the temperature field in the walls of buildings. Study [17] presents numerical modelling of the temperature distribution in samples of building thermal insulation materials. In [18] a mathematical modelling of the heat transfer process at the movement of fluid in a pipe is considered. Paper [19] discusses computer simulation in ANSYS of processes of mixing fluids with different temperatures. Investigation [20] presents a numerical simulation of the temperature field around gas pores in the structure of samples made by synthetic materials. In [21], a computer simulation in ANSYS of the velocity distribution of a fluid passing through a cone diffuser is presented. Paper [22] has proposed a mathematical model of the induction heating process of cylindrical pieces. Study [23] presents numerical modelling based on finite difference method of the thermal conditions of a LED driver.

The efficiency of heat consumers, representing various heat exchangers, has a significant importance in heat pump installations. The efficiency of these installations is also influenced by the parameters of the environment, as presented in [24].

This paper presents results from investigation the processes of heat convection in the ducts of a two-stage "water to air" heat exchanger, used as a heat consumer in a reversible heat pump installation. A comparison between the values of the heat convection coefficients, determined by numerical modelling of the processes in ABAQUS, and on the basis of criterion equations, has been performed.

2 Methodology

The investigation has been realized in order to design a new heat exchanger.

Figure 1 presents a block diagram of the heat pump installation. The heat pump is reversible, which allows research to be performed in "summer" and "winter" modes. Water buffer is used as the heat source. The heat consumer is a two-stage "water to air" heat exchanger (HE). In order to maintain a constant temperature of the water coming from the buffer into the heat pump, the stratification conditions in the buffer have been met. The block diagram in Fig. 1 presents valve positions of the system at investigation in "cooling" mode.

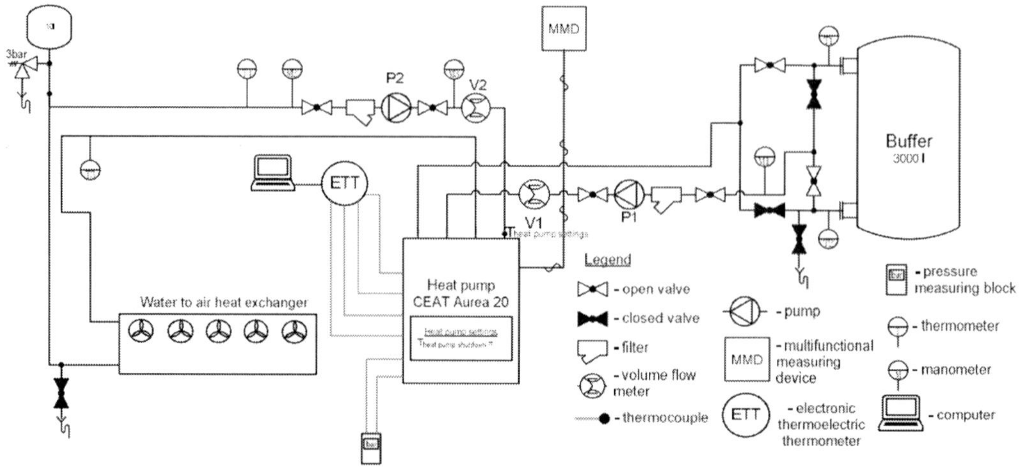

Fig. 1. Block diagram of heat pump installation.

Figure 2 presents a diagram of the "water to air" heat exchanger. The basic geometrical parameters in mm, necessary for implementation the thermal calculations of the apparatus, have been presented.

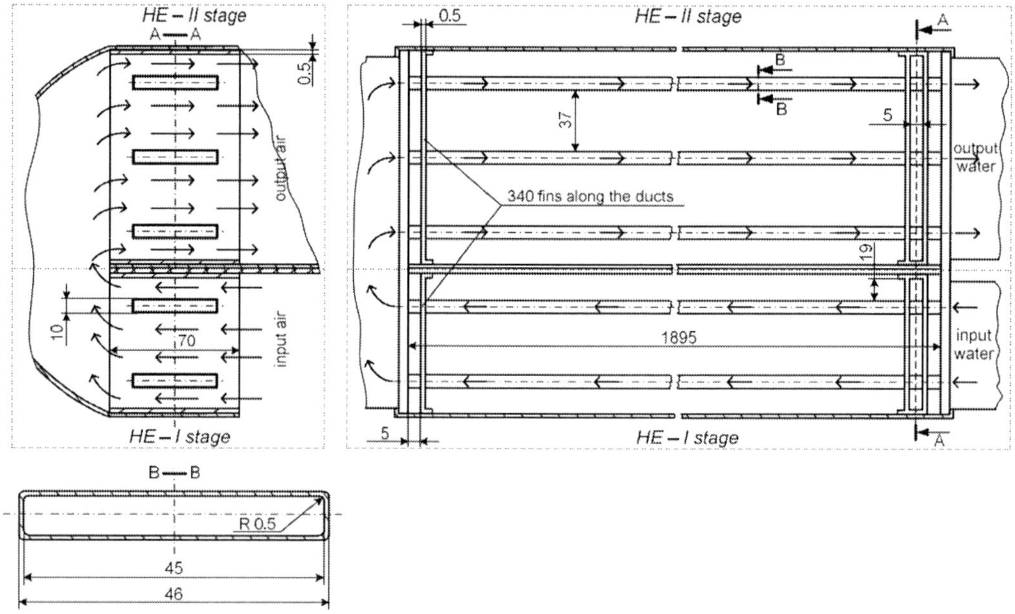

Fig. 2. A diagram of the heat exchanger.

The following equations have been used for calculation of the internal heat convection coefficients, by simulation of the heat transfer processes:

$$\bar{h}_{c,1-simulation} = \frac{Q_{1-duct-heating}}{A_{1-internal}\,\Delta\bar{T}_{1-internal}} \tag{1}$$

$$\bar{h}_{c,1-simulation} = \frac{Q_{1-duct-cooling}}{A_{1-internal}\,\Delta\bar{T}_{1-internal}} \tag{2}$$

where: $Q_{1-duct-heating}$ is the heat flow, exchanged between the water flowing in a duct from the heat exchanger, and its internal heat exchange surface, at "heating" mode of the installation, W;

$Q_{1-duct-cooling}$ - heat flow at "heating" mode, W;

$A_{1-internal}$ - the internal heat exchange surface area of one duct, m^2;

$\Delta\overline{T}_{1-internal}$ - average cross-section temperature difference of the duct, K.

The heat flows $Q_{1-duct-heating}$ and $Q_{1-duct-cooling}$ have been calculated by the basic heat equation.

The average temperature difference $\Delta\overline{T}_{1-internal}$ have been determined based on the simulated temperature field of the fluid, flowing in the duct. For this purpose, the computing mesh of the model has been verified on the basis of certain temperature values, by successively increasing the number of computing cells. The investigation has been accomplished in ABAQUS software environment.

For determination of the heat convection coefficients $\overline{h}_{c,1-criterion\,equation}$ by the criterion method, one of the following two criterion equations has been used, depending on the value of the obtained Reynolds number:

$$Nu_{fd,1} = 0.008\,Re_{fd,1}^{0.9}\,Pr_{f,1}^{0.43} \quad (3)$$

$$Nu_{fd,1} = 1.4\left(Re_{fd,1}\frac{d_{equiv.-duct}}{L_{duct}}\right)^{0.4} Pr_{f,1}^{0.33}\left(\frac{Pr_{f,1}}{Pr_{w,1}}\right)^{0.25} \quad (4)$$

$d_{equiv.-duct}$ is the equivalent internal diameter of the duct, m.

L_{duct} is the length of each duct from the heat exchanger, m.

The equivalent internal diameter has been calculated by the equation:

$$d_{equiv.-duct} = \frac{4f_{duct}}{P_{duct}} \quad (5)$$

where: f_{duct} is the area of the internal cross-section of the duct, m^2;

P_{duct} - perimeter of the cross-section, m.

The comparison between the heat convection coefficients, determined by the presented two methods, is done by calculating the percentage difference between them, according to the following equation:

$$Percentage\ Difference = \frac{\left|\overline{h}_{c,1-simulation} - \overline{h}_{c,1-criterion\,equation}\right|}{\dfrac{\overline{h}_{c,1-simulation} + \overline{h}_{c,1-criterion\,equation}}{2}}100 \quad (6)$$

3 Results and discussion

Table 1 presents the values of the boundary conditions used for computer simulation of the heat transfer processes. The boundary conditions are as follows: bulk fluid temperature of the duct input $T_{f,1-input}$, bulk fluid temperature of the duct output $T_{f,1-output}$, fluid velocity in the input cross-section of the duct $U_{1-input}$, manometric pressure of the fluid in the output cross-section of the duct $p_{m,1-output}$, heat flux in the duct q_{1-duct}.

Table 1. Boundary conditions for simulation of heat transfer processes.

Heat transfer process	$T_{f,1\text{-input}}$, °C	$T_{f,1\text{-output}}$, °C	$U_{1\text{-input}}$, m/s	$p_{m,1\text{-output}}$, Pa	$q_{1\text{-duct}}$, W/m^2
HE – I stage "Cooling" mode	9.9	11.9	0.216	90070	3577.91
HE – II stage "Cooling" mode	11.9	12.8	0.144	90000	1073.17
HE – I stage "Heating" mode	51.1	47.5	0.218	90050	6415.37
HE – II stage "Heating" mode	47.5	44.6	0.145	90000	3445.29

The used shape of the computing cells for simulation of the heat convection processes is "Tet". The verification of the computing mesh has been performed on the base of temperature results of a node, located at the output duct's cross-section. The selected mesh is with cells' size of 0.007 m.

Table 2 presents the obtained values of the heat convection coefficients and the percentage difference between them by apply of the two presented methods.

Table 2. Results for heat convection coefficients.

Heat transfer process	$\bar{h}_{c,1\text{-simulation}}$, W/(m^{2}K)	$\bar{h}_{c,1\text{-criterion equation}}$, W/(m^{2}K)	Percentage Difference, %
HE – I stage "Cooling" mode	553.92	933.42	51.03
HE – II stage "Cooling" mode	480.06	267.88	56.74
HE – I stage "Heating" mode	1058.93	1534.69	36.69
HE – II stage "Heating" mode	853.23	983.07	14.14

The heat convection coefficient $\bar{h}_{c,1-criterion\,equation}$ at HE - II stage, "cooling" mode, is determined on the basis of criterion equation (4). For determination the heat convection coefficients by the criterion method for the other three heat transfer processes, criterion equation (3) have been used.

It should be noted that the way of calculation of the determinative size of the heat exchange surface $d_{equiv.-duct}$ influences the determination of the heat convection coefficients, based on criterion equations. It is included in the equations for calculation of Reynolds number, in a separate expression in criterion equation (4), as well as in the equation for calculation of the heat convection coefficient, based on Nusselt number. The equivalent internal diameter $d_{equiv.-duct}$ is calculated by equation (5) which is valid for ducts with an arbitrary cross-sectional shape. This equation does not express the specific features

https://doi.org/10.1051/e3sconf/202128601003

of the investigated duct cross-section. This affects the accuracy of determining the heat convection coefficients by this method.

From the results presented in Table 2 can be concluded that in general the higher values of heat convection coefficients are obtained in "heating" mode, comparing to "cooling" mode. This is due to the set operating parameters of the heat pump, as well as the ambient temperature in the relevant experimental investigations.

It can be noted that in the first stage of the heat exchanger, the heat convection coefficients have higher values compared to the second stage. The reason for this is the temperature variation of the airflow, passing through the first stage of the heat exchanger, which leads to a smaller temperature difference in the second stage.

In "heating" mode, the percentage difference between the heat convection coefficients determined by the two methods is lower than in "cooling" mode.

Figure 3 presents the simulated temperature distribution and the simulated fluid velocity distribution in the central cross-section of the heat exchanger ducts.

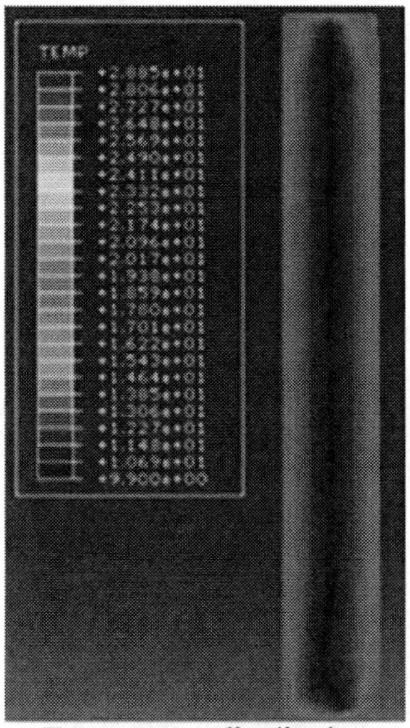

Temperature distribution
HE – I stage
"Cooling" mode

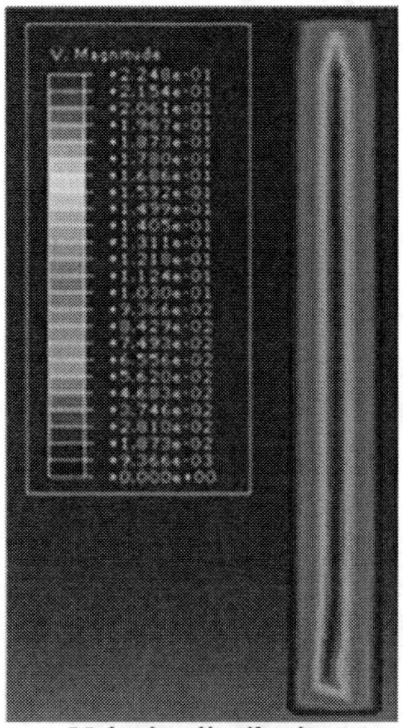

Velocity distribution
HE – I stage
"Cooling" mode

Temperature distribution
HE – II stage
"Cooling" mode

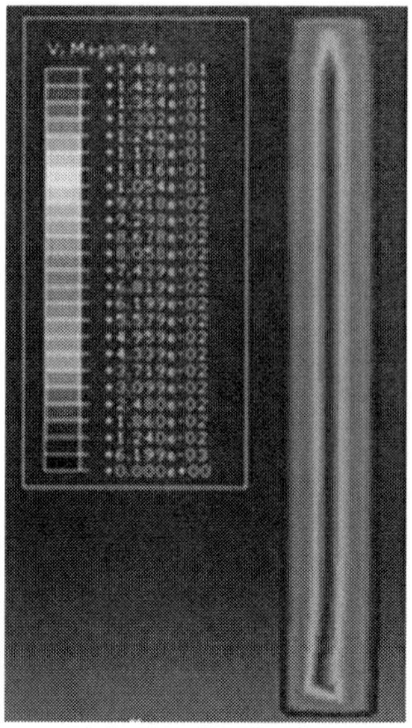

Velocity distribution
HE – II stage
"Cooling" mode

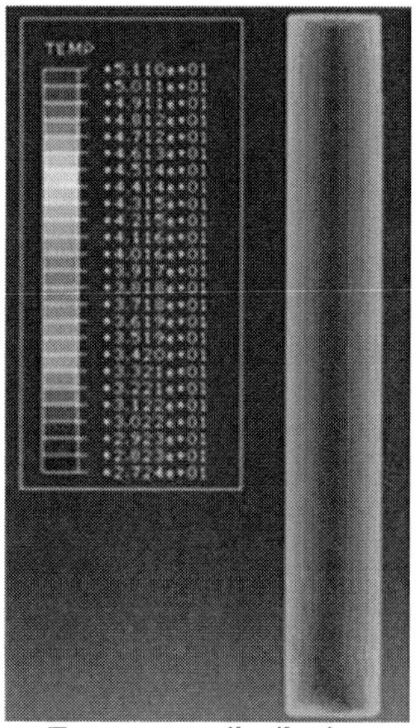

Temperature distribution
HE – I stage
"Heating" mode

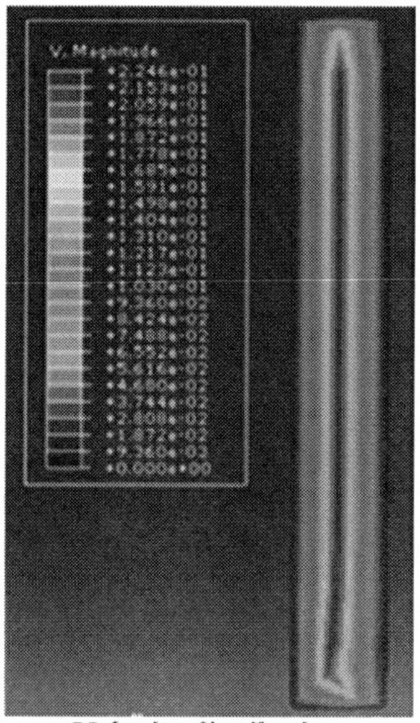

Velocity distribution
HE – I stage
"Heating" mode

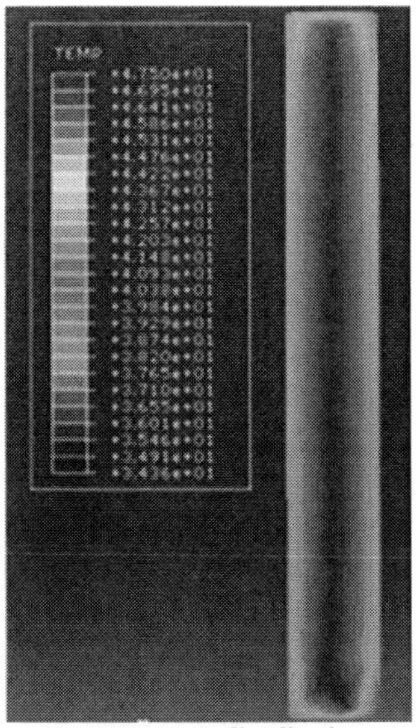

Temperature distribution
HE – II stage
"Heating" mode

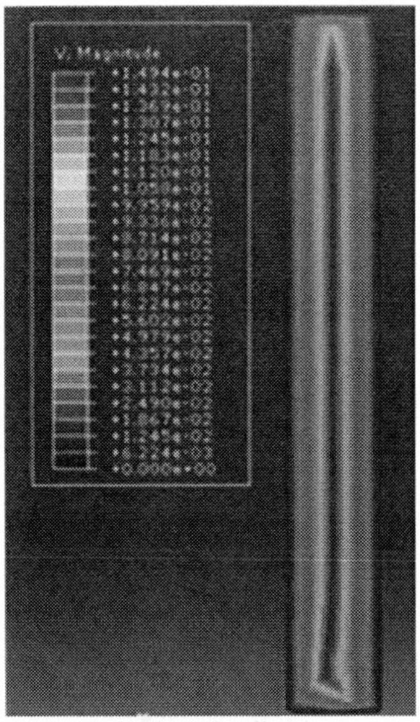

Velocity distribution
HE – II stage
"Heating" mode

Fig. 3. Simulation results for the temperature distribution and the fluid velocity distribution in the ducts' central cross-section.

The results presented in Figure 3 are used for determination of the heat convection coefficients by the method of simulation of the heat transfer processes in the ducts. From the simulation data for the temperature distribution of the fluid, it can be seen that the change in temperature near the internal surface of the duct is more significant. The results for the distribution of the fluid velocity present that near the internal surface of the duct the velocity has relatively low values, which is due to the higher forces of internal friction.

The influence of the specific shape and the geometric parameters of the ducts on the fluid temperature field has been noticed. This affects the values of the average temperature difference in cross-section, and respectively the values of the heat convection coefficients, calculated by equations (1) and (2).

In subsequent studies, the simulation results obtained for the temperature distribution on the internal surface of the ducts can be compared with experimental results. For this purpose, it is necessary to measure the temperature at certain points on the external surface, and the thermal resistance in the ducts' walls must be neglected.

4 Conclusion

Simulation results for the temperature and the velocity distribution of the fluid flowing in the ducts of heat exchanger have been obtained.

A methodology for determination of the heat convection coefficients in the heat exchanger ducts by using simulation data has been proposed.

Heat convection coefficients have been determined on the basis of simulation of the heat transfer processes and on the basis of criterion equations.

The percentage difference between the coefficients determined by the two methods has been calculated. According to the obtained results, the highest percentage difference between the coefficients is observed in the second stage of the heat exchanger, when the installation is operating in "cooling" mode. The lowest percentage difference is in the same stage of the apparatus, but in "heating" mode.

The influence of the method of calculation of the determinative size of the heat exchange surface on the accuracy of determining the heat convection coefficients by the presented criterion equations has been analyzed.

The influence of the investigation conditions including the system parameters and the environment parameters, and the variation of airflow parameters on the heat convection coefficients have been analyzed. A comparative analyze has been accomplished regarding the heat convection coefficient values in the two stages of the heat exchanger, and in the two modes of operation of the heat pump installation.

References

1. Z. Kadhim, M. Kassim, A. Hassan, Effect of integral finned tube on heat transfer characteristics for cross flow heat exchanger, International Journal of Computer Applications **(0975 – 8887)**, 139 (2016)
2. He Fa Jiang, Cao Wei Wu, Yan Ping, Experimental investigation of heat transfer and flowing resistance for air flow cross over spiral finned tube heat exchanger, Energy Procedia **17**, 741 (2012)
3. M.S. Baba, M. Bhagvanth Rao, A. V. Sita Rama Raju, Experimental study of convective heat transfer in a finned tube counter flow heat exchanger with Fe_3O_4 – water nanofluid, International Journal of Mechanical Engineering and Technology (IJMET), 500 (2017)
4. A.H. Dhumal, G.M. Kerkal , K.T. Pawale, Heat transfer enhancement for tube in tube heat exchanger using twisted tape inserts, IJAERS **4**, 89 (2017)
5. S. Albetel, A. Rus, E. David, V. Martian, The amplitude influence on the thermal and hydraulic performances for a wavy air fin in a compact heat exchanger used in agriculture applications, E3S Web of Conferences **180**, 01006, TE-RE-RD, (2020)
6. S. Tasheva, Z. Hodjeva, V. Rasheva, G. Valtchev, B. Milenkov, Investigation of the hydrodynamics of a floating head heat exchanger, Heat Engineering **2 (5)**, 11, (2013)
7. S. Valchev, I. Mihaylov, Analysis of energy efficiency of air handling unit with integrated air to air heat exchanger in heating mode, E3S Web of Conferences **207**, PE-PM, 01002 (2020)
8. M. Zaidan, A. Alkumait, T. Ibrahim, Assessment of heat transfer and fluid flow characteristics within finned flat tube, Case Studies in Thermal Engineering **12**, 557 (2018)
9. W. Osley, P. Droegemueller, P. Ellerby, CFD Investigation of Heat Transfer and Flow Patterns in Tube Side Laminar Flow and the Potential for Enhancement, Chem. Eng. Trans. J. **35**, 997 (2013)
10. K. Majewski, S. Grądziel, CFD Simulations of Heat Transfer in Internally Helically Ribbed Tubes, Chem. Process Eng. J. **37 (2)**, 251 (2016)
11. S. Singh, K. Sørensen, T. Condra, Multiphysics numerical modeling of a fin and tube heat exchanger, *Proceedings of the 56th SIMS*, 383 (2015)
12. D. Jung, D. N. Assanis, Numerical modeling of cross flow compact heat exchanger with louvered fins using thermal resistance concept, SAE Technical Paper Series, (2006)
13. S. Arena, E. Casti, J. Gasia, L. F. Cabeza, G. Cau, Numerical simulation of a finned-tube LHTES system: influence of the mushy zone constant on a phase change behavior, Energy Procedia **126 (201709)**, 517 (2017)

14. Lin Wei, Guorong Zhu, Zhijiang Jin, Numerical simulation of heat transfer in finned tube of heat recovery unit using fluid-solid coupled method, Advances in Mechanical Engineering, (2015)
15. Shubham Singh, Venkata Krishnan K., Spandana H., Mahesh Kumar N., P. S. Kulkarni, Numerical study of the heat transfer enhancement of circular tube bank fin heat exchanger with vortex generators, 20th Annual CFD Symposium – Bangalore, (2018)
16. R. Petkova-Slipets, K. Yordanov, P. Zlateva, A comparative thermal analysis of walls composed of traditional and alternative building materials, Civ. and Env. Eng. **16 (2)**, 388 (2020)
17. P. Zlateva, K. Yordanov, R. Petkova-Slipets, A study of the thermal properties of an alternative straw-containing building material, E3S Web of Conferences **207**, 01004, PE-PM, (2020)
18. C. Oon, H. Togun, S. Kazi, A. Badarudin, M. Zubir, E. Sadeghinezhad, Numerical simulation of heat transfer to separation air flow in an annular pipe, Int. Commun. Heat Mass. J. **39**, 1176 (2012)
19. E. Dimofte, F. Popescu, I. Ion, Numerical modelling of mixing fluids at different temperatures, *Proceedings of TE-RE-RD,* **35** (2016)
20. R. Petkova-Slipets, P. Zlateva, An analysis of the structure and thermal conductivity of hollow microsphere filled syntactic foams, Civ. and Env. Eng. **0 (0)**, (2019)
21. G. Popov, K. Klimentov, B. Kostov, R. Dimitrova, Determining the minor loss coefficient of cone diffusers, E3S Web of Conferences **207**, PE-PM, 04004 (2020)
22. B. Gilev, N. Hinov, H. Ibrishimov, Mathematical model of induction heating with heat transfer of cylindrical body for pressing treatment. *Proceedings of International Conference on High Technology for Sustainable Development HiTech*, (2019)
23. N.L. Evstatieva, B.I. Evstatiev, Modelling of the thermal conditions of a LED driver, *26th International Symposium for Design and Technology in Electronic Packaging (SIITME)*, (2020)
24. P. Zlateva, K. Yordanov, Experimental study of heat pump type air-water for heating system performance, E3S Web of Conferences **112**, 01007, TE-RE-RD, (2019)

Improving the lifetime of an electro-compressor for autoclave with optimized buffer tank

Iulian Vlăducă[1,*], *Claudia Irina* Borzea[1], *Carmen Gheorghița* Petre[1], *Romeo Dorin* Hrițcu[1], *Raluca Lucia* Maier[1], *Vicențiu Liviu* Ringheanu[1], and *Ramona Manuela* Stanciuc[1]

[1]Romanian Research and Development Institute for Gas Turbines COMOTI, 061126 Bucharest

Abstract. In order to increase the research capabilities in the field of low-weight intelligent materials and high structural performance, and also to increase energy efficiency, a study was conducted on improving the operating time of a hot air autoclave compressor with a buffer tank. The booster screw electro-compressor of 200 kW is designed to compress technological air to the installation's necessary parameters at 1 ppm air purity, and the pressurized buffer tank is designed to reduce the number of start-ups of the compressor unit and its motor, in order to increase its lifespan. The pressure can vary in the designed buffer tank from 7 to 26 bar (abs). The autoclave parameters record thermodynamic modifications as it heats up to 400 °C at the working pressure. Cool air is required during the cooling process. The air mass exchanged during operation for cooling down to atmospheric temperature was calculated. The data obtained was used for validating an optimally designed buffer tank within the required operational parameters limits, set and acquired by the custom-made automation system.

1 Introduction

An autoclave is basically a heating equipment combining an oven and a pressure chamber. Using an autoclave enables a good control of the pressure and temperature [1], and is used for curing. Autoclave moulding is widely used in the aerospace industry [2], for manufacturing composites of high quality [3, 4], based on prepregs.

The pre-impregnated laminae are laid onto the tool surface, on top of which a peel ply, bleeder ply and vacuum bag are placed. The entire assembly, once sealed, is placed in the autoclave. Vacuum operation has the role of removing the voids in the composite materials, thus consolidating the ply while volatiles that may form during the moulding process are continuously removed in order to reduce the incidence of porosity. Voids can cause structural damage which may significantly affect the structural integrity and remaining useful lifetime of the aircraft [3]. By combining vacuum and pressure applied on the assembly, a uniform pressure is results on the part surface. Thus, the autoclave process is able to provide very high-quality laminate composites in terms of material uniformity, dimensional tolerances. This technology, however, can be prohibitively expensive unless

* Corresponding author: iulian.vladuca@comoti.ro

the level of quality control and very narrow tolerances are imperatively imposed in the manufacturing field.

The autoclave discussed is used for the polymerization of composite materials for aerospace components [2]. The aim is to prolong the lifetime of the induction motor and its driven compressor by choosing the right installation parameters and buffer tank. The parameters are acquired by a specially designed automation control system with PLC and Panel PC distributed local data acquisition. The autoclave (Figure 1) was installed together with a compressor assembly and an air dryer.

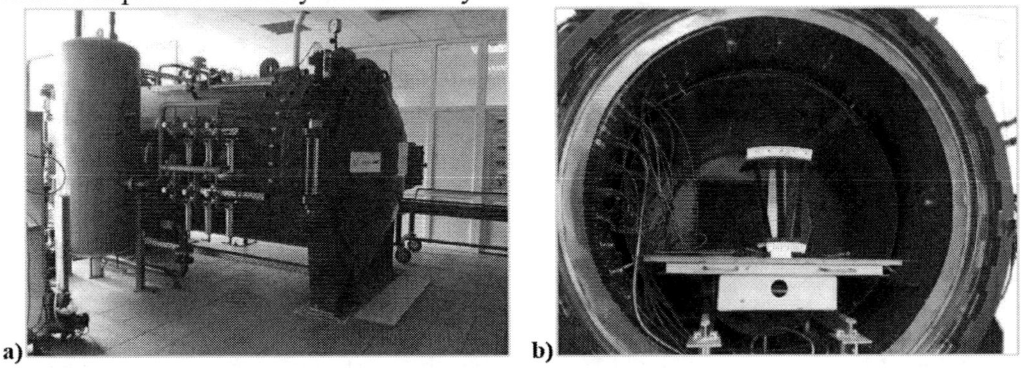

Fig. 1. a) Scholz autoclave, **b)** Interior with carbon fibre aerofoil

2 Twin-screw electro-compressor used

For the given application, a CU 128 [5] compressor unit [6] was chosen, having the characteristics in Figure 2. This one is driven by a 200 kW asynchronous motor. The minimum and maximum pressures of the CU64, CU90 and CU128 compressor units are constant, being: min. 6 bar (abs) and maximum 26 bar (abs). The characteristics are relatively similar, differing in the power consumed and circulated gas flow, depending on their construction size. The air inlet temperature range is -40 °C ÷ 50 °C. The compressor operating requirements of the autoclave installation are: working fluid: air; air flow: 780 Nm^3/h; suction pressure: 1 bar (atmospheric pressure); suction temperature: -20 °C ...+40 °C; discharge pressure: max. 20 bar; discharge gas temperature at autoclave inlet: 50 °C; residual oil content: 0.01 ppm; remanent particles: max. 1 μm.

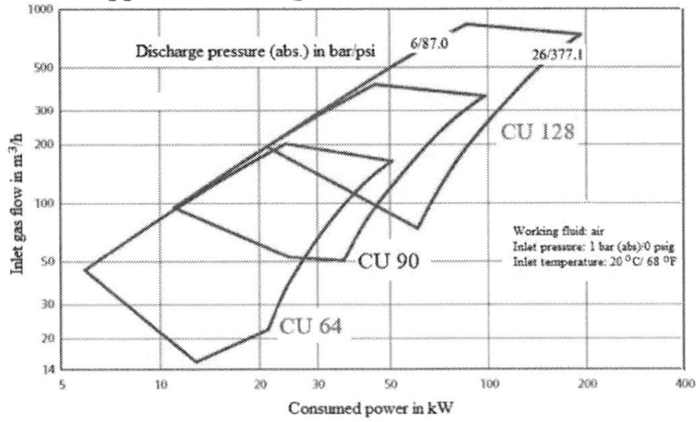

Fig. 2. Characteristics of the compressor units, showing CU 128 compressor used

The operation parameters of the installation (Figure 3a) were analysed, which conducted to the observation of the need for a buffer tank (Figure 3b) between the compressor and the autoclave, in order to ensure the air consumption during operation in the pressure-temperature curve of the autoclave. The buffer tank was sized, appreciating the exchange of air masses that takes place in the bearing stage, establishing an optimum volume of 3 m^3.

a) 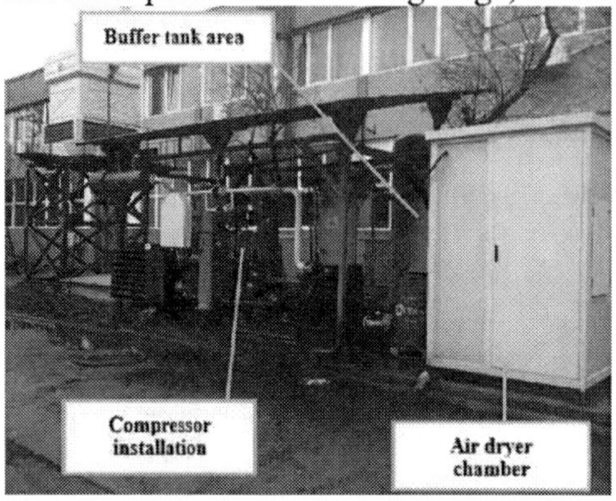

b) 

Fig. 3. a) Outdoor autoclave assembly with compressor, **b)** Buffer tank

The buffer tank has the role of limiting the starts/stops of the induction motor. Due to overcurrent of very high amperage occurring at the start-up of electric induction motors, the time required in order to accelerate loads with high inertia gives birth to sudden temperature rises. If the time interval between successive start-ups is very short, the motor's windings may overheat, which can cause insulation damage and thus reducing lifetime. Temperature is also influenced by the motor's capacity to dissipate heat. The useful lifetime of the motor depends on the life of the winding insulation, being affected by various factors, such as moisture, vibration, corrosion etc. Among all factors affecting electric motor lifetime, the most critical is the temperature of the insulation materials [7].

3 Tests computations

A maximum buffer tank volume of 3 m^3 was considered, due to placement constraints. The calculations [8] use the experimental data obtained during the commissioning tests with the compressor [9-11] without buffer tank. The tests were performed in the following conditions:

(1) Air pressure = 7 bar, working temperature = 140 °C.
(2) Air pressure = 15 bar, working temperature = 200 °C.
(3) Air pressure = 18 bar, working temperature = 400 °C.

The tests were conducted at relative pressure values, which were converted to absolute values. If not specified, the pressure unit [bar] generally means relative pressure [bar (rel)]. The input data is: autoclave volume: V_A = 12.5 m^3; buffer tank volume: V_{BT} = 3 m^3; compressor separator vessel volume: V_C = 0.5 m^3; compressor discharge air temperature: T_C = 50 °C. The buffer tank is dimensioned for 24 bar and 50 °C, considering the operating limits of the screw compressor. Three operation stages were considered:

A) Steady state at constant pressure and constant temperature.
B) Constant pressure and air cooling.
C) Evacuation of the air with decreasing of the pressure and temperature down to the atmospheric conditions. The autoclave filling time at required pressure is not considered.

3.1 First set of tests (7 bar, 140 °C)

A) <u>Steady state</u>

During the steady state period, there is an emptying-filling activity with pressure variations of -0.02/+0.01 bar at 1.5 minutes. The air mass required to fill up the autoclave is the difference between the air mass occupied at the pressure of +0.01 bar and the mass occupied at the pressure of -0.02 bar. An adiabatic process is considered, with the following initial values: $P = 7$ bar (rel), $T = 413.16$ K, $R = 287.14$ J/KgK. Two air states are assumed, at the absolute pressures of 8.01 bar (abs) and 7.98 bar (abs) respectively. At P_n pressure, the temperature T_n is calculated as:

$$T_n = T \cdot \left(\frac{P_n}{P}\right)^{\frac{k-1}{k}} \quad (1)$$

The corresponding air mass is:

$$M_n = \frac{P_n \cdot V_A}{R \cdot T_n} \quad (2)$$

Table 1. Data calculated for the two air states

State	P_n [°C]	T_n [°C]	M_n [kg]
1st state	$P_1 = 8.01$ bar (abs)	$T_1 = 413.3$ K	$M_1 = 84.371$
2nd state	$P_2 = 7.98$ bar (abs)	$T_2 = 412.9$ K	$M_2 = 84.145$

The air mass lost within the autoclave is $M_{loss} = M_1 - M_2 = 0.226$ kg. This amount of air is required for during the exhaust-intake cycles, for a period of time of $t_c = 1.5$ min. The calculations for the buffer tank are similar, with the observation that the pressure decreases from 24 bar (rel), namely 25 bar (abs), to a pressure of 7 + 3 bar or 8 + 3 bar (abs). A pressure of 3 bar is required, higher than the pressure in the autoclave. The temperature (T = 50 °C) decreases adiabatically. The initial conditions are: $P_{B1} = 25$ bar (abs); $P_{B2} = 11$ bar (abs); $T_{B1} = 323.16$ K. The mass and temperature in the buffer tank, states 1 and 2 are calculated with the following relations:

$$M_{BT_n} = \frac{P_n \cdot V_{BT}}{R \cdot T_n} \quad (3)$$

From the equation (1), we have $T_2 = T_1 \cdot (P_2/P_1)^{\frac{k-1}{k}} = 255.592$ K. Obtaining the air masses from equation (3) $M_{BT_1} = 80.829$ kg and $M_{BT_2} = 44.966$ kg, the mass loss in the buffer tank is hence $M_{BT} = MB_{T1} - MB_{T2} = 35.862$ kg. The buffer tank loses this mass during a pressure decrease of 14 bar, from 25 bar (abs) to 11 bar (abs). The number of cycles, $N_c \cong 159$, for autoclave operation with buffer tank, results from equation (4):

$$N_c = \frac{M_{BT}}{M_{loss}} \quad (4)$$

The maximum stationary time t_{st} of the autoclave is calculated with relation (5) below. If this time is longer than the calculated value, a new start-up of the compressor is required, for compensating the air mass lost by the autoclave.

$$t_{st} = N_c \cdot t_c = 3.97 \text{ h} \quad (5)$$

B) Constant pressure, air cooling at 7 bar

At the end of the steady state period, the air inside the autoclave is beginning to cool down, maintaining a constant pressure. Thus, at a constant pressure of 7 bar, the air mass corresponding to the temperature from Table 2 (showing 10 selected sample values) was calculated, as well as the differences between two successive temperatures.

Table 2. Temperature values at constant pressure of 7 bar

i = *1..55*	1	6	12	18	24	30	36	42	48	55
T [°C]	140.3	131.6	120.2	108.9	96.8	85	73.6	62	50.9	39.8

Due to the air cooling, it is necessary to supply an air mass in the buffer tank, in order to maintain a constant pressure inside the autoclave. The variation of the autoclave mass, M_{Ai}, with i = 1..55, is given by the equation below.

$$M_{Ai} = \frac{P_{1i} \cdot V_1}{R \cdot T_{1i}} \tag{6}$$

From the iterations, the total mass loss of the autoclave is $M_{loss} = M_A = M_{A55} - M_{A1} =$ 27.05 kg, which should be provided by the buffer tank. The mass lost by the buffer tank for this variation is M_{BT} = 35.862 kg. We conclude that a 3 m^3 buffer tank is sufficient for ensuring the installation's operation throughout the cooling cycle, $M_{BT} > M_A$. In this scenario (7 bar, 140 °C), mass lost can be supplied by buffer tank, without restarting the compressor.

C) Air evacuation

At the end of duty cycle, the air is evacuated from the autoclave, as it cools down to a temperature of 23 °C. The values for pressure, temperature and successive differences of the mass are presented in Table 3.

Table 3. Cooling process at the end of duty cycle

i = *1..37*	1	4	8	12	16	20	24	28	32	37
T [°C]	39.8	37.9	35.6	33.1	30.5	29.7	27.5	25.5	24.3	23.3
P [bar]	7	6.3	5.4	4.5	3.6	2.6	1.7	0.9	0.4	0
M_{Ai} [ND]	1.18	2.639	2.621	2.695	4.22	2.826	2.811	2.89	1.438	0

The calculation with (6) show that the mass differences discharged at successive temperatures are positive, see the table 3, and therefore we can conclude that during the period of autoclaved air evacuation and temperature drop, external air mass supply is not required, and compressor can be stopped.

3.2 Second set of tests (15 bar, 200 °C)

A) Steady state

Similarly to the first set of tests, an adiabatic transformation with the same emptying-filling pressure variations was considered, with the following initial values corresponding to a temperature of 200 °C: P = 15 bar, T = 473.16 K, R = 287.14 J/KgK. Two air states are assumed, at the absolute pressures P_1 = 16.01 bar (abs) and P_2 =15.98 bar (abs) respectively. Using equations (1) and (2), the autoclave air mass is calculated to be M_{loss} = 0.197 kg. Relying on equations (1) and (3), the air mass loss in the buffer tank is M_{BT} = 14.388 kg, for a pressure drop of 6 bar, from 25 bar (abs) to 19 bar (abs). The number of resulting operation cycles with buffer tank is calculated with relation (4) to be $N_c \cong 73$. A maximum stationary time t_{st} = 1.8 h is calculated with relation (5). As in the first case, if this time is longer than this value, a new compressor start-up is required, for compensating the air mass lost by the autoclave.

B) Constant pressure, air cooling at 15 bar

At the end of the steady state period, the air inside the autoclave is beginning to cool down, maintaining a constant pressure of 15 bar. The air masses corresponding to the temperatures in Table 4 were calculated, as well as the differences between two successive temperatures.

Table 4. Temperature values at constant pressure of 15 bar

***i* = 1..36**	1	4	8	12	16	20	24	28	32	36
***T* [°C]**	200	186.4	166	146.5	124.2	105.6	83.5	65.5	55.1	43.8

Applying relation (6) to iterations, the total air mass loss to be provided by the buffer tank is $M_{loss} = M_A = M_{A36} - M_{A1} = 72.547$ kg. The mass loss in the buffer tank is $M_{BT} = 14.388$ kg.

Hence, a 3 m^3 buffer tank alone cannot sustain the operation of the autoclave for the temperature drop during the entire cooling cycle, $M_{BT} < M_A$. This means that in this scenario (15 bar, 200 °C), the compressor start-up would be required during the autoclave duty cycle.

C) Air evacuation

At the end of duty cycle, the air is evacuated from the autoclave, simultaneously with its cooling, down 21.1 °C. The values in Table 5 are given for 10 iteration samples.

Table 5. Cooling process at the end of duty cycle

***i* = 1..36**	1	4	8	12	16	20	24	28	32	36
***T* [°C]**	43.8	37.9	33.7	29.8	26.8	24.9	23.7	23.2	21.7	21.1
***P* [bar]**	15	13	10	7	4.4	2.7	1.5	0.8	0.3	0
M_{Ai} [ND]	2.129	9.076	10.829	9.746	7.114	4.332	2.91	2.922	1.458	0.015

The computations with (6) show that the mass differences discharged at successive temperatures are positive see the table 5, which leads to the conclusion that during the period of autoclaved air evacuation and temperature decrease, it is no longer necessary to supply external air mass, the compressor being stopped (see subsequent figure 6 and figure 7).

3.3 Third set of tests (18 bar, 400 °C)

A) Steady state

The same calculation method and relations as described in the first two sets of tests were used for an adiabatic transformation, with the following initial values: $P = 18$ bar (rel), $T = 673.16$ K, $R = 287.14$ J/KgK. Two air states are assumed, at the absolute pressures $P_1 = 19.01$ bar (abs) and $P_2 = 18.98$ bar (abs) respectively. Relying on equations (1) ÷ (3), we calculated the air mass loss for autoclave $M_{loss} = 0.139$ kg, and for buffer tank $M_{BT} = 7.053$ kg, during a pressure drop of 3 bar, from 25 bar (abs) to 19 bar (abs). The number of resulting cycles for autoclave operation with buffer tank is calculated with relation (4) to be $N_c \cong 51$. Maximum stationary time is obtained with relation (5), to be $t_{st} = 1.27$ h. If this time is exceeded, compressor start-up is required.

B) Constant pressure, air cooling at 18 bar

The data regarding the temperature decrease is summarised in Table 6. The air inside the autoclave cools, keeping the pressure constant at 18 bar. The air mass corresponding to each temperature, and the differences between two successive temperatures have been calculated.

Table 6. Temperature values at constant pressure of 18 bar

i = 1..82	1	8	16	24	32	40	48	56	64	72	82
***T* [°C]**	400	368.2	327	286.8	246.8	207.7	167.6	127.6	86.8	59.5	38.8

From the iterations, using relation (3), the total mass loss which has to be compensated by the buffer tank is $M_{loss} = M_A = M_{A82} - M_{A1} = 142.271$ kg. The mass lost by the buffer tank is hence $M_{BT} = 7.053$ Kg. Therefore, a 3 m^3 buffer tank alone cannot sustain the operation of the autoclave for the temperature drop during the entire cooling cycle, $M_{BT} < M_A$. At cycle $N_C = 9$, the air mass loss is 7.188 kg, and at $N_C = 8$, it is 6.092 kg. The conclusion is that the buffer tank alone ensures the proper operation for only 9 cycles of 30 seconds each, namely for 4.5 min. In this scenario (18 bar, 400 °C), the compressor will function continuously during the whole autoclave duty cycle.

C) Air evacuation

At the end of duty cycle, the air is evacuated from the autoclave and cools down to 23.7 °C. The pressure, temperature and successive mass differences are summarized in Table 7.

Table 7. Cooling process at the end of duty cycle

i = 1..37	1	4	8	12	16	20	24	28	32	38
***T* [°C]**	38.8	35.5	33	30.7	28.9	27.8	26.8	26.1	25.5	23.7
***P* [bar]**	18	15.8	13.1	10.5	7.8	5.2	3	1.6	0.5	0
M_{Ai} [ND]	9.772	9.357	8.155	9.774	9.934	10.073	7.24	4.342	2.907	0.025

The computations show that the mass differences discharged at successive temperatures are positive, which means that during air evacuation, the compressor can be stopped.

Hereinafter, we present the graphical representations of the analytical data obtained. Figure 4 shows the variations of the evacuated air mass and the temperature for constant pressures discussed (7 bar, 15 bar and 18 bar respectively). A cycle duration is 1.5 min.

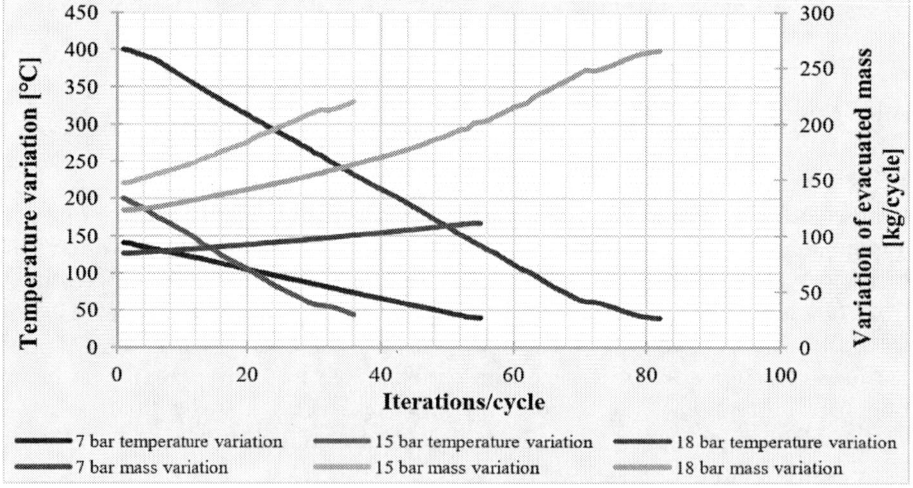

Fig. 4. Evacuated air mass and temperature at constant pressures

Figure 5 shows the computations results for the variation of negative air mass between two successive temperatures at constant pressure, during air cooling process.

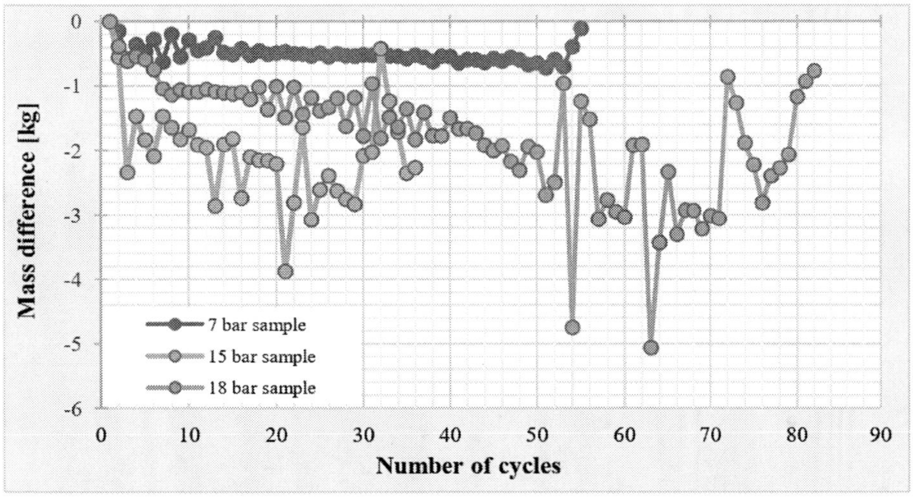

Fig. 5. Negative mass variation between two successive temperatures

Figure 6 shows the variation of the evacuated mass and temperature at the end of autoclave duty cycle.

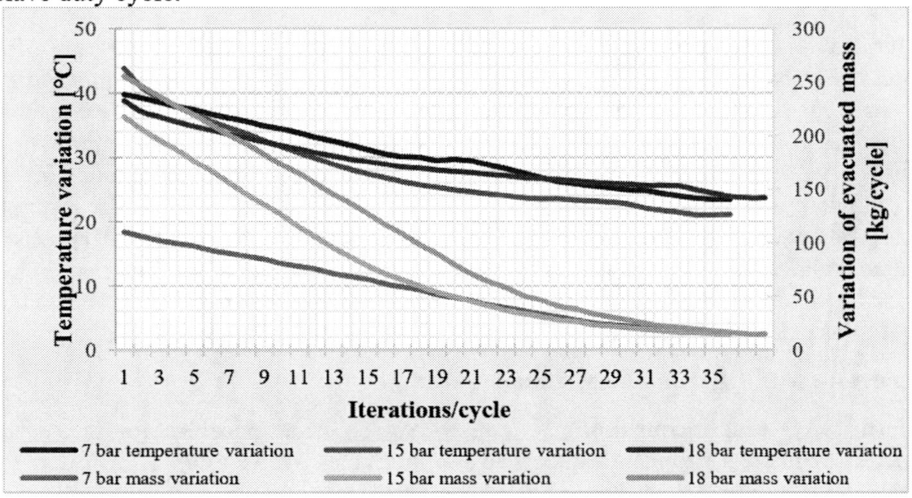

Fig. 6. Variation of the evacuated mass and temperature at the end of duty cycle

Figure 7 shows the computations results for the variation of positive air mass between two successive temperatures at constant pressure, during the air cooling process.

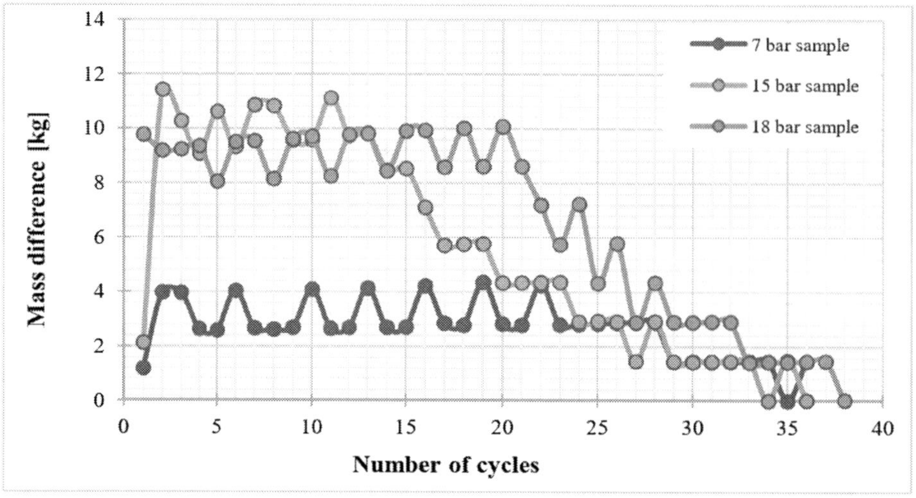

Fig. 7. Positive variation in mass between two successive temperatures at autoclave completion

4 Automation control system

The automation system has the purpose of driving, controlling and monitoring of an air compression equipment up to 24 bar (gauge). The compressed air is stored in a buffer tank and is used for supplying air mass to the autoclave. Depending on this one's consumption, the stored air volume decreases, being compensated by using the electro-compressor. If the pressure is within normal limits, the compressor enters in a standby mode, for saving electrical energy consumption. It only starts automatically when the pressure in the buffer tank falls below the limit value set in the PLC (Programmable Logic Controller) software.

Three architectures were considered for the implementation of the automation system, discussed hereinafter. For versatility, the software implemented in the PLC was realised to allow it to be extended for several other applications and functioning regimes.

A) Structure with PLC and local data acquisition

The local DAQ (data acquisition) is realised via connection cables for measuring the parameters of interest, from transducers to the interfaces of the PLC. The PLC is located remotely, in a control box situated in the control room. This supposes collecting the transducers' signal conductors in a junction box with terminal blocks, and gathering these ones in a joint cable to the electrical control box. Usually, the maximum distance between the transducer and PLC interface is between 10 m and 100 m. In case of exceeding the specified distance, the accuracy of the signal cannot be guaranteed. The advantages of this solution are: compact PLC, all data is in proximity, safe signal transmission. The main disadvantage is presented by the long cable comprising several conductors. The conductors have to be twisted-paired and shielded separately.

B) Structure with PLC and distributed data acquisition

The distributed DAQ is realised using modules mounted near the transducers. Thus, the data in the CPU (central processing unit) is obtained via communication between acquisition system near the measured process and PLC. The advantages are shortening connection length between transducers and DAQ module, and connection between local DAQ and PLC with communication cable with few conductors. However, a thermostatic enclosure is required for the acquisition equipment, and DAQ has to be doubled for safety systems.

C) PLC control with Panel PC and HMI software

Within the purchased Panel PC, we installed "Advanced HMI" which runs under "Microsoft Visual Studio Community" and "Microsoft .NET Framework version 4.8". In this panel PC, LabView can also be installed, for the monitoring and recording application. AdvancedHMI is an open-source software used for creating HMI (Human-Machine Interface) applications that communicate with the PLC or with the device. The result is a very fast and effective application. By using a Modbus TCP library, the visualisation can be realised in network for monitoring the compressor on the intranet or internet (using VPN for increased data safety).

For the practical implementation of the automation system (Figure 8), we chose an architecture with PLC and distributed DAQ, local commands and DAQ, and HMI configured in the Panel PC, within the front panel of the control box.

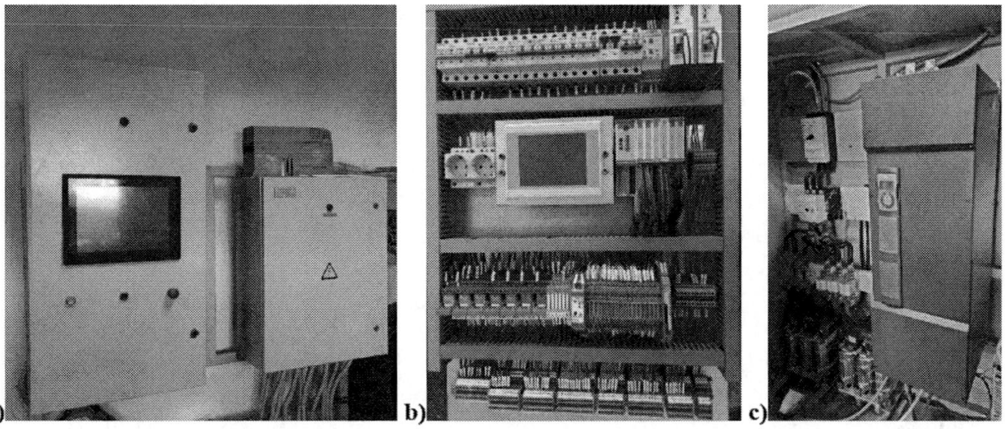

Fig. 8. Automation system: **a)** local automation cabinets; **b)** equipped counter-panel; **c)** drives cabinet with frequency converter, contactors and fuses

The digital signals are acquired in two ways: directly, if produced by their own power supply in contact with internal relay in the control room, or through a galvanic separator if the signals are generated in the field, using the PLC's power supply.

5 Conclusions

The paper presents the computations made for choosing the optimum operation parameters of an autoclave for developing aeronautical composite materials. Prolonging the lifetime of its electro-compressor and driving motor is realised by reducing the number of start-ups and shutdowns, and also avoiding these ones' continuous operation over the entire autoclave duty cycle. The results show that, for reducing the electricity demands and prolonging equipment lifetime, the optimum operation regime occurs for autoclave air pressure of 7 bar, at 140 °C temperature. Due to the fact that the autoclave requires a flow rate of 0.13 kg per second, a higher-pressure compressor (40 bar) and a higher volume buffer tank would be required for ensuring the air flow at 18 bar and 400 °C. However, the need for higher pressure and temperature occurs occasionally, only for certain special composites. In the main working range, the buffer tank suffices along with the present 20 bar compressor.

This research was funded within the project "Development of Research infrastructure for EMerging Advanced composite materials dedicated to innovative STator ogv technologies for aircrafts Engine noise Reduction", code MySMIS 132263, acronym REMASTER, contract no. 12/1.1.3H/28.04.2020.

References

1. V. Karbhari, Fabrication, quality and service-life issues for composites in civil engineering. *Durability of Composites for Civil Structural Applications*. pp. 13-30 (2007). DOI: 10.1533/9781845693565.1.13.
2. M. Condruz, C. Puscasu, L. Voicu, I. Vintila, A. Paraschiv and D. Mirea, "Fiber Reinforced Composite Materials for Proton Radiation Shielding", *Materiale Plastice* **55**, 1 (2018), pp. 5-8. DOI: 10.37358/mp.18.1.4952.
3. G. Seon, Y. Nikishkov, A. Makeev and L. Ferguson, "Towards a digital twin for mitigating void formation during debulking of autoclave composite parts", *Eng Fract Mech.* **225**, p. 106792 (2020). DOI: 10.1016/j.engfracmech.2019.106792.
4. A. Fomin, P. Khavanov, M. Kozhukhova, "Autoclave Composites Containing Ferriferous Industrial By-Products", *Materials Science Forum* **945**, pp. 1030-1035 (2019). DOI: 10.4028/www.scientific.net/msf.945.1030.
5. "Oil Injected Airends | GHH Rand", *ghhrand*, 2021, https://oem-solutions.ingersollrand.com/en-us/products/integrated-compression/oil-injected-airends.
6. "COMOTI - Romanian Research and Development Institute for Gas Turbines", *Comoti.ro*, 2021, http://www.comoti.ro/en/Compresoare-cu-surub.htm.
7. "Practical tips on how not to burnout an electric motor", *EEP - Electrical Engineering Portal*, 2021, https://electrical-engineering-portal.com/burnout-electric-motor.
8. S. Neacşu, *Termotehnică și Mașini Termice* (Bucureşti: Printech, 2009). ISBN: 978-606-521-375-3.
9. C. Eparu, S. Neacşu, A. Neacşa, A. Prundurel, "The comparative thermodynamic analysis of compressor's energetic performance", *Mathematical Modelling of Engineering Problems* **6**, 1 (2019), pp. 152-155. DOI: 10.18280/mmep.060120.
10. N. Stosic, I. Smith, A. Kovačević, *Screw Compressors Mathematical Modeling and Performace Calculation* (Berlin: Springer, 2005). ISBN 978-3-540-26228-2.
11. N. Seshaiah, R. Sahoo, S. Sarangi, "Theoretical and experimental studies on oil injected twin-screw air compressor when compressing different light and heavy gases", *Appl Therm Eng.* **30**, 4 (2009), pp. 327-339. DOI: 10.1016/j.applthermaleng.2009.09.010.

Cost analysis at different energy sources for heating of an air-supported dome for indoor tennis

Ventsislav Gavrilov[1], *Nina* Penkova[2,*], Penka Zlateva[3] and *Boian* Mladenov[2]

[1]University of Chemical Technology and Metallurgy, Department of Physical Education and Sport, 1756 Sofia, Bulgaria
[2]University of Chemical Technology and Metallurgy, Department of Silicate technology, 1756 Sofia, Bulgaria
[3]Technical University of Varna, Department of Thermal Engineering, 9010 Varna, Bulgaria

Abstract. Numerical analyses of the energy supply for heating of air supported structure for indoor tennis in Sofia via district heating, gas and liquid fuels are performed. The computational procedures are implemented through thermal balances of the structures, calibrated on the base of in situ measurements of the indoor temperature and the amount of the fuel, consumed at heating of an air supported balloon for tennis in University of Chemical technology and Metallurgy. The profitability of the different heating variants is estimated using the price of the energy and the systems.

1 Introduction

The wholly air-supported structures, also known as air supported domes (ASD), are lower cost alternatives to the traditional building, suitable for covering areas for indoor sports as soccer fields, large swimming pools, tennis and multisport facilities [1, 2]. They are mobile and semi-permanent as can be taken down and set up every season. The walls are made by fabric membranes (i.e. PVC coated polyester), anchored to the ground around their perimeter, without needs of supporting beams or columns. The dome access points include revolving doors, pedestrian airlocks for barrier free access, and vehicle airlocks for maintenance and lift equipment.

A higher air pressure inside the dome according to the local meteorological pressure supports the entire structure. This gauge pressure has to compensate the gravity force of the membrane, the wind loads and the snow loads during the winter periods. It is a function of the expected wind speed and snow intensity for the region at a given mass of the envelope. The practice shows that a gauge pressure of 250 Pa in the indoor space is enough to counteract the loads, mention above [3]. It is maintained by continuous fresh air supply into the structure in order to compensate the leakages and to ensure the necessary fresh air for the people in the dome [3, 4].

* Corresponding author: nina@uctm.edu

The supply airflows are provided by ventilation and heating/cooling units. Their temperature and volumetric flow are regulated by the automation system, based on a differential pressure gauge and temperature sensors. A recirculation of the return air can be maintained in order to prevent inlet temperatures higher than 40 °C. Temperatures above that value are dangerous for the polyester membranes and the people. However, the ventilation system have to ensure 100 % fresh air, if it is necessary. According to [3 and 4] the blower capacity of the ventilation unit should be designed to allow:

- a compensation of the maximal air leakage at the base, around the doors and through the vents. The experience indicates that the leakage losses at a gauge pressure 250 Pa in the dome can reach 57 m^3h^{-1} per linear meter of base perimeter, 340 m^3h^{-1} per door assembly, and 43891 x A m^3h^{-1} for vents (A is the total vent area in square meters);
- a fresh airflow of at least 60 m^3h^{-1} per person according to [5];
- a satisfactory initial inflation time specified by the design requirements.

The temperature and the relative humidity of the indoor air at sport halls have to be in the ranges of 14 - 16 °C and 30 - 60 % respectively, [6]. The optimal air velocity is 0.3 - 0.4 ms^{-1}. The Bulgarian regulation [5] imposes more précised parameters for covered sport halls in universities and schools: temperatures of 18 °C at the winter, 22 - 27 °C in the summer. The maintaining of these indoor parameters is connected with relatively high energy needs due to the low thermal resistance of the dome envelopes and the relatively high air exchange rate per hour.

The estimation of the heating and the cooling power at the air supported domes depends on solar gains. They are difficult for computation at the complex geometry of the walls, causing complicated solar ray tracing and infrared radiation heat transfer into the structure [7]. Comprehensive investigations of these related processes have not been published so far. The estimation of the energy needs and the relevant cost analysis at air supported domes are actual and open problems.

This paper presents a cost analysis at different energy sources for heating of an air-supported dome for tennis. The assessments are based on a calibrated thermal balance, based on a model for solar energy gains estimation.

2 Assessment of the energy inputs and outputs at heating

2.1 Object of investigation

The ASD for indoor tennis cover 3 tennis courts on the territory of University of Chemical Technology and Metallurgy (UCTM) in Sofia (fig. 1 and 2). Parameters of the structure are given in Table 1. The wall area is obtained based on the geometrical model of the construction (fig. 2). It is overestimated with 5 % for the thermal losses computation to reflect the surface relief, obvious from fig. 1.

Fig. 1. Air supported dome for indoor tennis at UCTM.

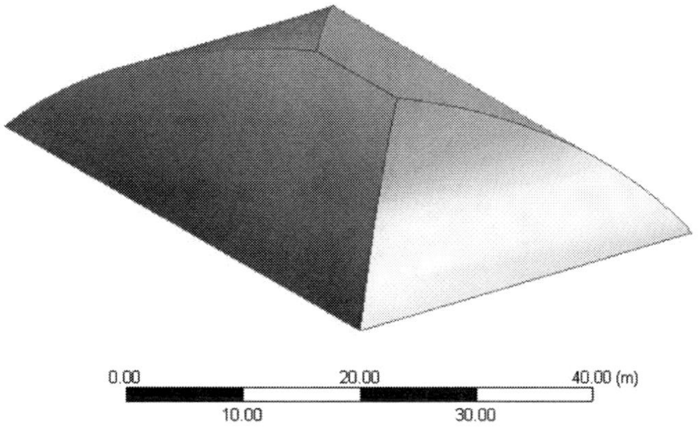

Fig. 2. Geometrical model of ASD.

The air-supported structure is set in the middle of December 2020 with an idea to be taken down in the end of the heating period in Sofia (23 April).

Table 1. Geometrical and thermal parameters of ASD.

Parameter	Dimension	Value
Width	m	37.10
Length	m	48.10
Maximal height	m	10
Floor perimeter	m	170.40
Floor area, A_f	m^2	1784.51
Wall area A_w	m^2	2259.59
Wall area + 5%	m^2	2372.46
Indoor volume V	m^3	10424
Supply fresh air flow $\dot{V}$ (over 90% of the time)	m^3h^{-1}	11400
Air exchange rate per hour (ARCH)	h^{-1}	1.09

The ventilation and heating unit (VHU) is type K-140 D, made by Gustav Nolting GmbH. It is consisted of a main blower (volumetric flow 11400 m^3h^{-1}, gauge pressure 300 Pa, power of 4 kW), a stand-by engine (volumetric flow 11000 m^3h^{-1}), and a burner coupled with a heat exchanger for heating of the supply airflow. The stand-by engine is used as a reserve power. The used fuel is diesel oil. The nominal heating capacity of the unit is 181 kW, the nominal heat output is 163 kW (10% thermal losses). The burner is switched on during the occupation periods of the ASD. The rest of the time the dome is not heated up – only the main blower is working.

There is a possibility for recirculation, but it is not used because of the relatively small capacity of the unit - it support the structure at a lower indoor gauge pressure than recommended. Additional blower with frequency inverter and a maximal power of 15 kW is used to supply additional fresh air in order to keep the gauge pressure at chilly nights and snow loads. The electrical energy consumption of the blowers (minimum 2976 kWh/month if the additional blower is not working) is not taken into account in the present study. It does not influence on the thermal balance for the heating as the blowers are outside the dome and the heat, released by the engines, is dissipating in the outdoor environmental. In addition, this energy consumption is the same at the different variants of the primary thermal energy.

2.2 Thermal balance of air supported domes for tennis

The thermal balance according to the system boundary on figure 3 is:

$$\dot{Q}_{in} = \dot{Q}_w + \dot{Q}_f + \dot{Q}_l - \dot{Q}_s - \dot{Q}_h, \text{ kW} \tag{1}$$

where $\dot{Q}_{in}$ is the input heat flow, $\dot{Q}_w$, $\dot{Q}_f$, $\dot{Q}_l$ are respectively the thermal losses through the wall, floor and the air leakage, $\dot{Q}_s$ represents the solar gains and $\dot{Q}_h$ are additional thermal sources (lighting, people, electrical equipment). The last are relatively small at the covered tennis courts and can be neglected.

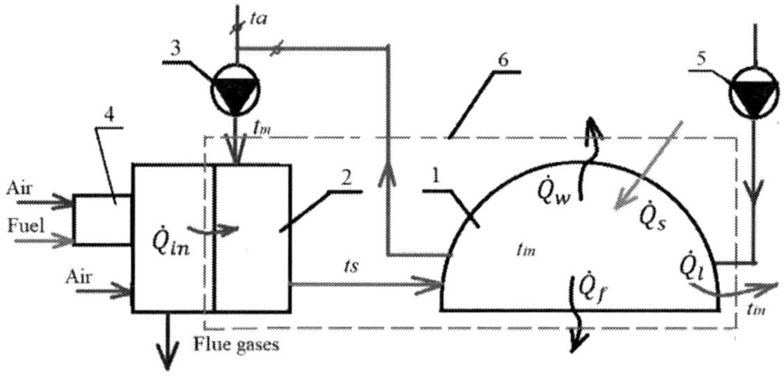

Fig. 2. System, consisted of an air dome for tennis and VHU. 1 - air dome; 2 - heat exchanger of VHU; 3 – main blower of VHU; 4 –burner of VHU; 5- additional brower; 6 - system boundary.

The input heat flow, necessary to heat the supply air with a mass flow $\dot{m}_l$ from the ambient temperature t_a to temperature t_s (fig. 1), is:

$$\dot{Q}_{in} = \eta_g \dot{Q}_g = \dot{m}_l c_{air}(t_s - t_a), \text{ kW} \tag{2}$$

where: η_g = efficiency of the heat generator; $\dot{Q}_g$ = the output heat flow of the generator, kW; c_{air} is the specific heat capacity of the air at constant pressure, $kJkg^{-1}K^{-1}$.
At a case of combustion, $\dot{Q}_g$ is obtained via the lower heating value LHV [$kJkg^{-1}$] and the mass flow $\dot{B}$ [kgs^{-1}] of the fuel:

$$\dot{Q}_g = \dot{B}.LHV, \text{ kW} \tag{3}$$

The mass flow of the fresh supply air is equal to the mass flow of air leakage. Usually it is higher than the necessary fresh air for the people (maximum 4 per court).

$$\dot{m}_l = \frac{ARCH.V\rho}{3600} \tag{4}$$

where ρ=air density at the indoor temperature, kgm^{-3}.

The thermal losses through the balloon walls are computed by the equation:

$$\dot{Q}_w = \frac{1.1 U_w A_w (t_{in} - t_a)}{1000}, \text{ kW} \tag{5}$$

where U_w = thermal transmittance of the walls, $Wm^{-2}K^{-1}$. It is computed by the thermal resistances of the membrane envelope R_m, internal and external thermal boundary layers R_{si} and R_{se} respectively, $[m^2KW^{-1}]$. U_w is increased with 10% in the equation above to reflect the thermal bridges [8].

$$U_w = \frac{1}{R_{si} + R_m + R_{se}}, \text{ Wm}^{-2}\text{K}^{-1} \tag{6}$$

The thermal resistance of one layer membrane is:

$$R_m = \frac{\delta_m}{K_m}, \text{ m}^2\text{KW}^{-1} \tag{7}$$

where δ_m=membrane thickness, m; K_m=thermal conductivity of the material, $Wm^{-1}K^{-1}$.
The heat flow, exchanged with the ground, can be obtained by the equation, used at the floor of greenhouses [9]:

$$\dot{Q}_f = 0.001 A_f \left[k_e (t_{in} - t_a) - \frac{1.2}{h_k} (t_{gw} - t_a) \right], \text{ kW} \tag{8}$$

where h_k = underground water depth, m; t_{gw}=underground water temperature, °C; k_e = equivalent coefficient of thermal transmittance, $Wm^{-2}K^{-1}$, function of floor area A_f and h_k [9].
The thermal losses due to the air leakage are computed according to:

$$\dot{Q}_l = \dot{m}_l c_{air} (t_{in} - t_a), \text{ kW} \tag{9}$$

The heat flow, expressing solar gains $\dot{Q}_s$ in equation (1) is non-zero at the daily time. It is difficult for computation due to the changeable and transient solar incident angle at the curved balloon envelopes (fig. 1 and 2). The polyester membranes are transparent, absorb part of the solar irradiation and have relatively high emissivity in the infrared spectrum [7]. Therefore, the solar heat gains are consisted of the directly transferred solar energy and the absorbed one, which is transferred by convection and radiation in the infrared spectrum to the indoor space. In the present paper the amount of the solar energy, utilized in the dome, is obtained by the monthly energy balance, knowing the fuel consumptions and the other amounts of heats:

$$Q_{in} = Q_w + Q_f + Q_l - Q_s, \text{ kWh/month} \tag{10}$$

The input amounts of heat are computed based on the known monthly fuel consumptions B [kg/month]:

$$Q_{in} = \eta_g B.LHV, \text{ kWh/month} \tag{11}$$

The monthly amounts of heat in the right side of equation (10) are computed with the averaged heat flows $\dot{Q}_w^{av}$, $\dot{Q}_f^{av}$ and $\dot{Q}_l^{av}$:

$$Q_w = \tau_m \dot{Q}_w^{av} = \tau_m \frac{1.1 U_w A_w (t_{in}^{av} - t_a^{av})}{1000}, \text{ kWh/month} \quad (12)$$

$$Q_f = \tau_m \dot{Q}_f^{av} = 0.001 \tau_m A_f \left[k_e (t_{in}^{av} - t_a^{av}) - \frac{1.2}{h_k} (t_{gw} - t_a^{av}) \right], \text{ kWh/month} \quad (13)$$

$$Q_l = \tau_m \dot{Q}_l^{av} = \tau_m \frac{ARCH.V\rho}{3600} c_{air} (t_{in}^{av} - t_a^{av}), \text{ kWh/month} \quad (14)$$

where τ_m=number of the hours of the correspondent month; t_a^{av} = monthly average ambient temperature, °C; t_{in}^{av} = time average indoor temperature, °C, computed by the part *d=n/24* of the operating hours *n* per day of the heating system:

$$t_{in}^{av} = d.t_{in} + (1-d) t_a^{av}, \text{ °C} \quad (15)$$

The monthly solar gains are accepted as part ***k*** of the real sky solar energy $Q_{s,m}$, incident on a horizontal plane with floor area A_f:

$$Q_s = k Q_{s,m}, \text{ kWh/month} \quad (16)$$

$$Q_{s,m} = q_{s,m} A_f D_m, \text{ kWh/month} \quad (17)$$

where $q_{s,m}$= specific amount of solar energy on a horizontal plane, $kWhm^{-2}day^{-1}$; D_m = number of the days per months at the heating period according to [8].
The coefficient k is obtained by equation (10). This approach is applicable as there is enough information about the solar energy, incident on the horizontal planes [10].

2.3 Calibration of the thermal balance

January and February are the first full months of the operation of the ASD. The average ambient temperature for these months, obtained via meteorological database [11] and the established indoor temperature during the time intervals with switched on heating are given in Table 2. The time duration of the heating correspond to the occupation periods that are initially smaller than the predicted one (12h/day) and increase gradually.

Table 2. Indoor and outdoor temperatures.

Month	t_a^{av}	**Heating time**	**Measured** t_{in}	t_{in}^{av}
-	°C	h/day	°C	°C
01. 2021	1.3	6	12	3.98
02. 2021	4.2	8	15	7.80

The daily amounts of the solar energy, incident on a horizontal plane at the location of the tennis courts is obtained by [10] as averaged values for 37 years period (1984-2019). They are used to compute the monthly solar gains in Table 3. The solar gains for January and

February are used in the energy balance (10) to obtain the part k of the utilized solar energy in the dome (Table 4).

Table 3. Incident solar energy on a horizontal plane

Month	January	February	March	April	October	November	December
$q_{s,m}$ kWhm^{-2}day^{-1}	1.706	2.476	3.414	4.263	2.840	1.801	1.448
D_m Days/month	31	28	31	23	17	30	31
$Q_{s,m}$ kWh/month	94397	123694	188886	174949	86148	96393	80075

The monthly amounts of heat, participating in the balance, are computed at the established monthly fuel consumptions and the parameters: LHV=9.7525 kWhl^{-1}; R_{si}=0.1 m^{2}KW^{-1}; R_{se}=0.043 m^{2}KW^{-1}; U_w= 6.51Wm^{-1}K^{-1}; η_g =0.1; k_e=0.5 Wm^{-2}K^{-1}; h_k = 6 m; t_{gw} =10°C. It is obvious that the thermal losses through the dome walls are the higher than the other, but they are compensate by the solar energy.

Table 4. Monthly amounts of heats.

Month	Diesel consumption	Q_{in}	Q_w	Q_f	Q_l	k	Q_s=$kQ_{s,m}$
-	l/month	kWh/month	kWh/month	kWh/month	kWh/month	-	kWh/mont h
01.2021	1000	8777	33830	-534	13561	0.345	32586
02.2021	1000	8777	41123	767	16260	0.346	42786
					Average	0.346	

The average k= 0.346 is the part of the solar energy, incident on a horizontal plane with floor area A_f, utilized in the investigated air supported dome. It is used in the next analyses at different energy sources.

3 Thermal and cost analyses at different energy sources

The 3-month experience show that the present heating capacity of the existing HBU is not enough to maintain the necessary indoor temperature at negative ambient temperatures. The solution of equation (1) in the absence of sunshine at minimal accepted temperature for Sofia (-16 °C) according to [12] gives a design (maximal) heat flow Q_{in}= 666.132 kW. It is several times higher according to the existing one. This fact and the established insufficient power of the main blower of VHU impose increasing of the capacity of the installation or a new one at the same or different primary energy.

Three options of energy sources for heating are analyzed: thermal energy from combustion of diesel oil (present variant), natural gas and from cogeneration power plants in Sofia (district heating). The monthly thermal balances are implemented at the average ambient temperatures in 7-th climatic zone (where Sofia is). A time occupation of the dome 12h/day and an indoor temperature of 15 are accepted. The necessary input amounts of heat are given in Table 6. The negative values of the necessary input energy for April and October show that the solar energy gains are enough to heat the dome to the desired temperature during these months.

Table 5. Technical and financial data at the investigated variants

Type	Thermal losses	LHV	Actual cost with VAT	Additional data
-	%	kWh/unit	BGN/unit	-
Diesel oil	10	11.75kWh/kg	1.82 BGN/l	Density: 0.83 kg/l
Natural gas	8	9.305 kWh/m^3	1.20 BGN/m^3	The m^3 are at Normal physical conditions
Thermal energy from district heating	8	-	0.1 BGN/kWh	The generator of the VHU is a heat exchanger

The primary energy, obtained at the thermal losses at the different generators, fuel consumptions and the subsequent actual financial costs for the period of November to March are systematized in Table 7.

Table 6. Monthly amounts of solar gains and input amounts on heat.

Month	D_m	t_a^{av}	Q_s=$kQ_{s,m}$	Q_{in}
-	day/month	°C	kWh/month	kWh/month
1	31	-0.4	32,619	89,317
2	28	0.2	42,743	63,149
3	31	4.6	65,270	17,865
4	23	10.4	54,408	-26,628
10	17	11.2	26,792	-9,707
11	30	5.1	33,309	43,326
12	31	0.4	27,670	88,663

Table 7. Necessary energy consumptions and costs.

Month	Diesel oil			Natural gas			District heating	
	Necessary amount of heat Q_g	Amount of fuel	Cost with VAT	Necessary amount of heat Q_g	Amount of fuel	Cost with VAT	Necessary amount of heat Q_g	Cost with VAT
-	kWh	l/month	BGN	kWh	m^3/month	-	kWh/month	BGN
1	99,241	10176	18520	97,083	10433	12520	97,083	9563
2	71,760	7358	13392	68,640	7377	8852	68,640	6762
3	20,301	2082	3789	19,419	2087	2504	19,419	1913

11	49,234	5048	9188	47,093	5061	6073	47,093	4639
12	100,754	10331	18803	96,373	10357	12429	96,373	9494
Sum	**341,289**	**34995**	**63691**	**328,608**	**35315**	**42378**	**328,608**	**32371**

The necessary thermal energy at the cases of natural gas and district heating are equal because of the same thermal losses. But their cost is different. The cost of the thermal energy for the heating of the ASD is maximal at the use of a diesel oil. It is approximatively twice in comparison with the heating option at district heating. The financial payback times of the energy costs per heating season at the investigated variants are given in Table 9. The computations are implemented at the assumption for a full occupation of the covered tennis courts 12 h/day and a cost of 30 BGN with VAT per hour per court.

Table 8. Financial payback time of the energy consumption for heating of ASD

Type	Payback time of the thermal energy cost	Payback time of the overall energy cost
-	Days per heating period	Days per heating period
Diesel oil	59	65
Natural gas	39	54
Thermal energy from district heating	30	36

If the cost of the electrical energy consumption of the blowers is taken into account (approximatively 30 MWh/year x 200 BGN/kWh), the payback time will increase (third column of Table 8).

4 Conclusions

The suggested approach for estimation of the solar energy gains at the heating of air-supported domes, based on inverse thermal balance at operating installations, is easy for implementation. It is tested and applied at balance assessment of the thermal energy inputs for heating of a covered tennis courts in Sofia at different energy sources. As results of the case study, it was found that:

- the solar energy cover significant part of the thermal losses at the chilly months and it is enough for the heating of the air dome during April and October;
- the energy supply by the district heating of Sofia is the more cost-effective variant for the heating.

Acknowledgements: This study has been financially supported by the project BG05M2OP001-1.001-0008: National Centre of Mechatronics and Clean Technologies, funded by Operational Programme for Science and Education for Smart Growth at EU Funds.

References

1. https://www.thefarleygroup.com/Products.htm
2. R. Torsing, K. Oosterman, J. Bakker, M. Hinssen, I. Bosveld, T. Huijsmans, Procedia Engineering **147**, 2848-853 (2016)

3. D. A Lutes, CBD-137. *Air-supported structures, Canadian Building Digest*, (1971)
4. Minimum Standards for Air-Supported Structures, *Air Structures Manufacturers and Suppliers Assoc.* (1961)
5. *Bulgarian regulation № РД-02-20-3 for design, implementation and maintenance of public service buildings in the field of education and science, healthcare, culture and arts* (2016)
6. *Bulgarian instructions № 34 for the hygiene of sports facilities and equipment* (2003)
7. H. Suoa, A. Angelottib, A. Zanelli, Energy and Buildings, **109,** 35–46 (2015)
8. *Bulgarian regulation № 7 of 2004 on the energy efficiency in* buildings (2017)
9. S. Stamov et all, *Handbook for heating, ventilation and air conditioning, II part, Tehnika* in Bulgarian (1991)
10. https://power.larc.nasa.gov: NASA Prediction Of Worldwide Energy Resources (POWER)
11. https://www.stringmeteo.com
12. *Bulgarian regulation № 15 of 28 July 2005 on technical rules and standards for design, construction and operation of sites and facilities for heat generation, transmission and distribution* (2016)

The effects of split direct injection on the operation of a tractor diesel engine fueled by biodiesel B20

Andrei Laurentiu Niculae[1,*], *Adnan Kadhim* Rashid[1], and *Radu* Chiriac[1,2]

[1]Faculty of Mechanical Engineering and Mechatronics, University Politehnica of Bucharest, Splaiul Independentei 313, 060042, Romania
[2]EA7341 CMGPCE of Conservatoire National des Arts et Metiers, Paris, France

Abstract. The use of biodiesel-diesel blends is a current solution to some important problems, such as the depletion of oil resources, global warming, and the pollutant emissions of smoke, carbon monoxide, and hydrocarbons of diesel engines. However, the use of this alternative fuel is characterized by a reduction in engine effective power and an increase in brake-specific fuel consumption and nitrogen oxide pollutant emissions. Using the AVL MCC zero-dimensional combustion model of the AVL BOOST simulation program, it was evaluated to what extent split injection strategies can improve the performance and fuel economy of a tractor diesel engine fuelled with biodiesel B20 at maximum brake torque condition considering noise and pollutant emissions limitation. Various pilot – main – post split injection strategies have been studied to establish the optimal injection characteristics in terms of performance and fuel economy. Subsequently, they have been adapted in terms of compliance with current emission standards. In this way, it has been emphasized that the split injection solution is a viable way to improve performance, economy, and pollutant emissions of a tractor diesel engine.

1 Introduction

The depletion of oil resources, global warming, and the harmful effects caused by pollutant emissions as nitrogen oxides (NO_x) and soot on the environment have led researchers to find solutions to overcome these problems.

Biodiesel is a fuel obtained by transesterification processes of animal or vegetable oils. The use of biodiesel – diesel blends allows slowing down the depletion of oil resources. A slight increase of biodiesel fraction in fuel will offer enough time to car manufacturers for adapting the internal combustion engines to the operation with higher percentages of biodiesel, this way the oil resources depletion phenomenon being less and less overwhelming. The use of vegetable oils in the process of biodiesel production and its subsequent use in blend with diesel as fuel for the compression ignition engines is an effective solution to combat the global warming phenomenon. This is because plants

* Corresponding author: andrei.niculae@upb.ro

consume carbon dioxide (CO_2) through the process of photosynthesis during their lifetime, and CO_2 is the main gas in the atmosphere responsible for producing this phenomenon.

The reduction of NO_x and soot emissions can be achieved by adopting split injection strategies associated with the presence of an electronically controlled injection system. These injection strategies involve the existence of pilot, main, and possibly post-injections coupled in different ways as number and timings.

By combustion of the amount of pilot injected fuel, there is a slight increase in pressure and temperature in the cylinder before the main injection. This reduces the ignition delay of the fuel amount delivered in the main injection, the combustion process developing more during the third stage, that of diffusive combustion with a lower peak temperature which causes a decrease in the formation of NO_x emissions [1].

A very effective and widely used solution for reducing soot emissions is the use of post-injections. A post-injection is a short injection that occurs after the main fuel injection [2]. The dwell time represents the duration between two consecutive injections. The amount of fuel injected during the pilot injection, as well as the dwell time, strongly influences the values of pollutant emissions, noise, and specific fuel consumption of a diesel engine [3].

D'Ambrosio and Ferrari [4] conducted an experimental study that aimed to evaluate the potential of triple injection (pilot – pilot – main) compared to the double injection (pilot - main). For medium loads and speeds of the Diesel engine, the use of triple injection reduces NO_x emissions, but the engine noise does not improve, and soot emissions increase.

Poorghasemi et al [5] have numerically studied the possibility of simultaneously reducing NO_x and soot emissions in a diesel engine by using split injection strategies. The results showed that the adoption of a post-injection with an adequate dwell time allows a significant and simultaneous reduction of these emissions.

Hiren et al [6] conducted a numerical and experimental study for the investigation of the influence of pilot injection on the combustion process and the formation of the pollutant emissions, keeping constant dwell time and main injection timing. A large amount of fuel introduced during the pilot injection reduces smoke and increases NO, HC emissions, and brake-specific fuel consumption. The injection strategy with only 15% pilot quantity is the optimal among all the considered pilot injection strategies (15%, 30%, and 45% pilot quantity) because it causes a reduction in brake-specific fuel consumption, NO, and HC emissions, while smoke emissions acceptably increase.

The authors of the paper [7] developed a single-zone phenomenological combustion simulation model for a diesel engine that uses conventional single injection or split injection strategies. Subsequently, they assessed the extent to which NO_x emissions can be reduced by using EGR or by adopting split injection strategies. They conclude that both methods can significantly reduce NO_x emissions, but with penalties in terms of engine performance when using EGR.

The researchers from reference [8] numerically and experimentally studied the effects of using various split injection strategies with different dwell times on the combustion characteristics and pollutant emissions. An increase of the dwell time determines the decrease of peak heat release rate, thus amplifying the diffusive combustion which leads to NO_x reductions and soot, HC, and CO increases.

The authors of work [9] investigated numerically and experimentally the effects of pilot injection on flame temperature and soot distribution for an optically internal combustion engine fuelled by pure diesel and biodiesel. The pilot injection improves combustion performance and increases fuel economy if it is positioned near the main injection. The combustion initiated by the pilot injection contributes to the development of the main fuel jet, resulting in a reduction of inhomogeneities in the combustion chamber. Thus, the fuel-rich local zones are reduced and the general temperature of the flame decreases, this having an improved distribution along the combustion chamber. These changes affect the formation of NO_x and soot emissions.

In paper [10] was highlighted the impact of injection rate shape on the performance, efficiency, and pollutant emissions of a tractor diesel engine using the AVL MCC zero-dimensional combustion model of the AVL BOOST software. The actual paper represents a numerical study performed with the AVL MCC zero-dimensional combustion model of the AVL BOOST software on the possibilities of improving performance and reducing NO_x and soot pollutant emissions of a tractor diesel engine fuelled with B20 (20% rapeseed biodiesel and 80% diesel) at full load and maximum torque speed condition by using split injection strategies. The model was calibrated against experimental data obtained for pure diesel and B20 fuelling with a single injection strategy.

2 Materials and Methods

The test bench is equipped with UTB 2404055 tractor Diesel engine. Its main specifications are presented in Table 1.

Table 1. Engine specifications

Description	Specification
Engine Type	UTB 2404055
Bore x Stroke	102 mm x 115 mm
Compression Ratio	17.5
Number of Cylinders	4
Displacement (L)	3.76
Piston Type	Omega
Rated Power	48 kW / 2400 rpm
Rated Torque	224 Nm / 1400 rpm
Air System	Naturally-aspired
Injection System	Delphi
Number of Nozzle Holes	5
Nozzle Hole Diameter	0.24 mm

Figure 1 shows the configuration of the test bench and the AVL BOOST symbolic model of the engine. The instant torque and speed are measured and controlled using the AVL Alpha 160 eddy current dynamometer. The fuel consumption is recorded using the AVL Dynamic Fuel Meter 733 S coupled with AVL Fuel Temperature Controller 753 C. The volumetric air consumption is measured using the airflow meter SCHLUMBERGER Fluxi 2000/TZ. The carbon dioxide (CO_2), carbon monoxide (CO), nitrogen oxides (NO_x), and total hydrocarbons (THC) pollutant emissions are evaluated using HORIBA Mexa 7170 D gas analyzer, while the smoke level in the exhaust gases is determined by the AVL Smoke Meter 415 S. The pressure traces are recorded for two engine cylinders (1 and 3) with a data acquisition system consisting of AVL GM 12 D quartz pressure transducers, AVL 3066A02 charge amplifiers and AVL 365 C crank angle encoder. All the equipment is connected to the data acquisition module AVL Indiset 620 provided with the AVL Indicom 1.6 software.

Numerous parameters were recorded, including engine brake power, fuel consumption, pressure trace, and pollutant emissions values when fuelling the engine with pure diesel (D100) and B20 for maximum brake torque test condition (full load, 1400 rpm).

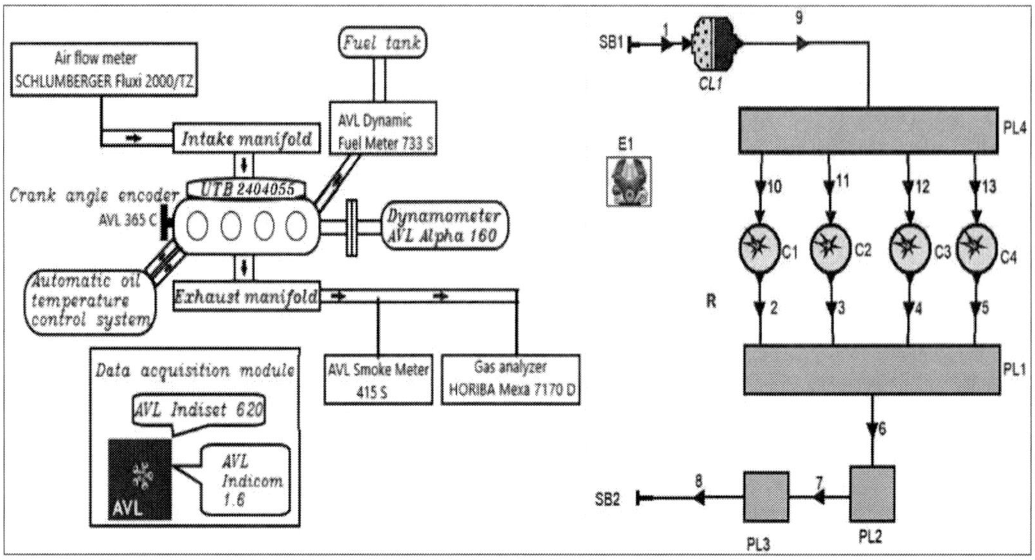

Fig. 1. Schematic diagram of the experimental set-up and the AVL BOOST symbolic model of the engine

Fig. 2 shows the values of the normalized rate of fuel injection (relative to peak velocity) depending on the crank angle for various injection strategies. Based on experimental data, the calibration of the simulation model was performed considering the single injection strategy with 16 °CA total duration. This injection starts at -5 °CA and ends at 11 °CA. Subsequently, to improve the engine performance without increasing the pollutant emissions level for B20 fuel, split injection strategies with 16 °CA and 10 °CA total duration were adopted. These strategies consist of: the pilot injection that delivers 10% of the total mass fuel, the main injection that delivers 80%, and the post injection that delivers 10%. The same value for dwell was considered between the pilot and main injections or between main and post. For split injection strategy with the same total duration like single injection (16 °CA), the start of injection (SOI) and dwell time were considered -11 °CA and 2 °CA (relative to Top Dead Center – TDC), while for the other one were set to -2 °CA and 4 °CA. These values have been chosen considering maximum engine effective power and pollutant emissions level closer to the experimental values obtained for B20 fuelling.

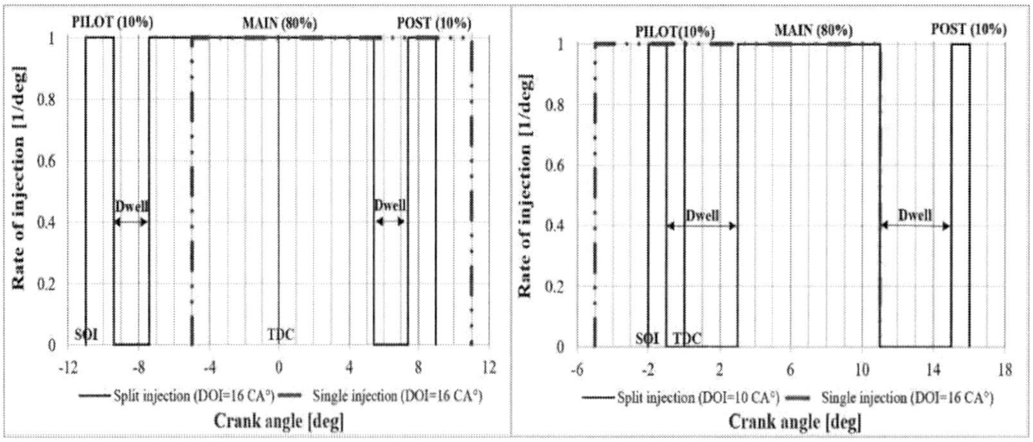

Fig. 2. The normalized split and single rate of injections studied with AVL BOOST software

3 Simulation Results

Table 2 shows the values obtained by experimental testing and by numerical simulation for brake power (BP), brake-specific fuel consumption (BSFC), peak pressure (p_{max}), peak pressure rise ($\dot{p}_{max}$), nitrogen oxides (NO_x), and soot when the engine was operated at maximum brake torque conditions (full load, 1400 rpm).

Table 2. The values of parameters considered for maximum brake torque test condition

	BP [kW]		BSFC [g/kWh]		p_{max} [bar]		$\dot{p}_{max}$ [bar/deg]		NO_X [ppm]		Soot [g/kWh]	
	D100	B20	D100	B20	D100	B20	D100	B20	D100	B20	D100	B20
Experimental	32.87	31.73	236.4	255.0	66.53	66.07	8.18	7.88	1082	1017	2.166	2.051
Single injection (16 CA°)	32.22	32.21	241.2	251.2	66.15	65.48	4.98	5.10	1083	1018	2.166	1.994
Split injection (16 CA°)	32.32	32.32	240.4	250.4	74.76	73.66	4.70	4.70	1422	1257	1.841	1.939
Split injection (10 CA°)	32.84	32.84	236.6	246.4	62.21	61.70	3.26	3.13	1094	1008	0.823	1.507

Table 3 displays the percentage differences of the AVL MCC combustion model results for the above-mentioned parameters when using single or split injection strategies relative to experimentally obtained values. Regarding the calibration step of the single injection strategy, it may be emphasised that the relative deviations of engine performance (ε_{BP} and ε_{BSFC}), pollutant emissions (ε_{NOx} and ε_{Soot}), and peak pressure ($\varepsilon_{p_{max}}$) considered relevant for simulation rest under 3% for both fuels. For peak pressure rise there are considerable errors ($\varepsilon_{\dot{p}_{max}}$), the calibration of this parameter requiring future improvements.

Table 3. Percentage differences relative to experimental data

	ε_{BP} [%]		ε_{BSFC} [%]		$\varepsilon_{p_{max}}$ [%]		$\varepsilon_{\dot{p}_{max}}$ [%]		ε_{NOx} [%]		ε_{Soot} [%]	
	D100	B20	D100	B20	D100	B20	D100	B20	D100	B20	D100	B20
Single injection (16 CA°)	-1.98	1.51	2.03	-1.49	-0.57	-0.89	-39.12	-35.28	0.09	0.10	0.00	-2.78
Split injection (16 CA°)	-1.67	1.86	1.69	-1.80	12.37	11.49	-42.54	-40.36	31.42	23.60	-15.00	-5.46
Split injection (10 CA°)	-0.09	3.50	0.08	-3.37	-6.49	-6.61	-60.15	-60.28	1.11	-0.88	-62.00	-26.52

Figure 3 illustrates the values of brake power (BP) and brake-specific fuel consumption (BSFC) obtained experimentally and by simulations considering different injection strategies. By adopting a split injection with 16 °CA total duration, the engine BP increased slightly from 32.21 kW (single injection) to 32.32 kW for both fuels. This increase is insufficient, so a split injection with a shorter total duration was adopted (10 °CA) because it allows obtaining a BP closer to that experimentally determined for pure diesel - D100 (32.84 kW for both fuels compared to 32.87 kW). In terms of BSFC, this injection strategy offers minimum values for both fuels (236.6 g/kWh for D100 and 246.4 g/kWh for B20 compared to 241.2 g/kWh for D100 and 251.2 g/kWh for B20 when using the single injection strategy).

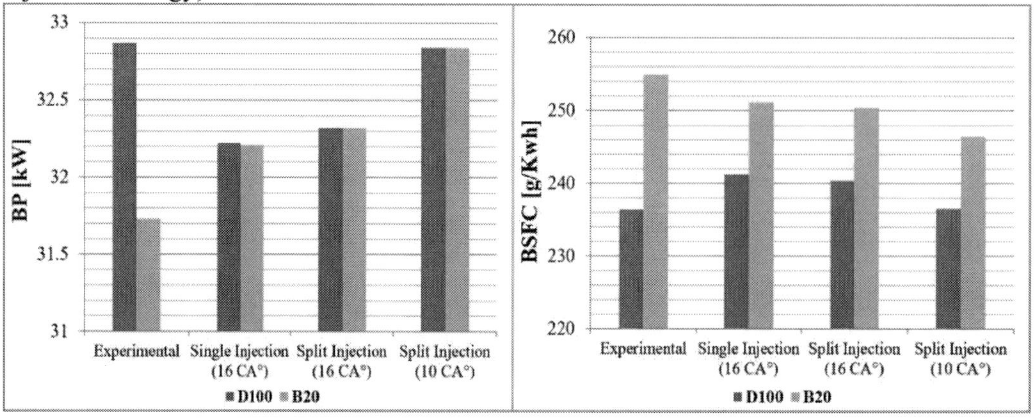

Fig. 3. Brake power (BP) and brake-specific fuel consumption (BSFC) variations by different injection strategies

Figure 4 displays the values of peak pressure (p_{max}) and peak pressure rise ($\dot{p}_{max}$) obtained experimentally and by simulations considering different injection strategies. The use of the split injection with a duration of 16 °CA determines a peak pressure increase of 12.4% for D100 and 11.5% for B20 and a peak pressure rise decrease of 42.5% for D100 and 40.4% for B20 compared to the experimental values. The split injection with 10 °CA total duration determines a simultaneous reduction of the peak pressure (6.5% for D100 and 6.6% for B20) and peak pressure rise (60.2% for D100 and 60.3% for B20), which determines lower stresses on the engine components and lesser operation noise.

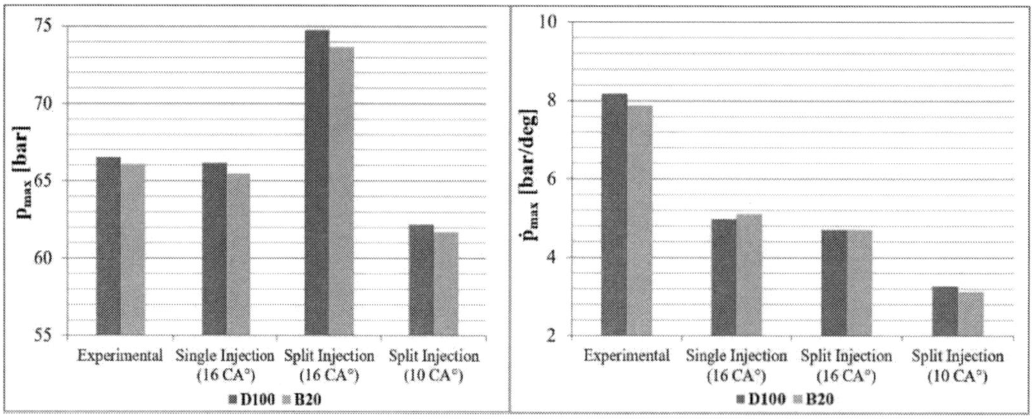

Fig. 4. Peak pressure (p_{max}) and peak pressure rise ($\dot{p}_{max}$) variations by different injection strategies

Figure 5 presents the values of nitrogen oxides (NO_x) and soot pollutant emissions obtained experimentally and by simulations considering different injection strategies. Using the split injection with a longer duration, NO_x emissions significantly increase by 31.4% for D100 and 23.6% for B20, while soot emissions decrease by 15% for D100 and 5.5% for B20 compared to the experimental values. The split injection with shorter duration allows obtaining a simultaneous reduction of these emissions for B20 (0.9% less NO_x and 26.5%

less soot) and slight growth in NO_x with a sharp decrease in soot for D100 (1.1 % more NO_x and 62% less soot).

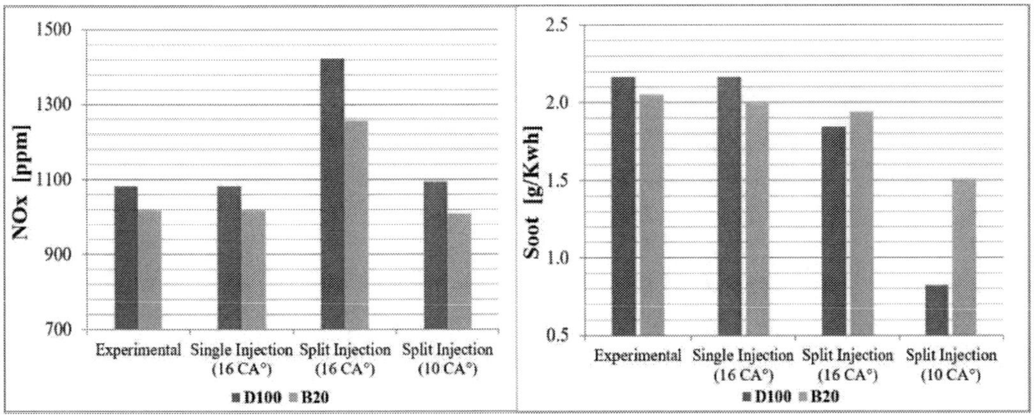

Fig. 5. Nitrogen oxides (NO_x) and soot emissions variations by different injection strategies

4 Conclusions

The main conclusions of this work can be summarized as follow:

- The use of a one-dimensional simulation model developed under AVL BOOST software allows the evaluation of the performance, efficiency, and pollutant emissions of a diesel engine when using various injection strategies with minimal computation resource consumption;
- The substitution of the original diesel fuel used in a compression ignition engine by biodiesel containing elevated fractions of biofuel is generally associated with a derating of the original engine power when no additional adjustment is performed; higher biofuel fractions are associated with lower engine performances and a similar effect was registered also in this case when for B20 use the decrease in brake power was 3.5%;
- A method to recover the engine power is to change the injection characteristic passing from the single injection to the split injection with multiple pulses (in this work with three pulses: pilot – main - post);
- The simulations performed showed that it is possible to find a suitable split injection characteristic for B20 and engine operating condition (full load and 1400 rpm speed) where power is entirely recovered but with lower peak pressure (61.7 bar relative to 66.53 bar), lower peak pressure rise (3.13 bar/deg relative to 8.18 bar/deg), lower NO_x (1008 ppm relative to 1082 ppm) and lower soot emission (1.507 g/kWh relative to 2.166 g/kWh);
- A slight penalty of engine economy in terms of BSFC outcomes, which increases by 4% from 236.4 to 246.4 g/kWh.

The model needs future refinements in the calibration stage but it emphasized the effectiveness of multiple pulses split injection technology to control the combustion process in diesel engines for lowering emissions while maintaining performances.

Acknowledgments

The authors of this paper acknowledge the AVL-AST team from AVL List GmbH for the special support offered under the University Partnership Program on performing the simulation activity.

References

1. S. d'Ambrosio, A. Ferrari, A. Mancarella, A. Mittica, Int. J. Automot. Technol., **Vol. 21**, No. 1, pp. 197-214 (2020)
2. J. O'Connor, M. Musculus, SAE Int. J. Engines, **Vol. 6**, Issue 1, pp. 400-421 (2013)
3. S. d'Ambrosio, A. Ferrari, Energy Convers. Manag., **Vol. 110**, pp. 457-468 (2016)
4. S. d'Ambrosio, A. Ferrari, Appl. Energy, **Vol. 155**, pp. 918-932 (2015)
5. K. Poorghasemi, F. Ommi, H. Yaghmaei, A. Namaki, J. Mech. Sci. Technol, **Vol. 26**, pp. 269-281 (2012)
6. D. Hiren, S. Bharatkumar, P. Brijesh, Clean Technol Environ Policy, **Vol. 21**, pp. 905-921 (2019)
7. R. Sindhu, G. Amba Prasad Rao, K. Mahdu Murthy, J Braz. Soc. Mech. Sci. Eng., **Vol. 38**, pp. 1865-1876 (2016)
8. P. Brijesh, S. Sreedhara, Clean Technol Environ Policy, **Vol. 18**, pp. 2325-2334 (2016)
9. J. Jeon, S. Park, Appl. Energy, Vol. 160, pp. 581-591 (2015)
10. A. L. Niculae, R. Chiriac, A. Racovitza, Manuscript in progress, Department of Thermotechnics, Engines, Heat and Refrigeration Equipment, University Politehnica of Bucharest, Romania

Improving energy performance of a Cement Manufacturing factory by using Waste Heat Recovery Systems, Estimated vs. Actual achievements

Mircea Scripcariu[1], *Cristian* Gheorghiu[2], *Miruna* Gheorghiu[3] and *Alexandra Gabriela* Dobrica[4*]

[1]University Politehnica of Bucharest, Energy Production and Use Department, 313 Splaiul Independentei, Bucharest, Romania
[2]University Politehnica of Bucharest, Energy Production and Use Department, 313 Splaiul Independentei, Bucharest, Romania
[3]ELSACO ESCO L.L.C.,500 Mihai Bravu, Bucharest, Romania
[4]University Politehnica of Bucharest, Energy Production and Use Department, 313 Splaiul Independentei, Bucharest, Romania

Abstract. Waste Heat Recovery (WHR) Systems are spreading more and more in cement factories and are essential in achieving the energy performance required by the European Directives, legislation, and standards. Using WHR Systems may assure an important percentage of the energy required by the manufacturing process, with no additional fuel and no additional greenhouse gas emissions. Using the waste heat as a power generation source, increases the energy efficiency of the process and decreases the thermal energy losses. As long as the kiln is functional, so is the WHR powerplant, generating the energy in an efficient manufacturing process with low operational costs and increased reliability. This paper aims at evaluating the actual technic and economic performance of a WHR System compared to the estimated performance determined in the feasibility study which was done prior to the investment in order to prove the viability of the technology in the cement manufacturing industrial sector. The paper proves that the WHR proved to be financially inefficient if the feasibility study input data was considered and correlated with the actual technical performance but lead to extremely attractive financial indicators when considering actual, updated capital expenditures and operational expenditures and technical performance.

1 Introduction

Currently there are numerous technical solutions available for a cement factory which can reduce the carbon footprint and improve the energy performance of the process. Using the best available technology may increase the energy performance of the process but, a high amount of energy is still demanded by the process. In 2020, the largest cement

*Corresponding author: alexandra.dobrica3@gmail.com

manufacturer worldwide reported a total energy consumption of 528 million of GJ. Fig. ***1*** shows the amount of thermal energy and electricity demand [1]. The purpose of the Organic Rankine Cycle System is to reduce the costs of energy extracted from the national electric power grid by having a reliable power source on-site. The first 1.5 MW ORC system was installed in Lengfurt in 1999 [2]. Since then, the technology had been used in many cement factories from 49 countries having 402 plants operating in 2021.
The cement manufacturers focused their attention on ORC technology because it has proven its energy performance and it has reduced the high amount of greenhouse gas emissions generated by this industry.

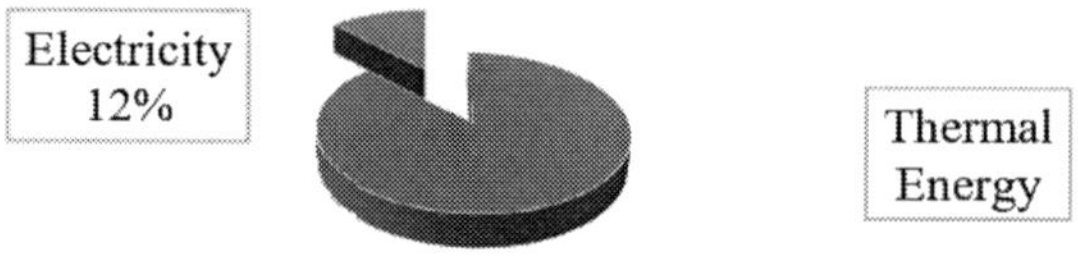

Fig. 1. Thermal energy and electricity use reported by the biggest cement manufacturer.

The cement industry has a significant environmental footprint due to the high amount of thermal energy required by the process, mainly coming from the burning of traditional fuels. In the last 25 years, the CO_2 emissions produced by the cement industry increased drastically, from 27 Gt in 2006 to 32 Gt in 2020. Without a sustainable strategy to reduce the greenhouse gas emissions, the quantity of CO_2 could achieve over 36 Gt after 2022, as shown in the Figure 2 [3]. 50% of CO_2 emissions are produced during the calcination process, in which the limestone is transformed into lime as an outcome of a chemical reaction [3]. About 40% of the total quantity of CO_2 are produced as a result of the burning processes required to achieve the 1,450 °C temperature of clinkerization. The rest of 10% of the greenhouse gas emissions are the result of electricity needed to run the plant [3]. To reach the target proposed by the Green Deal, a carbon neutral Europe by 2050 and reduce the CO_2 emissions in the cement sector with 30% by 2030, the industry has the opportunity to invest in kiln upgrades and waste heat recovery using ORC systems [4].

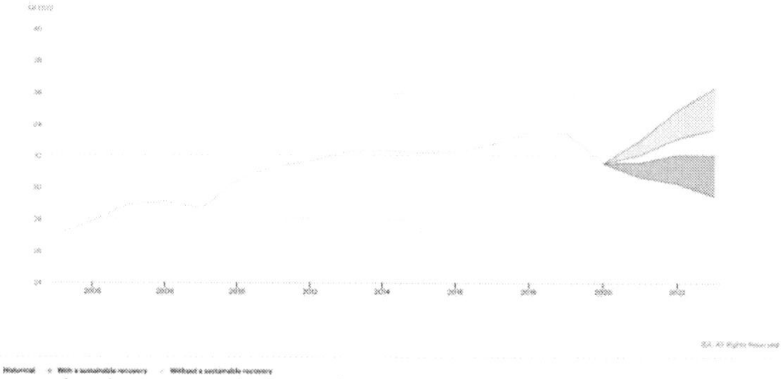

Fig. 2. Total energy related to CO_2 emissions with and without a sustainable recovery, 2005-2023 [3].

2 From Waste Heat to Power

In 2011, the first ORC system in Romania was operational. The system was installed on a 4,300 t/day rotary kiln. The ORC system has an installed power of 4.1 MW. In the first phase of the project, many key factors were taken into account, such as: temperatures from the process, waste heat energy potential, the working fluid, time allocated for the kiln inspection etc. [6]. Temperature is a high importance factor for the system because it has an influence on the power plant design, materials, heat exchangers for the system [7], and on the energy potential of the waste heat source. The main advantage of the system is its capacity to operate at relatively low temperatures of steam in comparison with the conventional steam cycle. The temperature at the preheater outlet is 360 °C, while at the clinker cooler the temperature is lower, the average temperature is around 250 °C, as shown in Figure 3.

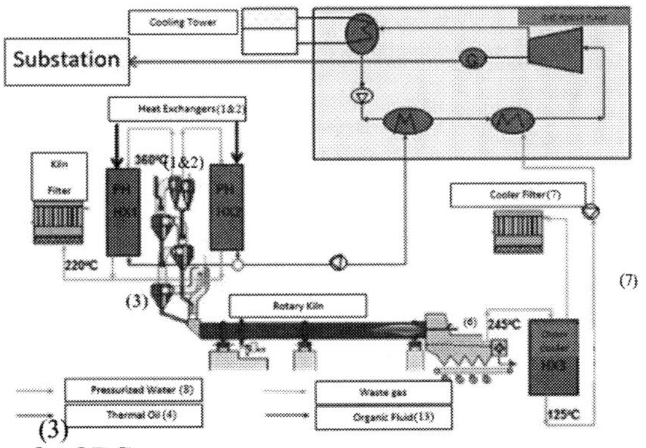

Fig. 3. Waste Heat Sources for ORC system (6)

The thermodynamic circuit is described in Figure 4. The preheater boilers (1 and 2) of the ORC system are extracting the waste gas after the superior stage cyclones (3), returning it before the fans after the gas has transferred the heat to the heat to the thermal oil (4) used as working fluid, while the clinker cooler boiler (5) is installed between clinker cooler (6) and the filter (7). The heat from the main heat exchanger from the kiln is extracted with two horizontal cross flow boilers (1 and 2), using thermal oil as heat transfer fluid and one vertical gas flow boiler with dust settling chamber, using pressurized water as heat transfer fluid (8). By using pressurized water, the boiler's "capacity" decreased with 20%. This combined solution of having thermal oil in the preheater boilers and hot pressurized water in the clinker cooler boiler as working fluids is a new technical solution for the ORC systems in the cement industry [7]. The ORC unit uses the working fluids to pre-heat and vaporize the organic working fluid in the evaporator. The organic fluid vapor powers the turbine (9), which is directly coupled to the electric generator (10), generating electricity. The exhaust vapor flows through the regenerator (11) where it preheats the organic liquid (12). The vapor is then condensed in the condenser (13) which has a water-cooling system (14). In the last transformation of the organic cycle, the organic fluid is pumped in the regenerator (15) and then to the evaporator (16).

During the estimation phase of the ORC project, some assumptions were made in order to investigate the performance of the system. To estimate the available heat for the plant, a production rate of 3,800 t/day and an average heat consumption during normal operation of 3,396 MJ/$t_{clinker}$. Because the power output of the ORC system depends on the preheater exit temperature, a couple of upgrades had to be implemented to the rotary kiln: Chlorine bypass and solid recovered fuel feeding to the calciner. The initial results, along with the implemented upgrades, had shown a waste gas flow of 1.6 $Nm^3/kg_{clinker}$ at the preheater, representing a waste heat source for the future ORC system. The maximum thermal energy output at the preheater was estimated at 36.5 MW, this output is obtained in ideal conditions during one year operation time at a 20 °C outside temperature as reference, an average temperature 366 °C at the preheater outlet and a gas flow of 253,328 Nm^3/h at the preheater.

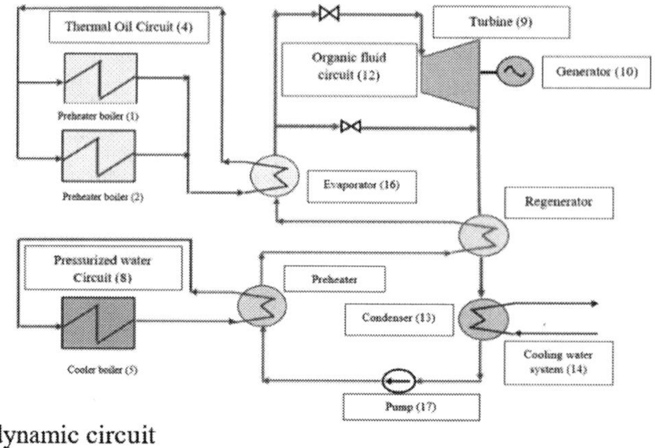

Fig. 4. ORC thermodynamic circuit

In the non-ideal case, during operation, the monthly thermal energy output it is expected not to exceed more than 21.6 MW because of the seasonal changes of the outside temperature. The same assumptions were made to calculate the maximum thermal energy output from the clinker cooler, resulting a cooler waste gas of 0.8 $Nm^3/kg_{clinker}$ at an average air temperature of 250 °C inside the cooler. The results had shown that waste air flow is 126,666 Nm^3/h generating a peak power of 10.7 MW. As in the preheater case, the waste thermal energy output in non-ideal conditions at the clinker cooler will not exceed 4.1 MW. The initial costs of the project were estimated in the feasibility study, the results are shown in the table 2. The operational costs and the maintenance ones were estimated at 100 kEUR/yr. In the financial evaluation of the project, the expected lifetime was considered 20 years and the availability of the plant at 97%, resulting the financial indicators from table 3.

Table 1. Natural Gas and Electricity consumption for each Brick Factory

Technical Characteristic	Measuring Unit	Value
Available thermal energy	MW	21.3
Gross electricity production	MW	3.8
Net electricity production	MW	3.5
Gross efficiency	%	18 [8]
Net electricity production at 95% availability of the plant	GWh/year	27.0

Table 2. Initial cost of the ORC system

Initial costs for the power plant				
Project stage		Currency	Amount	Percent
1	Land preparation and infrastructure	kEUR	100	1%
2	Engineering, supervision, commissioning	kEUR	500	4%
3	Equipment	kEUR	11,500	82%
4	Civil part	kEUR	600	4%
5	Contingencies	kEUR	1,300	9%
6	**Total Project Investment Cost**	**kEUR**	**14,000**	**100%**

Table 3. Estimated financial indicators.

Indicator	Value
NPV	4.7 million EUR
IRR	16.2 %
Dynamic Payback	12.5 years

3 Actual performance of the ORC system

After six years in operation, the Organic Rankine Cycle system has achieved its expected performance. The plant is still operational, covering 15% [9] of the electrical energy demanded by the cement factory. Because the system is designed to operate at low temperatures, the ORC is a much more suitable solution for cement industry in comparison with the conventional steam cycle [10]. After years of monitoring the main parameters of the plant, discrepancies were observed in comparison with the performances described in the initial study. In the estimations phase of the project, the availability of the plant was considered 95%, meaning 7,500 h/yr. The actual availability time of the plant had an average value of 7,200 h/yr, with 300 h/yr less than the expected time. It is very important to specify that the lower availability time is the came as a consequence of the direct interdependence between the ORC plant and the rotary kiln operation, and it should not be considered as a break-down time due to ORC system malfunction. Having an overview on the actual energy performance of the plant, the following were observed:

- The specific waste air flow from the clinker cooler has the actual measured value of 0.942 $Nm^3/kg_{clinker}$ at a temperature of 240 ˚C, representing an increased flow and a decreased temperature in comparison with the one given in the estimation phase.
- The specific waste gas flow from the preheater has a measured value of 1.090 $Nm^3/kg_{clinker}$, less than the 1.8 $Nm^3/kg_{clinker}$ expected in the estimation phase. The outlet temperature of the preheater is 397 ˚C instead of 360 ˚C which represents the calculated temperature of the waste gas coming from the preheater.
- The amount of thermal energy which can be recovered from the process is approximatively 18.5 MW, representing half of the thermal energy calculated in the ideal conditions of operation.
- The gross electrical power output has the actual value of 3.1 MW, resulting a a gross efficiency of 17%, representing a close performance in comparison with the estimated value of 18%.

Table 4. Actual energy performances of the ORC plant

Technical Characteristic	Measuring Unit	Value
Available thermal energy	MW	18.5
Gross electricity production	MW	3.1
Net electricity production	MW	2.9
Gross efficiency	%	17
Net electricity energy production at 95% availability of the plant	GWh/year	24.7

Considering the lower energy performance compared to those estimated in the feasibility study, the actual production of the power was lower than the electricity production forecast calculated in the feasibility study. The comparison between the actual production and the forecast is presented in figure 5. Regarding the economic aspects, the real maintenance cost did not exceed the estimated cost of 100 kEUR/yr, even though many revisions and inspections were done using external resources which were more expensive than expected in the estimation phase. The annually maintenance using internal and external resources cost are presented in table 5. In order to compare the forecasted financial viability of the project with the actual financial viability, an Actual Scenario analysis was done. The main criterions used in the technic and economic analysis were the Net Present Value – NPV (1), the Internal Rate of Return – IRR (2), the Simple Payback Period – SPP (3), determined by considering a variable annual net income and the Benefit – Cost Analysis – BCA (4).

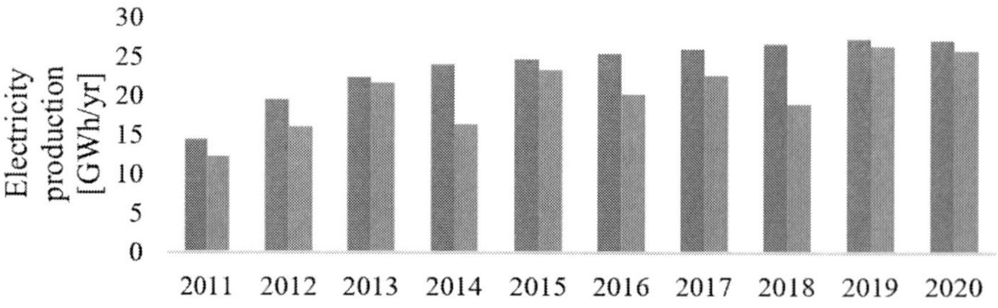

Fig. 5. Estimated (orange) vs. Actual (gray) production of the ORC

Table 5. Actual maintenance cost of the ORC plant

Year	Cost	I. Maintenance activities with own resources:	II. Maintenance activities using external resources:	Inspections and revisions	Pump fixing	Other repairing activities	TOTAL (I+II):
2013	[kEUR]	8.88	7.55	2.79	3.26	1.50	16.43
2014	[kEUR]	8.51	6.06	1.58	-	4.48	14.57
2015	[kEUR]	9.37	51.35	50.73	-	0.62	60.73

$$NPV = \sum_{t=1}^{tst} \frac{I_t - C_t}{(1+a)^t} - IC \; [EUR] \tag{1}$$

where t_{tst} is the analysis time-frame, in years, I_t is the yearly income in the tth year, in EUR/year, C_t are the yearly expenditures in the tth year, in EUR/year, a is the discount rate – 11.38%/year for this end-user and IC is the investment cost, in EUR.

$$NPV = \sum_{t=1}^{tst} \frac{I_t - C_t}{(1+IRR)^t} = 0 \; [EUR] \tag{2}$$

where the CAPEX can be included in the yearly expenditures as a depreciation cost.

$$SPP = \frac{IC}{\frac{\sum_{i=1}^{t} I_i - C_i}{t}} [years] \tag{3}$$

$$BCA = \frac{IC}{NPV} [-] \tag{4}$$

An average escalation rate for electricity prices of 5%/year was also considered, as determined in [11]. The technic economic analysis results will be presented in Table 6.
Because, as it can be observed in Table 6, the financial indicators of the project in the Actual Scenario turns out to be unfeasible for the company, a Present Day Scenario was also analyzed in order to quantify the differences generated by the maturity of the ORC technology. Because the project was implemented in 2011 (only 21 years after the first ORC was deployed in the cement industry sector), the technology was still new, and the benefit / cost forecast stage of the feasibility study was not accurate. The CAPEX of the

project was also considerably higher than it would be today (an average specific cost decrease of 914.63 EUR/kW from 3,414 EUR/kW in 2011 to 2,500 EUR/kW in 2021).

Table 6. Comparative Financial Analysis

Financial Indicator	Feasibility Study	"Actual" scenario	"Present day" scenario
CAPEX [EUR]	14,000.000.00	14,000.000.00	10,500.000.00
OPEX [EUR/year]	120,000.00	60,730.00	60,730.00
Average Annual Benefits [EUR/year]	N/A	1,817.969.89	3,304.934.66
NPV [EUR]	4,700.000.00	-2,884.775.69	9,984.237.28
IRR [%]	16.90	8.65	21.55
SPP [years]	N/A	12.02	6.31
BCA [-]	N/A	0.79	1.95

As it can be observed, the scenario in which the project was developed and deployed in 2021 would lead to more attractive financial indicators, even when properly quantifying the OPEX and the average annual benefits. This is mainly due to the maturity of the ORC technology and to the almost linear increase of the electricity price both in the 2011 – 2021 period and in the 2021 – 2040 forecast. In the feasibility study, the project may seem more attractive because the initial scenario is also quantifying the annual benefits of using the ORC system as a combined heat and power plant. The generated heat would have been used for administrative buildings and preparation of hot water. In other industries, the thermal energy produced by an ORC system is already valorized, improving the financial performance of the projects [12]. Even if the ORC system appears to be not feasible in the "actual" scenario, its contributions to reducing the annual CO2 emissions cannot be ignored. So far, the annually CO_2 reduction was approximatively 15,400 tCO_2/yr, economy which came as a result of the electricity generated without additional fuel. Regarding the differences in the OPEX between the feasibility study forecast and the actual registered costs, it was observed that the turbine of the power plant was the most reliable part of the system but, the most failures came from the generator break-down and leaks of the organic fluid, which is a high toxicity organic compound. The failure time has a strong impact on the power plant electricity production because of fewer running hours – some failures needed up to one week to be solved. Beside those unexpected events, another discrepancy between the availability time of 95% in the initial study and the actual operation of the plant is represented by the longer time needed for the rotary kiln inspection. Some inspections exceeded 2.5 months until completion, so the kiln was not functional during those months, making the ORC plant unavailable. The electricity production of the ORC system was also smaller in the "Actual" scenario compared to the forecast of the feasibility study. The main reason is that the plant is situated in the moderate-continental climate, the outside temperature during the cold season drops below 0°C (significantly lower than the average 20°C ISO reference temperature used for estimation). This, in turn, leads to a decrease in the waste heat extracted from the kiln, thus to a decrease of available thermal energy. Currently there are projects meant to increase the ORC capacity in the next few years. The cement manufacturer aims to increase the waste heat available for the plant with more upgrades on the rotary kiln. A higher daily production rate will produce more heat which can be recovered from de plant. An increased capacity will assure a high amount of electricity with no additional fuel and no greenhouse gas emissions which is a main issue nowadays.

4 Conclusions

In this paper, the energy performance of a 4.1 MW ORC plant was emphasized. The project meant to increase the energy efficiency was a success, despite the discrepancies between the initial estimation and the actual achievements. The system has proven its reliability, being a fully automated system with few break-down times during operation. Because of the similarities with the conventional steam cycle, there were no new additional knowledge requirements needed in order to assure the plant operation. Based on the lessons learned, t can be concluded that in order to make ORC technology more attractive and increase its penetration rate both in the cement industry and in other buildings materials industries, a number of adjustments have to be done in the feasibility stage:

1. The kiln (heat source) capacity must be better estimated – both in terms of operability and in terms of evolution over time,
2. The electricity production forecast must be done by considering the seasonal variation of various external factors, such as the outdoor temperature, as it strongly influences the efficiency of the system,
3. The operational costs of the system have to more accurately evaluated – both in terms of various parts and subsystem feasibility and in terms of actual costs for maintenance, capital works etc.,
4. The management of the company (both the Top Management and the Factory Management) have to get more involved in the feasibility study stage in order to properly evaluate various hypothesis made by the consultancy team and to mitigate the eventual misassumptions.

References

1. LafargeHolcim, Integrated Sustainability Report, (2020).
2. H. Legmann, "Recovery of Industrial Heat in the Cement Industry by Means of the ORC Process," 44th IEEE-IAS/PCS Cement Industry Technical Conference,5-9 May, pp 29-35, Florida, ISBN 0-7803-7254-9, (2002).
3. International Energy Agency, "Total energy related CO2 emissions with and without a sustainable recovery, 2005-2023," (2020).
4. European Commission, "Cement, waste and carbon markets. Problems related to waste incineration in cement kilns under the EU-ETS," (2020).
5. Cembureau – The European Cement Association, „Cementing the European Green Deal", (2020).
6. Institute for Industrial Productivity, "Waste Heat Recovery for the Cement Sector: Market and Supplier analysis Waste Heat Recovery for the Cement Sector", (2014).
7. M. Poerner and A. Rimpel, "Waste heat recovery," in "Fundamentals and Applications of Supercritical Carbon Dioxide (SCO2) Based Power Cycles", Woodhead Publishing, pp. 255–267, DOI: 10.1016/B978-0- 08-100804-1.00010, (2017).
8. S. Quoiilin, M. V. D. Broek, S. Declaye, P. Dewallef and V. Lemort, "Techno-economic survey of Organic Rankine Cycle (ORC) systems," "Renewable and Sustainable Energy Reviews", vol. IV, no. 22, pp. 168-186, (2012).
9. R. Menz, "From Waste Heat To Power," International Cement Review, no. 10, pp. 59-62, (2012).
10. S. Karellas, "Energetic and exergetic analysis of waste heat recovery systems in the cement industry," Energy, no. 58, pp. 147-156, (2013).

11. Ioan-Sevastian Bitir-Istrate, Cristian Gheorghiu and Miruna Gheorghiu, "The Cost of Indecision in Energy Efficiency. A Cost of Opportunity Analysis for an Industrial Consumer," 55th International Scientific Conference on Information, Communication and Energy Systems and Technologies (ICEST), DOI: 10.1109/ICEST49890.2020.9232894, (2020)
12. M. Scripcariu, I. Bitir-Istrate, M. Butu, R. Porumb, A. Pavel, M. Patrascan, "Heat recovery in metallurgy. Technical and economical aspects of using ORC installations in Romania," 6th International Conference on Thermal Equipment, Renewable Energy and Rural Development, June 8-10, Romania, ISSN 2359-7941, pp.: 149-154, (2017).

Experimental COP evaluation of a 65-litre household refrigerator running with R600a

Saleh J. Alqaisy[1,4], *Kamel S.* Hmood[1,2], *Mahdi Hatf Kadhum* Aboaltabooq[3] *Valentin* Apostol[*1], *Horatiu* Pop[1], *Jamal* Al Douri[1,4], and *Viorel* Bădescu[5]

[1]Dept. of Applied Thermodynamics, Engines, Thermal and Refrigeration Equipment, University Politehnica of Bucharest, Romania
[2]University of Babylon, Iraq
[3]Al-Furat Al-Awsat Technical University, Technical College Najaf, Iraq
[4]University of Technology, Iraq
[5]Candida Oancea Institute, Polytechnic University of Bucharest, Spl. Independentei 313, Bucharest 060042, Romania.Romania

Abstract: A household refrigerator is a major energy consumer in all homes. The paper aims to evaluate based, on experimental data, the COP of a 65-liter household refrigerator operating with R600a. The main components and instrumentation of the experimental setup are presented. A mathematical model based on the energy balance of the refrigerated enclosure is developed in order to evaluate the COP. Three experimental data sets have been obtained in quasi-steady-state operating conditions. The results point out that the compressor power absorbed from the grid was 60 [W]. The maximum value for the cooling capacity is 81.73 W while the minimum value is 80.24 W. In terms of COP, the maximum value obtained is 1.362, while the minimum one is 1.337. The COP values obtained in the present study are in good agreement with the ones displayed by the manufacturer in the compressor datasheet. Future development is discussed.

Keywords: COP, R600a, household refrigerator

1. Introduction

It is hard to imagine modern residential life without the advantages offered by artificial cooling. Artificial cooling in residential applications can be divided into two main categories, namely food preservation and air conditioning. The most affordable and widely used mean for food preservation is the household refrigerator (HR). The number of HR worldwide is 1 to 1.6 billion [1,2] consuming about 6% of the electricity generated worldwide [2]. The estimated lifetime of an HR is around 10 - 20 years [3] and it is accountable for about 32 to 50 % of the total energy consumption of a residence [4,5]. Given the high number of HRs worldwide and the associated energy consumption, improving their energy efficiency becomes of interest. The most commonly used indicator for the energy efficiency of HRs is the coefficient of performance (COP). As presented below, a key component in the struggle to achieve lower energy consumption is the type of refrigerant being used [6]. Since 1st of January 2015, in the field of household refrigerators

* Corresponding author: valentin.apostol@magr.ro

and freezers, refrigerants having a Global Warming Potential (GWP) higher than 150 are prohibited [7]. In this context, R600a, R290 and their mixtures have been proposed and used since, as alternative refrigerants. A brief literature survey regarding the use of R600a, R290 and their mixtures, correlated with the COP, is presented below.

The ambient temperature and the type of refrigerant are among the parameters that have a high influence on the COP of HR. In this regard, Raveendran and Sekhar [8] conducted a study, in a tropical area, on a 190 L refrigerator, considering an ambient temperature of 32 °C. The authors concluded that using a proper mixture of R600a and R290 with 54.8 % and 45.2 % mass fractions, respectively, leads to a 5.9 % reduction in energy consumption and an 8.9 % improvement in the COP, compared to the case when R134a is used as a refrigerant. Saran et al [9] concluded from their experimental research, that mixing R600a and R290, in 40 % and 60 % mass fractions, respectively, leads to higher COP and lower energy consumption, compared to the other mixing ratios. Hmood et al. [10] explained, in their analysis, that the R600a refrigerant leads to a 3% higher COP compared to R134a. The study was carried out for a small scale refrigerator. A drawback of R600a is its flammability, which determines the need to search for eco-friendly, energy-efficient and high-performance refrigerants [11]. Cho et al. [12] developed a simulation model and compared it with a practical experiment on household refrigerator-freezers working with R600a. The authors concluded that the length and the diameter of the hot-wall condenser pipe has a direct influence on the COP. The refrigerant charge, energy consumption and COP decrease with the reduction of the length and diameter of the hot-wall condenser pipe and vice versa. Abou-Ziyan and Fatouh [13] experimented on a household refrigerator comparing the case of using R134a and the case of using R600, R600a and R290 mixtures. The best mixing ratio proved to be 13.4 % R600a, 26.6 % R600a and 60 % R290. For this particular mixing ratio, the authors have modified the refrigerant charge in the range of 30, 40, 50, 60, and 70 g. Different capillary tube (CT) lengths have also been considered in the range of 4, 4.5, 5, 5.5 and 6 m. The authors concluded that, 5.5 m CT length and 70 g of refrigerant will lead to a 3% higher COP than in case of R134a.

In this context, the paper aims to evaluate experimentally, the COP of a small-scale refrigerator operating with R600a. The COP values are evaluated using data obtained from a specific experimental setup. A brief description of the experimental setup is given in Section 2. Section 3 presents the mathematical model used for COP evaluation. Experimental input data and results are presented in section 4. Finally, in Section 5, the main conclusions and development targets are presented.

2. Experiment setup

An overview image of the experimental setup is presented in Fig. 1. Basically, the experimental setup is comprised of a small refrigerator, measurement instruments and an automation/control panel.

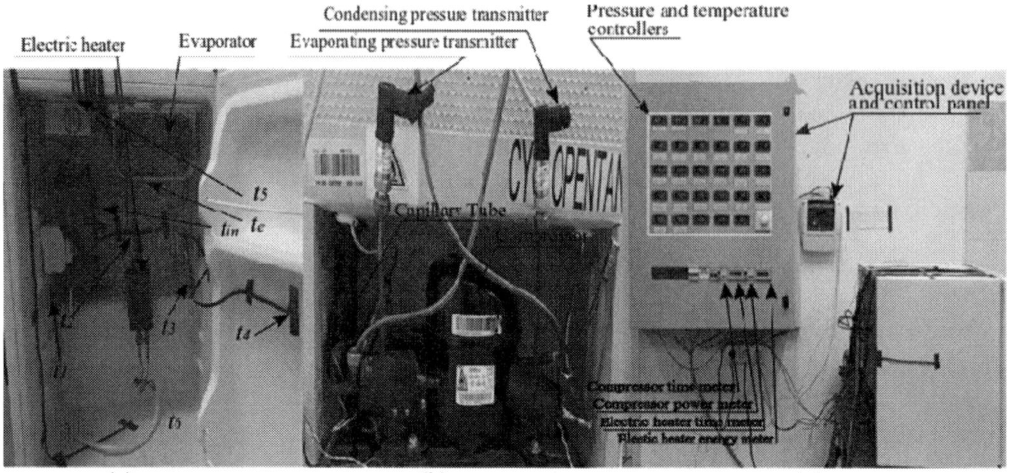

(a) (b) (c)

Fig.1. Overview image of the experimental setup: a. Front view, b. Back view, c. Side view

The refrigerator is a hot-wall condenser (HWC) MBM-67AP type, having 65 L. The dimensions of the refrigerator are 0.50 x 0.50 x 0.60 m. The refrigerant used is R600a. The evaporator is a roll bond type made of coated aluminum (see Fig. 1, front view). The compressor is a hermetic reciprocating compressor (HRC) (see Fig. 1 back view). Hot wall-condenser means, that the condenser is placed inside the right, upper and left walls of the refrigerator, being in contact with the outer metal sheet. From the outer metal sheet to the surroundings, the heat is transfer by natural convection. The CT has a length of 2.26 m, an inner diameter of 0.5 mm and an outer diameter of 1.2 mm (see Fig. 1, back view). The heat source is simulated using a 150 W adjustable static electric heater (see Fig. 1, front view).

The experimental setup is instrumented with temperature and pressure sensors. For temperature measurements, NTC thermistor sensors are used. In order to conduct the energy balance for the refrigerated enclosure, the temperature of each wall is measured. In Fig. 1, t_1 is the temperature of the left wall, t_2 is the temperature of the back wall, t_3 is the temperature of the right wall, t_4 is the temperature of the door wall, t_5 is the temperature of the top wall, t_6 is the temperature of the bottom wall. The evaporator wall temperature t_e, the temperature of the air inside the refrigerated enclosure t_{in} and the ambient temperature t_{amb}, are also measured. The condensing (p_{con}) and evaporating (p_{ev}) pressures are measured using pressure transducers (see Fig. 1 back view). The temperature sensors and the pressure transducers are connected to the corresponding controllers, mounted in the automation and control panels (see Fig. 1 side view).

The experimental setup is also fitted with a power adjustment device for the static electric heater, energy and time meters for the compressor and the static electric heater (see Fig. 1 side view).

3. The mathematical model for the COP evaluation

The mathematical model for COP evaluation is based on the energy balance of the refrigerated enclosure. A schematic of the refrigerated enclosure corresponding to the experimental setup is given in Fig. 2.

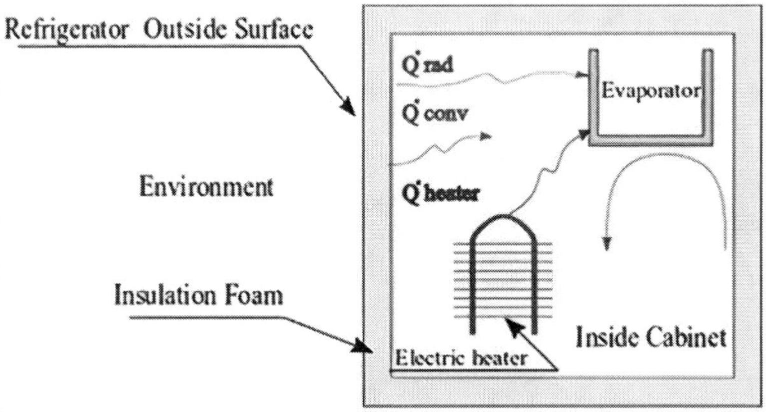

Fig. 2. Energy balance of the refrigerated enclosure

The energy balance of the refrigerated enclosure can be described by the following equation:

$$\dot{Q}_o = \dot{Q}_{rad} + \dot{Q}_{conv} + \dot{Q}_{Heater} \tag{1}$$

Where: $\dot{Q}_o$ is the cooling capacity of the refrigerator [W]; $\dot{Q}_{rad}$ is the heat transferred through radiation from the enclosure walls to the evaporator [W]; $\dot{Q}_{conv}$ is the heat transferred through free convection from the walls to the air inside the refrigerated enclosure [W] and $\dot{Q}_{Heater}$ is the heat transferred through convection and radiation from the electric heater to the air inside and then to the evaporator [W].

The heat transferred through radiation from the enclosure walls to the evaporator is the sum of the radiative heat fluxes rejected from each wall and received by each wall of the evaporator. The heat flux transferred through radiation can be computed based on the general formula presented by Welty et al, [14]:

$$\dot{Q}_{rad} = \frac{\sigma\left(T_w^4 - T_e^4\right)}{\frac{(1-\varepsilon_w)}{A_w\varepsilon_w} + \frac{1}{A_w F_{w-e}} + \frac{1-\varepsilon_e}{A_e\varepsilon_e}} \tag{2}$$

In eq. (2) $\sigma = 5.67 \cdot 10^{-8}\ W/\left(m^2K^4\right)$ is the Stefan Boltzmann constant, A_w [m^2] is the area of the enclosure wall for which the rejected radiative heat flux is computed, A_e [m^2] is the area of the evaporator wall which receives the radiative heat flux, T_w [K] is the temperature of the enclosure wall, T_e [K] is the temperature of the evaporator wall, ε_w is the emissivity factor of the enclosure walls (for the present study the value of ε_w has been considered to be 0.95 corresponding to white plastic materials), ε_e is the emissivity factor of the evaporator walls (or the present study the value of ε_e has been considered to be 0.88 corresponding to white painted aluminum materials) [15,16] and F_{w-e} is the view factor which takes into consideration the relative position of the enclosure wall to the evaporator wall. The view factor can be computed according to Lienhard [17].

The heat flux received by the air inside the enclosure from the refrigerator walls through free convection can be computed using the following relationship:

$$\dot{Q}_{conv} = h \cdot A_w \cdot (t_w - t_{air}) \tag{3}$$

In eq. (3), h [W/(m^2 K)] is the natural convective heat transfer coefficient, A_w [m^2] is the area of the refrigerator wall which rejects the convective heat flux, t_{air} [°C] is the temperature of the air inside the refrigerated enclosure and t_w [°C] is the temperature of the refrigerator wall. Eq. (3) is applied for each individual wall of the refrigerator.

The convective heat transfer coefficient can be determined as follows [18,19]:

$$h = \frac{Nu \cdot k}{L} \tag{4}$$

Where Nu stands for the Nusselt number, k [W/(m K)] is the thermal conductivity of the air inside the refrigerated enclosure [20] and L [m] is the characteristic length of the refrigerator wall rejecting the convective heat flux.

The heat flux transferred from the refrigerator wall to the air inside the enclosure takes place by means of natural convection. Natural convection can be characterized through the Rayleigh number:

$$Ra = Gr \cdot Pr = \frac{g \cdot \beta \cdot \Delta T \cdot L^3}{\alpha \cdot \nu} \tag{5}$$

Where: β [1/K] is the thermal expansion coefficient, g [m/s^2] is the gravitational acceleration, ΔT [K] is the temperature difference between the refrigerator wall and the air inside the refrigerator enclosure, α [m^2/s] is the thermal diffusivity and ν [m^2/s] is the kinematic viscosity.

The Nu number is computed accordingly, as presented next.

As it can be seen from Fig. 1 front view, the refrigerator has vertical and horizontal walls. For the present study, the refrigerator walls have been considered vertical and horizontal plates. For vertical plate walls, if the Rayleigh number $Ra \leq 10^9$, the Nu number can be computed as [14,17,21]:

$$Nu = 0.68 + 0.670 \cdot Ra^{1/4} \cdot \left[1 + \left(\frac{0.492}{Pr}\right)^{9/16}\right]^{-(4/9)} \tag{6}$$

For vertical plate walls, if the Rayleigh number $Ra > 10^9$, the Nu number can be computed as [22,23]:

$$Nu = \left[0.825 + \frac{0.387 \cdot Ra^{1/6}}{\left[1 + \left(\frac{0.492}{Pr}\right)^{9/16}\right]^{8/27}}\right]^2 \tag{7}$$

The heat transfer from the horizontal walls of the refrigerator has been calculated applying Eq. (8), as follows:

$$\alpha_{hor,down} = 0.27 . Ra^{0.25} . \frac{k}{L} \tag{8}$$

Eq. (8) is applied for 10^5<Ra<10^{10} and L is the characteristic length of the horizontal plate walls of the refrigerator.

The heat flux received from the static heater can be computed as:

$$\dot{Q}_{Heater} = \frac{E_{Heater}^{final} - E_{Heater}^{initial}}{\tau_{operation}^{final} - \tau_{operation}^{initial}} \tag{9}$$

In Eq. (9), $E_{Heater}^{initial}$ [kWh] is the electric heater energy meter reading at the beginning of the operation time period, E_{Heater}^{final} [kWh] is the reading of the electric heater energy meter at the end of the operation time period [kWh]; $\tau_{operation}^{initial}$ [h] is the reading of the time meter at the beginning of the operation time period and $\tau_{operation}^{final}$ [h] is the reading of the time meter at the end of the operation time period.

The real COP of the household refrigerated used in the present study is determined as [24]:

$$COP = \frac{\dot{Q}_o}{P} \tag{10}$$

Where P [kW] is the power absorbed by the compressor from the electric grid and it can be computed as:

$$P = \frac{E_{Comp}^{final} - E_{Comp}^{initial}}{\tau_{operation}^{final} - \tau_{operation}^{initial}} \tag{11}$$

In Eq. (11), $E_{Comp.}^{initial}$ [kWh] is the compressor energy meter reading at the beginning operation time period and $E_{Comp.}^{final}$ [kWh] is the compressor energy meter reading at the end of the operation time period.

Using the mathematical model described before, in the next section, experimental input data is presented. Results are presented in a tabular manner and commented on.

4. Experimental input data and results

The experimental data used in the present paper is presented in Table1 and Table 2 for three individual tests. Table 1 presents the experimental data regarding temperatures and pressures while Table 2 shows the energy consumption of the electric heater, energy consumption of the compressor and time meter readings at the beginning and the end of the corresponding test, respectively.

Table 1. Experimental input data for measured temperatures and pressures

	t_1 [°C]	t_2 [°C]	t_3 [°C]	t_4 [°C]	t_5 [°C]	t_6 [°C]	t_e [°C]	t_{in} [°C]	t_{amb} [°C]	p_{ev} [bar]	p_{con} [bar]
Test 1	9.6	10.4	9.1	8.5	8.6	7.4	-26.9	5.7	25	0.507	7.36
Test 2	10.2	11.3	9.4	9.1	9.1	8.0	-25.6	6.4	26	0.514	7.52
Test 3	12.5	13.8	12.9	11.5	12.9	10.3	-24.5	8.6	21	0.518	7.14

Table 2. Experimental data for energy consumption and time meter readings

	$E^{initial}_{Heater}$ [kWh]	E^{final}_{Heater} [kWh]	$E^{initial}_{Comp.}$ [kWh]	$E^{final}_{Comp.}$ [kWh]	$\tau^{initial}_{operation}$ [h]	$\tau^{final}_{operation}$ [h]	Operation total Time [h]
Test 1	34.46	34.57	16.63	16.75	504.2	506.2	2.0
Test 2	34.68	34.79	16.90	17.02	508.5	510.5	2.0
Test 3	35.50	35.61	17.85	17.97	425.6	527.6	2.0

The three experimental data sets presented in Table 1 and Table 2 correspond to a quasi-steady-state operation of the experimental setup. To reach the quasi-steady-state operation of the experimental setup, the compressor was programmed to turn off at *t_comp_off*=5 °C and turn on at *t_comp_on* = 12 °C while the electric heater was programmed to turn off at *t_heater_off*=10 °C and turn on at *t_heater_on* = 7 °C. The quasi-steady-state operation can be reached by adjusting the power of the static heater to match the cooling capacity of the compressor. Several experiments have been carried out to determine the quasi-steady-state condition, before the experiment tests presented in Table 1 and Table 2. As a general remark, at least three hours are required for each experimental data set, from which one hour to reach the quasi-steady-state and two hours to experiment with quasi-steady-state conditions. After a long operation time, the compressor will heat up and its efficiency will decrease leading to a reduction of its cooling capacity. After three hours, the cooling capacity of the compressor decreases and thus the power of the heat source should be adjusted accordingly. For the present experimental study, the ability to adjust the power of the heat source to match the cooling capacity of the compressor is limited and thus the maximum operation time in quasi-steady-state conditions has been limited to three hours.

Using the experimental data presented in Table 1 and Table 2, the mathematical model presented in Section 2 is applied. As it can be seen, the experimental data presented in Table 1 and Table 2 display close values for all three experimental tests.

The results obtained in the present study, for the three data sets, regarding the compressor power P [W], electric heater heat flux $\dot{Q}_{Heater}$ [W], heat flux transferred through radiation from the enclosure walls to the evaporator $\dot{Q}_{rad}$ [W], heat flux transferred through natural convection from the enclosure inside walls to the air inside $\dot{Q}_{conv}$ [W], cooling capacity $\dot{Q}_o$ [W] and COP are presented in Table 4.

Table 4. Results

	P [W]	$\dot{Q}_{Heater}$ [W]	$\dot{Q}_{rad}$ [W]	$\dot{Q}_{conv}$ [W]	$\dot{Q}_o$ [W]	COP
Test 1	60.0	55.0	12.56	13.39	80.95	1.349
Test 2	60.0	55.0	12.51	12.73	80.24	1.337
Test 3	60.0	55.0	13.22	13.51	81.73	1.362

From Table 4 it can be noticed that the compressor power absorbed from the grid, for the three tests, is constant and has the value of 60 [W]. The power absorbed by the static electric heater from the grid, which is equal to the heat flux released to the air inside the refrigerated enclosure and then to the evaporator, was 55 [W]. The maximum value, among the three data sets, obtained for the cooling capacity is 81.73 [W] while the minimum value is 80.24 [W]. The cooling capacity of the small scale refrigerator used in this study according to the manufacturer data sheet is 85 [W][25]. The maximum and minimum values obtained for the cooling capacity are very close. In terms of COP, the highest value obtained, among the three data sets, is 1.362 and the lowest one is 1.337. The highest and lowest values for the COP are close. The COP for the compressor displayed by the manufacturer in the datasheet [25] is 1.37. The values for the cooling capacity and COP displayed by manufacturer are in good agreement with the values obtained in the present work, especially that the three experimental tests have been conducted in non-standard conditions.

5. Conclusions

The paper presents the results obtained for the experimental COP evaluation carried out on a small scale, 65-liter refrigerator, operating with R600a. A brief description of the experimental setup is given. A mathematical model for the COP evaluation, based on the energy balance of the refrigerated enclosure, is presented. The mathematical model has been applied for three experimental data sets obtained for quasi-steady-state operating conditions. The results point out that the compressor power absorbed from the grid it is 60.0 [W]. The power absorbed from the grid by the electric heater is approximately 55.0 W. The maximum value, obtained for the cooling capacity is 81.73 [W] while the minimum value is 80.24 [W]. The cooling capacity of the small scale refrigerator used in this study according to the manufacturer data sheet is 85 [W]. The results obtained for the COP show that the maximum value obtained is 1.362, while the minimum one is 1.337. The COP for the compressor displayed by the manufacturer in the datasheet is 1.37. There is good agreement between the results obtained in the present study and the data given by the manufacturer.

Future development involves a modification of the experimental setup in order to better match the cooling capacity of the compressor with the power of the electric heater. This modification will enable tests conducted for a longer time period in quasi-steady-state operating conditions.

References

1. K. Liang, Energy **151,** 324–331 (2018)
2. L. Harrington, L. Aye, B. Fuller, Appl. Energy **211,** 346–357 (2018)
3. M. Fatouh, H. Abou-Ziyan, Appl. Therm. Eng. **145** 14–26 (2018)
4. A.E. Gürel, Ü. Agbulut, A. Ergün, I. Ceylan, Eng. Sci. Technol. an Int. J. 23365–372 (2020)
5. [H. Jin, Z. Sun, J. Gu, Y. Huang, X. Wang, Q. Zhang, X. Shen, Measurement **157** 107654 (2020)

   A. Pirvaram, S.M. Sadrameli, L. Abdolmaleki, Energy **181,** 321–330 (2019)

A. Maiorino, R. Llopis, M.G. Del Duca, C. Aprea, Int. J. Refrig. **117,** 132–139 (2020)

6. P. Saji Raveendran, S. Joseph Sekhar, Appl. Therm. Eng. **179,** 115684 (2020)
7. T.S. Pilla, P.K.G. Sunkari, S.L. Padmanabhuni, S.S. Nair, R.S. Dondapati, Energy Procedia. **109,** 113–121 (2017)
8. [K.S. Hmood, H. Pop, V. Apostol, A.Q. Ahmed, Am. Sci. Res. J. Eng. Technol. Sci. **35,** 251–265 (2017)
9. S. Choi, U. Han, H. Cho, H. Lee, Appl. Therm. Eng. **132,** 560–574 (2018)
10. W. Cho, D.S. Jang, S.H. Lee, S. Yun, Y. Kim, Int. J. Refrig. **117,** 295–306 (2020)
11. H. Abou-Ziyan, M. Fatouh, Appl. Therm. Eng. **129,** 446–462 (2018)
12. J.R. Welty, G.L. Rorrer, D.G. Foster, *Fundamentals of Momentum, Heat, and Mass Transfer*, (Sixth Edit. United States of America: John Wiley & Sons, 2015)
13. A.A. Imran, H.M. Jafal, J. Eng. Dev. **13,** 1–16 (2009)

14. [R.S. Espíndola, J. Boeng, F.T. Knabben, C.J.L. Hermes, Int. J. Refrig. **114,** 10–18 (2020)
15. J.H. Lienhard, *A Heat Transfer Textbook* (Fifth Edit. Phlogiston Press Cambridge, Massachusetts, U.S.A, 2019)
16. Y.A. Cengel, *Heat and Mass Transfer (SI Units) A Practical Approach* (Third Edit. United States of America: McGraw-Hill Education, 2006)
17. D. Azzouzi, M. Kelkouli, F. Amaryoucef, Appl. Therm. Eng. **122,** 528–534 (2017)
18. R.S. Espíndola, F.T. Knabben, C. Melo, C.J.L. Hermes, Int. J. Refrig. **110,** 262–267 (2020)
19. F.P. Incropera, D.P. DeWitt, T.L. Bergman, A.S. Lavine, *Fundamentals of Heat and Mass Transfer* (Sixth Edit. United States of America: John Wiley & Sons, Inc, 2007)
20. J.K. Gupta, M. Ram Gopal, Int. J. Refrig. **31,** 979–988 (2008)
21. Y. A. Çengel, A.J. Ghajar, *Heat and mass transfer fundamentals and applications* (Fifth Edit. United States of America: McGraw-Hill Education, 2015)
22. C.D. Pe´rez-Segarra, J. Rigola, M. So`ria, A. Oliva, Int. J. Refrig. **28,** 579–593 (2005)
23. www.GMCC-WELLING.com/Homeen/Product/index/classify/20.html (2018)

The Heat Transfer Performance of MWCNT, CuO, and Al_2O_3 Nanofluids in an Automotive Engine Radiator

Karaar Mahdi Al-Araji[1*], *M. A.* Almoussawi[1] and *Kareem* J. Alwana[2]

[1]Department of Power Mechanics Engineering, Engineering Technical Collage / Al-Najaf, Al-Furat Al-Awsat Technical University (ATU), 31001 Najaf, Iraq
[2]Al-Furat Al-Awsat Technical University, Najaf Technical Institute, 31001, Iraq

Abstract. The effect of enhancing heat transfer using three nanofluids, Multi-Walled Carbon Nanotubes MWCNT, Copper oxide CuO, and aluminum oxide Al2O3, have been experimentally studied on the automotive radiator with a concentration of 1% vol and different flow rates (4-8) l/min, air velocity of 3 m/s and inlet temperatures range (60-80) °C. The results showed that the use of nanofluids improved the thermal performance compared to the base fluid. Using the (MWCNT-Water) achieved 41.7% of Nusselt number where as, copper oxide (CuO) and aluminum oxide (Al2O3) have improved the Nussult number by 31.7 % and 12.3 % respectively. The Nusselt number showed an increase with the increase in the flow rate and the inlet temperature.

Keywords: Nanofluid, Al_2O_3, CuO, MWCNT, Automotive engine radiator, Nusselt number.

1 Introduction

The Environmental Protection Agency EPA imposes strict standards on automobile manufacturers to protect the environment from the influence of emissions of combustion engine exhausts. One of the essential things in reducing pollution and increasing engine performance is to improve the automotive cooling system. Nanofluids are new technology which consists of tiny particles (less than 100 nm) suspended in conventional cooling fluids (water and/or ethylene glycol) [1]. Many experiments on nanofluids performance as a coolant in automotive radiators in order to improve the performance and reduce the radiator's size have been carried out in the last decades. **Chaurasia et al.**[2] conducted an experimental study on a mixture of nanofluid (alumina with distilled water) at a concentration of 0.1%, 0.15%, and 0.2% with different flow rates and different temperatures. They found that a concentration of 0.2% of nanofluid has increased the heat transfer rate by 44.29%. **Topuz et al.** [3] carried out an experimental investigation on nanofluid containing aluminum oxide Al2O3 nano-particles mixed with water and ethylene glycol (50-50%), 0.5% concentration with different flow rates and fixed temperature 95 °C.

* Corresponding author: karaar.mahdi@gmail.com

Results showed an increase in cooling performance by 15% compared to the base fluid. **Ravisankar et al.**[4] applied an experimental and numerical study to reveal the performance of nanofluid (CuO with water) on a tractor radiator with a concentration of (0.025, 0.05%) and turbulent flow ($8000 \leq Re \leq 25000$). They found that the overall heat transfer coefficient has increased by 4% at a concentration of 0.05%. **Ali et al.**[5] studied experimentally the improvement in the heat transfer of automobile radiators using (MgO + water) nanofluid at different flow rates (8-16) l/min and different volume fraction. The study showed a 31% increase in the heat transfer rate at a concentration of 0.12%. **M'hamed et al.** [6] applied a nano-coolant fluid mix of MWCNT with water and EG at a 50-50% ratio on the Proton Kelisa 1000cc automobile engine. A volumetric concentration of 0.1 to 0.5% and a different flow rate with the laminar flow were suggested. Results showed an increase in the average heat transfer coefficient by 196.3% at a concentration of 0.5%. In the current study, nanofluids and their effect on automotive coolants are studied. The main aim is to enhance the average Nusselt number of the nanofluid as a coolant at the laminar flow conditions. Also, to overcome emissions of harmful gases and pumping power, to improve fuel consumption, and to reduce automotive radiators' size.

2 Materials and Methods

2.1 Preparation of Nanofluid

The nanoparticles used in the current study are multi-walled carbon nanotubes MWCNT, Copper Oxide CuO and Aluminum Oxide Al_2O_3, the characteristics of nano-particles are shown in Table 1. The nanoparticles were suspended at a volume fraction of 1% with pure water (pH 6.6). An ultrasonic for 3 hours with a frequency of 37 kHz and 100 Watts was used to produce the nanofluid. The particles' stability was monitored for 15 days and no sedimentation of nanoparticles was found during this period when a surfactant (gum Arabic GA) was added to increase the nanofluid stability [7]. The prepared samples are shown in Fig.1

Table 1. The characteristics of nanoparticles

Details	MWCNT	CuO	Al_2O_3
Appearance	Black powder	Dark brown powder	White powder
Purity	99%	99%	99.99%
Size (nm)	D: 30-60 nm L: 5-20 um	D: 30-50 nm	D: 20-30 nm
Thermal conductivity (W/m.K)	3000	76.5	46
Density (kg/m^3)	2100	6400	3890
Specific heat (j/kg.K)	769	531	778

Fig 1. A Photograph showing the effect of Arabic gum on nanofluids stability

2. 2 Experimental Work

Fig. 2 shows the rig test carried out to study the three nanofluids' thermal performance. It contains a coolant tank with 30 liters with a heater capacity of 3000 watts. The liquid's temperature is controlled by a thermostat placed on the heater with a range (40 - 80) °C. The centrifugal pump speed and the amount of flow are controlled by a frequency converter (AC driver), flow meter instrument, a pressure gauge to measure the inlet and outlet coolant pressure of radiators, insulated pipelines with diameter 0.5", two thermocouples (k-type) to measure the inlet and outlet temperature of the fluid, six thermocouples (k-type) attach to the surface of the tube are distributed as shown in Fig 3 to record the temperature of the tube surface. Automotive engine radiators with 60 vertical finned flat tubes, their specifications and dimensions are listed in Table 2. An axial fan with airflow speed 3 m/s was fixed to the radiator to simulate the radiator cooling engine. The study was conducted at different temperatures range 60, 70, 80 °C for three types of nanofluids at a concentration of 1%. The ambient temperature was 25 °C, and the flow rate of (4-8) liters per minute.

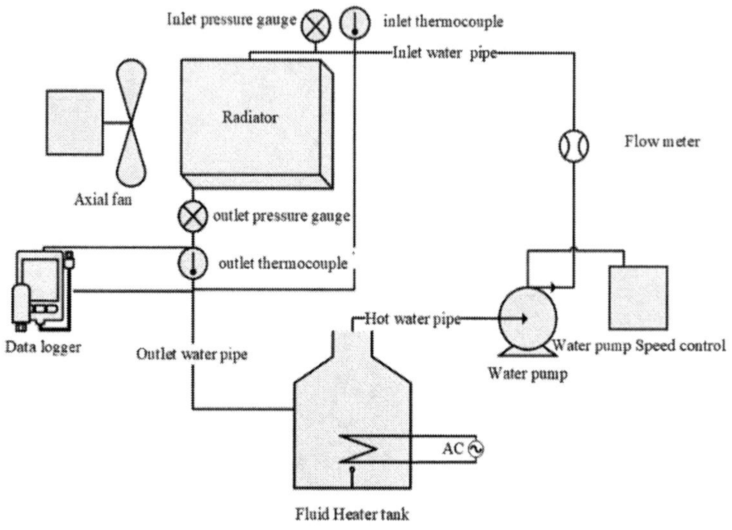

Fig 2. Schematic drawing of the experimental test rig.

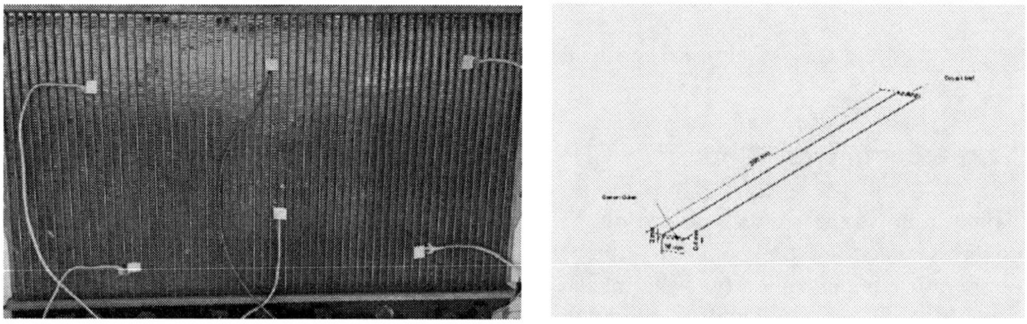

Fig 3. Thermocouple distribute on radiators and the dimension of the tube

Table 2. Dimension of the radiator in this study

Description	Specification
Tube length	350 mm
Tube height	16 mm
Tube width	8 mm
Thickness of tube	0.4 mm
Height of fins	16 mm
Width of fins	8 mm
Thickness of fins	0.09 mm
Material	Aluminum

2.3 Nanofluid Physical Properties

The following equation was used to evaluate the volumetric concentration of the nanoparticles and the amount of liquid. [8]. Nanofluid concentration is $\varphi = 1$ vol%, and the inlet temperature range of the coolants is 60-80 °C. where φ is the volume concentration

$$\varphi = \left[\frac{\frac{W_p}{\rho_p}}{\frac{W_p}{\rho_p} + \frac{W_{bf}}{\rho_{bf}}} \right] \times 100 \tag{1}$$

The following equations evaluate the effective thermal properties such as density, specific heat, viscosity, and thermal conductivity at various temperatures. Where (r) Density, kg/m^3, (Cp) Specific heat J/kg., (*bf*) base fluid, (*nf*) nanofluid, (*p*) particles.

$$\rho_{nf} = (1-\varphi)\rho_{bf} + \varphi\rho_p \tag{2}$$

$$Cp_{nf} = (1-\varphi)Cp_{bf} + \varphi Cp_p \tag{3}$$

By (Maxwell) model of the effective thermal conductivity for (Al_2O_3, CuO) nanofluids have approximate.

$$K_{nf} = \frac{K_p + 2K_{bf} - 2(K_p + K_{bf})\varphi}{K_p + 2K_{bf} - (K_p - K_{bf})\varphi} * K_{bf} \tag{4}$$

The (Hamilton and crosser) model of the effective thermal conductivity for (MWCNT) nanofluids have approximate. [9] where the (*k*) Thermal Conductivity, W/m. K.

$$K_{nf} = \frac{K_p + (n-1)K_{bf} + (n-1)(K_p + K_{bf})\varphi}{K_p + (n-1)K_{bf} - (K_p - K_{bf})\varphi} * K_{bf} \tag{5}$$

Thus, the effective dynamic viscosity is calculated by using the (Brinkman) model, where (μ) the dynamic viscosity Pa.s

$$\mu_{nf} = \frac{\mu_{bf}}{(1-\varphi)^{2.5}} \tag{6}$$

2.3 Data Extraction

According to Newton's cooling law Eq (7), the heat transfer rate can be measured by.[10]

$$Q = h\,As\,(T_b - T_w) \tag{7}$$

$$Q = \dot{m}\,Cp\,(T_{in} - T_{out}) \tag{8}$$

$$\dot{m} = \rho\,V\,A_{in} \tag{9}$$

Where (Q) heat transfer rate, (Cp) specific heat, (T_{in}) inlet temperature and (T_{out}) outlet temperature, ($\dot{m}$) mass flow rate and (ρ) density of fluid; (V) velocity of the liquid and (A_{in}) the inlet area of the flat tube.

The heat transfer coefficient and the Nusselt number were calculated using the following equation:[6]

$$h = \frac{\dot{m}\, Cp\, (T_{in} - T_{out})}{A_s(T_b - T_{w})} \quad (10)$$

$$Nu = \frac{h * D_h}{K} \quad (11)$$

$$D_h = \frac{4 \times [\pi d^2 (D - d) * d]}{\pi\, d + 2 * (D - d)} \quad (12)$$

Where (D_h) is hydraulic diameter, (k) the fluid thermal conductivity. (A_s) the flat tube surface area, (T_b) is the bulk temperature which is supposed to be the average temperature of the inlet, and (T_w) is the wall temperature.

3 Result and Discussion

3.1 Base Fluid (water)

Before starting experimental work on nanofluids in the radiator, it is necessary to conduct experiments on the same base fluid (pure water) and verify that the receiver operates under the same operating conditions of temperature and flow rate to compare results. It's also necessary to compare the Nusselt number measured from the experimental data with the predictions and other correlations for accuracy and reliability. There are three correlations; the first is suggested by (Dittus-Boelter) Eq.13 [11], the second developed by the researcher (Dehghandokht et al.) Eq.14 [12] and the last one was presented by (Sieder -Tate) Eq.15 [13]. Where the (Re) Reynolds number, (Pr) Prandtl number, (L) tube length.

$$\mathrm{Nu} = 0.0236 * \mathrm{Re}^{0.8} * \mathrm{Pr}^{0.3} \quad (13)$$

$$\mathrm{Nu} = 0.951 * \mathrm{Re}^{0.173} * \mathrm{Pr}^{\left(\frac{1}{3}\right)} \quad (14)$$

$$\mathrm{Nu} = 1.86 * \left(\frac{\mathrm{Re} * \mathrm{Pr}}{\mathrm{L}/\mathrm{D_h}}\right)^{0.3} * \left(\frac{\mu}{\mu_s}\right)^{0.14} \quad (15)$$

Fig 4. shows the experimental result for the base fluid (distilled water), which showed a good agreement with the (Sider-Tate) correlation at the fluid inlet temperature of 80 ° C. However, increasing the Reynolds number has led to an increase in the Nusselt number, where the error rate according to equation (15) was 6.7 %, while the error rate for the other correlation was 35% for Eq.13 and 39.8% for Eq.15.

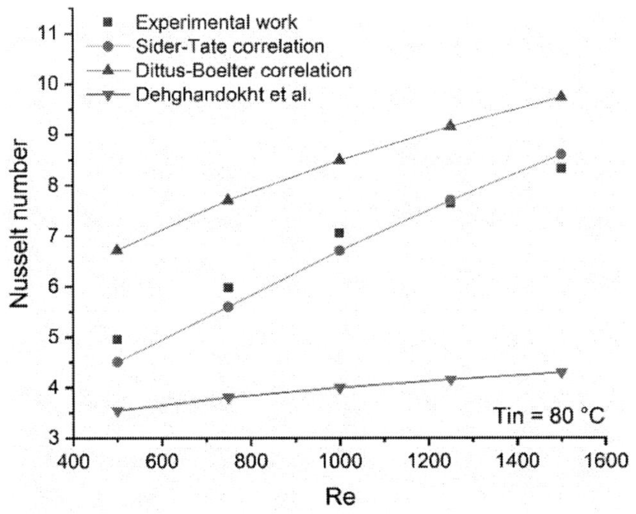

Fig 4. Comparison of the Nusselt number found from the experimental work with that resulted from different correlation

3.2 Nanofluid

Fig 5 shows the outlet temperatures of nanofluids compared to the base fluid at an inlet temperature of 80 ° C with different flow rates. The outlet temperature has decreased when using nanofluids; the highest enhance in the outlet temperature was achieved when MWCNT nanofluid has been used with an increase of 3.2% compared to the base fluid, copper oxide CuO and aluminum oxide Al_2O_3 showed lower values.

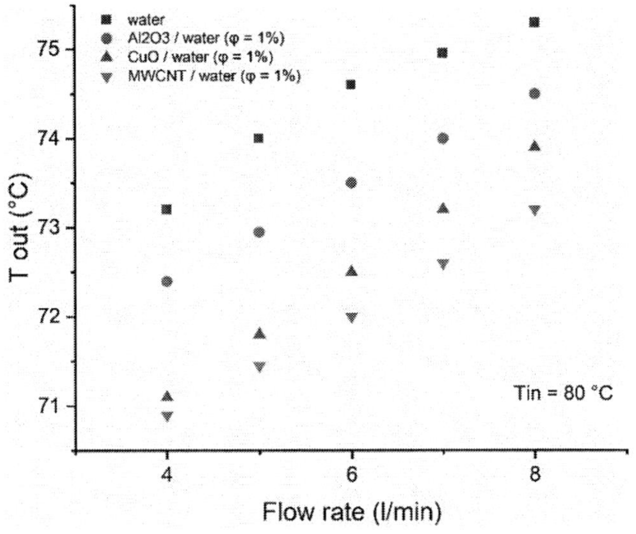

Fig 5. Comparison of the outlet temperature of the different nanofluids with the base fluid

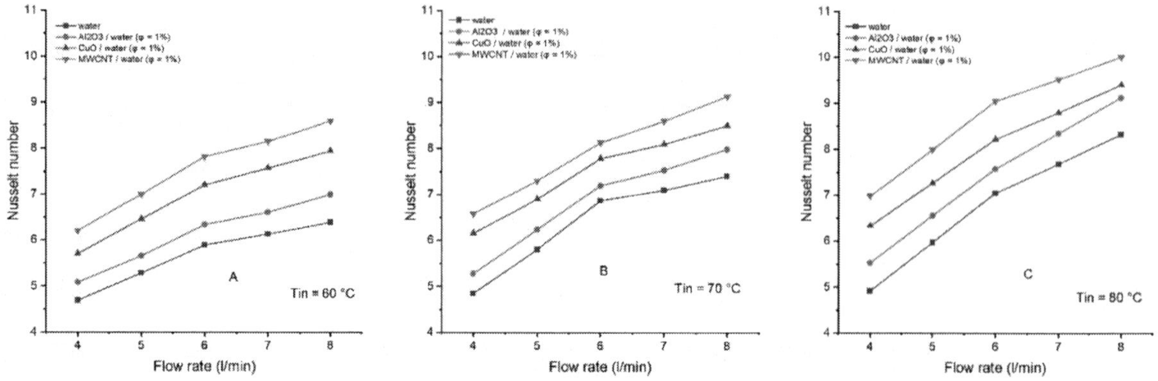

Fig 6. The effect of flow rate on Nusselt number at different inlet temperatures of nanofluids, A)60°C, B)70 °C, C) 80 °C

Fig 6. shows the enhancement of the nanofluids of (MWCNT, CuO, Al_2O_3) at different temperatures and flow rates. Also, increasing the flow velocity leads to an increase in the Nusselt number, Dattatraya et al.[10] found that Nusselt number increased by 12% when using a nanofluid at 80 ° C inlet temperature by increasing the flow rate from 7.63 to 8.82 l/min, this is the result of increasing the convection heat transfer rate, this also occur when increasing the nanofluid inlet temperature as a result of the decrease of specific heat of nanofluid which in turn lead to increase the rate of heat transfer[14].

Fig 7. shows the thermal performance enhancement represented by the ratio of the Nusselt number for the nanofluid to the Nusselt number for the base fluid of automotive radiator at an inlet temperature of 80 °C and a different flow rate. The highest improvement 41.8% was found in the flow rate of 4 l/min in the MWCNT nanofluid, while for the CuO nanofluid was 31.7 % and for the Al2O3 nanofluid was 12.3 %. The amount of heat transfer is more effective when the flow rate low with constant air velocity ambient temperature.

The main cause of enhancement of heat transfer rate can be attributed to the improve in the nanofluid's thermal properties (especially the thermal conductivity and specific heat) compared to the base fluid heat as showen in table 3. As well as, many research studies attributed this thermal improvement to Brownian motion, which causes random movement of particles in the fluid. Heris et al.[15] found that the thermal improvement rate was increased by 40% when using water-Al_2O_3 with a turbulent flow rather than laminar flow.

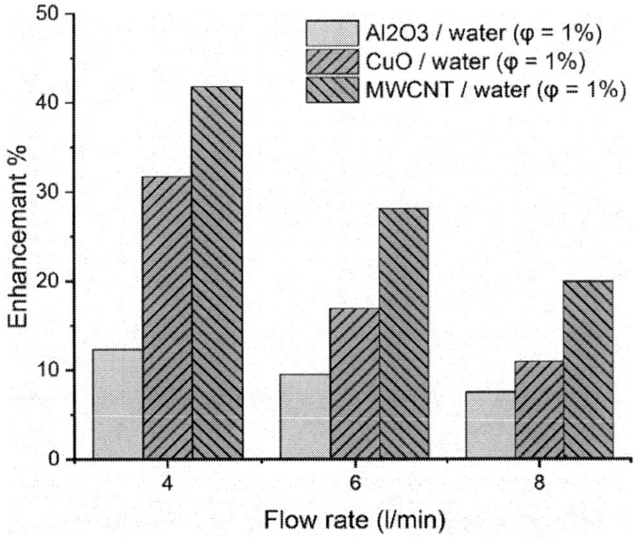

Fig 7. Variation of thermal enhancement in automotive radiators for different nanofluids at a volume concentration of 1%.

Table 3. The specific heat and thermal conductivity of nanofluids at a volume concentration of 1%.

Fluid	Volume fraction (%)	Specific heat (J/kg.K)	Thermal conductivity (W/m.K)
Water	0%	4200	0.667
Al_2O_3	1%	4162.81	0.684
CuO	1%	4160.34	0.688
MWCNT	1%	4154.96	0.707

4 Conclusions

In this research, the effect of using MWCNT, CuO, and Al_2O_3 nanofluids has been investigated in automotive radiators. A significant improvement in Nusselt number compared to the base fluid was achieved when using the mentioned nanofluids. The highest improvement was 41.8% for MWCNT mixed with distilled water at a concentration of 1%. The experimental results showed that the Nusselt number for base fluid and nanofluid has increased with the increase in the flow rate due to the increases in fluid turbulence inside the tube, resulting in more heat to transfer by convection. The increase in the inlet temperature has also resulted in the Nusselt number increase due to the change in the thermal properties. This is attributed to the increases in thermal conductivity and the the decrease in viscosity and specific heat of the fluids. Nanofluid can be a promising for coolants' radiators' future; this provides profitable methods for engineers to promote effective and compacted radiators to reduce the size and weight, thus; reducing the fuel consumption.

References.

1. U. S. Choi, *Am. Soc. Mech. Eng. Fluids Eng. Div. FED*, vol. 231, pp. 99–105, 1995.
2. P. Chaurasia, A. Kumar, A. Yadav, P. K. Rai, V. Kumar, and L. Prasad, *SN Appl. Sci.*, vol. 1, no. 3, 2019, doi: 10.1007/s42452-019-0260-7.
3. A. Topuz, T. Engin, B. Erdoğan, S. Mert, and A. Yeter, *Heat Mass Transf. und Stoffuebertragung*, 2020, doi: 10.1007/s00231-020-02916-8.
4. R. Ravisankar, V. S. K. Venkatachalapathy, and N. Alagumurthy, *Heat Transf. - Asian Res.*, vol. 46, no. 1, pp. 61–74, 2017, doi: 10.1002/htj.21198.
5. H. M. Ali, H. Ali, H. Liaquat, H. T. Bin Maqsood, and M. A. Nadir, *Energy*, vol. 84, pp. 317–324, 2015, doi: 10.1016/j.energy.2015.02.103.
6. B. M'hamed, N. A. Che Sidik, M. F. A. Akhbar, R. Mamat, and G. Najafi, *Int. Commun. Heat Mass Transf.*, vol. 76, pp. 156–161, 2016, doi: 10.1016/j.icheatmasstransfer.2016.05.024.
7. A. O. Borode, N. A. Ahmed, and P. A. Olubambi, *Phys. Fluids*, vol. 31, no. 7, 2019, doi: 10.1063/1.5105380.
8. S. S. Chougule and S. K. Sahu, *J. Nanotechnol. Eng. Med.*, vol. 5, no. 1, pp. 1–6, 2014, doi: 10.1115/1.4026971.
9. P. Estellé, S. Halelfadl, and T. Maré, *J. Therm. Eng.*, vol. 1, no. 2, pp. 381–390, 2015, doi: 10.18186/jte.92293.
10. D. G. Subhedar, B. M. Ramani, and A. Gupta, *Case Stud. Therm. Eng.*, vol. 11, pp. 26–34, 2018, doi: 10.1016/j.csite.2017.11.009.
11. F. W. Dittus and L. M. K. Boelter, *Int. Commun. Heat Mass Transf.*, vol. 12, no. 1, pp. 3–22, 1985, doi: 10.1016/0735-1933(85)90003-X.
12. M. Dehghandokht, M. G. Khan, A. Fartaj, and S. Sanaye, *Appl. Therm. Eng.*, vol. 31, no. 10, pp. 1588–1599, 2011, doi: 10.1016/j.applthermaleng.2011.01.035.
13. E. N. Sieder and G. E. Tate, *Ind. Eng. Chem.*, vol. 28, no. 12, pp. 1429–1435, 1936, doi: 10.1021/ie50324a027.
14. S. Z. Heris, M. Shokrgozar, S. Poorpharhang, M. Shanbedi, and S. H. Noie, *J. Dispers. Sci. Technol.*, vol. 35, no. 5, pp. 677–684, 2014, doi: 10.1080/01932691.2013.805301.
15. S. Z. Heris, S. G. Etemad, and M. N. Esfahany, *Int. Commun. Heat Mass Transf.*, vol. 33, no. 4, pp. 529–535, 2006, doi: 10.1016/j.icheatmasstransfer.2006.01.005.

Using CFD to improve the performance of a heat exchanger from a gasifier

Liliana Dumitrescu[2*], *Edmond* Maican[1] , *Ioan* Pavel[2], *Cătălin* Dumitrescu [2], *Iuliana* Găgeanu[3], and *Andrei* Pătruț[4]

[1]"Politehnica" University of Bucharest, Faculty of Biotechnical Systems Engineering, Spl. Independenței no. 313, Sector 6, Bucharest, 060042, Romania
[2]Hydraulics and Pneumatics Research Institute INOE 2000-IHP, Cuțitul de Argint no. 14, Sector 4, Bucharest, 040558, Romania
[3]National Institute of Research-Development for Machines and Installations designed for Agriculture and Food Industry - INMA, 6 Ion Ionescu de la Brad Blv., sector 1, Bucharest, 013813, Romania
[4]SC CALORIS SA, Șos. Berceni no. 8A, Sector 4, Bucharest, 041914, România

Abstract. In this paper, a Computational Fluid Dynamics (CFD) analysis is performed on a gas-liquid heat exchanger fitted on a gasification equipment. The flow and temperature patterns are preliminary investigated using SolidWorks Flow Simulation software, in order to gain insight into the involved physical processes, and to find the exchanger weak points before being manufactured and tested. The analysed equipment tranfers heat from the flue gasses generated by a gasification system, towards a liquid heat transfer medium. This is subsequently sent to a second liquid-liquid heat exchanger used to heat water from a boiler. As a result of the analysis, solutions aiming at performance improvement of the equipment are discussed and proposed.

1 Introduction

Full exploitation of biomass energy potential is a strategic part of the EU renewable energy policy, as stated within the Directive 2009/28/EC [1]. The emphasis is on supply chains and on a 20% improvement of the technology efficiency and greenhouse gas (GHG) savings. Unlike wind and solar energy, biomass can be continuously supplied regardless climatic conditions. In terms of heat production, small scale systems are seen as the most effective and viable option in harnessing the locally available biomass, due to transportation cost reduction. Higher efficiency alternatives to simple biomass burning are considered, such as thermo-chemical conversion through gasification, thus avoiding health threats such as relatively high levels of GHG emmissions (N_2O and CH_4), polycyclic aromatic hydrocarbons, or particulate matter [2, 3].

An important aspect of these type of equipment is the way heat is recovered from the flue gas. Depending on flue gas parameters, gasification system geometry, ease of maintenance, fluids, and performance requirements, various types of heat exchangers have been studied and used. The performance of heat exchangers is influenced over time in a

* Corresponding author: lilianad.ihp@fluidas.ro

negative way, mostly due to contaminants that can build up deposit layers onto the exchanger walls (tars, particulates, sulfur, and others), thus affecting heat transfer and also reducing the section of the flow area [4, 5]. Currently, computational fluid dynamics (CFD) is an usual tool that is used in performance assessment and improvement.

An optimized heat exchanger design can perform better for a longer periods of time, thus reducing the number of maintenance steps. This paper presents the use of CFD to evaluate the performance of a heat exchanger from an existing gasification system and then to assess a modified and improved design, and also to propose solutions for further improvements.

2 Methods

The initial design of the heat exchanger is presented in Figure 1 *b*. The total heat exchange surface of the five tubes is 0.25 m^2. Each tube has an inner diameter of 71 mm and a height of 225 mm. Heat is also transfered through the top and bottom walls of the exchanger, which translates into a total surface (including the tubes) of 0.31 m^2. In Flow Simulation, inlets and outlets for flue gas and water should be represented as solid surfaces (Figure 1 *b*).

a)

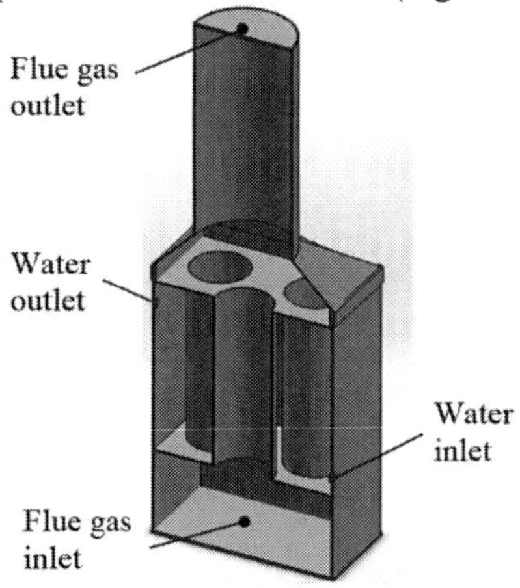

b)

Fig. 1 – Initial design of the heat exchanger: *(a)* before mounting into the evacuation assembly; *(b)* heat exchanger and flue gas admission and evacuation assembly (section view of a simplified CAD model for fluid dynamics simulation)

Based both on experimental data performed previously onto the existing gasification system, and on the requirements regarding the working regime of the entire heating assembly, the following initial data has been provided in the preprocessing step:

- Average temperature of gas at the inlet: 250 °C;
- Total gas flow at the inlet: 14.6 *l*/s;
- Water flow rate: 6 *l*/min and 3 *l*/min (both situations were simulated);
- Water temperature at the inlet: 20.05 °C;
- Water volume in the initial heat exchanger design: 5.9 *l*;
- Heat transfer coefficient: in order to consider heat losses through the exterior walls of the equipment and to assess the heat exchanger efficiency, the option of heat conduction in solids was activated. According to literature, for steel and gas-liquid exchangers, heat transfer coefficient ranges between 15...70 W/(m^2K) for liquid outside tubes and gas at atmospheric pressure inside tubes [6, 7]; simulations were made starting at 10 W/(m^2K), than a value of 20 W/(m^2K) was considered; these low values take into consideration the contaminant deposits that form onto the heat exchanger walls during operation.

 https://doi.org/10.1051/e3sconf/202128601010

3 Results and discussion

As mentioned earlier, the simulation was performed for two different water flow rates, namely 6 *l*/min and 3 *l*/min. These values were selected in accordance with the maximum (6 *l*/min) and minimum (2.76 *l*/min) flow rates of the pump that recirculates the water within the closed-circuit where the heat exchanger is installed. As expected, results were better for the lower flow rate value, but the overall average water temperature at the outlet can be considered unsatisfactory: just 24.7 °C. Figure 2 shows that one parameter that influence this result is the location of the water outlet under the layers of hot water.

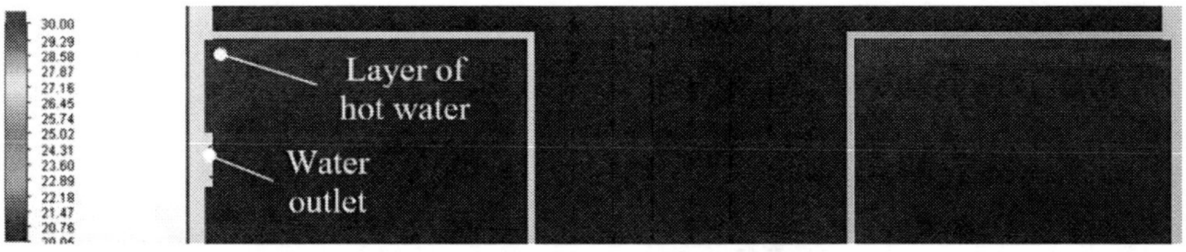

Fig. 2 – The water outlet is placed well below the layer of hot water

The other parameters that affected the heat exchanger performance are:

- Low water volume in the heat exchanger; considering a flow of 3 *l*/min, the entire volume of cold water is recirculated in less than two minutes;
- The exchanger tubes; their diameter is too large and they are too short in relation to the overall dimensions of the exchanger.

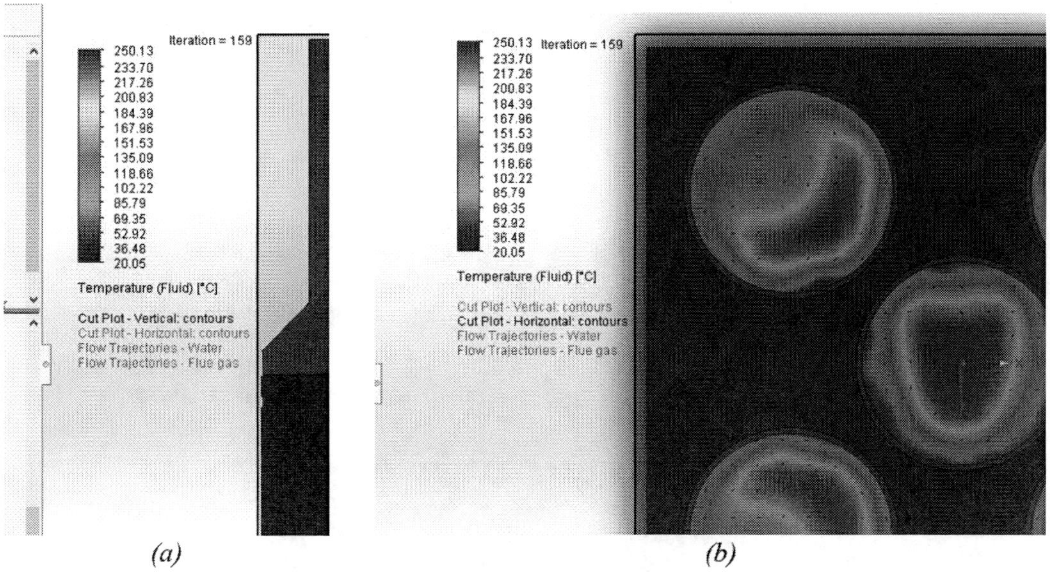

Fig. 3 – Flue gas temperature: a) vertical section view; b) cross-section at top of the exchanger tubes

Due to these two design parameters, flue gases leave the exchanger relatively quickly, without having enough time to efficiently transfer heat to the liquid thermal agent (Figure 3). The average temperature of the gases at the flue-gas outlet is about 180 °C, and the average temperature of the walls is 114 °C, meaning large heat losses through the walls. The average flue gas temperature at the outlet of the central tube is high (about 230 °C). To improve the performance of the heat exchanger, the following changes were made to the original design:

- The 5 pipes with an inner diameter of 70 mm were replaced with 9 cold drawn seamless pipes with an outer diameter of 48.3 mm (wall thickness: 2.93 mm, inner diameter of approximately 42.5 mm); keeping the cross section of the exchanger unchanged will result in a more compact design, characterized by a better heat transfer performance [8]

- The height of the heat exchanger was increased from 225 mm to 500 mm.

No change was made to the water outlet position, in order to asses the performance only due to changing the pipes number and dimensions. Simulation conditions have also been kept unchanged in terms of total gas and water flow rates and inlet temperatures. Heat transfer coefficient has been established to 20 W/(m^{2}K). Figure 4 shows a longitudinal section of the new design of the heat exchanger.

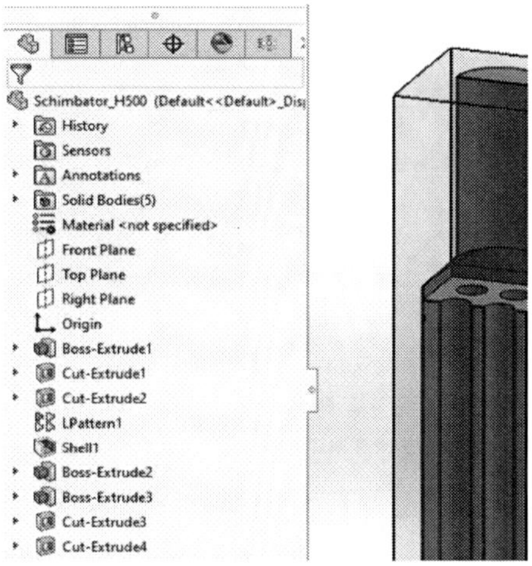

Fig. 4 – New heat exchanger design: 9 cold drawn seamless pipes (Φ = 48.3 mm)

Under the above conditions, the average water temperature at the outlet of the modified heat exchanger is 65.5 °C (Figure 5), and the average flue gas temperature at outlet is 155 °C.

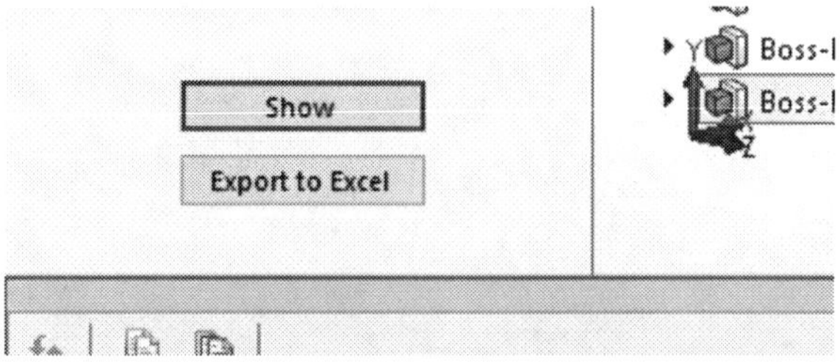

Fig. 5 – Water average temperature at the outlet increased to 65.5 °C in case of the new design

The layer of water above the outlet is thicker than in the previous design and has an average temperature in the range of the boiling point. The water outlet partialy captures this layer of hot water (Figure 6).

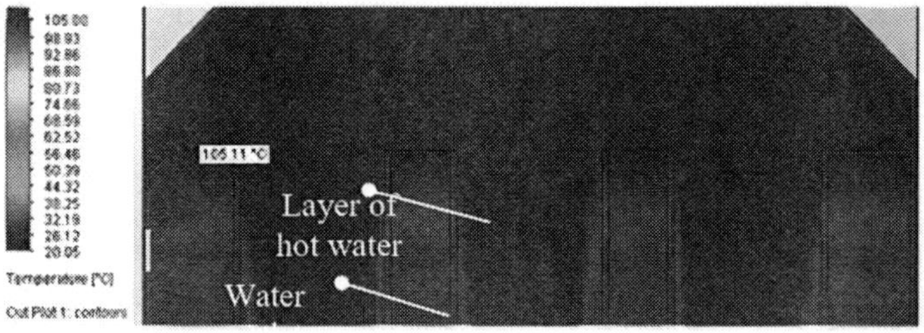

Fig. 6 – The layer of water above the outlet has a temperature in the range of the boiling point; placing the outlet higher will capture a higher volumic flow of high temperature water

This new design clearly shows performance improvement when compared to the original design. However, the overall performance can be further optimized by means of the following design changes:

- Insulation of the exterior walls with ceramic fiber wool (resistant to high temperatures), to limit heat loss due to heat transfer to the ambient environment by conduction through the walls. The entire housing should be insulated, except the chimney. In Figure 7, one can notice that the temperature of the gas near the walls (before entering the heat exchanger) drops just below 100 °C, due to large heat losses through the walls.

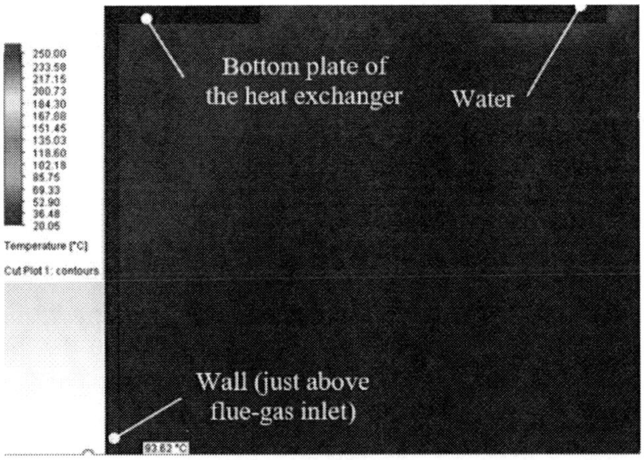

Fig. 7 – Flue gas temperature close to the exterior walls drops below 100 °C just before entering the heat exchanger, due to high heat conduction through the non-isolated exterior walls

- According to Figure 6, the water outlet takes a low volumetric flow from the upper layer of high temperature water. In order to take over a higher flow from this layer, one of the following three design changes should be adopted:
  a) Moving the water outlet with at least 15 mm above;
  b) The above design change can prove difficult due to the fact that the distance between the water outlet and the top plate of the exchanger is already relatively small (this area is "sensitive" due to existing welding seams); therefore, in an alternative design change, the water outlet position remains unchanged, but the top plate of the heat exchanger is inclined at an angle of 10° relative to the horizontal (Figure 8), to facilitate the formation of a thicker layer of hot water in the area of the outlet.

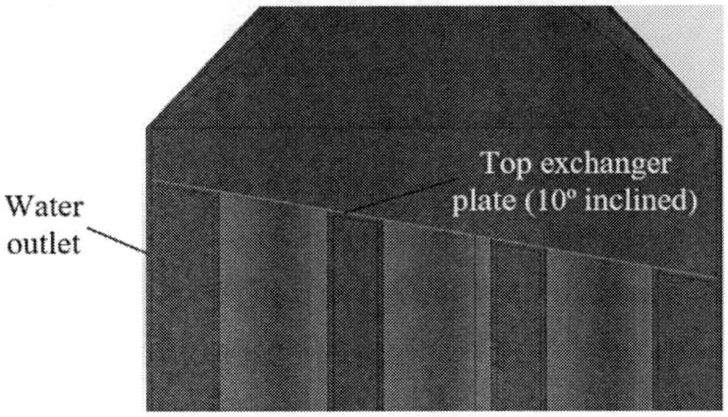

Fig. 8 – Inclining the top plate of the heat exchanger facilitates the formation of a thicker layer of hot water in the area of the water outlet

  c) Adoption of a combined solution, which consists in moving the water outlet higher with a distance less than the previously proposed (8 mm, instead of 15 mm) in order to avoid the nearby welding seams, together with the placement of the upper wall at an angle of 10° relative to the horizontal.

- Pressurizing the water circuit to prevent boiling at the top of the heat exchanger. A 3 bar pressure is recommended, with safety valve and expansion vessel.

4 Conclusions

After the evaluation of the original heat exchanger design, a 24.7 °C overall average water temperature at the outlet was determined. The average temperature of the flue-gas at the outlet was about 180 °C,and at the outlet of the central tube was about 230 °C. The average temperature of the walls was 114 °C, meaning important heat losses through the walls.
An improved design with 9 cold drawn seamless pipes and a height of the heat exchanger of 500 mm, was proposed and tested. The average water temperature at the outlet increased to 65.5 °C, and the average flue gas temperature at outlet decreased to 155 °C. Based on flow and temperature patterns, new improvements are proposed to further increase its performance: insulation of the exterior walls to limit heat loss through the walls; moving the water outlet 15 mm higher to catch the top high temperature water layer; inclining the exchanger top plate with 10° to increase the hot water layer thickness in the outlet area; pressurizing the water circuit to prevent boiling. Following the adoption of the above design changes, a new CFD analysis will be performed to asses the heat exchanger increase in performance.

This work was carried out under the project *Creating a high level proficiency nucleus in the field of increasing renewable energy conversion efficiency and energy independence by using combined resources*, project acronym: CONVENER, Financial agreement no. 37/02.09.2016, funded by the European Regional Development Fund through the Operational Program Competitiveness 2014-2020, Priority Axis 1: Research, technological development and innovation (RD&I) to support economic competitiveness and business development, Action 1.1.4 – Attracting high-level personnel from abroad in order to enhance the R&D capacity.

References

1. EU Parliament, Directive 2009/28/EC of the European Parliament and of the Council on the promotion of the use of energy from renewable sources, Official Journal of the European Union, Brussels (2009)
2. R. Devesa-Rey, X. Vecino, J.L. Varela-Alende, M.T. Barral, J.M. Cruz, A.B. Moldes, Valorization of winery waste vs. the costs of not recycling, Waste Management, Vol. 31, pp. 2327-2335, ISSN 0956-053X (2011)
3. E. Maican, E. Murad, G. Cican, I.C. Duţu, Analysis of a Top Lit Updraft Gasification System Designed for Greenhouses and Hothouses, Proceedings of the 16-th International Multidisciplinary Scientific Geoconferences (SGEM), vol. Renewable Energy Sources and Clean Technologies, Albena, Bulgaria, pp. 107-114, ISSN 1314-2704 (2016)
4. H. Lu, T. Ma, L. Lu, Deposition characteristics of particles in inclined heat exchange channel with surface ribs, International Journal of Heat and Mass Transfer, Vol. 161, 120289, ISSN 0017-9310 (2020)
5. Z. Zheng, W. Yang, P. Yu, Y. Cai, H. Zhou, S. K. Boon, P. Subbaiah, Simulating growth of ash deposit in boiler heat exchanger tube based on CFD dynamic mesh technique, Fuel, Vol. 259, 116083, ISSN 0016-2361 (2020)
6. J. Primo, Shell and Tube Heat Exchangers Basic Calculations, PDH Online Course M371, PDH Center, 5272 Meadow Estates Drive, Fairfax, pg. 12/32 (2012)
7. Engineering ToolBox, Heat Exchanger Heat Transfer Coefficients, Available at: https://www.engineeringtoolbox.com/heat-transfer-coefficients-exchangers-d_450.html (2003)

8. M. El-Shafie, M. Khalil Bassiouny, S. Kambara, S. M. El-Behery, A.A. Hussien, Design of a heat recovery unit using exhaust gases for energy savings in an absorption air conditioning unit, Applied Thermal Engineering, Vol. 194, 117031, ISSN 1359-4311 (2021)

Analytical model for estimating the production and composition of gas resulted through gasification

Lucian Mihăescu[1], *Ionel* Pîşă[1], *Iulia* Simion[2*] and *Gabriel Paul* Negreanu[1]

[1] Mechanics and Mechatronics Faculty, University Politehnica of Bucharest, Splaiul Independentei 313, 060042 Bucharest (Romania)
[2] Power Engineering Faculty, University Politehnica of Bucharest, Splaiul Independentei 313, 060042 Bucharest (Romania)

Abstract. For the modelling of gasification processes, several models have been developed over the years. It is remarked that gasification calculation models of very high complexity entail some complications. Therefore, simpler mathematical representations of gasification characteristics and process behavior are required as a first step in addressing such systems. The preliminary calculation simplicity is needed form two perspectives: First – the pre-sizing of gasification installations, and second – the estimation of experimental or functional results. For this kind of topics, an adequate simplified model should be defined. Further, to validate the results it will be necessary to use complex calculation models. The model proposed in this paper addresses gasification with distributive air in the air distribution current, considering general concurrent flow of air fuel. Previous successful investigations, conducted by the present research team, are taken into account within model definition stages. Thus, the work presented here provides useful advances in the field of mathematical modeling of gasification processes. The originality of the model consists in its easy computational accessibility, which allows the approach of technological optimizations, such as the variation of excess air and fuel composition.

Key words: biomass combustion, biomass gasification, energy recovery, modeling of gasification

1. Introduction

Performing or testing the functionalities of a gas generator depends on input obtained from the application of calculation models. The information thus obtained can be used in some cases.

Lately, the appearance of very complex calculation models which require time for the improvement of operations or for the previous processing of data is noticeable.

*Contact person: E-mail address julia.simion@gmail.com,

The existence of simple calculation models proves useful, even if the final accuracy shows some deviations. In many studies the predicted equilibrium compares reasonably well with the results observed in experiments and in other cases the deviation is small enough. The achievement of thermodynamic equilibrium for the multitude of reactions that take place in the gasifier, has a particular importance. The idle time in various temperature zones influences the thermodynamic equilibrium considered in fact reached by all calculation models.

Essentially, a calculation model for a gasifier, must pass through the steps inside the gas generator, namely [1] [2]:

- Drying (dehydration of fuel)
- Pyrolysis
- Combustion
- Reduction of the component CO_2 and H_2O in the hot carbon bed (embers).

Even if the gasifier as a whole is autothermal, the first two processes are endodermal.

The temperature level in the main zones of gas generator reflects the dynamics of the development of specific thermochemical processes.

In the gasification process, the biomass is heated to a high temperature, usually more than 700°C, producing a series of physical and chemical changes, which result in the development of volatile products plus solid carbonaceous residues. [3]

Gasification is one of the most economical and common applications for the use of biomass, which due to its abundant presence in different countries, is of great interest as a source of renewable energy. During gasification the mixture of biomass and gasifying agents (air, steam, carbon dioxide or oxygen) are bring to high temperatures, in order to obtain a fuel gas known as synthesis gas or simply syngas. [4]

The synthesis gas is composed of a set of gases and each of them is very important depending on the specific application of the synthesis gas. For this reason, there is a lot of research on modeling the composition of syngas obtained from both fixed bed gasifiers and fluidized bed gasifiers [5]. In a recent comprehensive review was reported that approximately 60% of all published gasification studies with models used only equilibrium models and just 40% used, or included, non-equilibrium deviations. [6]

The model developed in this research is for the gasification of biomass with air inlet only in the combustion zone.

2. Biomass gasification model

The information on the thermochemical processes in the gas generator, request the elementary analysis of the fuel in the form [7]:

$$C^i + H^i + S_c^i + O^i + N^i + W^i + A^i = 100 \tag{1}$$

Where the elements: carbon C^i, hydrogen H^i, sulfur S_c^i, oxygen O^i, nitrogen N^i, moisture W^i and ash A^i are in mass proportions. For biogas, the absence of sulfur is allowed (S_c^i=0).

The Figure 1 shows the physic-chemical model for the analyzed gas generator, with air supply in the combustion zone. [8]

 https://doi.org/10.1051/e3sconf/202128601011

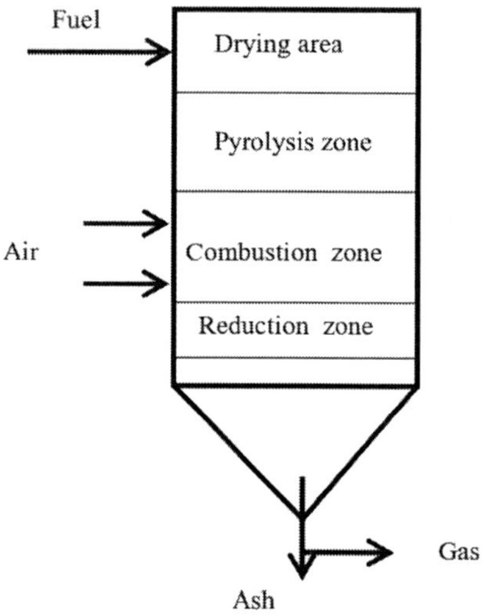

Fig 1. The thermochemical phases in a gasifier with air inlet

In the combustion zone the drying process reaches temperatures ranging between 70-200°C. Starting with 200°C for biomass, the release of volatiles begins. After drying, the fuel composition becomes (it was also considered the removal of the mineral mass):

$$C^{anh} + H^{anh} + O^{anh} + N^{anh} = 100[\%] \quad (2)$$

Participations from the anhydrous state are obtained from those of the initial state by the correction with the factor φ_1:

$$\varphi_1 = \frac{100}{100 - W^i - A^i} \quad (3)$$

$$C^{anh} = C^i \times \varphi_1, \qquad H^{anh} = H^i \times \varphi_1, \qquad O^{anh} = O^i \times \varphi_1, \qquad N^{anh} = N^i \times \varphi_1 \;\; [\%]$$

If the mass of the intake of fuel (flow) in gas generator is m_1, the mass of pyrolysis will become:

$$m_2 = m_1 \times \frac{1}{\varphi_1} \left[\frac{kg}{s}\right] \quad (4)$$

In the pyrolysis phase, volatile materials are released, which include the components: CO, CO_2, CH_4, H_2. Pyrolysis perform in the range of $200 - 700$°C. In this phase there is, as well, a release of liquid tar.

The amount of gas and tar released depends on the average temperature level in the pyrolysis zone [8] [9].

The mass quantity of gas m_3 released can be estimated from the mass m_2 by the function φ_2. The transformation functions in the pyrolysis process come from the literature [9].

$$m_3 = m_2 \times \varphi_2 \left[\frac{kg}{s}\right] \quad (5)$$

$$\varphi_2 = 0.01(1.12T^2 \times 10^{-4} - 0.058T + 30.77) \quad (6)$$

The mass amount of tar m_4 removed from the mass m_2 can be determined by the relation [9]:

$$m_4 = m_2 \times \varphi_3 \quad \left[\frac{kg}{s}\right] \tag{7}$$

$$\varphi_3 = 0.01(-1.38T^2 \times 10^{-4} + 0.12T + 12.64) \tag{8}$$

The composition of the released gas can be estimated with the relations [9]:

- mass proportion of CO:

$$g_{CO} = 0.01(-2.65T^2 \times 10^{-4} + 0.27T - 32.71) \quad [\%] \tag{9}$$

- mass proportion of CH_4:

$$g_{CH_4} = 0.01(6.69T^2 \times 10^{-5} - 0.037T + 4.28) \quad [\%] \tag{10}$$

- mass proportion of H_2:

$$g_{H_2} = 0.01(7.0T^2 \times 10^{-5} - 0.0371T + 5.111) \quad [\%] \tag{11}$$

The difference is obtained by the mass proportion of CO_2:

$$g_{CO_2} = 1 - \left(g_{CO} + g_{CH_4} + g_{H_2}\right) \quad [\%] \tag{12}$$

The massive amounts of gas will be:

$$m_{CO} = m_3 \times g_{CO} \quad \left[\frac{kg}{s}\right] \tag{13}$$

$$m_{CH_4} = m_3 \times g_{CH_4} \quad \left[\frac{kg}{s}\right] \tag{14}$$

$$m_{H_2} = m_3 \times g_{H_2} \quad \left[\frac{kg}{s}\right] \tag{15}$$

$$m_{CO_2} = m_3 - m_{CO} - m_{CH_4} - m_{H_2} \quad \left[\frac{kg}{s}\right] \tag{16}$$

After pirolysis process, the gasification process involves the oxidation of carbon with the air and the thermal reduction of carbon dioxide by carbon.
Mass of carbon participating in the reaction [10]:

$$m_5 = m_2 - m_3 - m_4 \quad \left[\frac{kg}{s}\right] \tag{17}$$

The amount of air m_6 depends on the excess air used λ. Usually, in order to control the mass of carbon participating in the reduction process, the excess air wiil be less than 1 ($\lambda < 1$). On burning carbon, the theoretical volume of combustion air is $9.03\frac{Nm^3}{kg}$.
For excess air λ, the volume of humid air introduced will be:

$$V_a = 9.03 \times \lambda \times m_5 \quad \left[\frac{Nm^3}{kg}\right] \tag{18}$$

Air mass corresponding to the volume $V_a$will be:

$$m_a = 6.4 \times \lambda \times m_5 \quad \left[\frac{kg}{kg}\right] \tag{19}$$

Excess air recommended value is in the range of $\lambda = 0.25 - 0.45$.
The mass of oxidized carbon will be:

$$m_6 = \lambda \times m_5 \quad \left[\frac{kg}{s}\right] \tag{20}$$

The emission of CO_2 at oxidation of the carbon unit results from the stoichiometry of combustion:

$$V_{CO_2} = 1.867 \quad \left[\frac{Nm^3}{kg}\right] \tag{21}$$

The emission of CO_2 from the amount of carbon existing in the reduction zone will be:

$$m_7 = 1.867 \times \lambda \times m_5 \quad \left[\frac{kg}{s}\right] \tag{22}$$

The nitrogen content in the gas will be:

$$V_{N_2} = 0.79V_a = 7.14 \times \lambda \times m_5 \quad \left[\frac{Nm^3}{s}\right] \tag{23}$$

The reduction reaction $C + CO_2 = 2CO$ depends on the system pressure and the temperature level. The reaction is endothermic.
According to Boudouard's law, the ratio of the transformation of CO_2 into CO is presented in table no. 1

Table 1. Relation CO/CO_2

t [°C]	CO [%]	CO_2 [%]
450	0.6	99.4
550	10.7	89.3
650	39.8	60.2
800	93	7.0
925	96	4.0

Analytically, the proportion of CO obtained by reducing the gas CO_2 is determined by the relation

$$k_p = \frac{4pa^2}{1 - a^2} \tag{24}$$

Where:

- $a = \frac{CO}{CO_2 + C}$
- p- reactor pressure $[bar]$
- k_p- the constant of Boudouard's law, which is determined using the data in table no. 2

Table 2. Variation of K_p with temperature

Temperature $[K]$	lgk_p
500	-4.603
1000	0.415
1500	2.240

The content of carbon monoxide after combustion and reduction with carbon will be:

$$m_8 = a \times m_7 \quad \left[\frac{kg}{s}\right] \tag{25}$$

The content of carbon dioxide in the bed will be:

$$m_9 = (1 - a)m_7 \quad \left[\frac{kg}{s}\right] \tag{26}$$

The reactions take place in the carbon bed:

$$C + O_2 = CO_2 + 34{,}096 \ [kJ] \tag{27}$$

$$C + {}^1/_2\,O_2 = CO + 10{,}276 \ [kJ] \tag{28}$$

$$C + CO_2 = 2CO - 13{,}544 \;\; [kJ] \tag{29}$$
$$C + H_2O = CO + H_2 - 9{,}909 \;\; [kJ] \tag{30}$$
$$C + 2H_2 = CH_4 + 6{,}285 \;\; [kJ] \tag{31}$$
$$CO + 3H_2 = CH_4 + H_2O + 17{,}004 \;\; [kJ] \tag{32}$$

Hydrogen is produced only by the reaction (30). Methane is produced by two reactions from equations (31) and (32). For reaction (30), the reaction rate constant can be determined by the relation [11] [9] [12]:

$$K_1 = e^{\left[\frac{4276}{T} - 3.961\right]} \tag{33}$$

For reaction (31), the reaction rate constant can be determined by the relation [13]:

$$\ln K_2 = \frac{7{,}082.842}{T} - 6.567 \ln T + \frac{7.467}{2} T 10^{-3} - \frac{2.167}{6} T^2 10^{-6} + \frac{0.702}{2T^2} + 32.541 \tag{34}$$

The relation can be reduced to form:

$$\ln K_2 = f(T) \tag{35}$$

with

$$K_2 = e^{f(T)} \tag{36}$$

For reaction (31) the speed constant will be [14] [15] [9] [16]:

$$K_3 = 1.198 \times 10^{13} e^{-\frac{26830}{T}} \tag{37}$$

Reaction (31) occurs when steam is used for gasification, in which case the volume of water in reaction (29) allows the emission of hydrogen, which allows the operation of the reaction around the values of the reaction constants K_2 and K_3.
In the study, not using the steam injection eliminates the reaction (32), because the estimated amount of water is consumed in the reaction (29), which has a much higher speed constant.
The stoichiometry of the reaction (29) requires the following amounts of substance [10]:

$$12\; kgC + 22.4\; Nm^3 H_2O = 22.4\; Nm^3 CO_2 + 22.4\; Nm^3 H_2 \tag{38}$$

The amount of water vapor V_{H_2O} in the carbon reaction zone imposed by that achieved in the drying process:

$$V_{H_2O} = 1.242 \frac{W^i}{100} \left[\frac{Nm^3 H_2O}{kg\; fuel}\right] \tag{39}$$

Replacing the relation (39) in (38) will result in a carbon consumption of 0.0066 W^i $\left[{}^{kg}/_{kg}\right]$ and a hydrogen emission of 1.242 $\frac{W^i}{100}\, {}^{Nm^3}/_{kg}$.
The total volume of hydrogen, formed in the pyrolysis phase expressed by relation (15) and by reducing the amount of water expressed by relation (30) will be:

$$V_{H_2}^* = \frac{m_{H_2}}{\rho_{H_2}} + 1.242 \frac{W_i}{100} \left[\frac{Nm^3}{s}\right] \tag{40}$$

According to relation (31) the emission of CH_4 with the maximum value $0.5 \times V_{H_2}$ will depend on the ratio of the reaction constants K_2/K_1, so that it becomes:

$$V_{CH_4} = 0.5\frac{K_2}{K_1}V_{H_2}^* \quad \left[\frac{Nm^3}{s}\right] \tag{41}$$

And the volume of hydrogen V_{H_2} is reduced to the quantity:

$$V_{H_2} = \left(1 - 0.5\frac{K_2}{K_1}\right)V_{H_2}^* \quad \left[\frac{Nm^3}{s}\right] \tag{42}$$

The gas from the gas generator, at the end of the thermochemical processes will have the composition:

$$V_{CO} = \frac{m_{CO}}{\rho_{CO}} + a\frac{m_7}{\rho_{CO}} \quad \left[\frac{Nm^3}{s}\right] \tag{43}$$

$$V_{CO_2} = \frac{m_{CO_2}}{\rho_{CO_2}} + (1-a)\frac{m_7}{\rho_{CO_2}} \quad \left[\frac{Nm^3}{s}\right] \tag{44}$$

$$V_{H_2} = \left(1 - 0.5\frac{K_2}{K_1}\right)\left[\frac{m_{H_2}}{\rho_{H_2}} + 1.242\frac{W_i}{100}\right] \quad \left[\frac{Nm^3}{s}\right] \tag{45}$$

$$V_{CH_4} = 0.5\frac{K_2}{K_1}\left[\frac{m_{H_2}}{\rho_{H_2}} + 1.242\frac{W_i}{100}\right] \quad \left[\frac{Nm^3}{s}\right] \tag{46}$$

$$V_{N_2} = 7.05 \times \lambda \times m_5 \quad \left[\frac{Nm^3}{s}\right] \tag{47}$$

The volume of gas from the gas formed by the thermochemical reactions will be:

$$V_g = V_{CO} + V_{CO_2} + V_{H_2} + V_{CH_4} + V_{N_2} \quad \left[\frac{Nm^3}{s}\right] \tag{48}$$

From this relation the participation of the gas in the gas generator can be determined:

$$CO = \frac{V_{CO}}{V_g} \times 100 \quad [\%] \tag{49}$$

$$CO_2 = \frac{V_{CO_2}}{V_g} \times 100 \quad [\%] \tag{50}$$

$$H_2 = \frac{V_{H_2}}{V_g} \times 100 \quad [\%] \tag{51}$$

$$CH_4 = \frac{V_{CH_4}}{V_g} \times 100 \quad [\%] \tag{52}$$

$$N_2 = 100 - (CO + CO_2 + H_2 + CH_4) \quad [\%] \tag{53}$$

The lower heating value of the gas of gasifier $H_{i,g}$ is determined by the relation:

$$H_{i,g} = 0.01\left(CO \times H_{i,CO} + H_2 \times H_{i,H_2} + CH_4 \times H_{i,CH_4}\right) \quad \left[\frac{kJ}{Nm^3}\right] \tag{54}$$

where $H_{i,CO}$, H_{i,H_2}, H_{i,CH_4} is the lower heating value for carbon monoxide, respectively for hydrogen and methane.
The efficiency of the gasifier will be defined by the ratio between the thermal power of the gasifier and that for the initial fuel:

$$\eta = \frac{V_g \times H_{i,g}^i}{m_1 \times H_i^i} \times 100 \tag{55}$$

where H_i^i is the lower heating value of the fuel subjected to gasification.

3. Thermal field estimation

A simple model can take in consideration:

- heat absorption for dehydration;
- heat absorption in the pyrolysis phase;
- heat emission on carbon oxidation, with the formation of CO_2;
- heat absorption when converting carbon dioxide to carbon monoxide;
- heat absorption when heating the mineral mass

the model did not consider the heat emission for the reduction reactions, considered in a small proportion and the heat loss to the external environment. The air was considered at reference temperature t_1.

In the design of the gas generator, the gradual air intlet was considered, starting with the ignition temperature of the carbon, the evolution of the temperatures in the case of the gas generator will be of the form:

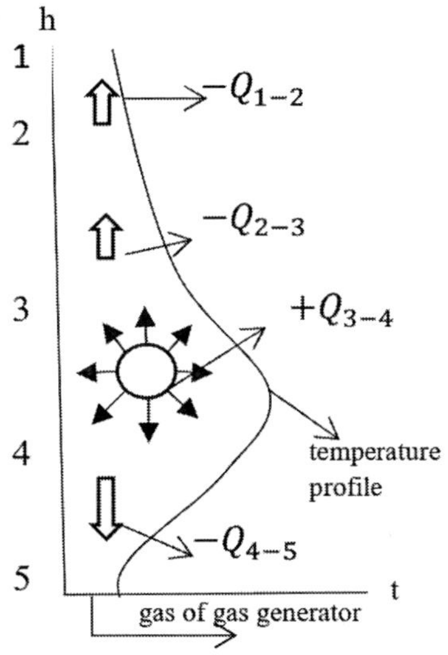

Fig. 2. The evolution of temperature in the gas generator body

Heat emitted by the combustion of fixed carbon (charred coal) will be:

$$Q_{3-4} = m_6 \times H_{i,c}^i \quad [kW] \tag{56}$$

Where $H_{i,c}^i$ is the lower heating value of carbon, measured in $\left[\frac{kJ}{kg}\right]$.

The output of this heat is:

- heating the vaporization of water, Q_{1-2} $[kW]$;
- heating for pyrolysis, Q_{2-3} $[kW]$
- cooling produced in the carbon reduction space, Q_{4-5} $[kW]$;

Finally, the temperature of the gas gasifer leaving the installation is obtained.
For biomass gasification, there is no need for external heating of the gasifier.

In this paper there is an appendix where a result of the composition of the gas obtained by the developed model is exemplified, for a certain temperature level existing in the specific areas of the gas.

4. Conclusions

- The model has taken in account the achievement of thermodynamic equilibrium for all existing stages into the gas generator processes. These hypotheses are sufficiently reached for the gasification of biomass, as shown by all the literature.
- The model led to a decrease in the amount of nitrogen (N_2) found in the gas within the gas generator compared to most other experimental data. The development of a thermal model for the thermochemical phases will contribute to the correction of these values. Gas flow rate influences both the thermal field and the reaction time.
- The gasification process was considered as a completely distinct succession of subprocesses, without any interference between them. This simplified the calculation model to the detriment of accuracy. However, the numerical application demonstrated its viability for the initial assumptions, related to the nature of the fuel and distinct phases of drying, pyrolysis and gasification by reducing some substances.
- The theoretical data obtained were compared with those from previous experiments and presented in the paper "Researches on biofuels gasification using the Lurgi process with homogeneous air inlet over the combustion space", the existing deviation will be used to optimize both directions - biomass gasification experiments, corrections on the calculation model. The model developed can be compared with the models developed in the paper "A mathematical model of biomass downdraft gasification with an integrated pyrolysis model" [17].

5. Appendix

The results of a numerical application in order to preliminary validation of the developed model.

Elementary analyze: $W^i = 20\%$; $A^i = 5\%$; $O^i = 25\%$; $N^i = 1\%$; $H^i = 5\%$, $C^i = 44\%$

Fuel flow: $m_1 = 1\frac{kg}{s}$; $\varphi_1 = 1.33$; $m_2 = 0.75\frac{kg}{s}$;

For $T = 700K$; $\varphi_2 = 0.45$; $\varphi_3 = 0.29$; $m_3 = 0.337\frac{kg}{s}$; $m_4 = 0.217\frac{kg}{s}$;$m_{CO} = 0.087\frac{kg}{s}$; $m_{CH_4} = 0.037\frac{kg}{s}$; $m_{H_2} = 0.043\frac{kg}{s}$; $m_{CO_2} = 0.17\frac{kg}{s}$;$m_5 = 0.196\frac{kg}{s}$; $\lambda = 0.4$; $m_6 = 0.078\frac{kg}{s}$; $V_a = 0.696\frac{m_N^3}{kg}$; $m_7 = 0.14\frac{kg}{s}$;

Temperature in bed 800°C; $a = 0.93$

For temperature of 1,000°K, $K_1 = 1.37$, $K_2 = 0.4$

$V_{CO} = 0.137\ \frac{m^3}{s}$; $V_{CO_2} = 0.09\ \frac{m^3}{s}$; $V_{H_2} = 0.51\frac{m^3}{s}$; $V_{CH_4} = 0.1\ \frac{m^3}{s}$; $V_{N_2} = 0.57\ \frac{m^3}{s}$;

$V_g = 1.503\ \frac{m^3}{s}$

The composition of gas of the gas generator is:

$CO = 11.5\%$; $CO_2 = 5.9\%$; $H_2 = 37.9\%$; $CH_4 = 6.6\%$; $N_2 = 38\%$

References

1. D. Baruah and D. Baruah, "Modeling of biomass gasification: a review," *Renew Sustain Energy Rev,* p. 39:806–15, 2014.

2. [A. Bridgwater, "The technical and economic feasibility of biomass gasification for power generation," *Fuel,* p. 74(5):631–53., 1995.

3. Elif Kirtay, "Recent advance in production of hydrogen from biomass," *Energy conversion and management,* vol. 52, no. pp.1778-1789, 2011.

4. P. Basu, " Biomass gasification and pyrolysis," vol. 1st ed. , 2010.

5. Jordan Amaro, Diego Jhovanny Mariños Rosadoa, Andrés Z. Mendiburu, Leila Ribeiro dos Santos and João A. de Carvalho Jr, "Modeling of syngas composition obtained from fixed bed gasifiers using Kuhn–Tucker multipliers," vol. 287, no. 119068, 2021.

6. Safarian S, Richter C. A and Unnþorsson R, "A review of biomass gasification modelling.," *Renew Sustain Energy Rev* , vol. 110, p. 378–91, 2019.

7. I. G. Carabogdan, A. Badea, L. Ionescu, A. Leca, I. Nistor and I. Cserveny, Industrial termical installtion, Bucharest: ET, 1978.

8. H.-J. Huang and S. Ramaswamy, "Modeling biomass gasification using thermodynamic equilibrium approach," *Appl Biochem Biotechno,* p. 154(1–3):14–25., 2009.

9. M. Trninić , A. Jovović and D. Stojiljković , "A steady state model of agricultural waste pyrolysis: a mini review.," *Waste Manage Res,* p. 34(9):851–65, 2016.

10. P. Nicolae, C. Cazacu, L. Mihaescu, L. Popescu and D. Biclesanu, Modernization of combustion installations for industrial boilers, Bucharest: Tehnical, 1993.

11. T. Patra and P. Sheth, "Biomass gasification models for downdraft gasifier: a state-ofthe- art review," *Renew Sustain Energy Rev,* p. 50:583–93, 2015.

12. Barman NS, Ghosh S and De S, "Gasification of biomass in a fixed bed downdraft gasifier – a realistic model including tar," *Bioresour Technol,* p. 107:505–11., 2012.

13. [Zainal ZA, Ali R, Lean CH and Seetharamu KN, "Prediction of performance of a downdraft gasifier using equilibrium modeling for different biomass materials," *Energy Convers Manage* , p. 42(12):1499–515., 2001.

14. M. Puig-Arnavat , J. Bruno and Coronas A, "Review and analysis of biomass gasification models," *Renew Sustain Energy Rev,* p. 14(9):2841–51, 2010.

15. V. Sikarwar, M. Zhao, P. Clough, J. Yao, X. Zhong and M. Memon, "An overview of advances in biomass gasification," *Energy Environ Sci,* p. 9:2939–77, 2016.

16. Bottino A, Comite A, Capannelli G, Di Felice R and Pinacci P, "Steam reforming of methane in equilibrium membrane reactors for integration in power cycles," *Catal Today* , p. 118(1–2):214–22., 2006.

17. T. Marta , S. Dragoslava , S. Øyvind , W. Liang, J. Aleksandar and M. Nebojsa , "A mathematical model of biomass downdraft gasification with an integrated," *Fuel* , vol. 265, p. 116867, 2020.

18. G. Lazaroiu, L. Mihaescu, G. Negreanu, I. Simion and D.A. Ciupageanu, "Researches on biofuels gasification using the Lurgi process with homogeneous air inlet over the combustion space," E3S Web of Conferences, vol 180/2020, DOI: https://doi.org/10.1051/e3sconf/202018001011.

E3S Web of Conferences **286**, 01012 (2021) https://doi.org/10.1051/e3sconf/202128601012
TE-RE-RD 2021

Review regarding defrosting methods for refrigeration and heat pump systems

Jamal Al Douri[1,3], *Kamel* S. Hmood[1,2], *Valentin* Apostol[1], *Horatiu* Pop[1*], *Saleh J.* Alqaisy[1,3] and *Elena Beatrice* Ibrean[1]

1Dept. of Applied Thermodynamics, Engines, Thermal and Refrigeration Equipment, University POLITEHNICA of Bucharest, Romania
2 University of BABYLON, Iraq
3 University of TECHNOLOGY, Iraq

Abstract: The paper presents a literature review regarding defrosting methods for household refrigeration and heat pumps. The literature review covers the period between 2008 and 2020. A number of 61 papers have been studied. The literature review was conducted by dividing the defrost methods into passive and active. Passive methods involve surface coating and off-cycle. The active defrost methods involve reverse cycle, ultrasonic vibration, hot–gas defrost, electric heating and hot fluid. Surface coating and ultrasonic vibration are recent defrost methods proposed in the literature. Also, recently, phase change materials have been used in household refrigerators to improve their efficiency. This is a topic that needs further investigation and is a subject of interest as it can lead to lower energy consumption during defrost cycles. From the present study, the research direction regarding the use of PCMs in defrost cycles can be highlighted.

Keywords: defrost, PCM, refrigeration systems, heat pumps

1 Introduction

Vapour compression refrigeration systems (VCRs) are the most widely used technical solution in the field of artificial cooling. Frost accumulation on the evaporator surface is an important problem. Frost appears when the moist air comes into contact with the evaporator surface having a temperature lower than 0°C [1]. When frost accumulates on the evaporator surface, it acts as a thermal insulator leading to lower performance of the refrigeration system and in consequence, to higher energy consumption [2, 3]. To avoid the low performance of the refrigeration system, the accumulated frost needs to be removed. The same observations can be applied in the case of heat pumps. In this context, the present work presents a literature review regarding defrosting methods used in VCRs and heat pumps. In this work, a defrost method refers to any method that prevents frost formation and also to frost removal after formation.

Section 2 of the paper presents the most widely used passive and active defrost methods, Section 3 shortly presents phase change materials (PCMs), Section 4 presents the application of PCMs in household refrigeration systems and discusses the possibility to use PCMs in defrosting cycles. Conclusions are presented in Section 5.

The literature review covers the period between 2008 and 2020. A number of 61 papers have been studied. The literature review was conducted by dividing the defrost methods into passive and active. For each method, the authors have struggled to cover the available

* Corresponding author: horatiu.pop@upb.ro

scientific literature and present the conducted work in a synthetic and easy to follow manner.

2 Passive and active defrosting methods

There are two main types of defrosting methods: passive and active. In this section, each of the two defrosting methods will be approached and discussed.

2.1 Passive defrosting

A passive defrost method is a method that does not involve direct energy consumption during the defrost process. The most commonly used passive defrost methods are surface coating and off-cycle [3].

2.1.1 Surface coating

In recent years, with the advance of material technology, evaporator surface coating has become a simpler and more effective anti-frosting technology. Wang et al [4] investigated three types of evaporator fins namely, hydrophilic, bare and superhydrophobic, as shown in Fig.1. They concluded that the superhydrophobic fin outperforms the hydrophilic and bare fins in terms of slowing droplet condensation time. Cai et al [5] investigated frost growth on hydrophobic and bare surfaces. Results indicated that hydrophobic surfaces can delay the growth of the frost layer.

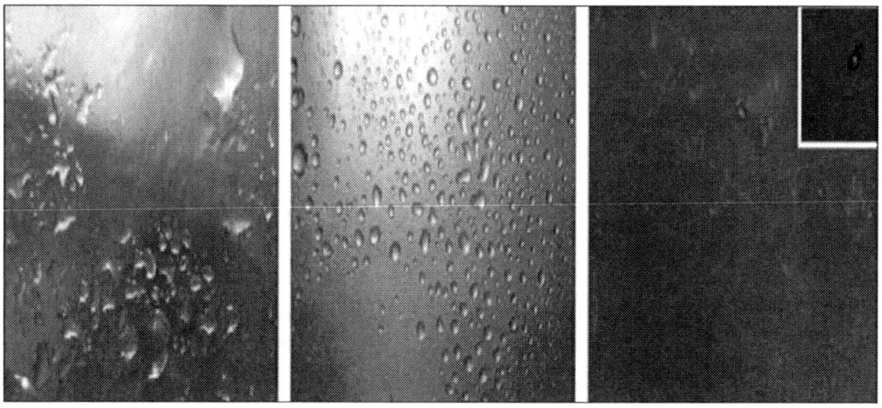

(a) (b) (c)

Fig.1. Water retention on three types of fins: (a) hydrophilic, (b) bare and (c) superhydrophobic.

Wang et al. [6] investigated the effect of the surface type on defrosting time. They compared the efficiency of superhydrophobic, hydrophilic and bare surfaces during defrosting. They concluded that in the case of superhydrophobic surfaces, the defrosting time and energy consumption have decreased compared to hydrophilic and bare surfaces. Jing et al. [7] investigated the phenomena of frosting and defrosting for a rigid superhydrophobic surface and a flexible superhydrophobic surface. The experiments show that the rigid superhydrophobic surface has the advantage of preserving its properties even after several cycles of frosting and defrosting while the flexible superhydrophobic surface did not.

2.1.2 Off-cycle defrost method

In off-cycle defrosting methods, the air surrounding the evaporator is used as a heat source. Therefore, this method can be applied for defrosting in environments where the temperature is higher than 1°C. When the refrigeration system needs to defrost, the compressor stops and with it the cooling process. The frost formed on the evaporator starts to melt in time. After the melting process ends, the operation of the refrigeration system is resumed. The advantages of the off-cycle defrost method involve simplicity and low cost [3]. The main disadvantage of the off-cycle defrost method is that it takes a lot of time because the required heat is taken from the surrounding air [8, 9].

2.2 Active defrosting

The active defrost methods are methods that involve direct energy consumption during the defrost process. The most widely used active defrost methods are: reverse cycle, ultrasonic vibration, hot-gas, electric heating and hot fluid [3].

2.2.1 Reverse cycle defrosting

Reverse cycle defrosting involves a four-way valve that allows changing the refrigerant flow path. At the moment when the defrost process is activated, the hot refrigerant coming from the compressors flows through the evaporator (which now acts as a condenser). In the evaporator, the refrigerant condenses and the heat released is used in the defrost process. After the evaporator, the refrigerant flows through a liquid receiver and undergoes an expansion process in an expansion valve before entering the condenser (which acts now as an evaporator). Finally, the refrigerant enters the compressor and the defrost process can continue until the frost formed on the evaporator is removed [10]. The reverse cycle defrost method is very effective leading to a short defrost time [3]. Minglu et al. [11] made a comparison for the defrosting system of an air source heat pump by using the thermal energy storage method and the reverse cycle defrost method. They concluded that in the case of the thermal energy storage method, the defrosting time shortened from 71.4 % to 80.5% and the energy consumption decreased by 65.1 % to 85.2% in comparison to the normal reverse cycle defrosting process. Wang et al. [12] experimentally tested two defrosting methods for a refrigeration system utilized in a cold room. The first method was electric heater defrosting and the second was reverse cycle defrost. They concluded that when an electric heater has been used for defrosting, the energy consumption was reduced by 20%. At the same time, the temperature inside the cold storage has been reduced by 7°C during the defrosting period. For the reverse cycle method, the time of defrosting has been reduced by 19% and the energy consumption was reduced by 27.2%. In the same operational condition, the reverse cycle defrost method saves more energy than the electric heater method.

2.2.2 Ultrasonic vibration

Ultrasonic vibration is one of the strategies applied for preventing the growth of the frost layer [13]. Cheng et al. studied the effect of low frequencies ranging from 100 to 200 Hz on the frost layer and found that it had little effect on the frost layer [14]. K. Adachi et al. [15] have studied the effect of 37 kHz high-frequency on an aluminium alloy plate in an environment with air at around 100% relative humidity and 2°C. The authors found that high-amplitude ultrasonic vibrations applied to an aluminium alloy plate can inhibit frost accumulation by about 60% [15]. Tan et al. [16] investigated the effect of using ultrasonic vibration to remove frost from a finned tube evaporator of an air source heat pump. The authors used an ultrasound rating power of 50 W and a resonance frequency of 40 kHz. The result showed that the consumption of energy for the defrosting process was 3.14% to 5.46% lower than without intermittent ultrasonic vibration while the heating capability improved by 2.2 % to 9.03%. The coefficient of performance improved by 6.51 % to 15.33 %. Li et al [17] have made an experimental study regarding the influence of 20 kHz ultrasound on frost development for a cold flat surface as shown in Fig. 2. The experiment shows that the frozen water droplets on the surface for which ultrasound effects have been applied became smaller and fragmented than those corresponding to the case without ultrasound effects. In the case of ultrasound effects, the freezing drops occupy less than 52 % in comparison to more than 65% in the case without ultrasound effect. The frost thickness can be reduce by 75 % during tests when ultrasound effect are applied compared to the case without ultrasound effects. Wang et al. [18] used ultrasonic vibrations to study the effect of frost release from finned evaporators. Experiments showed that ultrasonic vibrations cannot remove the primary ice layer from the fins, but the frost crystals and frost branches forming on the primary ice layer can be easily eliminated. Thus the increase of the frost layer can be suppressed.

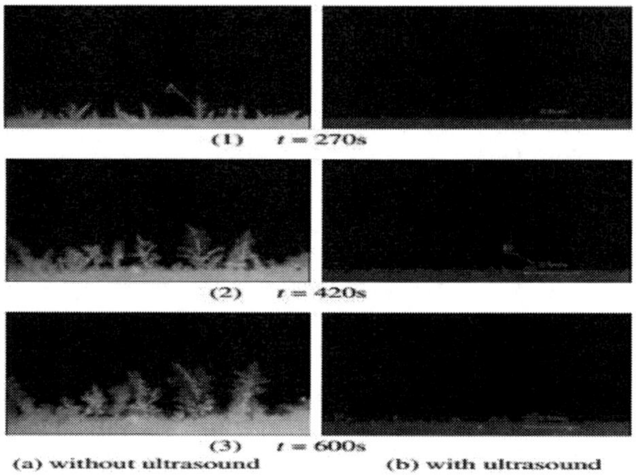

Fig. 2. Side-view of frost formation comparison [17].

2.2.3 Hot–gas defrost method

The hot-gas defrost method involves a pipeline connecting the compressor and the evaporator. The compressed refrigerant vapour reaches the evaporator first, then flows to the liquid-gas separator before entering the compressor [19]. During the hot gas defrost process, the high-pressure refrigerant enters the evaporator and warms the tube and thus, the frost is melting [19]. The advantage of hot-gas defrost is its efficiency because the heat comes from inside the evaporator during the defrost process [20]. The energy consumption of the hot-gas defrost method compared to the reverse-cycle method is smaller [20]. The disadvantage of this process is that during the defrost process the refrigerant condenses inside the evaporator and there is a risk of liquid entering the compressor. This risk can be removed by controlling the refrigerant flow in such a way that it can turn back to the vapour state on the suction line, by adding a liquid-gas separator before the compressor or adding a small evaporator before the compressor. Due to auxiliary valves, pipes and equipment the hot-gas defrost can generate supplementary costs.

Liu et al. [21] developed a bypass cycle defrosting technology that makes use of compressor casing thermal storage. The experimental results show that defrosting time was reduced by 65 % to 77 %. The overall energy consumption was reduced by 89 % to 92 % and also, the compressor noise was reduced by 18.5 %. Cho et al. [22] have studied the performance of the on-off and hot-gas bypass defrosting methods for three evaporators of a cabinet showcase refrigeration system. The results showed that, although it had comparatively more compressor power compared to the on-off cycling, the hot gas defrosting system had a better defrosting capacity and maintained a consistent storage temperature during defrosting.

2.2.4 Electric defrosting method

Electric defrosting methods use electric heating elements mounted in the evaporator. The heating elements are activated during the defrost cycle. Since the heat is applied from outside, only a portion of the heat emitted is used to remove the frost. The rest of the heat is transferred to the air inside the refrigerated space. When the refrigeration system resumes operation after a defrost period, this additional heat must be removed. The advantage of this method is that it is easy to install and control. The disadvantages of this method are: it is expensive and consumes a lot of energy [3, 8, 9].

Yoon et al. [23] focused on improving the defrosting efficiency of a household refrigerator by using three defrost modes. The first mode controls two heaters at the same time; the second mode controls each heater individually and the third mode uses a radiation heater in steps. The results showed that all modes were effective in preventing a temperature rise in the freezer and improving the defrosting efficiency. However, the best performance was obtained by the second mode. The temperature in the refrigerator drops from 11°C to 5°C at the time of defrosting and improves the defrosting efficiency by 15%. Some researchers such as Melo et al. [24] used three different types of electrical heaters, distributed, chlorine and glass. They discovered that all three types of heaters had the same defrosting efficiency. The glass tube heater had the highest efficiency of 48 %. Zhao et al. [25] added an extra heater of 60 W in the middle of the evaporator besides the original bottom one having 180 W. The aim was to observe the reduction of the temperature rise in the freezer cabinet during the defrosting process. The results showed that adding the heater enhanced the defrost by removing frost on both the lower and upper halves of the evaporator simultaneously. Therefore, the defrost duration was reduced by 3.3 minutes and at the same time, the temperature inside the freezer can be reduced by 1.1 °C during the defrost cycle. Yin et al. [26] studied a novel defrost method using air bypass circulation and an electric heater for cold storage. The results showed that during defrosting, circulating air was useful for enhancing the defrost speed. Compared with the traditional method (without air bypass), the defrost time of this new method was lowered by 62.1 %, defrost energy consumption decreased by 61.0 % and the fluctuation in storage temperature decreased by 70.1 %. Furthermore, the efficiency of defrosting raised up to 77.6 % being 2.93 times more efficient than the traditional method.

2.2.5 Hot fluid defrost method

The hot fluid defrost method involves the use of a fluid, usually water or brine, which is sprayed on the evaporator. The resulting mixture of water and frost is drained. During the defrost cycle the compressor of the refrigeration system does not work. This defrost method can be applied only if a hot fluid source is available [8, 9]. In Table 1, the advantages and disadvantages of the most widely used defrosting methods are summarized.

Table 1.The advantages and disadvantages of defrosting methods

Defrostmethods	Advantage	Disadvantages
Off-cycle defrost	Simple and economical [17], low cost, no reconstruction work [19], low maintenance costs, it is a safe method, easy to control	Not relevant in applications where temperatures are below freezing [19]. It takes a long time to melt the frost [20]
Hot fluid defrost	Applies hot liquid directly to the accumulated frost, able to achieve fast defrost	Limited for application, some fluid retained on the surface, not widely investigated or

		applied [19], not recommended for low-temperature applications
Hot - gas	Efficient in terms of defrosting time, mostly used in industry [19], widely used method.	Extremely high working pressures
Electric	Easy installation [19]	High energy consumption [19]
Reverse cycle	Time for defrosting shorter than hot –gas method [19]	Consumes more energy and is dangerous for use in refrigerators due to repeated reversing of the four-way valve, which can result in refrigerant leakage

2.2.6 Defrosting methods using phase change materials

There are other methods of defrosting which are using phase change materials (PCMs). The defrosting methods using PCMs are relatively new and are slowly gaining more attention from researchers. Next, Section 3 shortly presents an overview of PCMs and Section 4 presents the application of PCMs in household refrigeration systems and discusses the possibility to use PCMs in defrosting cycles.

3 Phase change materials

In recent years, the PCMs have attracted more interest from researchers because of:

- high storage energy capacity;
- low storage volume;
- isothermal behaviour during the charging and discharging phases;

The PCMs have high thermal energy storage capacity and have the ability to release and absorb a big quantity of energy at a constant temperature [27].

3.1 Classification of phase change materials

The PCMs can be classified according to the melting temperature into three groups:

- Low-temperature PCMs having a melting temperature below 15 °C.
- Medium temperature PCMs having a melting temperature between 15 °Cand 90 °C.
- High-temperature PCMs having a melting point above 90°C.

Fig. 3 shows the most important applications of PCMs depending on the melting point.

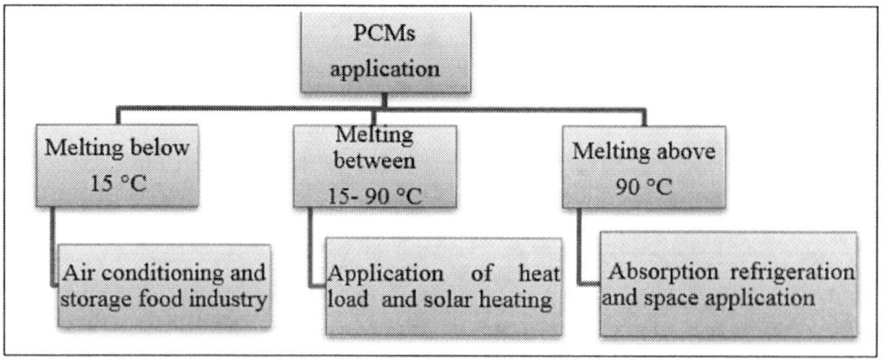

Fig.3.Classification PCMs according to the melting temperature

PCMs can be also classified according to the mode of phase transition: liquid-gas, solid-liquid, solid-gas and solid-solid, as shown in Fig 4 **[30].**

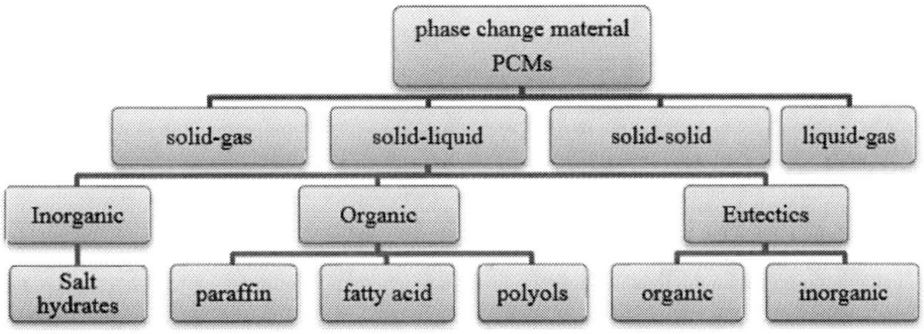

Fig. 4. Classification PCMs according to the mode of transition

3.2 Inorganic PCMs

Inorganic PCMSs have a melting temperature ranging from 10 °C to 900°C. This type of PCMs is widely used in solar energy applications where the required melting temperature is high. The thermal conductivity of inorganic PCMs is higher than the organic PCMs and so is the density. The main disadvantage of the inorganic PCMs, namely the salt hydrates, is that they exhibit great instability as they lead to water separation during the heating process. Most inorganic salts are not highly flammable.

Salt hydrates are the most studied type of PCM, consisting of water and salt. They have the general formula AB_nH_2O and melting temperatures ranging from 15 °C to 117°C [28-30].

The main advantages of salt hydrates are:

- Low cost;
- Availability;
- High melting temperature;
- High thermal conductivity;
- High heat of fusion per unit volume;
- Low volume changes;
- Compatible with plastic;

The main disadvantages of salt hydrates are:

- Separation means the production of additional hydrates or dehydrated salts which tend to decrease the available active volume for heat storage.
- Decrease in heat of fusion more than 73% for $Na_2SO_4.10H_2O$ after 1000 melt/freeze cycles.
- Salt hydrates show supercooling because the freezing point of other PCMs does not start to crystallize.
- Corrosion appears in metal containers [28–30].

3.3 Organic PCMs

Organic PCMs are the most widely used type. This group of PCMs includes the paraffin family, the fatty acids family and polyols. Among these organic PCMs, paraffin is the most common having a melting temperature ranging from 35 °C – 70 °C. Organic PCMs can store a large amount of energy in a small mass, are chemically and physically stable and are compatible with a wide range of materials. At the same time, the organic PCMs are flammable having flashpoints close to 200°C which is a value outside their range of application. Organic PCMs display low thermal conductivity [27-30].

Paraffin has the chemical formula CH_3-(CH_2)-CH_3. Its main advantages are:

- Large latent heat during phase change
- No tendency to separate;
- It is chemically stable;
- When exposed to oxygen, it oxidizes slowly;
- There is no change of thermal properties after frequent melting/solidification processes;
- High melting temperatures;
- Non-reactive and safe;
- It is compatible with all-metal containers and easy to integrate into heat storage systems.

The disadvantages of paraffin include:

- low thermal conductivity;
- when changing from solid to liquid, it takes a large volume;
- flammability [28].

3.4 Fatty acids

Fatty acids have the chemical formula CH_3 (CH_2) 2nCOOH and they have similar properties as paraffin. The main advantage of fatty acids is their sharper phase transformation. They are more expensive compared with paraffin and are mildly corrosive. However, after 1500 melting/solidification cycles, the fatty acids display thermal stability [28, 30].

3.5 Polyols

Polyols have the advantage of a small volume change. The main disadvantages involve high phase change temperature, low latent heat and a high cost [30].

3.6 Eutectics

They consist of two or more components and have the same melting and solidification temperature. Table 2 presents a list of PCMs and their properties. The corresponding references are also mentioned.

Table 2. The properties and characteristics of PCMs

Ref	Phase change materials (PCMs)	Melting Temperature [°C]	Heat of fusion [kJ/kg]	Thermal conductivity [W/(m K)]	Density [kg/m³]
[33]	E21	21	150	0.43	1480
[34]	Paraffin	26	82.7	0.2(s),0.15(l)	7870(s),750(l)
[35]	CaCl2·6H₂O	29.7	187.490	1.088	1710
[36]	C-18	-18.40	306	0.5-0.7	1300
[37]	n-octadecane	27.7	243.5	0.148	777
[38]	Na2CO3.10H2O	32-36	247	-	-
[39]	MgCl2 .6H2O	117	168.6	0.57(l,120°C)/0.694 (s,90°C)	1450(L,120°C)/1569 (s,20°C)
[39]	Mg(NO3)2.6H2O	89	162.8	0.49(l,95°C)/0.611(s ,37°C)	1550(l.94°C)/1636(s, 25°C)
[39]	Ba(OH)2.8H2O	48	265.7	0.653(l)/1.225(s)	1937(l)/2070(s)
[39]	Polyglycol E600	22	127.2	0.189	1126(l)/1232(S)

[39]	Palmitic acid	64	185.4	0.162	850(l)/ 989(S)
[39]	Capric acid	32	152.7	0.153(l)	878(l)/1004(S)
[39]	Caprylic acid	16	148.5	0.149	901(l)/981(S)
[40]	H20(salt hydrates)	0	333	0.6(20°C)/2.2(s)	917(0°C)/998(20°C)
[40]	KF.4H20	18.5	231	0.4	1455(18c)/1447(20°C)
[40]	Na2S04.10H20	32.4	254	1.93(s)	1485(s)
[40]	Na2Hp04.12H2	35	281	1.7(s)/1.95 (l)	1520(s)/1442(l)
[40]	Zn(N03)2.6H2	36.4	147	1.34(s)/2.26(l)	2065(14°C)
[40]	Na2S203.5H20	48	201	1.46(s)/2.39(l)	1730(s)/1670(l)
[40]	Ba(0H)2.8H20	78	267	1.17(s)	2180(s)
[41]	Wax(p116)	49	210	0.24(s)/0.15(l)	780(l)/860(s)
[41]	Calcium chloride	28	200	0.626(s)/0.45(l)	1500(l)/1680(s)
[42]	IGI 1230A	54.2	278.2	0.25 (s)/0.135 (l)	880 (sol)/770 (l)
[43]	Oleic acid	13	75.5	0.103(l)	871(l)
[43]	Capric acid	32	153	0.153(l)	1004(s)/878(l)
[43]	Lauric acid	44	178	0.147(l)	1007(s)/965(l)
[43]	Palmitic acid	64	185	0.162(l)	989(s)/850(l)
[43]	Stearic acid	69	202	0.172(l)	965(s)/848(l)

4 Application of PCMs in household refrigeration systems

Generally, in the case of household refrigerators, PCMs are used in order to improve their performance. In literature, work has been conducted to improve the coefficient of performance (COP) by using PCMs at the condenser level, evaporator level and inside the cold storage space [31]. In this regard, Table 3 presents a part of the literature which has as a subject the use of PCMs in household refrigerators. Table 3 covers the time period from 2008 to 2020.

Table 3. Literature regarding the use of PCMs in household refrigerators

Author	Main conclusion	Components where PCMs are used	Year
Azzouz et al [44]	The use of PCMs increases the energy efficiency and leads to enhanced heat transfer from the evaporator	The outer side of the evaporator	2008
Azzouz et al [45]	Adding PCMs shows a significant increase in efficiency compared to a conventional system	The backside of the evaporator	2009
Long Cheng et al [46]	Adding PCMs results in a lower condensation temperature, a higher evaporating temperature and much larger subcooling, as well as a 12% improvement in performance.	Condenser	2011
Khan and Afroz et al [47]	PCMs, decrease the fluctuation of the cabinet temperature.	Evaporator	2014
Khan and Afroz et al [48]	The effect of PCMs in compressor on-off cycle has been studied. The authors found that the number of compressors on-off cycles with PCMs is 5% to 30% lower. The efficiency increases and the temperature fluctuation inside the cabinet decreases.	Evaporator	2015
Sonnenrein et al [49]	Water, paraffin or copolymer compound has been used as PCM. The energy consumption has been reduced and the temperature lowered from 4°C to 0.5°C.	Evaporator	2015

Yusufoglu et al[50]	PCMs have been used to reduce the compressor on/off time, resulting in a 9.4% energy saving	Evaporator	2015
Reddy et al[51]	The authors used PCMs to cover the evaporator and the results show that the cooling effect inside the chamber increased. COP also increased from 3.29 to 3.5.	Evaporator	2016
Long Cheng et al [52]	PCMs have been used in condenser and evaporator and the result shows electrical consumption savings between 28 % and 32 %.	Condenser-evaporator	2017
Bakhshipour et al [53]	The authors point out that using PCMs increases the convection procedure and results in a 9.58 % improvement of the COP.	PCMs Heat exchanger after the condenser	2017
Elarem et al [54]	The experimental results indicate that energy consumption decreased by 12% and the COP increased by 8%, compared to the refrigerator without PCMs.	Evaporator	2017
Liu et al [55]	The refrigerator's energy consumption, with PCMs, decreased by 18.6 %, and the compressor ON-time ratio was reduced by 13.6 %.	Evaporator (free frost)	2017
Dandotiya and Banker et al [56]	The COP in the case of the PCMs-based condenser was 28% higher and the energy consumption was 15% lower.	Condenser	2017
Liu et al [21]	The authors proposed the use of a compressor casing thermal storage using PCMs for defrosting purposes. The results show that defrosting time can be reduced by 65% to 77 %, energy consumption can be reduced by 89 % to 92 % and also, the compressor operating noise level can be reduced by 18.5 %.	Compressor	2017
Liu et al [57]	The authors used a heat exchanger after the compressor for four different types of PCMs and modes of operation. As a result, defrosting speed was increased by 50% compared to the initial electric heating, while defrosting electric energy consumption was decreased by approximately 71%.	After compressor	2017
Zarajabad and Ahmadi et al[58]	The authors have used PCMs and fins. They noticed that the compressor's energy consumption decreased by 17.4 %.	Evaporator	2018
Tikudave and hole et al[46]	They used the heat storage condensers between the condenser and evaporator in the household refrigerator. The results show: the condensation temperature is lower, the temperature inside the cabinet is more stable, the on-time was improved, the COP increased by 21.5 % and the energy efficiency increased by 15%.	Evaporator	2018
Pirvaram et al [59]	The authors used two eutectic PCMs and the results show that the compressor work time percentage was reduced from 32.7 % to 27.6 %. The energy consumption was reduced by 13% when using one PCM and by 8% by using two PCMs.	Condenser	2019

Maiorino et al [60]	The use PCMs determined a noticeable decrease of the temperature gradient within the cabinet and extended the OFF time of the compressor.	Evaporator	2019
Abdolmaleki et al [61]	The energy consumption with PCMs was reduced by 8.37%.	Evaporator	2020

As it can be noticed from Table 3, PCMs have attracted attention as a solution to increase the efficiency of household refrigerators. As a remark, in the case of household refrigerators, the efficiency can be further improved by using PCMs during the defrost cycles. This is a topic that requires further investigation.

5 Conclusion

The paper presents a literature review regarding the defrost methods for vapour compression refrigeration systems and heat pumps. The literature review covers the period between 2008 and 2020. A number of 61 papers have been studied. One specialized in refrigeration systems and heat pumps can notice that the available scientific literature having as a topic defrost methods is scarce compared to other topics from refrigeration/heat pump technology like new types of refrigerants or efficiency enhancement.

There are two main types of defrosting methods: passive and active. Passive methods involve surface coating and off-cycle. The active defrost methods involve reverse cycle, ultrasonic vibration, hot–gas defrost, electric heating and hot fluid. Among these methods, surface coating and ultrasonic vibration are the most recent ones proposed in literature while the other methods could be classified as well know and widely used methods.

The present work shows that defrosting methods are an important source of direct energy consumption, except for the surface coating and off-cycle defrost. Reducing the energy consumption during the defrost cycle will lead to the higher efficiency of the refrigeration/heat pump system. In this regard, recently, phase change materials have been used in household refrigerators to improve their efficiency. The number of papers regarding this subject is constantly increasing. Related to phase change materials, the most recent work involves their use in the defrost cycles of the refrigeration systems. This is a topic that needs further investigation and is a subject of interest as it can lead to lower energy consumption during defrost cycles. Defrost cycles using phase change materials could be applied to both household and industrial refrigeration systems. In the next years, at least in Europe, the standards regarding household refrigerator energy consumption will become more strict. In this context and based on the present work, the research direction regarding the use of PCMs in defrost cycles can be highlighted.

References

[1] M. Qu, L. Xia, S. Deng, Y. Jiang, Appl. Energy **91**,122–129 (2012)
[2] W. Su, W. Li, X. Zhang, Energy Convers. Manag. **148**, 1157–1169 (2017)
[3] M. Amer, C.C. Wang, Renew. Sustain. Energy Rev. **73**, 53–74 (2017)
[4] F. Wang, C. Liang, M. Yang, X. Zhang, Exp. Therm. Fluid Sci. **61**, 113–120(2015)
[5] L. Cai, R. Wang, P. Hou, X. Zhang, Energy Build. **43**, 1159–1163(2011)
[6] F. Wang, C. Liang, M. Yang, X. Zhang, Appl. Therm. Eng. **113,** 229–237 (2017)
[7] T. Jing, Y. Kim, S. Lee, D. Kim, J. Kim, W. Hwang, Appl. Surf. Sci. **276,** 37–42 (2013)
[8] K. Nawaz, A. Elatar, B. Fricke, *A Critical Literature Review of Defrost Technologies for Heat Pumps and Refrigeration Systems*, (US DEPARTMENT OF ENERGY, 2018)
[9] M. Song, S. Deng, C. Dang, N. Mao, Z. Wang, Appl. Energy **211**,1150–1170

(2018)
[10] J. Shen, Z. Qian, Z. Xing, Y. Yu, M. Ge, Energy Procedia **160**, 491–498,(2019)
[11] M. Qu, T. Li, S. Deng, Y. Fan, Z. Li, Appl. Therm. Eng. **121**, 728–736 (2017)
[12] D. Wang, J. Jiang, L. Tao, Z. Kou, L. Yao, Appl. Therm. Eng. **127**, 1267–1273 (2017)
[13] H. Tan, X. Zhang, L. Zhang, T. Tao, G. Xu, Appl. Therm. Eng. **153**, 113–127 (2019)
[14] C.H. Cheng, C.C. Shiu, Int. J. Refrig. **26**, 69–78 (2003)
[15] K. Adachi, K. Saiki, H. Sato, T. Ito, Japanese J. Appl. Physics, Part 1 Regul. Pap. Short Notes Rev. Pap. **42**, 682–685 (2003)
[16] H. Tan, G. Xu, T. Tao, X. Sun, W. Yao, Appl. Energy **158**, 220–232,(2015)
[17] D. Li, Z. Chen, M. Shi, Exp. Therm. Fluid Sci. **34**, 1247–1252,(2010)
[18] D. Wang, T. Tao, G. Xu, A. Luo, S. Kang, Exp. Therm. Fluid Sci. **36**, 1–11(2012)
[19] N. Hoffenbecker, S.A. Klein, D.T. Reindl, Int. J. Refrig. **28**, 605–615(2005)
[20] J.H. Rainwater, ASHRAE J. 51 38–51 (2009)
[21] Z. Liu, F. Zhao, L. Zhang, R. Zhang, M. Yuan, Y. Chi, Appl. Therm. Eng. **130**, 1215–1223 (2018)
[22] H. Cho, Y. Kim, I. Jang, Energy **30**, 1915–1930 (2005)
[23] Y. Yoon, H. Jeong, K.S. Lee, Energy Convers. Manag. **157**, 511–516 (2018)
[24] C. Melo, F.T. Knabben, P. V. Pereira, Appl. Therm. Eng. **51**, 239–245 (2013)
[25] R. Zhao, D. Huang, X. Peng, H. Yang, Int. J. Refrig. **99**, 186–193 (2019)
[26] H.J. Yin, Z. Yang, A.Q. Chen, N. Zhang, Energy **37**, 623–631 (2012)
[27] A.F. Regin, S.C. Solanki, J.S. Saini, Renew. Sustain. Energy Rev. **12**, 2438–2458(2008)
[28] S.D. Sharma, K. Sagara, Int. J. Green Energy **2**, 1–56 (2005)
[29] J. Vadhera, A. Sura, G. Nandan, G. Dwivedi, Mater. Today Proc. **5**, 3411–3417(2018)
[30] S.F. Li, Z. hua Liu, X.J. Wang, Appl. Energy **255**, 113667(2019)
[31] K. Du, J. Calautit, Z. Wang, Y. Wu, H. Liu, Appl. Energy **220**, 242–273(2018)
[32] J. Shen, Z. Qian, Z. Xing, Y. Yu, M. Ge, Energy Procedia **160**, 491–498 (2019)
[33] F. Wang, G. Maidment, J. Missenden, R. Tozer, Appl. Therm. Eng. **27**, 2902–2910(2007)
[34] V. Antony Aroul Raj, R. Velraj, Int. J. Therm. Sci. **50**, 1573–1582 (2011)
[35] A.H. Mosaffa, F. Talati, H. Basirat Tabrizi, M.A. Rosen, Energy Build. 49, 356–361 (2012)
[36] E. Oró, L. Miró, M.M. Farid, L.F. Cabeza, Int. J. Refrig. **35**, 984–991 (2012)
[37] A. Ehsan, H. Singh, M. Saqlain, G. Farid, Sci.Int.(Lahore). **26**, 1435–1438(2014)
[38] M. Jaafer, A. Alatabe, Int. J. Sci. Res. Sci. Eng. Technol. **4**, 93–98(2018)
[39] M.M. Farid, A.M. Khudhair, S.A.K. Razack, S. Al-Hallaj, Energy Convers. Manag. **45**, 1597–1615 (2004)
[40] A. Abhat, Sol. Energy **30**, 313–332 (1983)
[41] H. El-Dessouky, F. Al-Juwayhel, Energy Convers. Manag. **38**, 601–617 (1997)
[42] R.J. Warzoha, R.M. Weigand, A.S. Fleischer, Appl. Energy **137**, 716–725(2015)
[43] H. Ge, H. Li, S. Mei, J. Liu, Renew. Sustain. Energy Rev. **2**, 1331–346 (2013)
[44] K. Azzouz, D. Leducq, D. Gobin, Int. J. Refrig. **31**, 892–901 (2008)
[45] K. Azzouz, D. Leducq, D. Gobin, Int. J. Refrig. **32**, 1634–1644 (2009)
[46] W.L. Cheng, B.J. Mei, Y.N. Liu, Y.H. Huang, X.D. Yuan, Energy **36**, 5797–5804 (2011)
[47] H.M.M. Afroz,Int. J. Recent adv. Mech. Eng. (IJMECH)**3**, 43–52(2014)
[48] M. Imran Hossen Khan, H.M.M. Afroz, Sci. Technol. Built Environ. **21**, 462–468 (2015)

[49] G. Sonnenrein, E. Baumhögger, A. Elsner, K. Fieback, A. Morbach, A. Paul, J. Vrabec, Int. J. Refrig. **60**, 166–173 (2015)
[50] Y. Yusufoglu, T. Apaydin, S. Yilmaz, H.O. Paksoy, Int. J. Refrig. **57**, 173–185(2015)
[51] R. Rahman, IOSR J. Mech. Civ. Eng. **10**, 08–16(2013)
[52] W. long Cheng, M. Ding, X. dong Yuan, B.C. Han, Energy Convers. Manag. **132**, 180–188 (2017)
[53] S. Bakhshipour, M.S. Valipour, Y. Pahamli, Int. J. Refrig. **83**, 1–13 (2017)
[54] R. Elarem, S. Mellouli, E. Abhilash, A. Jemni, Appl. Therm. Eng. **125**, 1320–1333(2017)
[55] Z. Liu, D. Zhao, Q. Wang, Y. Chi, L. Zhang, Int. J. Refrig. **79**,130–142 (2017)
[56] D. Dandotiya, N.D. Banker, Int. J. Air-Conditioning Refrig. **25**, 1–9 (2017)
[57] Z. Liu, A. Li, Q. Wang, Y. Chi, L. Zhang, Appl. Therm. Eng. **118**,256–265 (2017)
[58] O. Ghahramani Zarajabad, R. Ahmadi, Therm. Sci. Eng. Prog. **7**,115–124 (2018)
[59] A. Pirvaram, S.M. Sadrameli, L. Abdolmaleki, Energy. **181**, 321–330 (2019)
[60] A. Maiorino, M.G. Del Duca, A. Mota-Babiloni, A. Greco, C. Aprea, Int. J. Refrig. **100**, 255–264 (2019)
[61] L. Abdolmaleki, S.M. Sadrameli, A. Pirvaram, Renew. Energy **145,** 233–241 (2020)

Revaluation of bamboo as biomass

Nicoleta Alexandra Vanghele [1,*], *Andreea* Matache[1], *Mariana Mădălina* Stanciu[1], and *Dumitru Bogdan* Mihalache[1]

[1] INMA National Institute of Research - Development for Machines, and Installations designed to Agriculture and Food Industry, Bucharest

Abstract. Bamboo is a grassy plant that grows rapidly, is also a renewable natural resource and with a high yield. There are many applications and uses of bamboo, which led to its planting around the world. Among the industries in which bamboo has by now experienced extensive exploitation, we can remember the textile, construction, furniture ; even food, pharmaceutical or cosmetic industries.
Biomass is the plant material that is used as a renewable energy resource. This is achieved others by cultivating plants specially used for energy production or by using resistors from small industrial sectors.
The current paper highlights the potential for the recovery of bamboo as biomass, both in terms of the fact that bamboo is a plant that is believed to be an inexhaustible resource, but also for the wide amount of rest from its multiple uses.

Keywords : bamboo, biomass, renewable resource, residues

Introduction

In recent decades, there has been a tendency to exploit biomass sources to replace traditional fossil fuels. Such a material capable of producing large amounts of biomass is the bamboo [1].

Biomass (vegetable material) is a renewable energy source which becomes more competitive compared to traditional energy production sources in terms of system installation expenses and energy production costs [1]. Biomass stores the energy it receives from the Sun using the process called photosynthesis. In biomass processing, whether it is incineration, pyrolysis or gasification, there is a release of the energy captured from the Sun and its transformation into other forms of energy. From this perspective, biomass can be seen as an accumulator that can store solar energy for long periods. [2]. Unlike batteries, however, biomass is a sustainable source of energy storage, and if sustainable crop management methods are applied, it is unlimited. Generally, there are two main approaches to the use of plants to produce energy (1) the cultivation of plants specifically for the use of energy and (2) the use of residues resulted from agricultural practices [3]. The best

* Corresponding author : vanghelelrc@gmail.com

approach varies, depending on several factors such as climate, soils and geographical conditions [4].

Biomass may be considered an abundant source of energy because covers a large part of land surfaces and has a short life cycle, being an important candidate as an alternative fuel.

Energy crops are identified with the cultivation of annual or sustainable plants, at the industrial level, which are intended to provide biomass used to obtain various energy sources. Compared to other renewable sources of bioenergy such as wind powe or photovoltaic power, the energy produced from biomass is an intermitted energy source in nature, but can easily be stored in solid, liquids or gaseous forms. From this issue it can be concluded that bioenergy is a stable resource of auxiliary energy [5].

Bamboo is part of the *Poaceae* family, as well as corn, wheat, sugar cane, barley, oats, rye and rice. It can have up to 1250 species from 75 genres and is spread around the world [6].

The bamboo plants are fast-growing plants, representing a renewable and high-yield natural resource.

Bamboo is known as "the plant of a thousand uses", since plays an important socio-economic role, being also related to humanity since ancient times [7].

Bamboo has common characteristics with many other biomass raw materials in terms of the heating value and chemical composition. The commonly used conversion technologies of waste to energy involves the transformation of the row material into other forms of fuels [8]. The bamboo characteristics may differ depending on the species, plant section, the maturity of the plants, the season, or cultivation practices (e.g. fertilizer) and place of production [9].

This paper is a brief synthesis of the highlighter of the biomass potential of bamboo studied in literature.

Materials and methods

In the process of obtaining the carbohydrates that form the building blocks" of biomass, the CO_2 in the atmosphere, water and in the soil participate. If chemical energy is extracted (biomass is burned), oxygen from the atmosphere and carbon from plants react by forming carbon dioxide and water (fig.1). Since carbon dioxide can participate again in the biomass formation process, the process is cyclical [10].

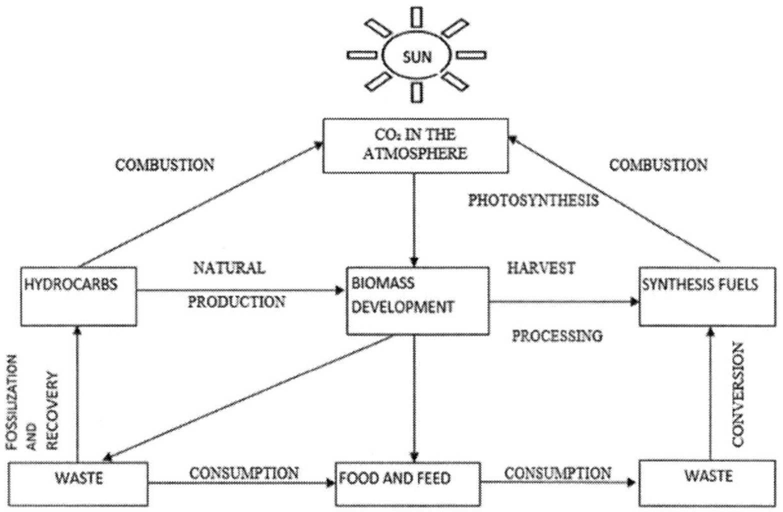

Fig. 1. Biomass formation [8].

The elemental composition of vegetable biomass is shown in Figure 2. The size of the boxes corresponded to the amount of each element in the composition of biomass.

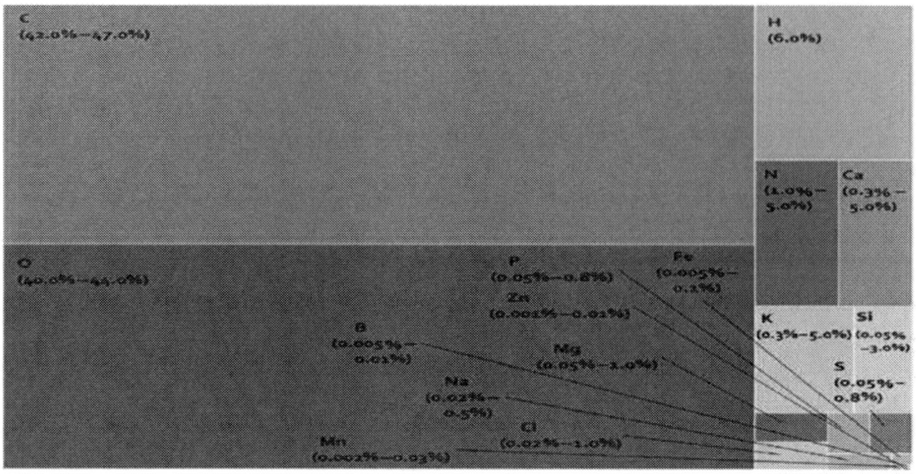

Fig. 2. Elemental composition of plant biomass. (%) [11].

The general composition and heating value of bamboo is found between the values of wood and the herbaceous material. The woody bamboo peaks, that are mature and dry, is considered the most suitable bio-fuel.

Bamboo presents several advantages compared to other lignocellulosic raw materials. The most important advantages are the higher crop yields and higher biomass densities - both being important features that may lead to lower production and transport costs. In addition, bamboo cropping doesn't need the utilization of seeds and plastics for baling. The total amounts of fertilizers required by the crops are also relatively low. These are key benefits in terms economy and of biomass value.

Table 1 shows the average characteristics of the plant compared to other biomass raw materials, that have been found in the literature [8].

Table 1. Biomass properties of bamboo compared to specific raw materials.

Raw		Bamboo culm	Cane bag	Wheat straw	Wood
Higher heating value (dry)	MJ kg^{-1}	17–20	18–20	16–19	17–20
Bulk density	kg m^{-3}	300–700	150–200	160–300	200–500
Crop yield	Ton ha/year	10–40	7–15	3–12	5–20
Cellulose	%	40	35	38	50
Hemicellulose	%	20	25	36	23
Lignin	%	20	20	16	22
Others	%	2–10	20	10	5

Bamboo affinity for energy pride.

Methods for producing energy from bamboo crops biomass

Apart from transesterification technique, transformation of waste biomass to energy is carried out using thermochemical and biochemical conversion [12]. There are many procedures to recover energy from bamboo biomass.
Thermochemical conversion can be performed:

- *By direct* combustion.

Direct combustion furnaces might be used for direct heat technologies, or for creating steam. Dry bamboo may be used for direct combustion in small incineration units or on industrial scale. The principle of combustion consists of burning the fuel, that is composed mainly of carbon and hydrogen atoms under controlled conditions.

- *By pyrolysis processes,*

Pyrolysis is the phenomenon of thermal decomposition of biomass into highly heterogeneous gaseous, liquid, and solid intermediates, in the absence of oxygen. Therefore, it is a thermal degradation of organic materials at a temperature of 350 to 600 °C. The main products produced from the pyrolysis process consist of coal (which is the solid phase of the obtained product), condensable oils (heavy hydrocarbons), tars (which is the liquid phase) and the condensable gases (which is called the gaseous phase). The pyrolysis is usually performed accordingly to the processes found in Figure 3 [13].

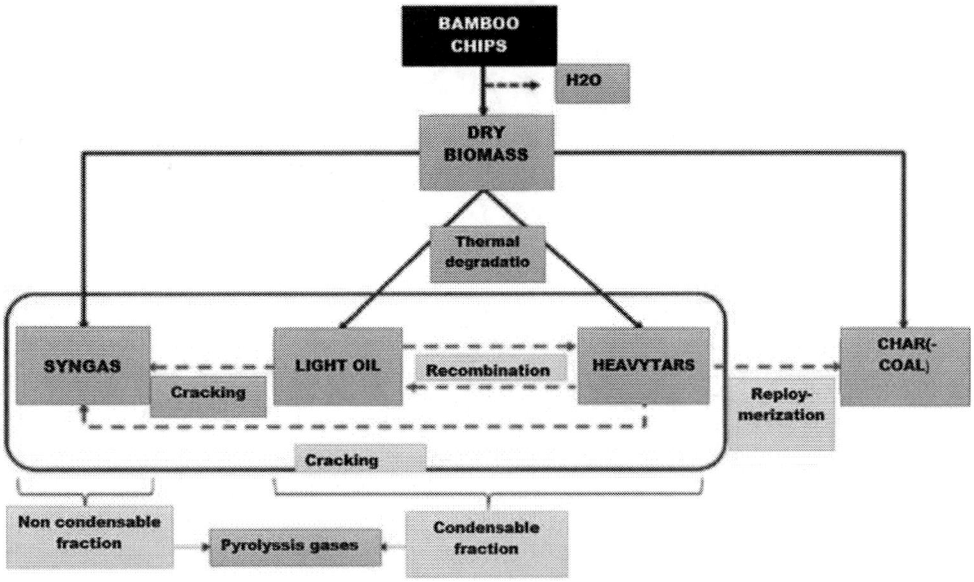

Fig. 3. The diagram depicting combustion process of bamboo biomass [13].

- *By gasification*

Gasification is the conversion of biomass- or fossil fuel-based carbonaceous materials into gases, such as: nitrogen (N2), carbon monoxide (CO), hydrogen (H2), and carbon dioxide (CO2). Gasification consists in a process of thermal and chemical conversion of organic material at increased temperature under special air intake conditions. The gasification process may include pyrolysis and partial combustion, occurring at a high temperature, usually between 750 °C and 1200 °C.

The gasification process will have as products syngas and ash. The gaseous products are usually named syngas, and can be either used directly in the burners, or are desulfurized and used in internal combustion engines. Syngas is a mixture of combustible gases and noncombustible gases. Approximately 40% of the volume of syngas can be used for generating electricity or heat. In practice atmospheric air is normally used since oxygen is quite expensive.

Biochemical conversion

In biochemical conversion, different strains of microorganisms that convert various products from biofuels are used. The principle of biochemical conversion is the fermentation of sugars and other substances contained in biomass. The processing is performed with the help of microorganisms which will produce different amounts of ethanol, methane and other fuels, chemical and thermal.

Bioconversion can be performed either in the presence of air or in its absence:

- *Anaerobic digestion* (Figure 4) is a process through which bacteria break down organic matter (animal manure, wastewater biosolids, and food wastes) in the absence of oxygen, producing biogas (methane) (60%) and CO2 (40%) [14].

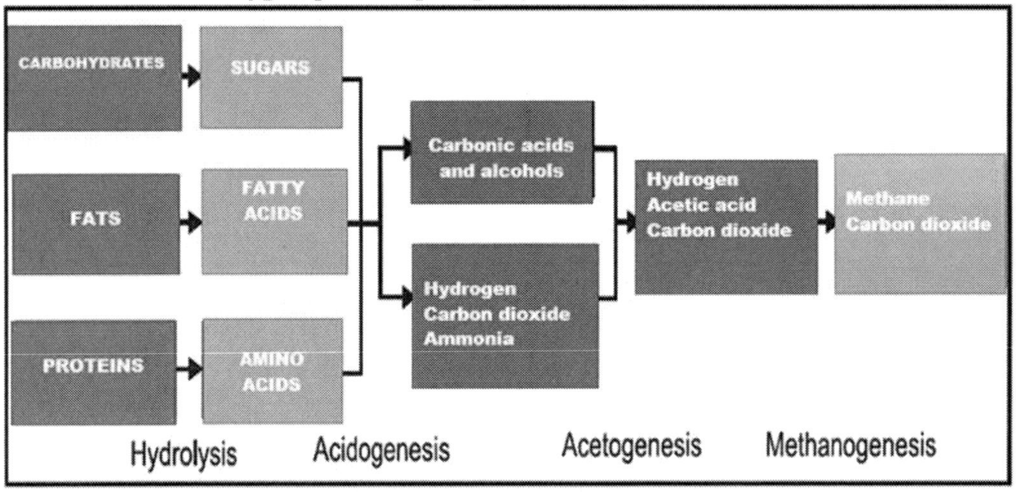

Fig. 4. Anaerobic digestion path [11].

- *Fermentation* usually occurs in the presence of oxygen being the decomposition of starch/sugar by microorganisms (yeasts and bacteria) to produce ethanol. By pretreatment with acid and enzyme at high processing temperatures, bamboo biomass turns into cellulose, which by hydrolysis becomes sugar which after fermentation turns into ethanol

  Bamboo → Biomass → Cells → Sugar → Ethanol.

Liquid fraction, also called the pyrolysis fuels is processed into a biorefinery to produce biofuels. Syngas is usually burned in cogeneration plants for the production of energy and electricity. In biochemical conversion, are usually used several different strains of microorganisms to convert biomass into biofuels. The technological flow involves the fermentation of sugars or other substances from bamboo biomass into (bio) ethanol, methane and other fuels.

Furthermore, bamboo has attractive fuel production characteristics, the most important parameters being the high heat values, the high volatile contents, the low ash and moisture content. This is why bamboo is considered an appropriate crop for production of bioenergy [15].

Results and discussions

Over the years, tests have been conducted to assess the energy potential of bamboo. Several research studies on this subject are highlighted in the following.

At Hasanuddin University, Tamalanrea, Makassar, Indonesia, the energy potential of bamboo leaf biomass has been verified.

In this study, bamboo leaves were used to conduct experiments according to temperature and heating time. Proximal analyses were used to determine first the volatile matter (VM). The ash content, the fixed carbon (CF) and moisture content (MC) have been also determined.

Have been used three prediction formulas to estimate the content of carbon, hydrogen and oxygen.

The dried bamboo leaves have been cleaned using distilled water, then a cutting process was applied, shredded and sifted at 500 μm until homogenized, after which they were heated to 250 °C, 300 °C, and 350 °C for 30, 60 and 90 minutes in - an environment devoid of oxygen.

The results of Table 2 [16] were obtained from the analyses.

Table 2. Main parameters obtained in percentages from bamboo leaves.

Temperature (°C)	**Time (min)**	**Moisture content (%)**	**Volatile matter (%)**	**Ash content (%)**	**Fixed carbon (%)**
250	30	2.72	32.21	34.15	30.57
	60	2.88	30.46	41.66	24.81
	90	4.40	38.22	32.27	24.63
300	30	6,63	29.71	36.64	27.35
	60	2.31	31.22	42.27	24.45
	90	3.50	28.33	45.70	22.26
350	30	2.40	25.94	50.10	21.40
	60	1.96	28.59	60.43	9.17
	90	2.87	22.32	71.37	3.89

Bambusa vulgaris has been found to have characteristics that make it possible to use it as a primary fuel. Biomass consisting of strains of *Bambusa vulgaris*, without branches, has a high quality as it progresses from base to top and is between 1 and 3 years old. This observation is mainly due to the energy density analysis, which has higher values in the upper position and at the age of three.

The quality of *the biomass Bambusa vulgaris* as an energy intake improves with increasing age and in the upper parts of the stems, producing a greater amount of energy per unit of volume.

The energy quality of *Bambusa vulgaris* was also investigated. Research has established that this species has a moisture content close to the maximum to generate energy (30%) 20.19%. In addition, the species under analysis can be used to increase the biomass base for energy production and to replace the wood that is used on large scale for this purpose [17].

Usually, the yield of bamboo biomass is no different within aspecies. However, following a fight at two locations in Thailand on two bamboo species, namely Bamboo *becheyana* and *Dendrocalamus membranaceus,* it was found that they may differ. As regards species *D. membraneaceus* the moisture content is lower while the calorific power is higher than that of species *B. beecheyana*, indicating that *D. membranaceus.* may be a more attractive option as a raw material for obtaining biomass [18].

Conclusions

Since bamboo is a very common plant used in different industries, this certainly leads to the production of numerous plant residues. They can be harnessed as biomass.

Over the years many researchers have studied the potential of bamboo to devein biomass and have shown that it lends itself to being a good raw material for the generation of plant biomass.

From the literature we can conclude that fossil fuels can be replaced with bamboo biomass, which is renewable and capable of producing different types of fuels, be they solid, liquid or gaseous.

For the transformation of bamboo biomass into another form of energy can be applied various technologies. These transformations provide new products such as coal, syngas, oil and ethanol.

Unlike wind and photovoltaic energy in addition to electricity production, bamboo biomass also produces biofuels.

As disadvantages we can say that bamboo biomass brings environmental risks by decreasing biodiversity, invasion of species and takes a long time to mature.

Although much research has been done in the field, in many countries energy projects from bamboo biomass are still being implemented.

To evaluate he sustainable mode of use and the capacity of this resource are necessary much more mute specialized studies.

This work was upheld by one establishing NUCLEU Program, carried out with the support of ANCSI, Project PN 5N/07.02.2019 "Research on the superior valorisation of some new plants species cultivated in Romania".

References

1. F. Nenciu and V. Vlăduţ, Studies on the perspectives of replacing the classic energy plants with Jerusalem artichoke and Sweet Sorghum, analyzing the impact on the conservation of ecosystems, IOP Conf. Ser.: Earth Environ. Sci. 635 012002, (2021);
2. F. Song, Bamboo Neosinolamus affinis-based thin film, a novel biomass material with high performances, (2014);
3. F. Nenciu, G. Găgeanu, V.N. Vlăduţ, M.G. Matache, I. Găgeanu, I. Voicea, G. Milian. Mircea C., Research on the use of technical plants and pomiculture waste in the production of biofuels, ISB INMA Teh (2020);
4. J. Piggot, Biomass supply economics: managing the supply cost of raw materials, TAPPI International Bioenergy and Bioproducts Conference, Memphis, TN; pp.14–16, (2009) pp.14–16;
5. ***Union of Concerned Scientists, How biomass energy works, (2009);

6. W.H. Chen et al., Progress in biomass torrefaction: Principles, applications and challenges, Progress in Energy and Combustion Science, **82**, January (2021);
7. M.F. Silva, et al., Bamboo as an eco-friendly material for food and biotechnology industries, Current Opinion in Food Science, (2020);
8. F. Nenciu, V. Vlăduţ, I. Voicea, I. Gageanu, Thermochemical conversion technologies for producing biofuels from agricultural biomass, Analele Universităţii din Craiova, seria Agricultură – Montanologie – Cadastru (Annals of the University of Craiova - Agriculture, Montanology, Cadastre Series), Vol. L (2020);
9. A. Gupta, A. Kumar, The potential of the bambas in the sustainable development of AsiaPac. In your sea. Rev., **4** (2008), pp. 100 – 107;
10. J.M.O. Scurlock, D.C Dayton, B. Hames, Bamboo: an overlooked biomass resource? , **19** Biomass and bioenergy (2000), pp 229-244;
11. A. Balázsi, Biomass as a source of renewable energy, analysis of energy technology from it, (2013);
12. J. Sánchez, et al. The Role of Bioenergy in the Bioeconomy, Biomass Resources, (2019) pp. 25–111;
13. D.R. Kerlero, J. Electrical Valorization of Bamboo in Africa; ENEA Consulting: Paris, France (2012);
14. J.E. Girard, Principles of Environmental Chemical, Jones & Bartlett Publishers, (2013);
15. R. Kumar, N. Chandrashekar, Fuel properties and combustion characteristics of some promising bamboo species in India. J. For. Res. **25**, (2014) pp.471–476;
16. A. Bidayatul, et al., Potentials of Biochars Rived from Bamboo Leaf Biomass as Energy Sources: Effect of Temperature and Time of Heating, Publishing International Journal of Biomaterials, (2019);
17. A.L. Balduino Junior et al. Energetic potential of bamboo culms for industrial and domestic use in Southern Brazil, Cienc. Rural **46** no. 11 Santa Maria, (2016);
18. A.Darabant, M. Haruthaithanasana et. al., Bamboo biomass yield and feedstock characteristics of energy plantations in Thailand, European Geosciences Union General Assembly, Energy Procedia **59** (2014) pp134 – 141;

Valorization on the energy potential of *Pistia stratiotes* in sustainable agriculture-review

Andreea Matache [1*], *Nicoleta Alexandra* Vanghele[1], *Mihai Gabriel* Matache [1] and *Ancuța Alexandra* Petre[1]

[1] The National Institute of Research - Development for Machines and Installations designed to Agriculture and Food Industry - INMA Bucharest

Abstract. Of late years, aquatic plants have received increased attention due to their energy potential in sustainable agriculture. *Pistia stratiotes* also belongs to this category. A vital aspect in sustainable agriculture, to which there is a growing tendency, is the use of environmentally friendly methods. Aquatic plants are used as raw materials for various applications, such as food industry, health, agriculture and aquaculture, ecosystem management, being an essential renewable resource. Due to population growth, both developed and developing countries are mainly facing problems related to future energy security and the proper use of natural resources. As the search for renewable energy resource intensifies, there is a growing drive to identify and develop promising, environmentally friendly, low-carbon and low-carbon bioenergy raw materials. Fuel demand is rising, while food costs and waste are gradually rising. It is therefore necessary to address waste management and reduce pollution caused by excessive energy use and industrialization. The current study focuses on the potential for harnessing renewable energy in the Nile lettuce (*Pistia stratiotes*), namely, highlighting the process of residual biomass, biofuel and compost.

Keywords: aquatic plants, sustainable agriculture, energy potential, *Pistia stratiotes*

Introduction

With the demographic development and increasing consumption, agricultural production is also constantly evolving. Classical agriculture uses numerous resources and various tillage operations that involve the use of significant amounts of water, fertilizers, pesticides [1]. As a major consumer of biodiversity, agriculture must constantly turn to new renewable resources. Special efforts are being made to implement efficient technologies, meant to support the evolution of classical agriculture toward a sustainable one [2, 3].

Sustainable agriculture offers the opportunity to exploit natural resources ecologically and analyze the principles of nature to develop their own systems of breeding crops and animals [4, 5]. The exploitation of renewable resources becomes essential in the context of the permanent decline of nonrenewable resources [6].

* Corresponding author: andmatache@yahoo.com

An important renewable resource is biomass, which includes residues from various activities: agriculture, forestry, animal husbandry, food and industrial waste. The energy potential of biomass is known, but being an intensively exploited resource, attention should be paid to the use of residual biomass available as an alternative energy source [7, 8].

Biomass is also a renewable source in the context of long-term exploitation of fossil fuels [9].

Fuels obtained using organic practices, such as agriculture and anaerobic digestion, are considered biofuels. The biofuel is formed by the biological fixation of carbon. While fossil fuels are considered nonrenewable, biofuels are a renewable source [10]. The category of biofuels includes ethanol, biodiesel, methanol, biobutanol [11]. Due to environmental requirements, there is a growing interest in the use of biofuels. Using them can also bring benefits by extending the life of engines, especially diesel [12].

Another way to use biomass is to use it to enrich the compost with nutrients, thus becoming biocompost. Composting is often presented as a form of recycling nature. Composting differs from natural decomposition by human control over the process of biological decomposition of organic matter. This results in a quality compost in a shorter time [13, 14].

Composting has become an alternative for treating organic waste to obtain a sanitized final material that can be used on any scale [15].

Knowing their energy potential in sustainable agriculture, aquatic plants have come to the attention of scientists interested in developing efficient and environmentally friendly techniques.

Although the *Pistia stratiotes* is known as an invasive aquatic plant, it also has multiple favorable uses. It is used as an ornamental species for aquariums, in phytoremediation of chemicals and metals and for wastewater treatment. Biomass from Pistia stratiotes can be exploited for feed production, compost, biogas [16].

By raising awareness of renewable energy resources, markets and scientific studies, the economy can become cleaner and more sustainable [17].

In this paper it is presented the energetic potential of the aquatic plant *Pistia stratiotes.*

Materials and methods – means of valorization

Nile lettuce (*Pistia stratiotes*) is considered an invasive floating macrophyte and therefore its control methods are sought. Despite all these negative aspects, several researchers demonstrate that P. stratiotes is an energy source.

Biomass

In Slovenia (Lake Sava), *Pistia stratiotes* biomass with chemical and physical parameters and seed resistance were analyzed to study the sustainability and spread of the species. Lake Sava has a special water temperature regime due to the underground thermal springs and the flow of hot water from the basins of a neighboring resort. The study area was divided into five sections (A, B, C, D, E) according to temperature differences.with application in the production of biofuel, bioenergy, fertilizer and compost [18].

Biofuels

In study [19], is presented the potential, by photofermentation, of the residual biomass of Pistia stratiotes to produce biofuel from hydrogen. This analysis was performed following the nitrogen removal treatment. A natural strain and a processed strain of Rhodopseudomonas palustris were used to determine the hydrogen. The use of Pistia stratiotes in the bioenergetic branch is largely based on its high productivity. The resulting composition was used for cultivating R. palustris. The natural strain adapted more easily, having a higher biochemical hydrogen yield than the processed strain.

The results obtained show that after the elimination of ammonium, fermentation and hydrolysis, the residual biomass of Nile lettuce (*Pistia stratiotes*) has considerable potential for the production of biofuels [19].

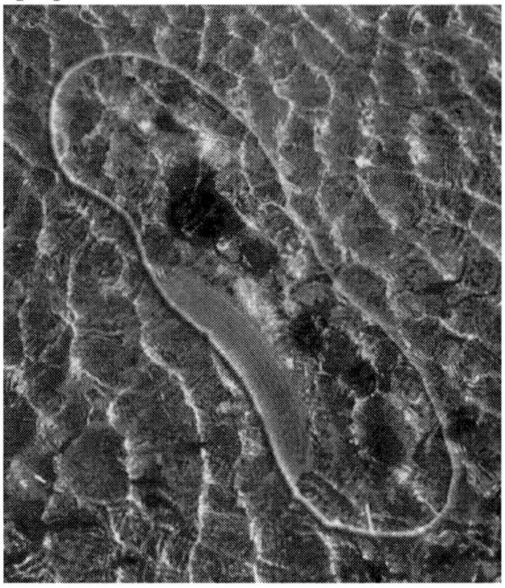

Fig. 1. Metabolic processes in *Rhodopseudomonas palustris* [27].

Biomass of *Pistia stratiotes* can be introduced as a substrate for bioethanol production due to the growth rate. Another advantage is the soft consistency of the plant, which facilitates its grinding process, to increase its biochemical reaction capacity. For enzymatic hydrolysis and fermentation of pretreated biomass of *Pistia stratiotes* to obtain ethanol, the researchers applied two fermentation methods, concomitant saccharification and fermentation (SSF) and separate hydrolysis and fermentation (SHF).

The fermentation microorganisms were *Escherichia coli* recombinant and *Saccharomyces cerevisiae*. Research results show that *Pistia stratotes* biomass has an increased yield in ethanol production when the fermentation process is improved [20].

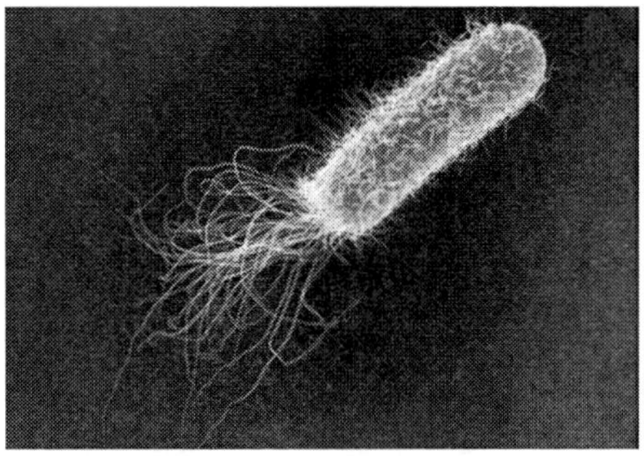

Fig. 2. *Escherichia coli* [28].

Fig. 3. *Saccharomyces cerevisiae* [29].

For the best possible management of fossil carbon emissions the focus is on new renewable sources. Therefore, in study [21] the biomass of Nile lettuce is analyzed, to produce bioethanol. Fresh and dried *Pistia stratiotes* biomass was administered with acid (H_2SO_4) and alkali (NaOH). These were mixed to pre-treat the free floating macrophyte biomass.

In the wild, *Pistia stratiotes* gathers starch-shaped food. After pre-treating the Nile salad with cellulose, ethanol (15.385 g / L), reducing sugar (4.773 ± 0.15%) and total sugar (11.350 ± 0.09%) were obtained. As a result, P. stratiotes has considerable potential in bioethanol production. In fig. 4 is presented graphically, the process of producing bioethanol from *Pistia stratiotes* biomass.

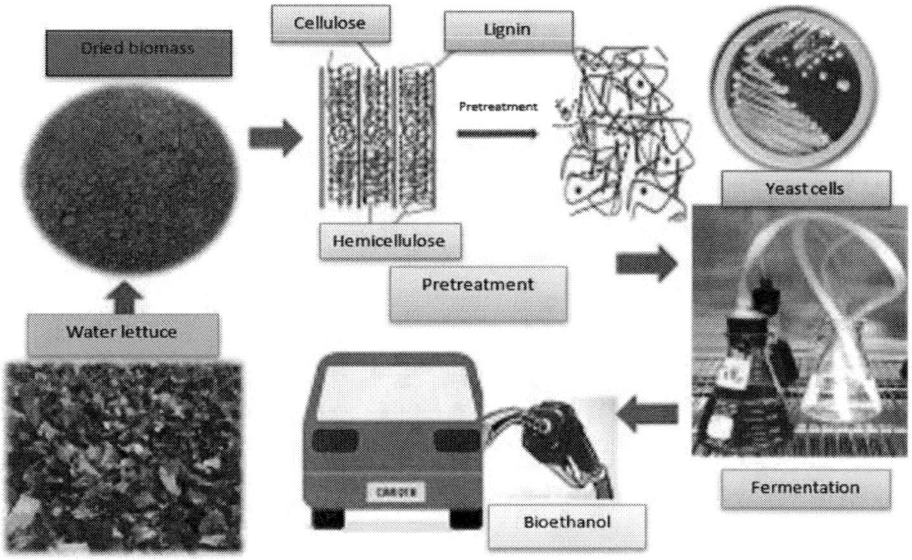

Fig. 4. Graphic presentation of bioethanol production from *Pistia stratiotes* biomass [21].

The following reserch paper analyzes *Pistia stratiotes* biomass to obtain biofuel [22]. The procedure was structured in 4 steps: 1. From the separated, washed and chopped leaves and roots resulted the juice; 2. The juice, then mixed with various concentrations of hydrochloric acid (HCL), was heated to 120 ° C for an average of 10 min. After cooling, the sample was filtered and collected; 3. At an average temperature of 35 ° C, fermentation began by adding dry yeast. 4. At the end of the fermentation process, distillation was conducted using steam and liquefaction. After that, the alcohol esterification process began with the use of a biodiesel separation impulse.

Energy (bioelectricity in hybrid system)

Increasing energy consumption from nonrenewable sources has led to the focus on new renewable energy sources. Paper [23], presents the ability of aquatic and terrestrial plants (*Ipomoea aquatica* and *Pistia stratiotes*) to simultaneously produce biomass and bioelectricity in a hybrid system.

Fig. 5. *Ipomoea aquatica* [30].

Fig. 6. *Pistia stratiotes* [31].

During photosynthesis, the microbial plant (PMFC) transmits part of what it emits to the roots. Then, the bacteria separate the nutrients causing electron transfer. Strategically positioned electrodes can receive electrons and thus produce electricity. A microbial plant produces current without impeding the development of the plant. Due to this, it can be used in agriculture to produce electricity and food at the same time.

The two systems were positioned in an area where the sun appears early in the morning and partial shade in the afternoon. These measures have been taken to protect plants from overheating. Microbial plants (PMFCs) were stacked individually in hybrid assemblies to amplify the power to simultaneously produce biomass and bioelectricity. The main assembly was made of ordinary glass with 9 individual cells (Fig.7a), each cell having a cotton cloth separator (Fig. 7b).

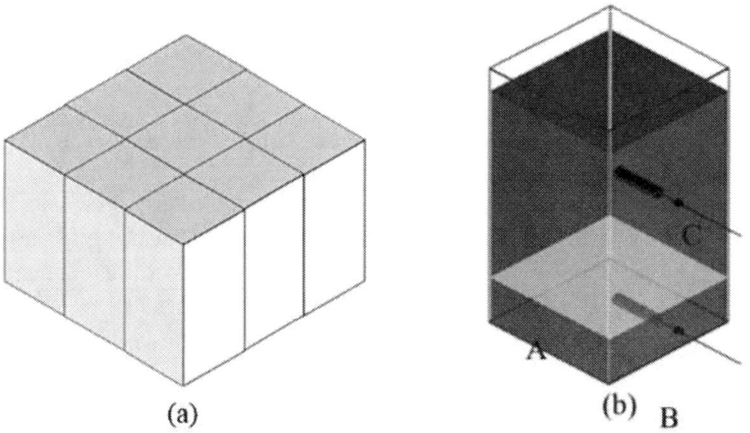

Fig. 7. (a) Design of stacked PMFCs (L x W x H: 18 in x 18 in x 12 in) and (b) Layout of the individual cells (A - cloth separator; B - cathode; C – anode) [23].

Energy by gasification

Study [24], analyzes the potential of transforming the residual biomass of Nile lettuce into energy through gasification technology. The technology consists of a system for handling ash and fuel and gasification (reactor, gas cooling, cleaning method).

Pretreated biomass prevents biological degradation and helps store for a long time. A good design of the gasifier facilitates the transformation of biomass into coal. Then, the coal is modified to produce CO and H_2 (fig.8).

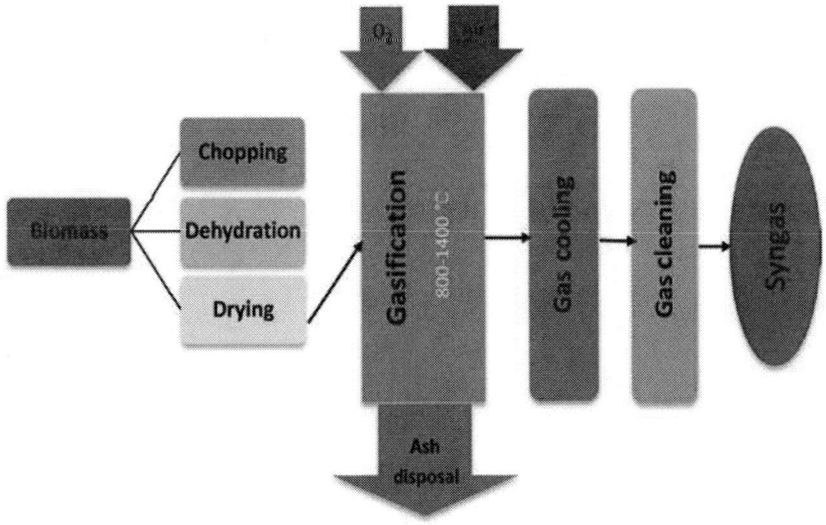

Fig. 8. Schematic diagram for syngas generation from biomass [24].

Fertilizer and compost in a combined system

Over time, scientists have conducted research to discover the most effective solutions for improving soil quality, with as few negative effects on the environment. Research has shown that the use of organic and inorganic materials as fertilizers supports plant growth and production.

Such an analysis was made in Indonesia, where farmers use biomass of aquatic plants as fertilizer. To determine the efficiency of using a combined treatment and with an improved technique, six treatments and three replicates were applied. The combination with P. stratiotes + Biotara + Decision Support System (DSS) was used to fertilize the rice crop. The results obtained showed that the application of the combined treatment had the highest productivity [25].

Another study was conducted, this time, to analyze the quality of compost by mixing rock phosphate with Nile lettuce (*Pistia stratiotes*) with and without effective microorganisms. *Pistia stratiotes* plants were dried for one week in the open field, after which the resulting material was shredded and sieved to reach the size of 5 cm. After the analysis performed, it was found that organic waste can be transformed into valuable organic biofertilizers by adding lower doses of chemical nitrogen fertilizer and inoculation with efficient microorganisms.

Applying a combination of treatment, with Nile salad (WL), rock phosphate (RP), effective microorganisms (EM) and urea has a positive effect on the nutritional growth of the compost. It also enriches matter with phosphorus, compared to other treatments. The time required for composting is also reduced by adding EM solution and urea. Thus, *Pistia stratiotes* plants become a useful product by adopting this solution [26].

Research results

Biomass

The research results show that P. stratiotes can also grow in temperate areas with thermal water. Being a perennial aquatic plant with a rapid propagation, *Pistia stratiotes* biomass can prevent light from entering the water, thus influencing the evolution of fauna. In fig. 9, dry biomass of *Pistia stratiotes* collected from the shores of Lake Sava in March 2010 (a), seeds glued together in formations on the underside of the oldest leaves (b), characteristic barrel-shaped seeds (c), germinating the seeds after a month in the rearing chambers (d).

Fig. 9. Dried plant of *Pistia stratiotes* [18].

Biofuels

When the fermentation process is optimized, the biomass of *Pistia stratiotes* has an increased yield in ethanol production. The results of the study show that the potential of *Pistia stratiotes* for ethanol production was higher in the fermentation method SSF, fig. 11 than in the SHF method, fig. 10.

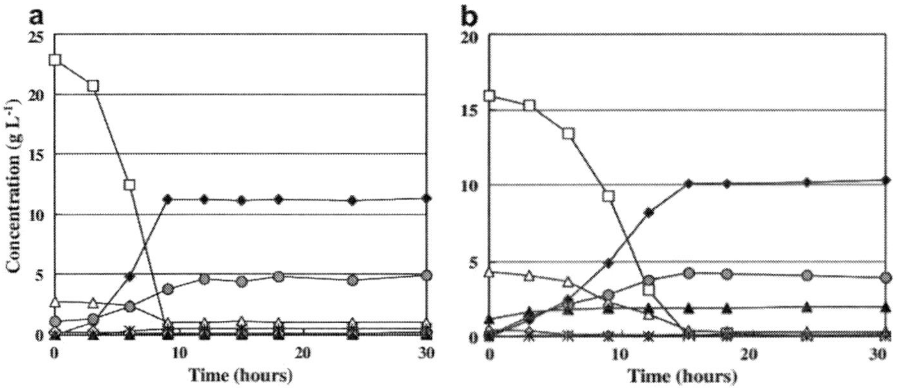

Fig.10. Ethanol production in SHF from water lettuce hydrolysate by *Saccharomyces cerevisiae* NBRC 2346 (a) and *Escherichia coli* KO11 (b). Symbols: □, glucose; Δ, xylose/ mannose/galactose; ◊, arabinose; *, glycerol; ▲, acetic acid; ●, biomass;♦ , ethanol.

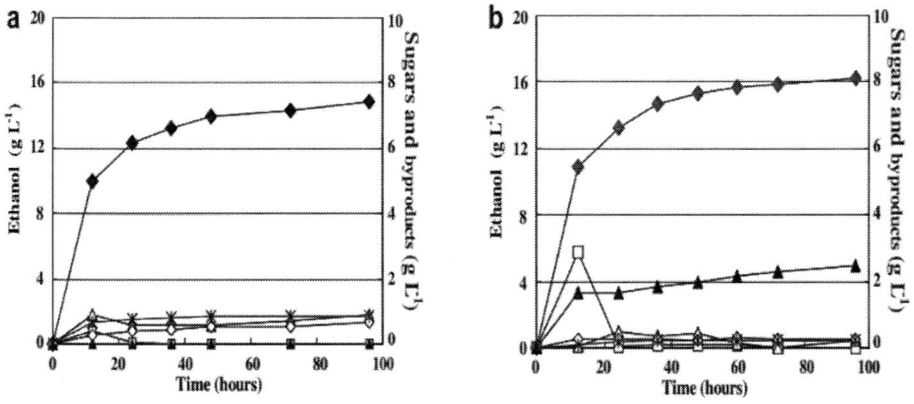

Fig.11. Ethanol and other byproduct concentrations in SSF fermentation of water lettuce leaves using *Saccharomyces cerevisiae* NBRC 2346 (a) and *Escherichia coli* KO11 (b). Symbols: □, glucose;Δ, xylose/mannose/galactose;◊ , arabinose; *, glycerol; ▲, acetic acid; ♦ , ethanol [20].

The results of obtaining biodiesel by using *Pistia stratiotes* biomass are presented in tables no. 1, 2, 3 şi 4 [22].

Table 1. Results of water lettuce (*Pistia stratiotes*) sample without addition of molasses (Only Leaf)

Sample	Total Reducing sugar (TRS)	Alcohol content	Biodiesel
Sample with 0.1% of HCL	0.12	0.15	
Sample with 0.5% of HCL	0.15	0.16	
Sample with 1.0% of HCL	0.2	0.19	0.16
Sample with 1.5% of HCL	0.29	0.32	

Table 2. Results of water lettuce (*Pistia stratiotes*) sample without addition of molasses (Only Radicle)

Sample	Total Reducing sugar (TRS)	Alcohol content	Biodiesel
Sample with 0.1% of HCL	0.2	0.24	
Sample with 0.5% of HCL	0.3	0.31	
Sample with 1.0% of HCL	0.5	0.55	
Sample with 1.5% of HCL	0.6	0.64	0.60

Table 3. Results of water lettuce (*Pistia stratiotes*) sample with addition of molasses (Only Leaf)

Sample	Total Reducing sugar (TRS)	Alcohol content	Biodiesel
Sample with 0.1% of HCL	0.29	0.3	
Sample with 0.5% of HCL	0.32	0.31	
Sample with 1.0% of HCL	0.35	0.4	
Sample with 1.5% of HCL	0.4	0.45	0.44

Table **4**. Results of water lettuce (*Pistia stratiotes*) sample with addition of molasses (Only Radicle)

Sample	**Total Reducing sugar (TRS)**	**Alcohol content**	**Biodiesel**
Sample with 0.1% of HCL	0.45	0.5	
Sample with 0.5% of HCL	0.7	0.71	
Sample with 1.0% of HCL	0.89	0.9	
Sample with 1.5% of HCL	0.98	1.09	1.02

Energy

Following the research, there were no significant differences in current voltage for the two terrestrial and aquatic plants, as shown in fig.12. Keeping the humidity, temperature and lighting at the same level, the different values observed were due to the physiological difference and the population of bacteria.

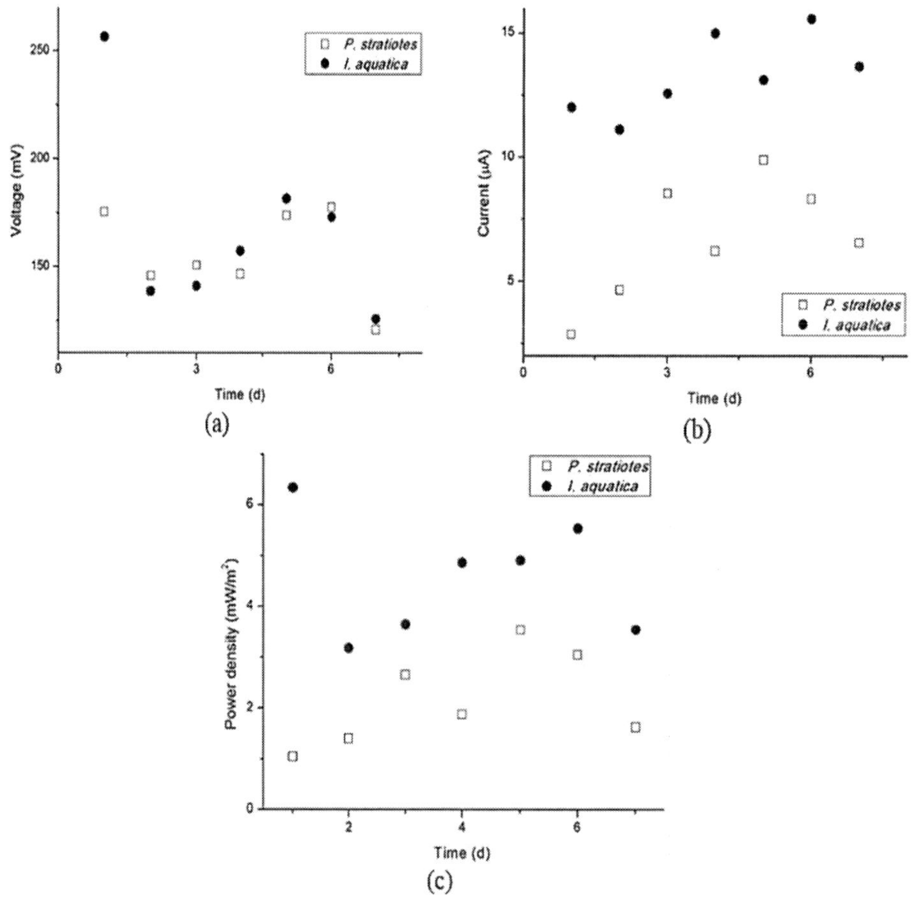

Fig. 12. (a) Average voltage readings of individual cells, (b) Average current readings of individual cells and (c) Average calculated power densities of individual cells [23].

Fertilizer and compost

The results of the experiment show that the application of the combined treatment of *Pistia stratiotes* (15 t / ha), bio-fertilizer Biotara/ha (25 kg), fertilization system DSS - K + B + DSS (50-kg urea / ha + 37 kg SP36 / ha + 24 kg KCl ha) had the highest potential compared to those obtained from local farmers and the control. The yield increased by 10.28% reaching 7.62 t/ha as shown in fig.13.

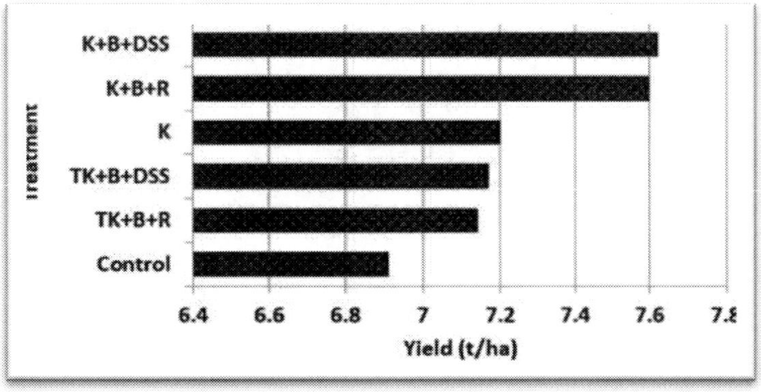

Fig. 13. Yield (t/ha) on treatment [25].

To obtain the biocompost, the following parameters were followed: maximum and minimum temperature, total precipitation and their duration, wind speed (table 5).

Table 5. Meteorological data during the composting period [26]

Weather parameters	April 2010	May 2010
Maximum temperature (°C)	33.8	36.9
Minimum temperature (°C)	16.4	19.2
Wind speed (Km/day)	51	74
Total rainfall (mm)	45.25	23.31
Rainfall days	5	7
Pan Evaporation (mm)	5.94	7.76
Relative Humidity (%)	60	52

Conclusions

Population development and continuous growth of consumption are two vital aspects in the management of nonrenewable and renewable resources. In this context, the discovery and use of renewable sources is of great interest in supporting sustainable agriculture.

Although aquatic plants are most often considered harmful to the habitat in which they grow, they also have many benefits. Nile salad is used to treat wastewater, as an ornamental plant for aquariums, in phytoremediation of metals.

To control the spread of invasive aquatic plants, researchers have shown that they can be used to improve soil quality, the plants that grow under its action, and the environment.

The aim of this paper was to present the aquatic plant, *Pistia stratiotes*, as a renewable energy source. The biomass of Nile Salad (*Pistia stratiotes*) has a diverse potential with applicability in obtaining biofuel, fertilizers and compost.

All these properties of *Pistia stratiotes* mean that in future research will continue to obtain new uses.

This work was supported by the Romanian Research and Innovation Ministry, through NUCLEU Programme, Project "PN 19 10 02 03: "RESEARCH ON THE INTENSIVE GROWTH OF FISH IN THE POLYCULTURE SYSTEM AND THE COMPLEX VALORIFICATION OF THE BIORESOURCES (PLANTS) OF AQUATIC", contract no. 5N / 07.02.2019.

References

1. R.M. Jadav, Current Science **118**, No.3, pp. 333–33 (2020);
2. P. Lloyd, Biodiversity,18(2–3), 124–125(2017);
3. I. Caba, D. Bordean et al., E3S Web of Conferences **112** (2019);
4. G. Ph. Robertson, Daedalus **144,** Issue 4, pp.76-89 (2015);
5. M. S. Kang, Journal of Crop Improv (2020);
6. A. Rodriguez, M. Brijaldo at al., Ciencia En Desarrollo, **11**(1), 63–80 (2020);
7. L. Hamelin et al., Renew. and Sust. Energ. Reviews**100**, pp.127-142 (2019);
8. J. J. Roberts et al., Renewable and Sustainable Energy Reviews, 41, 568–583 (2015)
9. G. Venturini, A. Pizarro-Alonso, & M. Münster, Applied Energy,**250,** 369–388 (2019);
10. M. A. Hanif et al., Env. Chem.: A Compr. Appr, pp.583–646 (2020);
11. F. Gerali, IEEE Engineering and Technology History Wiki. Hoboken, NJ: IEEE History Center (2020);
12. P. Găgeanu, I. Găgeanu, G. Bunduchi, and A. Zaica, E3S Web of Conferences **112** (2019);
13. M. Papale, I. Romano, I. Finore at al, Microorganisms, **9** (2), 218 (2021);
14. R. Kanissery, Citrus Ind. Magazine (2021);
15. T.Sayara, R. Basheer-Salimia, F. Hawamde, & A. Sánchez, Agronomy **10** (11) (2020);
16. D. Chapman, J.A.Coetzee et al., Bulletin OEPP/EPPO Bulletin **47**(3) (2017);
17. D. Beňová, K. Mareš, P. Hutla, Tatiana Ivanova, Jan Banout, M. Kolaříková, Agronomy, ***11***(1), 169 (2021);
18. M. Jaklič, Š. Koren, N. Jogan, Acta Bot. Croat **79** (1), 35–42 (2020);
19. E. Corneli, A. Adessi, E.J. Olguın, G. Ragaglini et al., Journal of Appl. Microbiology (2017);
20. D. Mishima, M. Kuniki, K. Sei, S. Soda, M. Ike, M. Fujita, Bio. Tech. **99**, Issue 7, 2495–2500 (2008);
21. K. Whangchaiet al., Bioresource Tech. Reports **14** (2021);
22. S. Bachhav et al., Production of Biofuel using Water Lettuces (*Pistia Stratiotes*), IRJET, vol. **6**, Issue 5 (2019);
23. K. R. S. Pamintuan et al., IOP Conf. Ser.: Earth Environ. Sci.**191** (2018);

24. S. Kathi, Bioremediation and Bioeconomy, pp. 111–128 (2016);
25. A. Hairani, M. Alwi, K. Anwar, BIO Web of Conferences **20** (2020);
26. S. Kanwal, S.Iram et al., African Journal of Microbiology Research **5** (14), pp. 1784–1793 (2011);
27. https://microbewiki.kenyon.edu/images/thumb/4/47/Rpalustris.png/300px-Rpalustris.png;
28. http://www.healthspectra.com/wp-content/uploads/2018/02/E-Coli-infection-HS02.jpg;
29. https://biologydictionary.net/wp-content/uploads/2020/06/Saccharomyces-Cerevisiae.jpg;
30.https://upload.wikimedia.org/wikipedia/commons/b/b4/Ipomoeaaquatica(Marsh_Glory)flower_W_IMG_0405.jp;
31.https://img.crocdn.co.uk/images/products2/pl/20/00/02/24/pl2000022461.jpg?width=940&height=940.

Investigation of the thermodynamic and thermal properties of clary sage (*Salvia sclarea* L.) essential oil and its main components

Stanislava Tasheva[1]*, *Vanya* Gandova[1], *Vanya* Prodanova-Stefanova[2], *Krasimira* Marinova[3], *Milen* Dimov[2], *Krasimira* Dobreva[2] and *Albena* Stoyanova[1]

[1]*University of Food Technologies, 4002 Plovdiv, Bulgaria*
[2]*Trakia University, Dept. Food Technology, 8600 Yambol, Bulgaria*
[3]*University 'Prof. Assen Zlatarov', 8010 Bourgas, Bulgaria*

Abstract: The some physochemical and chemical properties of the clary sage (*Salvia sclarea* L.) essential oil were determined. The main constituents in the clary sage essential oil (above 2%) were: linalyl acetate (34.62%), β-linalool (17.67%), α-muurolene (8.27%), β-caryophyllene (5.60%), α-ylangene (5.18%), α-terpineol (4.84%), n-docosane (3.00%), and neryl acetate (2.34%).The thermodynamic and thermal properties of essential oil and its main components were investigated.

1 Introduction

Clary sage essential oil is obtained by steam distillation of flowering tops and foliage of cultivated *Salvia sclarea* L. (Lamiaceae). The main compounds in the oil were linalyl acetate (56-78%) and linalool (6.5-24%) [1-8]. In the plant were identified polyphenolic compounds, sesqui- and diterpenes [9-12].

The thermal properties of borage (*Borago officinalis* L.) seeds were determined at temperatures ranging from 6 to 20°C and moisture contents from 1.2 to 30.3% [13].

Some authors were reported of viscosities and densities, at T = 298.15 K of the phases formed after diterpenation of bergamot, lemon and mint essential oils, by (liquid + liquid) extraction [14].

In the literature do not found data for determination of thermodynamic parameters of clary sage essential oil as well as its main components which is also the aim of the present study.

2 Material and methods

2.1 Materials. Clary sage oil was provided by Kateco Ltd, Bulgaria. The oil was kept in the refrigerator at 4 °C before analysis.

* Corresponding author: st_tasheva@abv.bg

2.1.1 Determination of refractive index. The refractive index (n_D^{20}) was determined according to ISO 280:1998 [15].

2.1.2 Determination of relative density. The refractive density (d_{20}^{20}) was determined according to ISO 279:1998 [16].

2.1.3 Infrared spectroscopy of clary sage oil. The infrared spectrum was recorded using a Nicolet iS 50 Thermo Scientific FT-IR spectrometer in the frequency region of 4000–400 cm^{-1}, with the samples embedded in KBr matrixes.

2.1.4 Chemical composition of the essential oil. GC analysis was performed using gas chromatograph Agilent 7890A [17]

2.1.5 Thermodynamic and thermal properties of the essential oil and its main compounds. Specific heat capacity is obtained by eq. (1) [18]:

$$C_p = \left(\frac{dH}{dT}\right)_p \tag{1}$$

where: C_p - specific heat capacity, $J.mol^{-1}K^{-1}$; H - enthalpy, $J.mol^{-1}$; T - temperature, ° K. Dependence between specific heat capacity, thermal conductivity and coefficient of thermal conductivity were determined according to [13, 18] by eq. (2).

$$a = \frac{\lambda}{\rho C_p} \tag{2}$$

where: a - coefficient of thermal conductivity, $m^2.s^{-1}$; λ - thermal conductivity, $W.m^{-1}K^{-1}$; C_p - specific heat capacity, $J.mol^{-1}K^{-1}$; ρ -density, $kg.m^{-3}$.

2.1.6 Statistical analysis. The experimental data were obtained after three replicates. The results are presented as a mean value of the individual measurements with the corresponding standard deviation (SD).

3 Results and Discussions

The clary sage essential oil presented light yellow with a specific odor and had followed properties: density 0.8623 ± 0.0 and a refractive index (n_D^{20}) 1.4635 ± 0.0. The obtained values didn`t exhibited differences with similar data from literature [1, 2].

The IR spectrum of clary sage essential oil is presented in Table 1. The analysis of FT-IR of studied oil represents mainly strong vibrations of the components in this oil. Intense line at 3423 cm^{-1} appears stronger connected with conformation in the *β*-linalool molecule. In this case the mixture OH-groups lose steric factor.

Another connection with the spectrum of pure *β*-linalool is the band at 2973 cm^{-1} and associated with CH_2-asymmetric vibrations. The probable cause of this behavior can be explained with antagonism with different bands from different components in the investigated oil. The other components does not show strong absorption bands because the limitations of FT-IR method which detects chemical compounds with concentration above 5%.

The experimental FT-IR spectrum of clary sage essential oil is the presented in Fig. 1.

 https://doi.org/10.1051/e3sconf/202128602003

Table 1. IR spectrum of clary sage essential oil

Characteristic bands (cm^{-1})		Group type
Experimental data	Reference data	
3423	3550 – 3400 3590 – 3420	γOH intramolecular H – bond
2973	2970 – 2950	linear γ_{as} CH_3
2928	2940 – 2915	γ_{as} $-CH_2-$
1735	1740 – 1720	R – CHO $\gamma_{C=O}$; characteristic band of carbonyl group
1640	1648 – 1640	non-conjugated C=C, low intensity. HRC=CH_2 end vinyl
1450	1465 – 1440	aromatic ring
1412	1435 – 1405	- CH_2–CO–; vibrations of ketones, δ CH_2
1375	1385 – 1370	a structural fragment of the type $C(CH_3)_2$
1250	1260 – 1240	a characteristic band for ether bond, of the type γ^{as} C–O–C
1172	1175 – 1165	a characteristic band for ether bond, of the type CH_3–CH–CH_3
1088	1150 – 1050	γ^{as} C–O–C, ether bond
1047	1070 – 1020	=C–O–C, γ^{s} C–O–C, aromatic and vinyl
921	925 – 9 05	C–H bond of the type HRC=CH_2
879	900 – 865	trisubstituted aromatic ring γ Ar – H in positions 1,2,4

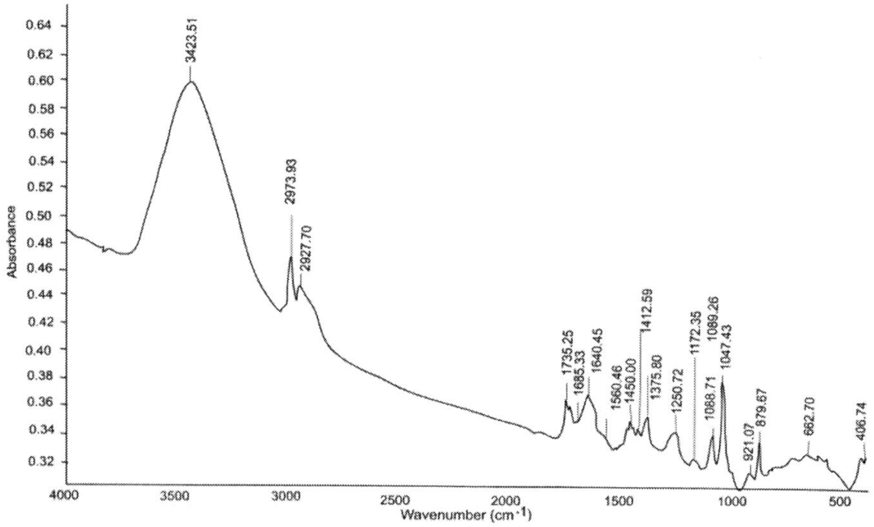

Fig. 1. IR spectrum of clary sage essential oil

Chemical composition of the clary sage essential oil is listed in Table 2.

From Table 2 as can see the oil contained 33 components representing 98.39% of the total content were identified. Fourteen of them were in concentrations over 1% and the rest 19 constituents were in concentrations under 1%. The main components in the essential oil (above 2%) from fruits were: linalyl acetate (34.62%), *β*-linalool (17.67%), *α*-muurolene (8.27%), *β*-caryophyllene (5.60%), *α*-ylangene (5.18%), *α*-terpineol (4.84%), n-docosane (3.00%), and neryl acetate (2.34%).

Table 2. Chemical composition of clary sage essential oil

№	Compounds	RI	Content, %
1.	(2E,4E)-Hexadienol	912	0.15 ± 0.0
2.	*β*-Myrcene	988	1.82 ± 0.01
3.	Limonene	1025	1.05 ± 0.01
4.	(Z)-*β*-Ocimene	1039	0.22 ± 0.0
5.	(E)-*β*-Ocimene	1049	1.59 ± 0.01
6.	*α*-Terpinolene	1084	0.56 ± 0.0
7.	*β*-Linalool	1090	17.67 ± 0.17
8.	*α*-Terpineol	1188	4.84 ± 0.04
9.	Linalyl formate	1215	0.18 ± 0.0
10.	Nerol	1230	0.15 ± 0.0
11.	Geraniol	1249	0.83 ± 0.0
12.	Linalyl acetate	1254	34.62 ± 0.33
13.	Geranyl formate	1297	0.10 ± 0.0
14.	Citronellyl acetate	1350	0.15 ± 0.0
15.	Neryl acetate	1361	2.34 ± 0.02
16.	*β*-Ylangene	1373	5.18 ± 0.05
17.	*β*-Cubebene	1386	0.33 ± 0.0
18.	*β*-Bourbonene	1389	0.38 ± 0.0
19.	*β*-Caryophyllene	1420	5.60 ± 0.05
20.	Aromadendrene	1439	0.27 ± 0.0
21.	*α*-Caryophyllene	1455	0.30 ± 0.0
22.	*γ*-Muurolene	1478	0.60 ± 0.0
23.	*α*-Muurolene	1496	8.27 ± 0.08
24.	Viridiflorene	1499	0.74 ± 0.0
25.	*γ*-Cadinene	1513	1.61 ± 0.01
26.	*δ*-Cadinene	1522	0.97 ± 0.0
27.	*β*-Cadinene	1537	1.56 ± 0.01
28.	Spathulenol	1577	0.72 ± 0.0
29.	(2-Z,6-E)-Farnesyl acetate	1821	1.12 ± 0.01
30.	*n*-Nonadecane	1900	0.53 ± 0.0
31.	*n*-Eicosane	2000	0.37 ± 0.0
32.	*n*-Heneicosane	2100	0.57± 0.0
33.	*n*-Docosane	2200	3.00 ± 0.03

The difference in the quantities of chemical composition of our essential oil and the reported data [1-8] may be due to environmental conditions under which the plant has grown as well as the variation in conditions of analysis.

Distribution of major groups of aroma substances in the essential oil is shown in Figure 2. The dominant groups in the oil were oxygenated monoterpenes (61.88%); then followed by sesquiterpene hydrocarbons (26.23%), monoterpene hydrocarbons (5.33%), and hydrocarbons (4.54%), oxygenated sesquiterpenes (1.87%), and oxygenated hydrocarbons (0.15%).

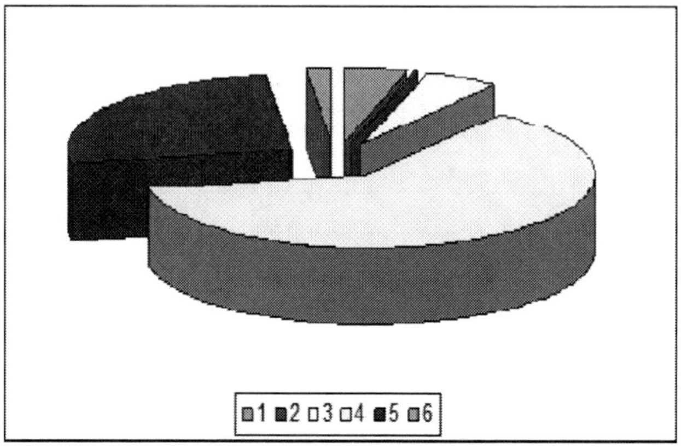

Fig.2. Group of components in clary sage essential oils, %.

1 - hydrocarbons; 2 - oxygenated hydrocarbons; 3 - monoterpene hydrocarbons; 4 - oxygen monoterpenes; 5 - sesquiterpene hydrocarbons; 6 - oxygenated sesquiterpenes.

In Table 3 are presented some chemical properties of the main nine aromatic substances, which formed 82.85% from total composition. It was determined scent and properties of the essential oil, too. The investigated substances were β-linalool, α-terpineol, linalyl acetate, and neryl acetate which are the main components from oxygen monoterpenes group, α-muurolene, β-caryophyllene, and β-ylangene – from sesquiterpene hydrocarbons group and n-docosane from hydrocarbons group. There distribution by functional groups and structure is: hydrocarbons (α-muurolene, β-caryophyllene, β-ylangene, and n-docosane), two alcohols (linalool and α-terpineol), and two esters (linalyl acetate and neryl acetate).

The calculations of thermo physical characteristics of the clary sage essential oil as well as thermal conductivity, specific heat capacity and coefficient of thermal conductivity are presented in Table 4. The values are obtained according to the chemical composition presented in Table 2.

The results compared with literature data of the same parameters, which is obtained experimentally [13] where: λ = 0.11-0.28 $W.m^{-1}\ K^{-1}$; Cp = 0.77–1.99 $кJ.mol^{-1}K^{-1}$; a = $2.32.10^{-7}$- $3.18.10^{-7}\ m^2.s^{-1}$.

Table 3. Thermo physical parameters of the main compounds [19]

Compounds	Ylangene	β-Caryophyllene	α-Muurolene	n-Docosane	β-Linalool	α-Terpineol	Neryl acetate	Linalyl acetate
Empirical Formula	$C_{15}H_{24}$	$C_{15}H_{24}$	$C_{15}H_{24}$	$C_{22}H_{46}$	$C_{10}H_{18}O$	$C_{10}H_{18}O$	$C_{12}H_{20}O_2$	$C_{12}H_{20}O_2$
Molecular Weight	204.35	204.32	204.35	310.60	154.25	154.25	196.29	196.29
ΔfG°, kJ.mol^{-1}	242.55	196.63	179.03	134.36	58.85	-55.88	-40.42	-21.41
ΔfH°gas, kJ.mol^{-1}	-125.10	-112.68	-164.97	-497.41	-177.85	-310.08	-320.95	-311.70
ΔfusH°, kJ.mol^{-1}	20.07	14.82	21.69	52.74	15.94	11.00	27.41	19.82
ΔvapH°, kJ.mol^{-1}	47.69	65.50	50.71	[111.90; 115.60]	65.00	54.62	51.54	49.53
*log*Potc/wat	4.27	4.73	4.58	8.83	2.67	2.50	3.24	3.24
Pc, kPa	2001.91	2027.23	1921.98	10000.00	2581.96	2950.48	2047.46	2054.89
Tboil, K	561.69	576.30	576.33	641.80	[468.70; 471.80]	490.70	407.20	493.20
Tc, K	775.53	802.06	791.72	786.00	697.57	740.92	748.06	740.67
Tfus, K	322.81	323.71	287.93	[313.20; 318.30]	244.90	308.15	259.08	278.78
Vc, m^3.kg^{-1}mol^{-1}	0.71	0.72	0.72	1.27	0.57	0.52	0.69	0.68
Cp, gas, J.mol^{-1}K^{-1}	503.35	501.77	503.29	962.02	347.03	353.52	426.65	428.80
Cp, liquid, J.mol^{-1}K^{-1}	-*	-	-	739.00	372.40	-	-	-
Cp, solid, J.mol^{-1}K^{-1}	-	-	-	563.60	-	-	-	-

$\Delta_c H°_{liquid}$ - standard liquid enthalpy of combustion (kJ.mol^{-1}); $\Delta_f G°$ - standard Gibbs free energy of formation (kJ.mol^{-1}); $\Delta_f H°_{gas}$ - enthalpy of formation at standard conditions (kJ.mol^{-1}); $\Delta_{fus}H°$ - enthalpy of fusion at standard conditions (kJ.mol^{-1}); $\Delta_{vap}H°$ - enthalpy of vaporization at standard conditions (kJ.mol^{-1}); P_c - critical pressure (kPa); T_{boil} - normal boiling point Temperature (K); T_c - critical temperature (K); $C_{p,gas}$ - ideal gas heat capacity (J.mol^{-1}K^{-1}); $C_{p,liquid}$ - liquid phase heat capacity (J.mol^{-1}K^{-1}); * no data.

Table 4. Thermal parameters of the clary sage essential oil

Thermal parameters	R_{cm}, J.mol^{-1}K^{-1}	E_a, kJ.mol^{-1}	C_p, J.mol^{-1}K^{-1}	λ.10^3, W.m^{-1}K^{-1}	a.10^6, m^2.s^{-1}
Clary sage oil	43.58	-1796.75	6026.32	330.31	1.49
Thermal Parameters	$M_{cm,}$ kg.mol^{-1}	ΔH, kJ.mol^{-1}	ΔS, kJ.mol^{-1}K^{-1}	ΔG, kJ.mol^{-1}	
Clary sage oil	190.77	1796.75	-14.34	6072.31	

R_{cm} - universal gas constant, J.mol^{-1}K^{-1}; E_a - activation energy, kJ.mol^{-1}; C_p - specific heat capacity, J.mol^{-1}K^{-1}; λ - thermal conductivity, W.m^{-1}K^{-1}; a - coefficient of thermal conductivity, m^2.s^{-1}; $M_{cm,}$ - molar mass of the main components, kg.mol^{-1}; ΔH - enthalpy, kJ.mol^{-1}; ΔS - entropy, kJ.mol^{-1}K^{-1}; ΔG - energy of Gibb`s, kJ.mol^{-1}.

The activation energy was calculated about -1796.75 kJ.mol^{-1}. The negative value specified that processes were non-spontaneous occurring as a result of heat and light. The molar mass of clary sage oil was determined experimentally using GC/MS analysis (190.77 kg.mol^{-1}). The thermodynamic parameters of Gibbs energy, enthalpy and entropy of the main components of sage essential oil were calculated. The thermodynamic parameters for salvia oil-ethanol-water solution were found in the literature [20], as the values obtained for the Gibbs energy are from -8968.87 to -10037.77 kJ.mol^{-1}, for the enthalpy (ΔH) -2490.24 kJ.mol^{-1} and for entropy (ΔS) from 21.71 to 24.83 kJ.mol^{-1}K^{-1}. The differences in the values obtained are probably due to the different composition of the studied essential oils. In pure oils the values of the thermodynamic parameters depend mainly on the chemical composition of the obtained essential oil, and when used in solution the values are also influenced by the solvents. Gibbs energy represents a thermodynamic potential and shows the possibility of the process. The positive values of Gibbs energy mean that the process that takes place in the main components of the essential oil is spontaneous and irreversible. Entropy as a thermodynamic parameter gives a connection about the mechanism of the reaction and the activation complexes. In this case the entropy values are negative, which means that the process is irreversible. The enthalpy values are positive too, which also means that the process is irreversible. In this case the enthalpy of the process is equal to the activating energy with the opposite sign. This means that the process will proceed with the adding of energy.

Table 5 presented values of heat capacity and enthalpy of main components of clary sage oil. Composition of the all components are summarised to 100% needs for calculations of the molar heat capacity and enthalpy for each one of main components. Comparison between results given in Table 3 and Table 5 are done for n-docosane and *β*-linalool. Differences which appeared between literature data [14, 19] (Table 3) and experimental values from this work probably connected for different composition of components in the essential oil. It didn`t found data for molar enthalpy of clary sage essential oil at 25^{0}C in literature and for this reason no comparison was made of the results obtained in this work.

Table 5. Specific heat capacity and enthalpy of main components of clary sage essential oil

Compounds	Ylan gene	*β*-Caryo phyllene	*α*-Muuro lene	*n*-Doco sane	*β*-Linalool	*α*-Terpi-eol	Neryl acetate	Linalyl acetate
Content,%	5.18	5.60	8.27	3.00	17.67	4.84	2.34	34.62
C_p, J.mol^{-1}.K^{-1}	229.87	248.50	366.99	87.59	1038.79	284.54	108.10	1599.38
$h_{i,}$ kJ.mol^{-1}	68.54	74.09	109.42	26.11	309.71	84.83	32.23	476.86

C_p - specific heat capacity, J.mol^{-1}K^{-1}; h_i - enthalpy, kJ.mol^{-1}.

4 Conclusions

Some thermodynamic and thermo physical properties of the industrial clary sage essential oil and its major compounds were investigated. The chemical characteristic was done with two experimental methods FT-IR and GC/MS. The FT-IR analysis of studied oil represented mainly strong vibrations of major components in the studied oil.

1. Thermal conductivity λ and coefficient of thermal conductivity were calculated as $330.31.10^{-3}$ $W.m^{-1}K^{-1}$ and $1.49.10^{-6}$ $m^2.s^{-1}$ respectively.
2. The thermodynamic parameters of sage essential oil were calculated on the basis of its main components, too: Gibbs energy (ΔG) - 6072.31 $kJ.mol^{-1}$, enthalpy (ΔH) - 1796.75 $kJ.mol^{-1}$, and entropy (ΔS) -14.34 $kJ.mol^{-1}K^{-1}$.
3. The specific heat capacity of sage oil and its main components are calculated, as well as the molar enthalpy of the main components.

References

1. K. Bauer, D. Garbe, H. Surburg, *Common fragrance and flavor materials, Preparation, properties and uses*, (fourth completely revised Edition, Weinheim, NewYork, Chichester, Brisbane, Singapore, Toronto, Wiley-VCH, 2001)
2. E. Georgiev, A. Stoyanova, *A guide for the specialistin the aromatic industry*. (UFT Acad. Publ. House, Plovdiv, Bulgaria, 2006)
3. E. Dimitrova, S. Ilieva, G. Zolotovitch, Compt Rend l'Acad Bul Sci, **8**, 3, 13-16, (1955)
4. S. Ilieva, Herba Hungarica, **18,** 3,197-203,(1979)
5. N. Kovatcheva, V. Zheljazkov, 7th International Symposium on Essential Oils, Vienna, (1996)
6. Y. Hristova, V. Gochev, J. Wanner, L. Jirovetz, E. Schmidt, T. Grigorova, A. Kuzmanov, J BioSci Biotechn, **2**, 1, 39-44 (2013)
7. T. Farukh, S. Sharopov, W. Setzer, Short Report, Rec. Nat. Prod., **6,** 1, 75-79, (2012)
8. A. Peana, M. Moretti, C. Juliano, Planta Med., **65**, 8, 752-754, (1999)
9. A. Caniard, P. Zerbe, S. Legrand, A. Cohade, N. Valot, J.-L. Magnard, J. Bohlmann, L. Legendre, BMC Plant Biology, **12**, 119, (2012)
10. B. Maurer, A. Hauser, Presented, in part, at the IXth International Congress of Essential Oils, Singapore, **66**, 7, 2223-2235, (1983)
11. Ł. Kuźma', M. Różalski, E. Walencka, B. Różalska, H. Wysokińska, Phitomedicine, **14**, 1, 31-35, (2007)
12. I. Jasicka-Misiak, A. Poliwoda, M. Petecka, O. Buslovych, V. Shlyapnikov, P. Wieczorek, Ecol. Chem. Eng. S., **25**, 1, 133-142, (2018)
13. W. Yang, S. Sokhansanj, J. Tang, P. Winter, Biosystems Engineering, **82,** 2, 169, (2002)
14. P. Florido, I. Andrade, M. Capellini, F. Carvalho, K. Aracava, C. Koshima, C. Rodrigues, C. Goncalves, J Chem Thermo-dynamics, **72**, 152, (2014)
15. ISO 280:1998. Essential oils - Determination of refractive index.

16. ISO 279:1998. Essential oils - Determination of relative density at 20 ^{O}C - Reference method

17. S. Tasheva, V. Gandova, K. Dobreva, I. Dincheva, V. Prodanova-Stefanova, A. Stoyanova, RJPBCS, **10,** 220-227, (2019).

18. F. Incropera, D. Dewitt, T. Bergam, A. Lavine, *Fundaments of Heat and Mass Transfer*,(John Wiley &Sons, 2006)

19. https://www.chemeo.com

20. S. Tasheva, V. Gandova, K. Dobreva, V. Prodanova-Stefanova, K. Marinova, I. Dincheva, A. Stoyanova, E3W Conf. 25th Sceintific Conference on Power Engineering and Power Mashines (PEPM'2020), **207**, 01021, (2020)

Investigation of the thermodynamic and physicochemical properties of lavender oil (*Lavandula angustifolia* Mill.)-ethanol-water system

Vanya Gandova[1], *Stanislava* Tasheva[1*], *Vanya* Prodanova-Stefanova[2], *Krasimira* Dobreva[2], *Ivayla* Dincheva[3] and *Albena* Stoyanova[1]

[1]University of Food Technologies, Plovdiv, 4002, Bulgaria

[2]Trakia University, Dept. Food Technology, Yambol, 8600, Bulgaria

[3]PlantGeneticResearch Group, Agrobioinstitute, Agricultural Academy, 1164, Sofia, Bulgaria

Abstract: The thermodynamic and physicochemical properties of lavender oil (*Lavandula angustifolia* Mill.)-ethanol-water system were measured. The chemical composition of solutions of the lavender oil in 70%, 75%, 80%, 85%, 90%, and 95% ethanol was determined using GC/MS analysis. The main compounds in the solutions are oxygenated monoterpenes (β-linalool, terpinen-4-ol, α-terpineol, and linalyl acetate), followed by monoterpene hydrocarbons ((Z)-β-ocimene and (E)-β-ocimene), and sesquiterpene hydrocarbons (β-caryophyllene and (E)-β-farnesene). The ternary phase diagrams were constructed. By multiply linear regression model for surface tension prediction was applied. The thermodynamic and kinetic parameters of lavender oil – ethanol - water system were calculated (energy of Gibb's, enthalpy, entropy, energy of activation, and coefficient of diffusion).

1.Introduction

Lavender essential oil is obtained by steam distillation of freshly cut, flowering tops of (*Lavandula angustifolia* Mill. (kind Lamiaceae). The solutions of the lavender essential oil in ethanol with concentration from 70% to 95% are used in different perfumery and pharmaceutical products. The major constituents of the oil are the oxygenated monoterpenes linalyl acetate, β-linalool, terpinen-4-ol, α-terpineol, and lavandulol acetate [1-3]. Influence of the temperature and the pressure on the thermodynamic and thermal properties of solution of different terpenic compounds, for example α- and β-pinene, linalool, borneol, *etc*., were investigated from different authors [4-11].

The essential oil content in the inflorescence of lavender cultivated in the mid hills of Uttarakhand was investigated. The quality of lavender oil produced in India was found to be comparable to that produced in Hungary, France, China, Bulgaria, Russia, and the USA [12].

*Corresponding author: st_tasheva@abv.bg

Different studies were provided for described the antibacterial properties of lavender essential oil. Besides, no data exist on its ability to activate human macrophages during the innate response against *Staphylococcus aureus* [13].

Lavender essential oil was used since antiquity as natural active constituents in products designed for dealing with a broad range of issues including those related to human hygiene, health and personal care, and those concerned with organic agriculture and pest management.

Lane and Mahmoud [14] report on the composition and insect repellent properties of the essential oils obtained from 12 varieties of the two most commonly cultivated lavender species (*L. angustifolia* and *L. intermedia*), grown in an experimental plot in the Okanagan Valley of British Columbia, Canada. The oil composition for three lavender oil samples was also analyzed. The results shown that the locally grown plants produce essential oils that are similar to those found in other parts of the world in terms of the overall olfactory properties, and in that they contain the same constituents found in comparable lavender oils.

The extracts obtained from lavender, which were grown in Burdur Örtülü locality, was determined using HPLC and GC-MS analysis. The anti-microbial effect of the essential oil was also investigated. The dried flowers were extracted and the essential oil was distilled from the remaining part. 31 different compounds were determined by GC-MS analysis: Linalool and linalyl acetate have the highest concentration. It was observed that the lavender essential oil could completely remove the contamination caused by the microorganisms as of the 14th day. The essential oil, can be used either directly or incorporated into the cosmetics without the necessity for any other extra preservative [15].

Lavender essential oil was approved by the European Medicines Agency as herbal medicine to relieve anxiety and stress. The capability of the oil to relieve other nervous system disorders such as neuropathic pain has never been established. Oral administration of lavender oil could represent a therapeutic approach in the management of neuropathic pain states [16].

The aim of this study is to be investigating some thermodynamic and thermal properties in the ternary lavender essential oil–ethanol–water system.

2.Material and methods

2.1.Material. For preparation of solutions was used distilled water, lavender essential oil was provided by Kateko Ltd, Bulgaria and 95% ethanol was provided by FILLAB, Bulgaria.

The lavender essential oil was exhibited light yellow color. The main oil has the following properties: density which is determined to 0.8623 ± 0.0 and a refractive index (n_D^{20}) which is determined to 1.4635 ± 0.0.

2.1.1.Solutions preparation.

To investigate the system 18 solutions between ethanol and lavender essential oil with different quantity of water were prepared. The solutions were divided in six series according the ethanol percent. First series were started with 70% and proceed with step of 5%. Last series solutions were prepared only with 95% pure ethanol.

The prepared solutions were investigated with the ratios between lavender oil and ethanol equals to: 1:5, 1:6 and 1:7. The primary composition of the samples is presented in Table 1.

Table 1. Molar ratio lavender oil: ethanol and prime composition in percent's of 18 solutions

N°	Lavender oil: ethanol	ethanol, %	Lavender oil, %	Water, %
1	1:5	70.0	3.3	26.7
2	1:6	70.0	2.5	27.5
3	1:7	70.0	2.5	27.5
4	1:5	75.0	2.5	22.5
5	1:6	75.0	1.7	23.3
6	1:7	75.0	1.7	23.3
7	1:5	80.0	3.3	16.7
8	1:6	80.0	3.4	16.6
9	1:7	80.0	3.4	16.6
10	1:5	85.0	4.2	10.8
11	1:6	85.0	3.4	11.6
12	1:7	85.0	3.4	11.6
13	1:5	90.0	4.2	5.8
14	1:6	90.0	3.4	6.6
15	1:7	90.0	3.4	6.6
16	1:5	95.0	5.0	0.0
17	1:6	95.0	5.0	0.0
18	1:7	95.0	5.0	0.0

2.1.2.Determination of surface tension.

The surface tension of prepared solutions was determined using maximum bubble pressure method. This method is connected with pressure P_{max} indicated as maximum pressure. In following to blow appeared a small bubble in solutions from the end of the capillary. The air bubble was flushed at a constant velocity through the capillary. Pressure was created by means of an aqueous pressure gauge. Surface tension is determined by equation (1)

$$\gamma = \frac{\mathrm{rg}}{2}(\Delta H\rho_{\mathrm{o}} - r\rho) \qquad (1)$$

where: r – capillary radius in (m), g – gravity in ($m.s^{-2}$), ΔH – maximum difference in the two gauges of the gauge, ρ_0, ρ – density of the manometric (water) and test liquid in ($kg.m^{-3}$).

2.1.3.Determination of density in solutions.

The density of liquids was determined by a pycnometer. A well-dried pycnometer was weighed. It is filled with the test liquid of which the density will be determined. After drying, the pycnometer is weighed with the test liquid. It is then emptied, rinsed with ethanol and filled with ethanol. Then, it is weighed again. The density is determined by the equation (2).

The density of the liquids was determined by a pycnometer according to the equation (2).

$$\rho = \frac{m_1 - m}{V} = \frac{m_1 - m}{\frac{m_1 - m}{\rho_1}} = \frac{m_1 - m}{m_2 - m}\rho_1 \quad (2)$$

where: ρ – the density of the test liquid in (kg.m^{-3}), ρ_1 – the density of ethanol in (kg.m^{-3}), m – mass of the pycnometer in (g), m_1 – mass of the pycnometer with the test liquid in (g), m_2 – mass of the pycnometer with ethanol in (g).

2.1.4.Determination of refractive index and molar refraction.

Measurements of the refractive index (n_D) were performed at a wavelength equal to 589.3 nm using an Abbe type refractometer. The measurements were used to calculated molar refraction by equation (3).

$$R_m = \frac{(n_D^2 - 1)(x_1 M_1 + x_2 M_2 + x_3 M_3)}{(n_D^2 + 2)\rho} \quad (3)$$

where: Rm – molar refraction in (m^3.mol^{-1}), ρ – density of the test liquid in (kg.m^{-3}), n_D – refractive index, x – mole fraction of components, M – molar mass of the component in (kg).

2.1.5.Gas Chromatograph (GC) analysis.

The GC analysis was provided for six samples at all ethanol concentrations in ratio 1:5, which is the highest content of essential oil. This series were contained the following ratio of lavender oil: ethanol = 1:5. The samples contained the largest amount of lavender oil and for this reason they were preferred as quantity of essential oil has an influence in the perfumery.

The GC analysis was performed using a GC Agilent 7890A, an HP-5 ms column (30 m x 250 μm x 0.25 μm), temperature: 35 °C/3 min, 5 °C/min to 250 °C for 3 min, total: 49 min; helium as a carrier gas at 1mL/min constant speed, and 30:1 split ratio.

The GC/MS analysis was carried out on an Agilent 5975C mass spectrometer, using helium as a carrier gas, and the same column and temperature as in the GC analysis.

The identification of chemical compounds was done by comparison to their relative retention indexes and library data. The identified constituents were arranged in order of retention time and quantity in percentage.

2.1.6.Linear regression model.

The surface tension prediction was provided by using of two pressures obtained after providing experiment. Maximum bubble pressure (ΔP) and Laplace pressure were included in the regression model. Their determinations were calculated by the equations (4-5):

$$P_{\max} = \rho g \Delta H \quad (4)$$

$$\Delta P = P_{\max} - P_L \quad (5)$$

where: P_{max} – maximum bubble pressure in (MPa), P_L – Laplace pressure in (MPa), g – gravity in (m.s^{-2}), ρ –density in (kg.m^{-3}).

2.1.6. Thermodynamic and thermal parameters calculations.

Some thermodynamic and thermal properties (Gibbs energy, entropy, enthalpy, thermal conductivity, thermal diffusivity and specific heat capacity) were calculated to determine the stability of the system of lavender oil-ethanol-water [17].

According to the transition state theory are determined activation thermodynamic parameters (equilibrium constant, activation free energy of Gibb's, activation enthalpy and entropy) by known equations [18-22].

Specific heat capacity, thermal conductivity and diffusivity were determined according [22-23].

The molecular diffusion coefficient was calculated according to the Wilke - Chang equation given in [20, 24].

2.1.7. Statistical analysis.

It should be noted that the experiments performed were repeated three times. The results were presented by OriginPro 7.0 software.

3. Results and discussion

Physicochemical properties of lavender oil (*L. angustifolia*)-ethanol-water system as density, surface tension and refractive index were determined experimentally. The obtained results are presented in Table 2. Density and refractive index were used to calculate molar refraction in the system.

Table 2. Density, surface tension, refractive index and molar refraction of 18 experimental solutions

N°	Density, ρ, 10^{-3} kg.m^{-3}	Surface tension, γ, mN.m^{-1}	Refractive index, n_D^{20}	Molar refraction, Rm, 10^{-6} m^3.mol^{-1}
1	0.853±0.02	18.555±0.17	1.3827±0.01	10.726±0.11
2	0.849±0.02	14.885±0.13	1.3800±0.01	10.282±0.11
3	0.845±0.02	13.665±0.12	1.3766±0.01	9.983±0.12
4	0.853±0.02	19.780±0.18	1.3820±0.01	11.243±0.12
5	0,852±0.02	14.881±0.13	1.3798±0.01	10.844±0.12
6	0.851±0.02	13.658±0.12	1.3760±0.01	10.565±0.10
7	0.834±0.02	18.578±0.17	1.3827±0.01	11.733±0.12
8	0.833±0.02	16.130±0.15	1.3810±0.01	11.463±0.12
9	0.831±0.02	14.907±0.12	1.3780±0.01	11.188±0.13
10	0.825±0.02	16.139±0.15	1.3820±0.01	12.313±0.12
11	0.822±0.02	14.918±0.13	1.3790±0.01	11.867±0.11
12	0.819±0.02	13.697±0.12	1.3771±0.01	11.607±0.11
13	0.816±0.02	19.825±0.18	1.3801±0.01	12.731±0.13
14	0.812±0.02	18.605±0.17	1.3791±0.01	12.327±0.12
15	0.808±0.02	16.160±0.15	1.3769±0.01	12.041±0.13
16	0.801±0.02	17.884±0.16	1.3808±001	13.208±0.13
17	0.799±0.02	16.171±0.15	1.3786±0.01	12.896±0.11
18	0.798±0.02	14.947±0.13	1.3772±0.01	12.677±0.12

The presented results in Table 2 have shown that all physicochemical dimensions decreased their values when ethanol in solutions increased as quantity. Surface tension and refractive index for pure ethanol are: γ = 22.80 mN.m^{-1} and ρ = 0.789 kg.m^{-3}[17]. The values of refractive index presented some small variations in the last solutions prepared with 90% and 95% ethanol. The refractive index in pure ethanol is 1.3617. In this reason probably decreasing of values of refractive index of solutions with increasing quantity of ethanol leaded to values closed of pure ethanol.

The surface tension of eighteen experimental solutions was measured. The obtained results are presented in Table 2. It was seen that with increased of quantity of ethanol in solutions, the surface tension decreases. Some of the highest surface tension values is approximately 19.825 mN.m^{-1}. Some low value is approximately13.697 mN.m^{-1}. The obtained experimental values is in agree with surface tension in pure ethanol equal to 22.80 mN.m^{-1} and water equal to 72.75 mN.m^{-1} [25]. It`s known that in perfumery and cosmetics are preferred solutions with high ethanol percent. These results presented ethanol and lavender essential oil as components as stabilized system.

The molar refraction Rm was calculated, too. The molar mass of lavender essential oil was determined experimentally by GC analysis. By application of equation 3, the mole fractions of three main components were calculated and used, too. The results which are shown in Table 2 for molar refraction it`s obvious that all values increased with increasing of ethanol percent in solutions. In solutions with 70% ethanol the values is 9.983.10^{6} m^{3}.mol^{-1} but in solutions with 95% pure ethanol is 13.208.10^{6} m^{3}.mol^{-1}. According [6] the molar refraction of mixtures exhibited values between 17-20.10^{6} m^{3}.mol^{-1}. The determination of molar refraction of pure water (3.88.10^{6} m^{3}.mol^{-1}) and ethanol (12.54.10^{6} m^{3}.mol^{-1}) and a mixture of ethanol-water solutions with different concentrations [26] compared with results obtained in this work (13.208.10^{6} m^{3}.mol^{-1}) and good agreement was observed.

Six from all solutions were investigated with GC analysis. The experimental results for chemical compositions of lavender essential oil: ethanol = 1:5 are presented in Table 3. The solutions contained the largest amount of lavender essential oil were preferred because the investigation is connected with the perfumery.

- In the solution prepared with lavender essential oil with 70% ethanol was identified 42 components representing 99.43% of the total content. Six of them were in concentrations above 1% and the rest 36 constituents were in concentrations under 1%. The main components in the solution (above 2%) were: β-linalool (48.30%), linalyl acetate (26.14%), terpinen-4-ol (5.40%), and α-terpineol (5.05%).
- In the solution prepared with lavender essential oil with 75% ethanol was identified 42 components representing 98.98% of the total content. Eight of them were in concentrations above 1% and the rest 34 constituents were in concentrations under 1%. The main components in the solution (above 2%) were: β-linalool (54.13%), linalyl acetate (20.81%), terpinen-4-ol (6.24%), and α-terpineol (3.82%).
- In the solution prepared with lavender essential oil with 80% ethanol was identified 42 components representing 98.39% of the total content. Six of them were in concentrations above 1% and the rest 34 constituents were in concentrations under 1%. The main components in the solution (above 2%) were: β-linalool (51.14%), linalyl acetate (23.80%), terpinen-4-ol (6.04%), and α-terpineol (3.52%).
- In the solution prepared with lavender essential oil with 85% ethanol was identified 42 components representing 98.67% of the total content. Seven of them were in concentrations above 1% and the rest 35 constituents were in concentrations under 1%. The main components in the solution (above 2%) were: β-linalool (49.07%), linalyl acetate (25.00%), terpinen-4-ol (5.89%), and α-terpineol (3.25%).
- In the solution prepared with lavender essential oil with 90% ethanol was identified 42 components representing 97.95% of the total content. Six of them were in concentrations above 1% and the rest 36 constituents were in concentrations under 1%. The main

components in the solution (above 2%) were: *β*-linalool (46.20%), linalyl acetate (28.30%), terpinen-4-ol (5.73%), and *α*-terpineol (2.96%).

- In the solution prepared with lavender essential oil with 95% ethanol was identified 42 components representing 97.37% of the total content. Seven of them were in concentrations above 1% and the rest 35 constituents were in concentrations under 1%. The main components in the solution (above 2%) were: *β*-linalool (38.75%), linalyl acetate (33.10%), terpinen-4-ol (4.92%), lavandulol acetate (2.40%), and *α*-terpineol (2.38%).

Table 3. Chemical composition of the six investigated samples between lavender essential oil and ethanol after GC analysis

Name	**70% ethanol**	**75% ethanol**	**80% ethanol**	**85% ethanol**	**90% ethanol**	**95% ethanol**
n-Hexanol	0.08±0.0	0.09±0.0	0.08±0.0	0.07±0.0	0.06±0.0	0.04±0.0
α-Pinene	0.08±0.0	0.17±0.01	0.14±0.01	0.12±0.01	0.18±0.01	0.13±0.01
Camphene	0.07±0.0	0.11±0.01	0.09±0.0	0.07±0.0	0.12±0.01	0.11±0.01
Sabinene	0.08±0.0	0.04±0.0	0.05±0.0	0.03±0.0	0.04±0.0	0.07±0.0
β-Pinene	0.06±0.0	0.06±0.0	0.05±0.0	0.04±0.0	0.07±0.0	0.13±0.01
n-Octan-1-en-3-ol	0.73±0.06	0.86±0.07	0.82±0.07	0.76±0.06	0.72±0.06	0.55±0.05
n-Octanone-3	0.54±0.04	0.59±0.05	0.57±0.05	0.60±0.05	0.56±0.04	0.60±0.05
Myrcene	0.18±0.01	0.16±0.01	0.19±0.01	0.22±0.01	0.28±0.02	0.35±0.03
n-Butyl butyrate	0.05±0.0	0.04±0.0	0.07±0.0	0.05±0.0	0.06±0.0	0.08±0.0
n-Octan-2-ol	0.12±0.01	0.13±0.01	0.15±0.01	0.11±0.01	0.12±0.01	0.14±0.01
n-Hexyl acetate	0.32±0.02	0.30±0.02	0.35±0.03	0.38±0.03	0.40±0.03	0.43±0.03
p-Cymene	0.05±0.0	0.07±0.0	0.04±0.0	0.06±0.0	0.07±0.0	0.16±0.01
Limonene	0.07±0.0	0.07±0.0	0.05±0.0	0.07±0.0	0.08±0.0	0.17±0.01
Eucalyptol	0.55±0.04	0.58±0.05	0.51±0.04	0.57±0.05	0.59±0.05	0.73±0.07
β-(Z)-Ocimene	0.80±0.07	0.52±0.04	0.69±0.05	0.86±0.07	0.94±0.08	1.81±0.17
β-(E)-Ocimene	0.27±0.02	0.18±0.01	0.24±0.02	0.30±0.02	0.33±0.02	0.57±0.04
cis-Linalyl oxide	0.38±0.03	0.52±0.04	0.41±0.03	0.39±0.03	0.34±0.03	0.28±0.02
trans-Linalyl oxide	0.24±0.02	0.32±0.02	0.26±0.01	0.23±0.01	0.20±0.01	0.17±0.01
β-Linalool	48.30±4.50	54.13±5.50	51.14±5.00	49.07±4.80	46.20±4.50	38.75±3.50
Ethyl heptanoate	0.94±0.08	0.85±0.07	0.97±0.08	1.08±0.09	1.24±0.10	1.42±0.12
Heptyl acetate	0.04±0.0	0.06±0.0	0.08±0.0	0.06±0.0	0.04±0.0	0.09±0.0
Camphor	0.35±0.03	0.38±0.03	0.37±0.03	0.35±0.03	0.34±0.03	0.30±0.02
Lavandulol	0.86±0.07	1.05±0.09	0.98±0.08	0.91±0.08	0.85±0.07	0.66±0.05
Borneol	1.51±0.13	1.83±0.17	1.68±0.15	1.61±0.15	1.48±0.13	1.17±0.10
Terpinen-4-ol	5.40±0.50	6.24±0.60	6.04±0.58	5.89±0.55	5.73±0.53	4.92±0.45
Cryptone	0.68±0.05	1.02±0.09	0.77±0.06	0.68±0.05	0.59±0.04	0.49±0.03
n-Butyl hexanoate	0.08±0.0	0.10±0.0	0.10±0.0	0.21±0.02	0.17±0.01	0.28±0.02

α-Terpineol	5.05±0.49	3.82±0.37	3.52±0.42	3.25±0.30	2.96±0.25	2.38±0.25
Ethyl octanoate	0.18±0.01	0.20±0.01	0.20±0.01	0.20±0.01	0.10±0.0	0.10±0.0
Nerol	0.53±0.04	0.56±0.04	0.60±0.05	0.50±0.04	0.50±0.04	0.30±0.02
Geraniol	0.29±0.01	0.31±0.02	0.30±0.02	0.20±0.01	0.20±0.01	0.20±0.01
Linalyl acetate	26.14±0.25	20.81±0.19	23.80±0.22	25.00±0.24	28.30±0.27	33.10±0.31

Name	**70% ethanol**	**75% ethanol**	**80% ethanol**	**85% ethanol**	**90% ethanol**	**95% ethanol**
p-Menth-1en-7-al	0.16±0.01	0.12±0.01	0.0	0.10±0.0	0.0	0.10±0.0
Lavandulol acetate	1.52±0.14	1.33±0.12	1.60±0.15	1.80±0.17	2.00±0.19	2.40±0.23
Linalool propanoate	0.26±0.02	0.39±0.03	0.20±0.01	0.20±0.01	0.10±0.0	0.24±0.02
Neryl acetate	0.23±0.02	0.20±0.01	0.20±0.01	0.30±0.02	0.40±0.03	0.41±0.03
Geranyl acetate	0.52±0.04	0.46±0.03	0.56±0.04	0.60±0.05	0.68±0.06	0.96±0.08
β-Caryophyllene	0.43±0.03	0.25±0.02	0.33±0.03	0.40±0.03	0.58±0.05	1.52±0.14
β-(Z)-Farnesene	0.11±0.01	0.14±0.01	0.08±0.0	0.10±0.0	0.16±0.01	0.51±0.04
β-(E)-Farnesene	0.14±0.01	0.10±0.0	0.06±0.0	0.08±0.0	0.09±0.0	0.12±0.01
Caryophyllene oxide	0.23±0.02	0.11±0.01	0.25±0.02	0.30±0.02	0.33±0.03	0.58±0.05
α-Cadinol	0.07±0.0	0.16±0.01	0.10±0.0	0.13±0.01	0.10±0.01	0.14±0.01
oxygenated hydrocarbons, %	3.81	4.26	4.21	4.35	4.21	4.40
monoterpene hydrocarbons, %	1.63	1.33	1.52	1.74	2.06	3.41
oxygenated monoterpenes ,%	93.52	93.58	93.41	92.82	92.38	89.10
phenyl propanoids%	0.05	0.07	0.04	0.06	0.07	0.16
sesquiterpene hydrocarbons, %	0.69	0.49	0.47	0.59	0.84	2.19
oxygenated sesquiterpene, %	0.30	0.27	0.36	0.44	0.44	0.74

In this study, the oxygenated monoterpenes and oxygenated hydrocarbons were the dominant groups of chemical components in the solutions according the Table 3.From obtained experimental data can be seen that low composition of hydrocarbons at small ethanol concentrations is connected with their difficult solubility. With increase of ethanol concentration, the oxygen monoterpenes content decreased but hydrocarbons increased.

GC analysis of six solutions the main components were determined. In the solutions they are oxygenated monoterpenes: β-linalool, terpinen-4-ol, α-terpineol, linalyl acetate with up to 5%. The results are used to constructed experimental phase diagrams. Two phase diagrams for solutions prepared with 75% and 80% ethanol are presented in Figs. 1-2. These two solutions were preferred because the main components β-linalool and linalyl acetate exhibited highest values in these solutions. Fig. 1 presented results for the solution with 75% ethanol. The symbols denoted the point of equilibrium between each tree component (oil-ethanol-water) total compositions of each sample. In the phase diagram four primary compounds were observed: β-linalool, terpinen-4-ol, α-terpineol, linalyl acetate and the equilibrium points they form with ethanol and water. The lines shown the tie-lines for the components connected with the equilibrium state in the system. Fig. 2 presented the same phase diagram at 80% ethanol and the components: β-linalool, terpinen-4-ol, α-terpineol, linalyl acetate and equilibrium points between these and ethanol. The lines presented again the tie-lines connected with the equilibrium state in the system.

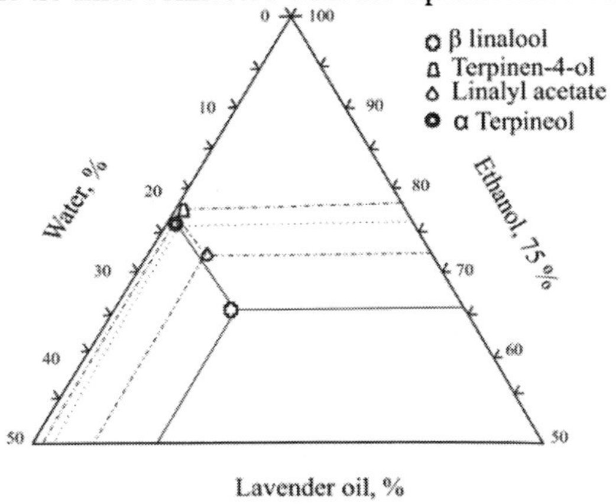

Fig. 1 Partial phase diagram for solution prepared with 75% ethanol in lavender essential oil-ethanol-water system

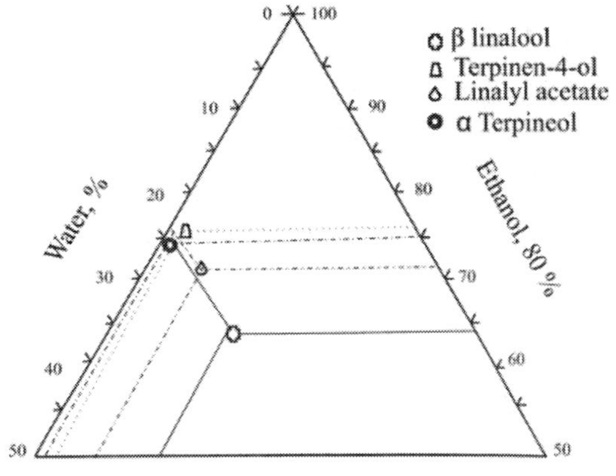

Fig. 2 Partial phase diagram for solution prepared with 80% ethanol in lavender essential oil-ethanol-water system

 https://doi.org/10.1051/e3sconf/202128602004

Between figures 1 and 2 appeared small differences in component *β*-linalool. At solution prepared with 75% the component is equal to 54.13% but at solution with 80% ethanol is 51.14%. With increasing of ethanol percent the *β*-linalool percent decreased. Opposite values for linalyl acetate increased with increasing of ethanol percent. At solution prepared with 75% is 20.81% but at solution with 80% ethanol is 23.80%. These differences in chemical compositions are connected with different properties and stability of the systems, probably.

Multiple linear regression. To estimate the surface tension from two pressures, a multiple linear regresased.in model was developed [14] using the values shown in Figs. 3 and 4. Fig. 3 presented values of surface tension with dependence of density and Laplace pressure. Fig. 4 presented values of surface tension with dependence of density and maximum bubble pressure. Almost linear dependence and preferable correlation has been observed. As can be seen from the figure, there is a preferable correlation between the two tensions. According to these results two equations were obtained. In the equations "x" interacts in equation (6) with a maximum bubble pressure and in equation (7) with Laplace pressure. A high coefficient of correlation (R^2 = 0.999 and R^2 = 0.905) indicates the excellent correlations between experimental and calculation values of surface tension.

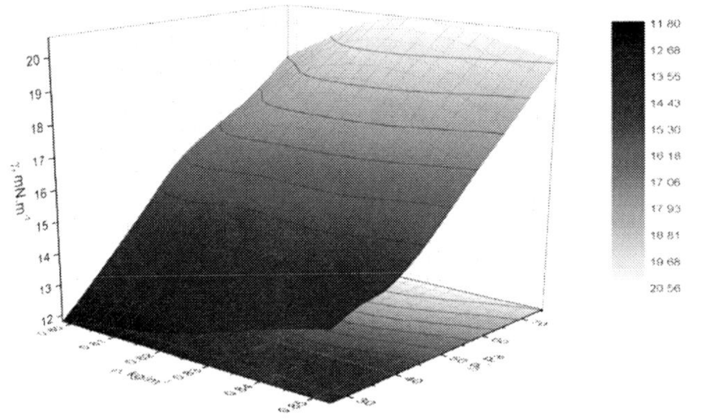

Fig. 3 Presents values of surface tension with dependence of density and Laplace pressure.

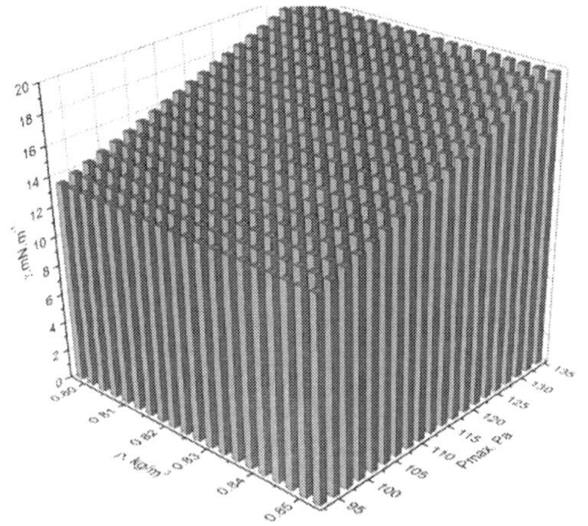

Fig. 4 Presents values of surface tension with dependence of density and maximum bubble pressure.

$$P_{max} = -0.371 + 0.150.x, R^2 = 0.999 \quad (6)$$

$$\Delta P = 9.00 + 0.433.x, R^2 = 0.905 \quad (7)$$

To estimate the surface tension from composition of oxygenated monoterpenes, a multiple linear regression model was developed using the values showed in Table 2 and Table 3. The model is shown in equation 8. Coefficient of correlation is $R^2 = 0.832$.

$$\gamma = - 211.878 +5.284.L. - 19.393.T\text{-}4\text{-}ol - 5.284.T + 3.970.La. \quad (8)$$

where: γ – surface tencion; L – β-linalool; T-4-ol – terpinen-4-ol; T - α-terpineol; La - linalyl acetate.

According to the transport theory, the values of the activating enthalpy, entropy and Gibbs energy for the transition state and for the final state of the investigated solutions of lavender oil-ethanol-water are calculated and are presented in Table 4. The value of the activating energy was calculated according to [19], which is 5726.07 kJ.mol^{-1}. As can be seen from the presented results in the transient state, the values of the activating thermodynamic parameters are negative and decrease with increasing lavender oil: ethanol ratio of the tested solutions.

Table 4. Thermodynamic parameters of experimental lavender solutions

№	$K^{\#}$	$\Delta G^{\#}$ kJ.mol^{-1}	$\Delta S^{\#}$ kJ.mol^{-1}K^{-1}	$\Delta H^{\#}$ kJ.mol^{-1}	ΔG kJ.mol^{-1}	ΔS kJ.mol^{-1}K^{-1}	ΔH kJ.mol^{-1}
1	3.052	-2705.956	-9.076	-2715.032	-2765.837	9.929	194.398
2	2.447	-2169.726	-7.277	-2177.003	-2217.740	11.767	1290.592
3	2.273	-1991.770	-6.680	-1998.450	-2035.847	12.377	1654.379
4	3.580	-3093.215	-10.375	-3103.590	-3161.666	8.601	-597.259
5	2.609	-2326.068	-7.802	-2333.870	-2377.542	11.231	970.987
6	2.412	-2134.816	-7.160	-2141.976	-2182.058	11.887	1361.956
7	3.588	-3098.340	-10.392	-3108.732	-3166.904	8.583	-607.737
8	3.032	-2689.772	-9.022	-2698.794	-2749.295	9.984	227.482
9	2.787	-2485.651	-8.337	-2493.988	-2540.657	10.684	644.758
10	3.255	-2862.425	-9.601	-2872.026	-2925.768	9.392	-125.465
11	2.976	-2645.158	-8.872	-2654.030	-2703.694	10.137	318.685
12	2.722	-2428.495	-8.145	-2436.640	-2482.236	10.880	761.601
13	4.637	-3720.434	-12.478	-3732.912	-3802.765	6.451	-1879.457
14	4.218	-3490.448	-11.707	-3502.155	-3567.690	7.239	-1409.307
15	3.490	-3031.359	-10.167	-3041.527	-3098.441	8.813	-470.811
16	4.326	-3551.819	-11.913	-3563.731	-3630.418	7.029	-1534.764
17	3.759	-3211.219	-10.770	-3221.989	-3282.281	8.197	-838.490
18	3.400	-2967.899	-9.954	-2977.853	-3033.577	9.031	-341.081

$K^{\#}$ - equilibrium constant of the transition state; $\Delta G^{\#}$- is the activation free energy or Gibb`s energy, kJ.mol$^{-1}$; $\Delta S^{\#}$ - is the activation entropy, kJ.mol$^{-1}$K$^{-1}$; $\Delta H^{\#}$ - is the activation enthalpy, kJ.mol$^{-1}$; $\Delta G$ - the free energy or Gibb`s energy, kJ.mol^{-1}; ΔS - is the entropy, kJ.mol^{-1}K^{-1}; ΔH - is the enthalpy, kJ.mol^{-1}.

This means that in addition to the temperature as a factor influencing the thermodynamic parameters, the ratio of raw material: solvent in the studied solutions is also important. According to the obtained results, the negative values of activation entropy $\Delta S^{\#}$ indicated the process was less random and non-spontaneous. The formation of activated complex due to the association mechanism is responsible for negative values of entropy. The negative values for Gibbs free energy change of the solution for ternary system specific that process was spontaneous, as shown by the values for the transient and final state. Which is probably due to better solubility of lavender oil in a solution with ethanol and water (better solubility of the main components of lavender oil in these solutions).

For the solutions of lavender oil-ethanol-water according to the Willie-Cheng equation, the internal molecular diffusion of the solutions at temperature (T = 298.15 K) and solvent ethanol was calculated, the value of the diffusion coefficient is $D = 56.27.10^{-9}\ m^2.s^{-1}$.

Based on the obtained values for the final state of enthalpy, the specific heat capacity of the lavender oil-ethanol-water solutions was calculated. The coefficient of thermal conductivity is calculated, which is a thermophysical characteristic of bodies and depends on many factors, but is usually determined by empirical dependencies. Using them [27], the thermal conductivity of a solution of lavender oil-ethanol-water was calculated.

The coefficient of thermal conductivity is also calculated, representing the ratio of the coefficient of thermal conductivity related to the specific heat capacity for the studied solutions.

The higher value of λ connected with the faster temperatures in the system equalize, or the faster given environment conducts heat. Thermal conductivity is a property of material [23]. According the experimental results the thermal properties of the system were calculated and presented in Table 5. The specific heat capacity increases with increasing ethanol concentration and with increasing lavender oil: ethanol ratio. The values for the specific heat capacity are similar to the values obtained with the linalyl acetate-ethanol-water solution presented in [22]. Due to the fact that linalyl acetate is one of the main components in lavender oil, probably. The difference between linalyl acetate and lavender solutions is based on lower values for specific heat capacity, which is probably due to the fact that lavender solutions involve the main components of lavender oil. The obtained results on the coefficient of thermal conductivity and thermal difusivity for solutions of lavender oil-ethanol-water are close to the values for solutions of linalool-ethanol-water presented in [17, 22], which is explained by the fact that the components involved in the solutions are oxygen-containing monoterpenes.

Table 5. Thermal parameters of lavender oils solution

№	c_p, $kJ.mol^{-1}K^{-1}$	$\lambda\ 10^3$, $W.m^{-1}K^{-1}$	$a\ 10^6$, $m^2.s^{-1}$
1	35.073	330.364	9.419
2	33.279	330.366	9.927
3	31.661	330.368	10.435
4	37.985	330.364	8.697
5	36.379	330.364	9.081
6	34.459	330.365	9.587
7	41.494	330.374	7.962
8	39.012	330.374	8.468
9	37.840	330.375	8.731
10	44.157	330.379	7.482
11	42.669	330.38	7.743
12	40.658	330.382	8.126
13	47.294	330.384	6.986
14	45.582	330.386	7.248
15	43.287	330.388	7.632
16	52.635	330.392	6.277
17	50.554	330.393	6.535
18	48.632	330.393	6.794

c_p - specific heat capacity $kJ.mol^{-1}K^{-1}$; λ - coefficient of thermal conductivity, $W.m^{-1}K^{-1}$; a - coefficient of thermal difusivity, $m^2.s^{-1}$.

Three thermal parameters according calculations obtained in this work: λ = 0.33 W. $m^{-1}K^{-1}$; Cp = 31.661 – 52.635 кJ.$mol^{-1}K^{-1}$; a = 6.277 – $10.435.10^{-6}$ $m^2.s^{-1}$. In literature are given results for same components of essential oil linalool and linalyl acetate, and the results compared with literature data for the same parameters, obtained experimentally [17, 22, 27].

4. Conclusions

Lavender oil-ethanol-water systems were investigated by different methods. Some physicochemical and thermal properties in the system were calculated. Density, surface tension and refractive index were determined experimentally. They are decreased when ethanol in solutions increased as quantity. These results presented lavender essential oil as component that stabilized system. Molar refraction of solutions was obtained with calculations. The values were between $9.98.10^6$ $m^3.mol^{-1}$ in solutions with 70% ethanol and $13.21.10^6$ $m^3.mol^{-1}$ in solutions with 95% ethanol. Two phase diagrams are presence at 75% and 80% ethanol in solutions. In the two phase diagrams four main components *β*-linalool, terpinen-4-ol, *α*-terpineol, and linalyl acetate are presence. Three thermal parameters according calculations obtained in this work presented values: λ = 0.33 $W.m^{-1}K^{-1}$; Cp = 31.661 – 52.635 кJ.$mol^{-1}K^{-1}$; a = 6.277 – $10.435.10^{-6}$ $m^2.s^{-1}$. The multiple linear regressions, preferable correlation between experimental surface tension values was obtained. A high coefficient of correlations was observed in three dependent and independent variables; R^2 = 0.999 and R^2 = 0.905, for pressures and R^2 = 0.832 for chemical composition of monoterpenes.

References

1. K. Bauer, D. Garbe, H. Surburg, *Common fragrance and flavor materials, Preparation, properties and uses*, (fourth completely revised Edition, Weinheim, NewYork, Chichester, Brisbane, Singapore, Toronto, Wiley-VCH, 2001).
2. K.H.C. Başer, G. Buchbauer, *Handbook of essential oils: science, technology, and applications*, (Taylor and Francis Group, 2010).
3. E. Georgiev, A. Stoyanova, *A guide for the specialistin the aromatic industry*. (UFT Acad. Publ. House, Plovdiv, Bulgaria, 2006).
4. A. Arce, A. Marchiaro, A. Soto. J. Sol. Chem., **33**, 561, (2004).
5. R. Clara, A. Gymez Marigliano, V. Del Campos, H. Sylimo, Fluid Phase Equilibria, **293**, 151, (2010).
6. S. Garcia-Abarrio, M. Torcal, M. Haya, J. Urieta, A. Mainar, J. Chem. Therm., **43**, 527, (2011).
7. S. Garcia-Abarrio, L. Viloria, L. Haya, J. Urieta, A. Mainar, Fluid Phase Equilibria, **308**, 78, (2011).
8. K. Li, K. Tamura, Fluid Phase Equilibria, **263**, 223, (2008).
9. V. Štejfa, M. Fulem, K. Ruzicka, C. Cervinka, J. Chem.Therm., **79**, 272, (2014).
10. V. Štejfa, M. Fulem, K. Ruzicka, C. Cervinka, J. Chem.Therm., **79**, 280, (2014).
11. M. Torcal, M. Teruel, J. Garcı, J. Urieta, A. Mainar, J. Chem. Eng. Data, **55**, 5332, (2010).
12. R. Verma, L. Rahman, C._Chanotiya, R. Verma, A. Chauhan, A. Singh, A. Yadav, J. Serbian Chem. Soc., **75,** 343, (2010).

13. D. Giovannini, A. Gismondi, A. Basso, L. Canuti, R. Braglia, A. Canini, F. Mariani, G. Cappelli, Immunol. Invest., **45**, 11, (2016).
14. W. Lane, S. Mahmoud, Nat. Prod. Commun., **3**, 1361, (2008).
15. A. Turgut, F. Emen, H. Canbay, R. Demirdöğen, N. Çam, D. Kılıç, T. Yeşilkaynak, JOTCSA, **4**, 283, (2017).
16. M. Sanna, F. Les, V. Lopez, N. Galeotti, Front. Pharmacol., **10,** 1, (2019).
17. Tasheva, V. Gandova, K. Dobreva, I. Dincheva, V. Prodanova-Stefanova, A. Stoyanova, RJPBCS, **10**, 220, (2019).
18. S. Sepida, Z. Abidin, R. Yunus, A. Muhammad, A. Amer, Jour. Applied Sci., **6,** 1390, (2009).
19. H. Topallar, Ü. Geçgel, Turk. Journ. Chemical, **24**, 247, (2000).
20. K. Anbalagan, M. Magesh Kumar, K. Ilango, R. Mohankumar, R. Lakshmi Pria, Ind. Crops Prod., **140,** 1, (2019).
21. D. Paunovic, S. Mitic, D. Kostic, M. Mitic, B. Stojanovic, J. Pavlovic. Advanced Technologies, **3**, 58, (2014).
22. V. Gandova, S. Tasheva, K. Dobreva, V. Prodanova-Stefanova, K. Marinova, A. Stoyanova, E3W Conf. 25th Sceintific Conference on Power Engineering and Power Mashines (PEPM'2020) **204**, 01022, (2020).
23. F. Incropera, D. Dewitt, T. Bergman, A. Lavine, *Fundaments of heat and mass transfer*, (John Wiley &Sons, 2006).
24. S. Tasheva, V. Gandova, V. Popova, T. Ivanova, A. Stoyanova, 2020 7th International Conference on Energy Efficiency and Agriculture Engineering (EE&AE), 1, (2020).
25. R. Belda, J. Herraez, O. Diez, Phys. Chem. Liq., **43**, 91, (2005).
26. J. Khan, M. Farooqui, S. Quardi, Ras. Jour. Chem., **4**, 944, (2011).
27. W. Yang, S. Sokhansanj, J. Tang, P. Winter, Biosystems Engineering, **82**, 169, (2002).

Performance of cylindrical solar still with hemispherical cover: CFD simulation study

Qahtan A Abed[1,*], *Dhafer M* Hachim[2], and *Wisam A* Abd Al-Wahid[2]

[1]Technical Institute/ Al-Rumaitha, Al - Furat Al- Awsat Technical University, 31001, Iraq

[2]Engineering technical college/ Najaf, Al - Furat Al- Awsat Technical University, 31001, Iraq

Abstract. Solar distillation collectors are the device that runs on freshwater products. The solar still is one solar thermal technology for purification of water. This technique is employed to convert saline water into freshwater. At this time, the human being increased freshwater required consumption because of the increasing population density. The thermal analysis of a new design of cylindrical solar still (CSS) with hemispherical cover has been studied. A computer simulation model was improved to investigate its performance. The 2D symmetry simulation models of cylindrical solar still have been validated. The simulation model can emulate the temperature value of different points inside the CSS. The simulation results confirm an acceptable agreement with the experimental result was reported in the literature. From this simulation, it was noted that the freshwater product from the cylindrical solar still obtains the maximum daily productivity of 6.1 kg/m^2 as compared to a traditional single slope solar still SSS as 3.18 kg/m^2 during the day.

1 Introduction

Renewable energy, especially solar energy, can be used to desalinate the saltwater by found the heat energy needed to carry out the evaporating and condensing process (phase change processes). There are two types of solar desalination processes, direct and indirect processes. In the first type of solar desalination processes (direct), the solar irradiance is received by the collector and converted into thermal energy to evaporate the saltwater. Whereas the second type (indirect system) used the dual separate systems, it combines solar collectors with conventional desalination systems. The lack of drinking water has long been linked with people who live in remote areas. Many people in rural areas of the world spend a large portion of their day getting the freshwater. The freshwater production process that consumes a lot of energy therefore, attempts have been done to improve the solar radiation for operating desalination process by various solar systems [1]. For the time being, various techniques are used to generate drinkable water from polluted water. There are a lot of previous studies of enhancing the desalination proses. Abu-Hijleh and Rababah [2] proposed new way to improve the solar desalination presses throw adding sponge cubes on the tope of water basin surface. This method increases the production of the freshwater

* Corresponding author: qahtan.abed@atu.edu.iq

from the solar still, it achieved more than 18% compared with single slope solar still without sponge cubes. There are many other studies in which applications of solar distillation have been conducted on the solar still collectors [3–6]. The freshwater production from any solar still types are increases with increase the water pressure saturation [7]. A theoretical model of steady-state heat transfer for tubular solar still type had been investigated by [8]. The results of FUKUHARA concluded that the suggested model could accurately predict the fresh water output of the solar still. Effects of many factors such as temperatures difference of glass, water vapour and water basin surface, depth of saltwater and glass cover slop etc. on solar still productivity are showed in [9].

Providing people with safe drinking water is getting the government's attention. Therefor many researchers have been doing to improve the productivity of solar distillation technology [10]. Many studies have been applied to improve the efficiency of solar stills collectors, through enhancing the solar still design [11]. The effects of the tilt angle of the glass cover and saltwater depth in the solar still basin on the distillation performance had been studied by [12]. The best daily freshwater productivity from the collector was with the tilt angle of the glass equal to the value of the latitude angle of the setup place. Also, it has been found that solar still productivity during the day decreases with higher saltwater depth.

To enhance the performance of different types of solar still, many previous researchers has been suggested coupling of a conventional solar still collector with flat plate solar collector [13–15]. A comprehensive review of the various types of solar desalination is provided by [16]. They are compared two solar stills models, the tubular type and triangular type. A unique study was done by [17] in a cylindrical solar distillation type. In this case, discrimination is done between the basic single slope solar still with new cylindrical solar still at the same basin area. They carried experiments to test the systems on the summer and winter days. Their conclusions indicated a 3% increase in the thermal efficiency of the cylindrical type. Nidal et al. [18] have presented a thermal analysis of the performance of a cylindrical solar collector with a curved absorber tube. It was noticed an enhanced heat absorption in the cylindrical solar collectors.

Reviews of the previous researcher show that a large number of experimental studies on a solar still. There are not enough previews studied done on cylindrical solar stills. Therefore, our goal of this simulation is to present a simple solar still building at a low price and can be operated by any person in the population.

Exactly the contrary, there is little numerical study. Therefore to cover this gap with the goal of reducing spent time and increase the accuracy of the results, a transient simulation model was applied in this simulation work. The main objective of this article is to create a numerical simulation model for the study of heat and mass transfer in cylindrical solar still. Presenting the varying temperatures and freshwater product during the day depending on real weather conditions. Also, Investigation the impacts of weather conditions and design to enhance the efficiency of cylindrical solar still.

2 Description of solar system

In order to investigate the performance of the TSS, a two-dimensional symmetry simulation model is improved using COMSOL multiphysics 5.4. Fig. 1 shows the suggested cylindrical solar still (CSS) system consists of transparent glass in a cylinder shape with 30 cm length, 15 cm diameter and has 0.4cm thickness. A coper trough with dimensions 13 × 2 × 0.8 cm was employed as a water basin to keep the saltwater. All sides of the trough are coated with an anti-corrosion paint, as well as it's coated with black paint to increase its absorptivity. To maintain a constant level of saltwater in the trough during day, a feed water tank is preparing for CSS. The model in all its dimensions was used in the previous experimental study by [17].

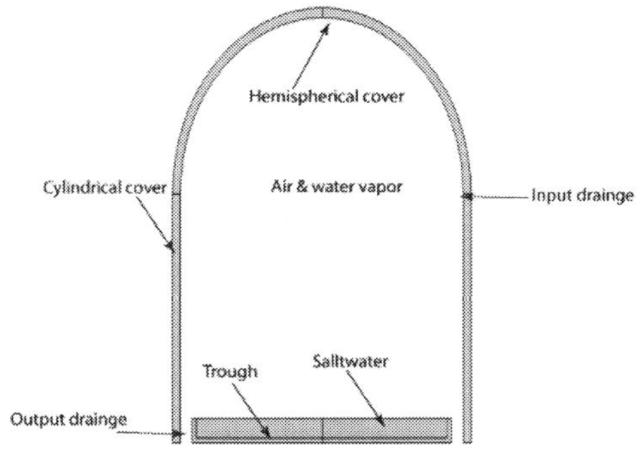

Fig.1. Schematic diagram of cylindrical solar still

3 CFD COMSOL Modeling

3.1. Create the geometry of CSS

Cylindrical solar still (CSS) is an uncomplicated solar desalination project, as it is employed to convert saline water to fresh water. The improved CSS used for the existing study consists of a transparent cover shaped of vertical tubular with hemispherical and a black water basin. A 2D symmetry geometry is applied for simulations in the current work. The geometry of the cylindrical solar still (CSS) has been modelled by COMSOL Metaphysics as CFD simulation software program. The two dimensions symmetry geometry of cylindrical solar still illustrates in figure 2. The geometric model of the cylindrical solar still consists of a hemispherical cover on the upper sided, a transparent cylindrical case on the front side and a black rectangular water basin containing salt water.

The solar radiation incident on the cylindrical cover surface, part of its path through it, and other parts are reflected to the sky. The basin is worked to absorb the solar irradiance, and the temperature will start to increase. The absorbed heat energy is transferred to the saltwater to increase its temperature. From the surface of saltwater there are three kinds of heat transfers processes which are radiative heat transfer (Q_{r,w_g}), convective heat transfer (Q_{c,w_g}) and evaporative heat transfer (Q_{e,w_g}) (see figure 2). The saturator vapour is accumulated up the water surface. Which begins to condense on the inner side of the cylinder cover after losing its latent heat. The drops of freshwater are falling below by gravity and it is accumulated in a freshwater tank. The number of drops of freshwater is increasing as the temperature differences between the saltwater temperature and the cylinder surface temperature increase. There are two types of heat transfers from cylinder cover to ambient, which is convective (Q_{c,g_a}) and radiation (Q_{r,g_a}) as losses. Other type of heat losses from the basin by convection to ambient (Q_{l,b_a}).

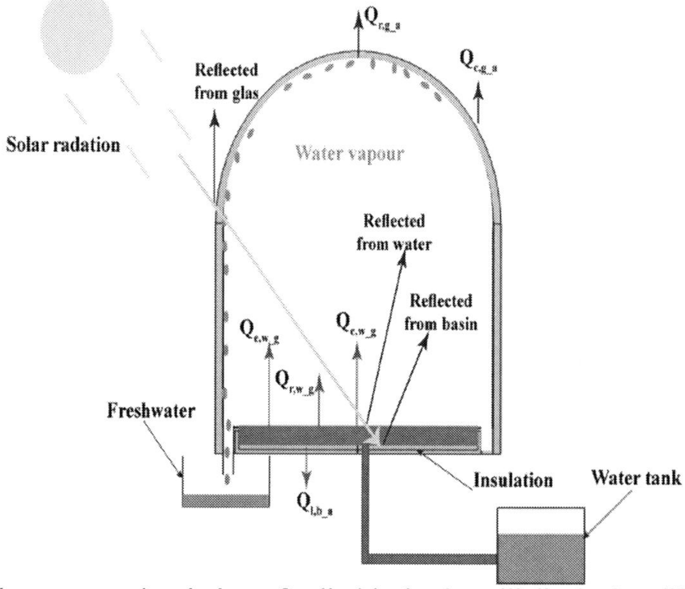

Fig. 2. The cross-sectional view of cylindrical solar still displaying different losses and gain heat

3.2. The model mesh

The two-dimension CSS Geometry (2D), properties and functions were defined in COMSOL Multi-Physics. Also the CSS grid has been created by COMSOL Multi-Physics. Given the difficult computational domain and the many intersections surface, the CSS is separated into different parts, which are divided by the mesh, and a system is applied to combine the meshes between all parts. To achieve accurate simulation results, the fine element size for the grid has been used. The 2D symmetry simulation model of cylindrical solar still with a grid is presented in figure 3. The geometry mesh was built from lots of small mesh elements. Which consists of 1500000 elements. The mesh independence test was performed to test the grid size of the simulation model. The mass transfer and heat energy were analysed in this simulation.

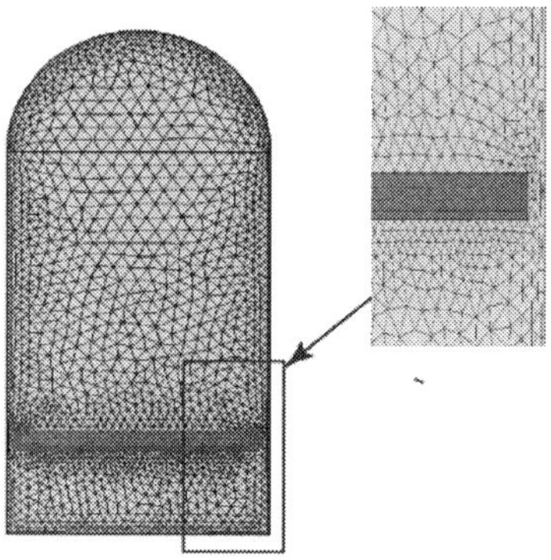

Fig. 3. Grid created at the CSS with different numbers of elements

3.3. Main governing equations

The mixture of air and vapour movement inside the cylindrical solar stills is decided to be a laminar flow moving according to the buoyancy force. The fluid in any state must be appropriate with the scent law of thermodynamics (conservation of mass) [19].

3.3.1 Continuity equation

$$\frac{\partial \rho}{\partial t}+\nabla.(\rho\vec{\bar{\upsilon}})=0 \tag{1}$$

Where $\vec{\upsilon}$ and ρ are the velocity vector and density, respectively.

3.3.2 Momentum equation

Convection heat transfer increase in the CSS as the temperature difference between the saltwater surface and the hemispherical cover surface. The momentum equation for two cases is expressed below:

For gas (mixture of air-water vapour)

$$\frac{\partial}{\partial t}(\rho\vec{\bar{\upsilon}})+\nabla.(\rho\vec{\bar{\upsilon}}\vec{\bar{\upsilon}})=-\nabla.(\bar{p})+\nabla.(\vec{\tau}_{eff})+\rho\vec{g} \tag{2}$$

For liquide (water)

$$\frac{\partial}{\partial t}(\rho\vec{\bar{\upsilon}})+\nabla.(\rho\vec{\bar{\upsilon}}\vec{\bar{\upsilon}})=-\nabla.(\bar{p})+\nabla.(\vec{\tau}_{eff})+\rho\vec{g} \tag{3}$$

where: $\vec{\tau}_{eff}$ shear stress effective, β thermal expansion coefficient (water)

3.3.3 Energy equations

The temperatures variation is achieved by solving the following energy equation in various parts of the CSS is expressed:

$$\frac{\partial}{\partial t}(\rho\bar{E})+\nabla.(\vec{\bar{\upsilon}}(\rho\bar{E}+\bar{p}))=\nabla.(\lambda_{eff}\nabla\bar{T}-\sum_j\bar{h}_{E.j}\vec{\bar{J}}_j)+S_h+S_L \tag{4}$$

For liquide (water)

$$\frac{\partial}{\partial t}(\rho E)+\nabla.(\vec{\upsilon}(\rho E+p))=\nabla.(\lambda\nabla T)+S_h+S_L \tag{5}$$

For solid (glass)

$$\frac{\partial}{\partial t}(\rho E)=\nabla.(\lambda\nabla T)+S_h+S_L \tag{6}$$

Where: $\bar{h}_{E.j}$ and $\vec{\bar{J}}_j$ are the diffusive flux and sensible enthalpy, λ_{eff} thermal conductivity effective, S_h is the latent heat of evaporation and S_L is the absorbed radiant energy.

3.4 The boundary and initial conditions

There are three physicals used in this simulation: heat transfer in solids, heat transfer in fluid and condensation physics. The solar irradiation is passing through the upper hemispherical cover. Then the transmissivities irradiance falls on the water surface, and absorbed by the water basin.

The governing equations for each of these elements are solved to simulate the system performance. The above governing equations were solved in accordance with expressions and assumptions that have been implemented for this simulation study:

1. The boundary symmetry conditions have been used in the current simulation model of CSS as given in figure 3.
2. Neglected any source for the thermal energy generation inside the solar still.
3. The considered the heat transfer from the top boundary were convection and radiation to the surrounding.
4. Forced heat transfer by convection from the still surface to the ambient was neglected.
5. The influence that the condensed water film has on the solar radiation that enters the collector have been neglected.
6. At the time of sunrise and sunset, the solar still trough was received the solar radiation through the cylinder wall.
7. The system leakages don't occur.
8. The estimated is no slip for all the walls of the cylindrical solar still

The initial temperature of the surrounding and inside the CSS are estimated to be 25 °C and 30 °C respectively. Additionally, the velocity was estimated equal to 0 m/sec in all initial states. The input data of weather conditions (solar radiation, wind speed and ambient temperature) are used according to the previous experimental work [17]. Fig. 3 presents the variation of solar irradiance and ambient temperature during the day for the CSS.

3.5 The model validation

The numerical simulations are performed in the cylindrical solar still. COMSOL multiphysics techniques are utilised to guess the amount of freshwater productivity and temperature distribution of different parts CSS. With the purpose of validating the simulation model, the same weather conditions among the present simulation and the previous experimental work have been implemented [17]. The productivity comparison is presented in fig. 4. It can be observed that the simulation results are very good to agreement with the experimental results. The formula of mean bass error (MBE) is used to show the simulation accuracy as follows:

$$MBE = \frac{1}{n}\sum_{i=1}^{n}(P_{i,simu} - P_{i,\exp}) \qquad (7)$$

where: $P_{i,simu}$ and $P_{i,\exp}$ are the simulation and experimental productivity, respectively, while n referring to the total number of values.

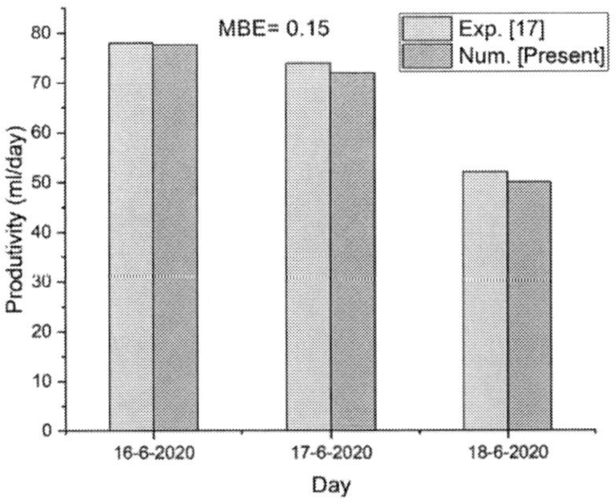

Fig. 4. Comparison of freshwater productivity between the simulation and experimental results

4. Results of the simulation

The CFD numerical simulations were performed by the COMSOL multiphysics program. The solar irradiance and surrounding air temperatures were measured during the experimental test (on Najaf- Iraq, latitude 32°02', longitude 44°20') have been plotted in figure 5. The meteorological data of weather have been selected based on the experimental test, and different five weather conditions days were used in the simulation (see figure 5). It presents the variations in ambient temperatures and solar irradiance during the day. This figure shows an hourly variation of solar irradiance and ambient air temperature for different days (22-5-2019, 3-10-2019, 30-12-2019, 18-2-2020 and 16-6-2020).

At the start of all days, the amount of solar irradiance was limited. After that, it's increasing to be the maximum value in the middle of the day, and then these values are returning to decrease. The air surrounding temperature (ambient) was the maximum value on 16-6-2020 compared with other days.

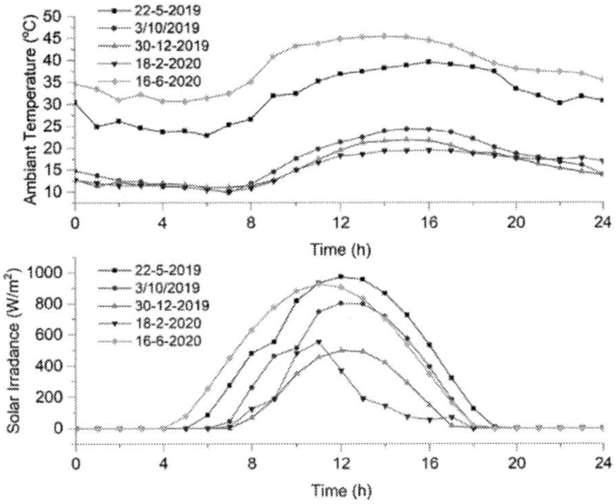

Fig. 5. Variation of the intensity of solar irradiance and ambient temperature during the day.

In this investigation, the CSS designed was simulated to show the influence of this new designed on the freshwater productivity. The contours of variation temperature on the front side of the CSS during the different hours of the day has been illustrated in figure 6. It shows the difference in temperature between on the cover saltwater surface. When this is increased, the productivity of freshwater will increase. Nevertheless, the difference in temperatures decreases about 15 °C at the time of 12 PM and 6 AM, which leads to decrease freshwater protect during the night.

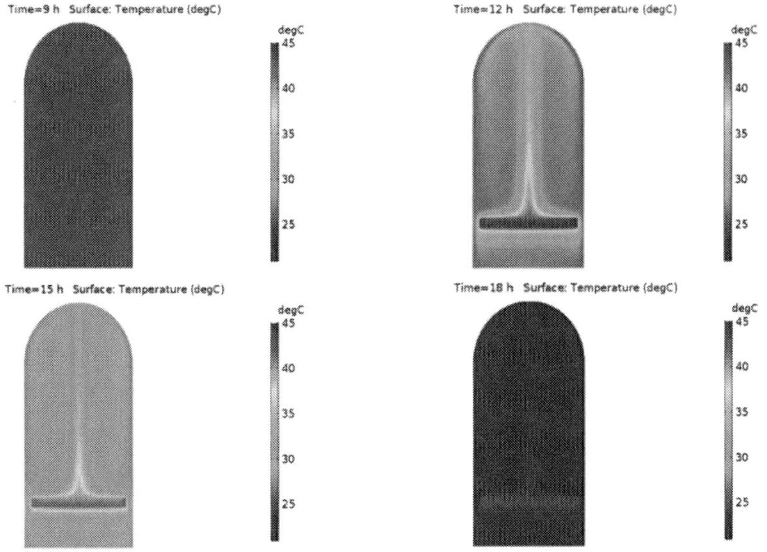

Fig. 6. Variation of the temperature contours of the CSS during a day at (a) 9AM (b) 12PM (c) 15PM (d) 18PM.

The contour of the various pressure on the symmetry cylindrical solar still is displayed in Fig. 7A. The non-uniform pressure distribution was seen for the inside of the symmetry plan CSS. The result showed there is an increase in vapour pressure in the upstream flow direction. A side view of velocity contours in running water vapour and air inside the CSS is shown in figure 7B. In the CSS there are no restrictions, so the evaporated water is moving up to stick to the cold glass surface and then dropped to exits out of the CSS.

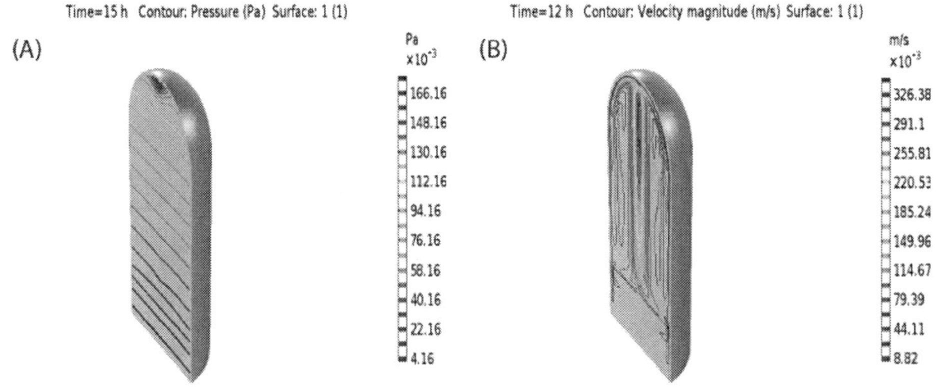

Fig. 7. (A) The static pressure contours for the optimized CSS model, (B) The vapour velocity contours inside the CSS model.

New simulation model for a conventional single slope solar (SSS) still have been developed to 0comparison with CSS. It is observed that the solar radiation incident during the day on the glass cover of SSS is less than it on the CSS. Due to geometry of solar still, it allows receiving the solar radiation at the front side of the SSS, while the CSS receives the solar radiation in all direction. The comparison results are illustrated in Figure 8. It shows that the fresh water product from of the present model CSS with hemispherical was higher than it in SSS.

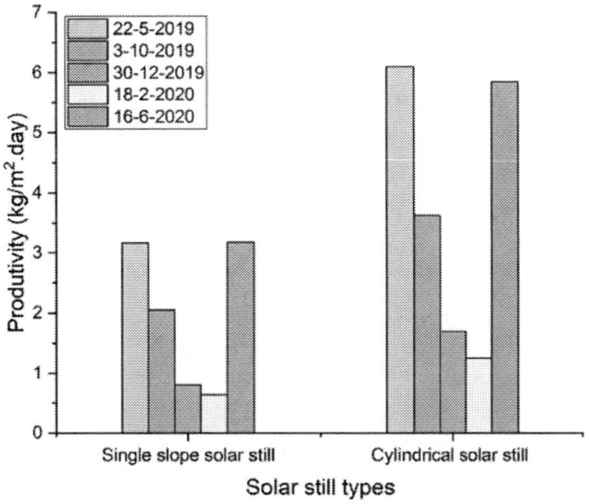

Fig. 8. Comparison of the hourly production for single slope solar still and cylindrical solar still.

5. Conclusion

Cylindrical solar still was successfully designed by using COMSOL multiphysics and the simulations were performed. The 2D with 2 phase simulation model was built. The simulation model was implemented for the water vapor mixture. The incident solar radiation was waking to vaporize the saltwater and the CSS was ready to display condensed water vapor accumulates on the inside glass surface. Also, this new design allows visible freshwater drops that drop down to the fresh water tank. Proposed new geometry of the cylindrical solar still in this study acceptable fresh water yield in compare with conventional single slope solar still. The daily fresh ware product was maximum with CSS and found as 6.1 kg/m^2, while from the traditional SSS is found as 3.18 kg/m^2. The maximum deviations between previous experimental and the present simulation results are found as 0.15.

References

1. A. Mahmoud, H. Fath, S. Ookwara, and M. Ahmed, Desalination **467**, 29 (2019)
2. B. A. Abu-Hijleh, Energy Conversion and Management **44**, 1411 (2003)
3. M. A. Sharaf, Journal of Solar Energy Engineering, Transactions of the ASME **134**, (2012)
4. S. Abdallah and O. O. Badran, Desalination **220**, 669 (2008)
5. A. S. Nafey, M. Abdelkader, A. Abdelmotalip, and A. A. Mabrouk, Energy Conversion and Management **43**, 937 (2002)
6. V. Velmurugan, S. Senthil Kumaran, V. Niranjan Prabhu, and K. Srithar, Thermal Science **12**, 153 (2008)
7. F. Muñoz, E. Barrera, A. Ruiz, E. M. Martínez, and N. Chargoy, Desalination **476**, 114241 (2020)
8. A. Ahsan and T. Fukuhara, Solar Energy **84**, 1147 (2010)
9. V. Velmurugan and K. Srithar, Renewable and Sustainable Energy Reviews **15**, 1294 (2011)
10. V. S. Vigneswaran, G. Kumaresan, B. V. Dinakar, K. K. Kamal, and R. Velraj, Journal of Energy Storage **26**, 101019 (2019)
11. S. Al-Kharabsheh and D. Y. Goswami, Desalination **156**, 323 (2003)
12. H. Ş. Aybar and H. Assefi, Desalination and Water Treatment 37 (2012)
13. R. S. Subramanian, G. Kumaresan, R. Ajith, U. Sabarivasan, K. K. Gowthamaan, and S. Anudeep, Materials Today: Proceedings (2020)
14. A. a. Badran, A. a. Al-Hallaq, I. a. Eyal Salman, and M. Z. Odat, Desalination **172**, 227 (2005)
15. O. O. Badran and H. A. Al-Tahaineh, Desalination **183**, 137 (2005)
16. H. Ş. Aybar and H. Assefi, Desalination and Water Treatment **7**, 35 (2009)
17. M. A. Al -Amir Khadim, W. A. A. Al-Wahid, D. M. Hachim, and K. Sopian, Smart Science 1 (2021)
18. N. H. Abu-Hamdeh, R. A. R. Bantan, M. Khoshvaght-Aliabadi, and A. Alimoradi, Renewable Energy **161**, 1260 (2020)
19. *COMSOL Multiphysics 5.4 Documentation* (2020)

Energy efficiency in the building materials industry. Case study: Brick manufacturing in Romania

Cristian Gheorghiu[1], *Mircea* Scripcariu[1], *Miruna* Gheorghiu[2*], and *Alexandra Gabriela* Dobrica[1]

[1]University Politehnica of Bucharest, Energy Production and Use Department, 313 Splaiul Independentei, Bucharest, Romania
[2] ELSACO ESCO L.L.C.,500 Mihai Bravu, Bucharest, Romania

Abstract. In this paper an overview of the construction materials industry, from an embedded energy point of view will be presented. A case study for four brick factories in Romania will also analyzed. The Energy Performance Indicators (EnPI) of each factory will be evaluated and compared with the global reference values and the most technically and economically feasible Energy Performance Improvement Actions (EPIAs) will be presented. The replicability of these EPIA's in different materials manufacturing industries will be also analyzed.

1 Introduction

As the European Construction Sector Observatory (ECSO) has shown in its latest report, even though the building construction market in Romania has grown by more than 173.4% from 2010 to 2018 [1], Romania still has some of the largest overcrowding (46.3%) and severe house depravation (16.1%) rates in the European Union [2]. The construction materials market growth continued throughout 2020, despite the COVID-19 pandemic and is expected to reach 1.5 trillion $ by the end of 2027 [3] and is currently responsible for 6% of the global share of energy use, respectively 11% of the Greenhouse Gases Emissions (GHG), of the entire building sector. Considering that the European Union's latest Directives regarding energy efficiency [4] and environmental impact reduction [5] increased the target for energy efficiency and set a target for carbon neutrality for 2050, the construction material industry has a new set of challenges which must be addressed. The most relevant challenges that need to be addressed in order to ensure a circular economy in the construction material industry are, as [6] shows: stimulating demand, training, innovation respectively energy efficiency and climate change. This paper analyzes the construction material manufacturing industry in Romania with emphasis on the brick manufacturing industry. As demonstrated by [7] and [8], the brick manufacturing sector is

* Corresponding author: miruna.gheorghiu.c@gmail.com

continuously trying to improve the technological process thus improving both the final product quality and the EnPI.
The main objective of the paper is to analyze the technic and economic viability of implementing large scale Energy Performance Improvement Actions (EPIAs) in brick factories and to evaluate the potential of environmental impact reduction. The second objective of the research is to identify the main static and variable factors that influence the EnPI and can be used to normalize the aforementioned indicators in order to properly develop a long-term energy performance analysis methodology.
Four Brick Factories (BF), owned and operated by the same company, will be analyzed from an energy performance and environmental impact point of view. Various EPIAs will be identified, presented, and analyzed from a technical and economic point of view to determine the Environmental Impact Reduction (EIR) potential of this building material manufacturing sub-sector.

2 Energy Boundary Description

In all four Brick Factories analyzed, electricity and natural gas are the main forms of energy used. Figure 1 shows the share of each form of energy of the total equivalent energy consumption.

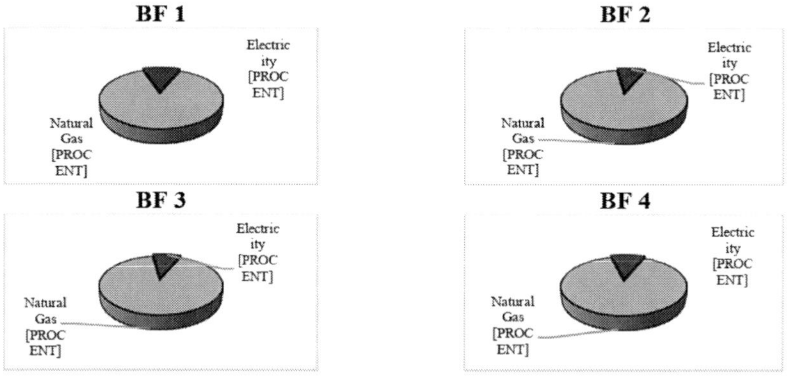

Fig. 1. The share of each form of energy of the total equivalent energy consumption

As it can be observed, in the brick production process natural gas has a higher share compared to electricity, by a factor of nine. Natural gas is used as fuel for the steam boiler, for the burners in the drying chambers and by the brick heat treatment oven, which ensures the ripening process and is the main natural gas end-user. Electricity is used by the industrial robots, air ventilation systems, lighting systems and the IT&C system.
In order to determine the Energy Performance Indicators (EnPIs), the Environmental Impact Indicators and to determine the energy baseline for the four analyzed Energy Boundaries, a brief energy use comparison is presented in Table 1.
Based on the data presented in Table 1, it can be observed that the natural gas and electricity use are directly proportional with the annual brick production in all four case study sites. In order to evaluate the energy performance of each site, to compare the results and to quantify the impact of the various proposed EPIAs, an in-depth analysis of specific EnPIs has to be performed.

Table 1. Natural Gas and Electricity consumption for each Brick Factory

Brick Factory	Natural Gas Use [MWh/year]	Electricity Use [MWh/year]	Annual Production [tons/year]
BF 1	80,216	10,713	237,626
BF 2	56,631	5,397	196,947
BF 3	70,523	7,281	234,516
BF 4	53,820	6,320	116,096

2.1 Energy Performance Analysis

To analyze and understand the energy performance related to energy use, it is necessary to identify the relevant Energy Performance Indicators.

Energy Intensity (EI) was selected as a global EnPI as it is also the most commonly used indicator in Energy Audits, and it is also the global EnPI that EU Member States have to annually report to the European Commission. EI was determined by using equation (1):

$$EI = \frac{W_{eq}}{PV}\left[\frac{t.o.e.}{EUR \cdot 10^3}\right] \tag{1}$$

where EE [t.o.e./year] is the annual equivalent energy use of the energy boundary, expressed in tons of oil equivalent (t.o.e.) and PV[thousand EURs/year] is the yearly production income.

In order to evaluate the Specific Electricity (W_{el}^{sp}) and Natural Gas (W_{ng}^{sp}) use with regard to the annual brick production, thus determining the electricity and natural gas baselines, equations (2) and (3) were used.

$$W_{el}^{sp} = \frac{W^e}{P}\left[\frac{MWh}{ton}\right] \tag{2}$$

where W^e [MWh/year] is the annual electricity use of the energy boundary, expressed in MWh/year and P[tons/year] is the yearly production of the BF.

$$W_{ng}^{sp} = \frac{W^{ng}}{P}\left[\frac{MWh}{ton}\right] \tag{3}$$

where W^e [MWh/year] is the annual natural gas use of the energy boundary, expressed in MWh/year and P[tons/year] is the yearly production of the BF.

To get an overall view of the energy performance of each site, the Specific Equivalent Energy (W_{eq}^{sp}) use was determined with equation (4).

$$W_{eq}^{sp} = \frac{W^{eq}}{P}\left[\frac{t.o.e.}{ton}\right] \tag{4}$$

where W^{eq} [t.o.e./year] is the annual equivalent energy use of the energy boundary, expressed in tons of oil equivalent (t.o.e.) and P[tons/year] is the yearly production of the BF.

The results of the unadjusted Energy Performance analysis will be presented in Table 2 and Figure 2.

Table 2. Energy Performance analysis results – EnPI baseline

Brick Factory	EI [t.o.e./EUR·10^3]	W_{el}^{sp}[MWh/ton]	W_{ng}^{sp}[MWh/ton]	W_{eq}^{sp}[t.o.e./ton]
BF 1	0.6616	0.0451	0.3376	0.0329
BF 2	0.5072	0.0274	0.2875	0.0271
BF 3	0.5892	0.0310	0.3007	0.0285
BF 4	0.9648	0.0544	0.4620	0.0444

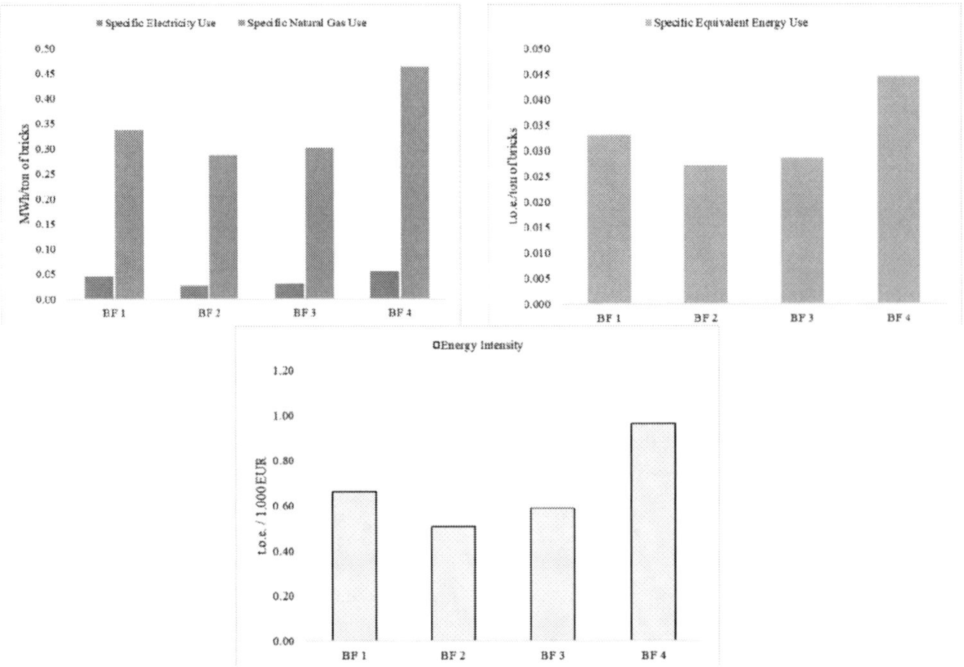

Fig. 2. EnPI Baseline – Comparison between the four factories

It can be observed that BF 1 and BF 4 have the highest values for all the analyzed EnPI, which leads to the conclusion that these two energy boundaries are the least energy efficient. The most energy efficient production site is BF 2. It obtained an Energy Intensity of only 0.5072 t.o.e./thousand EUR, representing only 52% of the EI registered by BF 4.
This difference can be explained by the fact that BF 4 was acquired by the company less than a year ago. The factory was still in the process of implementing the various good practice operational guidelines that the new owner already developed and successfully tested in the other production sites. In order to compare the energy performance of the analyzed energy boundaries with the world average, an industry specific EnPI has to be used. The specific equivalent energy use for each thousand bricks was determined by using equation (5):

$$W_{eq}^{sp,brick} = W_{eq} \cdot 41.868 \cdot \frac{P}{BW} \cdot 10^3 \left[\frac{GJ}{10^3\ bricks}\right] \quad (5)$$

where BW is the average brick weight of 15.5 kg for the four analyzed factories. The average value had to be considered because each factory produces at least 3 different types of bricks. As it can be observed from Table 3, some of the analyzed factories have better $W_{eq}^{sp,brick}$ values than the global average whilst others fall behind. This EnPI can be used by the Management of the owner-company to prioritize the EPIAs implementation in BF 4.

Table 3. EnPI comparison

Factory	Production		Specific Energy use [t.o.e./10^3 bricks]
	[tons]	[bricks]	
BF 1	237,626.44	15,330,739	21.36
BF 2	196,947.18	12,706,270	17.58
BF 3	234,516.22	15,130,079	18.52
BF 4	116,095.88	7,490,057	28.91
Clay brick average specific energy use in Canada [9]			22.046
Clay brick average specific energy use in S.U.A. [10]			9.3
Clay brick average specific energy use in Brazil [11]			55.211

2.2 Environmental Impact Analysis

Reducing the environmental impact is the most important national, European and international energy efficiency target. The Environmental Impact Indicators were calculated according to equations (5), (6) and (7).
In order to determine the annual quantity of equivalent greenhouse gases emission the following conversion factors for natural gas and electricity were used: 202 $gCO_{2,eq}/kWh$ (f_{ng}) and 345 $gCO_{2,eq}/kWh$ (f_{el}) [12].
The unadjusted overall equivalent CO_2 emissions (A_{CO2}) were determined, for each site, and will serve as an environmental impact indicator baseline.

$$A_{CO2} = W^{ng} \cdot f_{ng} + W^{el} \cdot f_{el} \left[\frac{tons\ CO_2}{year}\right] \tag{5}$$

As A_{CO2} cannot be used to compare the four production sites, specific environmental impact indicators have to be used. Thus, the specific CO_2 emissions reported to the production of each site was determined by using equation (6).

$$CO_{2,eq}^{sp} = \frac{A_{CO2}}{P}\left[\frac{tons\ CO_2}{tons\ of\ brick}\right] \tag{6}$$

Furthermore, a global environmental impact indicator was proposed in this paper. In order to properly quantify the environmental impact from an economic point of view, an analysis regarding the weight of CO_2 emissions in the global economic productivity was developed. The Environmental Impact Intensity (EII) was determined by using equation (7) and can be used to compare various industries.

$$EII = \frac{A_{CO2}}{PV}\left[\frac{tons\ CO_2}{EUR \cdot 10^3}\right] \tag{7}$$

The unadjusted results of Environmental Impact Analysis will be presented in Table 4.

Table 4. Energy Performance analysis results – EnPI baseline

Brick Factory	A_{CO2} [tons CO_2/year]	$CO_{2,eq}^{sp}$ [tons CO_2/ton of brick]	EII [tons CO_2/ EUR·10^3]
BF 1	19,899.93	0.0837	0.3450
BF 2	13,301.75	0.0675	0.2592
BF 3	16,757.63	0.0715	0.3024
BF 4	13,016.11	0.1121	0.4990

As expected, BF 4 has the worst EII value of 0.5 tons of CO_2/ton of brick. This means that for every ton of brick produced, BF 4 has to pay for 0.5 CO_2 certificates. In the context of the European Union's fourth phase of the EU-ETS mechanism [13], if the EII values will not be improved, the overall economic efficiency of the analyzed energy boundaries will decrease over time by a factor of 7%. The weight of CO_2 certificates in the overall operational expenditures (OPEX) is, as of 2021, approximatively 5.45%. When the CO_2 certificates will reach the maximum price of 100 EUR/certificate, the weight will increase to 13.63%. It is thus obvious that the owner company has to prioritize the implementation of EPIAs in order to maximize the potential for sustainability and decarbonization.

2.3 Static and variable factors

Over time the principle of brick production never changed, just the technology of production. Thus, the brick factories are more efficient, and the quality of products have been improved. A better knowledge of raw materials and their properties, of the equipment uses, of the factors that influence the production, the energy consumption and quality of products contribute to more advanced concepts for brick factories and a better quality of brick production [14].

There are six major ingredients for brick production. Silica (Sand) and Alumina (Clay), these two are the most prominent ingredients in brick clay.
The clay used must have certain properties and the most important are plasticity, that allows it to shape and homogenize with water, and calorific value according which, in the process of ripening the brick, more or less natural gas is consumed.

Because each factory is built near a clay quarry, there is no possibility to change the clay used in the production process. But a lot of tests have been done with different types of clay, and the amount of natural gas required for ripening was lower if the calorific value of the clay was lower and vice versa. CO_2 emissions varied similarly to natural gas consumption. Other static and variable factors that were identified during this case study are presented in Table 5.

Table 5. Factors influencing energy use

Factor	Type	Influences	Can be optimized?	How?
Sand granulation	Variable	Drying process duration / brick quality	Yes	Ensuring a 0.2 mm granulation
Sawdust humidity	Variable	Sifting process duration	Yes	Stored in controlled atmosphere environment
Outdoor temperature	Variable	Natural gas consumption	Yes	Optimizing heat flow in the heat treatment oven
Type of drying system	Static	Drying process duration	Yes	Replacing drying chambers with drying tunnels

3 Energy Performance Improvement Actions

3.1 Improving the energy monitoring system

In order to maximize the efficiency of identifying the relevant EPIA's, the implementation of an advanced energy monitoring system in each of the analyzed energy boundaries is mandatory. As [15] shows, by developing a system in accordance with [16] an overall EnPI improvement of up to 3% can be achieved with minimal investments and minor operational improvement actions. Based on the technological process diagram, a monitoring system architecture was proposed. The system will lead to an accurate measurement procedure for: global electricity use, individual major equipment electricity use, industrial water use, individual major equipment natural gas use, raw materials flows, intermediate product flows, final product flows, outdoor temperature and humidity etc. The monitoring system was particularized for each BF and the total Capital Expenditures (CAPEX), OPEX and the expected energy savings obtained by implementing the aforementioned no cost EPIA's and operational improvement actions are presented in Table 6.

Table 6. Input data for energy monitoring system EPIA

BF	CAPEX [EUR]	OPEX [EUR/year]	Energy savings [MWh/year]
BF 1	245,000	5,000	2.406,48
BF 2	212,000	3,500	1.698,93
BF 3	240,000	5,500	2.115,69
BF 4	208,000	3,500	1.614,60

3.2 Renewable energy sources

Considering that the environmental impact reduction is a main goal of the BF owner company, the use of renewable energy sources is an attractive mean of reaching the desired target of reducing the CO_2 by 14% by the end of 2021. As all the BFs are located in rural areas and occupy a significant surface, the implementation of photoelectric systems was proposed and analyzed from a technical point of view. By using a software simulation tool (RETScreen Expert) the Forecasted Electricity Productions were determined and are presented in Table 7. To mitigate the dusting of the PV modules, an automated washing system was also considered for each BF site (60 EUR/kWp).

Table 7. Photoelectric System simulation result

Factory	Available Surface [m^2]	Effective PV System area [m^2]	PV System Peak Power [kW]	CAPEX [EUR]	OPEX [EUR/year]	Forecasted Electricity Production [MWh/year]
BF 1	14,123.69	6,230.00	988.40	707,694.40	8,846.18	1,464.67
BF 2	12,596.58	4,414.00	600.25	429,779.00	5,372.24	810.38
BF 3	10,445.30	4,414.00	600.25	429,779.00	5,372.24	844.54
BF 4	9,347.48	2,522.00	400.05	286,435.80	3,580.45	534.22

* determined by considering an average standard CAPEX of 656 EUR/kWp in Romania.

3.3 Waste heat recovery

All four BFs have an available waste heat source in the form of exhaust gases at a temperature of 140 Celsius Degrees. As per [17], the specific Investment Cost of a small-scale Organic Rankine Cycle (ORC) system which can use the available heat source is estimated at approximatively 2,500 EUR/kW with an installed power of the ORC system of 150 kWe.

Thus, the estimated CAPEX for the EPIA is 375,000 EUR. As [18] has shown, an ORC System has an annual operation time of approximatively 7,800 hours/year with an average OPEX of 7,500 EUR/year. As per the Romanian legislative framework [19], an analysis period of 12 years was considered.

3.4 Global analysis results

The potential energy savings and the potential environmental impact reduction of each EPIA is presented in Table 8.

Table 8. EPIA Analysis

EPIA	Applicable to	Potential Energy Savings [MWh/year]	Potential Environmental Impact Reduction [tons CO_2/year]
Advanced Monitoring System	BF 1	4,026.84	813.42
	BF 2	2,842.88	574.26
	BF 3	3,540.25	715.13
	BF 4	2,701.76	545.76
Photovoltaic system	BF 1	1,464.67	505.31
	BF 2	810.38	279.58
	BF 3	844.54	291.37
	BF 4	534.22	184.30
ORC system	BF 1, BF 2, BF 3, BF 4	1,170.00	403.65

It can thus be concluded that by implementing the analyzed EPIAs, the overall potential energy savings could amount to 1,844.32 t.o.e./year, respectively 7.37% of the baseline energy use. The cumulated environmental impact reduction potential is 5,523.73 tons CO2/year, respectively 8.77% of the baseline environmental impact, as it can be observed from Table 9.

Table 9. The potential to reduce the environmental impact

Factory	Total Potential Equivalent Energy Savings [t.o.e./year]	Total Potential Environmental Impact Reduction [tons CO_2/year]
BF 1	572.89	1,722.38
BF 2	414.80	1,257.49
BF 3	477.71	1,410.15
BF 4	378.91	1,133.71

4 Financial Analysis

The main criterions used in the technic and economic analysis of the EPIAs were the Net Present Value – NPV (8), the Internal Rate of Return – IRR (9), the Simple Payback Period – SPP (10), determined by considering a variable annual net income and the Benefit – Cost Analysis – BCA (11).

$$NPV = \sum_{t=1}^{tst} \frac{I_t - C_t}{(1+a)^t} - IC\ [EUR] \tag{8}$$

where t_{tst} is the analysis time-frame, in years, selected as per [19], I_t is the yearly income in the tth year, in EUR/year, C_t are the yearly expenditures in the tth year, in EUR/year, a is the discount rate – 11.38%/year for this end-user and IC is the investment cost, in EUR.

$$NPV = \sum_{t=1}^{tst} \frac{I_t - C_t}{(1+IRR)^t} = 0\ [EUR] \tag{9}$$

where the CAPEX can be included in the yearly expenditures as a depreciation cost.

$$SPP = \frac{IC}{\frac{\sum_{i=1}^{t} I_i - C_i}{t}} [years] \tag{10}$$

$$BCA = \frac{IC}{NPV} [-] \tag{11}$$

An average escalation rate for electricity prices of 5%/year was also considered, as determined in [20]. The technic economic analysis results will be presented in Table 10.

Table 10. Financial Analysis

EPIA	Applicable to	NPV [EUR]	IRR [%]	SPP [years]	BCA [-]
Advanced Monitoring System	BF 1	123,390	21%	5.31	1.5
	BF 2	53,182	16%	6.15	1.25
	BF 3	83,511	18%	5.83	1.35
	BF 4	42,996	15%	6.32	1.21
Photovoltaic system	BF 1	1,120,265	26%	5.35	2.58
	BF 2	615,692	24%	5.74	2.43
	BF 3	661,285	25%	5.56	2.54
	BF 4	402,493	24%	5.79	2.41
ORC system	BF 1, BF 2, BF 3, BF 4	764,209	36%	4.03	3.04

It can be observed that all proposed EPIAs lead to attractive financial indicators. For BF 1 the most attractive EPIA is the installation of a photoelectric system. The difference between the economic viability of the PV system for the four BFs is mainly generated by the difference in the available rooftop surfaces and by the difference in solar potential. For all the other BFs the most attractive EPIA is the implementation of Organic Rankine Cycle electricity production system. It can also be observed that in all analyzed scenarios the SPP is less than or close to 5 years.

5 Conclusions

This paper demonstrated that the brick manufacturing industry, to achieve environmental sustainability, must be subjected to an in-depth analysis in order to properly identify specific EPIAs at process level. Even though, as proven in chapters 3 and 4, and shown in Table 11, the implementation of large scale, outside the process level, EPIAs can lead to a major EnPI and global EI improvement, the target set by the owner company cannot be reached in the desired timeframe.

As the carbon neutrality target implies the mitigation of all forms of CO_2 equivalent emissions, even though the electricity used by the BFs already include the EUA costs, an increase in the overall energy performance of the analyzed BFs will lead to a global decrease in the Environmental Impact generated by the company (electricity and natural gas). The various analyzed EPIAs include both forms of energy and, as a result, the EI improvement is a global rather than a natural gas specific one.

Table 11. Centralized results

Factory	Obtainable Energy Intensity [t.o.e./1,000 EUR]	EI Reduction	Obtainable Environmental Impact [tons CO_2/year]	Environmental Impact Reduction	Obtainable NPV [EUR]
BF 1	0.6132	7.32%	18,177.23	8.66%	2,275,427
BF 2	0.4678	7.77%	12,043.93	9.45%	861,309
BF 3	0.5471	7.14%	15,347.44	8.41%	984,436
BF 4	0.8939	7.35%	11,918.33	8.69%	1,392,581

As it was shown in chapter 2, there are certain variable factors which cannot be optimized. One such factor is the quality of the clay used in the process. As each factory has its own clay quarry located in the vicinity of the factory, the optimization of this variable factor is improbable. The only mean of optimizing this factor could be to find a new clay quarry, with better properties. This could, in turn, lead to an increase in the overall energy use, as the raw material should have to be transported from the quarry to the production site with an additional fossil fuel use. Further research is required in order to develop a normalization methodology for properly quantifying the influence of the static and variable factors on the EnPIs, thus facilitating the optimization of this interdependency. All the proposed EPIAs are easily replicable to other building materials industries such as the cement, lime or reinforced steel industries as they all have the same characteristics regarding the occupied land (PV System available surface), energy uses (electricity and heat) and process type (linear, first in – first out).An additional technic and economic analysis of the viability of retrofitting the existing heat treatment ovens with dual fuel burners and increasing the mix of green hydrogen in the brick manufacturing processes must be conducted. In accordance with [21] the share of hydrogen in Europe's energy mix is expected to grow from less than 2% to 14%. It is thus obvious that hydrogen will also play a key role in the transition towards environmental sustainability of the brick manufacturing industry sector.

References

1. https://ec.europa.eu/growth/sectors/construction/observatory_en, accessed on 28.02.2021.
2. Eurostat, "Housing Statistics," 2020, https://ec.europa.eu/eurostat/statisticsexplained/.
3. Research and Markets, "Construction Materials - Global Market Trajectory & Analytics," 2020.
4. The European Parliament , "Directive (EU) 2018/844 amending Directive 2010/31/EU on the energy performance of buildings and Directive 2012/27/EU on energy efficiency," 2018.
5. European Commission, "EU climate action and the European Green Deal," https://ec.europa.eu/clima/policies/eu-climate-action_en , 2019.
6. European Comission, https://ec.europa.eu/growth/sectors/construction_en, accessed on 12.03.2021.
7. Alaa Shakir, Ali Mohammed, "Manufacturing of Bricks in the Past, in the Present and in the Future: A state of the Art Review," International Journal of Advances in Applied Sciences, Vol. 2, No. 3, September 2013, pp. 145~156, ISSN: 2252-8814, 2013.
8. D. Sakthivel, R. Dharunkumarasamy, L. Karthikeyan, S. Praveen Kumar, B. Surya, "Experimental investigation on brick manufacturing by using sludge & pulp production residue," International Research Journal of Engineering and Technology (IRJET), Volume: 06, Issue: 03, pp 266 – 274, ISSN: 2395-0072, 2019.
9. Jeorge J. Venta, P.Eng. "Life cycle analysis of brick and mortar products", the ATHENATM Sustainable Material Institute, Canada, Ottawa, 1998.
10. Michael Chusid, Steven H Miller, Julie Rapoport. "The building bricks of sustainability", The Construction Specifier, Vol. 1. Pp. 30-40, 2009.
11. Dachamir Hotza and Bianca Goulart de Oliveira Maia, "Environmental performance and energy assessment of fired-clay brick masonry," Eco-efficient Masonry Bricks and Blocks: Design, Properties and Durability, pp 447-459, DOI: 10.1016/B978-1-78242-305-8.00020-6, 2014.
12. European Investment Bank, "EIB Project Carbon Footprint Methodologies: Methodologies for the Assessment of Project GHG Emissions and Emission Variations," (2020).
13. European Comission, https://ec.europa.eu/clima/policies/ets/revision_en#tab-0-0, accessed on 12.03.2021
14. www.gobrick.com, accessed on 11.03.2021.
15. International Standardization Organization, "ISO 50001:2018 Energy Management Systems," 2018
16. Efficiency Valuation Organization, International Performance Measurement and Verification Protocol (IPMVP), https://evo-world.org/en/products-services-mainmenu-en/protocols/ipmvp
17. Lorenzo Tocci, Tamas Pal, Ioannis Pesmazoglou and Benjamin FranchettiSmall, "Scale Organic Rankine Cycle (ORC):A Techno-Economic Review," Energies, 10, 413; doi:10.3390/en10040413, 2017
18. George Kosmadakis, Panagiotis Neofytou, "POTENTIAL AND COST EFFECTIVENESS OF A REVERSIBLE HIGH-TEMPERATURE HEAT PUMP/ORC UNIT FOR THE EXPLOITATION OF INDUSTRIAL WASTE HEAT," 5th International Seminar on ORC Power Systems, 2019
19. The Romanian Government, "Decision no. 2139/30.10.2004 for approving the normal period of use for capital goods," Official Gazette, Part I, No. 46/13.01.2005
20. Ioan-Sevastian Bitir-Istrate, Cristian Gheorghiu and Miruna Gheorghiu, "The Cost of Indecision in Energy Efficiency. A Cost of Opportunity Analysis for an Industrial

Consumer," 55th International Scientific Conference on Information, Communication and Energy Systems and Technologies (ICEST), DOI: 10.1109/ICEST49890.2020.9232894, (2020)
21. European Commission, "A hydrogen strategy for a climate – neutral Europe," 2020

Archimedean screw as fish-friendly turbines for harnessing hydropower potential

Cristian Purece[1*], and *Lilica* Corlan[1]

[1]INCDE ICEMENERG, Bucharest, Romania

Abstract. Hydraulic energy is one of the most important sources of renewable energy today. It is also a complementary source to other renewable energy sources, being the only one that offers an important non-polluting storage capacity (through pump storage facilities). Another essential quality of hydraulic energy is its ability to regulate the energy system to allow the integration of other sources, in which the process of generating energy depends on factors that cannot be controlled (sun, wind, etc.). However, hydropower plants can damage ecosystems, especially by affecting fish within their turbines. Usually, fish cannot pass unharmed through most of the turbines. The main exceptions generally include some low head (below 8 m) turbines, such as the Very Low Head (VLH) turbine and the screw turbine, but lately strategies have also been developed for high head hydropower plants such as: fish passage facilities and fish friendly turbines such as Alden turbine, Minimal Gap Runner turbine (MGR). However, this article only deals with screw-type turbines, low-head fish-friendly turbines and how to implement them in order to obtain sustainable green energy.

1 Introduction

Sustainable green energy production is one of the main global challenges today. Increasing CO_2 emissions, environmental pollution and global warming are leading researchers around the world to find new energy resources [1]. Renewable energies such as solar, wind, geothermal and hydropower, as popular types of renewable energy resources, can be found all over the world [2]. Among these resources, hydropower is widely used as high-head hydropower plants. Although almost all high power hydropower opportunities have been largely depleted, there is a strong potential for low head hydropower resources (small hydropower) potential to be exploited. Some of these resources are small rivers, irrigation systems, drinking water networks, wastewater networks, cooling systems, etc. In hydropower plants, the hydraulic power of water is converted into mechanical power using a turbine and then in electricity with a generator.

Recently, the reverse use of the conventional Archimedes screw pump is considered a popular technology to be used as a turbine to generate electricity from running water, called the Archimedes screw turbine. Traditionally, the Archimedes screw has been used to pump

* Corresponding author: cristianicemenerg@yahoo.com

water from a lower level to a higher level or to transport liquid or solid materials in a horizontal or inclined plane [3].

Currently there are different types of turbines for electricity generation, including Kaplan, Francis, Pelton type turbines, etc. Comparing the relative costs of an Archimedes screw turbine and a Kaplan turbine, a common turbine for a low head hydropower site, it was found that for an energy production of about 15% more, for an Archimedes screw turbine the cost is about 10% lower, and the annual cost capital is 22% cheaper [4]. In addition, Archimedes' screw turbine is an ecological and fish-friendly structure, it needs little civil works for installation, it can be installed even in existing structures [5].

From previous researches it can be seen that the efficiency of the screws decreases when the rotational speed increases, as the frictional forces become unexpectedly high [6]. Numerical researches on the actual performance of Archimedes screw turbine are limited. The main objectives of this paper are to perform numerical simulations to determine the performance of Archimedes screw turbines according to different parameters as well as the presentation of an example of implementation of Archimedes screw turbines.

2 Theoretical basis

It is generally assumed that the weight of water passing through the screw's blades drives the screw [7]. Assuming no losses, all potential energy contained in the water flow can be extracted, giving such a turbine maximum theoretical efficiency. However, most of the weight of the water in the Archimedean screw rests on the trough, which does not move (Figure 1). The power is equal to the force multiplied by velocity and, since the screw rotation speed vector acts tangentially on the screw, only a small part of the water weight (the part acting on the inclined outer section of the blade) contributes to the conversion of hydraulic energy into mechanical energy. Therefore, the contribution of gravity can be neglected.

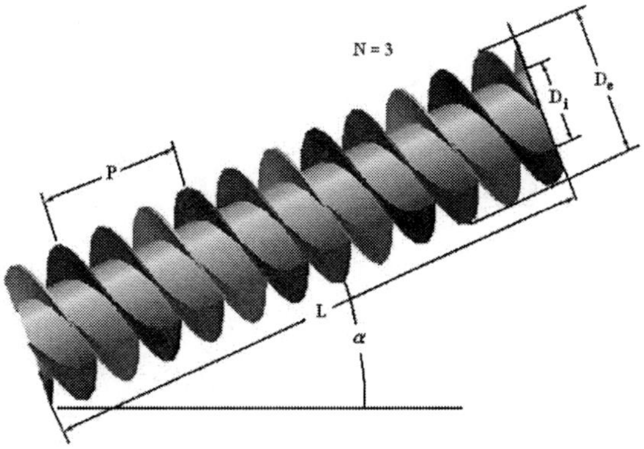

Fig. 1. The Archimedean screw [2]

We consider that the screw turbine has a head difference H, a screw length L, an inside diameter D_i, an outside diameter D_e, a pitch P, a number of blade N, a turns of the helix m, a slope α angle between the screw shaft and the horizontal. The longitudinal section of Figure 2 (let) shows the water level inside the screw; there is a difference in head between two blades, so that a horizontal hydrostatic force is developed. Figure 2 (right) indicates the hydrostatic pressure/force on inclined blades, with an upstream pressure higher than the downstream pressure acting normally on the blades. However, only the component in the direction of rotation of the screw will contribute to the rotation of the screw turbine blades.

 https://doi.org/10.1051/e3sconf/202128602007

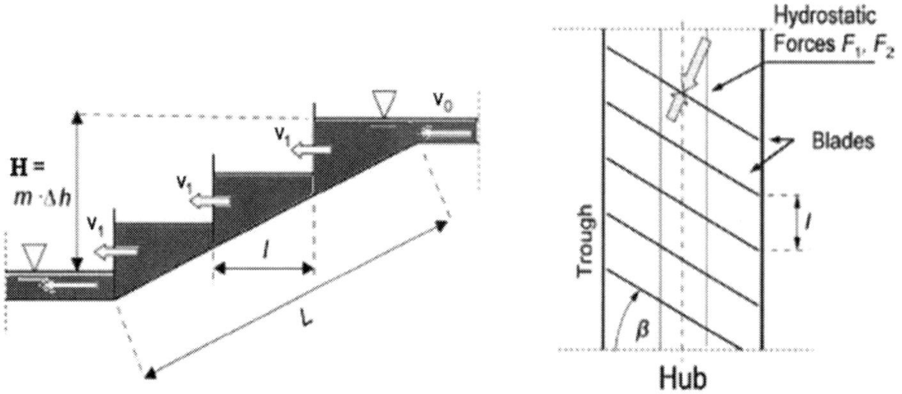

Fig. 2. Archimedean screw: longitudinal section (left), plan view of submerged blade (right)

A detailed view of a blade of Archimedes' screw is shown in Figure 3. The difference in water level between the ascending and descending parts of each blade produces a hydraulic force, F_{hyd}, which moves the screw with a velocity v_1, generating a power $P_{blades} = F_{hyd} \times v_1$. With an angle α between the screw shaft and the horizontal and a distance l between two individual blades, the water depth increases by:

$$\Delta h = l \tan \alpha = (L / m) \tan \alpha \text{ or } \Delta h = H / m \quad (1)$$

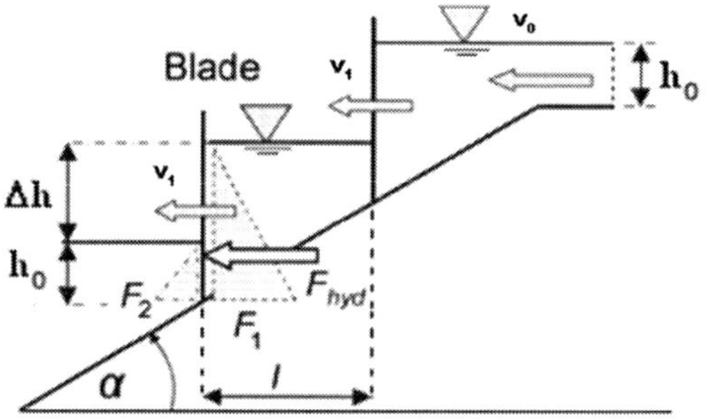

Fig. 3. Hydraulic forces acting on the screw blades

The specific hydraulic energy (E) changed in the hydraulic turbine is defined by the relation:

$$E = g \cdot H = \frac{(p_1 - p_2)}{\rho} + \frac{v_1^2 - v_2^2}{2} + g \cdot (z_1 - z_2) \quad (2)$$

where: H is the head or load available between the high pressure and the low pressure reference section of the hydraulic turbine, and g is the gravitational acceleration, ρ is the density of water, P_1 is the pressure in the high pressure reference section, P_2 is the pressure in the low pressure reference section, v_1 is the velocity in the high pressure reference section, v_2 is the velocity in the low pressure reference section.

In hydraulic turbines that convert only position-specific potential energy, equation (2) is reduced to the expression: $E = g\ (z_1 - z_2)$ and the degree of reaction is zero, because the pressure is constant and equal to that of the atmosphere ($p_1 = p_2 = p_{at}$), and the velocities in reference sections 1 and 2 are negligible, or have quasi-equal values, so the kinetic term in equation (2) is cancelled, $\left(\frac{v_1^2 - v_2^2}{2}\right) \equiv 0$. This category includes elevators (for example, Archimedes' screw turbine), gravitational water wheels, and hydraulic transformers for pumping used in antiquity (made by coupling a water wheel and an elevator) [8].

The mechanical energy produced by the rotation of the screw is transformed into electrical energy using a generator. The power supplied to the generator terminals is given by the following equation:

$$P = \eta \rho g H Q \tag{3}$$

and the hydraulic power is:

$$P_{hid} = \rho g H Q \tag{4}$$

where η is the efficiency of the hydro unit (turbine + generator), H is the head of the turbine, Q is the turbine flow [9].

Therefore, the higher will be the head turbine and the flow rate is, the higher will be the power generated by it. In order to obtain a very high energy, the head of the turbine should be increased to a substantial height by making a dam, but such a task is difficult to accomplish, it takes more time and a lot of money should be invested. Thus, hydropower plants should be without an accumulation basin or have a very small head. The Archimedes screw turbine can be used in such places with low head.

For the Archimedes screw turbine the head H can be written as:

$$H = m\,\Delta h \tag{5}$$

The mechanical power P_{mec} of the Archimedes screw turbine can be calculated by the relation:

$$P_{mec} = T\,\omega = T\,(2\pi n/60) \tag{6}$$

where: T is the torque, ω is the angular velocity, n is the turbine rotational speed.

The efficiency of the turbine η_{Tb} can be calculated with the relationship:

$$\eta_{Tb} = \frac{P_{mec}}{P_{hid}} \tag{7}$$

3 Calculation of screw turbine parameters

Numerical simulation was used to perform the calculation. The simulation of the Archimedes screw turbine model was performed using ANSYS FLUENT, Computational Fluid Dynamics (CFD) [10]. This method is important for industrial applications, the main advantage of using this technique is that it reduces the time and cost of design for industrial applications. By using this technique a reduction of the design cost is obtained by 40% compared to the experimental methods. The basic equations of this technique are the Navier Stokes equations, developed by Claude Louis Navier [11].

To model the fluid through the Archimedes screw turbine before applying the conventional Navier-Stokes equation there are necessary assumptions required. These assumptions are: fluid is in single phase; water is considered as incompressible fluid; properties of water are temperature independent. Numerical method used is the method of Reynolds-averaged Navier-Stokes equations (RANS).

The parameters of the Archimedes' turbine screw for which the numerical simulation was performed were: screw length L = 1000 mm, outer diameter D_e = 130 mm, inner diameter D_i = 70 mm, number of blades N = 1, pitch P = 70 mm, trough diameter Dt = 134 mm, slope $\alpha = 30^0, 35^0, 40^0, 45^0$, nominal turbine flow Q = 2.5 l/s.

The transient flow state was used in simulation as its boundary condition. The upstream water velocity (inlet) as well as the rotational speed of the of turbine screw, were taken from the experimental results, and the pressure at downstream (outlet) is equal to 0.

In order to obtain a good accuracy, a non-uniform grid was chosen. Convergence criteria were set to less than (10^{-6}) for continuity and momentum residuals. It has been used a number of iterations 100. After study, it was found that the best value for the size of the element is 5 mm, the large values from 5 mm did not give accurate results, and at small values from 5 mm, the results did not change, it was observed only an increase of the time of the analysis.

The results obtained from the numerical calculation are presented in Figures 4 - 9. Figure 4 shows the relationship between the mechanical power of the turbine and the flow.

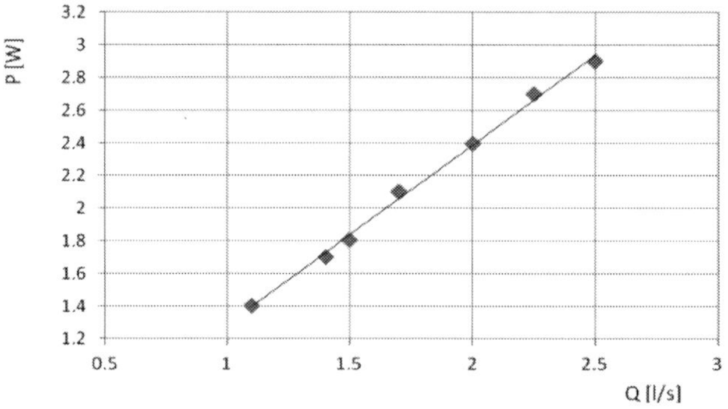

Fig. 4. The relation between flow rate and mechanical power

From the curve in Figure 4 it can be seen the increase of the mechanical power with the increase of the flow. The reason for the increase in mechanical power is that the flow increases the force acting on the blades and thus increases the rotation of the blade, which leads to increased mechanical power.

The relationship between flow and efficiency at an angle of inclination of Archimedes' screw of 35^0 is given in Figure 5. As the flow rate increases, the efficiency of the Archimedes screw increases until it reaches the maximum value corresponding to the optimal flow rate and then begins to decrease. This means that in order to obtain a high efficiency, the optimal flow must be found.

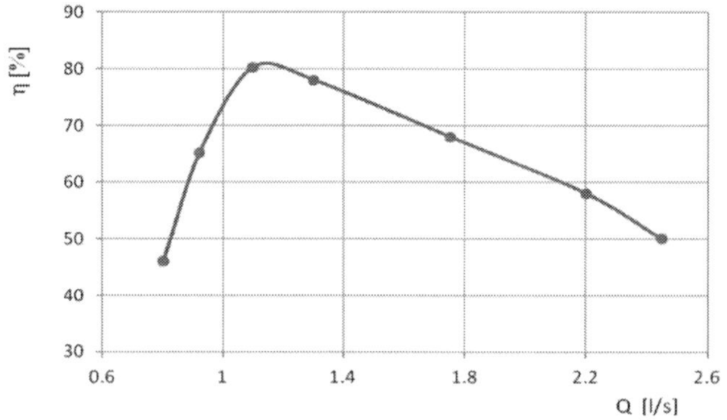

Fig. 5. The relation between efficiency and flow rate (35^0 Angle)

From Figure 6 it can be seen that as the flow rate increases, the rotational speed increases at each angle of inclination of the Archimedes screw axis (α), but when the value of the angle increases, the lines of rotation speed converge, which means that the change in rotation speed becomes small, this is due to the existence of energy losses caused by overflow.

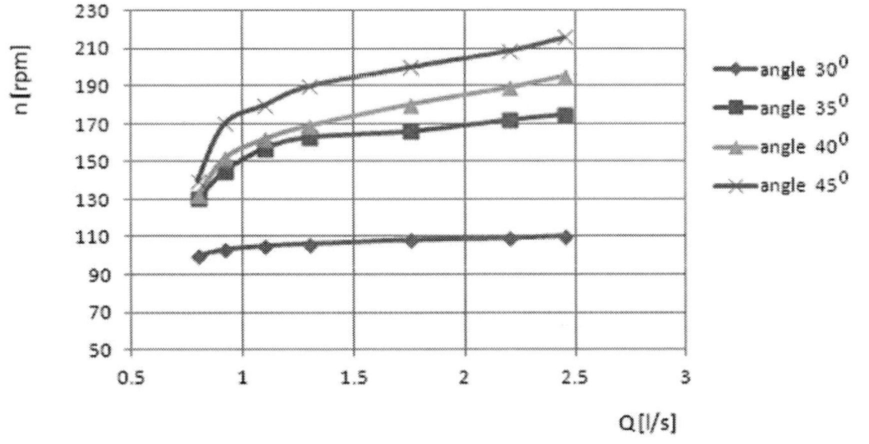

Fig. 6. Variation of the rotation speed and flow rate for different angles α

From the obtained results it was concluded that the optimal angle α between the turbine axis and the horizontal is the angle of 35^0. Figure 7 shows the relationship between efficiency and rotational speed for an angle of inclination of 35^0 between the screw turbine and the horizontal. We notice that the efficiency at the beginning increases with the increase of the rotation speed until it reaches the maximum value 80.2%, at a rotation speed of 157 rpm, after which the efficiency begins to decrease with the increase of the rotation speed.

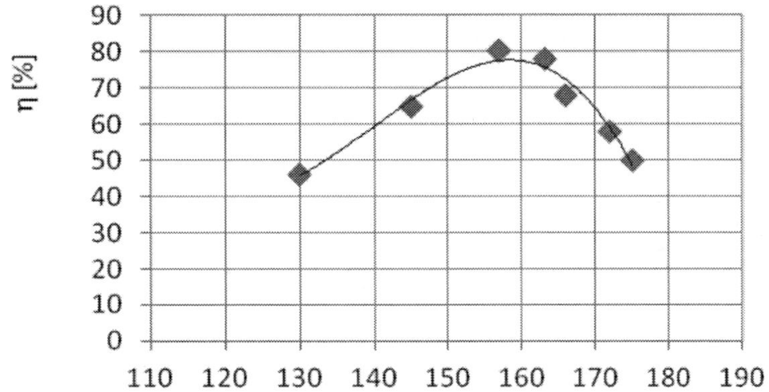

Fig. 7. The relation between efficiency and rotation speed (35^0 Angle)

The relationship between the torque of the screw turbine and the number of blades is shown in Figure 8. The higher the number of blades, the torque will increase, due to the increase of the water contact area with the blades, but with the continuous increase of the number of blades, the difference between the torque values becomes small, therefore you can choose the optimal number of blades, but with the larger the number of blades, the manufacturing process becomes more complicated and expensive.

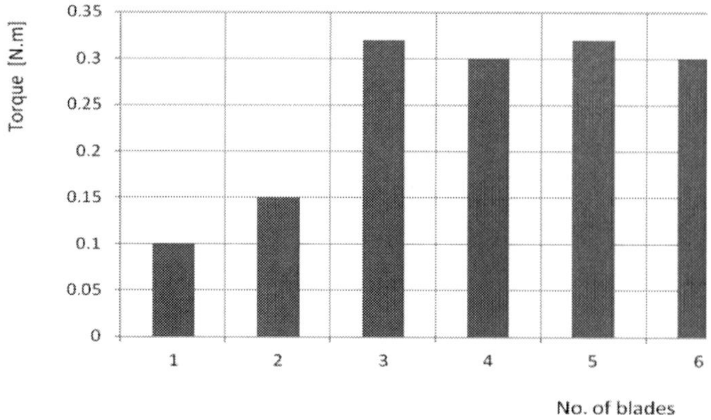

Fig. 8. The relation between torque and number of blades

Figure 9 shows the relationship between turbine efficiency and the number of blades for an Archimedes screw turbine. From the Figure 9, we notice that if the number of turbine blades is higher, the turbine efficiency is higher, and this is due to the higher number of blades, because of the larger contact surface with the turbine water, the couple will also increase. However, the increase in efficiency becomes small or oscillating if the number of blades continues to increase.

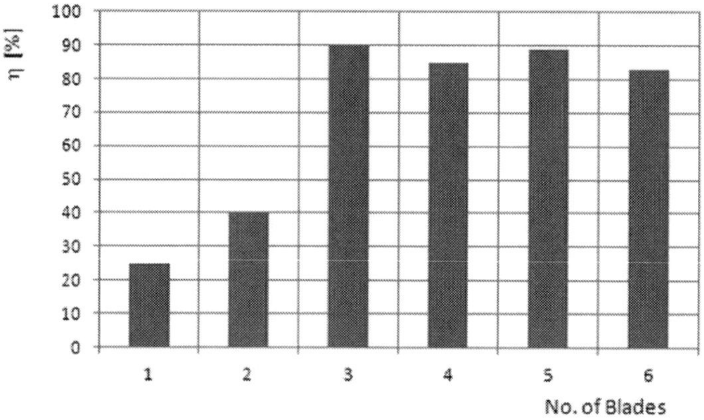

Fig. 9. The relation between turbine efficiency and number of blades

Analysis of results

Through the analysis resulting from Figures 4 to 9 we notice that there is a linear relationship between flow and mechanical power, with increasing flow increases and mechanical power. The efficiency of the turbine in relation to the flow does not necessarily increase in direct proportion to the increase of the flow, as can be seen from the efficiency curve as a function of flow (Figure 4). The efficiency of the turbine increases to a point where its value is maximum then begins to decrease, the reason for this decrease is due to the higher flow than the nominal flow of the turbine. The excess flow can no longer be transformed into mechanical energy.

The flow rate varies depending on the variation of the inclination angle of the turbine shaft (30^0, 35^0, 40^0, 45^0). It can be seen that as the flow rate increases, the rotational speed increases with the increase of the inclination angle of the turbine shaft, but if the inclination angle of the turbine shaft continues to increase, the relationship of the lines of rotation speed converges with the flow rate. This indicates that there is no need for an increase in the value of the angle of inclination of the turbine shaft because it gives the same value of the rotational speed (we get the same mechanical energy but with higher losses). The optimal inclination angle of the turbine shaft is 35^0. For this value of the angle the maximum value of the turbine efficiency is obtained.

From Figure 7, an increase in rotational speed can be observed with increasing flow. However, the efficiency of the turbine increases until it reaches a maximum value (80.2% for an angle of inclination of 35^0) and then it begins to decrease after reaching the optimum flow, due to the losses caused by the overflow.

In case of variation of torque and efficiency with the number of blades, it is observed that both torque and turbine efficiency increase with increasing number of blades due to increasing contact area between water and blades, but with the continuous increase in number of blades there is a convergence of the values of torque and turbine efficiency, which means that there is no need for an increase in the number of blades, because the torque and efficiency of the turbine have the same values, and increasing the number of blades means manufacturing difficulties but also an increase in costs.

4 Implementation of screw turbines at a small hydropower plant

4.1 Advantages of screw turbine implementation

Following the research carried out in recent years, it has been found that the implementation of such screw turbines at a small hydropower plant (SHPP), can bring many more benefits than in the case of other types of turbines like Kaplan, Francis, Propeller or Crossflow.

The installation costs of a hydraulic system with Archimedes screw turbines are much lower compared to those for Kaplan turbines. Following research by British researchers, it was found that a hydraulic turbine system with Archimedes' screw produces 15% more energy than a Kaplan system and the investment costs 10% less [4].

The results obtained can be seen in the Table 1.

Table 1. Results obtained with the Archimedes screw turbine compared to the Kaplan turbine

Criterion / turbine type	**Kaplan turbine**	**Archimedes screw turbine**
Annual energy generation	390 MWh	448 MWh
Installed capacity	95 kW	124 kW
The cost of the investment	414.000 euros	371.000 euros
Cost capital MWh/year	1.061 euro	828 euro
Internal rate of return (IRR, 20 years)	5 %	9 %

The data in Table 1 show that screw turbines are much more efficient than Kaplan turbines, both in terms of electricity production, but especially in financial terms, therefore the implementation of such turbines would be recommended as several regions. The costs of such hydropower facilities depend very much on the location of the work and cannot be considered universal, but they come with concrete evidence that such projects should be encouraged.

Archimedes' screw turbines are mainly used for flow rates between 0.1 m^3/s and 15 m^3/s and have been shown to be less harmful to fish.

The advantages of implementing screw turbines are:

- Changes in flow or water level in the river do not affect the efficiency of the hydraulic system.

- Studies have shown that fish do not suffer from the implementation of such a hydraulic system, they can cross from one end to the other of the turbine in maximum safety.
- Due to the screw shape of the turbine and the way it operates, branches or other debris can pass on without affecting the functionality of the system, thus reducing maintenance costs.
- The lifetime of such a system can be up to 40 years.
- They do not have a negative impact on the environment.
- The amortization period of the investment is relatively short.
- Maintenance costs are low.
- The installation time of such a system is short.
- The system is very solid, wear-resistant.
- It fits perfectly into the landscape due to the small size and elegance of the screwdriver mechanism.
- Fossil fuel is saved by capitalizing on the hydropower potential of a river.

4.2 Example of Archimedes' screw turbines at the Zăvoiul Orbului SHPP

An example of the implementation of Archimedes screw turbine is the one at Zăvoiul Orbului. Zăvoiul Orbului SHPP has the following characteristics:

- Arrangement execution period: 1985 ÷ 1988
- $V_{\text{reservoir}} \approx 12$ mil. m^3
- Side dams ≈ 7400 m
- Execution - stage I (cover level)
- $H = 4.5$ m (actual NRL)

The Zăvoiul Orbului SHPP has the following uses:

- Dâmbovița additional flow;
- water supply;
- irrigation.

Description of Zăvoiul Orbului SHPP:

- the hydropower arrangement has 2 low head steps - located on the right bank;
- SHPPs are equipped with Archimedes screw turbines (2 x 6);
- the arrangement produces basic, semi-peak and peak energy with delivery in the national energy system (NES).

Figure 10 shows the Zăvoiul Orbului SHPP.

Fig. 10. Overview and principle scheme of the Zăvoiul Orbului SHPP [12]

Also in Figure 10 we can notice the basic components of the hydropower arrangement: 1 – Water intake, 2 – Manual sluice gates, 3 – Steel culvert, 4 – Headrace channel for SHPP1, 5 – Hydraulically operated sluice gate for first SHPP1, 5A – Electrical equipment for SHPP1, 6 – SHPP1 with six Archimedes screws, 7 – Dissipater acting as headrace channel for SHPP2 screws, 8 – Reinforced concrete wall for mechanical equipment protection, 8A – Reinforced concrete wall, 9 – Hydraulically operated sluice gates in front of SHPP2, 9A – Electric equipment for SHPP2, 10 – SHPP2 screws, 11 – Concrete tailrace, 12 – Gabion mattress foundation.

Zăvoiul Orbului SHHP1 has the following technical characteristics:

Head $H_i \approx 4.5$ m
Installed flow rate $Q_i \approx 6 \times 6.50$ m^3/s
Installed power $P_i \approx 6 \times 195$ kW
Turbines: 6 turbines Archimedes screw
Turbine tilt angle screw (α) 22^0
Velocity flow ≈ 1.75 m/s
Figure 11 shows the Zăvoiul Orbului SHHP 1

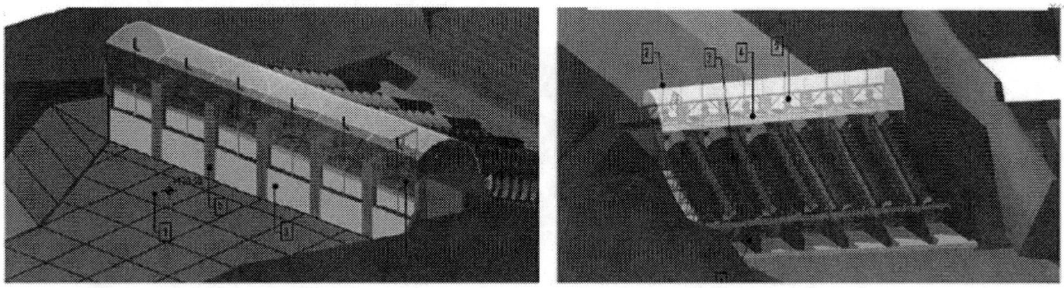

Fig. 11. Zăvoiul Orbului SHHP 1 [12]

Zăvoiul Orbului SHHP 2 has the following technical characteristics:
Head $H_i \approx 5.2$ m
Installed flow rate $Q_i \approx 6 \times 6.50$ m^3/s
Installed power $P_i \approx 6 \times 220$ kW
Turbines: 6 turbines Archimedes screw
Turbine tilt angle screw (α) 22^0
Velocity flow ≈ 1.75 m/s
Figure 12 shows the Zăvoiul Orbului SHHP2

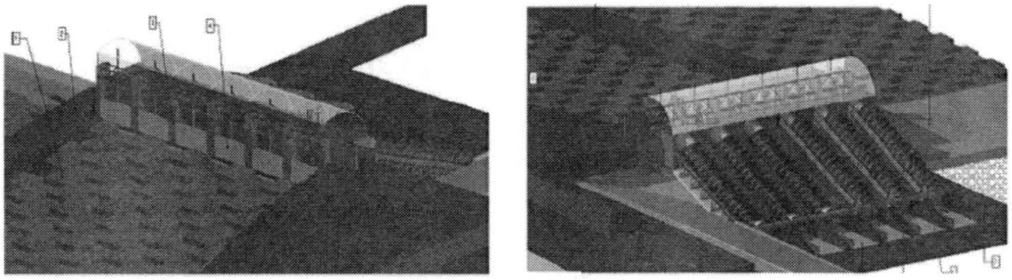

Fig. 12. Zăvoiul Orbului SHHP 2 [12]

The calculation of the energy produced in one year is done with the relationship:

$$E_{an} = P_i \cdot \zeta \cdot Ty \tag{8}$$

where: ζ is the utilization coefficient of the SHPP, $\zeta = 0,65$ % and $Ty = 8760$ hours/ year

For SHPP1 the annual energy produced is: $E_{anSHPP1} = 6662$ MWh. For SHPP2 the annual energy produced is: $E_{anSHPP2} = 7516$ MWh.

The annual income V_{an} is:

$$V_{an} = E_{an} \cdot P_{en} \tag{9}$$

where P_{en} is the price of energy and is approximate 100 € (we consider the price of one MW of energy of approximately 45 €/MW and a price of the green certificate of 28 €/MW).

For SHPP1 the annual income is: 728822.8 €/year. For SHPP2 the annual income is: 822250.4 €/year.

4.3. Locations for Archimedes' screw turbines

A possible location where a SHPP equipped with Archimedes' screw turbines can be located can be on the river Argeş, between Zeama-Rece and Catanele as in figure 13.

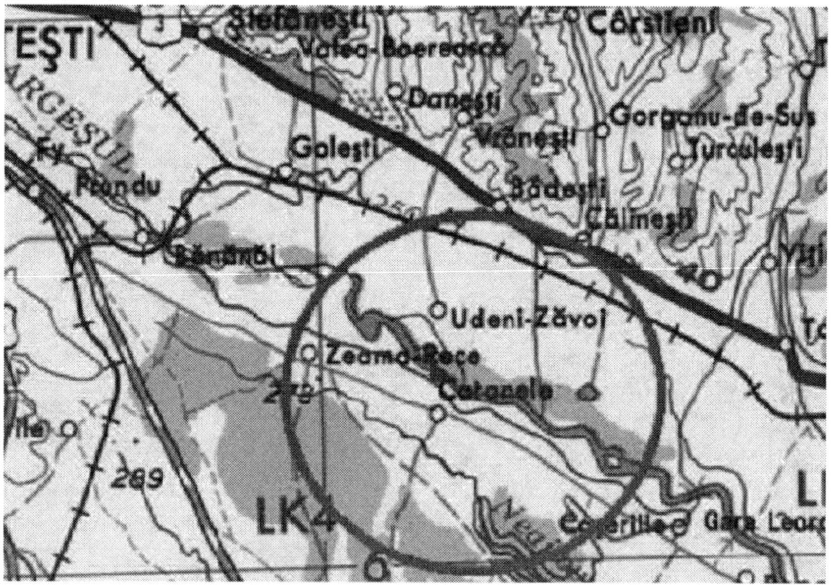

Fig. 13. Possible location for Archimedes' screw turbines

Another possible location for the location of Archimedes' screw turbines could be the irrigation canals. These are one of the best micro-hydropower resources, as they have many energy dissipation structures to reduce the hydraulic energy of flowing water. Some of these structures are the troughs, the settling tank etc., which dissipate excess water energy as heat loss. Excess energy from the water flowing in such structures can be converted into electricity. Given that Archimedes' screw turbines are efficient, fish-friendly, environmentally friendly devices and require little civilian work to implement in existing structures, they could be a good candidate for this purpose. There are also many new hydroelectric systems, including Archimedes' screw turbines that can be used in already existing structures (e.g. non-powered dams) to supply electricity.

5 The "fish friendly" character of the screw turbine

In the case of screw turbines it can be easily demonstrated that due to the slow rotational speed of the screw and the large space through which water passes, all categories of fish, even larger ones, can cross the turbine without suffering [13].

As a result, several environmental protection agencies have announced that it is not necessary to implement a protection screen.

Thousands of fish were monitored and video recordings were made using underwater cameras located both at the entrance to the turbine, inside the chambers through which the water passes, but also at the exit of the hydropower system, to evaluate the effect of the turbine Archimedes screw on some fish species of salmonids class. It was found that no fish were affected, they were crossing safely.

In the Figure 14 are presented fish species unaffected by screw turbines.

Fig. 14. Fish species unaffected by Archimedes' screw-type turbines [14]

A recent study by VisAdvies at the Hooidonkse Mill on the Dommel River in the Netherlands shows that these Archimende screw turbines are harmless to most fish species [14]. A total of 214 fish were monitored through the turbine. The size of the fish was between 4-15 cm and no fish was affected, they crossed safely.

Figure 15 shows the fish species that can pass through the screw turbine and the number of unaffected fishs.

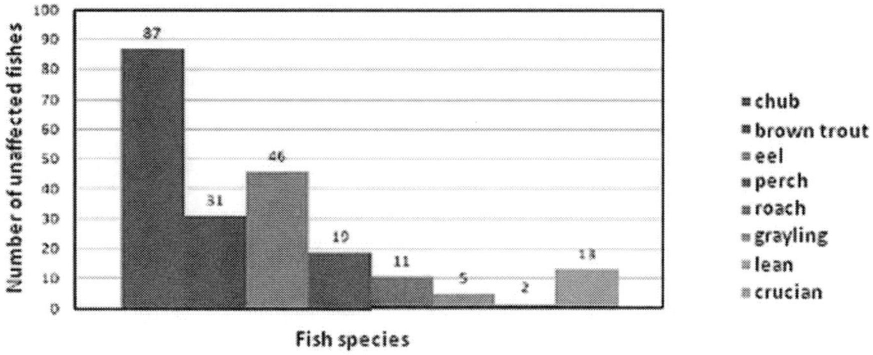

Fig. 15. The number of unaffected fishes by the Archimedes screw turbines [14]

6 Conclusions

As the world begins to understand the dangers of global warming, green energy is generating more and more interest. It is useful to look back at the great engineers of the past who first exploited the earth's natural energy, green/renewable energy. With a few design changes it can switch from screw pump to turbine screw. Archimedes screw turbine offers a lot of advantages over today's adjustable turbines. The lack of environmental effects, combined with a wide range of location possibilities as well as the robust, trouble-free nature of the screw turbine design, make this turbine offer sustainable renewable energy solutions for many locations.

Locations such as irrigation canals, or various existing energy structures could be used to locate Archimedes' screw turbines, e.g. non-powered dams. Archimedes' screw turbines can be successfully placed on different rivers, e.g. Argeş river between the localities Zeama-Rece and Catanele.

The implementation of Archimedes' screw turbines brings many more benefits than the implementation of other types of turbines, Kaplan, Francis, Propeller, Crossflow. According to the literature, the installation costs of a SHPP with Archimedes screw turbines are much lower compared to those for installing Kaplan turbines. At the same time, it was found that an Archimedes screw turbine hydraulic system produces 15% more energy than a Kaplan system and the investment costs 10% less.

Among the benefits of Archimedes turbines can be mentioned: the flat efficiency curve, which means that even dramatic changes in flow or fall do not significantly affect the efficiency of the system, high mechanical efficiency 85%. They can operate up to a flow rate of 5% of the design flow, are fish friendly turbines (many studies have been conducted and it has been shown that fish do not suffer any damage when crossing through the turbine Archimedes screw, in fact Archimedes screws are used in some areas as pumps to move fish from one basin to another), they have tolerance to dirty water due to their design and mode of operation, are robust due to their simple construction and no complex control/automation system is required.

Although remarkable progress has been made in recent years in the theoretical and practical study of Archimedes' screw turbines still further researches are needed to be able to optimize the design for the different ways of placing the turbine (sunken, inclined or horizontal), because each method generates electricity in a different way. Much of current design theory is based on pump design, not turbine design. Many of the design criteria can be used for any of them; however, there are key differences in operation, such as the direction in which water passes through the turbine.

Acknowledgments

This paper has been financed from the Nucleus Program funds that has been carried out with the support of the MEC project no. PN 19400401 contract MCI no. 27N /2019.

References

1. K. Kusakana, J. Munda, A. Jimoh, *Economic and environmental analysis of micro hydropower system for rural power supply*, Power and Energy Conference, 2008. PEC on 2008, IEEE 2nd International, pp. 441-444. IEEE, (2008)
2. J. Du, H. Yang, Z. Shen, J. Chen, *Micro hydro power generation from water supply system in high rise buildings using pump as turbines*, Energy **137** (2017), pp.431-440.
3. A. Kozyn, S. Ash, W. Lubitz, *Assessment of archimedes screw power generation potential in Ontario*, CCTC 2015, paper number 1570095585, in: Climate Change Technology Conference, Montreal (2015).

4. REN21, Renewables 2012: *Global Status Report, Renewable Energy Policy Network for the 21st Century*, (2012).
5. L. Lisdiyanti, Y. Hizhar, B. Yulistiyanto, *Effect of Flow Discharge and Shaft Slope of Archimides (Screw) Turbin on the Micro-hydro Power Plant*, (2012).
6. K. Kashyap, R. Thakur, S. Kumar, *Identification of Archimedes Screw Turbine for Efficient Conversion of Traditional Water Mills (Gharats) into Micro Hydro-power Stations in Western Himalayan Regions of India: An Experimental Analysis.* INTERNATIONAL JOURNAL of RENEWABLE ENERGY RESEARCH, Vol.**10**, No.3, (2020).
7. K. Brada, *Hydraulic screw generates electricity from micro hydropower stations.* Maschinenmarkt Würzburg, Mitteilung **14**, pp.52–56, (1999).
8. S. C. Georgescu, A. M. Gorgescu, *Hydraulics of pipelines and hydraulic machines.* Editura PRINTECH, (2007).
9. U. Kumar, P. Singh, A. C. Tiwari, *Suitability of Archimedes Screws for Micro Hydro Power Generation in India.* International Journal of Thermal Technologies, Vol. **6**, No. 3, pp. 273-278, (2016).
10. O.S. Abdullah, A.H. Kamel, W.H. Khalil, *Numerical and Experimental Modelling of Small Hydropower Turbine.* Journal of Engineering and Applied Sciences, Vol. **15**, No. 20, pp. 2244-2250 (2020).
11. S. Simmons, *A Computational Fluid Dynamic Analysis of Archimedes Screw Generators.* Presented to The University of Guelph In partial fulfilment of requirements for the degree of Master of Applied Sciences in Engineering Guelph, Ontario, Canada. (2018).
12. *** *https://www.hydroreview.com/*world-regions/adding-hydro-to-a-non-powered-dam-in-romania
13. A. YoosefDoost, W.D. Lubitz, *Archimedes Screw Turbines: A Sustainable Development Solution for Green and Renewable Energy Generation—A Review of Potential and Design Procedures.* Sustainability, 12, No (18) 7352; doi:10.3390/su12187352, (2020).
14. *** *http://www.ecoevolution.ie*

E3S Web of Conferences **286**, 02008 (2021) https://doi.org/10.1051/e3sconf/202128602008
TE-RE-RD 2021

The comparison of thermal comfort test results in selected traditional and modern buildings

Natalia Krawczyk[1,*], and *Justyna* Krakowiak[1]

[1]Faculty of Environmental, Geomatic and Energy Engineering, Kielce University of Technology, Poland

Abstract. The use of renewable energy sources in buildings is more and more common (both for heating and cooling purposes, as well as electricity generation). The paper focuses on the thermal comfort tests in two buildings: the traditional one (that uses non-renewable sources of energy) and the modern intelligent building "Energis" of Kielce University of Technology. The thermal sensations of students in these two buildings have been compared based on the questionnaire survey and conclusions have been drawn regarding the differences between the feelings of thermal comfort in those two buildings. Apart from subjective feelings of the volunteers, the measurements of indoor air parameters were conducted in both buildings and have been presented in the paper.

1 Introduction

Nowadays, the concept of thermal comfort is becoming more and more common. We spend more and more time indoors. Thermal comfort is a state in which a person is satisfied with the prevailing thermal conditions. This means that it does not feel cold or too warm. By providing the right thermal conditions, we feel good in the room, which affects our immune system and productivity. Therefore, in the existing and future intelligent buildings, the guarantee of thermal comfort has been increased. The main element of designing an artificial climate inside buildings is thermal comfort. It has a significant impact on human health and its safety in the building. Failure to provide good conditions can adversely affect our health, well-being and productivity. It makes us feel tired and makes us less productive, and consequently our effectiveness of the activities performed drops significantly. It is worth paying attention to the current situation in the world. The current coronavirus (Covid-19) pandemic has changed our private, social and economic lives. Most of us have changed our lifestyle to remote. This is due to the fact that when working in closed rooms, we want to feel comfortable in them. Overall, the feeling of thermal comfort significantly affects the work efficiency. Therefore, it is important to ensure appropriate conditions in a given room by using appropriate air-conditioning and heating devices. The supply of heat to the building can be produced with the use of conventional methods as well as alternative methods, including the use of renewable energy sources. Energy storage systems are important for renewable energy sources. A large part of the energy is used for thermal

* Corresponding author: nkrawczyk@tu.kielce.pl

comfort in buildings. It is important to ensure thermal comfort and thus reduce energy costs. A review of progress is essential in terms of thermal comfort. The main parameters influencing the determination of thermal comfort are, first of all, air temperature, relative humidity, air velocity, average radiation temperature and physical activity. The thermal comfort model was developed in the 1970s by O. Fanger on the basis of the applicable standards: ISO 7730 [1] together with PN-EN 16798-1: 2019 [2]. Fanger [3] determined two indicators: PMV - Predicted Mean Vote and PPD - Percentage of Dissatisfied people. According to the American standard ASHRAE 55 [4], the PMV index is expressed on a seven-point scale, from -3 to +3, where negative is cold and positive is hot. For the thermal environment to be acceptable, the PMV value must be between -0.5 and +0.5. The PMV index was expressed by the following equation according to [1]:

$$PMV = [0{,}303 \cdot exp\,(-0{,}036 \cdot M) + 0{,}028] \cdot \{(M - W) - 3{,}05 \cdot 10^{-3} \cdot [5733 - 6{,}99 \cdot (M - W) - p_a] - 0{,}42 \cdot [(M - W) - 58{,}15] - 1{,}7 \cdot 10^{-5} \cdot M \cdot (5867 - p_a) - 0{,}0014 \cdot M \cdot (34 - t_a) - 3{,}96 \cdot 10^{-8} \cdot f_{cl} \cdot [(t_{cl} + 273)^4 - (\bar{t}_r + 273)^4] - f_{cl} \cdot h_c \cdot (t_{cl} - t_a)\} \tag{1}$$

where: M - metabolic rate [W/m^2], W - effective mechanical power [W/m^2], p_a - partial pressure of water vapour [Pa], t_a - air temperature [°C], f_{cl} - clothing area ratio [-], t_{cl} - surface temperature of clothing [°C], $\bar{t}_r$ - average radiation temperature [°C], h_c - heat transfer coefficient [m^{2}K/W]. The projected percentage of dissatisfied people (PPD), describes the number of people who feel their environment is too warm or too cold. The PPD index is calculated according to the following equation [1]:

$$PPD = 100 - 95 \cdot exp\,(-0{,}03353 \cdot PMV^4 - 0{,}2179 \cdot PMV^2) \tag{2}$$

In recent years, an increasing number of scientists have studied the differences between PMV and thermal sensation voices. Biyik et al. [5] carried out research on the storage of energy from photovoltaic panels, which can then be used and thus maintain thermal comfort in the building. Aghniaey et al. [6,7] conducted research on thermal comfort on a university campus in the United States. They measured the temperature and air velocity in the room, the average radiant temperature, the CO_2 concentration and the relative humidity. The authors showed that the respondents preferred higher temperature ranges, especially women accept a warmer environment. The authors [8] analysed the thermal comfort by comparing the real feelings of people with the Fanger model. Xu et al. [9] studied the problem of electricity management for a shopping center. They showed that it is possible to simultaneously ensure thermal comfort and minimize operating costs by using renewable energy sources (photovoltaic and wind energy). In the article [10] a study on thermal comfort in a climatic chamber was carried out. 15 women and 15 men participated in the study. They concluded that thermal sensations are significantly influenced by air temperature and the body's thermal adaptability. Moreover, a significant correlation was observed between skin temperature and thermal sensations. Almeida et al. [11] analysed thermal comfort in school buildings. They showed a significant relationship between PMV and Mean Thermal Sensation (MTS) and the operating temperature. Vilcekova et al. [12] on the basis of research conducted in primary school, they showed a discrepancy between the results of PMV calculations and the actual feelings. They also observed an increase in CO_2 concentration during school activities. Homoda et al. [13] conclude that the air temperature and humidity have the greatest impact on the feeling of thermal comfort. The tests of the author and other co-workers [14-17] indicate that the thermal environment in modern buildings is largely influenced by the heating, ventilation and air conditioning systems installed there.

The article discusses the research on thermal comfort in two buildings: the traditional and the modern building "Energis" of the Kielce University of Technology (using renewable energy sources). The respondents' thermal sensations in the presented buildings were compared on the basis of the surveys carried out, and the parameters of the air in the rooms were examined.

2 Material and method

The research was carried out in June 2019 in the intelligent building "Energis" of the Kielce University of Technology and in the school building. The "Energis" building has mechanical ventilation, in which it is possible to control the operation of the ventilation. On the other hand, the school building is not ventilated in the rooms where the research was conducted. An intelligent building is equipped with solar collectors, heat pumps and photovoltaic cells that use solar energy to heat water and generate electricity to illuminate the building. On the roof of the building there are photovoltaic panels and solar collectors, thanks to which the building has the ability to accumulate heat and use energy. In terms of energy, the building is self-sufficient. The picture below shows a photo of solar collectors and panels.

Fig. 1. Solar collectors and photovoltaic panels on the roof of the "Energis" building.

The discussed Energis building has been designed to ensure adequate air parameters for people staying in it. The designed air-conditioning and heating systems should ensure appropriate parameters depending on the purpose of the room, because they significantly affect the conditions in the room. The second room does not have any kind of ventilation. The inflow of fresh air is possible by opening windows and doors, thanks to ventilation and by infiltration through leaks in the window and door joinery. This room is on the first floor and the windows were ajar during the research.

The study consisted of two research methods. The first was to measure the air parameters. A microclimate meter was used for the performed tests, on the basis of which air parameters in the rooms were determined. The temperature and air velocity, black sphere temperature and relative humidity were tested. Figure 2 below shows the Testo 400 gauge that was used for the tests.

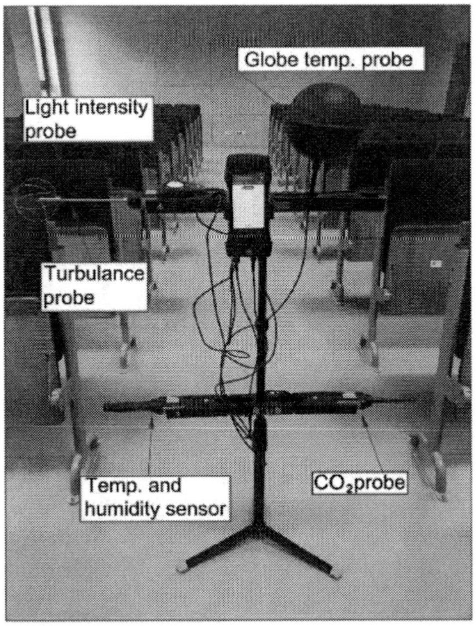

Fig. 2. Testo 400 microclimate measuring device.

The parameters were read from the meter after 15 minutes of stabilization of the measurements. During the tests performed, the measurements were characterized by high atmospheric air temperatures ranging up to 28°C, at an atmospheric pressure of about 985 hPa. Parameters indicated by the meter were read for both objects. Then the values in the given intervals were averaged and the average radiation temperature was determined. The table below shows the air parameters for the rooms in question.

Table 1. Air parameters for the tested rooms.

	Room 1: Intelligent building powered by renewable energy sources	Room 2: Traditional building
Air temperature, °C	27.60	27.70
Black ball temperature, °C	27.09	25.72
Air speed, m/s	0.01	0.05
Air humidity, %	53.15	47.60

The second method consisted in filling in the questionnaires by the respondents concerning the thermal impressions of the microclimate in which they stayed. The survey allowed for a clear assessment of thermal comfort, thermal preferences regarding the conditions in the room and thermal sensations. The questionnaire contained a question related to clothing, thanks to which it was possible to determine the average level of insulation of the garment. The thermal resistance of the office chair was added to the value of thermal resistance for clothes, which is 0.1 clo. The total value of thermal resistance for the first room was 0.49 clo, and for the second one - 0.62 clo. At the end of the questionnaire, there was a certificate which provided information on the sex, age, height and weight of the respondents. People aged 18 to 21 and one woman aged 52 participated in the study. Based on the height and weight, the BMI (Body Mass Index) was calculated. It is the ratio formed by dividing the body weight given in kilograms by the square of the height given in meters. The survey also included a question about health. When someone answered "yes" and "I do not know", such a questionnaire was rejected due to the thermal feelings of the sick, which are not reliable.

3 Results and discussion

The tests of the two rooms were carried out in June. In the first room (Energis intelligent building), 12 respondents took part, 4 of which were rejected, 3 due to ignorance about their health condition and 1 due to illness. Thus, there are 8 reliable questionnaires for the first room, including 3 women and 5 men aged 19 to 21. The first room has mechanical ventilation with an air parameters control panel. It was not used during the research. For the second room, 19 people were examined, 3 questionnaires were rejected, one due to selecting the answer "yes" to the question about the health condition, and 2 by selecting the answer "I do not know", therefore 16 questionnaires were taken into account for further analysis, including 15 women and 1 man. The subjects in the second room were 17 to 18 years of age and one woman was 52 years of age. The second object does not have any kind of ventilation, but the windows were ajar during the tests. The chart below shows the frequency of the given answer regarding thermal sensations vote experienced by the respondents in a room powered by renewable energy sources and in a traditional room (Fig. 3).

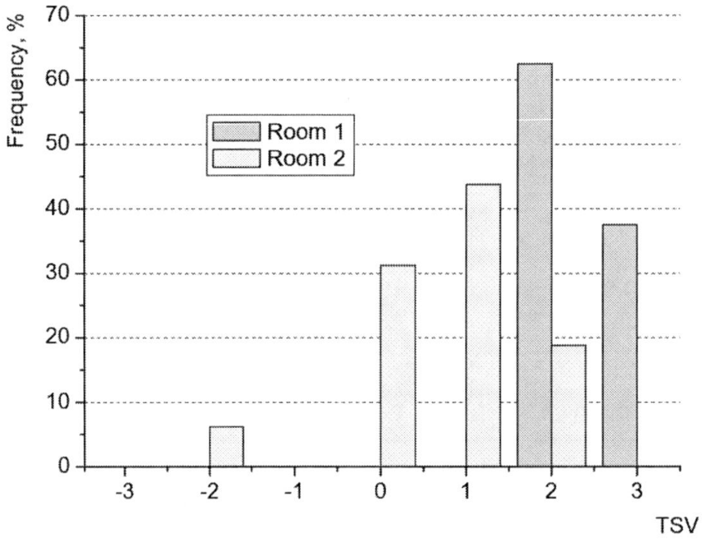

Fig. 3. The frequency of responses given by the respondents regarding thermal sensations vote (TSV): "-3" - too cold, "-2" - too cool, "-1" - pleasantly cool, "0" - comfortable, "1" - pleasantly warm, "2" - too warm, "3" - too hot.

The research on thermal sensation showed that the response "too cool" was 6.25% for a traditional building. The most frequently given answer was "too warm" for building 1 (Energis), which amounted to 62.50% for 5 responses out of 8. However, for a traditional building - "pleasantly warm" was the most frequently given answer, which corresponds to 43.75%. The answer "comfortable" for building 2 was marked 5 times which corresponds to 31.25%. The "too warm" option was repeated 3 times and amounts to 18.75%. For an intelligent building, the response "too hot" was 37.50%. Taking into account all the answers, it can be concluded that the respondents did not feel well in the rooms where they stayed. This means that there was no thermal comfort in the places where they stayed. The percentage of choice of answers -3, -2, +2, +3 was greater than 10.00% and amounted to 50.00%. The next figure shows the frequency of the answers given regarding the thermal acceptability vote.

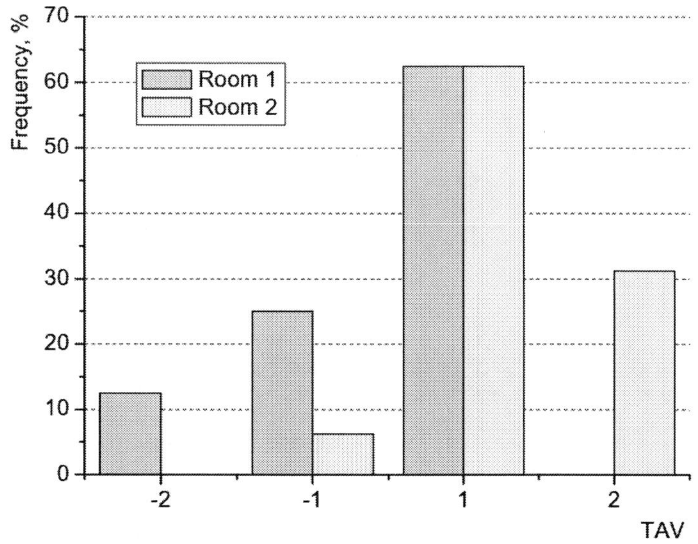

Fig. 4. The frequency of responses given by respondents regarding thermal acceptability vote (TAV): "-2" - definitely not acceptable, "-1" - no longer acceptable, "1" - still acceptable, "2" - definitely acceptable.

The above chart shows that for room one, the answer "definitely not acceptable" was 12.50%. In turn, for "no longer acceptable" it was twice as high, which corresponds to 25.00%. The most common answer for both rooms was "still acceptable", which is 62.50%. For the second room, 31.25% of respondents believe that the air is definitely acceptable. To sum up, the answers given by the respondents show that 62.50% of all respondents are satisfied with the conditions in the rooms. Only 16.67% of the respondents did not correspond to the temperature of the environment in which they were located. The next figure shows the frequency of the given answer regarding thermal preferences vote.

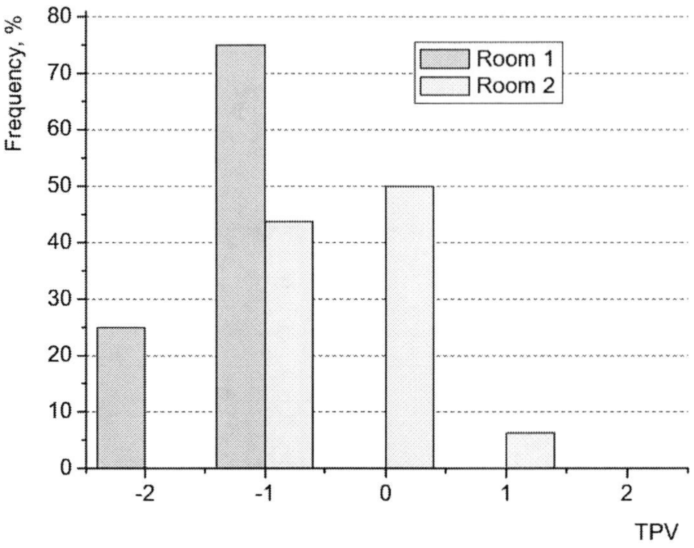

Fig. 5. The frequency of answers given by the respondents regarding thermal preferences vote (TPV): "-2" - definitely cooler, "-1" - cooler, "0" - unchanged, "1" - warmer, "2" - definitely warmer.

Of the respondents' responses to room one, the most common response was "cooler" and it was 75.00%, meaning 6 out of 8 want it to be cooler in the room. And 2 people (25.00%) selected "definitely cooler". However, for the second room, the situation is as follows: 50.00% of the respondents answered "unchanged", and 43.75% - wanted it to be "cooler" (7 people out of 16 chose this answer). The least frequently given answer for the second room was the answer "warmer" and it amounts to 6.25%. Only one person indicated that she wanted it to be warmer. It can be concluded that the respondents did not feel well in the examined rooms. When comparing all the results, 62.50% of respondents believe that the conditions in the rooms are poor. The figure below shows the frequency graph for the assessment of humidity for the tested rooms.

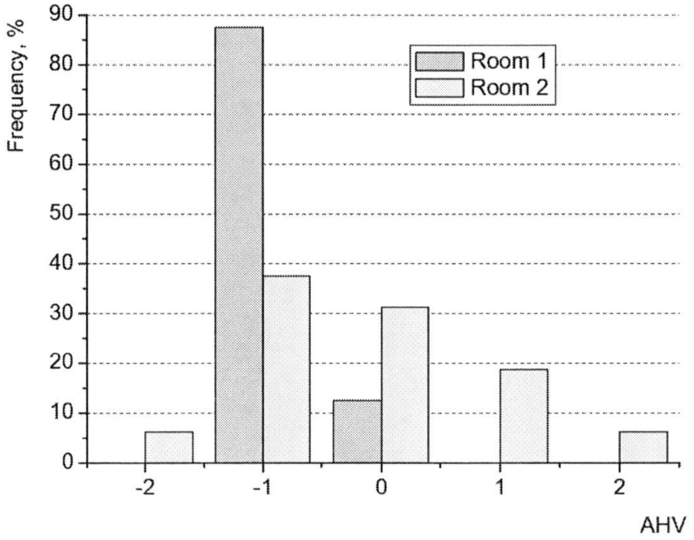

Fig. 6. The frequency of the answers given by the respondents regarding air humidity vote (AHV): "-2" – too dry, "-1" - fairly dry, "0" - pleasantly, "1" - quite damp, "2" - too humid.

From the chart above, it can be seen that for room 1, the most common answer was "fairly dry", which is 87.50%. This means that 7 out of 8 rate the air humidity as fairly dry. Only one person was satisfied with the air humidity in the tested room. Similarly, for the second room, the most frequently given answer was also "fairly dry" and it is equal to 37.50%. This answer was chosen by 6 out of 16 people, and 5 people chose the answer "pleasantly", which constitutes 31.25%. For 4 out of 16 people there was too much humidity in room 2. So, the percentage of "quite damp" and "too humid" responses is 25.00%. In general, it can be concluded that the tested rooms were quite dry. The next figure 7 shows the humidity preferences vote.

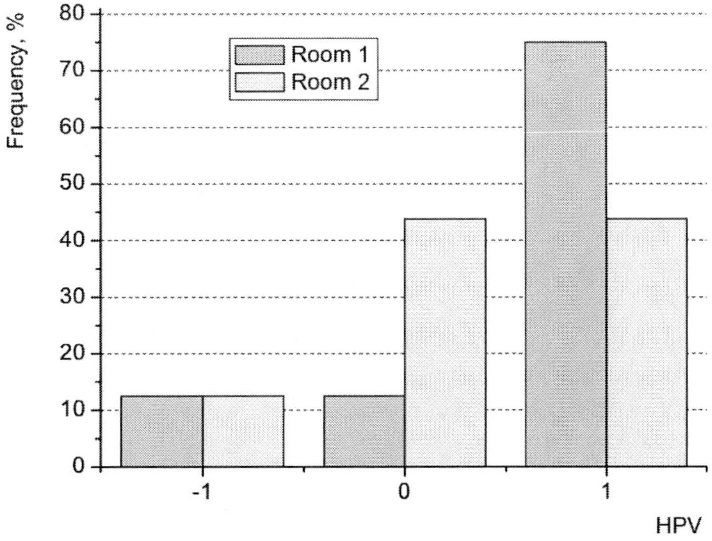

Fig. 7. The frequency of answers given by the respondents regarding the humidity preferences vote (HPV): "-1" - more dry, "0" - no change, "1" - more humid.

For the rooms studied, it can be seen that the respondents would like the air in the rooms to be more humid, which is 54.17%. The most frequently chosen answer for the first room was the answer "more humid", 6 of the 8 respondents chose this answer (i.e. 75.00%). The same number of responses was for "more dry" and "no change", it was 12.50% for both responses. The same percentage of the "more dry" answer was given to the second room. For the "no change" and "more humid" responses, the response rate was 43.75%. An analysis of the graph above shows that the room is quite dry. Figure 8 shows a comparison between TSV (Thermal Sensations Vote) and PMV (Predicted Mean Vote) for the rooms under study.

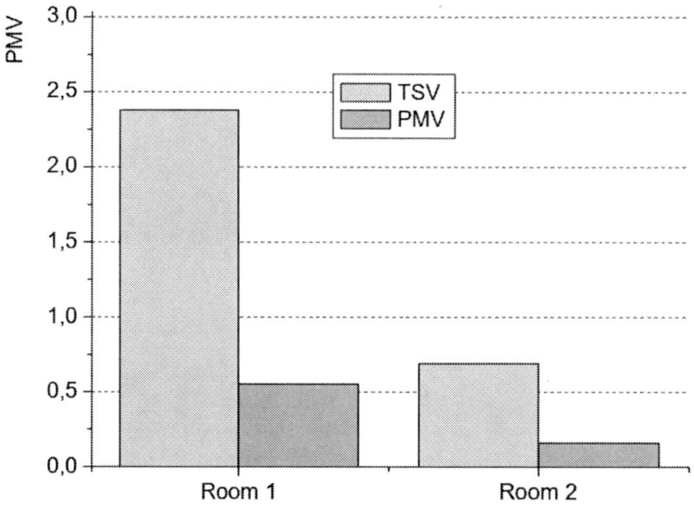

Fig. 8. Comparison of actual test results with calculated.

The above figure shows a comparison of the average thermal comfort rating from the questionnaires and that calculated according to the formula (1) from the standard [2] for the rooms under study. For the first room, the thermal sensations vote is 2.38. This value is inconsistent with the calculated one, i.e., PMV is 0.55. The values differ by 1.83. However, for the second room, TSV from the questionnaires is equal to 0.69. This answer is inconsistent with the standard, because PMV according to the standard is 0.16. These values are different and different from each other. Both tests were conducted at the same outside and room temperature. The ratio between TSV and PMV for the first and second room is as follows: for room 1 the ratio of TSV to PMV is 4.33, and for the second room 4.31. This means that both values are around 4.30. The thermal feelings of percentage of Dissatisfied people for the examined rooms are presented below.

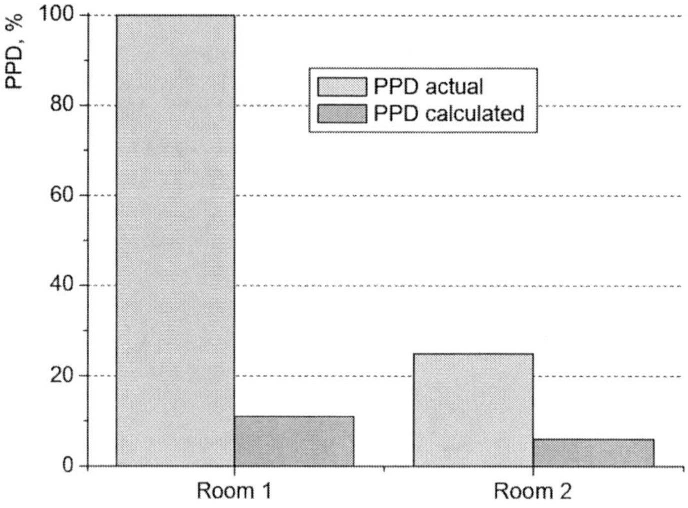

Fig. 9. Comparison of actual survey results with calculated results for PPD.

The parameters were read from the meter after 15 minutes of stabilization of the measurements. When analysing the above chart, a significant difference was noticed between the percentage of dissatisfied people according to the questionnaires and that calculated in accordance with the standard [2]. The percentage of dissatisfied with the prevailing conditions for the first room was analysed and the obtained value was 100%. You can see that all people find it too warm or too hot. According to the calculations from the norm, it is 11%. This is a significant difference in relation to the answers according to the respondents. According to the polls, for the second room, the percentage of dissatisfied people is 25%. PPD calculated according to the standard is 6%. There are 4 times more people dissatisfied for the second room than according to formula (2). The standard states that the percentage of dissatisfied people should not exceed 10%. The figure below presents the calculated BMI for the respondents and compares it with the thermal preferences voices.

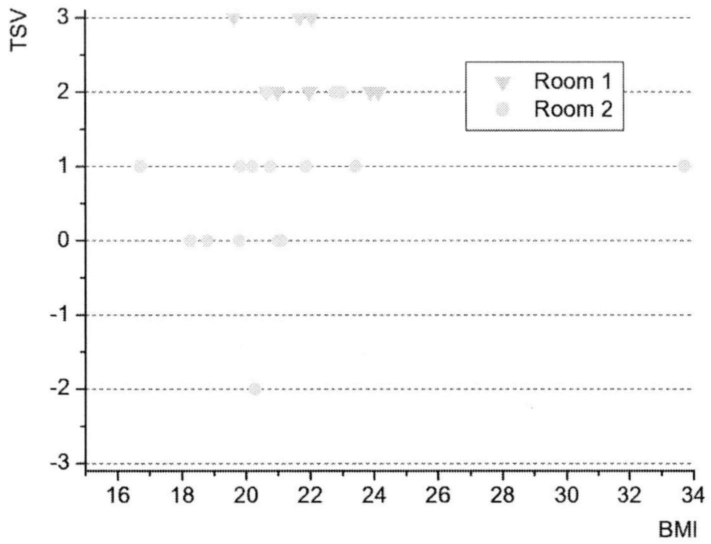

Fig. 10. Dependence of TSV on the BMI mass index for two rooms.

In the presented diagram, the first room (Energis intelligent building) is marked in blue, while the traditional building is marked in yellow. Figure 10 shows the calculated BMI for the respondents, which was compared with thermal sensations voices (TSV). Based on the results, it can be concluded that people with a higher BMI have higher thermal sensations. This means that they prefer lower indoor temperatures.

4 Conclusions

The measurement results from the microclimate meter were compared with the survey results. This showed that there was a significant difference between these results. There was a much smaller difference between the results from the microclimate meter and the results from the questionnaires for Room 2, which was 0.53. While for Building 1 it was 1.83. Both buildings had similar air temperature (for Room 1 - 27.6°C, Room 2 - 27.7°C). Research conducted in an Energis intelligent building with renewable energy sources and a traditional building showed that the number of people dissatisfied with the thermal conditions exceeded 10%. In an intelligent building, closed windows could have a significant impact. And in the traditional building, the windows were open, which made the number of dissatisfied people smaller. Thus, it might not be so important if the building is very modern and equipped with state-of-the-art technological solutions, but how these systems are used for the benefit of the people there. In addition, the survey showed that the respondents prefer the indoor environment to have higher relative humidity. This may be due to the time of year that the survey was conducted (summer). The presented survey will be continued to collect more data on thermal comfort. This will allow the current temperature settings in smart buildings to be changed to more accurately suit the occupants and will also serve to modify the Fanger model. It is important to ensure proper thermal conditions in educational spaces because it has a significant impact on productivity and will also reduce energy costs.

References

1. ISO International Organisation for Standardization, Ergonomics of the thermal environment – Analytical determination and interpretation of thermal comfort using calculation of the PMV and PPD indices and local thermal comfort criteria, International Standard ISO 7730 (2005)
2. PN-EN 16798-1:2019, Energy Performance of Buildings-Ventilation for Buildings-Part 1: Indoor Environmental Input Parameters for Design and Assessment of Energy Performance of Buildings Addressing Indoor Air Quality, Thermal Environment, Lighting and Acustics (2019)
3. P.O.Fanger., Komfort cieplny, Arkady, Warszawa (1974)
4. ASHRAE Standard 55-2017, Thermal Environmental Conditions for Human Occupancy (2017)
5. E. Biyik, A. Kahraman., Journal of Building Engineering, **25**, (2019)
6. S. Aghniaey, M. L. Thomas, T.N. Sharpton, S.P. Douglass, T. Oliver, M. Sutter, Building and Environment, **148** (2019)
7. S. Aghniaey, M. L. Thomas, **59** (4) (2017)
8. E.E. Broday, J.A. Moret, A.A. de Paula Xavier, R. de Oliveira, International Journal of Industrial Ergonomics, **69** (2019)
9. X. Xu, W. Hu, W. Liu, Y. Du, R. Huang, Q. Huang, Chen Zhe, Energy Conversion & Management, **230** (2021)
10. Q. Wu, J. Liu, L. Zhang, J. Zhang, L. Jiang, Building and Environment, **171** (2020)
11. R.M.S.F. Almeida, N.M.M. Ramos, V.P. de Freitas, Energy and Buildings **11** (2016)
12. S. Vilcekova, L. Meciarova, E.K. Burdova, J. Katunska, D. Kosicanova, S. Doroudiani, Building and Environment, **120** (2017)
13. R.Z. Homoda, K.S.H. Saharia, H.A.F. Almurib, F.H. Nagi, Building and Environment, **49** (2012)
14. N. Krawczyk, Z. Piasta, Structure and Environment, **11** (4) (2019)
15. N. Krawczyk, A. Kapjor, Structure and Environment, **12** (3) (2020)
16. J.Zb. Piotrowski, Ł.J. Orman, X. Lucas, E. Zender – Świercz, M. Telejko, D. Koruba, EPJ Web of Conferences, **67**, 02095 (2014)
17. G. Majewski, M. Telejko, Ł.J. Orman, E3S Web of Conferences, **17**, 00056 (2017)

A study of the transient processes in the physical model of the electric power systems with predominant part of the renewable energy sources

Ivaylo Nedelchev[1*], *Hristo* Zhivomirov[1], and *Yoncho* Kamenov[2]

[1]Technical University of Varna, Department of Theory of Electrical Engineering and Measurements, Bulgaria

[2]Technical University of Varna, Department of Power Engineering, Bulgaria

Abstract. The renewable energy take part in the most of the electric power systems in the modern world. The part of this type of energy in the global electric power system, as well as in the local scale, increases with the setting the stricter requirements for decreasing the level of the carbon dioxide emissions. This is the result of the newest international conventions and decision for saving the nature. By these conditions, the electric power systems are forced to work with more different types of energy sources: wind power, photovoltaic, biomass plants etc. Switching of such miscellaneous power sources, leads to complicated transient processes, which are developed due to specific electrical parameters, especially harmonic components, of the synchronous generators, photovoltaic and wind power plants. This paper represents data from measurements of the transient processes into the physical model of the electric power system with predominant part of renewable energy and assesses the applicability of the model. For conducting this study, the multichannel DAQ measurement system is used.

1 Introduction

During the last few decades, the scale of renewable energy sources (RES) integrated into the electric power system (EPS) has been rapidly increasing. In the global EPS, there is a real possibility to appear a case with a predominant part of the renewable energy in a large region or during a long time period [1, 2, 3]. In the global EPS there exist the regions with predominant part of RES [2]. In those regions, one or more types of RES could exist. The most often cases of RES types are: wind power turbines and photovoltaic modules [4]. All of RES electrical schemes incorporate inverter, with switch-mode power supply (SMPS) module, which creates a high level of high frequency powerful harmonics, decreasing the power factor in the circuit, quality of the power energy, sustainability and reliability of the system [1, 4, 5]. These effects increase the risk of unpredictable and

[*] Corresponding author: iynedelchev@abv.bg

unprovoked failures and collapse of the system. That type of effect was the reason for the failure of the system in large regions in Western Europe in the beginning of 2021.

This paper represents a study of the processes in a small-scale physical model of the EPS with predominant part of the RES [6, 7, 8]. The purpose of this study is to represent the verification and applicability of the constructed physical model, as well as the possibility for monitoring of the transient processes during system mode adjusting, changing of the level of the load of the power lines, and quality of the generated electric energy from different types of RES.

2 Short description and connecting diagram of the physical model

In every EPS, there exist different types power units of RES. The most used are: wind power turbines, photovoltaic plant, water turbine generators, biomass power plants. The constructed physical model is designed from the team of the power-engineering department in the Technical university of Varna [6]. For the simulation of the wind power and sun radiation power, the real inverters with DC voltage source are used. Simulating the RES is performed via adjustable DC voltage source (ADCV). With the help of controllable three-phase power rectifiers the inverters was sourced with DC voltage at the input side. Thus, the energy regime of this type of power units can be changed via adjusting the mode of the rectifiers. In the model, one of the inverters represents a photovoltaic generator (PVG) and the other one – wind power generator (WPG). For representing of the water turbine generator in the system, the direct coupled motor synchronous generators are included (SG1, SG2, SG3). Multi-stage resistive loads are used as consumers of the lines.

On the Fig.1 is shown the diagram of the physical model. The main studied power units in the diagram are marked as: synchronous generators (SG1, SG2, SG3), photovoltaic generator (PVG) and Wind power generator (WPG). The terminals of the resistive loads are marked as L1 - L7.

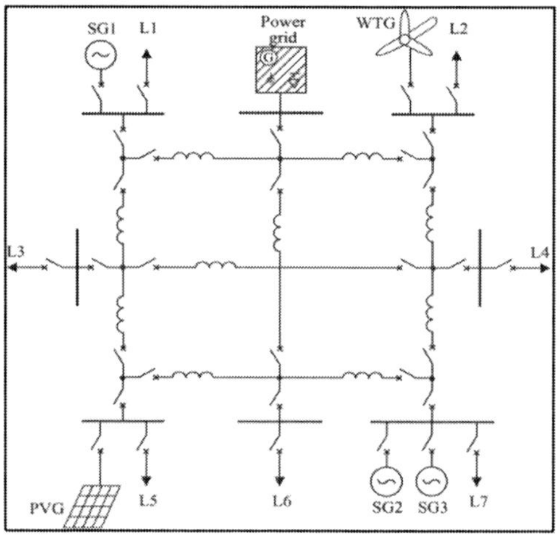

Fig. 1. Single line diagram of the physical model as an object of the measurement (*cf* Fig.1 in [6]).

The inductors in the diagram represent the inductance of the transmission lines. The synchronization of the SG1-3 is implemented by controllers and all requirements for the synchronization to the global power net, concerning voltage frequency, phase angle, phase sequence and waveform of the generated voltage are fulfilled. Two types of controllers are used: for internal synchronization between the different units in the model structure and for external synchronization during the connection to the global power net. The model is capable for working in so-called "island mode". In this mode, the synchronization of the SG 1-3 can be done manually or automatically. Detailed diagram of the power units and description of the physical model are given in [6].

3 Measurement equipment. Connecting diagrams, transducers and software

The experiments was conducted via measurement set, which consist of: DAQ module NI-6211, supplement primary transducers and software. The connecting diagram for the cases of measurements including RES power units or synchronous generators are given on the Fig.2 and Fig.3.

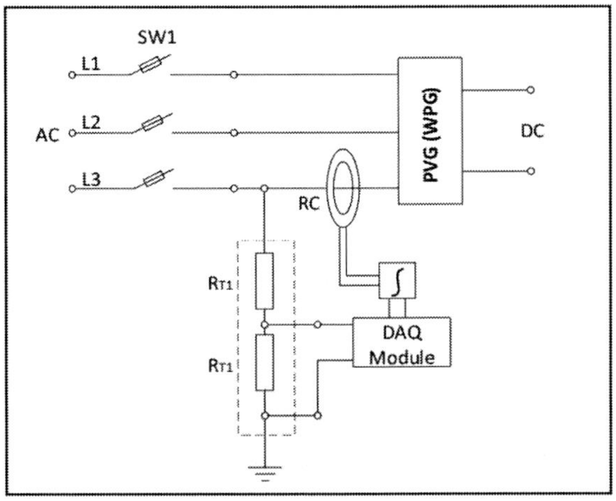

Fig. 2. Measurement the current and voltage in the output circuit of the simulated photovoltaic generator (PVG) or wind power generator (WPG).

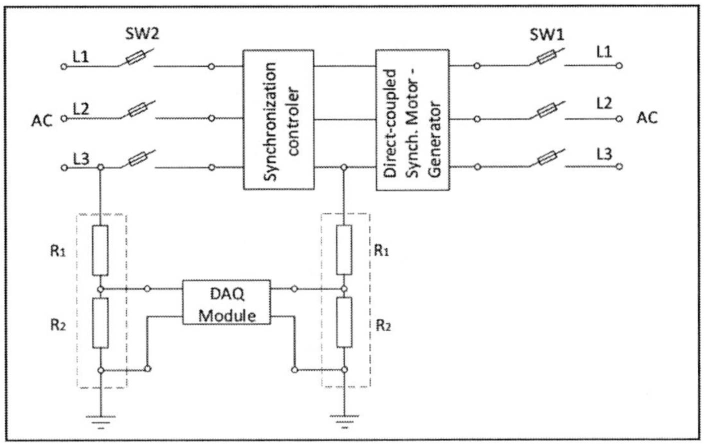

Fig. 3. Measurement of the input and output voltage in the circuit of the synchronous generator (SG).

Due to the symmetric type of the load, it was measured the output voltage and current in a single phase of the three-phase type of EPS. In other cases, the input or output voltages of the single power unit was observed. In this way, for measurements two of the channels of the DAQ module was involved. For the measurement of the voltage it was used a high-voltage divider with a large input resistance [9]. As a current transducer, the Rogowski coil with incorporated integrator was used [10]. For the purpose of the experiment, the LabVIEW software was involved and as a result, was obtained the time and frequency domain graphs.

The main parameters and conditions of the experimental set are:
- Sampling rate of the DAQ system – 5000 Hz;
- Input resistance of the high-voltage divider – 1 GΩ;
- Frequency band of the Rogowski coil measurement module – 100 kHz;
- Current range of the Rogowski coil – 100 A;
- Load of the power units – the load was gradually changed through three levels: 10%, 50%, 100%; For measuring of the biggest effect of the transient process, a step changing of the load was used from 10 % to 100 %;
- The resistive load – the rate of the involvong the resistive load in the energy transmission lines was executed via IGBT switch regulator. The levels of the resistive load were adjusted from 10 % of the nominal load with the step of 10 % up to the nominal capacity of the load.
- Coefficient of transformation for the Rogowski coil is [10]:

$$K_{RC} = \frac{I_1}{U_2} = 2000\ \frac{A}{V}, \tag{1}$$

and for the high-voltage divider is:

$$K_d = \frac{U_2}{U_1} = 2000, \tag{2}$$

On the Fig.4 is shown the WPG power unit box of the physical model and DAQ system during the measurements [6].

Fig. 4. Common view of the part of the physical model (WPG power unit) and measurement equipment.

4 Parameters of the transient process of the studied physical RES model

During the operation of the RES into the EPS system, the energy generation passes through three stages [11]:
- synchronization, of the power units with the main EPS,

- establishment of the frequency and amplitude – energy transfer into the EPS;
- disconnecting the power units from the EPS.

It could be distinguished one more stage case – the changing of the operation mode of the RES.

Usually switching on the RES and disconnecting from the EPS requires a few seconds in order to set stable parameters of generation [11]. From the other hand, the quality of the generated energy also is important factor for system stability. For example, the output voltage shape of the inverters and noise in the system are one of the strongest factors for system failures [1]. Having in mind upper reasons and facts it will be correct the parameters of the model transient process to be defined as follows [11]:

• time for synchronisation with the EPS frequency - t_s, *s.*

t_s is the time between switching on the power units into the EPS and establishment of the frequency and in-phase mode of the module in the system;

• time for disconnecting - t_d, *s*.

t_d is the time between switching off the power units from the EPS and attenuation of their output voltage to the U_m/e level, where U_m is the amplitude of the nominal output voltage before involving the system onto the transient mode.

• time for switching the mode – t_{sm}, *s*.

t_{sm} is the time for switching of the power between two different modes. Considering the stable levels of operation, the largest difference in working modes are by load level between 10 % to 100 % of the nominal regime.

The parameters t_d, t_s are most applicable for monitoring the output voltage of the synchronous generators (SG1, SG2, SG3), while the parameter t_{sm}, is more useful for PVG and WPG. All of them are important transient mode parameters, which characterize flexibility of the EPS with respect to outer RES impact.

• total harmonic distortion of the power units – ***THD***;

The expression for THD is given by:

$$THD = \frac{\sqrt{\sum_{n=2}^{N} U_n^2}}{U_1}, \quad (3)$$

where n is the highest harmonic of the measurement set, U_1 is the RMS value of the fundamental harmonic voltage in the system, U_n are the voltage harmonics with the numbers higher than fundamental one (RMS value).

• the impact of the higher harmonic (percentage from the basic one) – δU_n, % (U_n – the biggest RMS voltage value from the sequence of the higher harmonics of the RES or synchronous generator; n is the number of the certain harmonic)

$$\delta U = \frac{U_n}{U_1}.100, \%, \quad (4)$$

5 Measurement results

5.1 Parameters of the transient process

The measured parameters of the transient process, which characterize the behaviour of the power units and resistive load regulator in the physical model, are given in the Table 1. The quality of the generated energy depends on the level of the THD and δU_n. They are obtained by 50% of the nominal load level of the system. The biggest distortion were obtained by SG power unit – 10,97% and the lowest level by PVG – 2,54%. In the Table 1, *n* is the number of the largest higher harmonic. The other frequency components also have the similar RMS level such, as the measured biggest one higher harmonic. The impact of the higher harmonics is assessed with the THD parameter.

Table 1. Observed parameters of the transient process of the studied power units and resistive load.

	SG	PVG	WPG	LA
THD, %	0,83	0,91	1,95	23,55
δU_n, %	10,97	7,73	2,54	35,9
n	5	5	7	3

Via spectral analysis of the output power unit voltages, their higher harmonics level is obtained. In the Table 2 are represented further calculations of the times of the transient process for synchronization and disconnecting (t_s, t_d) of the SG1 from the circuit. Depending on the defined power units also, the switching mode times t_{sm} of input (DC) voltages for the both inverters representing PVG and WPG are found to be 0,5 s. The AC switching times mode of the inverters' voltages have approximately the same level – 0,54 - 0,55 s. During the synchronization with the grid and disconnecting the SG1, this module shows different behavior. The time for synchronization t_s, is approximately three times more than disconnecting time t_d.

Table 2. Synchronization periods of power units by switching between different modes.

Synchronization periods	SG	PVG (AC)	PVG (DC)	WPG (AC)	WPG (DC)
t_s, s (by load level of 50 %)	7,7	-	-	-	-
t_d, s (by load level of 50 %)	2,9	-	-	-	-
t_{sm} , s (by swithching from 10% to 100 % of the load level)	-	0,55	0,5	0,54	0,5

5.2 Time domain graphs of the measured transient process

The most detailed information of the studied transient processes in the physical model, can be represented via the graphs across the time domain and frequency domain. Fig.5 - Fig.7 show the time domain graphs for the SG1 power unit.

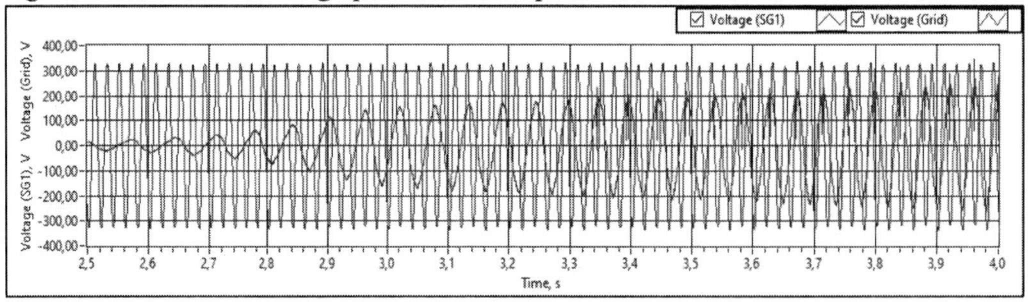

Fig. 5. The process of the switching the SG1 power unit into the grid.

On the time domain graphs, the defined transient process times, shown in the Table 2 can be measured. The synchronization of the SG module occurs, when synchronization controller (Fig.3), detect matching in phase the both voltages – the output of the SG1 and the grid voltage. Then the switch SW2 is closing. It can be mentioned, the beginning of the long period for synchronization (more than 7s.), which is calculated from the software but prolongs after the diagram on the Fig. 5. Unlike, by disconnecting, the level of the output voltage fades away within 2,5 s. pseudo periodically.

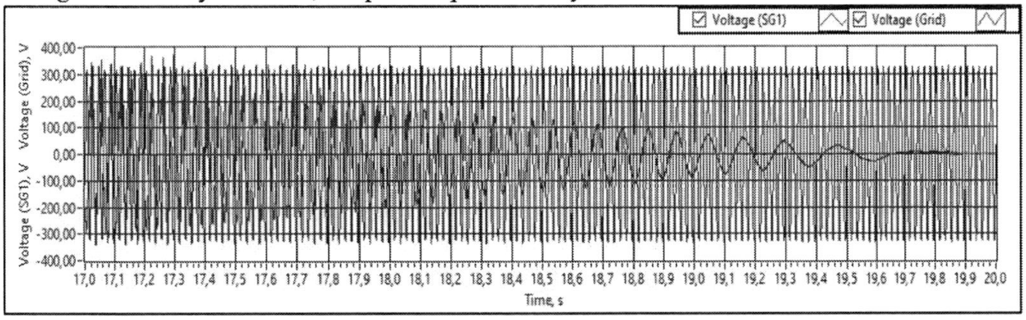

Fig. 6. The process of disconnecting the SG1 from the grid (common view).

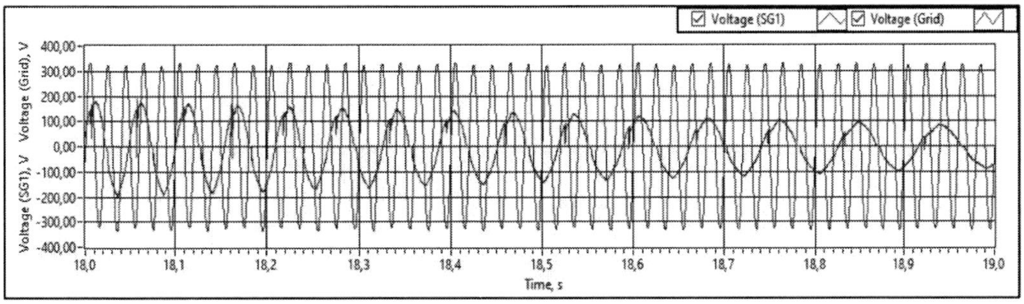

Fig. 7. The process of disconnecting the SG1 from the grid (larger scale). Fading the oscillations of the SG1.

The process of the changing the load level from 10% to 100% of the nominal mode for the AC and DC side of the WPG unit is represented on the Fig.8 and Fig.9.

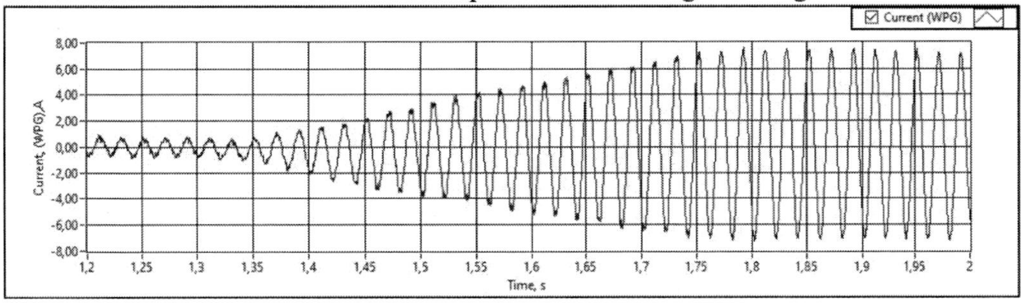

Fig. 8. The process of switching the WPG from 10% to 100% of the load (AC current).

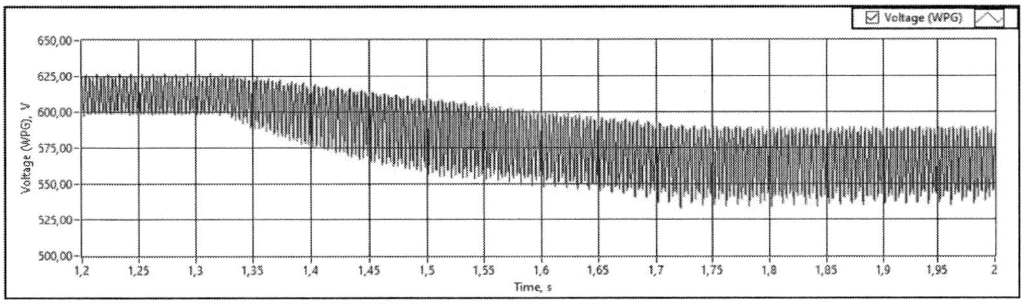

Fig. 9. The process of switching the WPG from 10% to 100% of the load (DC voltage).

Sharply increasing of the current load, can be seen on the Fig.8. It can be observed establishing the level of the raised current of the AC side of the WPG unit and on the other hand, the decreasing of the source voltage in the DC part of the inverter (Fig. 9). The different period times of the transient process can be distinguished easily.

Another dynamic process in the model can be observed during the changing the level of the load of the power line. Fig. 10 and Fig.11 show graphs, which represent the operation of the current output circuit in the IGBT regulator and its frequency response. It can be mentioned the rich spectral set and the big level of the higher harmonics. The spectral analysis shows a high level of distortion. THD of the output current exceed 23 %.

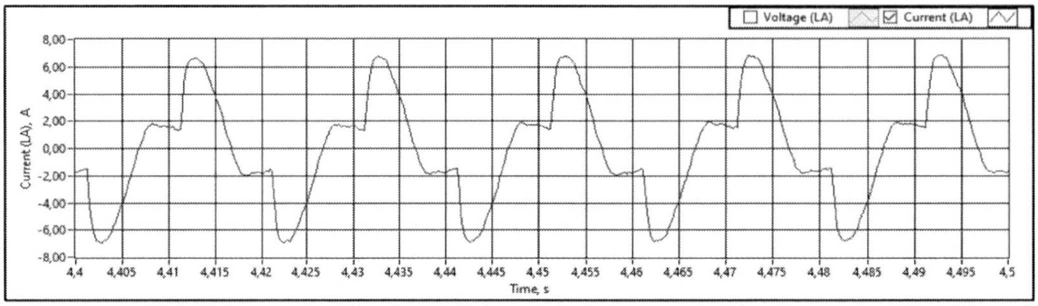

Fig. 10. The time domain of the current by 70% level of the resistive load, regulated by IGBT switch.

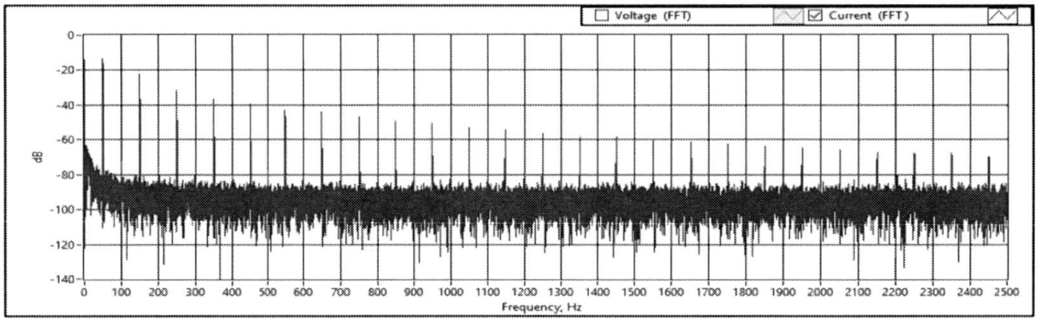

Fig. 11. The frequency domain of the current by 70% level of the resistive load, regulated by IGBT switch.

The moment of switching between different modes of the active loads or initially turning the power generation on, are shown in the Fig.12. The drift of the output current is obvious and its appearance is due to the saved energy in the parasitic elements in the output circuit. Both capacitance of the switching elements and stray inductance parameters of the power line generate a pseudo periodic process of the output current circuit.

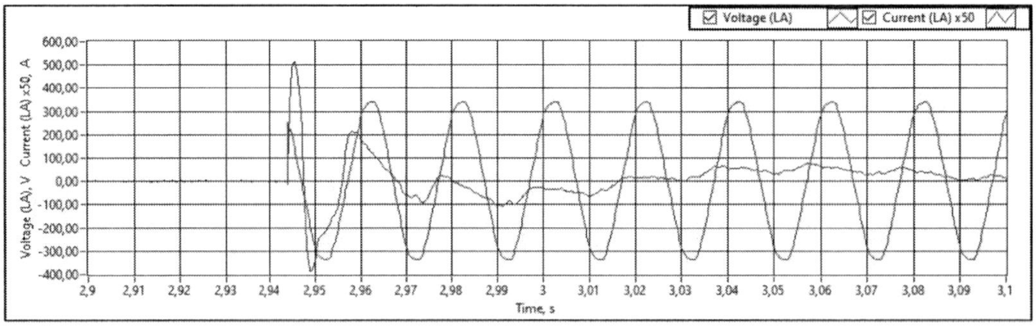

Fig. 12. The time domain graph of the current and voltage by the switching moment in the power line.

5.3 Frequency domain graphs of the measured transient process

In the studying physical model of RES, about the considering power generating units, as it is known, the rich composition of frequency components have to be present. On the frequency domain graphs are shown the spectral sequences of the higher level harmonics, which modulate the output quantities of the power units. The predominant part of them, consists of odd harmonics. Fig. 13 - Fig. 15 represents normalized frequency domain graphs of the output SG1 voltage, WPG and PVG currents (the biggest impact of the higher harmonics are enclosed on the graph area).

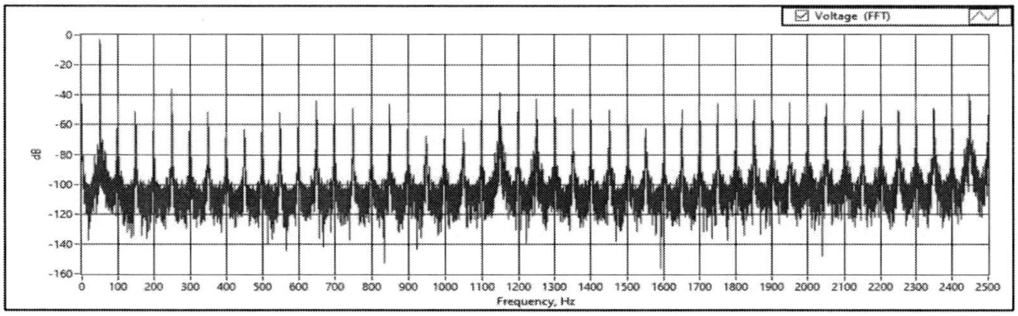

Fig. 13. The spectral analysis of the SG1 voltage, synchronized with the grid voltage.

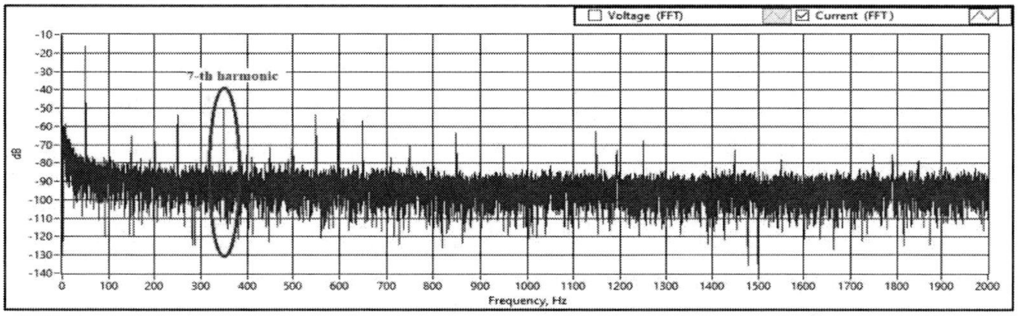

Fig. 14. The spectral analysis of the WPG current, synchronized with the grid voltage.

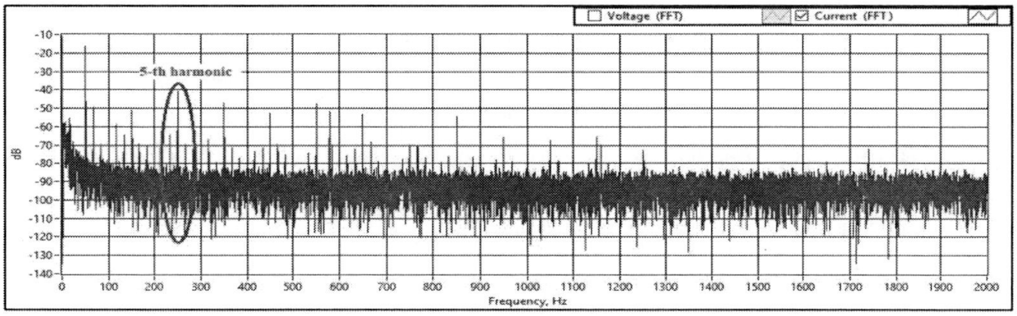

Fig. 15. The spectral analysis of the PVG current, synchronized with the grid voltage.

6 Conclusions

The experimental results of the physical RES model, leads to some general conclusions, which can affect the next scientific researches.

• The constructed physical model of the RES system and the simulation of the different power units, represent in an adequate way the processes in the real EPS;

• The synchronization period and disconnecting times are clearly distinguishable and can be measured despite of the type or model of the used inverter or synchronous generator.
• The spectral analysis of the signals measured by this DAQ measurement set, shows the distribution of the frequency components in a real time, generated from the RES power units in the energy transmission line and EPS;
• By this model and software, can be assessed the influence of the predominant RES part of the energy in the EPS;
• The method for monitoring with DAQ system could be applied over asymmetric loaded three-phase system by involving more measurement channels of the DAQ system.

Along with undisputed utilities of the studied physical model, some disadvantages can be mentioned, which can be a base for future work.
• The switching of the load, using IGBT power switches inputs a lots of higher frequency harmonics, which have to be filtered;
• It will be better, the source DC voltage for simulation of the WPG and PVG to be executed with stabilization, filtering and more flexible way for regulation.
• The validation of the constructed physical model by the present measurements could be useful for studying different types of RES inverters furtherly included in the model.

References

1. L. Meegahapola, A. Sguarezi, J.S. Bryant, M. Gu, E.R. Conde, R.B.A. Cunha, *Energies* 2020, **13**, 3441, doi.org/10.3390/en13133441
2. G.V.B. Kumar, R.K. Sarojini, K. Palanisamy, S. Padmanaban, J.B. Holm-Nielsen, *Energies* 2019, **12**, 1996, doi.org/10.3390/en12101996
3. T. Markvart, L. Castaner, *Practical Handbook of Photovoltaics: Fundamentals & Aplications*, (2003)
4. L. G. Acuña, R.V. Padilla, A. Santander-Mercado, *Renewable Energy*, **106**, 68-77, (2017)
5. L. Gang, *Renewable and Sustainable Energy Reviews*, **31**, 611-621, (2014)
6. Y. Rangelov, N. Nikolaev, Y. Kamenov, K. Gerasimov, *IITI 2017*, **680**, 396-402, (2018), doi:10.1007/978-3-319-68324-9_43
7. P. Yankov, P. Popov, N. Nikolaev, Y. Rangelov, *SIELA 2018*, (2018), doi:10.1109/SIELA.2018.8447075
8. R. Stanev, *BulEF 2020*, (2020), doi:10.1109/BulEF51036.2020.9326034
9. J.G. Webster, *Measurement, Instrumentation, and Sensors Handbook*, (2014)
10. http://www.heyiele.com/Catalogue/Flexible_Rogowski_Coil_With_G1_Integrator-633.html
11. I. Nedelchev, H. Zhivomirov, Y. Rangelov, N. Nikolaev, R. Vasilev, *E3S Web of Conferences* **180**, 485-490, (2018), doi.org/10.1051/e3sconf/202018002006

E3S Web of Conferences **286**, 02010 (2021) https://doi.org/10.1051/e3sconf/202128602010
TE-RE-RD 2021

Study of regime parameters of the fermenter in the production of biogas from animal liquid waste materials

Penka Zlateva[1,*], *Angel* Terziev[2], and *Krastin* Yordanov[1]

[1]Department of Thermal Engineering, Technical University of Varna, Varna, Bulgaria
[2]Department of Power Engineering and Power Machines, Technical University of Sofia, Sofia, Bulgaria

Abstract. The focus of the present study is a small biogas power plant for anaerobic fermentation of several types of animal waste raw materials used for biogas production. The impact of some of the characteristics of substances such as composition, temperature, humidity, and pH of the mixture in the bioreactor has been considered. The above is vital for optimizing the fermentation process, and also to improve the biogas production process. The plant is located in Northeastern Bulgaria and the raw liquid manure is supplied by several neighboring small farms. The annual quantities of raw waste are as follows: cow manure - 1252 t / a; chicken manure - 427 t / a and pig manure - 639 t / a. The manure is collected in a preliminary tank and then pumped to the bioreactor. The fermenter itself is a hermetically sealed and thermally insulated tank where constant temperature is maintained. It is equipped with a stirring system, which helps the mixing and homogenization of the substrate. The tests were performed during three charges of the installation. The fermentation takes approximately 23 up to 25 days. The experiments were performed during the summer and autumn seasons when the ambient air temperature varies from 28 to 45 °C. The biogas can be used as an energy carried as the obtained organic fertilizer is suitable for agriculture purposes.

1 Introduction

Owners of small farms face the problem of manure utilization on yearly basis. One way of efficient utilization is to combine several neighboring farms and use eco-technologies that allow the extraction of biogas from manure [1]. The production of biofuels from various natural sources starts in the late 18th century and developed successfully in the 19th century. The first bioenergy plants were established in the 1940s. Manure biogas plants are most common in India, China, and Japan. Biotechnology has long been used in many European countries such as Germany, Austria, Denmark, Sweden, and others [2-3]. Due to the significant negative environmental impact because of the worldwide fast growing

* Corresponding author: pzlateva@tu-varna.bg

industry and increasing energy price, it is necessary to switch to alternative sources of energy and heat production [4].

The main part of the biogas plant is the fermenters. This is the place where the well-homogenized fertilizer mass is heated to a temperature of 30 up to 58 °C. Depending on the temperature, mesophilic or thermophilic fermentation takes place. Most biogas and the highest percentage of methane gas is produced at a temperature between 45 and 55 °C. The biogas released from the fertilizer mass has a different composition depending on the percentage of dry matter, the content of carbohydrates, proteins, and the temperature of the fermentation process [5-6]. Biogas production requires maintaining the homogeneity and constant density of the substrate in the fermenters. This is done through stirrers, that prevent the formation of crust in the fermenter, which further could prevent the release of gas. It is especially important to maintain the homogenous temperature of the substrate in the fermenter. For this purpose, heat exchangers are installed on the walls and the floor, where hot water heats up the substrate.

The substrate in the fermenter, in addition to liquid and solid manure, may contain plant and animal waste, waste from slaughterhouses and dairies, distilleries and breweries, chopped straw or corn, chopped green mass, as well as many other food wastes [7]. In recent years, the produced biogas is burned in cogeneration units where both thermal and electric energy is produced. The electricity produced is used to cover the farm own needs as the rest is supplied to the network at a preferential feed in tariff. Biomass after the fermentation process is suitable for fertilization and has much better performance than ordinary manure [8-9]. Such plants are suitable for all types of farms (poultry farms, cow farms, sheep farms, pig farms, etc.). The use of biogas plants from manure has economic and social effect and leads to protection and enrichment of the environment [10-11].

Bulgaria has significant quantities of waste wood, which can be used both by direct combustion to produce heat and by gasification to produce synthetic gas. In addition, there is significant waste biomass from agriculture, which can be used for the production of biogas through fermentation plants. In the last 20 years, the focus of recovery has been large scale power plants with a capacity of 2 to 10 MW, which mainly use waste wood. In the last decade, in order to increase the energy efficiency of farms, small (local) power plants have been used for the recovery of vegetable and animal waste.

The work focuses on the experimental study of biogas production with animal and poultry manure from several neighboring farms with a small capacity of 70 cows, 6 500 chickens and 350 pigs.

2 Experimental set up

The process of decomposition of organic substances during anaerobic fermentation is strongly influenced by the characteristics of the substances such as composition, temperature, humidity, pH of the medium, stirring of the sludge, etc. This requires measuring the values of temperature, humidity and pH of the medium in the bioreactor, using calibrated measuring instruments. The measurements were made at three charges with substrate in the fermenter.

2.1 Preliminary set up for biogas production

The liquid manure is collected in a receiving tank, after which it is pumped to the bioreactor via a pipeline system. The fermenter itself is a hermetically sealed and thermally insulated tank where a constant temperature is maintained. It is equipped with a stirring system to help mix and homogenize the substrates both during loading and fermentation process. The substrates are heated via water type heat exchanger. This creates the proper conditions for microorganisms to start decomposing organic matter during their stay in an air-free environment (anaerobic environment) and their conversion into methane (CH_4), carbon dioxide (CO_2) and organic fertilizer. The fermentation time cycle is 23 days at temperatures of 33 up to 34 °C in mesophilic mode and 53 up to 55 °C in thermophilic mode. Since the measurements were performed during the summer and autumn seasons, when the external air temperature is from 33 ° C to 43 °C, no further heating of the substrate was required. The produced biogas has different concentration of methane, which depend mostly on the composition of the raw materials. As the concept of the installation was developed for the use of mainly liquid fertilizer, a methane content of about 62% can be expected.

The biogas produced by the wet fermentation method is burned in a cogeneration unit to produce simultaneous electricity and thermal energy. The heat produced is partly used for own needs and partly for heating purposes. An emergency gas flare is also installed to burn biogas during overproduction, as well as during maintenance and shutdown of the installation. The gas tank (gas storage), designed as a low pressure tank, serves for intermediate storage of biogas and to balance production fluctuations.

The substrate of the fermentation residues is separated by a solid and liquid phase separator. The liquid phase is stored intermediate in two open liquid manure storages and then exported to agricultural land. The solid phase is stored on an open paved area. Figure 1 is a representation of the proposed biogas plant.

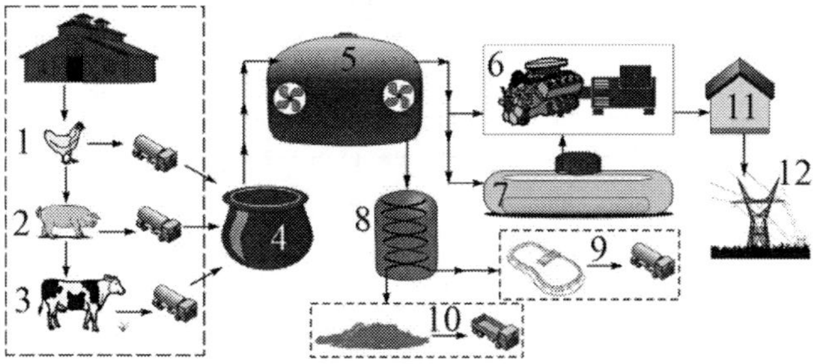

Fig. 1. Scheme of the biogas installation: 1 - poultry farm; 2 - pig farm; 3 - cow farm; 4 - receiving tank (liquid fertilizer); 5 - fermenter; 6 – cogeneration unit; 7 - gas storage; 8 - separator; 9 - liquid fertilizer; 10 - solid fertilizer; 11 - transformer; 12 - electrical network

2.2 Raw material specifics

The experimental studies were performed in a fermenter with the following raw waste materials: cow manure, pig manure and chicken manure. The quantities of the concerned raw waste in question are given in Table 1.

Table 1. Amount of the used inlet raw materials

	Raw materials			
	Animals	**Substrate amount**		
	number	m [kg]	m [kg/d]	m [kg/a]
Cow manure	70	49,00	3430	1 251 950
Chicken manure	6 500	0,18	1170	427 050
Pig manure	350	5,00	1750	638 750
Total	**6 920**	**54,18**	**6350**	**2 317 750**

The maximum quantities of raw waste for loading the fermenter are distributed and shown in Table 2. To calculate the parameters for liquid fertilizers from cows, pigs and chickens, the average dry matter content CB and organic dry matter content (ORS) is accepted. The results are shown in Table 2.

Table 2. The annual amount of manures and dry matter content

	Manures				
	Raw material			Dry material	
	[t/a]	[t/m^3]	[m^3/d]	TS [% d. FM]	OTS [% d. FM]
Cow manure	1 252	1,10	3,12	10 %	8,5%
Chicken manure	427	1,10	1,06	31%	26,35%
Pig manure	639	1,02	1,72	6%	5,1%
sum/average/t liquid	2 318	1,08	5,90	-	-

3 Results and Discussion

As a result of the on-site tests performed on the fermentation process, and also the tested fertilizers from cows, pigs and chickens, the following results for biogas (methane) production were obtained, shown in Table 3.

Table 3. Biogas (methane) production from animal waste

	Manures				
	Dry material			Methane production	
	TS [t/a]	OTS [t/a]	OTS [t/d]	[Nm^{3}CH$_4$/kgOTS]	[Nm^{3}CH$_4$/a]
Cow manure	125	106	0,29	0,22	23 411
Chicken manure	132	113	0,31	0,22	24 756
Pig manure	38	33	0,09	0,23	7 493
sum/average/t liquid	296	252	0,69	0,22	55 660

Table 4 is a representation of the total biogas production. A gas analyzer is used to determine the composition of the produced biogas. The results of the experimental study cover a period of one year.

Table 4. Total biogas production

Substratum	Methane	Energy	Power
	[Nm^{3}CH$_4$/a]	[kWh/a]	[kWel]
Total animal substratum	55 660	556 600	30

Values of the measured characteristics from several charges of the substances of the substrates in the fermenter for 23 up to 25 days are given in Table 5 as average values of 3 measurements per day.

Table 5. Substances characteristics in the feremnter

Fermenter	Type of the process	Temperature	Humidity	pH
		°C	%	
1 filling	Mesophilic	33	75	6,6
	Thermophilic	53	75	6,8
2 filling	Mesophilic	34	75	6,7
	Thermophilic	55	75	6,9
3 filling	Mesophilic	33	75	6,6
	Thermophilic	53	75	6,9

The analysis of the results shows good repeatability at the measured temperature, humidity and pH of the medium in the bioreactor. In the mesophilic process the temperature is about 33 °C, and in the thermophilic process it is about 53 °C. Humidity is 75% and pH 6.6 up to 6.9. For all feedings, the most stable and consistent biogas production was during the retention period. Biogas production started on the 8th day in the afternoon, gradually increasing and reaching its peak on the 15th day. There was no production on the 20th day. This continued for several more days until 23-25 days, when there was an unexpected production of biogas. It should be noted that all three studied fermenter charges contain poultry manure, which is responsible for the production of a larger volume of biogas.

Given the amount of animal material used, the annual amount of nitrogen generated can be divided as insignificant - 156 kg / year.

The gas analyzer is located in the cogenerator room so that after taking into account the H_2S readings the sulfur amount can be regulated at any time by supply of air via fan. In addition, this location provides good protection from the weather and good accessibility.

4 Conclusions

Based on the performed experimental studies in the fermenter of the biogas power plant with anaerobic fermentation of waste raw materials from cows, chickens and pigs and their analysis, the following outputs can be summarized:

- The on-site measurements were performed for one-year period. Gas analyzer is used to define the content of the produced biogas as flowmeter is installed for the amount of the produced biogas. By using analytical relations, the amount of nitrogen generated is equal to156 kg / a. Air fan is used to control the amount if air, and hence the total amount H_2S into the system.
- The measured substrate temperatures in the fermenter show good repeatability and the average values of the measurements in mesophilic regime is about 33 °C and about 53 °C in thermophilic regime during fermentation. The measured and averaged values of humidity and pH of the substrate in the fermenter during the fermentation process are 75% and 6,6 up to 6,9, respectively;
- The annual amount of methane produced by 2 318 t / a of raw material is 55 660 Nm^3CH_4 / a. The larger is the production of methane from chicken manure 24 756 m^3CH_4 / a and the lowest in pig manure 7 493 Nm^3CH_4 / a. Poultry manure is also responsible for the production of a larger volume of biogas;
- The obtained energy is 556 600 kWh / a and is used for cover own needs and partly for heating purposes in the farms. The produced fertilizer can be successfully used for agricultural purposes.

The investigations in that paper have been implemented with the financial support of National Program "Young scientist and postdoctoral students", funded by Bulgarian Ministry of Education.

References

1. T. Al Seadi et al., *Biogas Handbook* (Univ. of South. Denmark Esbjerg, 2008)
2. European Commission, *General and regional statistics* (Eurostat Database, 2020)
3. M. Mouftahi, N. Tlili, N. Hidouri et al., Processes, **9**, 48 (2021)
4. E. Sosnina, A. Shalukho, L.Veselov, Proc. of UralCon, art. no.9216233 (2020)
5. M. Temirbekova, M. Aliyarova, I. Iliev et. al., E3S Web of Conf., **180**, 02019 (2020)

6. B. G. Carp, G. Mocanu, I. Ion, F. Popescu, E3S Web of Conf., **207**, 02001 (2020)
7. E. Pop, L. Mihăescu et al., IOP Conf. Ser.: Mat. Sc. Eng., **595**, art. no. 012051 (2019)
8. D. Y. Damara, M. D. Dermawan, Proc. of ICT-PEP, 9249898 (2020)
9. Z. Kolev, S. Kadirova, E3S Web of Conf, **112**, 01008 (2019)
10. N. Komitov, N. Shopov, V. Rasheva, E3S Web of Conf., **207**, 02006 (2020)
11. G. S. Lakshmi, O. Rubanenko, G. Divya, V. Lavanya, INDISCON (2020)

Renewable energy as an integral part of the Bulgarian energy mix

Todorka Stankova[1,*], and *Daniela* Toneva[2]

[1,2] Technical University of Varna, Department Ecology and Environmental Protection, 9010 Varna, Studentska str., Bulgaria

Abstract. An economy with net-zero greenhouse gas emissions is the new EU 2050 is a main long– term strategic objective, which follows the EU RES promotion policy. Investment in renewable energy is a key factor for reduction of the conventional energy sources' carbon footprint. The transition from conventional to green energy is an on-going process. The renewable energy becomes an integral part from the energy mix of EU countries, including Bulgaria. Current research on RES component of Bulgarian energy mix covers the period from 2007 to 2019. It reveals the internal renewable energy subsector structure. The preconditions for solar and wind energy project are analysed, together with the levels of RES energy production in Bulgaria. The Bulgaria RES sector incoherence is highlighted. Some trends of RES development in the context of gross final energy production and consumption are revealed. Some gaps in Bulgarian practices regarding RES promotion policy implementation are identified and discussed.

1 Introduction

By 2050, Europe aims to become the world's first climate-neutral continent [1]. Renewable energy share in the total EU energy mix was doubled between 2004 and 2019 but at the same time the European environmental agency reported that around 90 % of city dwellers in Europe are exposed to pollutants at concentrations higher than the air quality levels deemed harmful to health [2].These facts unambiguously confirm that the anthropological impact on the environment develops faster than it can be counteracted. According to Eurostat over the past decade (2008-2018), the trend in primary energy production was generally negative for solid fossil fuels, oil, natural gas and nuclear energy. The production of natural gas and oil and petroleum products accounted for the biggest decreases (with 46.4 % and 35.3 % respectively) while solid fossil fuels production fell by 27.9 % [3]. However, there was a positive trend in the production of renewable energies (RE) over the same period (with an exception in 2011), with a 49.2 % increase, as well as for waste (non-renewable) with a 46.0 % increase [3,8]. At the same time the data shows that the exports are much lower than imports which make the EU still energy dependence.

* Corresponding author: stankova.todorka@gmail.com

2 Discussions

Due to the progress in renewable energy field Bulgaria declared that the 20/20/20 targets had been achieved and 18,6% is the share of the gross final energy consumption produced by renewable sources [4].The gross final energy consumption from renewable energy sources in 2017 and 2018 was 2,037.8ktoe and 2,226.1ktoe, and compared to the previous reporting period (2015-2016) it increased by an average of 9.2%. Important for the achieved increase is the larger amount of solid biomass used in 2018 for the production of heat and/or electricity in thermal power plants, as well as the increase in the use of hydropower. In 2018, the solid biomass used for energy production (398.5ktoe) is almost 9 times more than in 2017 (47.8ktoe) [4].

In the period 2017-2018 there is an increase in the installed capacity for the production of electricity from biomass, which in 2018 reached 195MW. Compared to 2017, the installed capacity of this type of power plants has increased almost 4 times. A change is also reported in hydropower plants (HPPs), where the change in 2018 compared to 2017 is mainly due to the commissioning of rehabilitated facilities. The new HPPs introduced in 2017 and 2018 are 5.2MW, respectively. In wind and photovoltaic power plants, the installed capacity is almost unchanged [4].

The production of electricity from renewable energy sources in the period 2017-2018 increased by 10.5%, which is due to the significant increase in electricity produced from solid biomass and in lower growth rate in electricity production from HPPs and HPPs [4].

Table 1. Calculation table for the renewable energy contribution of each sector to final energy consumption, ktoe

	2011	2012	2013	2014	2015	2016	2017	2018
Gross final consumption of RES for heating and cooling	1 043.3	1 122.0	1 142.5	1 087.3	1 146.6	1 203.0	1 229.2	1 345.2
Gross final consumption of electricity from RES	420.7	514.7	586.7	595.8	614.9	620.8	636.2	729.8
Gross final consumption of energy from RES in	6.0	5.1	105.7	114.9	149.9	168.6	172.4	151.1
Gross total RES consumption	1 470.0	1 641.8	1 834.9	1 798.0	1 911.5	1 992.4	2 037.8	2 226.1

A National energy plan and climate plan of Republic of Bulgaria for the period 2021-2030 was already accepted by the Ministry of Energy and the Ministry of Environment and Water. In order to achieve the set new national target of 27.09% share of renewable energy in the gross final energy consumption by 2030, existing as well as additional policies and measures will be applied [5]. As an appropriate form of support, the possibility of conducting tenders for the provision of capacity for the production of electricity from renewable energy sources and the provision of a supplement in the form of a premium to the market price for the electricity sold on the electricity market is also considered.

In the period 2020-2030, an increase of 2,645 MW in the net installed capacity of power plants using renewable energy is expected. This growth was accompanied by an increase in the net installed capacity of photovoltaic power plants (PV) by 2,174 MW and of wind power plants only by 249MW[5].The photovoltaic energy production was the most promoted in the country during the past years and is planned to be the most developed source till 2030. Noting that the electricity productivity per unit area compared by wind and solar energy is in favour of the wind energy, the planned till 2030 by the Bulgarian government 8,73 times more photovoltaic than wind energy is incomprehensible. Creating a map which provides information about the potential, the grid balance need and capacity can support the renewable energy sources development in complex, avoiding predominance of some sources over another and insuring grid balance in the production.

Table 2. Forecast – RES development in Bulgaria, 2020-2030, MW

Forecast for the net installed capacity of renewable power plants for the period 2020-2030, MW											
	2020	**2021**	**2022**	**2023**	**2024**	**2025**	**2026**	**2027**	**2028**	**2029**	**2030**
Hydro power plants	2508	2508	2508	2508	2508	2508	2508	2508	2508	2508	2508
Wind energy	699	709	719	729	739	749	788	828	868	908	948
PV	1042	1191	1339	1488	1636	1785	2071	2357	2643	2930	3126
Biomass	80	114	149	184	219	253	263	273	282	292	302
Total	**6349**	**6543**	**6737**	**6932**	**7126**	**7320**	**7656**	**7993**	**8329**	**8667**	**8914**

Another gab which is visible from the Bulgarian national action and climate plan is the obligatory of the EU due to coal and lignite Thermal Power Plants (TPP) cessation activity. Nowadays Bulgaria relay on 4555MW TPP installed capacity. The determination of the CO_2 emissions in the European Union is a role of both renewable and not renewable energy [6]. Till 2025 the Bulgarian government has to adopt an action plan of cancelling that conventional and pollutant energy production.

Table 3. Installed capacity in % by type of course, Bulgaria 2019

Type of power plant	**Installed capacity in %, Bulgaria 2019**	**Share in %**
Nuclear power plant	2000	15,7
Thermal power plant-lignite	4199	32,2
Termal power plant-coal	356	2,8
Termal power plant-gas	1235	9,7
Hydropower plant	3211	25,2
Wind power plant	701	5,5
Photovoltaic power plant	1059	8,3
Biomass power plant	77	0,6

Data of the Energy system operator (ESO) about the share in the energy mix in Bulgaria is provided in the following table. About 50 % of the electricity production is done by state own companies as the nuclear power station Kozloduy and some of the Thermal power plans (coal) and Hydro power plant (and pump hydro power plant). All renewable energy projects are private as well as some of the Thermal power plant.

According to the Bulgarian national action and climate plan by 2030 only 2645MW renewable energy sources are expected to be erected when 4555MW (from the thermal lignite and coal TPP) has to be cut off the grid. This make concerns about the energy security of the state at all. An adequate and immediate energy policy changes have to be adopted and implemented in order to insure the Bulgarian population and business with enough energy sources.

In 2017 and 2018, the Republic of Bulgaria achieved a share of energy from renewable sources equal to 18.7% and 20.49% of the gross final energy consumption in the country, respectively. The established values show exceeding the intermediate targets of the National Action Plan for Renewable Energy for the indicated years by 5 and 6.79 percentage points, and overfulfillment of the national mandatory target for 2020 of 16% share of renewable energy in gross final energy consumption.

The progress made is due to higher energy consumption from renewable sources in the heat and cooling and electricity sectors [4]. The state continually report the all wood used for household purposes as a biomass which doesn't respond to the sustainable criteria of paragraph 1 of Article 5 of the Directive, detailed reviewed in a previous science work of our team which accurse another concern about the real achieved share in the RES gross final energy consumption of the state [7].

In line with the EC's Third Liberalization Package [10], Bulgaria has taken steps towards full liberalization of the electricity market. As a result of legal changes from the beginning of 2018, the entire amount of electricity produced for the free market is traded only on the trading platforms of the IBEX. The liberalization of the electricity market is in force since 2007. The process was done step by step. Till October 2020 all industrial/non household consumers had to choose their own supplier with own price and till 2021 also all households will have the right to do it. Customers are not more obliged to choose the grid operator of their region of living as it was before. The Power Exchange Operational Rules (PEOR) regulate the organization of the activities, rights and obligations of the power exchange operator and all trading participants on the organized power exchange market for electricity. The Power exchange operator administrates three market segments- Intraday, Day-ahead and Bilateral contracts.

Pursuant to changes to the Energy Act published in State Gazette, 79/2019, the Energy and Water Regulatory Commission (EWRC) shall approve operational rules for organized power exchange market for electricity (art. 21, par. 1, item 42 EA). In this regard, on 19th June 2020, in issue 55 of State Gazette Power Exchange Operational Rules were approved for the first time by EWRC [5].

Wholesale baseload electricity prices 4-th quarter of 2019, Euro per MWh, Bulgaria 46,6Euro/MWh. Compared to Netherland, Germany, Belgium, France. The price in Bulgaria is on the top margin. For the whole 2019 the average price for day-ahead by IBEX on the base /1-24h/ was 47,37 €/MWh [3,9,11].

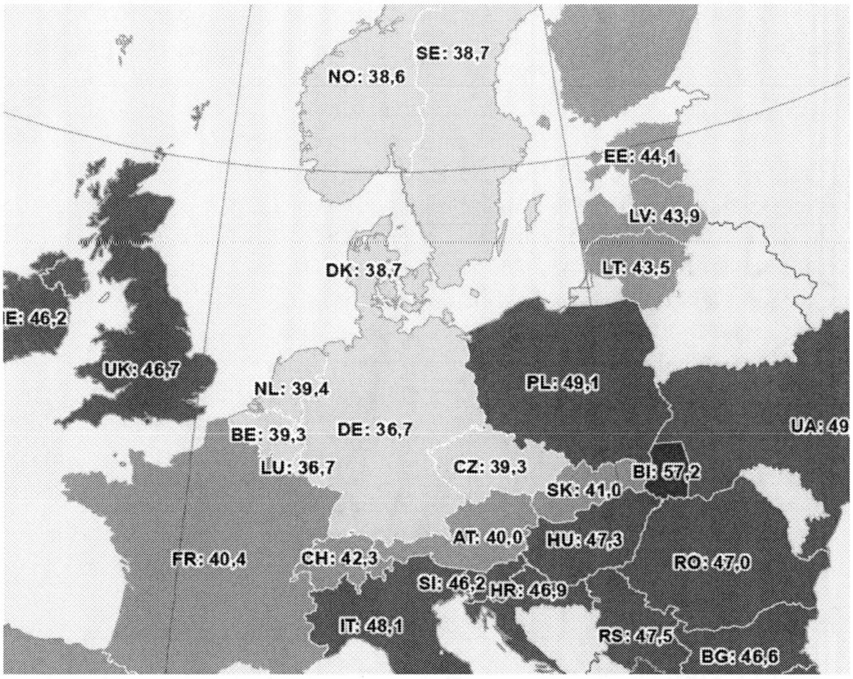

Fig. 1. Wholesale baseload electricity prices 4-th quarter of 2019, Euro per MWh.

Despite the fact that new renewable energy courses were installed during the past years, Bulgaria still have very high percentage of electricity energy dependency which is clearly visible from table 4. Having all the above mentioned fact the state still didn't run into operation the statistical transfer methods means that the Bulgarian end consumer is obliged to consume all RES energy and all bordering country can import only energy from the TPP or the Nuclear power plant [6,7]. This allows the government artificially to control the energy mix and the energy consumption.

Table 4. Energy dependence, Bulgaria 2019, %

	Energy dependency				**%**
	2014	**2015**	**2016**	**2017**	**2018**
Energy dependency	35,3	36,5	38,6	39,5	36,5

3 Conclusions

The photovoltaic energy production is the most promoted in the country. More than 70 % of the planned renewable energy installations have to use the solar irradiation as a source till 2030 [5]. There is a categorical imbalance in the forecast, noting the area interaction of solar compared to wind. If on 500sq.m plot of land a 3MW wind turbine could be installed, than for the same installed capacity of standing PV power plant approximately 60000sq.m will be needed together with service areas included.

From the Forecast in RES development in Bulgaria it is absolutely clear that even till 2030, the planned increase of 2 645MW by renewable sources isn't enough for the energy security of Bulgaria [5,6,7]. The option of change of the raw material coal or lignite with gas has a very low efficiency- up to 25 % and isn't an appropriate solution for investors, so 4555MW TPP installed capacity have to cut off the grid within less than 5 years. This is

35% of the whole energy mix in the country and the lack of an action plan of cancelling and replacing that conventional and pollutant energy production of the state make the energy security risk higher.

The possibility of the government for artificial control of the share in RES energy consumption by the Bulgarian citizens and entities by counting all household used woods in the part of biomass as well as the lack of statistical methods for transferring renewable energy to other counties confirms that major change in the energy policy is imperative.

The lack of balance between the renewable energy sources erection and the promotion of one above another is discriminatory and once more confirmed by the written in the 5-th RES progress report and the National action and climate plan, data.

All that mentioned above facts make concerns about the Bulgarian energy development market up to 2030 and the energy security especially after 2025 when the TPP have to cease their activity. Nevertheless the European Union have an action plan by 2050 which is missing in Bulgaria.

Part of the results of the present research were obtained under the National program "Young scientists and postdoctoral students", financed by the Ministry of Education and Science of Bulgaria.

References

1. The European Green Deal, communication from the commission COM (2019) 640 final, Brussel (2009) https://eur-lex.europa.eu/legal-content/EN/TXT/?qid=1596443911913&uri=CELEX:52019DC0640#document2
2. S. Angheluta, S. Burlacu, A. Diaconu, C.Curea, *European Journal of Sustainable Development*, Vol.**8**, №5; Special issue, p.57-65, (2021)
3. Eurostat, Renewable energy in the EU in 2018, news release **17**/2020-(2020) https://ec.europa.eu/eurostat/documents/2995521/10335438/8-23012020-AP-EN.pdf/292cf2e5-8870-4525-7ad7-188864ba0c29
4. *Fifth Renewable energy progress report*, European Commission, at all (2020)
5. *National energy and climate plan of Bulgaria* 2021-2030, p.68-70 (2020)
6. E. Dogan, F. Seker, Determinants of CO_2 emissions in the European Union: The role of renewable and non-renewable energy, *Renewable energy*, Vol. **94**, p.429-439 (2016)
7. T. Stankova, D.Toneva 2021 IOP Conf. Ser.: *Material Science and Engineering* 1032 012027 (2020)
8. J. Brodny, M.Tutak, Analyzing Similarities between the European Union Countries in Terms of the Structure and Volume of *Energy Production from Renewable Energy Sources, Energies*, Vol. **13**, Issue 4/10.3390, p.3-37 (2020)
9. M.Pacesila, S. Burcea, S. Colesca, Analysis of renewable energies in European Union, *Renewable and Sustainable Energy Reviews*, Vol.**56**, p.156-170, (2016)
10. P. Eikeland, The Third Internal Energy Market Package: New Power Relations among Member States, EU Institutions and Non-state Actors. *Journal of common market studies*, Vol.**49**, Issue 2, p. 243-263 (2011)
11. S. Djorup, J. Thellufsen, P. Sorknas, The electricity market in a renewable energy system, *Energy*, Vol. **162**, p. 148-157 (2018)

Improving the efficiency of crystalline silicon solar cell through regulating their temperature using thin films of polyvinyl alcohol

Ali Kadhim Naser [1,*], *Dhafer Manea.*Hachim[1], and *Qahtan A* Abed[2]

[1]Engineering Technical College / Najaf, Al-Furat Al-Awsat Technical University, 31001 Najaf, Iraq.
[2]Technical Institute/ Al-Rumaitha, Al - Furat Al- Awsat Technical University, 31001 Najaf, Iraq.

Abstract. Rising temperatures significantly affect the PV module, decreasing its voltage and lowering output power. Furthermore, temperature rises have been linked to several PV module failures or degradation modes. The purpose of this study analyzes polyvinyl alcohol PVA on crystalline silicon solar cells as a thermal insulation thin film. PVA thin films were prepared by dip-coating technique with a thickness of 1.15μm. The films exhibit suitable solar cell temperature controlling though it's an effect on masking the ultraviolet wavelength. The maximum temperature variation on the coating solar cell's surface was 4.5 °C as a comparison to a bare solar cell with irradiate exposure time 1800sec, and maximum efficiency obtained 18.99% in which Voc = 0.566 v and Isc = 330.2 mA when compared with bare solar cell 15.07% with benefit efficiency +3.92%. The concentrations of the polyvinyl alcohol influences on thin films and their effect on solar cell parameters are discussed.

Keywords: UV-mask, polycrystalline solar cells, surface temperature controlling system.

1 Introduction

Because of the rising cost of oil-based electricity, the termination of fossil fuel resources, and the contamination of the atmosphere, renewable energy sources have recently attracted a lot of interest. For these reasons, countries have been encouraged to replace fossil fuels with new energy sources, such as solar energy. Crystalline silicon solar cells are the earliest solar cell technologies to be developed, and it is still the most widely used type [1]. Via photovoltaic processes, Solar cells convert sunlight directly into electricity. In photovoltaic cells, semiconductors convert light to electricity since the semiconductor absorbs enough energy from light photons. The absorbed energy aids in the Excitation and transition of an electron from its valance to its conduction band; thus, an electron-hole couple is created (exactions) [2]. The generated electrical current is carried through a P/N junction in the solar cell. Consequently, only photons of the same to or beyond the bandgap energy are involved. When the photon's energy exceeds the bandgap energy, electrons are freed, and the excess is radiated as heat, elevating the solar cell temperature. Consequently, The Excitation of the semi-conduction of light electrons causes the generation of electrical current and heat in solar cells. You can achieve maximum

* Corresponding author: alikadhim558@gmail.com

power efficiency by consuming the entire solar spectrum within the band of that range without loss of power. In today's technological world, UV irradiation protection is a significant concern. The action of photonic energy in non-UV protective materials can cause undesired breakage of chemical structures, and it may be toxic to living organisms; Photo-aging of the skin can occur under intense ultraviolet light [3]. When it comes to UV radiation, it is usually divided into UV-A, UV-B, and UVC radiation, with wavelengths ranging from 315–380nm to 280–315nm and from 100–280nm, respectively. UVC light rays with energy levels of approximately 4.43-4.4eV cause more harm than UVA and UVB rays of approximately 3.94-4.4.43eV. Ultra-violet-blocking substances are thus required to help in reducing damage. Via strengthening inorganic fillers into a transparent polymer matrix, these materials can be customized to achieve a bandgap in the desired region that may help block UV rays even further [4]. Solutions or fine powder polymeric films are needed to make realistic use of these materials. Since reinforced polymers have the added advantage of being UV-protective shields, inorganic nanoparticles (which can cut UV absorption) can be incorporated into a polymer. (Enhancing their properties), rendering them significantly more effective in day-to-day life. A small amount of inorganic matter can improve the polymer matrix's versatility in thermo-mechanical, optoelectronic, and mechanical properties [5].

The cost efficiency and biocompatibility of polyvinyl alcohol (PVA) have made it an appealing alternative to other polymer film-forming additives. PVA's chemical formula is [CH_2-CH (OH)] n, with n indicating the degree of polymerization [6].

Aslam M, Kalyar MA study the effect and investigate the impact of an organic semiconductor. Polymer –polyvinyl-alcohol PVA has been widely used for Ultraviolet protection due to its characteristic nature, i.e., UV scattering and absorbing nature, broad-band UV absorption, and excellent stability against high temperature and Ultraviolet rays [7]. Polyvinyl alcohol is a renowned semi crystalline polymer that comes from the OH group and hydrogen bonding feature. It has a high aspect ratio, a high ultimate tensile power, a high modulus of elasticity, and is chemically compatible [8]. Polymers have the property of absorbing ultraviolet-B radiation in general of solar photons [9]. Polymeric nanocomposites produced from polyvinyl alcohol) (PVA) have become a prime candidate for hydroxylation, which permits the usage of polymerization through hydrogen bonding to Nano well composition to the molecular level because they possess a vast number of hydroxyl groups adjoined to their carbon backbone [10,11]. The structural, transparent, thermally stable , optical, UV shielding , mechanical and good forming film characteristics of polyvinyl alcohol make it a good insulation film and fit for different application in general and solar cells in especial [12,13].

PVA thin films were prepared on a silicon substrate in this work using a dip-coating technique. UV-Vis Spectroscopy, solar cell analyzer module, and thermo-couples K-type with data-logger PC-link were used for optical properties and thermal studies. These films are excellent UV-BLOCK and thermal insulators for solar cells, and they extend the life of the solar cell.

2 Experimental set-up

The 4 cm × 2cm polycrystalline silicon solar cells used in this work illustrated in Figure 1; the samples first washed with ethanol used ultrasonic water bath type for 1 hour.

PVA (polyvinyl alcohol) had a polymerization frequency of 1700-1800, which was purchased from Sigma-Aldrich with molecular weight (Mw) from 2000 to 200000 g/mol. The solubility of PVA in water is influenced by the degree of hydrolysis and polymerization. The molecular weight of a polymer increases as the degree of polymerization rises, and it has been shown that the solubility of a polymer reduces as the molecular weight rises at a given temperature [12].

Sodium nitrate (Nano3) and Sodium dihydrogen phosphate (NaH_2PO_4) with concentrations 0.2M and 0.01M respectively with a limited ratio of distillation water (DI) were used as a solvent to PVA at room temperature.

Fig .1. Mc-Si (4cm × 2cm) solar cell.

3 Experimental procedures

The 125ml distillation water (DI), 2.125 g (NaNo3), and 0.145 g (NaH_2 PO_4) were agitated in a magnetic stirrer for 1hr to get the best polymer-solvent at room temperature. Many concentrations of polyvinyl alcohol (1.5g, 1 g, 0.75g, 0.5g, and 0.125g) for every 25ml of the solvent were prepared prior. The mixture stays on a magnetic stirrer 300 RPM for 3hr at room temperature. Leave the solution to sit until foaming is finished and sealed tightly to avoid bacteria formation. Repeat the procedure five times to prepare all samples.

Deposition of the PVA GEL as a film on Si substrate via the dip-coating technique is one of the best and easiest techniques to control obtaining a uniform thin film. Industrial blower to dyer the sample used for 10-15 min at 1500 °C.

Fig.2. 125ml distillation water+ Poly vinyl alcohol+ Nano3 + NaH2PO4.

3.1 Thin-film thickness measurement

The most common approach is using an electronic balance. The substrates were weighed before and after deposition. Differences in substrate thicknesses make up the dissimilarity. Thinness has been accurately measured using the new equation [$t=(\Delta m/\rho.a.b)$] [13]. Where (a) and (b) are the substrate dimensions, Δm is the substrate's mass variance, ρ is the material's density, and (t) is the thickness of thin-film deposited.

Table .1. Substrate's dimensions, PVA density, and mass' variance

Density (PVA)	1.19.g/cm3
a: substrate's length	4 cm
b: substrate's width	2 cm
Δm: substrate's mass variance prior and after deposition	0.0011 gram

After applying the substrate's parameters, the PVA thin-film thickness was measured to be (1.15μm). The other approach to measuring thin-film thickness is; use an (INSIZE) micrometre to measure the thickness of the prepared films. The less thickness that can measure is 1μm. The device is calibrated before Initiation to measure the thickness of the layer. By measured before and after the coating process and by knowing the difference between the two measurements, Δms were obtained. Several readings of several positions are taken on the base's surface to get a more accurate thickness and take the average for those readings. The average thin-film thickness obtained was (1.156 μm).

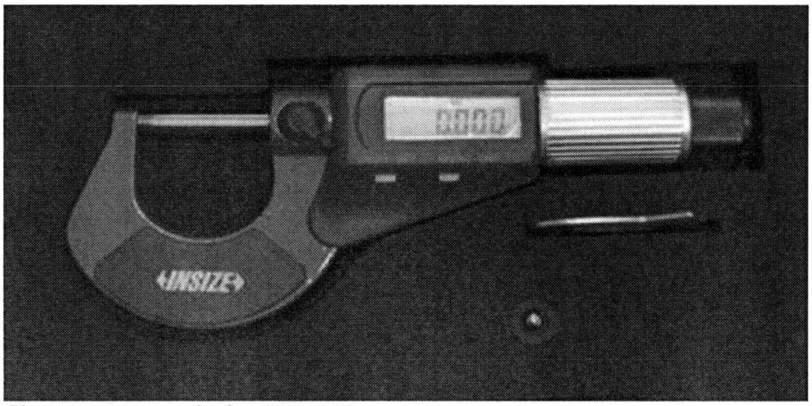

Fig.3. thin-film measurement device (INSIZE).

4 Results and Discussion

4.1 Optical characteristics

As the simplest component of matter, the atom or molecule has energy levels. The atom's electrons are in a stable state at room temperature within a level called the lower energy level. Electron's Excitation and this transition of the atom to high energy level are associated with whether the incident photon's energy is equal to the difference in energy between the two levels (the lower and the excitation levels) [14]. If the rays fall on a transparent part, part of this incident rays passes out from the other side. An absorption ray is a difference between incident and transmitted rays. The atoms absorb a specific spectrum of UV-Vis. The higher the concentration quantity represents, the more significant the absorption intensity. Figure 4 shows the absorption spectrum of a polymer with different concentrations of VS intensities. PVA has been presented as a promising optoelectronic candidate. Optical properties studies by UV-Vis's spectroscope indicated the absorption broad in the range 250-350 nm, as shown in Figure 4. Absorption intensity increase with increasing the PVA concentrations maximum peak occurs at 0.2 g at 300nm. The conclusion from the data is clear: there is no absorption in the visible region in PVA film. PVA can absorb UV light, transmit visible light, and filter out UV light.

Polymer PVA is an Organic semiconductor with a wide energy gap of 3 eV. Molecules are bound to each other by weak van der Waals forces, whereas conjugated π-bonds bond atoms within an organic semiconductor. This molecular structure gives organic semiconductors their flexibility, lightweight, and low sublimation point. The HOMO and LUMO molecular orbitals are like the valence band and conduction band, respectively [3].

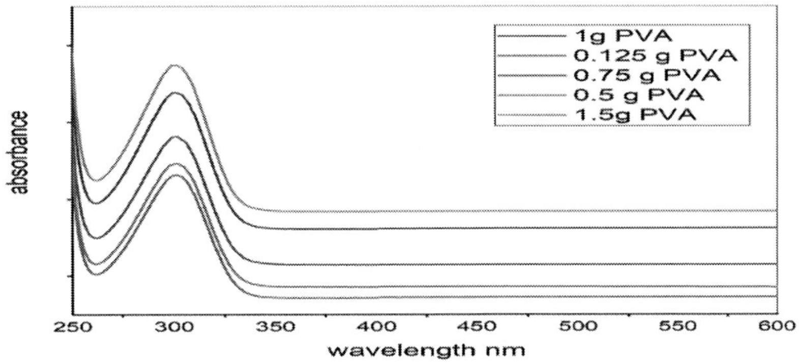

Fig. 4. UV-Vis represent the absorption of PVA for different concentrations.

4.2. Thermal studies and temperature effect

The temperature has a significant effect on solar cells. This is evidence of the loss of photovoltaic (PV) efficiency and energy yield because of the rise in panel temperature [16]. The Solar cells operating in harsh sunlight will experience a panel temperature of 50–70°C, which will cause a reduction in efficiency of 10–25 percent [17].

The molecule contains, in addition to levels of electronic energy; there are levels of vibrational energy and rotation. As a result of the molecular bonding with chemical bonds, the absorbed energy leads to the vibration of these bonds and thus the vibration of the atoms concerning each other. As for the rotational energy levels, they appear due to the rotation of the molecule around its axis. These transitions are closely related to the intensity and energy of the incident rays, wherein in the ultraviolet region, all the aforementioned transitions appear. The excited electron returns to a stable state due to the loss of energy that is absorbed in the form of emitted radiation or lost as a result of a collision of atoms with each other, which results in a rise in temperature [15,18].

This work used a data logger and thermo-couples K-types and halogen lamp with normal incident rays, its intensity 1000 W /m^2 for 1800 secs. It is clear from the figure below Fig.5 that the cell with a higher concentration has a large temperature variation than a bare solar cell because of the UV wavelength effect, which has high frequency, high energy. The excess energy from raising the electron from the valence beam to the conduction beam and the flow of current would be dissipated in the form of heat, causing the cell surface temperature to rise.

Table.2. Temperature regulation with time for mc-Si coating and without coating solar cells.

samples	Concentrations	Temperature variations	Time
S1	Bare	-----	1800 sec
S2	1.5 g	4.5 °C	1800 sec
S3	1 g	3.8 °C	1800 sec
S4	0.75 g	2.3 °C	1800 sec
S5	0.5 g	1.6 °C	1800 sec
S6	0.125 g	0.9 °C	1800 sec

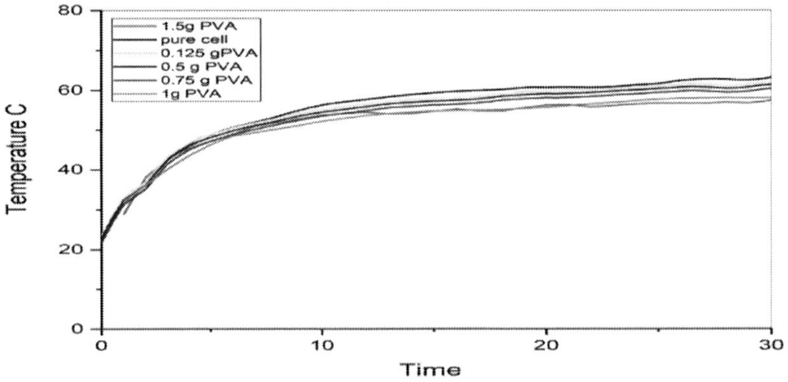

Fig.5. temperature versus time curves for bare and coating cells with different concentrations.

4.3 I-V curve characteristics

One of the most significant parameters affecting solar cell efficiency is rising temperature. Because Increasing the operating temperature of the PV module has significantly impacted the PV module by reducing its voltage [14,18]. The experimental data in table.3 was extracted from the solar cell module type (PROVA-200A); after the end time test 1800 secs, and reach the solar cells steady-state condition, all the solar cells have various temperatures.

This data observes the impact of temperature on the efficiency of solar cells. The lower efficiency was for a pure solar cell with a high-temperature surface which was 15.07 %, and the best efficiency was obtained with a lower temperature surface and the best concentration of polyvinyl alcohol, which was 18.99%. Fig.6 show the I-V characteristics curves for crystalline silicon solar cells. Observe the effect of different PVA concentrations on the solar cell's parameters .that test is done after the solar cell reaches the steady-state.

Table.3. I-V characteristics for mc-Si solar cells with and without coating,

samples	Concentration	Voc (V)	I_{SC}(mA)	FF	η (%)
S1	Pure	0.545	279.8	0.692	15.07 %
S2	1g (PVA)	0.533	376.8	0.633	18.16
S3	1.5 g (PVA)	0.566	330.2	0.711	18.99%
S4	0.75g(PVA)	0.556	321.5	0.702	17.94%
S5	0.5g(PVA)	0.592	0.592	0.535	17.18%
S6	0.125g(PVA)	0.556	305.4	0.670	16.26%

Table.3. I-V characteristics for mc-Si solar cells with and without coating,

samples	Concentration	Voc (V)	I_{SC}(mA)	FF	η (%)
S1	Pure	0.545	279.8	0.692	15.07 %
S2	1g (PVA)	0.533	376.8	0.633	18.16
S3	1.5 g (PVA)	0.566	330.2	0.711	18.99%
S4	0.75g(PVA)	0.556	321.5	0.702	17.94%
S5	0.5g(PVA)	0.592	0.592	0.535	17.18%
S6	0.125g(PVA)	0.556	305.4	0.670	16.26%

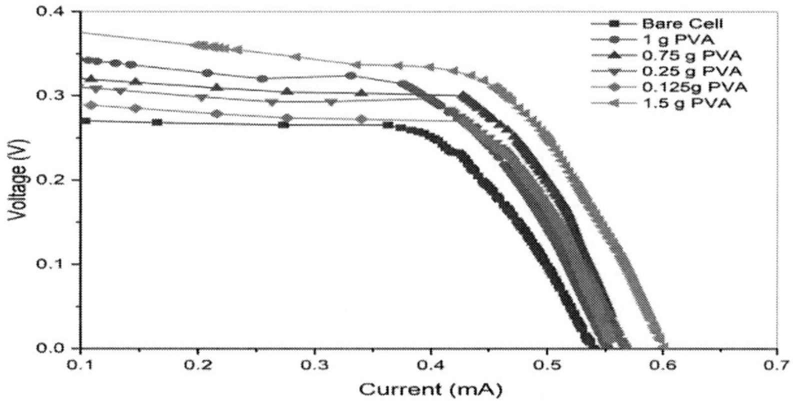

Fig.6. I-V characteristics for Bare and coating solar cells.

5 Conclusions

In this research, polyvinyl-alcohol was used as a coating thin film on silicon substrate by dip-coating technique. The result obtained was these thin films made as ultraviolet masking and, thus, it made an excellent solar cell's temperature regulation. Because of the apparent effect of ultraviolet rays on the temperature rise of the solar cell's surface, the photon's energy within this wavelength range of the ultraviolet rays possesses high frequencies with high photon energy that exceeds the mc-si cell's energy gap. Thus this difference in energy harms raising the temperature of the solar cell. Temperature controlling was positively influenced the improvement of the solar cell's performance. Because increasing the operating temperature of the PV module has significantly impacted the PV module by reducing its voltage. 1.5 g was the best concentration of the polymer polyvinyl-alcohol to creating the best thin-film as so on. The best temperature difference was 4.5 °C between the coating and un-coting solar cells.

References

1. E. Radziemska, Renewable Energy **28**, 1 (2003)
2. A. Mohammad Bagher, American Journal of Optics and Photonics **3**, 94 (2015)
3. R. P. Chahal, S. Mahendia, A. K. Tomar, and S. Kumar, Applied Surface Science **343**, 160 (2015)
4. H. Takele, S. Jebril, T. Strunskus, V. Zaporojchenko, R. Adelung, and F. Faupel, Applied Physics A **92**, 345 (2008)
5. V. Kaler, U. Pandel, and R. K. Duchaniya, in *Materials Today: Proceedings* (2018), pp. 6279–6287
6. R. Singh, S. G. Kulkarni, and S. S. Channe, Polymer Bulletin **70**, 1251 (2013)
7. M. Aslam, M. A. Kalyar, and Z. A. Raza, Polymer Engineering & Science **58**, 2119 (2018)
8. H. Karami and A. Mohammadi, Int. J. Electrochem. Sci **10**, 7392 (2015)
9. M. T. Ramesan, P. Jayakrishnan, T. Anilkumar, and G. Mathew, Journal of Materials Science: Materials in Electronics **29**, 1992 (2018)
10. M. T. Ramesan, P. Jayakrishnan, T. K. Manojkumar, and G. Mathew, Materials Research Express **5**, (2018)
11. T. A. M.T. Ramesan, Chippy Jose, P. Jayakrishnan, Polymer Composites **39**, 38 (2018)
12. S. Mahendia, A. K. Tomar, R. P. Chahal, P. Goyal, and S. Kumar, Journal of Physics D: Applied Physics **44**, 205105 (2011)
13. S. S. Samanta, M. S. Shinde, and R. S. Patil, **57**, 53 (2016)
14. K. Rakesh Tej Kumar, M. Ramakrishna, and G. Durga Sukumar, International Journal of Energy Research **42**, 2305 (2018)
15. D. M. Hachim, Q. A. Abed, and F. Kamil, Energy Sources, Part A: Recovery, Utilization, and Environmental Effects **1**, 1 (2020)
16. Q. A. H. Al-naser, N. M. A. Al-barghoothi, and N. A. S. Al-ali, International Journal of Engineering , Business and Enterprise Applications (IJEBEA) 108 (2013)
17. P. Singh and N. M. Ravindra, Solar Energy Materials and Solar Cells **101**, 36 (2012)
18. S. Chander, A. Purohit, A. Sharma, Arvind, S. P. Nehra, and M. S. Dhaka, Energy Reports **1**, 104 (2015)

Experimental researches on poultry manure combustion in co-combustion with biomass

Gheorghe Lăzăroiu[1]*, *Gabriel Paul* Negreanu[2], *Ionel* Pîșă[2], *Rodica Manuela* Grigoriu[1], and *Dana-Alexandra* Ciupăgeanu[1]

[1] Power Engineering Faculty, University Politehnica of Bucharest, Splaiul Independentei 313, 060042 Bucharest (Romania)
[2] Mechanics and Mechatronics Faculty, University Politehnica of Bucharest, Splaiul Independentei 313, 060042 Bucharest (Romania)

Abstract. Combustion of pure avian waste is strongly affected by its humidity content. According to the results of composition analysis, the initial humidity reaches up to 50%, inhibiting direct combustion initiation and development. Drying of poultry waste is an alternative, but simple relatively long-term storage or thermal pre-processing, complicates the technological process of energy recovery and increases the associated costs. Co-combustion represents a promising solution to enrich the heating value of raw poultry manure. Experiments using biomass (in a mass fraction of up to 30%) led to positive results in terms of efficiency. It is highlighted that the process depends on the quality of the woody biomass used for the mixture, but also on its bulk density, which causes an accelerated diffusion of air and influences the burning speed. This paper presents the experimental investigations on the co-combustion of poultry manure and woody biomass, performed on a 55 kW pilot boiler equipped with a post-combustion grate. The focus of the analysis is on the influence of the biomass bulk density and its heating value on the co-combustion process. The results obtained favor the development of a technology that is easy to apply and has a reduced cost. The technology investigated here is suitable for onsite applications in poultry farms, enabling meeting the energy demand based on co-combustion of resulting poultry waste.

1 Introduction

For the past decades, research interest in the energy field focuses on harmful emissions abatement and renewable energy technologies development and integration [1], [2]. It is remarked as, according to statistics released by the International Renewable Energy Agency, renewable energy sources (RES) almost doubled their cumulated generating capacity over the last decade [3].

One important but still less exploited RES is represented by the renewable waste [4]. For instance, waste mixtures resulting from bird houses operating cycles are characterized by increased soil restoration potential, so an alternative for their management is by spreading them as fertilizers [5], [6]. However, limitations regarding nitrogen content in agricultural areas imposed by 91/676/CEE Directive emphasize the need to intensively consider other management approaches such as treatments with energy recovery potential [7]. Combustion of poultry manure on site of poultry farms can be a solution for heating the halls, even to a certain extent, together with another fuel [8], [9].

The research developed in this paper aims to evaluate the efficiency of poultry manure co-combustion in combination with different types of solid biomass, such as forestry or agricultural. The approach presented meets also the objective of reducing the consumption of hydrocarbons (liquid or gaseous), according to the Paris Agreement. Specifically, the research refers only to the combustion of wet poultry manure, thus reducing the drying phase, usually naturally (which may be a polluting source by prolonged exposure to soil). The energy valorization of dry poultry manure has been previous research and achievements [4], [5], [10]. A mass participation of a maximum of 30% for poultry waste, which is fed above the burning biomass layer (only the fixed layer combustion technology for solid biomass) is envisaged.

The rest of the paper is structured as follows. Section 2 details the energy characteristics of poultry manure and the biomass which serves as thermal support.

* Corresponding author: glazaroiu@yahoo.com

2 Energy characteristics of poultry manure

Solid biomass is a category of fuels characterized by a high volatile content, which involves a high ignition capacity, a low ash content, a very high oxygen content and a strong dependence of calorific value on moisture content [11], [12]. Usually, the humidity is required to be below 30%, in which case the calorific value becomes higher than 12 000 kJ/kg, superior to lignite [13].

In the present research, the solid biomass considered is represented by wood waste, while the agricultural biomass included vine ropes and tree branches.

The energy characteristics are assessed based on the elementary analysis, which includes the mass participations for carbon (C), hydrogen (H), sulfur (S), oxygen (O), nitrogen (N), water (W) and ash (A), according to eq. (1). The index *i* refers to the initial state, and the humidity is the total (W_t^i).

$$C_i + Hi + S_c^i + Oi + N_i + W_t^i + A_i = 100 \quad (1)$$

The calorific value can be calculated based on the elementary analysis. Flue gas emissions and air demand are determined by the stoichiometric relationships of combustion reactions. If only the lower calorific value H_i^i is in MJ/kg, the carbon content C, employing eqs. (2-4):

$$C = 0.0548929 + 0.023736\, H_i^i \quad [kg/kg] \quad (2)$$

$$V_g^o = 1.57933 + 0.217974\, H_i^i \quad [m_N^3/kg] \quad (3)$$

$$V_{CO_2} = 0.101619 + 0.043991 H_i^i \quad [m_N^3/kg] \quad (4)$$

Table 1 presents the energy characteristics of the biomass considered for the experimental tests.

Table 1. Energy characteristics of some solid biomass varieties.

Biomass	**Ci**	**H_i**	**O_i**	**N_i**	**W_t^i**	**A_i**	**H_i^i**
	%	%	%	%	%	%	MJ/kg
Dry wood	46.7	5.1	33.6	0.6	10.5	3.5	17.15
Wood with medium humidity	32.9	4.6	28.6	0.7	30	3.2	12.01
Sunflower stems	37÷42	4.1÷5.2	35÷40	1.2÷1.7	11.1÷17.1	3.3÷5.5	13.6÷15.2
Branches	36.7	5.2	30.8	0.6	25.2	1.5	13.80
Vine ropes	40÷43	3.7÷4.9	35÷42	0.6÷2.5	10÷26	2.5÷4.5	14.7÷15.8
Corn cobs	36÷39	3.6÷5.1	35÷44	1.4÷3.8	11.5÷17.1	3.7÷5.7	13.2÷15.0

In the experimental investigations, the poultry manure is pure or mixed with solid biomass, according to the operating conditions in the breeding halls.

The elementary analysis for biomass bed manure resulted within the ranges:

$C_i = 12.4 \div 22.6$ %
$H_i = 4.4 \div 5.4$ %
$S_c^i = 1.8 \div 2.1$ %
$O_i = 34.8 \div 37.6$ %
$N_i = 1.4 \div 2.4$ %
$A_i = 6.2 \div 12.8$ %
$W_t^i = 33.8 \div 38.8$

Subsequently, the lower calorific value is equal to $H_i^i = 3\,840 \div 5\,950$ kJ/kg

The elementary analysis of pure manure resulted within the ranges:

$C_i = 12.4 \div 22.6$ %
$H_i = 4.4 \div 5.4$ %
$S_c^i = 1.8 \div 2.1$ %
$O_i = 34.8 \div 37.6$ %
$N_i = 1.4 \div 2.4$ %
$A_i = 6.2 \div 12.8$ %
$W_t^i = 33.8 \div 38.8$

The lower calorific value is evaluated in this case at $H_i^i = 3\ 750 \div 5\ 870$ kJ/kg. It is noticeable that these values are slightly lower (less than 2%) than the ones calculated in the previous case.

It is remarked that, according to the lower calorific value for poultry manure, it is required their combustion with a thermal support. In the present research, the thermal support comes from biomass, which has a calorific value about three times higher.

3 Solid fuels combustion in fixed layer

Fixed layer combustion is characterized by a concentration of it in a very small volume compared to the general size of the hearth. The result is a difficult penetration of air into the fuel layer, which leads to an excessive increase in pyrolysis - gasification processes, the resulting combustible gases continuing to burn in the rest of the hearth area [14]. Consequently, the overall combustion process depends outside the energy characteristics of the fuel and the porosity of the layer (the volume of voids in the fuel layer) [11].

However, the porosity responds positively to particles of a shape that can be approximated with a sphere and a strict particle size distribution. The higher the granulation, the lower the porosity of the fuel layer, as follows (the porosity was denoted by m).

- Particle size 2 ÷ 25mm, m = 0.46 ÷ 0.48
- Particle size 25 ÷ 50mm, m = 0.42 ÷ 0.45

For biomass with shapes and sizes which are very far from spherical in shape (such as branches, vines, wood in the form of perches), it is proposed to replace the air intake criterion represented by porosity with the criterion "fuel layer density", ρ_c in kg/m^3.

An oxygen-deficient fuel layer leads to a two-stages combustion mechanism, the first stage comprising the phase of gasification pyrolysis, with strong CO emission, the second stage leading to the final oxidation, with the formation of CO_2 [15]. Figure 1 shows the physical model for two-stage combustion, the avian waste being attached to the second stage of combustion.

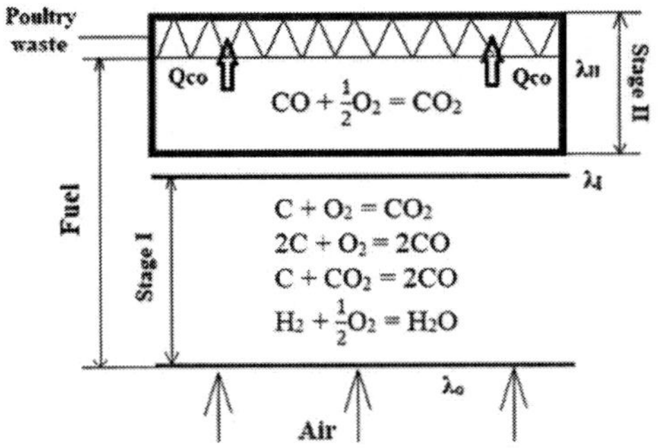

Fig. 1. Combustion scheme in a fixed layer of fuel with low porosity (high density), with two combustion steps

In stage I, the reactions take place:

$$C + O_2 = CO_2 \qquad (5)$$
$$2C + O_2 = 2CO$$
$$C + CO_2 = 2CO$$
$$H2 + \frac{1}{2}O_2 = H_2O$$

The reactions with formation of CO are a consequence of the air deficit. It can be considered that at the end of the first combustion stage the excess air is λI ≈ 0.5λo (λo being the total excess air for the combustion in the layer). As for the combustion in the layer, the excess air λo ≈ 2, we can consider λI = 1 (stoichiometric combustion). An estimative calculation for biomass having the lower calorific value equal to H_i^i = 14MJ/kg, indicated the following combustion results:

- The volume of CO_2 will be:

$$V_{CO_2} = 0.5 \cdot (0.101619 + 0.043991 \cdot 14) = 0.358 \quad [m_N^3/kg] \tag{6}$$

- Volume of stoichiometric flue gas (λ = 1) emitted by burning the entire quantity of fuel:

$$V_g^o = 1.57133 + 0.217974 \cdot 14 = 4.63 \quad [m_N^3/kg] \tag{7}$$

- The concentration of CO_2 in the stoichiometric volume of carbon gas is:

$$CO_2 = \frac{V_{CO_2}}{V_g^o} = 0.077 \frac{m_N^3}{m_N^3} = 7.7\% \tag{8}$$

- The excess air λo allowed will be self-distributed to the second combustion stage. The concentration of CO in the second stage depends on the degree of oxidation of carbon to CO_2, and analytically for 1 kg of carbon is represented by the relation:

$$CO = 34.7 - 1.65CO_2 \quad [\%] \tag{9}$$

- The heat released by burning CO in the second combustion stage will be:

$$Qco = \frac{CO}{100} \cdot H_{i,CO}^i \; [\frac{kJ}{m_N^3}] \tag{10}$$

where $H_{i,CO}^i$ is the lower calorific value of CO, resulting in:

$$Qco = \frac{34.7-1.65CO_2}{100} \cdot H_{i,CO}^i \quad [\frac{kJ}{m_N^3}] \tag{11}$$

- For a fuel with a C_i/100 carbon content, the equation becomes:

$$Q_{CO}^r = \frac{C^i}{100} \cdot \frac{34.7 - 1.65CO_2}{100} \cdot H_{i,CO}^i \qquad [\frac{kJ}{m_N^3}], \tag{12}$$

Q_{CO}^r represents the actual amount of CO released when a solid fuel is burned.

The low heat input released by the second-stage combustion of CO, to support the combustion of poultry waste, leads to heavy ignition and entry into a gasification regime, with an excessively large extension of the total combustion time.

For the concentration of CO_2 = 7.7% achieved in combustion stage I, the heat released by burning CO will be equal to:

$$Q_{CO}^r = (0.0548929 + 0.023736H_i^i)\,\frac{34.7 - 1.65 \cdot 7.7}{100} \cdot 12644 = 995.8\,\frac{kJ}{m_N^3}$$

This heat obtained by combustion of CO and is released at the same time as the combustion of the upper layer of poultry waste, but representing an insufficient thermal support. Thus, stoichiometrically, for the biomass considered having the lower calorific value equal to H_i^i = 14MJ/kg, the volume of flue gases having the value of $4.63 m_N^3/kg$, the total heat accumulated by combustion will be 3 023.7 kJ/m_N^3. The thermal participation of CO combustion in the second combustion stage results $\frac{995.8}{3023.7}$ = 0.32 = 32%, supporting the combustion of poultry manure.

The second stage combustion solution involves a low density of fuel in the layer (ρ_c < 250kg/m_N^3). If the correspondence with the porosity coefficient m is considered by the relation:

$$m \approx \frac{\rho_c}{\rho_o} \tag{13}$$

where ρ_o= 1 000 kg/m^3 is the absolute density for biomass, the porosity results in this case less than the value m = 0.25 enabling an easy penetration of air. The physical model of combustion is shown in Figure 2.

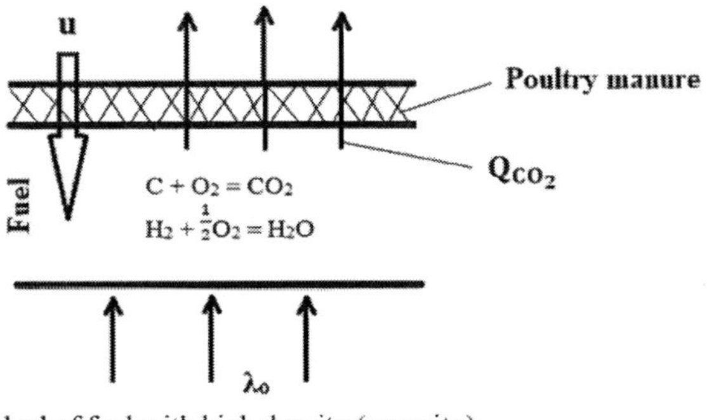

Fig. 2. Combustion in a fixed bad of fuel with high density (porosity).

The combustion rate u, in m/s, represents the rate of decrease of the fuel layer height. The higher it is, the more the heat released by combustion is transmitted directly to the upper poultry manure layer.

The combustion rate is determined by the lower calorific value H_i^i of the fuel and the density of the fuel layer ρ_c, as given in eq. (14):

$$u = f(H_i^i, \rho_c) \qquad [m/s] \tag{14}$$

Based on this observation, the law of heterogeneous combustion velocities q, can be expressed as in eq. (15):

$$q = \frac{C_o}{\frac{1}{k}+\frac{r_o}{D}} \qquad [m/s] \tag{15}$$

As a physical interpretation, the lower calorific value is attached to the combustion rate constant k [m/s] and the molecular diffusion density D [m^2/s]. With r_o [m] was noted the radius of the fuel particle and C_o represents the initial concentration of the oxidizing medium. The corresponding variation of the combustion rate u results, compared to the combustion parameters, H_i^i and ρ_c, is shown in Figure 3.

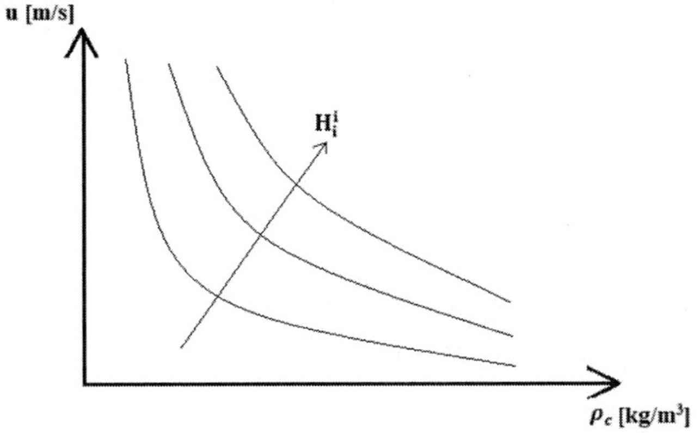

Fig. 3. Law of variation of combustion rate in a fixed layer of fuel

4 Experimental investigation on poultry manure – solid biomass co-combustion

The experiments conducted in this research aim to determine the efficiency of the co-combustion process depending both on the biomass quality and its bulk density. In reference to the moisture of the manure, it is highlighted that the investigations are conducted at very high humidity content of poultry manure. The combustion speed and the air diffusion behavior depending on the bulk density is observed.

4.1 Equipment and methodology of research

The experiments are performed at laboratory scale, employing a pilot installation comprising a boiler equipped with a fixed grill of 55 kW, set in the Thermotechnical Laboratory of the Faculty of Mechanics and Mechatronics, Polytechnic University of Bucharest.

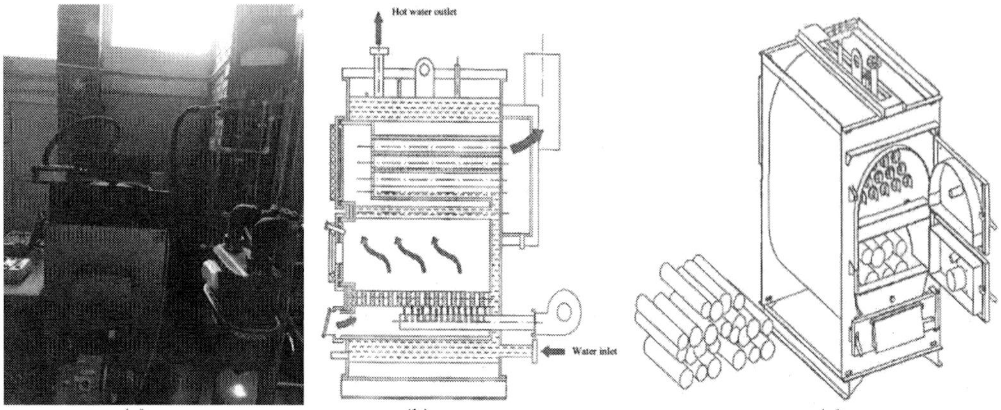

(a) (b) (c)

Fig. 4. Pilot installation, 55 kW boiler: (a) Overview; (b) Cross-sectional diagram; (c) Front diagram.

The 55 kW pilot boiler has the following dimensions of the hearth:

- length: L_{hearth} = 750 mm; width: l_{hearth} = 550 mm; height: h_{hearth} = 600 mm;
- hearth volume: V_{hearth} = 0.25 m^3;

The hearth is equipped with a fixed bars grill with the following dimensions:

- length: L_{grill} = 520 mm;
- bars width: l_{grill_bars} = 15 mm;
- space between bars: s = 15 mm;
- grill surface: S_{grill} = 0.286 m^2;
- active surface of the grill: S_{active_grill} = 0.19 m^2;

4.2 Experimental cases

The tests are performed in three sets of conditions. The fuel used consists of a mixture of poultry manure (collected directly from the poultry house bedding), which is spread over the fixed layer of solid biomass (wood or agricultural), in a controlled proportion. The energy characteristics of the solid biomass used for the experimental tests are analyzed according to the relevant standard methods and presented in Table 1.

4.2.1 Case 1

The fuel mixture used is represented by:

- 70% forest biomass, wood mixture of different species (beech, oak, hornbeam) in the form of boards of size (2 x 5.5 x 25)cm^3, with humidity 25%, lower calorific value 13 200 kJ/kg and mass 5.3 kg;

Fig. 5. Solid forest biomass in the form of planks – 1.

- 30% poultry manure, with 50% humidity, calorific value 4 100 kJ/kg and mass 1.59 kg. In this scenario, the lower calorific value results:

H_i^i = 0.7 · 13 200 + 0.3 · 4100 = 10 470 kJ/kg

The volume of the fuel mixture placed in a fixed layer on the hearth grill was (40 x 30 x 25) cm^3, the fuel layer being characterized by a low density, ρ_c = 226.6 kg/m^3.

(a) (b) (c)

Fig.6. Experiment 1 - Images of the burning process inside 55 kW pilot boiler at different times of the process
a) the 3th minute; b)the 12th minute; c) the 17th minute.

The burning time of the sample was 1080 s, leading to a burning speed of 0.231 mm/s.

4.2.2 Case 2

The fuel mixture used is represented by:

- 80% forest biomass, wood mixture of different species (beech, oak, hornbeam) in the form of boards of size (3.5 x 7 x 25) cm^3, maximum humidity of 25%, the calorific value 13 400 kJ/kg and mass 5.0 kg.
- 20% poultry manure, 50% humidity, calorific value 4 250 kg/m^3 and mass 1.0 kg.

Fig. 7. Solid forest biomass in the form of planks – 2.

The calorific value of the biomass-poultry manure mixture result in this case:

$H_i^i = 0.8 \cdot 13400 + 0.2 \cdot 4250 = 11\ 570$ kJ/kg

The volume of the fuel mixture placed in the fixed layer on the hearth grill is (40 x 30 x 25) cm^3, the fuel layer being characterized by a lower layer density than in Case 1, specifically ρ_c= 200 kg/m^3. Figure 8 shows the images of the combustion process.

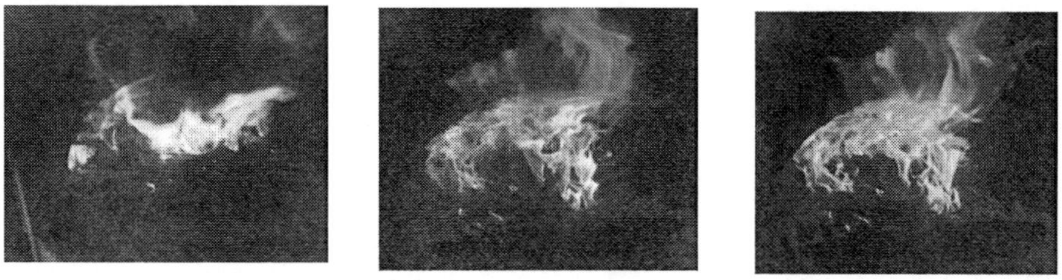

(a) (b) (c)

Fig. 8. Experiment 2 - Images of the burning process inside
55kW pilot boiler at different times of the process a) first minute; b) the 7th minute; c) the 9th minute.

The burning time of the sample was 1050s, leading to a burning speed of 0.238mm/s.

4.2.3 Case 3

The fuel mixture used in this case is represented by:

- ≈ 85% agricultural biomass, mixture of vines and branches with tree bark, in the form of furrows 25 cm long, and different thicknesses between 4 and 17 mm, maximum humidity of 20%, calorific value 13700kJ/kg and mass 4 kg;
- ≈ 15% poultry manure, 50% humidity and mass 0.75 kg.

The calorific value of the biomass-poultry manure mixture results:
H_i^i = 12270 kJ/kg

Fig. 9. Solid agricultural biomass in the form of firewood and twigs

The volume of the fuel mixture placed in the fixed layer on the hearth grill is (40 x 30 x 30) cm^3, with the density of the fuel layer much reduced compared to the first two experiments, which facilitated rapid ignition and very intense combustion. The density of the fuel mixture was ρ_c = 132 kg/m^3.

(a) **(b)** **(c)**

Fig. 10. Experimentul 3 - Images of the burning process inside 55 kW pilot boiler at different times of the process; a) first minute; b) the 3th minute; c) the 18th minute.

The burning time is reduced to 980s, and the burning speed increased to 0.306 mm/s.

4.3 Results and discussions

Table 2 specifies the experimental parameters for the fuel mixture used in each of the three experimental tests.

Table 2 – Experimental parameters of the fixed layer fuel mixture

Parameters / No. exp.	Fuel mixture composition Solid biomass [%]	Poultry manure [%]	Fuel mixture composition Solid biomass [kg]	Poultry manure [kg]	Fuel layer volume [m^3]	Burning time [s]
Experiment 1	70 Wood	30	5.3 Wood	1.59	0.3	1080
Experiment 2	80 Wood	20	5.0 Wood	1.0	0.3	1050
Experiment 3	85 Agricultural	15	4.0 Agricultural	0.75	0.36	980

Variables / Exp. No.	u [mm/s]	ρ_c [kg/m^3]
1	0.231	226.6
2	0.238	200
3	0.306	132

The graph in Figure 11 shows the influence of the burned fuel density (ρ_c) containing a certain proportion of poultry waste in the technology of fixed layer combustion on the combustion rate (u). Poultry waste is fed to the top of the fuel layer.

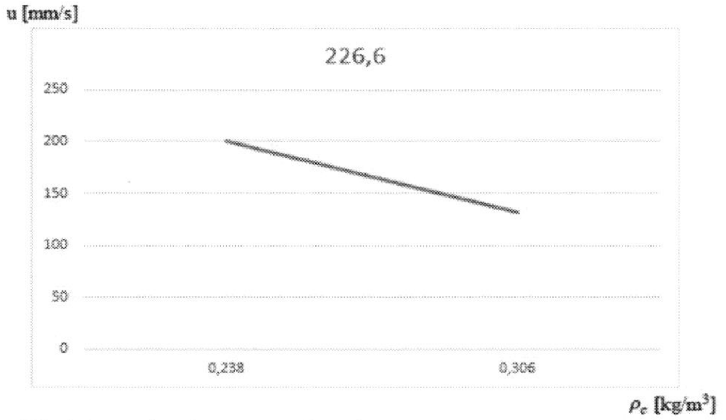

Fig. 11. The influence of fuel layer density on combustion rate

5 Conclusions

The co-combustion process shows great complexity, being influenced of several factors, such as: time, spatial distribution, flow currents, flue gases, (controlled) combustion rate and convection conditions. In all three experimental test conditions, the combustion (without prior mixing of fuels), is stationary, three-dimensional, with turbulent flow and unfolds in a heterogeneous environment.

The combustion of volatiles, including CO, took place in the volume between the biomass pieces and at the surface of the layer, as shown in Figures 6, 8 and 10. The time of drying and release of volatile matter reaches only 10% of the biochar burning period. The amount of heat released in the combustion process is largely determined by the combustion of the biochar residue. The normal development of the combustion process is still conditioned by the design of the air supply in the biomass co-combustion layer.

A comparative analysis between the three experimental tests highlights the decrease of the burning speed with the increase of the wood pieces dimensions (Case 2), but also an increase with the decrease of the fuel density in the layer. There are no significant differences between the use of wood or agricultural biomass.

The experimental test conducted in this research highlighted the possibility of energy recovery from poultry waste by means of combustion with solid biomass.

This work was supported by a grant of the Romanian Ministry of Research and Innovation, CCCDI-UEFISCDI, project number PN-III-P2-2.1-PED- 2019-2350, within PNCDI III (contract number 322PED/2020) and project number PN-III-P1-1.2-PCCDI-2017-0404/31PCCD/2018, within PNCDI.

References

1. O. Udrea and G. Lazaroiu, "RES INTEGRATION IN ROMANIA," *U.P.B. Sci. Bull., Ser. C*, 2015.
2. E. Brizio, G. Genon, F. Becchis, and D. Russolillo, "Renewable energy plants: Environmental compatibility and external costs assessment at global, regional and local scale," *Renew. Energy Power Qual. J.*, vol. 1, no. 8, pp. 1015–1019, 2010.
3. International Renewable Energy Agency, "Renewable Energy Statistics 2017," 2017.
4. G. Lazaroiu, D. A. Ciupageanu, L. Mihaescu, M. Grigoriu, and I. Simion, "Energy recovery from poultry manure: A viable solution to reduce poultry industry energy consumption," *Renew. Energy Power Qual. J.*, vol. 18, no. 18, pp. 202–206, 2020.
5. G. Lazaroiu, L. Mihaescu, R.-M. Grigoriu, D.-A. Ciupageanu, and I. Simion, "EFFICIENT POULTRY INDUSTRY WASTE MANAGEMENT APPROACH IN THE BIOECONOMY FRAMEWORK," in *8th International Conference on Advanced Materials and Systems (ICAMS)*, 2020, pp. 409–415.
6. E. Brizio and G. Genon, "Environmental compatibility of renewable energy plants," *WIT Trans. Ecol. Environ.*, vol. 136, pp. 149–159, 2010.
7. F. Cotana, V. Coccia, A. Petrozzi, G. Cavalaglio, M. Gelosia, and M. C. Merico, "Energy valorization of poultry manure in a thermal power plant: Experimental campaign," *Energy Procedia*, vol. 45, pp. 315–322, 2014.
8. L. Mihaescu, G. Lazaroiu, I. Pisa, D.-A. Ciupageanu, and V. Berbece, "Energy valorization of poultry manure through products resulted in a pyrolysis process," in *18th International Multidisciplinary Scientific GeoConference SGEM2018*, 2018, pp. 349–355.
9. G. Petrecca and R. Preto, "The use of poultry manure to produce clean electric power," *3rd Int. Conf. Clean Electr. Power Renew. Energy Resour. Impact, ICCEP 2011*, pp. 676–681, 2011.
10. L. Mihaescu, G. Lazaroiu, I. Pisa, G.-P. Negreanu, V. Berbece, and D.-A. Ciupageanu, "Thermal conditions required for poultry manure combustion in furnace boilers," in *19th International Multidisciplinary Scientific Geoconference SGEM2019, Volume 19(4.1)*, 2019, pp. 569–576.
11. G. Lazaroiu *et al.*, "Experimental Investigations of Innovative Biomass Energy Harnessing Solutions," *Energies*, vol. 11, no. 12, p. 3469, 2018.
12. R. Amirante, S. Bruno, E. Distaso, M. La Scala, and P. Tamburrano, "A biomass small-scale externally fired combined cycle plant for heat and power generation in rural communities," *Renew. Energy Focus*, vol. 28, no. 00, pp. 36–46, 2019.
13. G. Lazaroiu, *Holistica impactului surselor regenerabile de energie asupra mediului si climei. Vol. 2 Biomasa solida sursa de energie curata*. Bucharest: Politehnica Press, 2018.
14. G. Lazaroiu, L. Mihaescu, G. Negreanu, I. Simion, and D. A. Ciupageanu, "Researches on biofuels gasification using the Lurgi process with homogeneous air inlet over the combustion space," *E3S Web Conf.*, vol. 180, pp. 1–10, 2020.
15. I. Pîşă, G. Lăzăroiu, C. Rădulescu, and L. Mihăescu, "Experimental studies on the corrosion occurrence during biomass combustion process," *Environ. Eng. Manag. J.*, 2012.
16. E. Pop, L. Mihăescu, G. Lăzăroiu, I. Pîşă, G. Negreanu and V. Berbece, Energy characteristics of the char obtained by two steps pyrolysis of chicken manure, XXIInd National Conference on Thermodynamics with International Participation, NACOT 2019, May 22-24, 2019, Galați, Volume 595, Issue 120 September 2019 Article number 012051, pp. 1-8, ISSN 17578981.

Defining of criteria for flue gas decarbonization efficiency in methanation reactors with membrane technology

Gheorghe Lăzăroiu[1], *Lucian* Mihăescu[2], *Dana-Alexandra* Ciupăgeanu[1]*, *Rodica-Manuela* Grigoriu[1], and *Dana-Andreya* Bondrea[1]

[1] Power Engineering Faculty, University Politehnica of Bucharest, Splaiul Independentei 313, 060042 Bucharest (Romania)
[2] Mechanics and Mechatronics Faculty, University Politehnica of Bucharest, Splaiul Independentei 313, 060042 Bucharest (Romania)

Abstract. The paper presents an investigation on the conditions for implementing a methanation membrane decarbonator coupled to an energy installation that generates flue gases. The retention of the carbon dioxide content in the flue gases and its conversion to methane is envisaged. For start, low thermal power installations, employing natural gas as main fuel supply, are considered. Internal combustion engines (also working with natural gas fuel) are taken into account for the testing of the carbon dioxide retention process. For this, a classification of the flue gas composition by fuel categories is initially carried out. The decarbonation efficiency is defined and clarifications are made withal regarding the connection between the decarbonation installation and the energy plant. The first practical achievements are also presented, resulting from a decarbonator with a volume of 940 cm^3 (having the inner diameter of 12 cm and a height of 50 cm). The results prove that the proposed solution has great potential for practical applications, further research being however necessary. In terms of operating costs (including hydrogen consumption), it is remarked that they can be reduced by exploiting the methane production and eliminating the carbon tax, extending the integration perspective form economic point of view.

1 Introduction

At first, carbon dioxide (CO_2) methanation processes targeted exclusively methane (CH_4) production, for chemical or energy uses. As the interest on the subject of CO_2 conversion grew constantly, several methanation technologies have been developed, in addition to the chemical ones. In particular, gasification, plasma employment or even pyrolysis are investigated or used nowadays [1], [2].

Today, depollution requirements in reference to classical power plants highlight the need to reduce CO_2 emissions from flue gases. In this context, new questions about integrating methanization technologies on the exhaust path of classical power plants arise [3]. The feasibility and efficiency of CO_2 emissions reduction in the flue gases depends primarily on their composition. The thermal level of the flue gases plays an important role as well, corrections only in methanation reactors, for large amounts of flue gases involve high costs [4], [5]. Generally, the flue gases contain low amounts of CO_2, the main components in the exhaust stream being represented by nitrogen (N_2) and water vapor (H_2O). The amount of water vapors within the flue gas enable a reduction (up to full replacement) of the water vapor requirement in the methanation reactor, as displayed in Figure 1 [6]. Moreover, the composition of the flue gas depends on the nature of the fossil fuel used, if we accept the conclusion of the Kyoto Protocol, on the total zero or partially zero emission of biomass combustion [7].

A promising solution in order to reduce operational related costs is remarked in the possibility of producing hydrogen from renewable energy sources (RES) [8]. Solar energy and biomass are of in great interest, showing potential also for enabling the reaction conditions, supplying the energy for heating the reactants or cooling for water vapor condensation [9], [10].

In this paper, the level of costs is not addressed, the present research focusing on the possibility of technical application of the catalytic methanation technology to flue gases produced in the energy field. It is highlighted that the proposed research addresses exclusively the issues related to energy installations where adequate decarbonization facilities (in particular through methanation) must be coupled taking into account the flue gas characteristics, according to the prior fuel combustion.

* Corresponding author: dana_ciupageanu@yahoo.com

The second section presents the analysis of flue gases composition, including the equations serving to evaluate the emissions. Section 3 addresses the operation of a catalyst methanizer, while the following section introduces the decontamination of CO_2 as CH_4 production technology, with two-folded advantages. Conclusions are exposed in the last section.

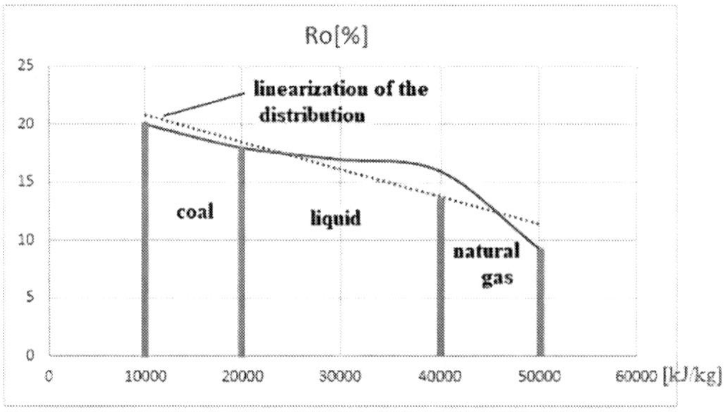

Fig. 1. Variation of CO_2 participation in the flue gases ($\lambda = 1$).

2 Flue gases composition analysis

The emission of flue gases, including CO_2, is determined based on the stoichiometry of the combustion reactions. Accurate calculations require basic knowledge of fuel analysis in order to evaluate the CO_2 emission level [11]. In the case when only the lower calorific value H_i^i is known (measured in *MJ/kg* for solid and liquid fuels in MJ/m_N^3 for gaseous fuels) in the initial conditions characterizing the state of the fuel, empirical calculation relations can be employed, providing acceptable accurate results [12], [13]. Thus, depending on the lower calorific value, the equations presented forward allow assessing the CO_2 emission in the flue gas flow, taking into account as input only the lower calorific value of the fuel [14].

- Emission of CO_2:

Solid fuels, V_{CO_2}= 0,101619 + 0,043991H_i^i $[m_N^3/kg]$

Liquid fuels, V_{CO_2}= 1,26601 + 0,0075569H_i^i $[m_N^3/kg]$ (1)

Gaseous fuels, V_{CO_2}= -0,142711 + 0,0325796H_i^{an} $[m_N^3/m_N^3]$

- Stoichiometric flue gas emission ($\lambda = 1$):

Solid fuels, V_g^o = 1,57933 + 0,217974H_i^i $[m_N^3/kg]$

Liquid fuels, V_g^o = -0,728139 + 0,292333H_i^i $[m_N^3/kg]$ (2)

Gaseous fuels, V_g^o = 0,618368 + 0,277279H_i^{an} $[m_N^3/m_N^3]$

The presence of excess air ($\lambda > 1$), increases the actual flue gas content, V_g, calculable with eq. (3):

$$V_g \approx V_g^o + (\lambda - 1)\, V_{aum}^o \qquad [m_N^3/kg]; [m_N^3/m_N^3]; \tag{3}$$

where, V_{aum}^o is the amount of stoichiometric air for burning the unit of fuel.

The participation of CO_2 in the flue gas is defined by the ratio R_{CO_2} is evaluated according to eq. (4a) and (4b):

$$R_{CO_2}^o = \frac{V_{CO_2}}{V_g^o}, \text{ for stoichiometric combustion} \tag{4a}$$

$$R_{CO_2} = \frac{V_{CO_2}}{V_g}, \text{ for burning with excess air} \quad (4b)$$

According to Figure 1, the participation of CO_2 in the flue gases (excess air $\lambda = 1$) decreases with increasing calorific value. According to the equations previously presented, the participation of CO_2 in the flue gas is calculated for different values of the lower calorific power, the results being listed in Table 1.

Table 1. Stoichiometric participation of CO2 in the flue gases.

R_o[%]	H^i_i[kJ/kg]
20	10000
18	20000
17	30000
16	40000
9.3	50000

It is mentioned that the calorific value of natural gas is considered equal to 53 190k J/kg, resulting a stoichiometric CO_2 participation in the flue gases of $R_o = 9.34$ %. Moreover, it is highlighted that gaseous fuels show the lowest value for CO_2 participation in the flue gases.

To exploit the potential in the flue gas flow and reduce the CO_2 emissions level, a decontamination solution is proposed. Specifically, the layout of a depollution methanizer is shown in Figure 2. It is not mandatory that the entire amount of flue gas is taken over by the methanizer, the difference being discharged through the chimney of the installation (this is the case of high power energy installations).

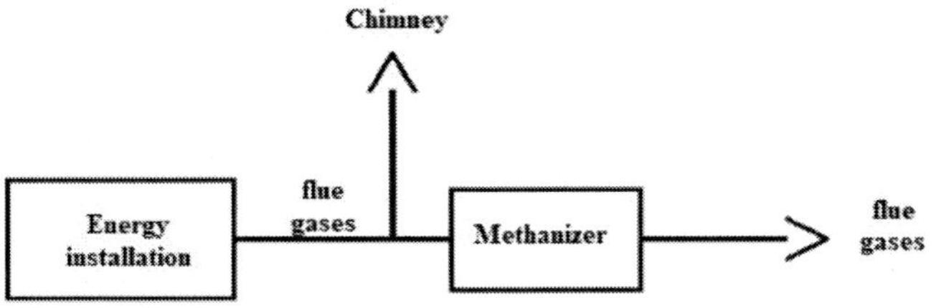

Fig. 2. CO_2 decontamination facility based on methanation.

3 Operation of a catalyst methanizer

There is a lot of effort currently investigating CH_4 production from CO_2 and H_2, by means of catalytic technology, as a solution with high practical potential among many methanation approaches [15], [16]. The efficiency of the catalytic methanation depends on the pressure and temperature in the reactor. The efficiency increases with pressure and temperature. The recommended temperature level in the reactor is 150-550°C, the pressure being usually limited to 30 bar [17]. In these conditions, the reaction environment also contains water vapors. The catalytic methanation installation is schematically presented in Figure 3.

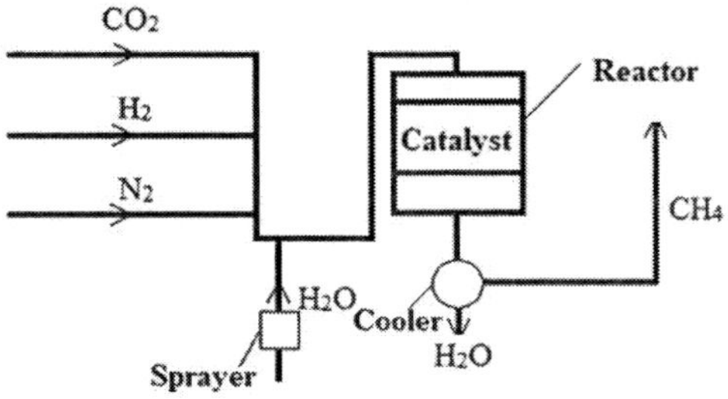

Fig. 3. Catalytic methanation scheme for CO_2

The methanation reaction unfolds at temperature within the range of 150 - 550 °C, following eq. (5):

$$CO_2 + 4H_2 \rightarrow CH_4 + 2H_2O - 165 \text{ kJ/mol} \quad (5)$$

As practical implementation, at the National Research - Development Institute for Electrochemistry and Condensed Matter Timisoara (INCEMC Timisoara) a reactor with Ni Raney catalyst is developed. The reactor consists of a cylinder with a diameter of 120 mm and a height of 500 mm. The maximum working temperature is between 200 and 300 °C, depending on the sealing gasket used.

The material balance shows that the methanation for 1 m_N^3 of CO_2 results in 1 m_N^3 of CH_4 and 2 m_N^3 of water vapors. As water vapors are captured and removed as condensation, the resulting volume of CH_4 is equal to that of CO_2. The stoichiometric reaction assumes a yield of 100%. For sub-unitary yields, as in real cases, the equivalence of the transformation is given by eq. (6):

$$CO_4 = 0{,}01\ \eta\ CH_2 \quad (6)$$

For the Ni Raney catalyst the methanation reactions comprise two steps, according to eqs. (7a) and (7b):

$$CO_2 + H_2 \rightarrow CO + H_2O \quad (7a)$$

$$CO_2 + 4H_2 \rightarrow CH_4 + H_2O \quad (7b)$$

H_2 consumption is similar to the global eq. (1), as well as the resulting amount of water.

4 CO_2 decontamination solution for CH_4 combustion in energy installations

The depollution efficiency is achieved for a combustion plant marked with IA_1, which hypothetically burns $1m_N^3$ CH_4. The flue gas balance, after the coupling of a membrane methanizer is shown in Figure 4, considering the stoichiometric combustion process (λ = 1).

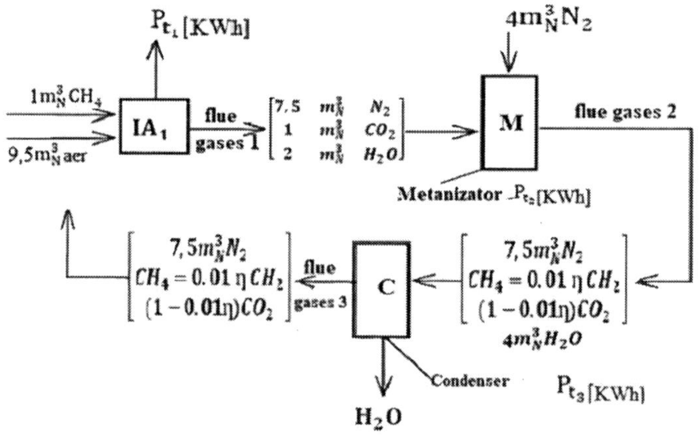

Fig. 4. Flue gas balance for combustion of $1m_N^3$ CH_4

It must be highlighted that, for CH_4 combustion, considering the stoichiometric process does not introduce a significant deviation from reality, given that the excess air is very low anyway for such fuels (λ = 1.02 - 1.03).

The yield η, in %, for the methanizer also taken into account. If the energy produced exclusively through combustion in the plant is P_{t_1}, the operation of the methanizer involves an energy consumption of ($-P_{t_2}$). If operation with CO_2 retention is considered, a condenser for water vapor needs to be inserted, a capacitor that will produce the energy P_{t_3}.

The flue gas balance indicates the presence of CH_4 at the output of the installation, which is a much more harmful gas than CO_2 if released the atmosphere. As a result, the combustion of flue gases 3 is taken into account in a new energy installation marked with IA_2. Figure 5 shows the proposed cascade energy production facility.

Further analysis are performed considering a methanizer efficiency of 90% (η = 90%).

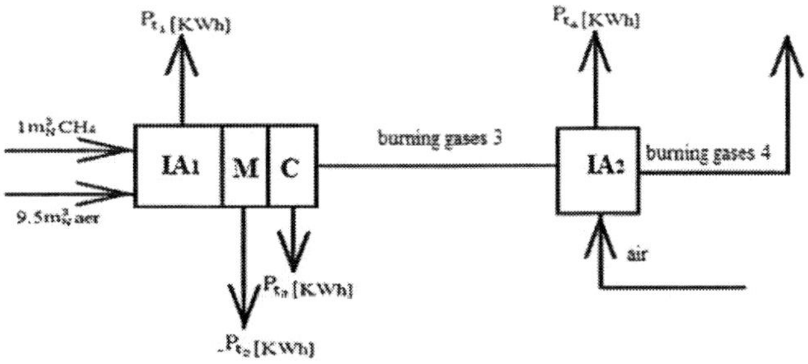

Fig. 5. Cascade combustion plant layout

The combustion plant IA_2, will produce the energy P_{t_4}, consuming the amount of air resulting from the stoichiometric calculation. The composition of the exhaust gases 3, for η = *90%*, in stoichiometric conditions results as:

$$\begin{bmatrix} 7{,}5 & m_N^3 & N_2 \\ 0{,}9 & m_N^3 & CH_4 \\ 0{,}1 & m_N^3 & CO_2 \end{bmatrix} = 8{,}5\ m_N^3 \text{ flue gases 3}$$

As a percentage, the flue gases 3, have the composition:

CH_4 = 10,6%, CO_2 = 1,17%, N_2 = 88,23%.

For this composition the calorific value of the flue gas is equal to H_{i3}^{i} = 3800 kJ/m_N^3.

After combustion, the volume and composition of the flue gases discharged from IA_2 into the atmosphere shall be:

$$\begin{bmatrix} 8{,}29 & m_N^3 & N_2 \\ 0{,}21 & m_N^3 & H_2O \\ 0{,}18 & m_N^3 & CO_2 \end{bmatrix} = 8{,}68\ m_N^3$$

There is a decrease in CO_2 concentration, from R_o = 9.34% at the exit of IA_1 to the value of 2.07% at the exit of IA_2.

The ratio to the power (energy) produced by combustion leads to the values:

-after IA_1 for the combustion of $1 m_N^3$ CH_4, it results:

$$E_{CO_{2_1}} = \frac{V_{CO_{2_1}}}{P_{t_1}} = \frac{m_{CO_{2_1}}}{P_{t_1}},$$

where m_{CO_2} is the mass amount of CO_2 emitted in g CO_2.

Numerically:

$$E_{CO_{2_1}} \approx \frac{2000\ g}{9{,}94KWh} \approx 200\text{g/kWh}$$

-after IA_2, the CO_2 emission will be:

$$E_{CO_{2_1}} \approx \frac{0{,}18\ m_N^3 CO_2}{P_{t_1} + P_{t_2} + P_{t_3} + P_{t_4}}\ [\frac{m_N^3 CO_2}{kWh}]$$

if it is assumed that $P_{t_2} + P_{t_3} \approx 0$, the CO_2 emission becomes:

$$E_{CO_{2_1}} \approx \frac{0{,}18\ m_N^3 CO_2}{P_{t_1} + P_{t_4}} = \frac{360\ g}{(9{,}94+1{,}05)kW} = 32{,}75\ [\frac{g}{kWh}]$$

There is a significant reduction in CO_2 emissions from cascade combustion, the reduction being of about 6 times.

5 Conclusions

The paper addressed some issues that need to be considered when using methanizers as a solution for decarbonating flue gases. To this aim, starting from the composition of the flue gases discharged by an energy installation, the analysis is performed taking into account natural gas fuel (CH_4). The calculations are performed for stoichiometric combustion ($\lambda = 1$), a situation very close to the real combustion conditions for gaseous fuels.

It is remarked that higher efficiencies for the methanizer are necessary, so that the CH_4 content in the flue gas flow after the methanizer is enough for a general combustion in the second stage. For an efficiency of 90% in the methanation process, the volume evacuated at the output of the methanizer is characterized by a lower calorific value. For the model introduced in this paper, there is noticeable a significant reduction of the CO_2 emissions after the second stage of combustion. Thus, this technology can be successfully applied to natural gas combustion.

As first applications, natural gas engine installation show increased potential, as the second combustion stage can be achieved in a simple way for the production of thermal energy. Regarding the application of the proposed technology to other fuels, it is highlighted that it must be treated selectively, representing a future research perspective. The present work is inceptive in the field of flue gas methanation, the calculations presented here being made for theoretical stoichiometric combustion of different fuels. Further research developed by the Authors aims investigating the excess air problem for all fuel types.

This work was supported by a grant of the Romanian Ministry of Research and Innovation, CCCDI-UEFISCDI, project number PN-III-P2-2.1-PED- 2019-2350, within PNCDI III (contract number 322PED/2020) and project number PN-III-P1-1.2-PCCDI-2017-0404/31PCCD/2018, within PNCDI.

References

1. I. Balcu et al., "Decarbonization of Low Power Applications through Methanation Facilities Integration," 2019 IEEE PES Innov. Smart Grid Technol. Eur., pp. 1–5, Sep. 2019.
2. W. Wang and J. Gong, "Methanation of carbon dioxide: An overview," Front. Chem. Eng. China, vol. 5, no. 1, pp. 2–10, 2011.
3. H. Blanco, W. Nijs, J. Ruf, and A. Faaij, "Potential of Power-to-Methane in the EU energy transition to a low carbon system using cost optimization," Appl. Energy, vol. 232, no. July, pp. 323–340, 2018.
4. International Renewable Energy Agency, "International Renewable Energy Agency: Renewable Energie: A Key Climate Soluation," 2017.
5. M. da Graça Carvalho, M. Bonifacio, and P. Dechamps, "Building a low carbon society," Energy, 2011.
6. L. Mihaescu, T. Prisecaru, and I. Oprea, Cazane si turbine. Bucharest: Politehnica Press, 2002.
7. G. Lazaroiu, Holistica impactului surselor regenerabile de energie asupra mediului si climei. Vol. 2 Biomasa solida sursa de energie curata. Bucharest: Politehnica Press, 2018.
8. B. Castellani et al., "Thermal integration of a high-temperature co-electrolyzer and experimental methanator for Power-to-Gas energy storage system," Energy Convers. Manag., vol. 171, no. 4, pp. 140–155, Apr. 2018.
9. F. Salomone, E. Giglio, D. Ferrero, M. Santarelli, R. Pirone, and S. Bensaid, "Techno-economic modelling of a Power-to-Gas system based on SOEC electrolysis and CO2 methanation in a RES-based electric grid," Chem. Eng. J., no. xxxx, pp. 1–20, 2018.
10. M. Ancona et al., "Renewable energy storage system based on a Power-to-gas conversion process," Energy Procedia, vol. 101, 2016.
11. L. Mihaescu, Cazane de abur si apa fierbinte. Bucharest: PERFECT, 2007.
12. [12] G. Lazaroiu et al., "Experimental Investigations of Innovative Biomass Energy Harnessing Solutions," Energies, vol. 11, no. 12, p. 3469, 2018.
13. M. Prisecaru, Constructia si exploatarea cazanelor de abur si apa fierbinte. Bucharest: Politehnica Press, 2009.
14. G. Lazaroiu, Holistica impactului surselor regenerabile de energie asupra mediului si climei. Vol. 3 Spre o energetica circulara. Bucharest: Politehnica Press, 2020.
15. F. Meylan, F. Piquet, and S. Erkman, "Power-to-gas trough CO2 methanation: Assessment of the carbon balance regarding in directives," J. Energy Storage, vol. 11, 2007.
16. M. Thema, F. Bauer, and M. Sterner, "Power-to-gas: Electrolysis and methanation status review," Renew. Sustain. Energy Rev., vol. 112, 2019.
17. G. Emanuele, A. Lauzini, M. Santarelli, and P. Leone, "Synthetic natural gaz via integrated high-temperature electrolysis and methanation: Part I – Economic performance," J. Energy Storage, vol. 1, 2015.

Evaluation of the influence of the windbreak trees on the change of wind shear in weakly complex terrains

Angel Terziev[1*], *Yancho* Panteleev[1], *Iliya* Iliev[2] and *Hristo* Beloev[2]

[1]Department of Power Engineering and Power Machines, Technical University of Sofia, 8 Kliment Ohridski blvd., 1000 Sofia, Bulgaria
[2] University of Ruse, Department of Thermotechnics, Hydraulics and Ecology 7017 Ruse, 8 Studentska Str., Bulgaria

Abstract. The turbulent nature of the wind above the earth's surface depends on both the topology of the terrain and the presence of natural obstacles along the way such as low grasses and shrubs, as well as medium-tall trees. When the wind passes through the indicated obstacles, detachment is observed i.e. formation of large eddies, which are carried away by the main flow, after which they dissipate. The size of the vortices, as well as the period of dissipation, depends on the wind speed, as well as the type of obstacle. The presence of windbreak trees significantly changes the wind shear over the surface, and hence the energy potential of the wind in the vicinity of trees. In present work, the influence of the tree belt on the wind shear at the adopted prevailing wind direction is investigated. The degree of deformation of the speed profile after the obstacle in weakly complex terrain is shown. Relevant prescriptions for the location of wind turbines in the vicinity of windbreak trees are presented in view of minimum shading and maximum energy output.

1. Introduction

Studying the wind behaviour over terrain is essential for determining the energy potential. Every terrain specific such as hills and elevations, as well as natural and artificial obstacles, affect the distribution of the speed profile in height, which changes the potential of the energy winds. The above factors of varying severity affect the change in velocity distribution on the earth's surface. High hills and elevations lead to the generation of large scale vortices, which, depending on the size, can affect both the distribution of the velocity profile near the earth's surface and at a significant height. Tall vegetation also represents wind resistance, the main purpose of which is to significantly reduce the speed in the surface boundary layer in order to prevent the transfer of planting material from agricultural land. However, this vegetation significantly affects the speed profile in the vicinity of the

* Corresponding author: aterziev@tu-sofia.bg

windbreak trees, which is a prerequisite for reduced energy output. Installing a wind turbine in this area would reduce production by up to 20% depending on the technical characteristics (including tower height) of a wind turbine [2].

Measurements carried out in the vicinity of the tree belt can give a clear idea of the distribution of the wind shear, especially when the terrain is flat. In [1] a comparison was made between experimental and numerical results for the wind speed profile in height, where a good match is observed, especially at a height of up to 50 m. Above this height, however, the differences between measured and predicted results can reach up to 10%.

The use of reference measurements to analyze wind parameters at a considerable distance from the terrain point also does not provide the necessary information on the influence of woody vegetation [2, 3]. In [4] an analysis of the influence of the terrain and surrounding vegetation on the wind speed profile in height is presented. The roughness modeling was performed using a WAsP software package. The roughness factor is 0.2.

Numerical modelling tools can also be used to predict the effect of small obstacles on wind flow turbulence [5]. In this work, a CFD code is presented that can be successfully used for the micro-sitting of wind turbines where the wind flow is affected by these obstacles. A similar study where the numerical procedures are used to study the behaviour of the wind when passing different types of trees is presented in [6]. K-ε turbulent model is used to predict the impact of the surface roughness on the main wind flow by using z_0 function [7, 8].

2. Site specifics and on-site wind data

2.1. Site specifics

The spot, object of the analysis is located in the northeastern part of Bulgaria, where the terrain is weakly complex. The terrain is open in all directions, and there are only windbreak trees with an average height between 15 up to 20 m. Tree species are deciduous trees that represent some significant resistance to wind especially in the vicinity of the surface boundary layer. The nearest settlement Prolez village is located 2.9 km southwest, and about 2 km in a northwest direction from the mast is the village of Tvarditsa (source Google map view). Fig. 1 is a representation of the location of the measuring mast, the wind belts (windbreak trees), and the nearest settlements (google maps view). The distance between the mast and the nearest tree is 80 m.

Fig. 1. Google maps view of the location of mast and windbreak trees

2.2. Raw wind data analysis

A tall meteorological mast (60 meters) equipped with calibrated and certified equipment is used to determine the wind potential [3]. Four wind speed sensors were used: two at 60m, one at 50m, and 1 sensor at 40m height. Two sensors at 60 and 50 m were used to measure the prevailing wind direction (wind vanes). There is also a temperature sensor installed for measuring the air temperature. Data logger was used to collect wind data - wind data for a three-year period (about 145 000 records with 10-mitues mean intervals) were registered. The data processing was done carefully and the freeze periods were eliminated. Table 1 presents the average wind speed at the specified mast heights.

Table 1. Summarized average wind data from on-site measurements

Sensor	Installation Height, m	Parameter	Value
Anemometer	60	Wind speed, m/s	5.88
Anemometer	60	Wind speed, m/s	5.92
Anemometer	50	Wind speed, m/s	5.66
Anemometer	40	Wind speed, m/s	5.34
Wind vane	60	Wind direction, °	201.2
Wind vane	50	Wind direction, °	202.2
Temperature sensor	10	Temperature, ° C	13.2

The data from the field measurements show that there are two prevailing wind directions: north-northwest and south-southwest (fig. 2). Further in the analysis, the following two prevailing wind directions and how the wind shear is affected when passing through the tree missives are adopted.

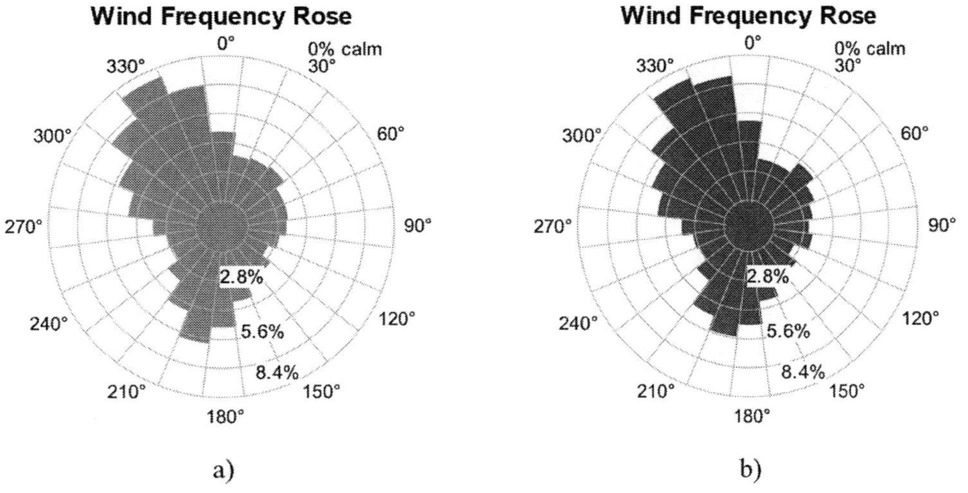

Fig. 2. Prevailing wind direction data a) 60m height; b) 50 m height

3. 3D topography modeling

A three-dimensional model of the terrain surface was prepared on the basis of a topographic map. No significant changes in elevation were observed within the considered terrain.

Fig. 3a shows the isolines of the terrain surface. It can be seen that the altitude varies from 20 to 80 m. Fig. 3b is a representation of a three-dimensional model with the location of the meteorological mast. In addition, the prevailing wind directions are indicated.

As mentioned above, the main purpose of the work is to show the influence of the existing tree belts on the deformation (distortion) of the wind shear over the terrain. It is known that the topography of the terrain, despite the flat nature, has little effect on the speed profile. For the purpose of the study in fig. 3 c and d are also shown the terrain profiles along the prevailing wind direction. Fig. 3c shows that the wind speed will be slightly decreased when the prevailing direction is from the north-northwest, and slightly increased when the prevailing wind direction is from the south-southwest.

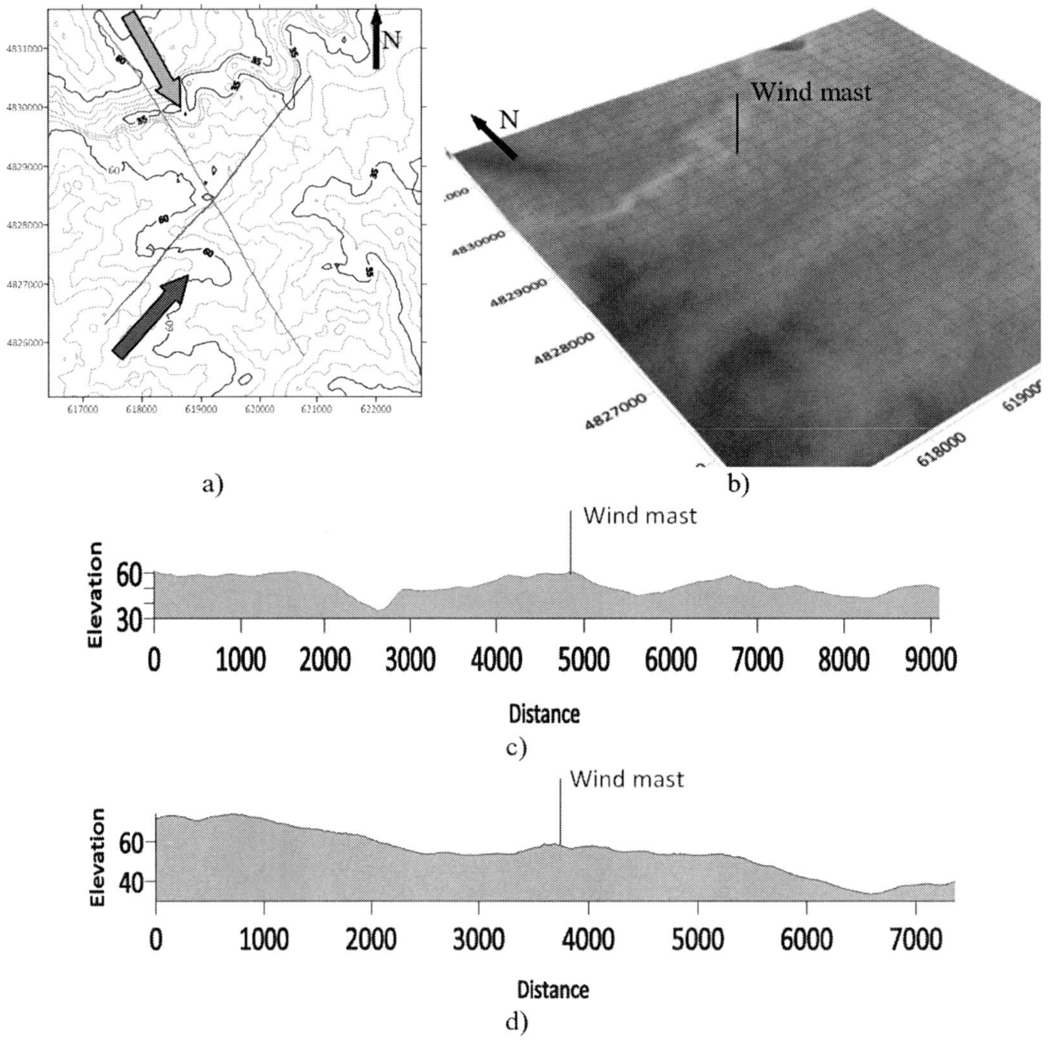

Fig. 3. Terrain specifics: a) Isolines of the terrain with prevailing wind directions; b) 3D terrain orography; c) terrain profile along the noth-norwest prevailing wind direction; d) terrain profile along the south-southwest prevailing wind direction.

4. Wind shear analysis

When the prevailing wind direction is from the north, before the wind reaches the mast, it passes through three windbreaker groups of trees, which are at a distance of about 500 m from each other. When the prevailing wind direction is from the south, the distance between tree belts and tall tower is about 1 300 m (Fig. 1). An analysis of the wind speed profile in height for the two prevailing wind directions was made, using the data from the measuring instruments at the three heights.

Figure 4 shows the wind speed distribution at the point of the mast after the wind passes through windbreaker trees with a height of 20 m at a distance of 1.3 km from the mast. The performed analysis shows that the power-law exponent alpha is 0.314 and the surface roughness factor is 2.0m (data obtained with Windographer software). The wind shear distribution shows that the wind profile is not affected by the trees located 1.3 km away from the mast. The increased velocity in the surface boundary layer is because of the speed-up of the flow in terms of site specifics.

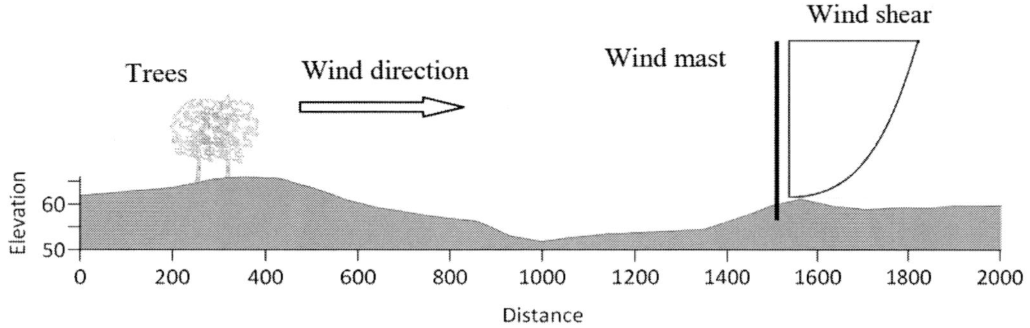

Fig. 4. Wind shear over the terrain at the point of the mast when prevailing wind direction from south-southwestern direction

Figure 5 shows the wind speed distribution at the point of the mast after the wind passes through a couple of windbreaker trees with a height of 20 m at a distance of 0.5 km between the separate rows of trees and 80m distance between the closest tree row and the mast. The analysis performed shows that the power law exponent alpha is 0.267 and the surface roughness factor is 1.14m. The wind shear distribution shows that the wind profile is affected significantly close to the surface boundary layer - it can be seen that the wind shear is steeper. Still away from the surface boundary layer and above 50m height the wind shear stays unaffected.

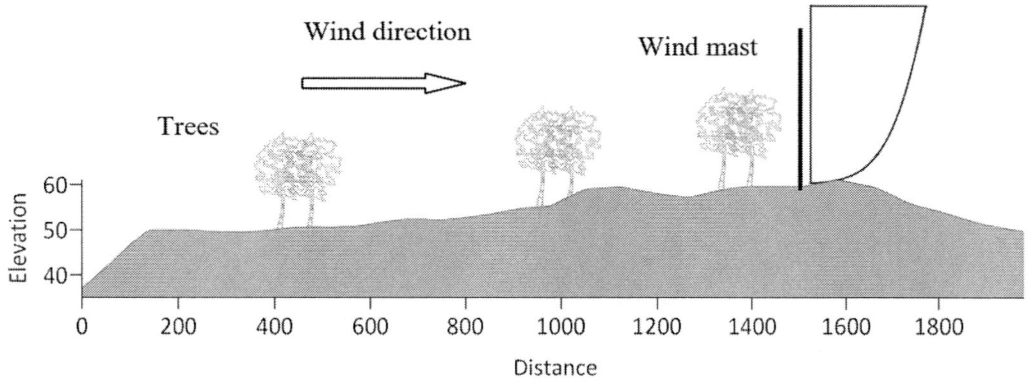

Fig. 5. Wind shear over the terrain at the point of the mast when prevailing wind direction from North-Northwestern direction

A comparison between the two velocity profiles in height is presented in fig. 6. The graph shows that in the range of the windbreaker trees will be affected (distorted) low-power wind turbines, respectively with a small tower - up to 50 m. In addition, the nature of the flow will be disturbed at a distance of up to 200 m after the respective obstacle (at an obstacle height of 20 m).

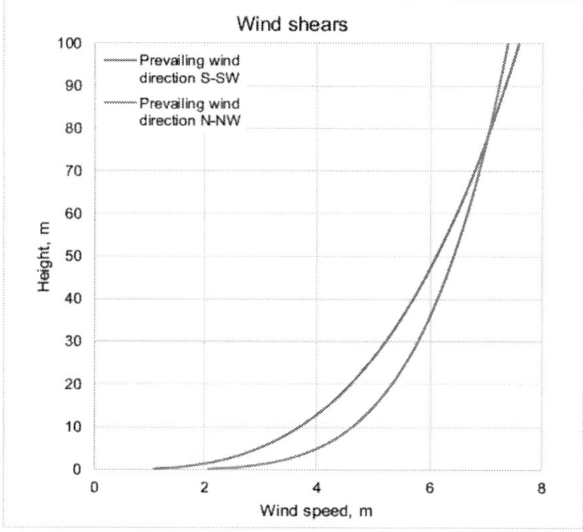

Fig. 6. Wind shears in the vicinity of mast in terms of prevailing wind directions

5. Conclusion

The paper examines the influence of low windbreaker trees (20 m high) on the change in wind speed profile in height. The performed long-term measurements with a tall meteorological mast show two predominant wind directions – N-NW and S-SW. With the adopted direction from the south-southwest, the wind passes through the windbreaker trees at a distance of about 1.5 km before reaching the mast. At the specified distance, no influence of the speed profile from the tree shield is observed. With north-northwest prevailing direction, the wind passes unaffectedly through three tree missives, which are at a distance of about 500m from each other, the latter being at a distance of 80m from the mast. The deformation of the profile is observed, especially in the area of the surface boundary layer, which leads to a lower wind speed in height. However, above 50 m above the ground, the influence of windbreaker trees can be assessed as insignificant. Thus, in cases where the turbines are less than 50 m high or are located near the wind belt, their productivity will be reduced.

The study was supported by the Research and Development Sector at the Technical University of Sofia under project 202ПД0015-02.

References

1. R. Kumar, T. Stallard, P. K. Stansby, Assessment of WRF prediction of velocity profile and turbulence intensity by comparison to field measurement, (2017), *Proceedings of the International Offshore and Polar Engineering Conference*, pp. 215-221

2. A. Terziev, Analysis of the possibilities of using reference wind measurements in the calculation of energy production from wind farms on low complex terrains, (2019) *E3S Web of Conferences*, **112**, pp. 1-11.
3. A. Terziev, Y. Panteleev, M. Ivanov, Possibility analysis for the use of remote reference wind, measurements, in the evaluation of energy production from low complex terrains, (2020) 2020 7th International Conference on Energy Efficiency and Agricultural Engineering, *EE and AE 2020 - Proceedings*, art. no. 9278994
4. P. Sharma, A. Gautam, P. Baredar, V. Warudkar, J. L. Bhagoria, S. Ahmed, Analysis of terrain of site Mamatkheda Ratlam through wind modeling tool ArcGIS and WAsP, Materials Today: Proceedings, (2020).
5. J. Vermeir, M. Runacres, T. De Troyer, CFD modelling and measurements of the atmospheric boundary layer for micrositing of small wind turbines, (2012) European Wind Energy Conference and Exhibition 2012, *EWEC 2012*, **2**, pp. 716-725.
6. M. Salim, K. Schlünzen, D. Grawe, Including trees in the numerical simulations of the wind flow in urban areas: Should we care?, *Journal of Wind Engineering and Industrial Aerodynamics*, **144**, (2015), p.p. 84-95
7. Yu Lun, A. Mochida, S. Murakami, H. Yoshino, T. Shirasawa, Numerical simulation of flow over topographic features by revised k–ε models, *Journal of Wind Engineering and Industrial Aerodynamics*, **91**, Issues 1–2, (2003), p.p. 231-245
8. L. Kang, X. Zhou, T. van Hooff, B. Blocken, M. Gu, CFD simulation of snow transport over flat, uniformly rough, open terrain: Impact of physical and computational parameters, *Journal of Wind Engineering and Industrial Aerodynamics*, **177**, (2018), p.p. 213-226

A theoretical study on the mass input and output and energy of the biomedical waste incinerator

Sari Jaber[1], *Alaa Aldin* Aljawad[1], *Tudor* Prisecaru [1] and *Elena* Pop[1,*]

[1]University Politehnica of Bucharest, Faculty of Mechanical Engineering and Mechatronics, Romania,

Abstract. Medical waste production has increased with many government and private hospitals and health centres due to the increase in the population, especially in the spread of Coronavirus Covid-19. This increase in medical waste is treating in medical incinerators. One of the advantages of incineration is the reduction of volume, weight and energy recovery. This paper examines the medical waste incinerator of one of the Medical City hospitals in Baghdad, as this incinerator internally divided into primary and secondary chambers. The medical waste for yellow bags placed in the primary chamber is burned with the help of air and a burner of 178-356 kW, resulting in waste burned. Then the combustion products pass through the secondary chamber, wherewith air and another burner of 178-356 kW, the combustion of gases and volatile materials from the waste completed. Air is then supplied to the flue gases to dilute the emission concentrations and reduce the flue gas temperature. This paper presents a study on the inputs and outputs of waste, air, fuel and the advantage of the heat quantity generated from the combustion of biomedical waste. As a result of this theoretical work, this incinerator provides an acceptable and durable solution to waste disposal problems and the risks of spreading viruses today.

1. Introduction

Medical waste has increased due to the increase in the population [1][2], especially in the spread of Coronavirus Covid-19 as the quantities of waste increased due to the frequent use of protective clothing and for all people. In the incineration process, waste treated to reduce the volume of waste by about 90%, weight by about 70%, disinfection, energy recapture, detoxification, and resource recapture [3][4][5][6]. Incineration is a practically reliable method of waste treatment. It is currently the best treatment method at present, especially from the third month of 2020. Incinerators have proven the efficiency of their work and the rapid disposal of medical waste [7]. The advantages of incineration treating become clearer. Medical waste completely burned at high temperatures of 1200°C, and high temperatures

* Corresponding author: elena.pop@upb.ro

must be under controlled to ensure complete combustion to reduce environmental and health pollution. The results of waste incineration are flue gases and residual ash.

2. Materials and Methods

2.1. Medical waste incineration

There are three types of incinerators for treating biomedical waste: Multiple-chamber [8], Controlled air, Rotary kiln [9], and Mobil incinerator [10].

2.2. Specifications of the medical waste incinerator

The incinerator used in this study is the air-controlled type of incineration to treat a mixture of yellow bags of medical waste with a capacity of 100 kg per hour, shown in fig 1.

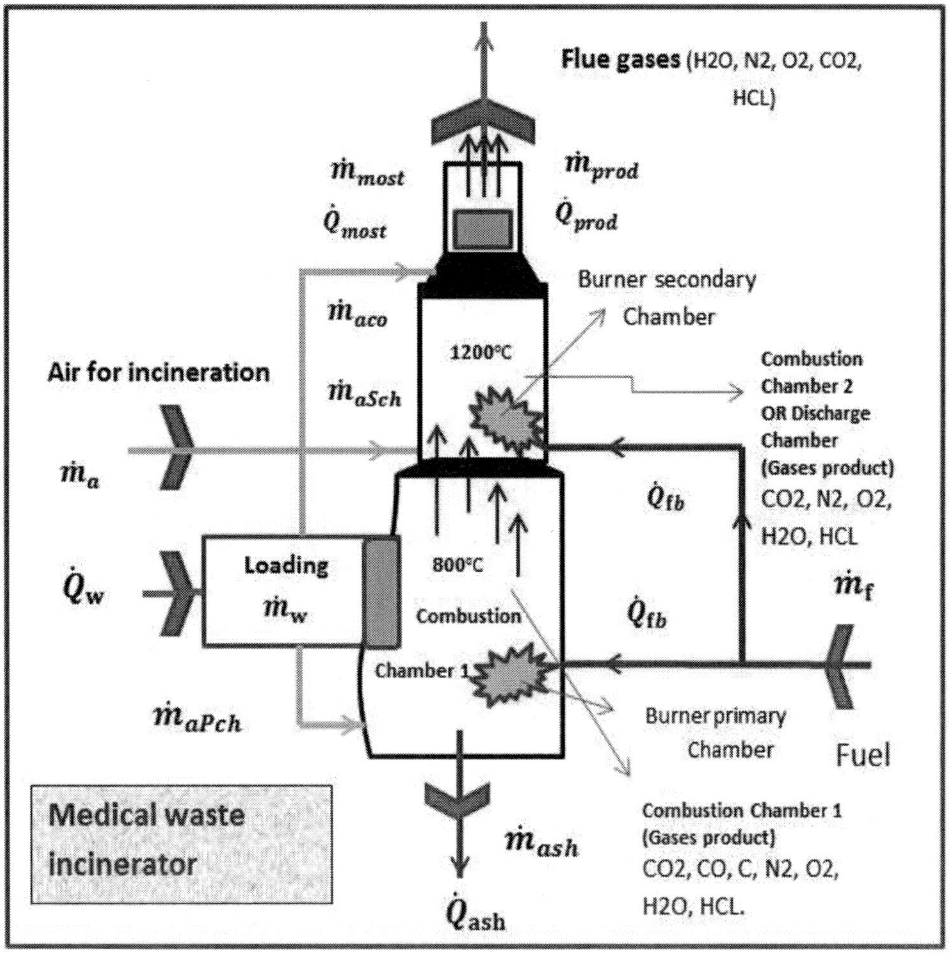

Fig. 1. System medical waste incinerators and the heat generated from incineration waste.

Technical parameters:

- Pch (Primary chamber) temperature reaches 800°C and Sch (Secondary chamber) temperature reaches 1200°C.

- The residence time for gases resulting from the combustion in the Sch is two seconds.
- The (Pch) is fed with 30% air of the required air.
- The air in the (Sch) supplied 150% of the air required at the peak combustion.
- Incinerators are made of refractory materials, and the combustion chambers are well insulated to maintain temperatures [11].
- Burners in the (Pch) and the (Sch) are fuelled with diesel fuel at 15 kg per hour.
- There are no air pollution control devices and no filter bags in the chimney.

2.3. Thermal processes inside a medical waste incinerator

The Medical Waste (MW) incineration process divided into:

Drying. When the heat vaporizes, superficial water and inherent water in the combustion chamber, the drying stage divide to: conduction drying, convection drying and radiation drying achieved by heat transfer. Whereas the higher the water content of the MW, accompanied by longer the drying phase will be consumed more thermal energy, and the temperature will be reduced and subsequently affecting the burning process.

Decomposition and thermal volatilization of combustible materials in MW generates many volatile hydrocarbons and carbon capture products. The pyrolysis reaction consists of endothermic and exothermic reactions. The thermal decomposition speed is related to the combustible components' composition, heat and mass transfer rate, and the organic solids' particle size.

Combustion. Gaseous and solid combustible substances resulting from drying and thermal decomposition, insufficient contact with air in the incinerator become a flame and start combustion at high temperatures. Therefore, MW burning is a mixed gas-phase combustion process and heterogeneous combustion, which is more complex than gaseous fuels and liquid fuels [12].

2.4. Heat and Material Balance

Heat and material balance calculation is an integral part of designing and evaluating incinerators. This technique involves a detailed estimation of the input and output conditions of the incinerator and to determine the combustion air and auxiliary fuel requirements for incinerating a given medical waste and determining an existing incinerator's limitations when charged with a known waste [12][13]. The characterization of biomedical waste is described extensively in [13]. The figure below (Fig 2) shows the different types of waste and the properties of hospitals biomedical waste in yellow bags. The steps to calculate the heat and material balance sample are presented below.

2.4.1. Heating values of input material

The material flow per hour into the incinerator is 100 kg/h. Based on an input waste, yellow bags waste is assumed to have the composition, according to the Table 1 and as shown in Figure 2 as well to consist of dry tissue, polyethylene, polyvinylchloride, cellulose, water, and ash represents 10, 45, 3, 30, 2, 10 kg/h respectively.

A; primary combustion chamber B; Secondary combustion chamber
C; Gate waste loading. D; Bottom ash exit gate. E; yellow bags of biomedical waste

Fig. 2 The shape of incinerator and biomedical waste

Table 1. Higher heating values and total heat of the combustible medical waste

Compound	Chemical Composition	Molecular Weight	Fraction assume	Generation (kg/h)	Calorific Value (HHV) (kJ/kg)	Total heat (kJ/h)
Tissue	$C_5H_{10}O_3$	118.1	0.10	$0.10 * 100 = 10$	20,471	204,710
Plastic (Polyethylene -96%)	$(C_2H_4)_x$	28.1	0.45	$0.45 * 100 = 45$	46,304	2083680
PVC	$(C_2H_3CL)_x$	62.5	0.03	$0.03 * 100 = 3$	22,630	67,890
H_2O	H_2O	18.0	0.02	$0.02 * 100 = 2$	0	0
Ash	-	-	0.10	$0.10 * 100 = 10$	0	0
Total heat generation from yellow waste bag			100		2913320	

2.4.2. Determination of stoichiometric oxygen for combustible medical wastes and combustion air rates

The total theoretical amount of oxygen required to burn the waste determined by the chemical equilibrium equations of the biomedical waste's components from laboratory analysis, the theoretical oxygen to burn the medical waste's combustible component (100 kg/h). We can calculate the total requirements for the theoretical amount of oxygen required to burn (oxidize) the waste [3].

Table 2. Material chemical composition

Material	Chemical Composition
Tissue	$C_5H_{10}O_3$ + $6O_2$ = $2\ CO_2$ + $5\ H_2O$ 118.1 6(32) 2(44) 5(18) 1.0 1.63 1.84 0.76 10 16.3 18.6 7.6
Polyethylene (as fired)	$(C_2H_4)_x$ + $3O_2$ = $2\ CO_2$ + $2\ H_2O$ 28.1 3(32) 2(44) 2(18) 1.0 3.43 3.14 1.29 45 154.35 141.3 58.05
PVC	$2(C_2H_3CL)_x$ + $5O_2$ = $4\ CO_2$ + $2\ H_2O$ + 2 HCl 2(62.5) 5(32) 4(44) 2(18) 2(36.5) 1.0 1.28 1.41 0.29 0.58 3 3.84 4.23 5.22 1.74
Cellulose, Swab.	$C_6\ H_{10}\ O_5$ + $6O_2$ = $6\ CO_2$ + $5H_2O$ 162.1 6(32) 6(44) 5(18) 1.0 1.19 1.63 0.56 30 35.7 48.9 16.8
Diesel	$4\ C_{12}\ H_{23}$ + $71\ O_2$ = $48\ CO_2$ + $46\ H_2O$ 4 (167.31) 71(32) 48(44) 46(18) 1.0 3.65 3.39 1.33 30 101.7 94.8 37.2

Total combustible components of waste (kg/h) = $10 + 30 + 3 + 45 = 88$ (kg/h).

The total amount of oxygen required for stoichiometry of waste incineration generated per hour = $16.3 + 154.35 + 3.84 + 35.7 + 101.7 = 311.89$ (kg/h).

The stoichiometric air is calculated = $311.89 * (100/23) = 1356.04$ (kg/h)

Air is supplying two combustion chambers, where the first is supplied, with 30% of the air required for complete combustion of waste.

Air required primary chamber = $(0.30 * 311.89) + 311.89 = 1762.85$ (kg/h)

In the secondary chamber for combustion, excess air supplied to 150% of what is theoretically required [14].

Air excess secondary chamber = $(1.5 * 311.89) + 311.89 = 3390.12$ (kg/h)

2.4.3. Material balance for combustion chambers

A) Total mass input $\dot{m}_i$ (kg/h)

Mass Input, Waste Input Rate ($\dot{m}_W$) $= 100$ kg/h

Dry air ($\dot{m}_{asch}$) $= 3390.12$ (kg/h)

Moisture in air ($\dot{m}_{amoist}$) $= 0.0132 * 3390.12 = 44.75$ (kg/h)

$\dot{m}_{diesel} = 15 + 15 = 30$ (kg/h)

Total Mass input $\dot{m}_i = \dot{m}_w + \dot{m}_{asch} + \dot{m}_{amoist} + \dot{m}_{diesel}$

$\dot{m}_i$ (kg/h) = $100 + 3390.12 + 44.75 + 30 = 3564.86$ (kg/h)

B) Total Mass Output (under the condition of complete combustion) $\dot{m}_o$ k, h.

1- Dry Products from Waste.

Air requirement for Waste = 3390.12 (kg/h)

Subtracting Stoichiometric air requirement = 1356.04 (kg/h)

Total excess air ($\dot{m}_{a.exces}$) $= 3390.12 - 1356.04 = 2034.07$

The nitrogen from stoichiometric air (77%) $= 0.77 * 1356.04 = 1044.15$ kg/h
Total production $= 2034.07 + 1044.15 = 3078.22$ kg/h.
The total CO_2 from the combustion to determine the theoretical air required only
CO_2 formed from $C_5H_{10}O_3$ = 18.6 kg/h.
CO_2 formed from $(C_2H_4)_x$ = 141.3 kg/h.
CO_2 formed from $2(C_2H_3Cl)_x$ = 4.23 kg/h.
CO_2 formed from $C_6H_{10}O_5$ = 48.9 kg/h.
CO_2 formed from 4 $C_{12}H_{23}$ = 94.8 kg/h
Total CO_2 generated = 307.83 kg/h.
Total Waste Dry products ($\dot{m}_{prod}$) = 3078.22 + 307.83= 3386.05 kg/h
The CO_2 will be released to the ambient after complete combustion

2- Moisture (H_2O)
H_2O Moisture in the waste = 2 kg/h.
H_2O Moisture generated from Combustion reactions = 124.87 kg/h.
H_2O Moisture in combustion air = 44.75 kg/h.
Total mass moisture ($\dot{m}_{moist}$) $= 2 + 124.87 + 44.75 = 171.62$ kg/h.

3- Ash from combustion waste.
Total Ash output ($\dot{m}_{ash}$) = 10 kg/h.

4- HCl generation from Waste (PVC).
HCl from $2(C_2H_3Cl)_x$ = 1.74 kg/h.
Total Mass Out = Total waste dry products + Total moisture + Total CO_2 from combustion +HCl formed from waste + Ash Output
Total Mass Output $\dot{m}_o$ (kg/h) $= 3386.05 + 171.62 + 10 + 1.74 = 3569.41$ kg/h
Net mass balance $= (\dot{m}_i - \dot{m}_o) = 3564.86 - 3569.41 = 4.55$ kg/h

2.4.4. Arithmetic balancing of the incinerator energy.

Total heat generation from medical waste incineration (Q_i kJ/h)
Total heat generation from the yellow waste bag ($Q_w = 2913320$ kJ/h) from table 1
Add heat burner primary chamber and secondary chamber from medical waste incinerator
The calorific value of diesel fuel is 45.5 MJ/kg = 45500 kJ/kg
$\dot{m}_{fbpch}$ = 15 kg/h
$\dot{m}_{fbsch}$ = 15 kg/h
Total heat generation from 2 burners Q_{tfb} kJ/h $= 30 * 45500 = 1365000$ kJ/h
$Q_{tfb} = 1365000$ kJ/h
Q_i (kJ/h) $= Q_w + Q_{tfb} = 2913320 + 1365000 = 4278320$ Kj

Fig. 3 Characteristics of the burner used in the two combustion chambers

a) Radiation losses = 5% (of the heat produced from the yellow waste bag burning +heat produce from 2 burners)[13]. Radiation heat loss generation from waste:
$Q_{Rad} = 0.05 * 4278320 = 213916$ kJ / h

b) Ash heat rate:

$$\dot{Q}_{ash} = \dot{m}_{ash} Cp\, \Delta T \ [15] \tag{1}$$

Where:
$\dot{m}_{ash}$ − mass of ash = 10 kg / h
Cp − for ash = 0.831 kJ / kg °C
ΔT − Temperature difference = (1200 °C − 20 °C) = 1180°C
$\dot{Q}_{ash} = 10 * 0.831 * 1180 = 9805.8$ kJ /h

Fig. 4 Sample of bottom ash remaining from waste incineration.

c) Heat to dry combustion Products:

$$\dot{Q}_{prod} = \dot{m}_{dpro} Cp \Delta T \tag{2}$$

Where:
$\dot{m}_{dpro} = (\dot{m}_{a.exces} + \dot{m}_{tco2})$ kg/h
$\dot{m}_{dpro}$; mass waste dry products kg/h = 2034.07 + 307.83 = 2341.90 kg/h
$\dot{m}_{a.exces}$; Total excess air kg/h = 2034.07 kg/h
$\dot{m}_{tco2}$; Total CO2 generated kg/h = 307.83 kg/h
Cp − for dry products = 1.086 kJ / kg °C
ΔT - Temperature difference = (1200 − 20) °C = 1180 °C
$\dot{Q}_{prod} = 2341.90 * 1.086 * 1180 = 3001098.01$ kJ / h

d) Heat to moisture

$$\dot{Q}_{moist} = \dot{m}_{moist} Cp \Delta T + \dot{m}_{moist} Hv [12]\ [13] \tag{3}$$

Where:
$\dot{m}_{moist}$ − Total mass moisture = 171.62 kg /h
Cp − Water heat capacity const. = 2.347 kJ/kg. °C
ΔT - Temperature difference = (1200 − 20) °C = 1180 °C
H_v − Water vapor latent heat const. = 2460.3 kJ/kg
$\dot{Q}_{moist} = 171.62 * 2.347 * 1180 + 171.62 * 2460.3 = 897531.4$ kJ /h

Total Heat output:

$$\dot{Q}_o = \dot{Q}_{Rad} + \dot{Q}_{ash} + \dot{Q}_{prod} + \dot{Q}_{moist} \tag{4}$$

$\dot{Q}_o = 21396 + 9805.80 + 3001098.01 + 897531.40 = 3929831.21$ kJ/h
Net Heat Balance $= (\dot{Q}_i - \dot{Q}_o)$ kJ/h $= 4278320 - 3929831.21 = 348488.79$ kJ/h

2.4.5. Efficiency of the Machine.

$$\eta_{incinerator} = \frac{Energy\ output}{Energy\ input} = \frac{\dot{Q}_o}{\dot{Q}_i} \quad (5)$$

$$\eta_{incinerator} = \frac{3929831.21}{4278320} = 91.85\ \%$$

3. Conclusion

The experiment calculates the heat, mass and energy from the incineration of medical waste using a double-stage diesel furnace and the effect of recovering the positive thermal energy generated from the incineration of the waste component. Could use the heat generated from waste burning to heating water that supplies the hospital; this system is more valuable with a large waste quantity.

The theoretical calculations for an incinerator located in one of Baghdad hospitals, in the Medical City, and the temperatures were constant in the burning: 800°C-1200°C for the primary and secondary chamber. Our reading was depending on actual waste which daily collected from the hospital and diesel consumption.

The inputs and outputs of waste, air, and fuel studied were close to the standard range. The heat generated is compared to the energy produced by the burners in the combustion chambers the energy of medical waste. The result was a very small difference due to the amount of diesel fuel consumed for the burners or heat loss in the insolation area.

As per the experiment, medical waste incinerators are appropriate for treating waste, reducing its volume and weight, purifying it, and recovering useful energy.

4. References

1. World Health Organisation et al., "*Safe management of wastes from health-care activities*," p. 329, 2014, [Online]. Available: http://apps.who.int/iris/bitstream/10665/85349/1/9789241548564_eng.pdf.
2. M. S. Arti Pamnani, "*Muncipal solid waste management in India: A review and some new results*," I.J. Civ. Eng. Technol., vol. **5**, no. 2, pp. 1–8, (2014).
3. E. Okechukwu, C. Chukwutoo, and E. Chibundo, "*Theoretical design of a non-energy recovery incinerator for awka municipality*," JEAS, vol. **13**, pp. 154–166, (2018).
4. H. A. A. Salam, "*Operating and emission characterstics of a novel design four chambers infectious meat incinerator*," IJRET pp. 322–328, (2013).
5. D. T. Lup, A. M. Stroe, P. M. Chezan, E. Pica, "*The importance of waste incineration*," Studia Universitatis Babeş-Bolyai Ambientum, 63. pp. 43–48, (2018), doi: 10.24193/subbambientum.2018.1.04.
6. R. O. Adu, S. F. Gyasi, D. K. Essumang, and K. B. Otabil, "*Medical Waste-Sorting and Management Practices in Five Hospitals in Ghana*," J. Environ Public Health, vol. **2020**, (2020).

7. Y. Yang, H. Wang, K. Chen, J. Zhou, S. Deng, and Y. Wang, "*Shelter hospital mode: How do we prevent COVID-19 hospital-acquired infection?*," I. C. H. Ep., vol. **41**, no. 7, pp. 872–873, (2020), doi: 10.1017/ice.2020.97.
8. S. Request, T. O. Fogarty, P. Feel, F. To, D. To, and P. Interested, "*Small-scale Medical Waste Incinerators : Evaluation of Risks and Best Management Practices*," no. October 2003, pp. 55–65, (2004).
9. F. Lombardi, E. Lategano, S. Cordiner, and V. Torretta, "*Waste incineration in rotary kilns: A new simulation combustion tool to support design and technical change*," Waste Manag. Res., vol. **31**, no. 7, pp. 739–750, (2013), doi: 10.1177/0734242X13484187.
10. J. Wajs, R. Bochniak, and A. Golabek, "*Proposal of a mobile medical waste incinerator with application of automatic waste feeder and heat recovery system as a novelty in Poland*," Sustain., vol. **11**, no. 18, (2019), doi: 10.3390/su11184980.
11. O. O. Olanrewaju and R. J. Fasinmirin, "*Design of Medical Wastes Incinerator for Health Care Facilities in Akure*," J. Eng. Res. Reports, vol. **5**, no. 2, pp. 1–13, (2019), doi: 10.9734/jerr/2019/v5i216919.
12. J. Shiju Easo and C. Nanjunda Swamy, "*Design of incinerator for the treatment of bio-medical solid waste in chikmagalur city*," J. Ind. Pollut. Control, vol. **27**, no. 2, pp. 173–179, (2011).
13. R. Ganguly, P. Vasistha, and A. K. Gupta, "*Design of an incinerator to treat combined biomedical wastes generated from four major hospitals in Chandigarh and Shimla City, India*," Res. J. Pharm. Biol. Chem. Sci., vol. **8**, no. 3S, pp. 201–212, (2017).
14. R.Walter and S.P. Niessen, "*Combustion and incineration processes*," 4th Edition**.,** (2010).
15. R. K. Patel, "*Design of Biomedical Waste Incinerator*," Int. J. Res. Appl. Sci. Eng. Technol., vol. **V**, no. IX, pp. 436–442, (2017), doi: 10.22214/ijraset.2017.9064.

The environmental situation of the ash medical waste in Baghdad city, Iraq

Sari Jaber[1], *Alaa Aldin* Aljawad[1], *Tudor* Prisecaru[1] and *Elena* Pop[1,*]

[1]University Politehnica of Bucharest, Faculty of Mechanical Engineering and Mechatronics

Abstract. The paper presents and discusses medical waste issues (including the hazardous ones 15%) and their proper management in Iraq. Hence, improper handling of infectious medical waste endangers human health and the environment. Incineration is considered one of the effective ways to dispose of medical waste, especially in the time of Coronavirus (COVID-19) can eliminate all pathogens, reduce waste weight by more than 60-75% by weight and reduce volume accordingly. But that goes with ♀ a large quantities of heavy metals and reactive, toxic materials produced, which are throws in municipal in the landfill area. That causes pollution of the soil, groundwater and air as a result of its light flying ash. The only disposal method available in Iraq is incineration in purpose-built incinerators. This paper examined three incineration processes for three hospitals in Bagdad; Medical City Complex, Al-Amal National Hospital and Al Shifa coronavirus Hospital, and we worked on burning waste and analyzing ash. The results were that most of the waste after incineration had high heavy mineral concentrations and a small amount of other materials.

1. Introduction

Medical waste can define as any solid wastes generated in the diagnosis, treatment or immunization of human beings or animals, research about it, or the production or testing of biological [1]. Because of increasing quantities and inadequate management system, medical waste management and selection of appropriate management systems are also important global problems worldwide. Incineration and disposal of the resultant, such as ash, could be determined as a proper application for managing medical waste. However, it is possible to obtain a 70% mass and 90% volume reduction in medical waste using the incineration process [2]. Considerable amounts of solid residuals, such as bottom and fly ashes containing heavy metals, are produced during this process [3, 4]. Furthermore, medical waste defined as any solid waste generated in the diagnosis, treatment or immunization of human beings or animals in related research, biological production or testing. The hazards to human health and the environment posed by medical waste justify a high level of concern with its management. Medical waste classified according to its source, typology and risk factors associated with handling, storage and ultimate disposal. Treatment by incineration and disposal of the resultant ash by landfilling is the most widely

* Corresponding author: elena.pop@upb.ro

used treatment process for managing medical waste. A properly operated incinerator reduces the waste volume by about 90%, decreasing final residues to a relatively small amount.

Suppose correct practices of main tenancy and operation used during the incineration process. In that case, ashes can be innocuous, being possible to landfill them [5, 6]. According to the World Health Organization (WHO), 80 wt % of the waste generated at health units present similar properties with the corresponding of municipal solid wastes. The remaining 20% has been classified as hazardous, appearing properties such as infectivity, toxicity, carcinogenicity, radioactivity, thus requiring special treatment [7, 8]. This typically contains many kinds of potentially infectious and toxic materials [9]. Medical waste that is not properly handled and disposed of represents a high risk of infection or injury to healthcare personnel and lesser risk to the general public through the spread of micro-organisms from healthcare facilities into the environment [10]. Medical waste is mainly composed of 26.69% organic garbage, 25.10% paper, 35.29% medical plastic waste, 8.13% glass, 0.44% metal, 1.19% textile fibre, and 3.17% wood timber [11], which has characteristics of high energy-dense and high volatiles since it contains a high proportion of organic matters [12], and found that the evolved volatile products from medical waste pyrolysis included 2-butanone, benzaldehyde, formic acid, acetic acid, hydrocarbon, carbon dioxide, carbon monoxide, and water [8, 11].

Baghdad, Iraq capital, the yearly amount of medical waste-collecting form 294 health institutions is approximately 1902 tons. Medical wastes collected separately from health institutions transported with special vehicles to medical waste site incineration located in said or outside the hospital. The capacity of the incineration is different depending on the number of beds. Some have a one or two-stage combustion system consisting or a rotary kiln or a secondary combustion chamber; at a temperature between 400 and 1000°C, the ash transferred to landfill sites. Volume and mass reductions of incineration facility are 95% and 75%, respectively [3].

2. Waste characteristics suitable for burning

According to the specifications of the World Health Organization (WHO):

- The waste treated by burning must have a calorific value of no less than 14650 kJ / kg;
- The proportion of combustible materials is more than 60%;
- The percentage of non-combustible materials is less than 5%;
- The moisture content is less than 30%.

As for the waste not suitable for burning, it represented by the following:

- Compressed gas containers;
- Imaging waste (silver salts) and radiography;
- Plastic waste as a PVC material;
- Large quantities of active chemical waste;
- Closed ampoules or ampoules containing heavy elements;
- Waste containing large amounts of mercury and cadmium, such as a broken thermometer and used batteries.

2.1 The factors that enhance the efficiency of the burning process

The efficiency of the burning process lies in the occurrence of a case of complete combustion of waste, that is, the conversion of organic materials in the incinerator to neutral gases such as carbon dioxide gas, water vapour, sulphur oxides and nitrogen, and the factors that join this case:

- The characteristics of the burned waste and the feed rate of the incinerator, as the presence of large quantities of plastic, prevents the completion of the burning process well, as this material, when exposed to high temperatures, softens and kneads and surrounds the rest of the waste, preventing it from burning completely, in addition to that burning the plastic and the presence of the resulting water vapour generates. One of the main causes of emission of dioxin is hydrochloric acid, which causes erosion of the liner of the incinerator over time, the burning of plastics, especially polyethylene vinyl chloride (PVC).
- Moisture content: the higher the moisture content, the more time the burning process needs.
- Temperatures: the temperatures must be in a manner that ensures the combustion process's completion. For medical waste, the temperatures at which the combustion process takes place ranges between (900-1200) °C.
- The presence of a mechanism for turning the waste inside the incinerator (Baffles), where the burning efficiency increases with the degree of stirring to ensure that it exposed to oxygen in quantity sufficient to complete the burning.

2.2 Waste incineration products

The waste incineration products, in general, can be divided into two main parts:

- Emission and releases to the external environment or the atmosphere, which include multiple gases and volatile particles, and these gases, are usually carbon oxides, sulphur oxides, chlorides and fumes of organic compounds of halogens or hydrocarbons in addition to dioxins and furans whose concentration increases, with an increase in the percentage of burnt plastic, especially PVC. The volatile includes fly ash, which forms a percentage of the total ash content. It contains a percentage of heavy elements in its standard form or its salts (zinc, cadmium, nickel, lead, mercury, manganese, etc.);
- Ash left in the incineration chamber (bottom ash), and some materials are not combustible. The ash left inside the combustion chamber constitutes (75-90%) of the total ash generated from the burning process. It contains a percentage of the aforementioned heavy metals, which makes the ash of hazardous waste due to the effect of its high toxicity on the groundwater if it disposed of by burying it in the soil without pre-treatment, such as placing it in special containers before burying it and preparing the site correctly.

2.3 Bottom and fly ashes characterization

There are two types of ashes produced from medical waste burning:

- The fly ash derived from the incinerator waste gases de-dusting system (fabric filter dust collector);
- The bottom ash produced from the heavier particles of the agglomerated remaining matter precipitated and accumulated usually in the combustion chamber.

The coarser bottom ash appeared black, opaque and glassy, while the finer fly ash, grey in colour (figure 1).

Fig. 1. Incineration ashes, produced during the combustion of medical wastes [2]

Most heavy metals migrate or concentrate in the fly ash and bottom ash depending on the compounds of heavy metals and their physicochemical properties during incineration. The migration of heavy metals during incineration is showing in figure 2.

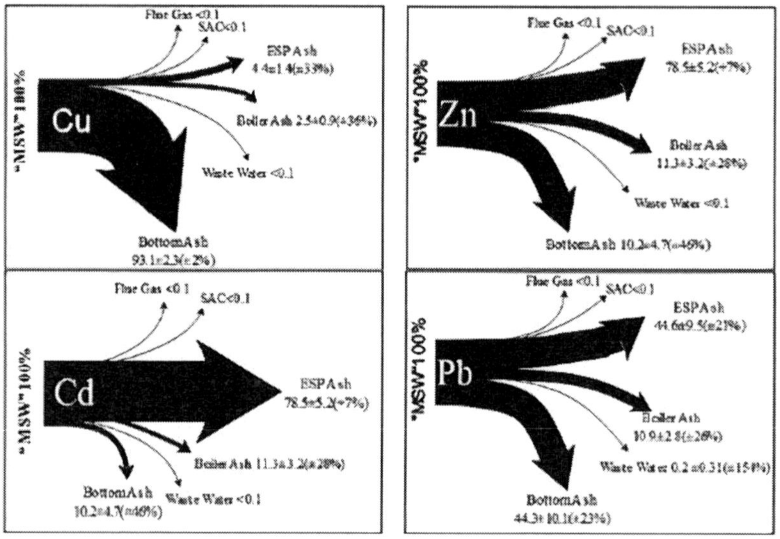

Fig. 2. Partitioning of Cu, Zn, Cd, and Pb during medical waste incineration [13]

3. The study & experiments

3.1 Area of study

The medical services sector is responsible for providing health care to the citizen through the services it provides in diagnosis, examination and treatment. This important vital sector includes health institutions such as government and private hospitals, health centres, health laboratories, blood banks, etc. In Iraq, the total number of health institutions in Baghdad city was 294, the total number of government hospitals was reached 49, while the total numbers of private hospitals were 37 and the total number of other health institutions reached was 208 as shown in figure 3.

The typical medical waste is sampled from three health institutions in Baghdad city to compare the results in this study.

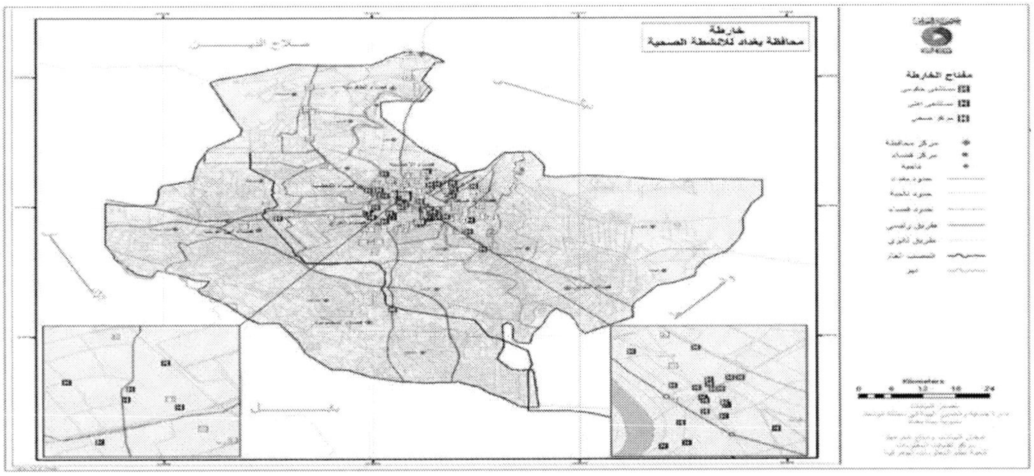

Fig. 3. The health institutions' sites, (government hospital - private hospital – health centre) in Baghdad [15]

The total and infectious healthcare waste generation rates by the number of hospital beds for a high-income country, as shown in the table 1.

Table 1: The total and infectious healthcare waste generation rates by The number of hospital beds [14]

United state	Total waste		Infectious waste	
No. of hospital beds	Kg/bed a day	Kg/patent a day	% of total waste	Kg/day(calculated)
<100	2.59	5.13	13.3	0.34
100-299	4.70	7.16	15.0	0.70
300-499	5.67	8.63	14.9	0.84
>500	5.83	0.69	14.9	0.87
total	4.18	6,93	15.0	0.63

Baghdad produces approximate 1901.8 tons of medical waste per year. The government hospital produces 1614 ton/ year, the Private hospital produces 103.7 ton/year and the health centre produces 184.1 ton/year, as shown in figure 2. After the pandemic of COVID-19 the medical waste has doubled in Baghdad city. For example, in December 2019 medical wastes were 82537 kg, whereas in January 2021 the medical waste reached 174385 kg.

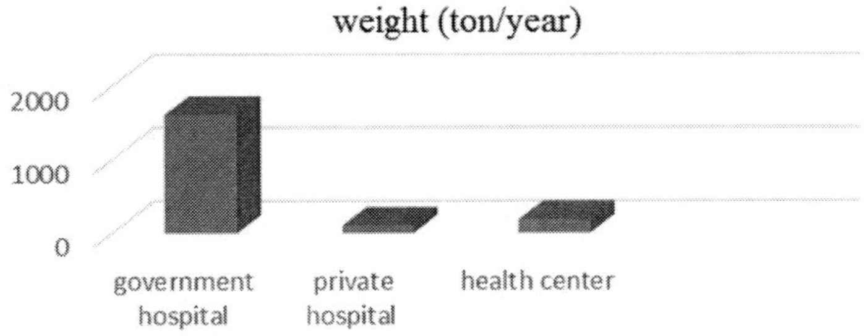

Fig. 4 The total weight of all medical waste (ton/year)

This study is a comparison between three waste incineration operations in three types of hospitals in Baghdad and the resulting ash study:

- Medical City Complex.
- Al-Amal National Hospital.
- Al Shifa Corona Virus Hospital

3.2 The work steps

We weighed the required amount of 30 kg of medical waste, then loaded it in the incinerator and ran the incinerator for an hour to monitor the combustion temperature. On the second day (the hospital safety department instruction), we open the chambers to collect the ash sample.

Fig. 5 Ash samples after incinerating medical waste

3.1. Medical City Complex

A medical city complex is a group of hospitals that contain 2715 beds. The medical waste production average per hospital is between (0.5-1) kg for each bed per day. The hospitals complex incinerator has two combustion chambers. Each of them has one diesel burner, which raises the temperature of the primary chamber to 400°C, and the secondary chamber raises the temperature to (600-700)°C. The tested waste weight is 30 kg, and after burning, the remaining ash is 9 kg, whereas the sampled ash weight is only 6.7 grams.

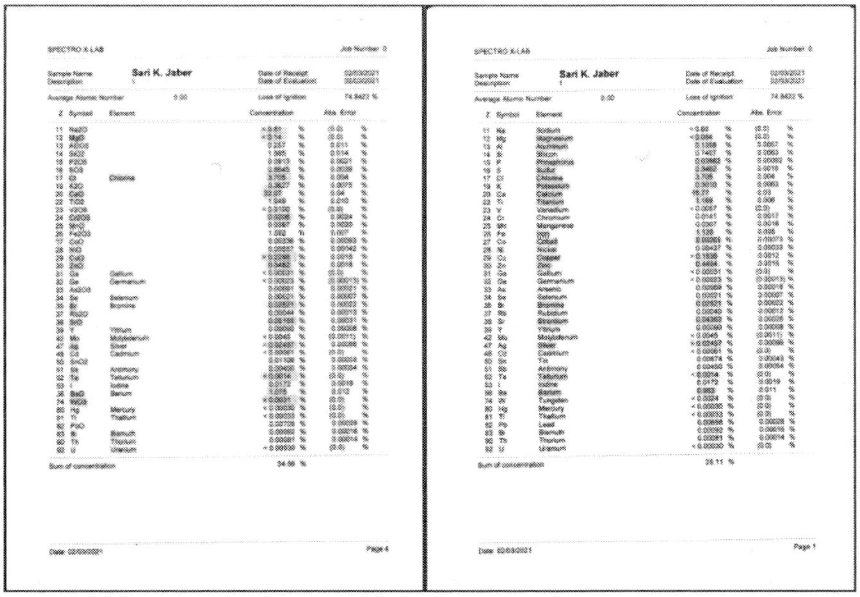

SPECTRO X-LAB

Sari K. Jaber

SPECTRO X-LAB

Sari K. Jaber

Fig. 6 The result of the laboratory test of the elements and oxides in Medical City Complex.

Note the element and oxides, yellow-marked are above the standard safe level for use in sanitary landfills.

Table 1. Concentrations of heavy elements present in the ash resulting from burning MW

Heavy elements	PPM	Permissible PPM Ministry of Environment in Iraq	
(Pb) Lead	65.8	530	
(Fe) Iron	11200	6000	+
(Cu) Copper	1836	190	+
(Zn) Zinc	4404	720	+
(Cd) Cadmium	6.1	12	
(Cr) Chromium	141	380	

Table 4. Concentration of another elements present in ash resulting from burning MW

Author's elements	PPM	Permissible PPM Ministry of Environment in Iraq	
(Ca) Calcium	157700	12	+
(Hg) Mercury	245.7	10	+
(Ni) Nickel	43.7	210	
(Sb) Antimony	45	15	+
(Se) Selenium	2.1	100	
(Ag) Silver	245.7	15	+
(Ti) Thallium	3.3	15	
(As) Arsenic	6.9	55	

3.2 Al-Amal National Hospital.

This hospital is specialized in cancer disease treatment and it contain 200 beds. Each bed produces between (0.5-1) kg of waste daily. The hospital incinerator contains two combustion chambers. The primary chamber has two diesel burners raise the temperature to 400°C, and the secondary combustion chamber has one diesel burner that raises the temperature to 600°C. The tested waste weight is 30 kg and the remaining ash after the burning is 15 kg whereas the sample ash weight is 5.25 grams.

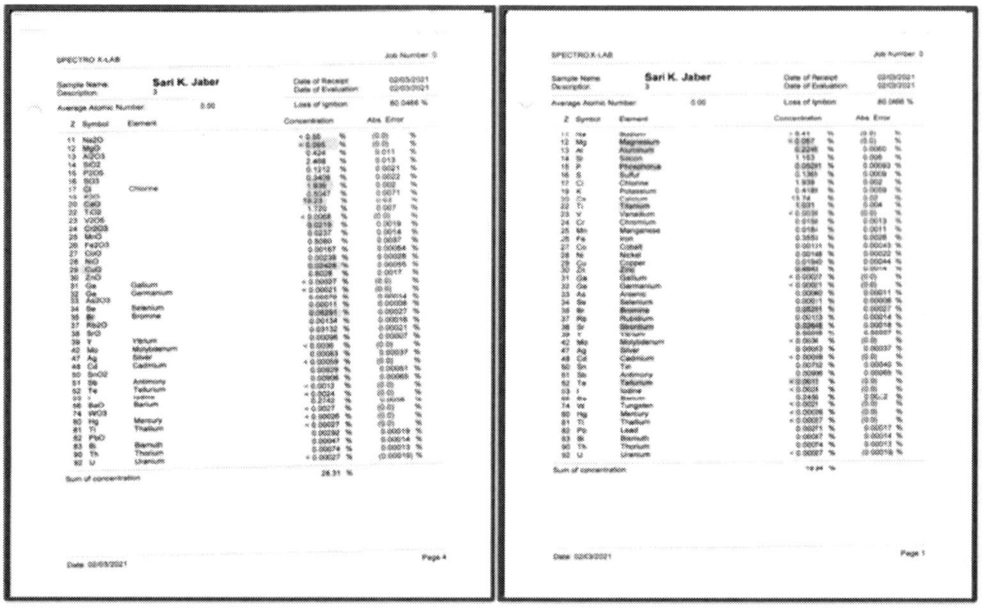

Fig. 8. The result of the laboratory examination of the elements and oxides in the Al-Amal National Hospital.

Note the element and oxides, yellow-marked are above the standard safe level for use in sanitary landfills

Table 5. Concentrations of heavy elements present in the Ash resulting from burning MW

Heavy elements	PPM	Permissible PPM Ministry of Environment in Iraq	
(Pb) Lead	27.1	530	
(Fe) Iron	3553	6000	
(Cu) Copper	194	190	
(Zn) Zinc	4843	720	
(Cd) Cadmium	5.9	12	
(Cr) Chromium	150	380	

Table 6. Concentration of another elements present in ash resulting from burning MW

Author's elements	PPM	Permissible PPM Ministry of Environment in Iraq	
(Ca) Calcium	137400	12	
(Hg) Mercury	2.6	10	
(Ni) Nickel	18.8	210	
(Sb) Antimony	90.6	15	
(Se) Selenium	1.1	100	
(Ag) Silver	8.3	15	
(Ti) Thallium	2.7	15	
(As) Arsenic	6.0	55	

3.3 Al Shifa coronavirus Hospital

This hospital is a healing center for the (COVID-19) patients, as it contains 250 beds and produces (5-6) kg of medical waste per bed /day. The hospital incinerator contains two combustion chambers. The primary chamber has a diesel burner that raises the temperature to 400°C, and the secondary combustion chamber also has a diesel burner that raises the temperature from 400 °C to 800 °C. The tested material weight is 30 kg, and the remaining ash weight after burning is 11 kg, whereas the sampled ash materials are 3.5 grams.

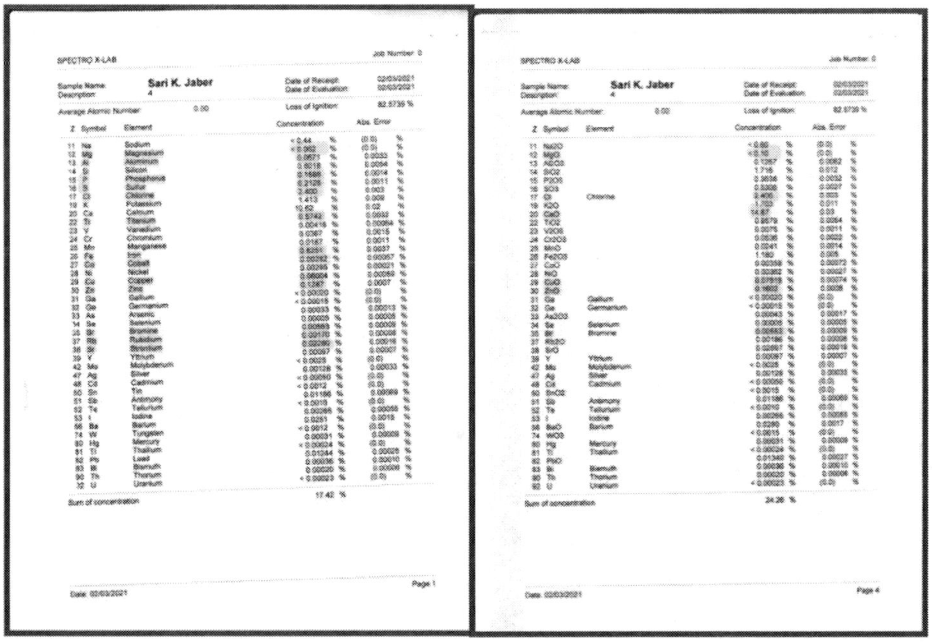

Fig.9. The result of the laboratory examination of the elements and oxides in the Al-Amal National Hospital.

Note the element and oxides, yellow-marked are above the standard safe level for use in sanitary landfills.

Table 7. Concentrations of heavy elements present in the ash resulting from burning MW

Heavy elements	PPM	Permissible PPM Ministry of Environment in Iraq	
(Pb) Lead	124.4	530	
(Fe) Iron	8251	6000	
(Cu) Copper	6004	190	
(Zn) Zinc	1287	720	
(Cd) Cadmium	5	12	
(Cr) Chromium	367	380	

Table 8. Concentration of another elements present in ash resulting from burning MW

Author's elements	PPM	Permissible PPM Ministry of Environment in Iraq	
(Ca) Calcium	106200	12	
(Hg) Mercury	3.1	10	
(Ni) Nickel	28.5	210	
(Sb) Antimony	118.6	15	
(Se) Selenium	55.3	100	
(Ag) Silver	12.8	15	
(Ti) Thallium	2.4	15	
(As) Arsenic	1.5	55	

4 CONCLUSIONS

We worked on burning medical waste according to its arrival from the hospital, and the experience was in each hospital and its incinerator separately. To analyses the ash, we found that the proportions of heavy minerals and amount of another materials is high. One of the main reasons is that the process of isolating waste before incineration, even if it is present, is imprecise and efficient. Then the ashes are thrown in the municipality dumps and not in certain areas for this purpose, which causes them to spread in the air. It causes serious diseases in the respiratory system. The penetration and oxidation of minerals into landfill soil is a major risk to the soil. It may also reach the groundwater. For this reason, it is necessary to define areas designated for burying medical waste that differ from those used for municipal waste. Or mixing heavy metals with bricks and cement is an excellent idea for fixing the heavy metals in the cement and are useful in construction operations.

References

1. L. WJ, Li J, Wu BL, Yi JQ Xie R, "*Emissions investigation for a novel medical waste incinerator. J Hazard Mater.*," J. H. M., vol. **166**, no. 1, pp. 365-371, Jul (2009).
2. S. Papamarkou, P. Tsakiridis, G. Bartzas, K. Tsakalakis, C. Tsakalou, "*Characterization and leachability evaluation of medical wastes incineration fly and bottom ashes and their vitrification outgrowths*," J. Env. Che. Eng., vol. **6**, no. 1, pp. 367-376, (2018).
3. E.Tınmaz Köse, A. Yıldız. A. Akyıldız, "*Compressive strength and heavy metal leaching of concrete containing medical waste incineration ash.*," C. B. Mat., vol. **138**, no. 1, pp. 326-332, (2017).
4. A.Ferraz, S. Afonso, M.M, "*Incineration of different types of medical wastes: emission factors for gaseous emissions*," At. Env., vol. **37**, pp. 5415-5422, (2003).
5. A. Ferraz, B. Cardoso, S. Pontes M.M, "*Concentration of Atmospheric Pollutants in the Gaseous Emissions of Medical Waste Incinerators*," J. A. & WMA., vol. **50**, pp. 131-136, (2011).
6. WHO, "*Guidelines for the safe disposal of unwanted pharmaceuticals in and after emergencies, Report No. WHO/EDM/PAR/99.2.*," Geneva, WHO, (1999).
7. W.H.O, "*Health care waste management rapid assessment tool (RAT), 2nd ed.*," Geneva, (2010).
8. J.Han, B. Zhao, Y. Wang, W. Chen, F. Xing L. Qin, "*Thermal degradation of medical plastic waste by in-situ FTIR, TG-MS and TG-GC/MS coupled analyses*," J.A. and Ap. Pyr., vol. **136**, pp. 132-145, (2018).
9. Qiushuang Li, Yuxi Ge, Chen Jiang, Zhiyuan Ren, Ren, "*Use Incentive Approach to Promote BAT/BEP for Centralized Incineration Facilities of Medical Waste in China*," Pro. Env. Sci., vol. **31**, pp. 457-464, (2016).
10. T. Chen, J.Wu, J. Wu, J.Zhang, "*TG-MS analysis and kinetic study for thermal decomposition of six representative components of municipal solid waste under steam atmosphere,*" J. Waste Manag., vol. **43**, pp. 152-161, (2015).
11. N. Deng, Y. Zhang, Y. Wang, "*Thermogravimetric analysis and kinetic study on pyrolysis of representative medical waste composition*," J. Waste Manag., vol. **28**, pp. 1572-1580, (2008).
12. M. Su-Ling, Brooks, E. Windfeld, "*Medical waste management – A review,*" J. Env. Man., vol. **163**, pp. 98-108, (2015).

13. H. Liu,Guo-XiaWei Fang Liu, "*Characteristics and Treatment Methods of Medical Waste Incinerator Fly Ash: A Review*," Tianjin, September 26,vol.**10,** pp.173, (2018).
14. F.Zhang, Z. Hao, H.Wang. L. Zhao, "*Levels of polycyclic aromatic hydrocarbons in different types of hospital waste incinerator ashes*" Sc. Tot. Env., vol. **397**, pp. 24-30, (2008).
15. Iraqi Ministry of Environment. Technical department.

Selective, weighted morphological research, by dividing the equipment for biocompost production by submorphologies

Mihai Olan[1], *Anişoara* Păun[1], *Paul* Găgeanu[1], *Nicoleta* Ungureanu[2*], *George* Bunduchi[1] and *Alexandru* Zaica[1]

[1] National Institute of Research – Development for Machines and Installations Designed to Agriculture and Food Industry – INMA Bucharest, Ion Ionescu de la Brad no. 6, sector 1, Romania
[2] University Politehnica of Bucharest, Department of Biotechnical Systems, Splaiul Independentei no. 313, sector 6, Romania

Abstract. This paper describes the use of the method of selective, weighted morphological research by dividing by inventics submorphologies to analyze the equipment for biocompost production. The selective morphological research method was used to find the optimal constructive solution which is then developed in the research, design and testing of an innovative piece of equipment. Using Bellman's principle and the method of morphological analysis, the known constructive solutions will be studied, their weighting will be achieved by dividing by submorphologies and the Bellman diagram will be constructed to determine the weighting coefficient of the optimal solution.

1 Introduction

The determination of the optimal AoE solution, maximal, is made starting from Bellman's principle according to which "an optimal policy has the property that whatever the initial state and initial decision are, the remaining decisions must constitute an optimal policy with regard to the state resulting from the first decision."

The increasing amount of organic waste generated in cities is a major problem for the environment and human health, and one of the solutions to solve this problem is the composting technology. The recovery of organic waste produced in kitchens in the production of compost is very important [1]. The first research in this regard was carried out in 1926 in Austria, where an industrial station was built for the transformation of urban organic waste into compost.

Composting is a biological, aerobic process of controlled decomposition of organic waste with the help of microorganisms [2], through which compost, a product similar to humus and extremely useful for horticulture and agriculture, is obtained [3, 4].

* Corresponding author: *nicoletaung@yahoo.com*

Composting reduces or prevents the release of methane into the atmosphere. It is estimated that methane released by the decomposition of organic matter is 26 times stronger than carbon dioxide [5], both being greenhouse gases that affect the environmental sustainability.

Composting technology involves several stages and takes place under controlled conditions, as it requires the regulation of pH, temperature and moisture [6], as well as a correct mixture between materials rich in carbon and nitrogen [7].

Bian et al. (2019) [8] studied the composting technology on a pilot scale, using several mixed waste recipes (agricultural waste, including manure, vegetable leaves and rice husks in a mass ratio of 6:3:1). Their study aimed to determine the influence of temperature and bioconversion time on composting performance, evaluated by the content of moisture, nutrients and carbon, respectively the C/N ratio of compost. The optimal parameters were obtained at 75±5°C and 18 hours, and the obtained compost met the requirements of the quality standard.

If the compost pile contains manure, this particular organic waste can favor the heating of the compost due to the activity of decomposing microorganisms. It is recommended that the moisture in the compost pile be 60% and the temperature sould be between 65 and 70°C [9, 10]. To obtain a quality compost, the aerobic fermentation technology at high temperatures requires periodic analysis of the maturity of the compost before its application as an organic fertilizer [11].

The extraction of mono-fertilizers from the compost as well as the odours capture technique can contribute to the improvement of the technical characteristics of the composting [12]. The addition of activators to the raw materials can improve the nutritional quality of the compost [13].

A methodology for the design and manufacture of portable organic waste shredders that can be used in compost production, respectively the factors that influence the shredding process of organic waste was analyzed by Ajinkya et al. [14].

To create an efficient equipment to loosen the compost, deductive logical methods of technical creation can be used by researching the known systems by the method of weighted, selective morphological research, with division by submorphology.

The aim will be to increase the degree of objectivity in choosing the optimal variant and thus an ABCD morphology is divided into the morphologies AB, BC, CD, and to combine biasassemblies from each submorphology, each biasamblation is denoted by a subunit coefficient so that the sum of the coefficients on a line be equal to the number 1, a technique similar to the decision technique imposed in value engineering [15].

In the case of equipment for biocompost production, there is the following classification in forming assemblies:

- A- the basic structure of the equipment;
  - A_1- fixed equipment (Fig. 1);
  - A_2 - mobile equipment.
- B - movement system of the equipment;
  - B_1- towed equipment;
  - B_2 - self-propelled equipment with diesel engine;
  - B_3 - self-propelled equipment with electric drive;
  - B_4 - self-propelled composting equipment.
- C - drum construction:
  - C_1 - with fixed knives:
  - C_2 - with removable knives;
  - C_3 - with drum welded supports and removable knives.
- D - adjustment of working depth:
  - D_1- side leg with height-adjustable wheel and fixed drum;

D_2- side leg with fixed height wheel and height adjustable drum.

A_3 - construction of the drum with knives.

a. According to the type of actuation - the forming assembly B:

A_1 - towed equipment;

A_2 - self-propelled equipment with diesel engine;

A_3 - self-propelled composting equipment.

A_1 - Fixed equipment for compost aeration. The Big Dutchman system (Fig. 1) is an innovative system that allows easy handling, automatic process control, easy adaptability to the amount of material to be composted, quick and easy installation. The final product is a valuable fertilizer that can be used in horticulture and agriculture.

Fig. 1. Big Dutchman equipment [16]

B_1 - Towed equipment. SNDBERGER (Germany) manufactures towed equipment for compost aeration, the transition of the system from the transport position to the working position is achieved by folding the mixing drum frame with a hydraulic system (Fig. 2).

B_2 - Self-propelled equipment with diesel engine. The Crawler Type Compost Turner equipment (Fig. 3) has the direction made with two levers with which it can turn left or right and can return to 180°, the height of the drum with knives being adjusted hydraulically, so that the equipment can loosen piles of compost of max. 1.5 m, the width of the resulting compost pile being 6 m.

Fig. 2. SNDBERGER equipment [17]

Fig. 3. Crawler Type Compost Turner equipment [18]

B_3 - Self-propelled equipment with electric gearmotor. The CMC SF 200 equipment (Fig. 4) made by the Austrian company Compost Systems Gmbh has electromechanical actuation for moving and rotating the drum with knives and achieves a good loosening of the compost material on fixed platforms.

B_4 - Self-propelled composting equipment. WILDCAT has developed a new type of equipment - self-propelled, a construction machine equipped with hydraulic lifting and loading systems. The WILDCAT compost turner model type LS117A (Fig. 5) has a folding side wheel and the drive of the drum with a separate heat engine, being moved by a tractor with its own heat engine.

Fig. 4. CMC SF 200 equipment [19]

Fig. 5. WILDCAT compost turner equipment, model LS117A [20]

2 Material and method

To increase the degree of objectivity in selecting the optimal solution, the morphology ABCD is subdivided into submorphologies AB, BC, CD. To compare the biasassemblies in each submorphology, each assembly is estimated with a subunit coefficient, so that the sum of the coefficients of on a line to be equal to 1, a technique similar to the decision technique imposed by value engineering [21].

The appreciation criterion for establishing the weighting coefficients is the ratio between the productivity and the cost price of the analyzed equipment.

	B_1	B_2	B_3	B_4
A_1	0	0	1	0
A_2	0,1	0,3	0,4	0,2

	C_1	C_2	C_3
B_1	0,1	0,4	0,5
B_2	0,2	0,4	0,4
B_3	0,1	0,3	0,6
B_4	0,3	0,3	0,4

	D_1	D_2
C_1	0,5	0,5
C_2	0,4	0,6
C_3	0,7	0,3

The decision variables for each section are noted as x_0, x_1, x_2, x_3:

X_0: A_1, A_2, A_3
X_1: B_1, B_2, B_3, B_4
X_2: C_1, C_2, C_3
X_3: D_1, D_2

The following notations are made:

$V_I(x_0, x_1)$ - the share of biasassemblies $A_0A=1$
$V_{II}(x_1, x_2)$ - the share of biasassemblies AB,
$V_{III}(x_2, x_3)$ - the share of biasassemblies BC,

$V_{IV}(x_3, x_4)$ - the share of biasassemblies CD,

The total share of tetrasemblies will be:

$$F(x_0, x_1, x_2, x_3)=V_I(x_0, x_1) * V_{II}(x_1, x_2) * V_{III}(x_2, x_3) * V_{IV}(x_3, x_4)$$

3 Results and discussion

In the first stage, the starting point is considered in A_0 and the maximum weight of the AB bias assemblies will be calculated according to the formulas:

$F_1(B_1) = MAX [(f_1 (x_1) * V_{II} (x_1, B_1)],$ $f_1(x_1) = 1$
$F_1(B_2) = MAX [(f_1 (x_1) * V_{II} (x_1, B_2)],$ $x_1=A_1A_2A_3$
$F_1(B_3) = MAX [(f_1 (x_1) * V_{II} (x_1, B_3)],$
$F_1(B_4) = MAX [(f_1 (x_1) * V_{II} (x_1, B_4)],$

By replacing the values it will be obtained:

$F_1(B_1) = MAX (-\infty \times 1 \quad 0,1 \times 1) = 0,1 \quad x_1=A_2$
$x_1=A_1 \quad x_1=A_2$
$F_1(B_2) = MAX (\; (-\infty \times 1 \quad 0,3 \times 1) = 0,3 \quad x_1=A_1$
$x_1=A_1 \quad x_1=A_2$
$F_1(B_3) = MAX (1 \times 1 \quad 0,4 \times 1) = 1 \quad x_1=A_1$
$x_1=A_1 \quad x_1=A_2$
$F_1(B_4) = MAX (-\infty \times 1 \quad 0,2 \times 1)=0,2 \quad x_1=A_2$
$x_1=A_1 \quad x_1=A_2$

The maximum weight of the ABC triasemblies is noted as $f_{I, II} (x_2)$ and it will be obtained:

$f_{I, II} (C_1) = MAX [f_{I, II} (x_2) \times V_{III} (x_2C_1)]$
$x_2=B_1, B_2, B_3, B_4$
$f_{I, II} (C_2) = MAX [f_{I, II} (x_2) \times V_{III} (x_2C_2)]$
$f_{I, II} (C_3) = MAX [f_{I, II} (x_2) \times V_{III} (x_2C_3)]$
By replacing the numeric values, it will be obtained:

$f_{I, II} (C_1)=MAX [0,1x0,1 \quad 0,3x0,2 \quad 1x0,1 \quad 0,2x0,3]=MAX [0,01 \quad 0,06 \quad 0,1 \quad 0,6]=0,1$
$x_2=B_1 \quad x_2=B_2 \quad x_2=B_3 \quad x_2=B_4 \quad (x_2=B_3)$
$f_{I, II} (C_2)=MAX [0,1x0,4 \quad 0,3x0,4 \quad 1x0,3 \quad 0,2x0,3]=MAX [0,04 \quad 0,12 \quad 0,3 \quad 0,06]=0,3$
$x_2=B_1 \quad x_2=B_2 \quad x_2=B_3 \quad x_2=B_4 \quad (x_2=B_3)$
$f_{I, II} (C_3)=MAX [0,1x0,5 \quad 0,3x0,4 \quad 1x0,6 \quad 0,2x0,4]=MAX [0,05 \quad 0,12 \quad 0,6 \quad 0,08]=0,6$
$x_2=B_1 \quad x_2=B_2 \quad x_2=B_3 \quad x_2=B_4 \quad (x_2=B_3)$

In the case of tetraassemblies ABCD, it will be obtained:

$f_{I, II, III} (D_1)=MAX[f_{I, II, III} (x_3) \times V_{IV}(X_3D_1)]$
$f_{I, II, III} (D_2)=MAX[f_{I, II, III} (x_3) \times V_{IV}(X_3D_2)]$
By replacing the numeric values it will result:

$f_{I, II, III} (D_1)=MAX[0,1x0,5 \quad 0,3x0,4 \quad 0,6x0,7]=MAX [0,05 \quad 0,12 \quad 0,42]=0,42$
$x_3=C_1 \quad x_3=C_2 \quad x_3=C_3 \quad (x_3=C_3)$
$f_{I, II, III} (D_1)=MAX[0,1x0,5 \quad 0,3x0,6 \quad 0,6x0,3]=MAX [0,05 \quad 0,18 \quad 0,18]=0,18$

$x_3=C_1$ $x_3=C_2$ $x_3=C_3$ $(x_3=C_2=C_3)$

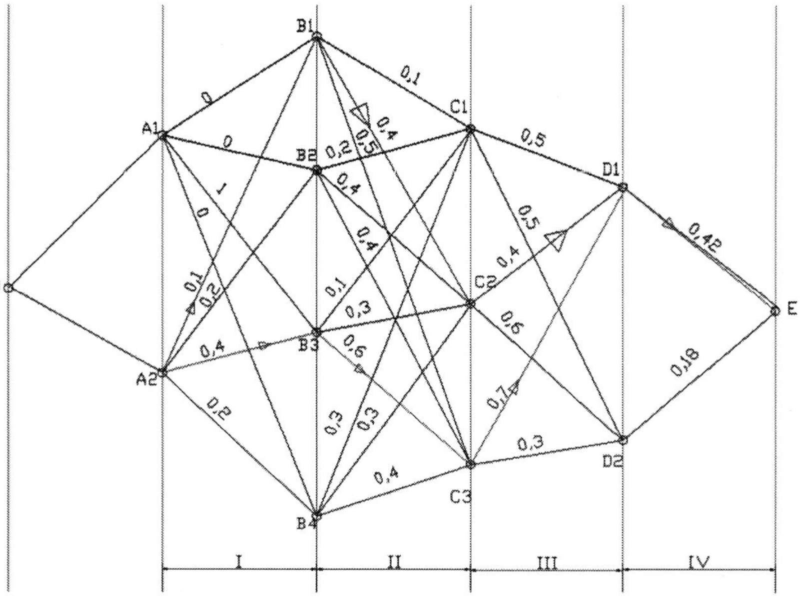

Fig. 6. Dots and arrows diagram

From these matrices will be constructed the diagram of "points and arrows" through the graph in Figure 6 [21].

Analysis of possible constructive-functional variants

After removing the incompatible assemblies, the representative assemblies are calculated:

1. $A_2B_3C_3D_1$= 0,4x0,6x0,7x0,42=0,07056 - Self-propelled mobile equipment with electromechanical drive (gear motors), the drum has welded supports and removable knives.

2. $A_2B_1C_3D_1$=0,1x0,5x0,7x0,42=0,0147 - Mobile towed equipment, the drum has welded supports and removable knives.

Based on a research contract, the two types of equipment were designed and a type of choice is to be executed and tested.

Variant 1 - Self-propelled mobile equipment with electromechanical drive (gear motors), the drum has welded supports and removable knives

This equipment consists of a support (1) with a drum with removable knives (2) and the drive of this drum is performed with a conical-cylindrical electric gearmotor (30). The movement of the system is done with two rubber tracks (4) and (5) driven by two electric gearmotors (6) and (7) which are controlled by an electrical panel (8). To increase the moisture of the compost to the set value, water pipe with spray nozzles (9) is used (Fig. 7).

This system has the following advantages:

- working autonomy – it does not require a tractor to move and drive the drum with knives;
- it has a drum with removable knives that can be replaced and resharpened in case of wear;
- it is an efficient, functional constructive solution and at a lower cost price than the imported equipment.

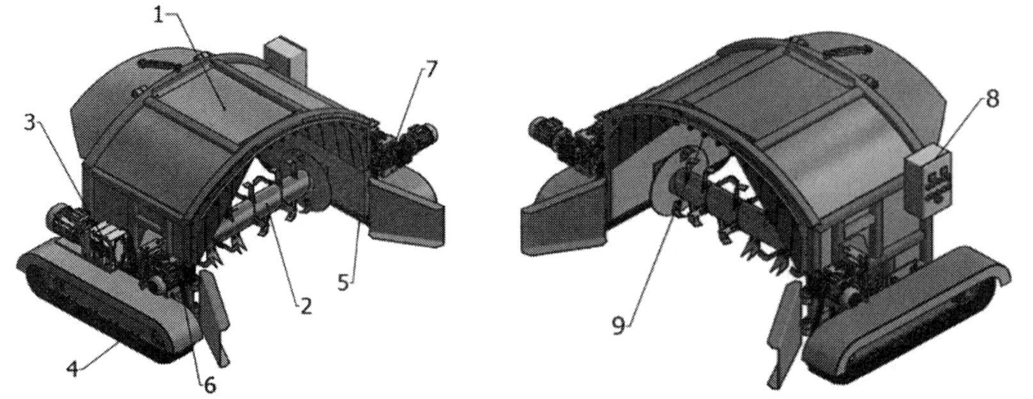

Fig. 7. Self-propelled equipment for biohumus production

Variant 2 - Mobile towed equipment, the drum has welded supports and removable knives

The system consists of a welded frame (1) on which is mounted a drum (2) with welded supports and removable knives. The equipment has a side wheel (3) that can be folded to 90^0 while moving and an articulated frame (4) stiffened with two tie rods (5). The coupling of the equipment to the tractor is made with the lugs (6) and the drive is made with the cardan shaft (7) that transmits the rotational movement to the drum (2) by means of a Gall chain transmission (8), a conical reducer (9) and a cardan shaft (10). For directing the compost from the edges this equipment is provided with two orientation paddles (11) and for wetting the compost the system is provided with a pipe (12) with spray nozzles for water (Fig. 8).

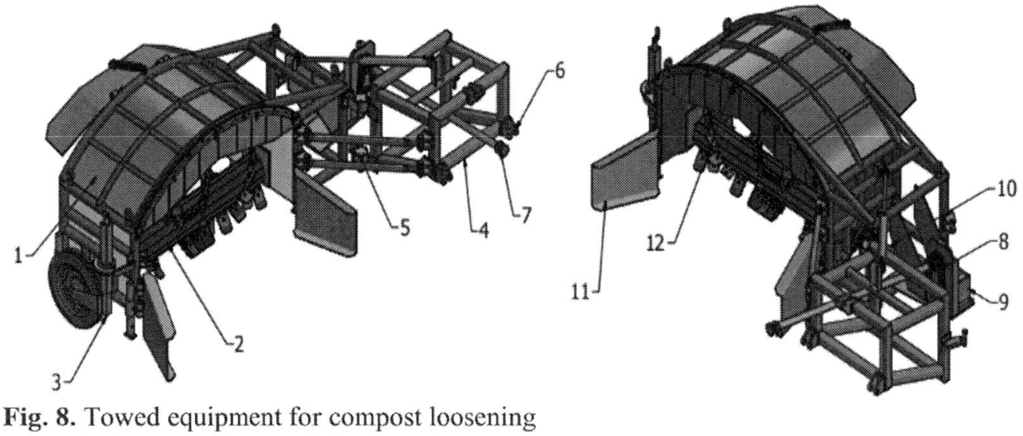

Fig. 8. Towed equipment for compost loosening

This system has the following advantages:

- low cost price, considering the work capacity offered and the degree of aeration;
- operational safety;
-the possibility of folding the entire system during transport and rotating the side wheel support that provides support during transport and during transport.

4 Conclusions

The analysis study of the technical solutions made by the companies with achievements in this field shows that there is a wide range of different equipment from a constructive point of view, but also from the production capacity.

Following the analyis of the main representative technical solutions available on the market and using the method of weighted, selective morphological research, with division by submorphology to analyze existing solutions and find new constructive-functional variants that will be then developed through the design-research activity, in the two representative types of equipment for biohumus production.

Using these results, two pilot installations were designed and executed, and the results obtained from the use of this equipment in compost production confirm the technical data provided from the design stage and demonstrate the usefulness of using this method of scientific research for research in this field.

This work was supported by a grant of the Romanian Research and Innovation Ministry, program Nucleu, PN 19 10 02 02 - CERCETĂRI PRIVIND REALIZAREA UNEI TEHNOLOGII DE PRODUCERE A BIOHUMUSULUI (VERMICOMPOST).

References

1. M. Diacono, A. Persiani, E. Testani, F. Montemurro, C. Ciaccia, Sustainability **11(14)** (2019)
2. S. Singh, L. Nain, Proc. Indian Natl. Sci. Acad. **80** (2014)
3. C.O. Ijagbemi, S. Adepo, IOSR **4** (2014)
4. E.A. Laza, I.L. Caba, V. Vlăduţ, D.M. Bordean, At. Atanasov, Research People and Actual Tasks on Multidisciplinary Sciences, Lozenec, Bulgaria (2019)
5. C. Cristescu C., AGIR Publishing, Bucharest (2008)
6. V. Vlăduţ, E. Marin, I. Voicea, L. Apostol, I. Găgeanu, M. Olan, D. Cujbescu, A. Pruteanu, Gh. Matei, N. Ungureanu, R. Oprescu, M. Dincă, B. Zăbavă, M. Catană, International Symposium ISB-INMATEH – Agricultural and Mechanical Engineering, Bucharest, Romania (2018)
7. I. Voicea, V. Vlăduţ, M. Matache, D. Dumitru, S. Isticioaia, E. Trotuş, V. Arsenoaia, The 15th Annual Meeting "Durable Agriculture – Agriculture of the Future", Craiova, Romania (2019)
8. B. Bian, X. Hu, S. Zhang, C. Lv, Z. Yang, W. Yang, L. Zhang, Bioresource Technol. **287** (2019)
9. S. Sutripta, Procedia Environ. Sci. **34** (2016)
10. A. Ameen, J. Ahmad, S. Raza, Int. J. Sci. Res. Publ. 6 (2016)
11. M. Zhang, S. Aiping, A. Muhammad, Y. Lihua, A. Muhammad, Biomass Convers. Biorefin. **3** (2021)
12. M. Toledo., J. Siles, M. Gutiérrez, M. Martín, M. Waste Manag. **76** (2018)
13. M.S. Ayilara, O.S. Olanrewaju, O.O. Babalola, O. Odeyemi, Sustainability **12(11)** (2020)
14. S. Ajinkya, A. Hande, A. Deshpande, IJIRST **1(7)** (2014)
15. V. Belous, Gh. Doncean, Performantica Publishing, Iasi (2001)
16. https://www.bigdutchman.com
17. https://www.i-bidder.com
18. https://compost-turner.net/
19. www.compost-systems.com
20. www.https://urcrecycle.com
21. T. Stanciulescu T., B. Vitalie, I. Moraru I., Performantica Publishing, Iasi (1998)

Desorption capacity of raw energy bars with fruits of *Zizyphus jujuba* Mill. - Bulgaria

Albena Durakova[1,*], *Adelina* Vasileva[2], *Tzvetana* Gogova[3], *Milena* Temelkova[4] *and* *Rumyana* Krasteva[5]

[1*] Department of Process Engineering, University of Food Technologies – Plovdiv, 26 Maritza Blvd. BG-4002 Plovdiv, Bulgaria
[2] Department of Mechanical and Instrument Engineering, Technical University - Sofia, Plovdiv Branch, 25 Tsanko Duistabanov Str., Plovdiv 4000, Bulgaria
[3] Department of Technology of Grain, Fodder, Bread and Confectionery Products, University of Food Technologies – Plovdiv, 26 Maritza Blvd. BG-4002 Plovdiv, Bulgaria
[4] Language Training Centre, University of Food Technologies – Plovdiv, 26 Maritza Blvd. BG-4002 Plovdiv, Bulgaria
[5] Centre of Physical Education and Sport, University of Food Technologies – Plovdiv, 26 Maritza Blvd. BG-4002 Plovdiv, Bulgaria

Abstract. The current study presents a mixture for raw energy bars of fresh fruits of the exotic fruit Zizyphus jujuba Mill. with origin - Bulgaria. The physico-chemical indices of the ready mixture are analyzed. A high content of carbohydrates - 24.45% and of nourishing fiber - 22.08% is reported. The desorption capacity of the exotic mixture is examined for temperatures of 10°C, 25° and 40°C in eight watery activities from 11% to 85%. The analysis shows that the equilibrium moisture lowers with the raise of temperature in constant watery activities. The desorption isotherms are of type III according to the classification of Brunauer et al. The modified model of Halsey is recommended for the isotherms description according to the estimation criteria distribution, and the model of Oswin - as a second modified model. The values of the monolayer moisture content in a_w <0.5 for temperatures 10°C - 5.46% dry basis (d.b.), 25°C - 5.72% d.b. and 40°C - 4.05% d.b. are calculated Brunauer-Emett-Teller model.

Keywords: *Zizyphus jujuba* Mill, sorption isotherms, raw bars.

1 Introduction

The modelling of the mass transfer processes - sorption characteristics, provides the opportunity the regimes and the ways of treatment, preservation and packaging of the nourishing products to be examined [1-3]. The sorptions isotherms allow to be examined the character and the quality of the different types of connected water [4-5]. The monolayer moisture content (MMC), an important sorption characteristic, influencing the products

* Corresponding author: aldurakova@abv.bg

microbiological stability, is calculated through the Brunauer-Emmett-Teller (BET) model [6-8]. More and more studies are directed to the prospecting of natural plants with a high content of biologically active components in the last years [9] According to Jinous Asgarpanah and Elaheh Haghighat the Zizyphus jujuba Mill (the Chines fruit of the date plant) is between the five most useful medical plants in the world [10]. The fruits of the *Zizyphus jujuba* Mill contain a lot of valuable nutrients for the human organism: sugars, proteins, acids and fats. They are extremely rich in cobalt, iodine, iron, high content of vitamin C [11-12]. The fruits are used except for food, as a medical treatment for disorders of the respiratory and the digestive systems. The fruits have the capacity to lower the blood pressure and to regulate the cholesterol. *Zizyphus jujuba* Mill contains kaempferol and myricetin, flavonoids, characterized with strong antioxidant, cardio protective and anticancer activity. The fruits of *Zizyphus jujuba* Mill contain the glycoside neomirtilin, which has the capacity to regulate the glucose levels with people suffering from diabetes mellitus. A daily intake of a glass of fresh fruits is recommended, as the researchers think that the levels of the blood sugar will normalize in a week in that way [13-14].

The proper and responsible choice of food can be a real challenge. The discovery of food providing energy, both nourishing and healthy, is one of the main goals of the contemporary person. The raw bars prepared in home conditions are a good alternative for satisfaction the organism's needs from the necessary quality of useful materials.

No details for the desorption capacity of a mixture for raw energy bars of fresh fruits of *Zizyphus jujuba* Mill. with Bulgarian origin were found in the carefully made literature survey, which gave us grounds for fulfilment of the present study.

2 Materials and methods

2.1 Materials

The object of the present study is jujube (*Zizyphus jujuba* Mill.) of Bulgarian origin supplied by an experimental facility in the town of Assenovgrad, Bulgaria.

Oatmeal, flax seeds with mulberry, raw sunflower seeds, dried plums and Rozhkov flour – purchased in Bulgaria by "Internet café-BG" ltd, packed by "Zoya bg Organic Shop".

2.2 Methods

Rules for taking of samples and methods of examination – according to Bulgarian State Standard BDS 754:1980/Amendment 4: 2003 [15]

pH – Bulgarian State Standard (BDS) EN ISO 11688:1993 [16];

Dry matter, in weight - BDS EN ISO 12145:2000 [17];

TTA (total titratable acidity) BDS EN ISO 6996:1993 [18];

Fibre - BDS EN ISO 11374:1986 [19];

Fat content, (%) – Soxhlet method via solvent extraction with petroleum ether. BDS EN ISO 6997:1984 [20];

Carbohydrate - BDS EN ISO 7169:1989 [21];

Protein – Direct Kjeldahl method analysis (determination of nitrogen content/ nitrogen determination method) – Regulation (EC) №152/2009 [22];

Energy value - Regulation 1169:2011, Appendices IVX ; XV [23].

Moisture, (%) of the products studied - expressed in dry basis 5 standard method via drying of 5g of flour at 105°C to constant weight, according to AOAC, 1990 [24].

The resulting products dosage is made manually on a scale „Mettler Toledo".

The dough mass kneading is done manually.

The examination of desorption capacity of the ready mixture for raw energy bars is made through the static gravimetrical method [25]. For the description of the sorption are used the modified three-parametrical methods of Oswin, Chung-Pfost, Halsey and Henderson. The coefficients (A, B and C) are determined by a program for nonlinear regression "Statistica" (procedure „Nonlinear estimation"). MMC values are calculated using the model of Brunayer-Emmett-Teller with program Excel by the method of the smallest squares [8]. The whole methods is fully described and presented by Durakova et al., [26-28].

3 Results and discussion

A mixture for raw energy bars of fresh fruits of *Zizyphus jujuba* Mill of Bulgarian origin is prepared. The mixture consists of fresh fruits of *Zizyphus jujuba* Mill – Bulgaria - 41%, oatmeal – 23%, flax seed with mulberries – 9%, dried plums – 22%, raw sunflower seeds – 4%, Rozhkov flour – 1%. The physico-chemical indices of the composed mixture are analyzed and introduced in Table 1.

Table 1. Physico-chemical indices of a mixture of fresh fruits of *Zizyphus jujuba* mill for raw energy bars of 100 gr ready product.

Name of the index	Quantity units	Results	Standard deviation	Conditions
Dry quantity, weight	%	68.27	± 0.69	105 ± 1°C
PH		4.86	± 0.05	25 ± 2°C
Acidity (titratable)	%	0.52	± 0.10	25 ± 2°C
Fats	%	7.96	± 0.33	25 ± 2°C
Carbohydrates	%	24.45	± 0.20	25 ± 2°C
Proteins	%	5.94	± 0.07	25 ± 2°C
Fibre	%	22.08	± 0.53	25 ± 1°C
Energy value	Kcal/KJ	193.2/807.8		

The high percentage content in 100 gr product of dry quantity – 68%, carbohydrates – 24.45%, nourishing fibre – 22.08% make impression from the results. The carbohydrates are of high importance for the functioning of the muscle system, the brain and the central nerve system. The nourishing fibre accelerates the food passage through the gastrointestinal tract, cleaning it from toxins and other unfavourable matters. It is considered that the products with high content of fibre have extremely healthy effect [29].

The analyses for the examination of the sorption characteristics of the mixture for raw energy bars are made for three temperatures - 10°C, 25°C and 40°C in an environment created by eight saturated solutions of salts, sustaining above their surface watery activities (a_w) in the range between 0.11 to 0.85. The initial moisture of the examined product is 42.27%, due to which the moisture migration from the product to the surroundings is interesting, and the examination is hold only for the process of desorption. It is observed that the equilibrium moisture content lowers with the increase of temperature for constant water activity, a behavior relevant to a row of nutrients, studied by many authors [30-31]. As for example for a_w = 0.431 the equilibrium moisture content from 40°C to 10°C has lowered with 3.19% d.b.

Table 2. Equilibrium moisture content EMC, % d.b. of a mixture of fresh fruits of Zizyphus jujuba mill for raw energy bars

Sel	10°C			25°C			40°C		
	a_w	M*	sd**	a_w	M*	sd**	a_w	M*	sd**
LiCl	0.113	8.77	0.05	0.113	5.99	0.11	0.112	4.46	0.06
CH3COOK	0.234	9.48	0.13	0.225	6.78	0.16	0.201	4.54	0.09
$MgCl_2$	0.335	10.28	0.28	0.328	8.02	0.14	0.316	4.85	0.12
K_2CO_3	0.431	10,66	0.08	0.432	9.35	0.09	0.432	7.47	0.09
$MgNO_3$	0.574	16.91	0.14	0.529	10.32	0.15	0.484	8.61	0.08
NaBr	0.622	18.86	0.13	0.576	12.16	0.14	0.532	8.93	0.15
NaCl	0.757	18.98	0.24	0.753	18.11	0.18	0.747	15.30	0.08
KCl	0.868	42.59	0.15	0.843	27.01	0.23	0.823	21.74	0.21

* Average of three repetitions, ** Average deviation of three repetitions

Figure 1 shows the isotherms of the test product. Isotherms are of type III according to the classification of Brunauer et al., 1940 [32].

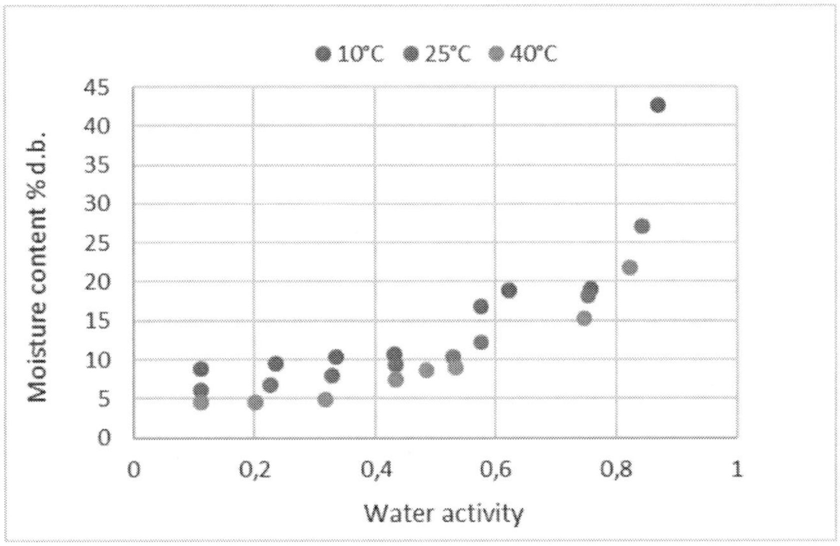

Figure 1. Desorption isotherms for 10°C, 25°C and 40°C.

The resulting coefficients from the present study for three parametrical models (A, B, C) and the relevant values of average relative error P, % and standard deviation SEM are presented in table 3. The graphical analysis of the analyzed sample residues is presented on figure 2.

Table 3. Coefficients of the models (A, B, C), average relative error P, % and standard deviation SEM

Model	A	B	C	P	SEM
Oswin	15.519	-0.172	0.540	15.59	2.49
Halsey	**4.1610**	**-0.029**	**1.604**	**10.01**	**2.17**
Henderson	0.00047	19.392	1.445	17.65	3.75
Chung-Pfost	162.309	0.1460	19.120	18.12	4.88

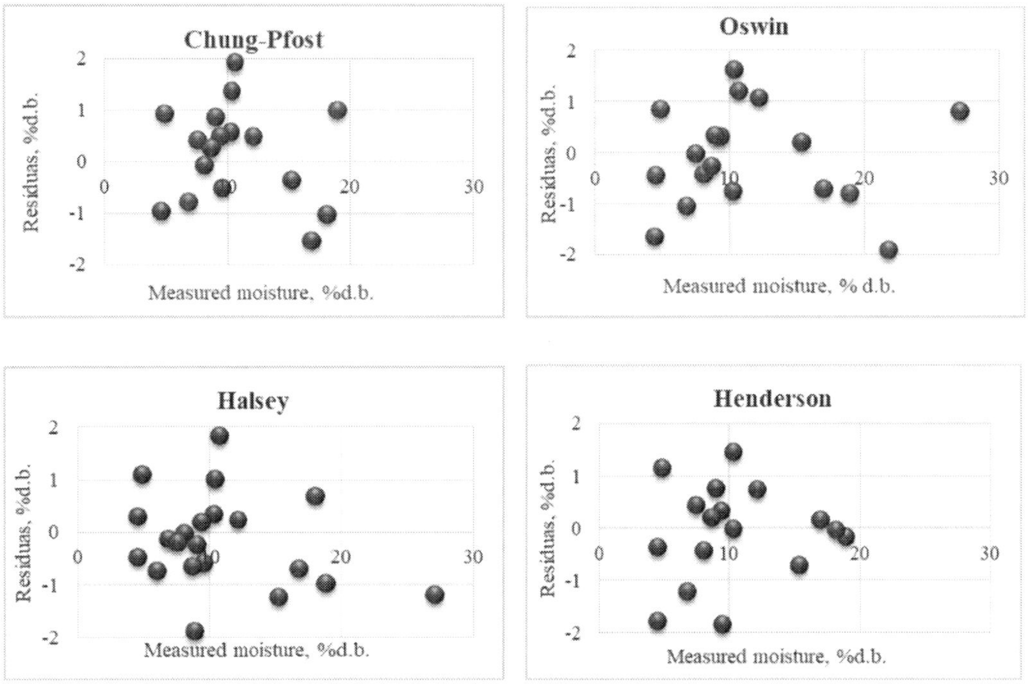

Figure 2. Residues graphical analysis.

The data in table 3 show that the lowest values of P and SEM are taken with the modified model of Halsey. The residues deviation is with random character by the modified model of Oswin from the graphical analysis. As a result, according to the criteria for estimation and usefulness of the models, we can recommend for description the desorption isotherms of a mixture for raw energy bars of fresh fruits of *Zizyphus jujuba* Mill of Bulgarian origin, as the most useful - the modified model of Halsey, and as a second – the modified model of Oswin.

For calculation the MMC the equation of Brunauer-Emmett-Teller is put in a linear type [8]. The linearization with trial data for $a_w < 0.5$ is presented on figure 3.

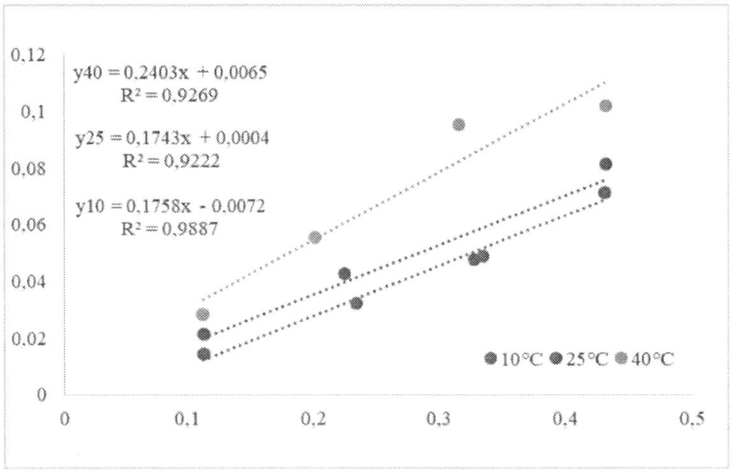

Figure 3. Linearization of the BET model

The values of MMC are calculated from the received coefficients of the linear equations and are presented in table 4.

Table 4. Monolayer moisture content, % d.b.

t (°C)	Desorption
10	5.46
25	5.72
40	4.05

It is obvious from the data in the table that the temperature does not affect MMC. The highest value is calculated for temperature 25°C, and the lowest for 40°C. In relation to the received results we can assume that in moisture of raw energy bars' mixture in the range from to 4.05% d.m. to 5.72% d.m., the microorganisms' growth will be lowered to minimum for a more prolonged period of time [25]. It is confirmed in a surveyed article of Syamaladevi et al., 2016, that flour products' pathogenic microbiological safety is provided through considerable reduction of the product's moisture [33].

Conclusions

1. A data base of the physico-chemical composition of the ready mixture for raw energy bars is received, as the content of carbohydrates -24.45% and nourishing fibre -22.08%, is most highly reported.
2. A database of desorption isotherms of for temperatures 10°C, 25°C and 40°C and aw in the range between 11% to 85%.
3. The desorption isotherms of the finished mixture are of type III according to the classification of Brunauer et al.
4. It is established that the temperature affects the desorption capacity of the ready mixture for raw energy bars of fresh fruits of Zizyphus jujuba Mill.
5. The equilibrium humidity of the studied product decreases with increasing temperature at constant water activity.
6. Throughout the statistic treatment appropriate models responding to the commonly accepted criteria for description of sorption isotherms are recommended – most useful – the modified model of Halsey, and second – the modified model of Oswin. The models can be used for calculation of the equilibrium moisture for the conditions of the experiment.
7. MMC for the process of desorption are calculated with trial data for $a_w < 0.5$ - 10°C - 5.46% d.b., 25°C - 5.72% d.b. and 40°C - 4.05% d.b.

Acknowledgements

This study was part of the Fund Scientific Research "FS N11/20-H", director project: Assoc. Prof. PhD Albena Georgieva Durakova, University of Food Technologies – Plovdiv, Bulgaria.

References

1. R. Moreira, F. Chenlo, J. Sineiro, M. Sánchez, S. Arufe, J. Appl. Phycol. **28** (1), 609-618, (2016).
2. Y. Suhag, G. A. Nayik, V. Nanda, J. Food Meas. Charact. **12** (4), 2553-2560, (2018).

3. R. Kulchan, W. Boonsupthip, P. Suppakul, J. Food Eng. **100** (3), 461-467, (2010).
4. T. P. Labuza, B. Altunakar, Water activity prediction and moisture sorption isotherms. Water activity in foods: fundamentals and applications, 161-205, (2020).
5. M. Dupas-Langlet, J. Dupas, S. Samain, M.-I. Giardiello, V. Meunier, L. Forny, J. Food Eng., **184**, 53-62, (2016).
6. S. Brunauer, P.H. Emmett, E. Teller, JACS, **60** (2), 309-319, (1938).
7. A. Nascimento, M. E. Cavalcanti-Mata, M. E. Martins Duarte, M. Pasquali, H. M. Lisboa, J. Food Process Eng., **42** (6), e13228, (2019).
8. M. Naderi, Surface Area: Brunauer–Emmett–Teller (BET). In Progress in filtration and separation (pp. 585-608). Academic Press, (2015).
9. A. Pereira, M. Maraschin, J. Ethnopharmacol., **160**, 149-163, (2015).
10. J. Asgarpanah, E. Haghighat, J. Pharm. Pharmacol., **6** (31), 2332-2339, (2012).
11. A. M. Pawlowska, F. Camangi, A. Bader, A. Braca, Food Chem, **112** (4), 858-862, (2009).
12. J. Li, L. Shan, Y. Liu, L. Fan, L. Ai, Int. J. Biol. Macromol., **49** (3), 255-259, (2011).
13. S. M. Al-Reza, J. L. Yoon, H. J. Kim, J. S. Kim, S. C. Kang, Food Chem. Toxicol. **48** (2), 639-643, (2010).
14. A. Wojdyło, A. Carbonell-Barrachina, P. Legua, F. Hernández, Food Chem, **201**, 307-314, (2016).
15. B.S. Standard BDS 754:1980/Amendment 4: 2003
16. B.S. Standard, pH - BDS 11688: 1993
17. B.S. Standard, Dry matter - BDS EN ISO 12145:2000
18. B.S. Standard, TTA – BDS EN ISO 6996:1993
19. B.S. Standard, Fibre - BDS EN ISO 11374:1986
20. B.S. Standard, Fat content - BDS EN ISO 6997:1984
21. B.S. Standard, Carbohydrates - BDS EN ISO 7169:1989
22. B.S. Standard, Proteins, in Reglement (EO) №152/2009
23. B.S. Standard, Energy value - Regulation 1169:2011, Appendices IVX ; XV
24. AOAC, 960.39 (Washington, DC), 15 ed. AOAC. (1990)
25. L. Bell, T. Labuza, AACC, 33-36, (2000)
26. A. Durakova, A. Bogoeva, V. Yanakieva, T. Gogova, K. Choroleeva, E3S Web of Conferences, **180**, 03007, EDP Sciences, (2020).
27. A G Durakova, A L Bogoeva, A P Krasteva and M L Temelkova, IOP Conf. Series: Materials Science and Engineering, (2020).
28. Albena Durakova, E3S Web of Conferences **180**, 03008 (2020).
29. L. E. Rojo, C. M. Villano, G. Joseph, B. Schmidt, V. Shulaev, J. L. Shuman, I. Raskin, J Cosmet Dermatol., **9** (3), 185-195, (2010).
30. K. Muzaffar, P. Kumar, Powder Technol., **291**, 322-327 (2016).
31. R. Muangrat, C. Nuankham, Food Sci. Nutr., **6**(3), 585-593 (2018).
32. S. Brunauer, L. S. Deming, W. E. Diming, E. Troller, JACS, **62**, 1723-1732, (1940)
33. R. M. Syamaladevi, J. Tang, R. Villa-Rojas, S. Sablani, B. Carter, & G. Campbell, Comprehensive Reviews in Food Science and Food Safety, **15** (2), 353-370, (2016).

Sorption characteristics of ready-made mixtures containing oatflakes, apples, and cinnamon

Albena Durakova[1,*], *Adelina* Vasileva[2] *and Kornelia* Choroleeva[3]

[1*] Department of Process Engineering, University of Food Technologies – Plovdiv, 26 Maritza Blvd. BG-4002 Plovdiv, Bulgaria
[2] Department of Mechanical and Instrument Engineering, Technical University - Sofia, Plovdiv Branch, 25 Tsanko Duistabanov Str., Plovdiv 4000, Bulgaria
[3] Language Training Centre, University of Food Technologies – Plovdiv, 26 Maritza Blvd. BG-4002 Plovdiv, Bulgaria

Abstract. The present paper studies the sorption characteristics of ready-made mixtures containing oatflakes – 89.5%, apples – 10%, and cinnamon – 0.5 % at the temperatures of 10°C, 25°C and 40°C and for eight water activities in the 11% - 85% range. The results obtained demonstrate that at a constant water activity, the increase in temperature is coupled with a decrease in equilibrium moisture content. The sorption isotherms of the analyzed ready-made mixture belong to Type III according to the classification of Brunauer et al. In order to describe the isotherms following the evaluation criteria, that is mean relative error, standard deviation, and distribution of residuals, the study recommends the three-parameter modified Oswin for adsorption and the Henderson model for desorption. Via linearization of the Brunauer-Emmett-Teller model, we calculate the values of monolayer moisture content (MMC) at aw < 0.5 for both processes. For adsorption, the MMC falls within the 2.51% - 3.57% d.b. range, whereas for desorption it is within the 3.41% - 3.71% d.b. range. Temperature affects the MMC since its increase leads to a decrease in the MMC for both processes. For adsorption, at a temperature between 10°C and 40°C the MMC decreases by 1.06%, and for desorption - by 0.3%.

Keywords: ready-made mixtures, oatflakes, apple, cinnamon

1 Introduction

Latest years have witnessed the increasing application of ready-made mixtures. Their usage helps every household since it is not necessary to mix ingredients or to homogenize and store them separately. On the market, there are many ready-made mixtures for the preparation of various food products – pancakes, muffins, bread and bakery products, cereals, etc. [1-3].

* Corresponding author: aldurakova@abv.bg

The ready-made mixture made of oatflakes, apples, and cinnamon is healthy food due to a number of properties of all three ingredients which are beneficial to human health. Oatflakes are ancient food which humanity has been consuming for more than 4 000 years. They are rich in phosphorus, magnesium, potassium, copper, iron, and B5 vitamins and when consumed raw they contain great amounts of B1 and B9 vitamins. They have a low glycemic index, which means that they are slow to digest and do not affect considerably blood sugar or insulin. Oatflakes can be consumed by people suffering from Type 2 diabetes. What is more, oat-containing products contain beta-glucans, soluble fiber, which have been proven to decrease actively cholesterol levels, which may lower the risk of heart disease [4-7].

Apples are fruit which can be consumed throughout the year. They are tasty and high in nutrients which are beneficial in many respects to the human organism and human health. The main beneficial properties of apples are due to the great amounts of antioxidants and pectin. Apples protect against the harmful effects of free radicals and they simultaneously enhance the natural process of body detoxification. Owing to their high potassium amounts, apples constitute a natural diuretic. Their soluble fiber can lower the blood sugar levels of patients who are highly prone to acquire Type 2 diabetes [8-10].

The medicinal uses of cinnamon are due to the three active components in the essential oils in its bark: cinnamyl acetate, cinnamic aldehyde, and cinnamyl alcohol. Cinnamon is also an excellent source of manganese, fiber, iron, and calcium which are the most important nutrients in the human body [11-14]. A number of studies prove that cinnamon decreases the proliferation of leukemia and cancer cell lymphoma [15-18]. Other studies show that cinnamon contains some polyphenols which may help Type 2 diabetes patients to regulate their blood sugar levels [19]. A study published in the Diabetes Care journal in 2003 discovers that Type 2 diabetes patients, who have taken in on a daily basis between 3 and 6 grams of cinnamon for 40 days, lower their blood sugar levels by 18 - 29% [20].

The modelling of the sorption processes in organic food is very important with respect to choosing regimes of processing and storage. Major sorption characteristics of food products are equilibrium humidity and monomolecular moisture. The moisture of the products affects directly their texture, taste, flavor, quality, and safety during consumption . The excess or deficiency in water content may lead to earlier deterioration, worse quality parameters and effects. This is a basic parameter which is analyzed in different stages and with a greater or lesser intensity. Food samples are regularly tested in order to determine if water activity is maintained within the safety levels established by HACCP. The term "water activity - aw" was suggested for the first time by Scott in 1952 and is widely used in the monitoring of the Critical Control Points in Hazard Analysis Critical Control Point (HACCP) [21].

The in-depth literature review has not revealed data on any research concerning the sorption characteristics of ready-made mixtures containing oatflakes, apples, and cinnamon, which has encouraged us to carry out the present analysis.

2 Materials and methods

2.1 Materials

A ready-made mixture containing oatflakes, apples, and cinnamon purchased in Bulgaria from "Internet café-BG" Ltd, packed by "Zoya bg Organic Shop", certified by BG – BIO 18.

2.2 Sorption isotherms methods

Moisture, (%) of the products studied - standard method via drying of 5g of flour at 105°C to constant weight, according to AOAC, 1990 [22].

For the purposes of the present study, we used the static gravimetric method recommended by Project COST 90 and revised by Bell and Labuza, 2000 [23-24]. The analysis of sorption characteristics – adsorption and desorption – was conducted for temperatures of 10°C, 25°C and 40°C at eight different relative air humidities maintained via the saturated solutions of the following salts: LiCl, CH_3COOK, $MgCl_2$, K_2CO_3, $Mg(NO_3)_2$, NaBr, NaCl, KCl. The samples were dehydrated in advance in a desiccator above diphosphorus pentoxide (P_2O_5) for 20 days, so that adsorption could be studied. For desorption, the samples were hydrated above distilled water (H_2O). In order to conduct the analysis, we used aluminum weighing vessels in which we weighed samples being 1±0.01g in mass. The vessels were placed in hygrostats in which we prepared in advance saturated solutions of salts maintaining above their surface specific constant water activities in the 0.11 - 0.85 range. In the hygrostats, at water activity of $a_w > 0.6$, we placed thymol crystals in order to prevent the microbiological deterioration of the product for the whole duration of the experiment. The hygrostats were placed in thermostats at the respective temperatures. Sample weighing was conducted every three days using an analytical balance with a precision of up to 0.0001g until three equal values were reached (reaching the level of equilibrium moisture content).

The equilibrium moisture content M (% dry mass) was determined via a dehydration method – for 24h at 105°C, the moisture content being calculated by the respective formula [22]. All tests were carried out in triplicate runs.

2.3 Sorption characteristic modeling

To describe sorption isotherms, we used the modified Oswin, Chung-Pfost, Halsey and Henderson three-parameter models. The mean relative error, the standard deviation, and the distribution of residuals were applied as criteria for the evaluation and comparison of the methods.

The Brunayer-Emmett-Teller (BET) model was transformed linearly in order to calculate the values of monolayer moisture content (MMC) [25]. Sorption characteristic modeling has been presented and described in detail by Durakova et al., 2020 [26].

3 Results and discussion

The ready-made mixture, purchased on the market, was characterized by the following percentages of the three components: wholegrain gluten-free oatflakes – 89.5%; apples – 10%; cinnamon – 0.5 % (bioingredients). The nutritive value of 100g of product was: energy value - 1486kJ/352kcal; protein – 11.6g; carbohydrate – 59.7g, of which sugars – 6.7g; fat – 5.5g, of which saturated – 1.2g; fiber – 8.6g, and salt < 0.01g.

In order to analyze sorption characteristics, we dehydrated the ready-made mixture with an initial moisture content of 8.00% to 3.91% for adsorption and hydrated it to 17.60% for desorption. The equilibrium moisture content values for adsorption and desorption are shown in Tables 1 and 2.

Table 1. Equilibrium moisture content *M, %* d.m. for adsorption at different temperatures *t (°C)* and water activities a_w

Sel		10°C			25°C			40°C	
	a_w	M*	sd**	a_w	M*	sd**	a_w	M*	sd**
LiCl	0.113	3.66	0.17	0.113	2.55	0.05	0.112	1.92	0.04
CH3COOK	0.234	4.39	0.40	0.225	3.55	0.17	0.201	2.50	0.14
$MgCl_2$	0.335	4.87	0.02	0.328	4.14	0.10	0.316	3.14	0.02
K_2CO_3	0.431	6.32	0.17	0.432	5.56	0.09	0.432	4.13	0.11
$MgNO_3$	0.574	8.35	0.48	0.529	7.36	0.02	0.484	5.43	0.36
NaBr	0.622	9.88	0.31	0.576	8.33	0.08	0.532	5.74	0.32
NaCl	0.757	9.97	0.27	0.753	8.39	0.39	0.747	7.87	0.65
KCl	0.868	14.22	0.46	0.843	14.09	0.28	0.823	9.34	0.48

* Mean of three runs, ** Mean deviation of three runs

Table 2. Equilibrium moisture content *M, %* d.m. for desorption of different temperatures *t (°C)* and water activities a_w

Sel		10°C			25°C			40°C	
	a_w	M*	sd**	a_w	M*	sd**	a_w	M*	sd**
LiCl	0.113	2.59	0.28	0.113	2.32	0.11	0.112	1.89	0.07
CH3COOK	0.234	4.16	0.09	0.225	3.00	0.07	0.201	2.39	0.13
$MgCl_2$	0.335	4.67	0.94	0.328	4.03	0.09	0.316	3.57	0.60
K_2CO_3	0.431	6.03	0.20	0.432	5.91	0.02	0.432	4.98	0.13
$MgNO_3$	0.574	7.80	0.36	0.529	7.26	0.21	0.484	5.13	0.22
NaBr	0.622	9.51	0.08	0.576	7.33	0.17	0.532	5.33	0.09
NaCl	0.757	10.16	0.25	0.753	8.96	0.07	0.747	8.01	0.58
KCl	0.868	13.77	0.50	0.843	13.45	0.53	0.823	9.14	0.42

* Mean of three runs, ** Mean deviation of three runs

For both processes, we observed a behavior typical of many food products under study, namely that the increase in temperature at constant water activity is coupled by a decrease in equilibrium moisture content [27-31].

Figure 1 shows the comparison of isotherms at a temperature of 10°C. It is evident that the isotherms belong to Type III according to the classification of Brunauer et al., 1940 [32].

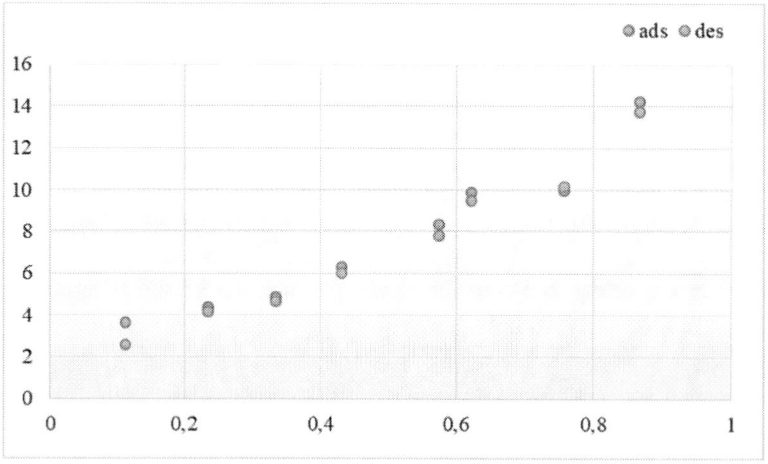

Figure 1. Comparisons of isotherms at 10°C

In the present study, the obtained coefficients of the three-parameter models (*A, B, C*) and the corresponding values of mean relative error P,%, standard deviation SEM and distribution of residuals are shown in Table 3 for adsorption and in Table 4 for desorption.

Table 3. Coefficients of the models (A, B, C), average relative error P, % and standard deviation SEM for adsorption

Model	*A*	*B*	*C*	*P*	*SEM*	Residuals
Oswin	**7.7084**	**-0.0615**	**0.3996**	**9.87**	**0.89**	**arbitrary**
Halsey	2.7094	-0.0186	1.4806	13.61	1.92	non-arbitrary
Henderson	0.0007	31.865	1.5606	11.38	0.99	non-arbitrary
Chung-Pfost	217.48	0.2621	35.536	12.94	0.97	non-arbitrary

Table 4. Coefficients of the models (A, B, C), average relative error P, % and standard deviation SEM for desorption

Model	*A*	*B*	*C*	*P*	*SEM*	Residuals
Oswin	7.2853	-0.0521	0.4115	9.66	0.70	non-arbitrary
Halsey	2.4055	-0.0131	1.4024	14.62	2.02	non-arbitrary
Henderson	**0.0006**	**44.672**	**1.5399**	**8.35**	**0.69**	**arbitrary**
Chung-Pfost	268.60	0.2606	52.471	9.96	0.77	non-arbitrary

The data show that the lowest values of P and SEM and arbitrary distributions of residuals for adsorption were obtained by the modified Oswin, whereas for desorption – by the modified Henderson. As a result, we can recommend both models. For the description of the adsorption isotherms of the ready-made mixture containing oatflakes, apples, and cinnamon, we recommend the modified Oswin, whereas for the description of the desorption isotherms of the mixture we recommend the modified Henderson.

In order to calculate MMC, we had to linearize the Brunauer-Emmett-Teller equation. The linearization using test data for $a_w < 0.5$ in the case of adsorption and desorption is shown in Fig. 2 and 3 [25].

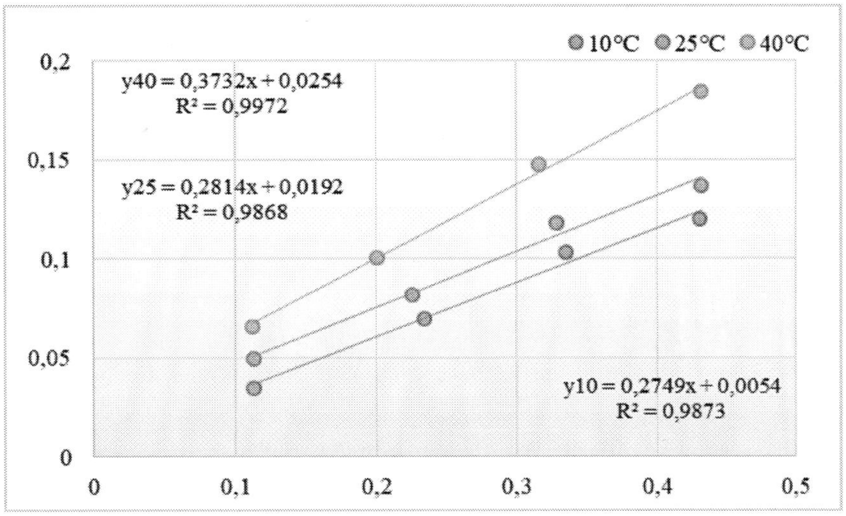

Figure 2. Linearization of the BET model at different temperatures for adsorption

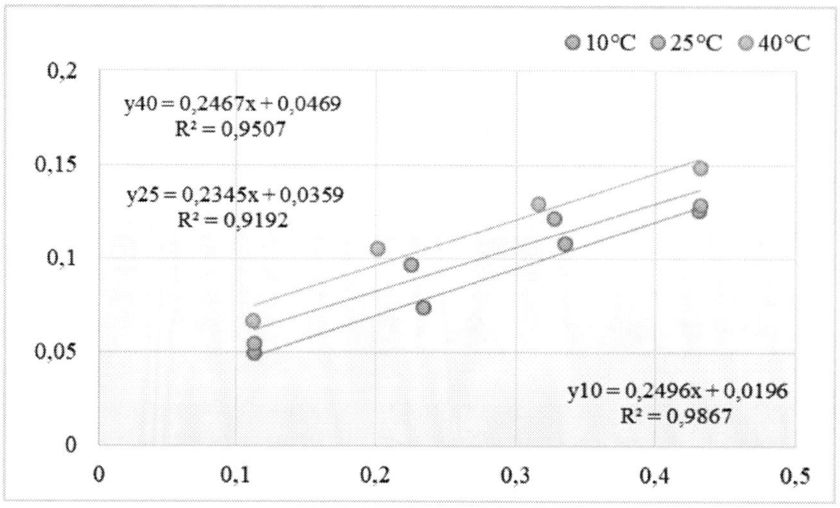

Figure 3. Linearization of the BET model at different temperatures for desorption

On the basis of the coefficients obtained from the linear equations, we calculated the MMC values for both processes and the three temperatures, as shown in Table 5. The value of monomolecular moisture (MMC) is a sorption characteristic which affects the stability of the food product. A number of studies prove that when the product under storage reaches a value corresponding to MMC, it preserves its qualitative parameters [33].

Table 5. BET monolayer moisture content, % d.m. at different temperatures (*t, °C*) for adsorption and desorption

t (°C)	**Adsorption**	**Desorption**
10	3.57	3.71
25	3.33	3.70
40	2.51	3.41

The data show that temperature affects MMC. The increase in temperature is coupled by a decrease in MMC values for both processes. In the case of adsorption, MMC diminishes by 1.06% from 10°C to 40°C, while with desorption it is lowered by 0.3%. On the basis of the obtained results, we can predict that in the case of mixture moisture content in the 2.51% d.m. - 3.71% d.m. range there will be microbial load for a greater period of time [24].

Conclusions

1. We obtained the equilibrium moisture content data for adsorption and desorption of a ready-made mixture containing oatflakes, apples, and cinnamon at temperatures of 10°C, 25°C and 40°C and relative air humidities in the 0.11 - 0.85 range.

2. We established that the increase in temperature leads to a decrease in equilibrium moisture content at constant water activity.

3. We discovered that the isotherms of the product under study belong to Type III according to the classification of Brunauer et al.

4. For the description of the adsorption isotherms of the ready-made mixture containing oatflakes, apples and cinnamon, we recommended the modified Oswin, while for the description of desorption isotherms we suggested the modified Henderson.

5. The modified models of Oswin and Henderson, suggested for the description of the sorption characteristics of the ready-made mixture of oat flakes, apple, and cinnamon can be used to make predictions concerning the equilibrium humidity relative to the experimental conditions.

6. On the basis of the linearization of the Brunauer-Emmett-Teller model, we calculated the MMC values for adsorption and desorption using test data for $a_w < 0.5$ in the case of adsorption within the 2.51% d.m. - 3.57% d.m. range and in the case of desorption – in the 3.41% d.m. - 3.71% d.m. range.

7. We established that, as regards the ready-made mixture containing oatflakes, apples, and cinnamon, temperature affects MMC values. With the increase in temperature, the values are decreased from 10°C to 40°C by 1.06% for adsorption and by 0.3% for desorption.

Acknowledgements

This study was part of the "FS N11/20-H" Scientific Research Fund, project director: Assoc. Prof. PhD Albena Georgieva Durakova, University of Food Technologies – Plovdiv, Bulgaria.

References

1. I. Y. Reznichenko, D. M. Borodulin, A. M. Chistyakov, N.Y. Ruban, S. S. Komarov, IOP Conf Ser Earth Environ Sci, **613** (1), 012124, IOP Publishing. (2020).
2. M. H. Mughal, BJSTR, **18** (4), 13804-13811, (2019).
3. A. Bogoeva, A. Durakova, R. Vrancheva, A. Slavov, BJAS, **26** (4), 906-912 (2020).
4. S. Al-Hooti, J. S. Sidhu, J. Al-Otaibi, H. Al-Ameeri, H. Al-Qabazard, Plant Foods Hum Nutr., **51** (2), 125-135, (1997).
5. X. - Z. Hu, J. - M. Zheng, X. - P. Li, C. Xu, Q. Zhao, J. Cereal Sci., **60** (2), 297-301, (2014).
6. H. G. Liljeberg, Y.E. Granfeldt, I. M. Björck, J Nutr, **126** (2), 458-466, (1996).
7. B. D. Singu, P. R. Bhushette, U.S. Annapure, Food Bioproc Tech., 1-12, (2020).
8. M. Lauková, Z. Kohajdová, J. Karovičová, Potr. S. J. F. Sci., **10** (1), 506-511, (2016).
9. R. C. Skinner, J. C. Gigliotti, K. M. Ku, J. C. Tou, Nutr. Rev., **76** (12), 893-909, (2018).
10. M. O. Al-Shawwa, Classification of Apple Fruits by Deep Learning, (2020).
11. K. Singletary, Nutr. Today, **54** (1), 42-52, (2019).
12. F. Askari, B. Rashidkhani, A. Hekmatdoost, Nutr Res, **34** (2), 143-148, (2014).
13. T. Bandara, I. Uluwaduge, E. R. Jansz, Int. J. Food Sci. Nutr, **63** (3), 380-386, (2012).
14. P. Ranasinghe, S. Pigera, G. S. Premakumara, P. Galappaththy, G. R. Constantine, P. Katulanda, BMC Complement Altern. Med., **13** (1), 275, (2013).
15. J.-E. Kim, J. E. Son, H. Jeong, D. J. Kim, S. G. Seo, E. Lee, T. G. Lim, J. R. Kim, Y. R. Kimbung, H. Chen, A. M. Bode, K. W. Lee, Z. Dong, Cancer Res., **75** (13), 2716-2728, (2015).

16. J. Gruenwald, J. Freder, N., Armbruester, Crit Rev Food Sci Nutr, **50** (9), 822-834, (2010).
17. S. A. Almatroodi, M. A. Alsahli, A. Almatroudi, S. Anwar, A. K. Verma, K. Dev, A. H. Rahmani, Gene Rep., 100966, (2020).
18. B. Goel, S. Mishra, Eur J Med plants, 10-16, (2020).
19. V. Assadollahi, K. Parivar, N. H. Roudbari, A. R. Khalatbary, M. Motamedi, B. Ezatpour, G. R. Dashti, (2013). Adv. Biomed. Res., **2**, (2013).
20. A. Khan, M. Safdar, M. M. A. Khan, K. N. Khattak, R. A. Anderson, Diabetes care, **26** (12), 3215-3218, (2003).
21. E. Sandulachi, Technical University of Moldova, 42-43, (2012).
22. AOAC, 960.39 (Washington, DC), **15** ed. AOAC, (1990).
23. W. Wolf, W. Spiess, G. Jung, G. In Properties of water in foods, 661-679, Springer, Dordrecht, (1985).
24. L. Bell, T. Labuza, AACC, 33-36, (2000).
25. S. Brunauer, P.H. Emmett, E. Teller, JACS, **60** (2), 309-319, (1938).
26. A. Durakova, A. Bogoeva, V. Yanakieva, T. Gogova, K. Choroleeva, E3S Web of Conferences, **180**, 03007, EDP Sciences, (2020).
27. A. Durakova, A. E3S Web of Conferences, **180**, 03008. EDP Sciences, (2020).
28. A. Bogoeva, JCEA, **21** (3), 609-617, (2020).
29. T. P. Labuza, B. Altunakar, Water activity in foods: fundamentals and applications, Chapter **7**, 161-205, John Wiley & Sons, Inc, (2020).
30. A. W. Gichau, J. K. Okoth, A. Makokha, JFST, **57** (3), 962-970, (2020).
31. S. Al-Ghamdi, Y. K. Hong, Z. Qu, S. S. Sablani, J. Food Eng., 273, 109820, (2020).
32. S. Brunauer, L. S. Deming, W. E. Diming, E. Troller, JACS, **62**, 1723-1732, (1940).
33. S. Kaya, & T. Kahyaoglu, J. Food Eng., **71** (2), 200-207, (2005).

An experimental investigation of working with ordinary and electromagnetic nozzles

Dimitar Kehayov[1], *Atanas* Atanasov[2] and *Dan* Cujbescu[3*]

[1]Agricultural University-Plovdiv, Department of Agricultural Mechanization, 4000 Plovdiv, Bulgaria
[2] University of Ruse, Agricultural Machinery Department, 7012, Ruse, Bulgaria
[3]INMA Bucharest, 6 Ion Ionescu de la Brad Bld, Bucharest, Romania

Abstract. The great variety of plant protection equipment, the shape, length and slope of the cultivated area, the type of sprayers used and the professional skills of the operator require an individual approach to each sprayer and cultivated area. The aim of this study was to compare the area of the double-treated sections when working with ordinary and electromagnetic nozzles at different slopes of the treated area. For the analysis, actual field coverage data were used when working with the Amazone self-propelled sprayer - Pantera 4502, with a boom width of 24 m, half of which is equipped with electromagnetic nozzles and the other half - with ordinary nozzles. It was found that when working with ordinary nozzles, the area of overlap increases with increasing degree of overlap and the angle of inclination of the section boundary. The area of overlap when using ordinary nozzles is almost 3 times larger than when using electromagnetic at 100% overlap of the treated areas. Regression models, surfaces and lines were obtained, with the help of which the area of overlap in each section of the treated field can be determined.

Introduction

Mechanized technologies in plant protection increase labor productivity, reduce relative costs, improve working conditions for workers, improve the quality of work and contribute reduce adverse effects on the environment.

More and more new techniques for growing crops are entering in our agriculture. This requires a very good knowledge of machines and technologies. Today's agriculture is inconceivable without the use of various chemical plant protection products, as well as biological and mineral fertilizers to stimulate the growth and development of crops. The great variety of machines for performing these operations imposes the need to know the capabilities of each of them, for their application in individual techniques and technologies. Reducing unintentional pesticide application errors can improve profitability and reduce the loss of pesticides and nutrients to the environment from agricultural operations. The potential for saving pesticides and fertilizers by automatically based control of spray nozzles with use ZYNX X15 control console with a 30-channel electronic sprayer is present in [5]. Other authors [4] evaluate the use of ultrasonic sensors for variable speed automatic sprayers. An important condition for quality work is to the strength calculus of the nozzles from the sprayers [1]. Uneven spraying of the working solutions during the plant protection

* Corresponding author: *dcujbescu@yahoo.com*

operations leads to overconsumption of the working solution and failure to achieve the desired effect. In this regard, an interesting study was performed by [2, 3, 8] using a scanning system to assess the distribution of the working solution, both in the laboratory and in specific cultures in real time. The uniform spreading of the working solution on the treated area largely depends on the type of nozzle used, the angle and the spray pressure considered by [7, 9], [10]. How to optimize the solution distribution and reduce the spray diversion is suggested by [6].

The search for optimal techniques and approaches for the implementation of an even distribution of plant protection products on the soil, plants or storages, in order to destroy or suppress weeds, diseases and pests for a certain term is the basis of the research. The great variety of plant protection equipment, the shape, length and slope of the cultivated area, the type of nozzles used, the professional skills of the operator and other influencing factors require an individual approach to each sprayer and cultivated area.

The aim of the present work is to compare the area of the double-treated places when working with ordinary and electromagnetic nozzles at different slopes of the treated area.

Materials and methods

In the present work the object of study is the self-propelled sprayer of the company Amazone - Pantera 4502. The width of the spraying system is 24 m. It consists of 7 sections - the middle 3 have 8 nozzles with 4 m working width, the final 4 have 6 nozzles with 3 m working width. For the purpose of our research we have equipped half of the boom system with electromagnetic nozzles, and the other half - with ordinary ones. Observations were made for the overlap to the middle boom. Both types of nozzles are equipped with a sprayer IDK-120-03 from Lechler. The solenoid valves are manufactured by Arag (s/n 140467) The experiment was conducted in 2020 in the experimental field of the Agrarian University-Plovdiv. The operating speed of the sprayer in the cultivated area was 20 km / h. The influence of the field slope at 5^0, 35^0 and 65^0 and the degree of overlap at 0%, 50% and 100% on the overlap area when working with ordinary and electromagnetic nozzles was reported.

The area of overlap is defined as follows:

$$S = \frac{ab}{2} \tag{1}$$

$$\frac{b}{a} = tg\alpha \tag{2}$$

$$b = a.tg\alpha \tag{3}$$

$$S = \frac{a^2 tg\alpha}{2} \tag{4}$$

Where: a- length equal to the length of the section, m.
b - length equal to the boundary of the untreated field
α - the angle locked between the section and the boundary of the untreated field °.

When operating the sprayer with conventional sprayers with 0% overlap untreated areas remain colored green. This is due to the fact that when the outer part of the section reaches the field boundary, the section crane switches off the operation of the whole section.

In the second variant of the test when operating the sprayer with ordinary nozzles with 50% overlap we have double-treated areas - shown in the figure in red and those that remain untreated - shown in green. During the test, 3 sections work - the first with a width of 4 m, and the two end - with a width of 3 m each. The size of the double-treated sections is determined according to formula (4) at a width for side "a" for the first section is 2 m, and for the second and third respectively 1.5 m.

For the third test variant with 100% overlap the sectional crane switches off the section only when its entire working width has entered the already treated field. The double-treated areas are coloured red. The size of the double-treated sections is determined according to the formula (4) at "a" on the first bar 4 m, "a" on the second and third respectively 3 m.

When working with an electromagnetic nozzle, an experiment was performed with two variants - without overlap (0%) and with full (100%) overlap. What is special here is that the working width for each nozzle is 0.5 m. When working without overlapping, the nozzle is switched off when it reaches the field boundary. At 100% overlap, the nozzle is switched off after its torch has fully entered the treated field.

Results and discussion

For working with ordinary nozzles, the following data were obtained from the calculations made in Table 1.

Table 1. Overlap area when working with ordinary nozzles, m^2

Overlap	**Slope at the field boundary**		
	5°	35°	65°
0%	0	0	0
50%	0,568	2,975	9,052
100%	1,479	11,9	36,448

It was found that with increasing degree of overlap and slope of the field boundary, the area of the double-cultivated part of the field increases. As a result of the regression analysis the following model was obtained:

The model has the form:

$$Z = 0{,}005442\, X\, Y \tag{5}$$

Where: Z - double cultivated area, m^2;
X - slope of the field boundary,
Y - degree of overlap, %.

The change of the double-treated area and the lines of equal response are presented in Fig. 1. and Fig. 2. From Fig.1. it is seen that with increasing the slope of the field the double treated area increases.

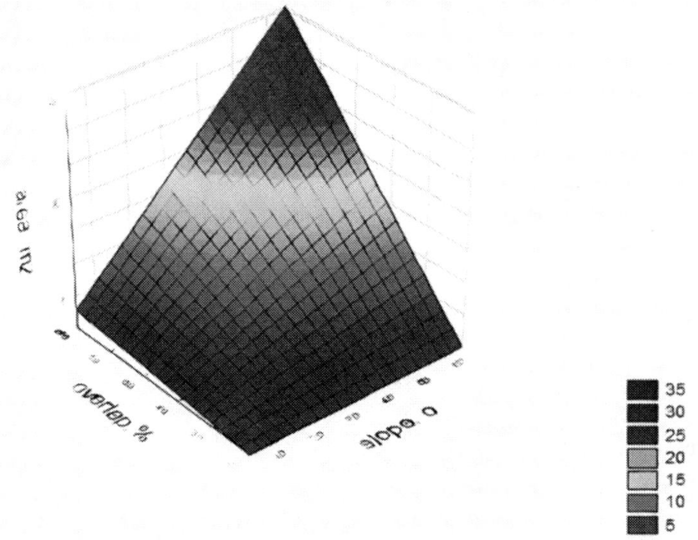

Fig.1. Variation of the double treated area

The lines of uniform response make it possible, knowing the slope of the field and the degree of overlap, to easily and quickly determine the double-treated area. For this purpose, from the point of the abscissa corresponding to the slope, we raise vertical lines. From the point of the ordinate corresponding to the degree of overlap, we draw horizontal lines. The intersection of the two lines usually lies on one of the family curves. From the scale to the right of the coloring, the double-treated area is reported.

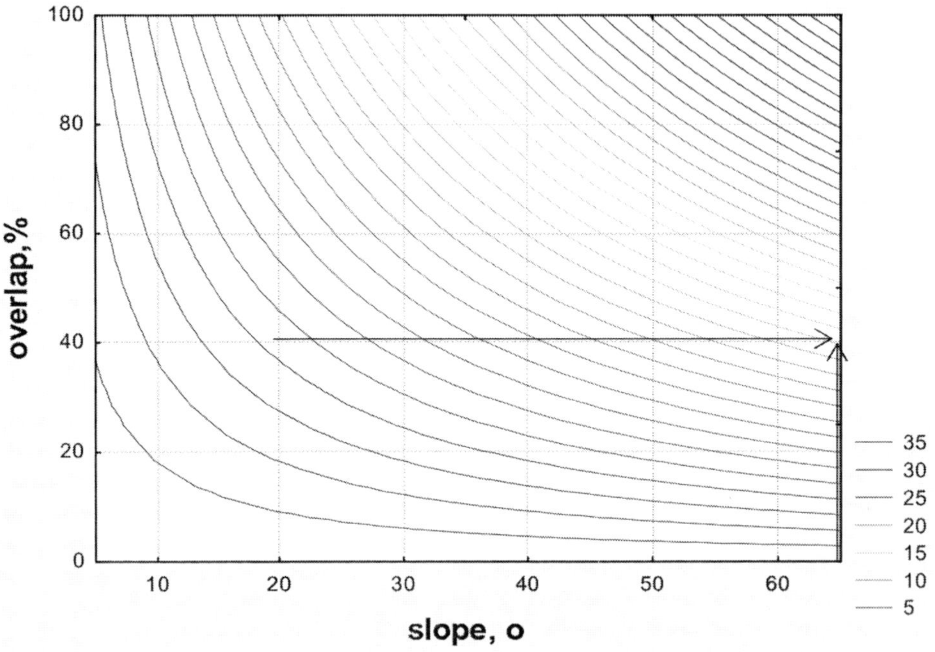

Fig.2. Lines of equal response

Example: Let the slope be 50^0 and the overlap be 40%. After baking the horizontal and vertical lines from the respective points of the ordinate and the abscissa, their point of intersection lies on the curve corresponding to 15 m^2 double treated area.

Table 2 Overlap area when working with electromagnetic nozzles, ^{m2}

Overlap	Slope at the field boundary		
	5°	35°	65°
0%	0	0	0
100%	0,435	3,5	10,72

From the obtained results Table 1 and Table 2 at 100% overlap it was found that at different slopes of the field boundary the overlap is almost 3 times greater when working with ordinary nozzles. This is due to the fact that when working with ordinary nozzles, the entire section is switched off when it fully enters the treated field. Parameter "a" (formula 4) varies from 3 to 4 m depending on the working width of the section. When working with electromagnetic nozzles, each is switched off when it fully enters the already treated field, one after the other. In this case, parameter "a" has a constant size of 0.5 m. The double-treated area for the whole section is obtained according to formula 4 by summing the areas for each nozzle.

The following regression models were obtained for the area of overlap of the two nozzles:

$$S\ o.n = 1{,}5026X^{2{,}9261} \quad (6)$$

$$S\ e.m.n = 0{,}4417X^{2{,}9267} \quad (7)$$

Where: S o.n - Overlap area when working with ordinary nozzles
S e.m.n – Overlap area when working with electromagnetic nozzles

Overlap area at 100% overlap, Series1 - with electromagnetic nozzle, Series2 - with ordinary nozzle is shown in Fig. 3.

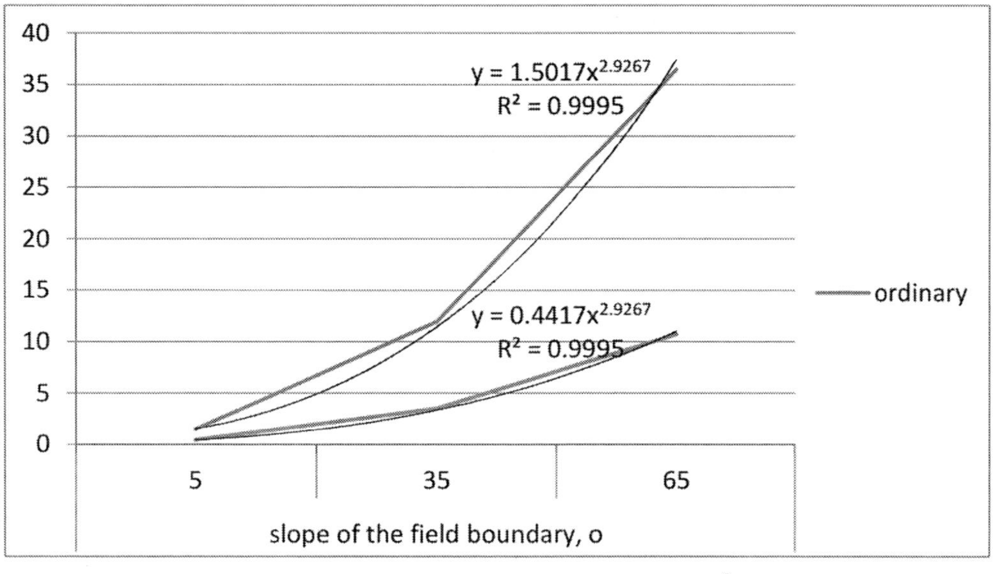

Fig.3. Overlap area with 100% overlap, m^2

Conclusions:

1. When working with ordinary nozzles, the area of overlap increases with the increase of the degree of overlap and the angle of inclination of the section boundary;
2. The area of overlap when using ordinary nozzles is almost 3 times larger than when using electromagnetic at 100% overlap of the treated areas;
3. Regression models, surfaces and lines are obtained, with the help of which the area of overlap in each section of the treated field can be determined.

References

[1] Bungescu S., Stahli W., Biriş S., Vlăduţ V., Imbrea F., Petroman C. - Cosmos programm used for the strength calculus of the nozzles from the sprayers, proceedings of the 37 International symposium on agricultural engineering "Actual Tasks on Agricultural Engineering", Opatija - Croaţia, 2009, pp. 177-184.

[2] Emanuele Cerruto, Sabina Failla, Domenico Longo, Giuseppe Manetto, Simulation of water sensitive papers for spray analysis, AgricEngInt: CIGR Journal, Vol. 18, No. 4, 2016, pp.23-29.

[3] Heping Zhu, Masoud Salyani, Robert D. Fox, A portable scanning system for evaluation of spray deposit distribution, Computers and Electronics in Agriculture, vol.76, 2011, pp. 38–43

[4] Hong Y. Jeon, Heping Zhu, Richard Derksen, Edal Ozkan, Charles Krause, Evaluation of ultrasonic sensor for variable-rate spray applications, Computers and Electronics in Agriculture vol.75, 2011, pp. 213–221.

[5] J.D. Luck, S.K. Pitla, S.A. Shearer, T.G. Mueller, C.R. Dillon, J.P. Fulton, S.F. Higgins, Potential for pesticide and nutrient savings via map-based automatic boom section control of spray nozzles, Computers and Electronics in Agriculture, vol.70, 2010, pp.19–26.

[6] Jan C. van de Zande, J. F. M. Huijsmans, H. A. J. Porskamp, J. M. G. P. Michielsen, H. Stallinga, H. J. Holterman A. de Jong, Spray techniques: how to optimise spray deposition and minimise spray drift, Environmentalist, 2008, vol. 28, pp.9–17.

[7] Nasir .S. Hassen, Nor Azwadi .C. Sidik, Jamaludin .M. Sheriff, Effect of nozzle type, angle and pressure on spray volumetric distribution of broadcasting and banding application, Journal Mechanical Engineering Research, vol.5 (4), pp.76-81.

[8] Qamar Uz Zaman, Travis J. Esau, Arnold W. Schumann, David C. Percival, Young Ki Chang, Scott M. Read, Aitazaz A. Farooque, Development of prototype automated variable rate sprayer for real-time spot-application of agrochemicals in wild blueberry fields, Computers and Electronics in Agriculture vol.76, 2011, pp. 175–182.

[9] Roşu (Niţu) M., MatacheM., Vlăduţ V., Căsăndroiu T., Bungescu S. - Influenceof the jet's angle size on the spraying process, proceedings of the 43 International symposium on agricultural engineering "Actual Tasks onAgricultural Engineering", 2015 pp. 275-286.

[10] Vlăduţ V., Matache M., Voicea I., Găgeanu P., Bungescu S.,Biriş S., Mihailov N., Popescu S., Savin L. - Comparison of a sprinkler's transverse distributionwith used and new nozzles, proceedings of the 39 International symposium on agricultural engineering "Actual Tasks on Agricultural Engineering",, Opatija – Croaţia, 2011, pp. 307-312,

Experimental research on obtaining pelletized feed recipes with applicability in semi-intensive aquaculture systems

Iulian Voicea[1*], *Corina Ioana* Moga[2], *Sorin-Ştefan* Biriş[3], *Gabriel-Alexandru* Constantin[3], *Sorin* Boruz[4], and *Mihai Gabriel* Matache[1]

[1]National Institute of Research – Development for Machines and Installations Designed for Agriculture and Food Industry – INMA Bucharest, 6, Ion Ionescu de la Brad, sector 1, Romania

[2]DFR Systems S.R.L., 46, Drumul Taberei, Bucharest 061392, Bucharest, Romania

[3]Politehnica University of Bucharest, Faculty of Biotechnical Systems Engineering, Romania

[4]University of Craiova, Romania

Abstract. Romanian aquaculture can be economically sustainable and viable if the farm is profitable, the farm's income is secure, and the products are accepted by customers. In many cases, improving environmental sustainability can lead to optimizing economic sustainability. For example, a more efficient use of food and nutrients or reduced water and energy consumption are not only beneficial for the environment, but can also reduce costs. Growth in polyculture system, which represents the growth of several species of fish in the same space (pond, pool, basin, etc.), and diversification of supply can increase consumer confidence and thus economic sustainability. The diet must provide the fish with the nutrients necessary for growth (proteins with a plastic role) and provide the energy necessary for physiological processes. Quality food ensures good development, facilitates reproduction and prevents extremely harmful diseases in an aquaculture system. The paper presents a series of experimental recipes for pelletized feed made using an experimental model of small capacity. Nutrient pellets made for feeding fish from the Cyprinid family were analyzed in terms of gross energy value, ash content, moisture, mechanical durability over time, bulk density etc.

1 Introduction

Aquaculture, in the broadest sense of the word, means the technology of growing organisms in the aquatic environment, useful to humans, in which the predominant role belongs to fish, algae, molluscs and crustaceans. Aquaculture is currently considered as the science that deals with the study of biological bases and the development of biotechnical methods based on which aquatic organisms multiply, grow, develop in natural or anthropogenic

* Corresponding author: voicea_iulian@yahoo.com

aquatic basins, [1,2]. Aquaculture is increasingly contributing to the world's production of aquatic food, given that, for most wild fish stocks, sustainable exploitable limits are now almost reached or even exceeded. Semi-intensive growth of fish or other aquatic organisms is based on natural productivity and / or enriched by fertilization of anthropogenic ecosystems, respectively on the administration of additional food. In fact, semi-intensive cultivation involves obtaining a fish biomass with low production costs due to the use of inexpensive inputs. For the aquaculture in our country, this system is widely used, contributing to obtaining food-safe products. In this case, it is obvious the intervention of anthropogenic factors to create favourable conditions for fish farming, such as: own system of supply and evacuation of water from the ecosystem (allows the replacement of 10-15% of water), application of treatments to improve physical-chemical quality of water and soil (application of organic and inorganic fertilizers, amendments, etc.) and the use of aeration systems to maintain optimal concentrations of dissolved oxygen in water, [3,4].

In these growing systems only one species can be raised (monoculture) but it is agreed to grow several fish species that capitalize on complementary trophic niches (polyculture). Taking into account the cost / benefit ratio (the ratio between inputs / outputs) it can be can appreciated that this system can be applied in the case of farms with fish production on a smaller scale (but higher than> 500 kg / ha), [5,6,7].

Raising different species of fish together in polyculture is a widespread system in many parts of the world. It is estimated that this system leads to obtaining productions clearly superior to the monoculture system, as a result of the maximum exploitation of all trophic niches. In order to maximize fish production in polyculture systems, it is essential to know the relationships established both between fish species and between fish and the environment. The farmer's choice of the most appropriate species of fish or other aquatic organisms, age classes, population densities, fertilization regime, feed ensures the success of such a production system, [8,9]. Usually, the efficiency of using the resources from the same pond increases as the fish species have different nutritional spectrum. Increasing production efficiency is achieved when the metabolic wastes of one species are used as food by another. Choosing the most suitable species for polyculture will minimize the antagonistic relations and maximize the synergy of the pond relations, leading to the increase of the availability of food resources for the cultivated species and at the same time the improvement of the environmental conditions. The specialized literature describes a synergistic relationship in the polyculture of plankton-eating fish - carp that appears as a result of the activity of the benthos agitation carp, thus facilitating the efficient use of organic and inorganic matter. This process significantly reduces the accumulation of ammonia and nitrites, as well as the organic load at the bottom of the water. Through this movement, nutrients are more available for macroscopic algae which, in turn, can be used in fish feed. More intense photosynthesis also leads to better oxygenation, increased pH and efficient nitrification.

Feeding has the most important impact on the results in fish farms, because the quantity and quality of the fish stock is directly proportional to its diet.

But fodder also has the highest production cost, having the greatest influence on profit, which is why the correct choice of fodder is crucial. Carp can be fed both complete and incomplete combined feed.

In salmon farms, the use of combined feed is an old practice, due to the special economic and ecological impacts. Increased productivity in this sector is constantly demanding better and better feed, as the results of intensive aquaculture (health, fish growth and development) are closely correlated with feed quality. In addition, environmental legislation requires that these feeds have as little impact on water quality as possible, especially since in some more developed countries each aquaculture farm is required to provide a certain amount of feed that can be used in one season, and the farmer is forced to make the most of this amount (sometimes quite limited).

In carp farms, there are extended reserves concerning increasing productivity. The majority of farms in the central and eastern Europe have 50 years old technologies, that are based on large areas of water coverage, n natural productivity of ponds and on additional feeding.

Pelleting / granulating is one of the best treatments for combined animal feed. Pelleting translates as a process of compressing materials that have been grinded to the intended sizes, by feeding them into an equipment and, by means of compaction rollers, forcing them to pass through the cylindrical orifices of a plane or ring dies. By pelleting the combined feed, bulk density increases, the feed is homogenous, harmful organisms are destroyed, and better handling and distribution are ensured. [10,11]

2 Material and method

The paper studies four different recipes of pelleted feed for aquaculture, especially for the cyprinid family. Pelleting was performed using an experimental model installation model for obtaining granulated / pelleted feed for aquaculture. The main technical characteristics of the pelleting / granulating press are the following: supply voltage: 230 V, electric motor power: 2800 W, maximum speed: 1400 rpm, production capacity: 100 kg/h, die orifice diameter: 3-8 mm.

Fig.1. Pellets resulted from the four feed recipes for cyprinids

The basis for the four pelleted feed recipes was represented by cereal material: sun flower seeds, corn, wheat, soybean. For an adequate binding and in order to increase the energy intake, additive materials were added, such as: fish flour, vitaminized calcium, gum arabic colouring, vitamin C, roasted hemp, krill flour, sweet corn starch with integrated starch, yeast extract to increase digestibility. Therefore, the composition of recipe 1 included: 50 % sunflower, 25 % corn, 10 % wheat, 5 % soybean, 10 % additives (fish flour, vitaminized calcium, black gum arabic colouring, roasted hemp, yeast extract, starch); recipe 2 included: 65 % corn, 15 % sunflower, 5 % wheat, 15 % additives (fish flour, vitaminized calcium, yellow gum arabic colouring, vitamin C, sweet corn starch with integrated starch, yeast extract, starch). In the composition of recipe 3, we have the following ingredients: 20 % sunflower 50 % corn, 10 % wheat, 10 % soybean, 10 % additives (fish flour, vitaminized calcium, red gum arabic colouring, krill flour, yeast extract, starch); and recipe 4 included the following: 10 % corn, 10 % wheat, 10 % soybean and 60 % additives (20 % was represented by fish flour + vitaminized calcium + green gum arabic colouring + vitamin C + roasted hemp + yeast extract, and 40 % starch + algae extract + spirulina + floatability extract).

For a faster determination of the moisture of the 4 types of pelletized feed recipes for aquaculture (cyprinids), a KERN thermobalance was used. The device uses the method of drying materials by heating the sample by means of electromagnetic waves with infrared wavelengths, produced by a special lamp. During drying, water is removed from the sample

 https://doi.org/10.1051/e3sconf/202128603005

in the form of vapour, the sample being permanently weighed by an analytical balance existing in the apparatus. This infrared drying method is a thermogravimetric method for rapidly determining moisture.

The ash content from the 4 types of pellets was determined using the calcination furnace [2, 5]. Thus, the samples were subjected to the calcination process according to the following stages:

- Weighing the samples using an analytical balance (the mass of the sample should be more than 1 g);
- Establishing the parameters of the program, namely increasing the temperature inside the calcination furnace at a rate of 5 °C after the following program: raising the temperature to 250 °C in a period of 30 - 50 minutes, followed by maintaining the temperature for 60 minutes, then an increase of temperature from 250 ° C to 550 ° C in a period of 30 minutes, followed by keeping it at the same temperature for 120 minutes and lowering it to a temperature below 30 ° C from where the samples are taken with a special tong and placed in the desiccator;
- After the samples were cooled for 24 hours in the desiccator, the ash content is determined for each pellet sample.

Fig. 2. Determining moisture and ash content

The determination of the gross energy value of the feed pellets was performed with the help of a CAL 3K-F calorimetric system, and the durability of the pellets was calculated using a TUMBLER 1000 laboratory system and the method used was 500 complete rotations during 10 min.

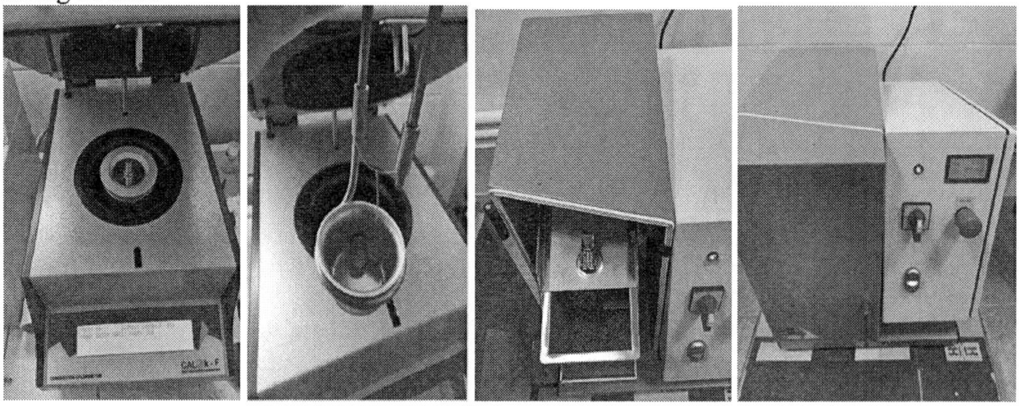

Fig. 3. Determination of calorific value and durability of feed pellets

3 Results

By using the laboratory equipment and test methods described above, the ash content, the moisture of the feed pellets, the gross energy value, the bulk density and the mechanical durability of the 4 types of pellet recipes were determined. The results obtained for the ash content are presented in figure 4, for each feed recipe being performed 4 distinct measurements.

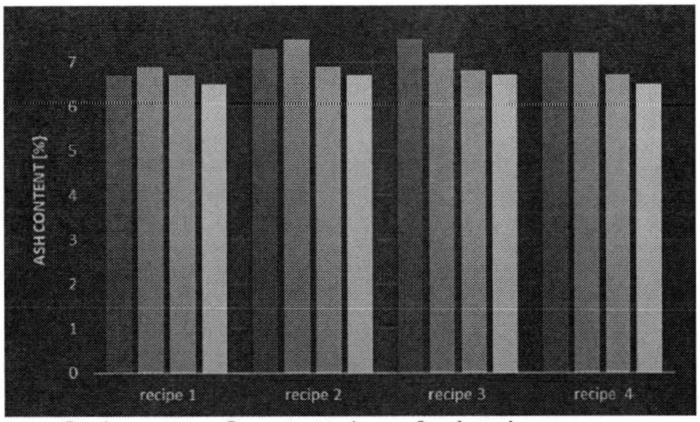

Fig. 4. Percentage of ash content for aquaculture feed recipes

In Table 1 are presented the results obtained for bulk density, moisture and gross energy value.

Table 1. Results obtained for the four types of pelleted feed

Pelleted feed recipe	Bulk density [kg/m^3]	Moisture [%]	Gross energy value [MJ]
Recipe 1	596.28	6.6	20.6
	597.36	7.1	19.5
	598.12	7.4	19.7
Recipe 2	588.12	7.2	19.5
	586.27	7.5	19.2
	587.56	7.6	18.8
Recipe 3	496.47	7.3	19.1
	498.76	7.1	18.6
	504.12	7.9	18.7
Recipe 4	85.32	6.9	17.8
	79.12	6.5	16.9
	84.29	6.8	17.5

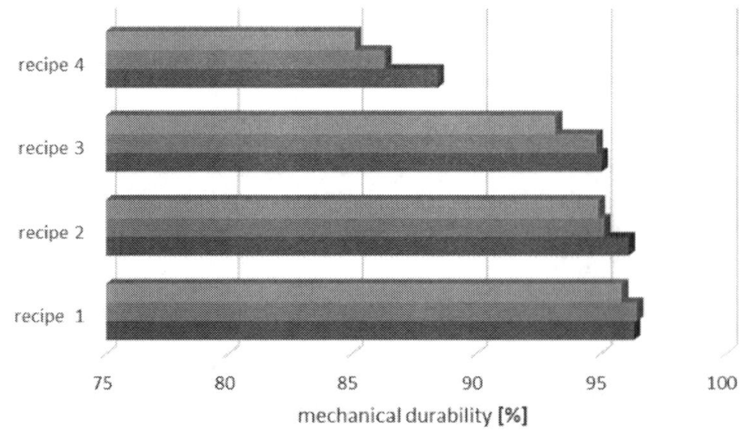

Fig. 5. Mechanical durability for the four aquaculture feed recipes

Figure 5 shows the mechanical durability of the 4 types of feed pellets obtained during the experiments.

4 Conclusions

Following the analysis of the experimental data presented above, the following conclusions can be drawn:

1. The average ash content for the four types of pellet recipes was between 6.7 (recipe 1) and 7.1 (recipe 2).
2. The average bulk density varied between 82.91 kg/m^3 (recipe 4) and 597.25 kg/m^3 (recipe 1).
3. The gross energetic value for recipe 1 was 19.93 MJ, for recipe 2: 19.19 MJ, recipe 3: 18.8 MJ and 17.4 MJ for recipe 4.
4. The mechanical durability of the feed pellets was appropriate, also due to the fact that each recipe had in its composition starch as a food additive binder. The maximum mechanical durability was found in recipe 1 being 91.26% and the minimum in recipe 4: 86.64%.
5. It can be stated that the 4 types of pellet recipes presented laboratory results appropriate as values with the pellets currently found on the market at this time, being used separately experimentally in feeding cyprinids for the study of digestibility and optimal nutritional assimilation of fish.

Acknowledgement: This work was supported by the Romanian Research and Innovation Ministry, through NUCLEU Programme, Project "PN 19 10 02 03: " Researches on the intensive polyculture growth of fish and the complex valorisation of aquatic bioresources (plants)", contract no. 5N / 07.02.2019.

References

1. K. Liua, J. Frost, T. Welker, F. Barrows, Animal Feed Science and Technology, **273**, (2021).
2. I. Gageanu, D. Cujbescu, C. Persu, Gh. Voicu, Proceedings of 17th International Scientific Conference Engineering for Rural Development – Jelgava, **17** (2018).

3. M.G.Jayaram, H.P.C.Shetty, Aquaculture, **23** (1981).
4. W. Y.Moa, Z. Chenga, W. M. Choia, Y. B. Man, Y. Liu, M. H. Won, Marine Pollution Bulletin, **85**, 2 (2014).
5. I. Găgeanu, C. Persu, D. Cujbescu, G. Gheorghe, Gh. Voicu, Proceedings of 18th International Scientific Conference Engineering for Rural Development, **18** (2019).
6. P.Chukwulozie, I. Chigoziri., J. Lekwuwa, C. Jude, Heliyon, **5**, 6 (2019).
7. J. Baea A. Hamidoghlia, S. Wona, W. Choia, W. Kim, Aquaculture, **525** (2020).
8. I. Găgeanu, P. Cârdei, M. Matache, Gh. Voicu, INMATEH Agricultural Engineering Journal, **59**, 3 (2019).
9. T. Synnøve, M. Oehmeace, M. Sørensen, G. Hee, Aquacultural Engineering, **44**, 1 (2011).
10. P. Cardei, I. Găgeanu, Gh. Voicu, M. Matache, INMATEH Agricultural Engineering Journal, **58**, 2 (2019).
11. G. P. Salze, A. Davis, *A*quaculture, **437**, 1 (2015).

Determination of qualitative working indices of harrows with elastic fangs

Dan Cujbescu[1], Gheorghe Matei[3], *Dimitar* Kehayov[4*], and *Atanas* Atanasov[5]

[1]National Institute of Research – Development for Machines and Installations Designed to Agriculture and Food Industry – INMA Bucharest, Ion Ionescu de la Brad nr. 6, sector 1, Romania
[2]University Politehnica of Bucharest, Department of Biotechnical Systems, Splaiul Independentei nr. 313, sector 6, Romania
[3)]University of Craiova, Fac. of Agronomy, Alexandru Ioan Cuza st., No 13, Craiova, Romania
[4]Agricultural University-Plovdiv, Department of Agricultural Mechanization, 4000 Plovdiv, Bulgaria
[5]University of Ruse, Agricultural Machinery Department, 7012, Ruse, Bulgaria

Abstract. In modern agriculture, grinding, levelling, loosening the soil surface layer and weed destruction, with application in the vegetable and technical crops sector, are carried out by a wide range of agricultural machines generically called harrows. These operations must be carried out by splitting on the natural lines of cohesion and not by hitting, cutting, compressing, breaking, when everything turns to dust. The paper presents the field tests for a harrow with elastic fangs in order to determine the following qualitative work indices: degree of soil grinding, degree of plant damage, degree of weed destruction, degree of levelling, working depth and operating energy indices for the maintenance of corn and sunflower crops.

1 Introduction

The mechanical processing of topsoil is one of the most important land preparation operations for the cultivation of future crops [1].

Equipment for the primary soil processing, such as ploughs, chisels, rippers, etc., are widely used, but don't produce adequate loosening and fragmentation of topsoil. Secondary processing tools are generally towed by agricultural tractors and can be divided into two main categories: passive and active equipment. Passive tools, such as harrows or vibratory cultivators, only fragment the soil by absorbing power from the tractor's drawbar, while the active equipment is additionally operated, by the tractor's power take-off [2, 3]. Active and passive secondary processing equipment have been shown to have similar soil fragmentation efficiency [4, 5, 19].

The main objectives of the harrowing work are to achieve the control of weeds, the incorporation of vegetable residues and organic manure to improve soil moisture and fragmentation [6, 7, 8].

* Corresponding author: *dkechajov@mail.bg*

Harrowing work (tillage) has also been identified as one of the operations that requires high costs of resources and management, such as the need for high-power tractors that involve high fuel consumption [9].

In modern agriculture, aspects related to energy conversion and mechanical efficiency are very important, because most agricultural machinery manufacturers focus on streamlining their products by designing new optimized shapes and using innovative materials to reduce wear and fuel consumption [10].

From an economic and environmental point of view, soil tillage operations in modern agriculture are also mandatory [11, 12].

Thus, experimental researches were undertaken in order to highlight the importance of working stability for soil tillage aggregates, in terms of quality of work and fuel consumption [13, 14].

Studies have been conducted on the automation of harrowing work [15]. Experimental researches have also been conducted, focusing on the physical and mechanical properties of soil tillage aggregates in order to define what parameters influence the germination and emergence of future plants [16].

In order to obtain uniform and healthy crops with minimal impact on the environment, aggregates with homogeneous dimensions should be made for soil tillage [17, 18].

The paper presents the field tests for a harrow with elastic fangs in order to determine the qualitative working indices as well as the exploitation energy indices during the maintenance of corn and sunflower crops.

2 Material and method

The harrow with elastic fangs is a machine carried on the tractor (driven from the power take-off), being intended for the maintenance of crops of straw cereals, corn, sunflower, beets, peas, beans, potatoes, rapeseed, vegetables, pastures, before and after conducting the sowing work.

Figure 1 shows the diagram of the harrow with elastic fangs used for experimental research.

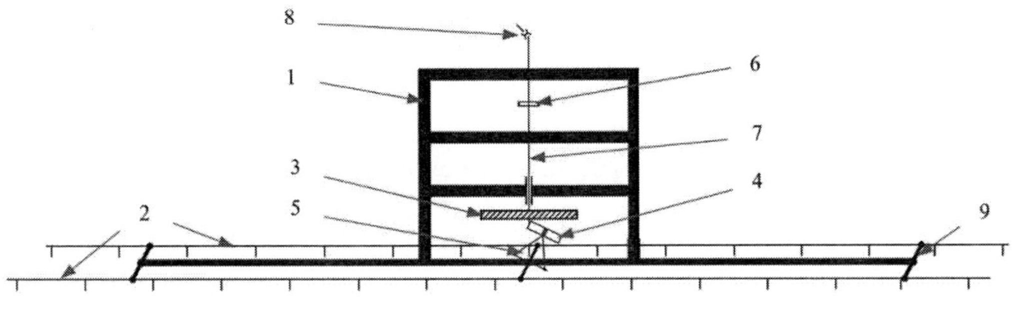

Fig. 1. Diagram of the harrow with elastic fangs
1 – frame; 2 – bank with elastic fangs; 3 – flywheel; 4 – oscillating washer; 5 – oscillating fork; 6 – flexible coupling; 7 – shaft; 8 – universal coupling; 9 – oscillating levers

The machine consists of an 80 mm square pipe frame that is attached on the tie rods of the tractor's hydraulic lift.

The batteries with elastic fangs are attached to the rear cross member of the frame by means of three oscillating levers.

The elastic fangs are springs that have the active part of 145 mm, and the spiral part inserted in a "U" profile provided with channels for the passage of the active part. The channels are long enough to move back the elastic fangs in case of meeting the plants, also allowing a strong lateral vibration.

The angle of attack of the elastic fangs can be adjusted in the range of 5-90° by an indexing system with bolts (in 6 positions).

The machine transmission includes the flywheel shaft, the elastic coupling, the oscillating washer, the oscillating swingarm that imprints the oscillating movement of the batteries with elastic fangs.

In the working position, the angle of attack of the elastic fangs is adjusted according to the stage of vegetation of the plants and the degree of soil moisture.

Table 1 presents the conditions for conducting the experimental research.

Table 1. Conditions for conducting experimental research

Specifications	**Determinations in the corn sown plot**	**Determinations in the sun flower sown plot**
Previous crop	corn	corn
Soil type	brown-red forest soil	brown-red forest soil
Previous works	autumn ploughing, harrowing and sowing	autumn ploughing, harrowing and sowing
Degree of grinding	without clods larger than 50 mm	without clods larger than 50 mm
Soil moisture between 0 – 10 mm	20.5 %	16 %
Terrain slope	max. 3.5 %	max. 4.2 %

The experiments were performed using an aggregate consisting of an 80 HP tractor and a harrow with elastic fangs for the maintenance of corn and sunflower crops.

3 Results and discussion

The determination of the qualitative working indices of the harrow with elastic fangs required the following steps:

3.1 Determining the degree of soil grinding

The experimental researches were conducted in repetitions, on a 0.5 x 0.5 m^2, with a maximum angle of attack of the elastic fangs. The results are presented in Table 2.

Table 2. Determining the degree of soil grinding

Crop	**Repetition**	**Degree of grinding [%]**	
		between 0 – 10 mm	**between 10 – 50 mm**
Corn	R1	89.6	10.4
	R2	78.3	21.7
	R3	83.2	16.8
	Average	83.7	16.3
Sun flower	R1	88.1	11.9
	R2	81.2	18.8
	R3	92.7	7.3
	Average	87.3	12.7

From the analysis of Table 2 it results that the harrow with elastic fangs achieves, in appropriate moisture conditions, a good grinding of the soil.

Also, it was noticed that it achieves a good breaking of the crust formed after a torrential rain, if it is put to work about four days after the rain.

3.2 Determining the degree of plant damage

The experimental researches were conducted in three repetitions and at two working speeds (4 and 6 km/h).

The results are presented in Table 3.

Table 3. Determining the degree of plant damage

Crop	Repetition	Degree of damage [%]	
		4 [km/h]	6 [km/h]
Corn	R1	1.1	1.3
	R2	0.8	1.5
	R3	0.4	1.2
	Average	0.7	1.3
Sun flower	R1	0.6	0.9
	R2	0.3	1.0
	R3	0.7	1.1
	Average	0.5	1.0

The results in Table 3 show that the degree of plant damage has values according to agrotechnical norms.

It should be noted that some of the plants recover in time if they are not uprooted.

3.3 Determining the degree of weed destruction

The experimental researches were conducted in three repetitions and at two working speeds. The results are presented in Table 4.

Table 4. Determining the degree of weed destruction

Crop	Repetition	Degree of weed destruction [%]	
		4 [km/h]	6 [km/h]
Corn	R1	43.5	61.5
	R2	58.9	66.6
	R3	51.7	57.3
	Average	51.4	61.9
Sun flower	R1	48.1	56.6
	R2	37.9	50.1
	R3	55.8	59.3
	Average	47.3	55.3

The results presented in Table 4 show that the degree of weed destruction is relatively low, as the destruction of weeds that are at a high stage of development would increase the degree of plant damage.

It would be advisable to use the machine when the weeds are in an early stage and to repeat the work whenever necessary, but this has negative effects on fuel consumption, as well as on the degree of soil compaction.

3.4 Determination of the degree of levelling and the degree of loosening

The experimental researches were conducted (in two repetitions) in order to determine the influence of the work performed by the harrow with elastic fangs.

The determination surface has the dimensions of 20 x 20 m.

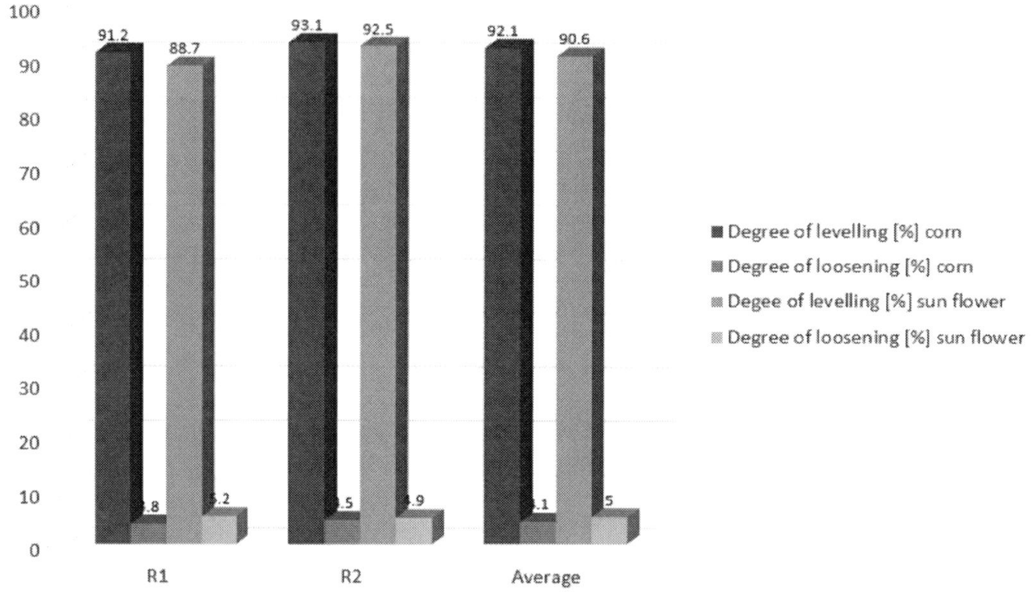

Fig. 2. Determination of the degree of levelling and the degree of loosening

From the analysis of the data presented in Figure 2 it results that the harrow with elastic fangs achieves a degree of soil levelling between 90.6 - 92.1%, being considered appropriate and a degree of loosening between 4.1 - 5.0%.

3.5 Determining the working depth of the machine

The working depth was set at 10 points along the entire working width of the machine, in two positions of attack of the elastic fangs: maximum and minimum.

The results are presented in Table 5

Table 5. Determining the working depth of the machine

Crop	Angle of attack	Working depth [mm]										Average
Crop	minimum	25	28	32	40	21	29	37	30	35	31	30.8
	maximum	47	50	42	45	51	42	46	38	55	41	45.7
Sun flower	minimum	12	19	22	23	17	18	18	20	22	27	19.8
	maximum	34	37	30	30	42	47	45	38	32	45	38.0

From the results obtained in Table 5 it results that the working depth has minimum values between 19.8 - 30.8 mm, the maximum values being between 38 - 45.7 mm.

3.6 Determination of energetic indices

Energetic indices were determined on a flat terrain, sown with corn in the state of the second node.

Experiments were conducted at three speed gears, in three repetitions, the result being presented in table 6.

Table 6. Determining the energetic indices

Indices name	M.U.	Values for gears		
		I	II	III
Effective working speed	km/h	3.78	6.15	8.57
Hourly fuel consumption	l/h	5.49	6.23	7.7
Skidding	%	3.35	3.47	3.67
Tensile strength of the machine	daN	231	212.5	224
Power required to operate the unit	CP	3.23	4.84	7.11
Engine power utilization coefficient	%	14.7	14.9	16.5

From the data presented in Table 6is observed that the values of hourly fuel consumption are situated between 5.49 l/h (1st gear) and 7.7 l/h(3rd gear).

Skidding has low values, situated around the value of 3.47%.

The tensile strength of the machine also has small vales, between 212.5 and 231 daN.

The maximum power necessary for driving the aggregate is 7.11 HP.

The coefficient of using the engine's power is very small (maximum 16.5%), leading to the idea of using the harrow in aggregate with small power tractors (<65 HP).

4 Conclusions

The experimental field researches of the harrow with elastic fangs were performed in order to determine the qualitative work indices:

- Degree of soil grinding,
- Degree of plant damage,
- Degree of weed destruction,
- Degree of levelling,
- Working depth,
- Energetic indices.

From the analysis of the data, it resulted that:

- the degree of soil grinding had values between 83.7 and 87.3% for the 0-10 mm fraction and between 12.7 and 16.3 for the 10-50 mm fraction;
- the degree of plant damage has a value of 0.76% for corn and 0.5% for sun flower at the speed of 4 km/h and of 1.3 % for corn and 1.0% for sun flower at the speed of 6 km/h;
- the degree of weed destruction has value between 47.3 – 51.4% at the speed of 4 km/h and between 55 – 61.9 % at the speed of 6 km/h;
- the degree of soil levelling has values between 90.6 – 92.1 %, and the degree of loosening between 4.1 - 5.2 %
- the working depth was between 19.8 – 30.8 mm at the minimum angle of attack of the elastic fangs and of 38 – 45.7 mm at the maximum attack angle of the elastic fangs;
- the hourly fuel consumption had the value of 7.71 l/h (3rd gear);

- the coefficient of using the engine's power was 14.9 % for the 2nd gear and 16.5 % for the 3rd gear.

Through the mechanical control of weeds, the machine leads to the decrease of the quantity of herbicides introduced in the soil.

Acknowledgement

This work was supported by the Romanian Research and Innovation Ministry, through the project entitled "Researches on achieving integrated systems for the bioeconomy field according to the concept of intelligent agriculture" – PN 19 10 01 01 – Ctr. 5N/07.02.2019.

References

1. P. Balsari, A. Biglia, L. Comba, D. Sacco, L. Eloi Alcatrao, M. Varani, M. Mattetti, P. Barge, C. Tortia, M. Manzone, P. Gay, D. Ricauda Aimonino, Performance analysis of a tractor - power harrow system under different working conditions, BIOSYSTEMS ENGINEERING **202**: 28-41 (2021)
2. S. H. Aday, Theory of agricultural machines. Basrah, Iraq: Algadeer Co. for Printing and Publishing (2015)
3. P. P. Navalade, V. M. Salokhe, T. Niyamapa, P. Soni, Performance of free rolling and powered tillage discs, Soil and Tillage Research **109**: 87-93 (2010)
4. K. M. Adam, D. C. Erbach, Secondary tillage tool effect on soil aggregation, Transactions of the American Society of Agricultural Engineers **35**: 1771-1776 (1992)
5. R. Berntsen, B. Berre, Soil fragmentation and the efficiency of tillage implements, Soil and Tillage Research **64**: 137-147 (2002)
6. E. Nafi, H. Webber, I. Danso, J. B. Naab, M. Frei, T. Gaiser, Interactive effects of conservation tillage, residue management, and nitrogen fertilizer application on soil properties under maize-cotton rotation system on highly weathered soils of West Africa, Soil and Tillage Research **196**, 104473 (2020)
7. N. Garcia-Franco, J. Albaladejo, M. Almagro, M. Martinez-Mena, Beneficial effects of reduced tillage and green manure on soil aggregation and stabilization of organic carbon in a Mediterranean agroecosystem, Soil and Tillage Research **153**: 66-75 (2015)
8. B. S. Acharya, S. Dodla, L. A. Gaston, M. Darapuneni, J. J. Wang, S. Sepat, et al., Winter cover crops effect on soil moisture and soybean growth and yield under different tillage systems, Soil and Tillage Research **195**, 104430 (2019)
9. V. Damanauskas, A. Velykis, A. Satkus, Efficiency of disc harrow adjustment for stubble tillage quality and fuel consumption, Soil and Tillage Research **194**, 104311 (2019)
10. M. Mattetti, M. Varani, G. Molari, F. Morelli, Influence of the speed on soil-pressure over a plough, Biosystems Engineering **156**: 136-147 (2017)
11. A. Calcante, R. Oberti, A technical-economic comparison between conventional tillage and conservative techniques in paddy-rice production practice in Northern Italy, Agronomy **9**, 886 (2019)
12. E. Uceanu, Gh. Bolintineanu, V. Vladuţ, Researches regarding the identification of qualitative characterisitcs at the soil work, INMATEH Vol. **24** (1): 67-79 (2008)
13. G. Paraschiv, E. Maican, I. Paraschiv, D. Bucur, Study on stability at work for tillage aggregates, Cercetări Agronomice în Moldova Vol. **2** (130): 5-13 (2007)

14. N. Constantin, D. Irimia, C. Persu, Testing the multifunctional aggregate for soil tillage works – MATINA, INMATEH Vol. **36** (1): 5-12 (2012)
15. G.G. Peteinatos, V. Rueda-Ayala, R. Gerhards, D. Andujar, Precision harrowing with a flexible tine harrow and an ultrasonic sensor, Precision Agriculture **15**: 579–586 (2015)
16. A. A. Tagar, J. Adamowski, M. S. Memon, M. C. Do, A. S. Mashori, A. S. Soomro, et al., Soil fragmentation and aggregate stability as affected by conventional tillage implements and relations with fractal dimensions, Soil and Tillage Research, **197** 104494 (2020)
17. I. Hakansson, A. Myrbeck, A. Etana, A review of research on seedbed preparation for small grains in Sweden, Soil and Tillage Research **64**: 23-40 (2002)
18. D.I. Vlăduţ, V. Vlăduţ, I. Grigore, S. Biriş - Experimental research on qualitative indices of work for equipment for seedbed preparation in conservative system, ENGINEERING FOR RURAL DEVELOPMENT **16**: 1174-1179, (2017)

 https://doi.org/10.1051/e3sconf/202128603007

Experimental research for obtaining functional ingredients from Jerusalem artichoke and hemp seeds

Iuliana Găgeanu[1], *Cătălin* Persu[1], *Gabriel* Gheorghe[1], *Diana* Popa[2] and *Liliana* Dumitrescu[3*]

[1]INMA Bucharest, Ion Ionescu de la Brad nr. 6, sector 1, Bucharest, Romania
[2]Agricultural Research-Development Station Secuieni, 377 Principala, Secuieni, Romania
[3]Hydraulics and Pneumatics Research Institute INOE 2000-IHP, 14, Cuțitul de Argint, Bucharest, 040558, Romania

Abstract. Functional ingredients have gained a large-scale development determined by the increased interest shown by consumers worldwide. Functional ingredients are bioactive compounds that can be used in the manufacture of functional food products. These bioactive compounds can be obtained from accessible materials. The use of these materials aims to transform affordable raw materials at low prices into functional ingredients concentrated in proteins, amino acids, minerals and fibres, which can be used to obtain foods with a beneficial role in health. The paper presents a series of experimental researches for obtaining and characterizing functional materials obtained from Jerusalem artichoke and hemp seeds. The samples obtained had low moisture content, between 5.11 – 5.43 % for Jerusalem artichoke and 4.8 – 5.13 % for hemp, ideal for storage and use as powder. Jerusalem artichoke registered high sugar contents (between 21.2 – 27.2 %), while hemp flour had high fat content (between 18.56 – 23.52 %), making both products suitable to be used as functional ingredients in the food industry.

1 Introduction

Functional ingredients can be described as bioactive compounds that have the potential to beneficially influence health above the basic nutritional value of normal foods. Foods containing one or more functional ingredients (also called functional foods) can be defined as foods having health properties over and above their original nutritional value [1 – 5].

Bioactive compounds / functional ingredients can be obtained from a variety of sources (primary produce, marine sources, microorganisms, inorganic raw materials). They can also result as a by-product of food processing, bringing new economic value to food processors.

The use of these primary produce aims to transform cheap raw materials into functional ingredients concentrated in proteins, amino acids, minerals and fiber, which can be used to obtain foods with a beneficial role on health, due to a synergism between the density of nutrients and the presence dietary fiber with proven benefits in regulating cholesterol, regulating intestinal transit, increasing satiety and weight management, improving glucose

* Corresponding author: lilianad.ihp@fluidas.ro

tolerance and insulin response, reducing hyperlipidaemia, hypertension and risk factors for coronary heart disease, reducing the risk of developing cancers [7-9].

In order to obtain functional ingredients, the materials used as raw materials must go through a conditioning stage, ensuring the maintenance of nutrients and all their qualities.

Cereal crops occupy more than half of crops in the world, grains being the main food item in agriculture [10]. Hemp has been used from ancient times both for its fiber, and also for its nutritional benefits. Jerusalem artichoke has been widely used both for its biomass potential (used for obtaining biofuels) and for its food potential, its inulin content having benefits in diabetes diets [10-19].

The food supplement obtained from these two types of plants contribute both to the secretion of gastric juices and to the increase of the capacity to contract the intestinal walls (thus stimulating the intestinal peristalsis).

In addition, bioactive compounds separated by supercritical extraction at normal temperature (30-35 °C) from hemp cakes and Jerusalem artichoke may contribute to the binding (in the form of insoluble compounds) to substances harmful to the body (free radicals, heavy metals, radionuclides) thus contributing to their elimination, regulating blood composition (especially the number of red blood cells and leukocytes) and improving liver function [20-22].

Cakes resulting from the processing (cold pressing) of hemp seeds and Jerusalem artichoke flour can be used to create food supplements ennobled with natural antioxidants, which contribute to the normalization of lipid and carbohydrate metabolism, and to maintain the intestinal flora in optimal conditions.

The paper presents a series of experiments for the preliminary processing of Jerusalem artichoke and hemp cakes in the view as using them as functional ingredients and further obtaining by-products and dietary supplements.

2 Material and method

The materials used for obtaining functional ingredients were represented by three varieties of Jerusalem artichoke, grown within the premises of INMA Bucharest, Figure 1, and three cake samples from three varieties of hemp (resulted from the extraction of oil from seeds), Figure 2.

Fig. 1. Jerusalem artichoke samples used for the experimental research. a – Rareş variety; b - Dacic variety; c – Olimp variety

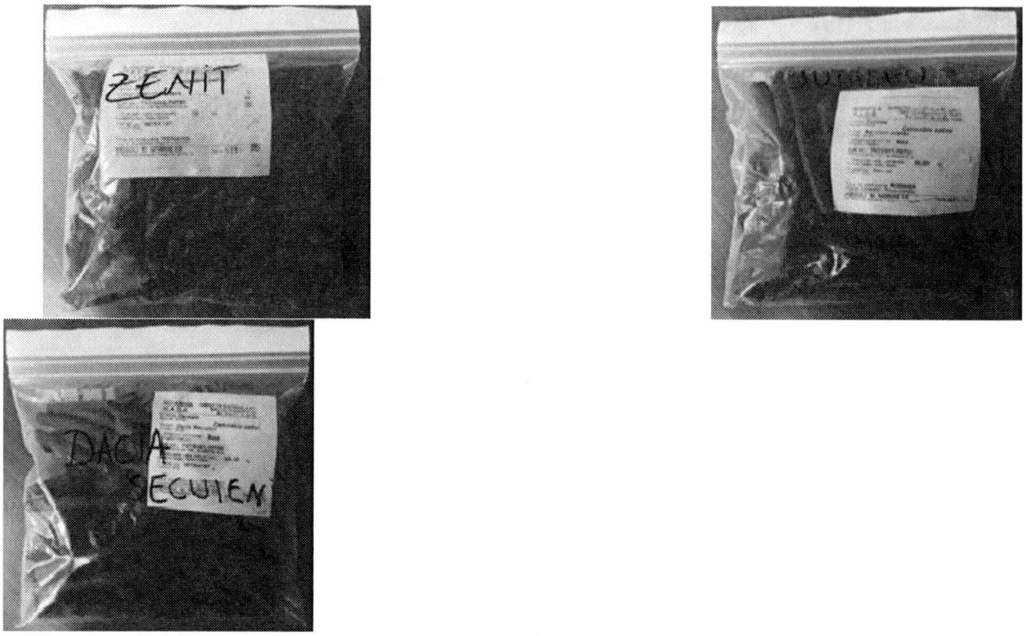

a b c

Fig. 2. Hemp samples used for the experimental research. a – Zenit variety; b - Jubileu variety; c – Dacia Secuieni variety

Table 1. Initial samples used for experiments.

Sample number	**Variety**		**Sample weigh [g]**
1	Jerusalem artichoke	Rareş	5000
2		Dacic	5000
3		Olimp	5000
4	Hemp	Zenit	1000
5		Jubileu	1000
6		Dacia Secuieni	1000

In order obtain functional ingredients, the materials had to first pass through a drying phase. The drying was achieved using a technical drying equipment (Figure 3). It is designed to act on the product, depending on the quality requirements imposed by the beneficiary, respectively on its physical characteristics, offering the possibility to dry samples at low temperatures, thus maintaining intact their properties.

The drying of all samples in the oven was done at 40 °C and the process was constantly monitored.

Fig. 3. Drying of samples used for the experimental research.

The grinding of samples was achieved using a mill for the fine grinding of dry food products. The granulation of the flour obtained varies depending on the orifice diameter of the sieve used. For the hemp and Jerusalem artichoke samples grinded was used a 2 mm orifice diameter sieve.

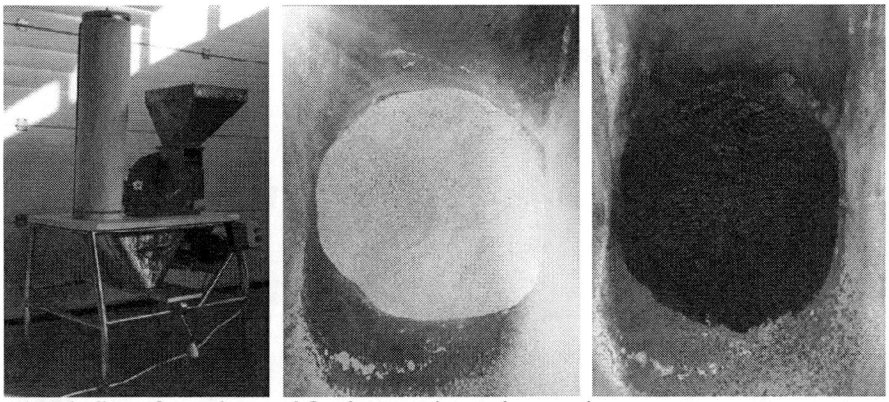

Fig. 4. Grinding of samples used for the experimental research.

The moisture of the samples before drying and after drying and grinding was determined by drying in an oven at 105 °C and calculating the difference in mass,

The bulk density of the samples was determined by filling a stainless-steel cylinder of known volume and subsequently weighing the quantity in the cylinder.

For each sample of Jerusalem artichoke and hemp, before conditioning by drying and grinding, the percentage of fat was calculated using the Soxhlet method by extraction until depletion with organic solvents.

3 Results

The results obtained after dying the samples of Jerusalem artichoke and hemp seed cakes are presented in Table 2.

Table 2. Results obtained after drying the samples

Sample number	Initial moisture [%]	Weight after drying [g]	Moisture after drying [%]

1	73.41	1329	5.43
2	66.95	1652	5.11
3	75.46	1227	5.34
4	15.27	907.9	5.13
5	16.02	899.1	4.8
6	15.18	904.2	5.21

After grinding the samples, a series of tests were conducted, the results being presented in Table 3 and Figure 5.

Table 3. Results obtained after grinding samples

Sample number	Bulk density [kg/m^3]	Fat content [%]	Sugar content [%]
1	711.14	-	21.2
2	726.87	-	26.4
3	729.87	-	27.2
4	815.33	18.56	1,52
5	812.78	20.52	1,47
6	894.14	23.52	1,51

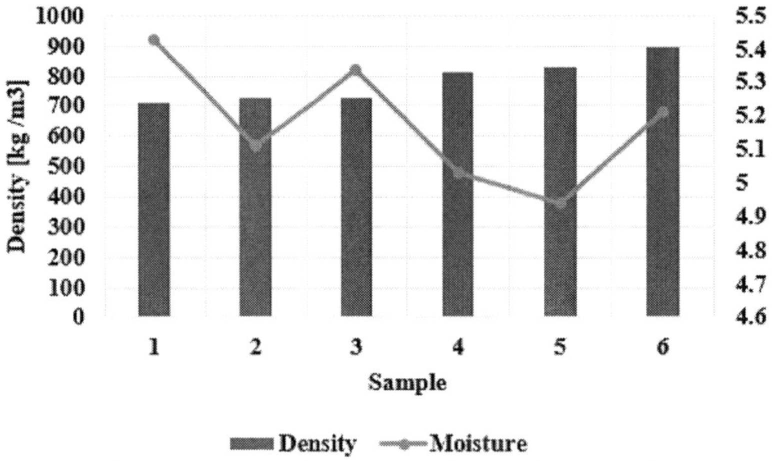

Fig. 5. Variation of density with the moisture content for samples analyzed.

From the results in Tables 2 and 3 and Figure 5 it can observed that Jerusalem artichoke flour had a lower bulk density than hemp flour, mainly due to the high lipid content of hemp. All dried samples had moisture content close to 5%, making them ideal for storage and use as dry ingredients.

4 Conclusions

Following the experiments of conditioning the Jerusalem artichoke samples in order to obtain functional ingredients that will be subsequently included in food products, the following conclusions can be drawn:

1. the moisture before drying had values between 66.95 - 75.46% for Jerusalem artichoke, the highest values being recorded for the Olimp variety, and between 15.18 – 16.02 % for hemp, the highest values being recorded for Jubileu variety;
2. Following the drying process of samples, Jerusalem artichoke reached moistures between 5.11 - 5.43 % and hemp between 4.59 – 5.13%, ideal for storage in the view of future use;
3. Hemp samples retained a fat content between 18.56-20.52 % and a sugar content between 1.57 – 1.52 %, being indicated for consumption;
4. Jerusalem artichoke samples had a sugar content between 21.2 – 27.2 %, the high content making the samples valuable ingredients.

Acknowledgement: This work was supported by a grant from UEFISCDI, project "Complex system of integral capitalization of agricultural species with energy and food potential", code PN-III-P1-1.2-PCCDI-2017-0566, Contract no. 9PCCDI /2018, P4 – Capitalizing on the food potential of selected agricultural crops (Jerusalem artichoke, sorghum seeds, cakes resulting from obtaining oil from cold-pressed hemp seeds).

References

1. C.L. Kruger, S.W. Mann, Food and Chemical Toxicology J., **41**, Issue 6, 793-805, (2003)
2. S. Cozzano, A. Fernández, M. D. del Castillo, A. Medrano Fernández, Ency. of Food Security and Sustain., **1**, 351-359, (2019)
3. I. Ferreira, L. Barros, *Functional Food Ingredients from Plants,* Vol. **90**, 1st Edition, ISBN: 9780128165676, (2018)
4. Y. H. Lee, C. Choo, M. Watawana, N. Jayawardena, V. Y. Waisundara, J. of the Sc. of Food and Agr., **95**, 2956–2964, (2015)
5. N. Belc, L. Apostol, G. Vlasceanu, C. Mosoiu, T. Manasia, V. Vladut, S.A. Martinez, Romanian Biotech. Letters, **25**, Issue 1, 1178-1185, (2020)
6. L. Apostol, N. Belc, L. Gaceu., V. Vladut, O.B. Oprea, Chemistry J., **70**, Issue 6, 2048-2053, (2019)
7. C Popescu., C. Popescu (Popiniuc), Ş. Manea, O Calin., C. Morgovan, V. Vlăduţ, A. Dune, S. Ardelean, Annals of the Uni. of Craiova – Agr., Montanology, Cadastre, Series, **46**, No. 2, 214-223, (2016)
8. S.F. Isticioaia, L. Apostol, G. Matei, O. Mîrzan, P. Pintilie, E. Trotuş, A. Leonte, V. Vladut, O. Cristea, B. Oprea, B. Furdui, C.M., Marinciu, M. Bularda, Romanian Agr. Res., **38**, (2021)
9. O-M. Dumitru, S Iorga., N.V. Vladut, C. Bracacescu., INMATEH Agr. Eng. J., **62**, No. 3, 133-146, (2020)

10. G. Matei, V. Vladut, S. Isticioaia, R.L. Panzaru, D. Popa, Scientific Papers-Series A-Agronomy, **63**, Issue 1, 387-393, (2020)
11. A. Kowalczyh-Jusko, K. Kupryaniuk, T. Oniszczuk, A. Wojtowicz, D., Janczak D., A. Smurzynska, K. Jozwiakowski, M. Czechlowski, *Proceedings of the 3rd Int. Conf. on Energy and Environment (ICEE 2017)*, 89-94, (2017)
12. V. Vlăduţ, S. Şt. Biriş, N. Ungureanu, I. Voicea, D. Cujbescu, L. Apostol, C. Moşoiu, S. Boruz, S. Isticioaia, Gh. Stroescu, A. Matache, D.N. Dumitru, Annals of the Uni. of Craiova – Agr., Montanology, Cadastre Series, **48**, No. 2, 451-457, (2018)
13. Gh. Matei, V. Vlăduţ, M. Ştefan, E Constantinescu., C. Sălceanu, *Int. Symp. ISB-INMA TEH' 2018 Agr. and Mechanical Eng.*, 1041-1046, (2018)
14. R. Oprescu, S.Şt. Biriş, I. Voicea, V. Vlăduţ, Acta Tech. Corviniesis – Bull. of Eng., Tome **XII**, Fascicule 1, 85-88, (2019), ISSN: 2067-3809
15. D. Popa, S. Isticioaia, G. Teliban, V. Vlăduţ, A. Nedelcu, Gh. Matei, O. Mîrzan, A. Panda, P. Pintilie, C. Melucă, A. Agapie, Sc. Papers, Horticulture Series, **63** (1), Agronomy series, 65-68, (2020)
16. L.D. Popa, S.F. Isticioaia, V. Vlăduţ, G.C. Teliban, Gh. Matei, A. Leonte, A. Zaica, O. Cristea, Matache M., *Int. Symp. ISB INMA TEH' 2020 Agr. and Mechanical Eng.*, 854-860, (2020)
17. R. Oprescu, S.Şt. Biriş, I. Voicea, V. Vlăduţ, *Int. Symp. ISB-INMA TEH' 2018, Agr. and Mechanical Eng.,* 325-330, (2018)
18. L-D. Popa, S. Isticioaia, V. Vlăduţ, G. Teliban, G. Matei, A. Nedelcu, *Int. Symp. ISB-INMA TEH' Agr. and Mechanical Eng.,* 838-845, (2019)
19. F. Sîrbu, I. Dumitru, D. Iuga, I. Voicea, V. Vlăduţ, R. Oprescu, Acta Tech. Corviniensis – Bull. of Eng., Fascicule **2**, 97-100, (2019)
20. V Vlăduţ., I. Găgeanu, L. Apostol, Gh. Matei, D. Popa, S. Isticioaia, N. Ungureanu, S.Şt. Biriş, T. Zelazinski, M. Epure, C. Dumitru, S.I. Biriş, *Int. Symp. ISB-INMA TEH" Agr. and Mechanical Eng.*, 958-965, (2019)
21. V. Vlăduţ, A.I. Grigore, I Găgeanu., M. Neagoe, N. Vanghele, L. Apostol, Gh. Matei, D. Popa, S. Isticioaia, N. Ungureanu, S.Şt. Biriş, G Paraschiv., M. Epure, C. Dumitru, S.I. Biriş, *Int. Symp. ISB-INMA TEH' 2019 Agr. and Mechanical Eng.*, 952-957, (2019)
22. A-G Burnete, A-M Lazăr, M. Catană, L. Catană, G. Stamatie, R.I. Teodorescu, N. Belc, V. Vlăduţ,, *Int. Symp. ISB-INMA TEH' 2019 Agr. and Mechanical Eng.,* 817-825, (2019)

Strength analysis and modal analysis for the load-bearing structure of the equipment for opening and compartmentalizing watering furrows, using 1D

Petru Cardei[1,*], *Raluca* Sfiru[1] , *Marius* Oprescu[2], *Vergil* Muraru[1], and *Cornelia* Muraru[1]

[1]INMA Bucharest, RDI, Bv. Ion Ionescu de la Brad no 6, Romania

[2]INMA Bucharest, DITRMA-RENAR, Bv. Ion Ionescu de la Brad no 6, Romania

Abstract. The paper proposes the analysis of the load-bearing structure of the equipment for opening and compartmentalizing watering furrows (EOCFW), using a 3D structural model, built with 1D finite element. Based on information from the experimental results of the EOCFW equipment, the load-bearing structure is supported and loaded. The results of the linear static analysis of the structure consist of the distributions of the relative displacement fields and the equivalent voltage in the structure. Also, the own frequencies of the structure and the deformed forms of the structure are obtained when it vibrates in its own ways with the lowest own frequencies. It shows how these results can be used. The field of relative displacements is used to assess the effects on the quality of the soil processing performed. The equivalent stress field is used to estimate the safety factor of the structure, by reference to the flow stress of the material from which the structure is built. The first four or five own frequencies are important for the prognosis of possible vibration regimes with resonance, their explanation and their amelioration. It emphasizes the ease with which the model can be modified to obtain improved or even optimized variants.

1 Introduction

Structural analysis is a modern tool, but already normal in the research of physical phenomena and the design of industrial, civil or other products. The tool is modern today because it constantly addresses new topics and deepens the already classic ones.

In the field of agricultural machinery, structural analysis has gained intense use in the last fifty years. In addition to the hesitant beginnings, in the last twenty years, there have been many works related to the design of equipment used in agriculture or the structural modelling of physical processes in agriculture, [1]. A review of the use of MEF in the field of agricultural mechanization is, for example [2, 5]. Most of the literature of the last 20 years, in this field, proposes models with solid (three-dimensional) finite elements, [3, 4, 6, 7-13].

* Corresponding author: petru_cardei@yahoo.com

3D structural models, built with 1D bar-type finite elements, are older than those built with 3D finite elements. Mathematical models of the straight bar (Euler or Tymoshenko, for example) were used long before the appearance of the finite element method. The finite element method integrated these mathematical models. 3D structural models built with 1D finite elements are also well known in the literature in the field of agricultural machinery, [1, 14-17]. Structural models built with 1D finite elements are well suited to large structures formed with relatively thin bars, such as the load-bearing structures of bridges, vehicles or agricultural machinery [18-20]. The literature still abounds with examples of the use of 3D structural models, mashed with 1D or mixed finite elements, hybrid models that contain 1D elements (of the bar or beam type). This category also includes the model presented in this paper. The advantages and disadvantages of this model are presented in relation to the 3D models mashed with 3D elements. On the other hand, the usefulness of such a model in the research - design activity for certain structures, signalled above, is also argued, their structural models simplifying the calculation procedures and the interpretation of the results. Research on the testing of agricultural equipment in real field conditions was carried out by [27, 29, 33], simulation by finite element analysis (FEM) [24, 28, 31, 32], on special installations/test stands in regime simulated and accelerated [26, 30, 31] or theoretical research [25], to identify and develop the most modern constructive solutions [34], similar to those carried out worldwide.

2 Material and method

The subject of the analysis, the results of which are presented in the article, is the load-bearing structure of EOCFW, fig. 1.

Fig. 1. Rotor version of the load-bearing structure of the EOCFW.

The structural model (geometry, fixing conditions and loads, elements and nodes) of the load-bearing structure of EOCFW, is presented in fig. 2. There is a difference between the load-bearing structure of the EOCFW and the entire structure of this machine. The working organs (the double mouldboard and the palette) are represented in the model only by the forces with which they act on the load-bearing structure. Fixing the structure or the conditions on the border means cancelling the relative linear displacements in the articulation points to the tractor. The loading is done with the force of 4500 N from the double mouldboard side and 700 N from the blade side, in the direction of movement in the opposite direction to the direction of movement of the unit. A force of 100 N acts on the resistance frame (the load-bearing structure) in the same direction and in the same sense as the other two forces, as a result of the resistance of the copy wheel, fig.2. The material of the structure, for linear static analysis and modal analysis, is characterized by the linear elasticity modulus, $E = 2.0 \cdot 10^{11}$ Pa, the Poisson's ratio $\nu = 0.29$ and the density $\rho = 7900$ kg / m^3. The model is mashed with ninety-eight nodes and ninety BEAM3D elements included in the finite element library of the COSMOS / M 1.75 program, [21].

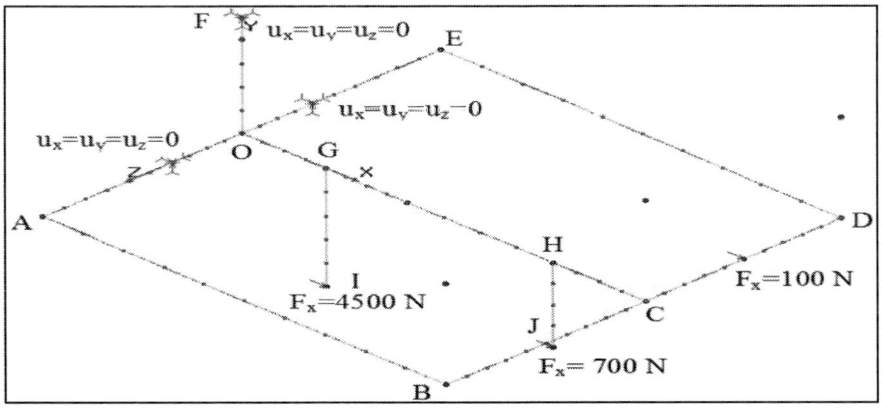

Fig. 2. Rotor or cam version of the load-bearing structure of the EOCFW.

The characteristics of the cross-sections of the bars used in the construction of the structural model of the load-bearing structure of the EOCFW are given in Table 1.

Table 1. Characteristics of cross-sections of EOCFW structural model profiles.

Beam, section, mm	**Area, mm^2**	**Moment of inertia, Ix,mm^4**	**Moment of inertia, Iy, mm^4**	**Depth, mm**	**Width, mm**
AO,OE, BC, CD, Caisson, 60x60x6	1296.0	63.76	63.76	84.852 (60)	84.852 (60)
AB, OC, ED, GI, HJ 55x30	1650	12.375	41.594	55.0	30.0
OF	1440	1.728	43.2	60	84

3 Results

The main results of the two analyzes are relative displacement field (values in nodes), the tensor field of Cauchy stresses (average values per element), reaction values, deformed shape of the structure (from static linear elastic analysis), respectively: the lower first five eigenfrequencies and the deformed shapes of the structure when it vibrates in the corresponding eigenmodes.

3.1 The main results of the linear elastic analysis

The vector field of relative displacement or the deformation field as it is called in engineering language has three components, corresponding to the three axes of the absolute reference system. Therefore there are three scalar fields of relative displacements in the directions of the three coordinate axes. A scalar field of the relative deformation, which reflects the overall displacement, is that of the resultant relative displacement, whose distribution in the structural model is given in fig. 3.

The stress field has in this model two main components, the axial stress field and the equivalent stress field (Von Mises), [21]. In fig. 4 represents the map of the average equivalent stress field distribution for each element of the model.
The maximum value of the relative displacement is 4.96 mm and corresponds to the component along the vertical axis at the ground plane (Oy). The maximum value of the equivalent stress in the structure is 123 MPa, is located at the upper part of the double mouldboard support and at the lower part of the beam with which the structure is coupled to the upper link of the tractor.
Reaction values at the tractor points of linked to the load-bearing structure is given in Table 2.

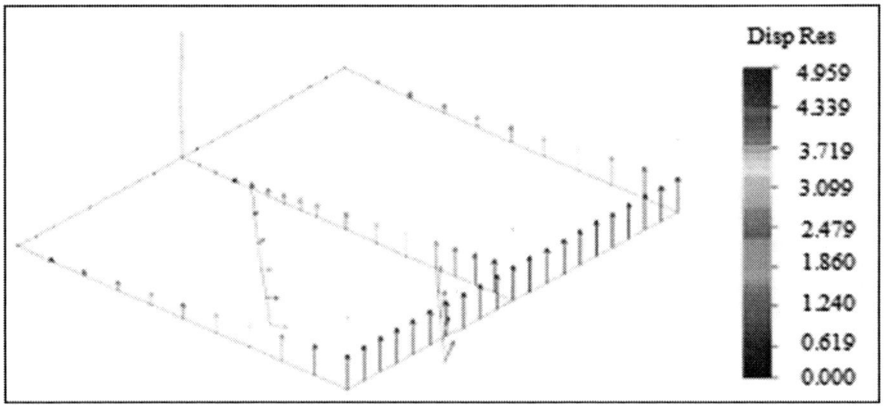

Fig. 3. Distribution of the resultant relative displacement field in the load-bearing structure of the EOCFW, in mm.

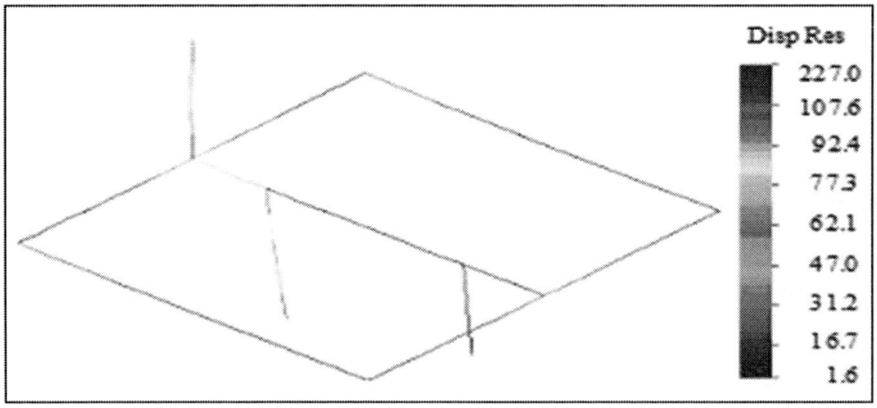

Fig. 4. Distribution of the Von Mises stress field in the load-bearing structure of the EOCFW, in MPa.

Table 2. Reaction values at tractor points for load-bearing structure an EOCFW.

Place	Ox reaction, N	Oy reaction, N	Oz reaction, N	Resultant, N
Right link	-5115.0	-843.1	5.7	5184
Central link	5072.0	168.6	0.0	5345

Left link	-5258.0	-843.1	-5.7	5325
Total	-5300.0	0.0	0.0	5300

3.2 The main results of the modal analysis

The first five eigenfrequencies of the structure, in Hz, have, in ascending order, the values: 17.57, 24.60, 42.44, 107.48, and 126.62. The deformed shapes of the structure when it vibrates on one of its own frequencies, help to determine the components of the assembly, which most likely causes a possible movement with high amplitude or even resonance.

4 Discussion

In addition to the results presented in the previous chapter, the FEA program that solves the analysis of the structural model, COSMOS / M 1.75, [21], also provides other results. These results include the inertial characteristics of the entire structure, for example, the mass of 70.07 kg, the length, the area and the volume of the model bars, the geometric and mass moments of inertia usable in calculations of the aggregate dynamics, the coordinates of the mass centre and radius of gyration. The mass given by the program is smaller than the mass of the EOCFW because it does not include the working organs and the additional elements for their support or guidance (in total, around 110 kg). The presented model is the simplest possible with natural loads in forces derived from experimental data, [22]. Obviously, the model can be complicated by the introduction of working organs and additional guiding elements.

For ordinary steels, the yield strength (at the end of which irreversible deformations begin to occur), has the value of about 220 - 230 MPa. Taking into account the maximum value of the equivalent stress in the structure (fig. 4), a safety factor of 1.8, [23], that required in the standards for agricultural machinery and equipment, results. If the structure is built with steel whose minimum yield stress value is 351,571 MPa, then the minimum safety factor of the structure is 2.86.

5 Conclusion

The results presented in the paper show that a mashed 3D structural model with 1D finite elements (Tymoshenko bars), is particularly useful in the rapid design of a load-bearing structure if sufficiently precise loading assumptions and boundary conditions are known. Changing the bars of the structure is very simple because the design does not change but only the numerical characteristics of their cross sections change. The changes, purely numerical, are made simply in the database. In this way, the bars are replaced by probing until a safety factor is reached in accordance with the standards and a negligible deformation for the working process. In the same way, some vibrations that can affect the work process can be largely removed. This optimization process can be also operated on the experimental model after detecting some deficiencies in the tests carried out in the field.

The use of mashed 3D structural models with 1D type elements is useful at a first start design, it is simple and fast compared to the direct calculation on mashed structural models with 3D finite elements (SOLID). 3D CAD-CAM models are models prepared especially for execution documentation. Their use in the structural analysis must go through the stage of transformation into a CAD - CAE model (elimination of gaps and interferences, mashing). Modifications by probing beams or other components are difficult, often involving redesign. For this reason, a mashed 3D structural model with 1D elements (possibly also 2D) is useful, the pre-design made with its help greatly limiting the calculation on models with 3D finite elements or even eliminating it.

Acknowledgement

This work was supported by a grant of the Ministry of Research, Innovation and Digitization (MRID), CCCDI – UEFISCDI, project number PN-III-P1-1.2-PCCDI-2017-0560 "Eco-nanotechnologies and intelligent equipment for soil properties mapping and evaluating the dynamics of the plant in order to improve agricultural production and environmental protection, within PNCDI III and was done by "NUCLEU" Programme, developed with the support of MRID project PN 19 10 01 02.

References

1. P. Cardei, and others, *Structural analysis and new materials focused on mechanics, mechatronics, maintenance and operation of technical equipment for agriculture and food industry*, Terra Nova Publishing Iasi, (2012)
2. N. S. Velloso, A. L. Costa, R. R. Magalhães, F. L. Santos, E. T. D. Andrade, The Finite Element Method Applied to Agricultural Engineering: A review, Current Agriculture Research Journal, **6**(3), 286-299, (2018)
3. H. K. Celik, R. Cinar, G. Kunt, A. E. W. Rennie, M. Ucar, I. Akinci, Finite Element Analysis of a PTO Shaft Used in an Agricultural Tractor, Ergonomics International Journal, **2**(3), (2018)
4. M. Boryga, P. Kolodziej, Stress analysis of the sugar beet lifter with the finite element method (FEM), Agricultural Engineering, **23**(3), 29-39,(2019)
5. Y. Shen, *Soil-Machine Interactions A Finite Element Perspective*, CRC Press, (1998)
6. P. Bazyar, A. Jafari, R. Alimardani, V. Mohammadi, J. Grichar, Finite Elements Analysis of Small-scale Head of Combine Harvester for Harvesting Fine-Grain Products, International Journal of Advanced Biological and Biomedical Research, **8**(4), 339-357, (2020)
7. N. Nankali, M. Namjoo, M. R. Maleki, Stress Analysis of Tractor Tire Interacting with Soil using 2D Finite Element Method, International Journal Advanced Design and Manufacturing Technology, **5**(3), 107-111, (2012)
8. F. Tian, Y. Chen, Z. Song, Y. Yan, L. Fade, Z. Wang, B. Xiong, Finite Element Simulation and Perforance Test of Loading and Mixing Characteristics of Self-Propelled Total Mixed Rotation Mixer, Journal of Engineering,15 pag., (2020)
9. E. P. da Silva, F. M. da Silva, R.R. Magalhães, Application of Finite Elements Method for Structural Analysis in a Coffee Harvester, Engineering, 6, 138-147, (2014)
10. V. Vladut, E. Maican, L. Apostol, N. Ungureanu, I. Dumitru, R. Oprescu, Verification of Stress by FEM Analysis Mechanical Testing of Agricultural Mobile Aggregates Coupling Devices, INMATEH Agricultural Engineering, **54**(1),39-46, (2018)
11. S. Şt. Biriş, Ungureanu N., Cujbescu D., Modelling of soil compaction under heavy-duty tractors, INMATEH Agricultural Engineering, **57**(1), 261-270, (2019)
12. G.A. Constantin, Gh. Voicu, B. Olac, F. Ilie, G. Paraschiv, V. Ştefan, G. Musuroi, Structural analysis with finite elements of a subsoiler working part, INMATEH Agricultural Engineering, **61**(2), 61-26, (2020)

13. G. Ipate, G. A. Constantin, S. Şt. Biriş, Gh. Voicu, C. O. Rusănescu, V. Ştefan, Numerical 3D analysis a mini wind turbine with horizontal axis, for implementation in agricultural farms, INMATEH Agricultural Engineering, **61**(3), 325-333, (2020)
14. V. Niţescu, P. Cardei, V. Muraru, I. Cojocaru, R. Sfîru, Modalităţi de dimensionare a cilindrului hidraulic pentru mecanismul de reversare al unui plug reversibil, folosind cinematica şi dinamica procesului de reversare, INMATEH Agricultural Engineering, **16**(1), 163-168, (2006)
15. V. Muraru, N. Constantin, P. Cardei, R. Sfîru, Structural Analysis of MATINA soil working machine, INMATEH Agricultural Engineering, **31**(2), 17-26, (2010)
16. I. Cojocaru, P. Cardei, M. Neacşu, *Utilisation of elasto-plastic models of the soil in order to optimise the resistance structures of the chisel ploughs and to estimate their effects in soil*, Proceedings of the International Conference "Soil Compaction and Crop Production, Gödölo, Hungary, (1998)
17. X. Wang, B. Li, Z. Yang, Finite Element Analysis and Lightweight Optimization Design on Main Frame Structure of Large Elecrostatic Precipitator, Advances in Materials science and Engineering, (2018)
18. M. H. Mat, A. R.Ab. Ghani, *Design and Analysis of 'Eco' Car Chassis*, Procedia Engineering 41, International Symposium on Robotics and Intelligent Sensors, 1756-1760, (2012)
19. S. Babak, A structural investigation to develop guidlines for the finite element analysis of the Mini-Baja vehicle, Electronic Theses and Dissertations, (2012)
20. I. Vayas, T. Adamakos, A. Iliopoulos, Three dimensional modeling for steel-concrete composite bridges using system of bar elements - modeling of skewed bridges, International Journal of Steel Structures, **11**(2), 157-169, (2011)
21. Structural Researches and Analysis Corporation (SRAC), *COSMOS/M Finite Element Analysis System, Version 1.75. Manuals: User Guide*, Basic FEA System, Advanced Modules, Cammand Reference, (1995)
22. M.R. Oprescu, I. Gageanu, M. Cristea, Determining the qualitative working indices for various forms of the active body of a rut soil compartmentation equipment, E3S Web of Conferences 180, 03011 (2020)
23. I. I. Ştefănescu, C. Spânu, *Organe de maşini*, vol. **I**, Editura Europlus, Galaţi, (2009)
24. S.Şt. Biriş, V. Vlăduţ, N. Faur, A. Cernescu, O. Kabaş, M. Matache, I. Voicea, S. Bungescu, C. Popescu, *FEM analysis / testing resistance of a tractor seat*, Proceedings of the 43 International Symposium On Agricultural Engineering "Actual Tasks on Agricultural Engineering", Opatija, 189-200, (2015)
25. S.Şt. Biriş, N. Ungureanu, N.E. Gheorghiţă, E. Maican, V. Vlăduţ, *Theoretical research on enhancing the performance of cultivators working bodies,* Proceedings of the 45 International Symposium On Agricultural Engineering "Actual Tasks on Agricultural Engineering", Opatija, 121-130, (2017)
26. D. Cujbescu, S.Şt. Biriş, Gh. Voicu, M. Matache, G. Paraschiv, V.Vlăduţ, M. Bularda, N. Ungureanu, *Determination of sowing precision in simulated laboratory conditions*, INMATEH – Agricultural Engineering, **61**(2), 209-216, (2020)
27. D. Cujbescu, N. Ungureanu, V. Vlăduţ, C. Persu, R. Oprescu, E. Gheorghiţă, *Field testing of compaction characteristics for farm tractor universal 445*, INMATEH – Agricultural Engineering, **59**(3), 245-252, (2019)
28. D.I. Vlăduţ, S. Biriş, V. Vlăduţ, D. Cujbescu, N. Ungureanu, I. Găgeanu, *Verification of stress by FEM analysis / mechanical testing of agricultural mobile aggregates couplig device*, INMATEH – Agricultural Engineering, **54**(1), 37-462, (2018)

29. D.I. Vlăduţ, V. Vlăduţ, I. Grigore, S. Biriş, *Experimental research on qualitative indices of work for equipment for seedbed preparation in conservative system*, 16th International Scientific Conference "ENGINEERING FOR RURAL DEVELOPMENT", 1174-1179, Jelgava, Latvia, (2017)
30. V. Vlăduţ, V. Gângu, I. Pirnă, S. Băjenaru, S. Biriş, S. Bungescu, *Complex tests of the resistance structures in simulated and accelerated regime on hydropulse installation*, Proceedings of the 35 International Symposium On Agricultural Engineering "Actual Tasks on Agricultural Engineering", 393-404, Opatija – Croaţia, (2007)
31. V. Vlăduţ, Matache M., Faur N., Biriş S., Paraschiv G., Bungescu S., *Traction bar testing of a 161 kW tractor in simulated and accelerated regime and its analysis by finite elements method for determining its rupture resistance*, Proceedings of the 37 International Symposium On Agricultural Engineering "Actual Tasks on Agricultural Engineering", 77-84, Opatija – Croaţia, (2009)
32. V. Vlăduţ, S. Biriş, S. Bungescu, N. Faur, A. Cernescu, M. Matache, O. Kabaş, G. Paraschiv, At. Atanasov, Gh. Ivan, *The verification of stress by MEF analysis/ mechanical testing of a traction bar*, Proceedings of the 43 International Symposium On Agricultural Engineering "Actual Tasks on Agricultural Engineering", 141-152, Opatija – Croaţia, (2015)
33. V. Vlăduţ, E. Marin, I. Grigore, S.Şt. Biriş, N. Ungureanu, G. Gheorghe, M. Matache, C. Persu, I. Voicea, D. Cujbescu, S. Bungescu, I. Kiss, C. Vlad, *Field and laboratory wear testing of integral seedbed implement's chisel tines point*, Proceedings of the 45 International Symposium On Agricultural Engineering "Actual Tasks on Agricultural Engineering", 131-140, Opatija – Croaţia, (2017)
34. L. Vlăduţoiu, P. Cârdei, V. Vlăduţ, L. Fechete - *Modern trends in designing and selecting the machine / equipment for soil deep tillage*, 16th International Scientific Conference "ENGINEERING FOR RURAL DEVELOPMENT", 1415-1420, Jelgava, Latvia, (2017).

Analysis of the resistance frame of the equipment for opening and compartmentalizing watering furrows, using 3D finite elements

Petru Cardei[1,*], *Marius* Oprescu[2] , *Nicolae* Constantin [1], *Ştefan-Sorin* Biriş[3], and *Sebastian* Muraru[1]

[1]INMA Bucharest, RDI, Bv. Ion Ionescu de la Brad no 6 Bucharest, Romania

[2]INMA Bucharest, DITRMA, Bv. Ion Ionescu de la Brad no 6 Bucharest, Romania

[3]Politechnic University of Bucharest,

Abstract. The proposed article sets out the results of the finite element structural analysis for the open and compartmentalised watering furrow equipment (OCWFE). It uses a 3D structural model with 3D finite elements. The analysis set out in the article is made in order to determine the field of relative displacement and equivalent stress in the load-bearing structure of the OCWFE. The structural model is generated in CAD-CAM. For the structural analysis, it is necessary the CAE model, which is obtained from the CAD-CAM model, mainly by eliminating gaps and interferences, but also by the techniques of realizing the contact between the components of the assembly and a careful mashing of the structure. The structural model thus created is supported and loaded in accordance with the experimental results from the literature. The relative displacement field and the equivalent stress field within the resistance frame of the OCWFE is obtained following the linear static analysis. The field of relative displacements is used to assess the effects on the quality of the work performed. The equivalent stress field is used to estimate the safety factor of the structure, by reference to the flow stress of the material from which the structure is built.

1 Introduction

The complexity of agricultural machines generates very varied requirements on their resistance frames (load-bearing structures). In addition to the supports of the working bodies, the resistance frames must also support ancillary assemblies with a functional role, sometimes real installations with an important role in carrying out the work process or in carrying out additional work. In order to meet these requirements, a high-level design is necessary, leading to a resistant structure but also sufficiently supple, optimized, so as not to produce excessive energy consumption. Also, the structure must be designed so as not to occur in the operation process, resonant vibration regimes that lead to poor quality of the

* Corresponding author: petru_cardei@yahoo.com

work, premature wear and failures. Also, a very important problem of load-bearing structures is the achievement of the high security of human personnel working on agricultural machinery, its protection even in case of serious accidents. For these reasons, the design of load-bearing structures of agricultural machinery is difficult enough to require the help of specialized tools for drawing, calculation and simulation. The study, research, and design of load-bearing structures is a field deeply involved in many disciplines of science and technology: civil engineering, [1-2], automobiles, aircraft or water vehicles, medicine, etc. For agricultural machines through load-bearing structures we understand, most of the time, resistance frames [3-14]. In general, the works [3-14] use 3D finite element analysis for linear-elastic static analysis, implicitly elementary analysis of the strength of structures. There are also works that address the dynamic working regime of assemblies or subassemblies of agricultural machines, problems of modal analysis related to the avoidance of working regimes with resonances, analysis of structural failures using linear and nonlinear analysis, [3]. Also addressed in many papers is the problem of optimizing the strength of agricultural machinery, also using structural models with 3D finite elements, for example [3, 7, 11, and 15]. The analysis presented in this paper is only a simple application of linearly elastic static analysis for the frame of a small dimension machine with precise loading points. The first result presented is the resistance analysis, a common result in the field, the second presenting the possibility to control the quality of soil processing, more precisely, the control of the depth of the irrigation channel created. In the studied literature, we did not find references to resistance studies for agricultural machines of this type, OCWFE. Research on the testing of agricultural equipment in real field conditions was carried out by [22, 24, 28], simulation by finite element analysis (FEM) [19, 23, 26, 27], on special installations / test stands in regime simulated and accelerated [21, 25, 26] or theoretical research [20], to identify and develop the most modern constructive solutions [29], similar to those carried out worldwide.

2 Material and method

The material of this research consists of the load-bearing structure of the OCWFE (fig. 1) and the boundary and loading conditions. The working method is the structural analysis using finite element method with 3D finite elements.

Fig. 1. The open and compartmentalised watering furrow equipment (OCWFE).

The structural model (geometry, boundary conditions and loads), and the meshed structural model, for OCWFE, are given in fig.2. The CAD-CAM model is an extension of the structural model in fig. 2 because it also contains the working members, the double mouldboard and the blade, as well as the copy wheel and the drive mechanism of the blade. The CAD-CAE model only takes over the load-bearing structure (resistance frame) with the supports of the double mouldboard and of the blade. To obtain the geometry of the CAD-CAE model from fig. 2, from the CAD-CAM model were eliminated the working organs and the annexes (the copying wheel and the drive mechanism of the blade). In addition, gaps and interferences have been eliminated from the CAD-CAM model. The effects of the working organs on the structure were substituted with the loads in forces, fig. 2. The conditions on the border consisted in the rigid fixing of the structure in the points of connection to the tractor (cancellation of all degrees of freedom on the borders of the holes where the fastening bolts are found).

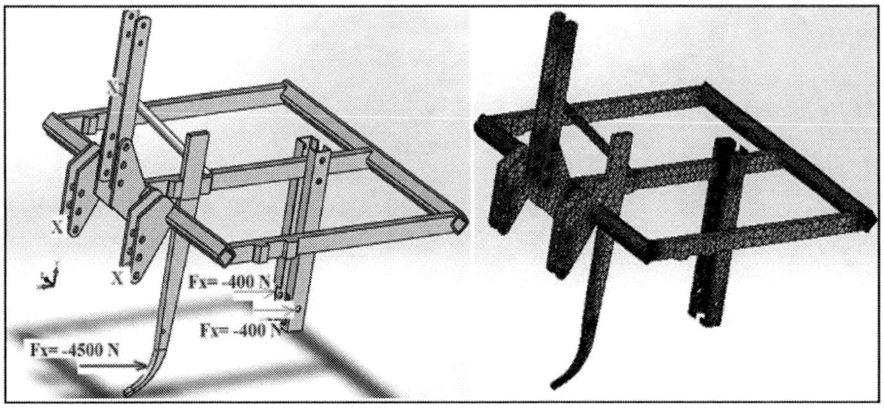

Fig. 2. The structural model (geometry, boundary conditions and loads), left, respectively the meshed structural model, right for OCWFE.

The load corresponding to the double mouldboard (4500 N) is applied in the centre of its contact area with its support. The load corresponding to the blade (2 x 400 N) is distributed in two equal components, on symmetrical areas in relation to the longitudinal axis of the structure, areas located on the connection support to the central resistance beam. The structure is meshed by probing, until the calculation program accepted the total meshing, fig. 2.

The only important material constants in the linear-elastic calculation are: the modulus of longitudinal elasticity $E = 2.0 \cdot 10^{11}$ Pa, the Poisson's ratio $\nu = 0.29$. In addition, to estimate the safety factor of the structure, the yield stress with the value 351.571 MPa (steel with AISI code 1020 from the material library of the Solid Works program, [16]) is retained.

3 Results

The important results retained from the elastic-linear static analysis of the OCWFE structural model, using the finite element method, are the fields of the relative displacement (deformation), the strain, and Cauchy stress. Based on these results, the first conclusions are drawn, which refer to the strength of the load-bearing structure in the work process and to the quality of the soil processing performed. For the two evaluations about the behaviour of the equipment in work, it is sufficient to consider in discussion only the equivalent stress field (Von Mises), fig. 4, and the resulting relative displacement field (deformation), fig. 3. The maximum value of the resulting relative displacement field is 4.58 mm (fig. 3), located on the model boundary at the end of the double mouldboard support. The maximum value of the equivalent stress is 186.642 MPa, located at the top of the double mouldboard support and at the connection points to the central tie rod of the tractor (fig. 4). The component with the highest intensity of the resulting relative displacement is the one along

 https://doi.org/10.1051/e3sconf/202128603009

the Ox axis. The distribution of the values of this component on the border of the structural model is given in fig. 5. In fig. 6 is given the map of the distribution of the component values along the Oy axis (the vertical to the ground surface) of the relative displacement, on the border of the structural model of OCWFE.

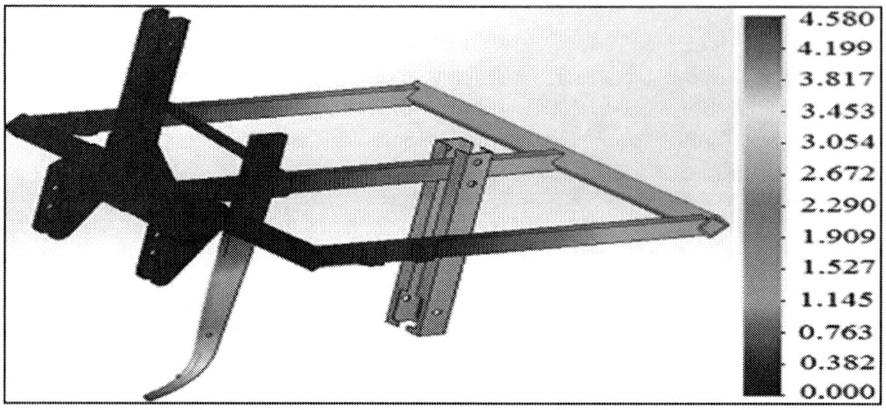

Fig. 3. Resultant displacement field on the model border, in mm.

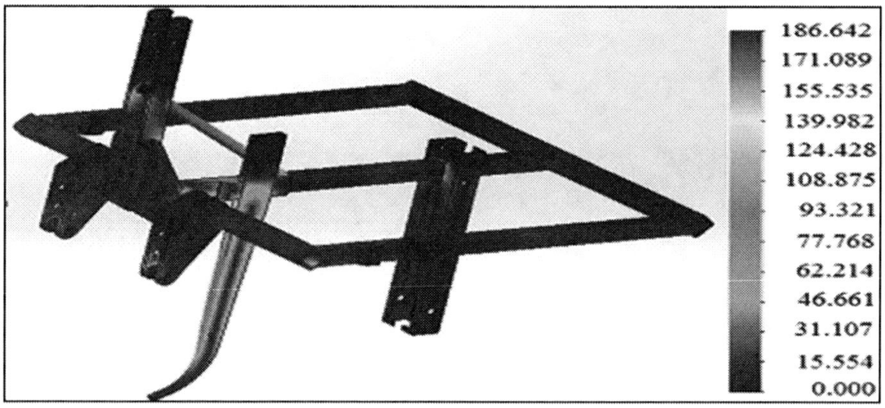

Fig. 4 Map of equivalent stress values on the model boundary, in MPa.

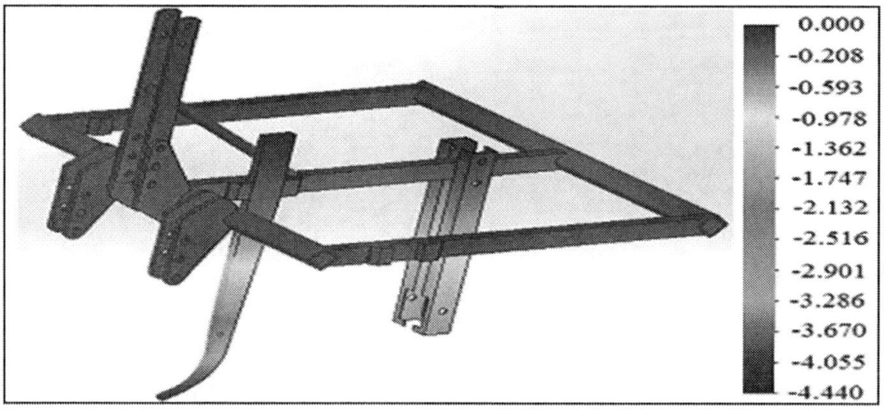

Fig. 5 Map of the distribution of the component values along the Ox axis of the relative displacement field, on the border of the structural model of OCWFE, in mm.

Regarding the relative displacement of the structure along the Oz axis (transverse direction to the direction of movement in the horizontal plane) it varies, on the boundary of the structure, between -0.194 and 0.312 mm. In general, this component is used to detect possible asymmetries in motion, phenomena that can give major effects in working process.

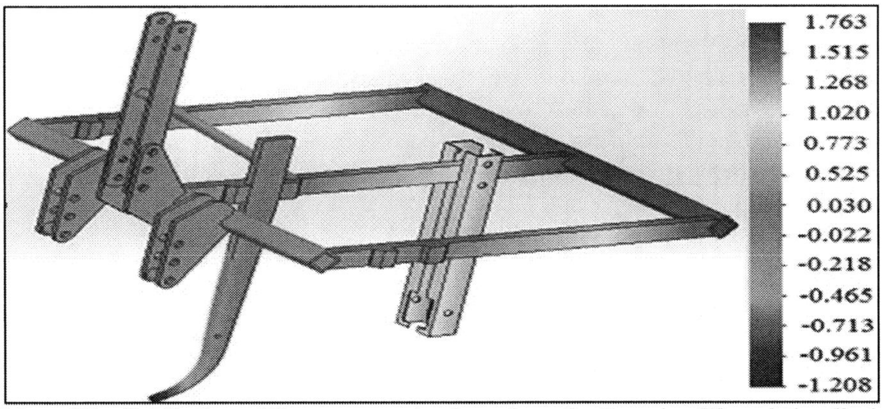

Fig. 6 Map of the distribution of the component values along the Oy axis of the relative displacement field, on the border of the structural model of OCWFE, in mm.

4 Discussion

The maximum value of the resulting relative displacement shows that the maximum deformation of the structure compared to the rest position is located at the base of the support of the double mouldboard that causes the same displacement of the double mouldboard, therefore, there is the possibility of a maximum vertical deviation of 2.29 - 2.69% of the channel depth which must be produced by passing the machine (depth of 170-200 mm). The deviation is negligible considering that the soil can itself produce such deviations.

The maximum value of the stress registered in the upper area of the double mouldboard support is not dangerous, ensuring a safety factor with a minimum value of 1.88, sufficient in relation to the standards for agricultural machines, [17]. In addition, the double mouldboard support has been simplified in the structural model used in this analysis, in reality being much more rigid. High values of equivalent tension in the tractor attachment area are negligible because the bearing is excessively rigid compared to reality (joints with large clearances).

In fig. 5 can be observed the distribution of the component values along the Ox axis of the relative displacement vector (deformation). This component has the dominant contribution to the resulting relative displacement. The analysis of the relative displacement on components shows that the absolute maximum value is concentrated on the Ox axis (direction of advance of the aggregate), in the opposite direction to the direction of advance (4.44 mm, at the area where the double mouldboard is attached to its support). The component of the relative displacement on the Oy direction has the maximum value of 1.763 mm, in the positive sense, i.e. the structure rises or tends to come out of the ground. This value is small than half the maximum of the resultant displacement value. Thus the error of achieving the working depth is much smaller than the one calculated above. In addition, the map of the distribution of the relative displacement on the border shows that the vertical lift is maximum at the rear of the load-bearing structure, and the minimum value (which is negative) is registered in the area where the double mouldboard is attached to its support. The negative sign of this component at the top of the double mouldboard support shows that the working body tends to enter deeper into the soil.

The values between which the transverse component varies with the direction of movement in the ground plane of the relative displacement (Oz axis), show that, although the structure is not symmetrical, the asymmetry in displacement is still negligible.

All the above considerations are valid in the hypothesis that the support of the structure (connection to the tractor) corresponds to reality. In reality, however, attaching the structure to the tractor is more complicated, including play joints, which makes it very difficult to model boundary conditions as a contact problem. For design needs, however, the information obtained in the case of rigid boundary conditions is sufficient.

The efficiency of a complex structural analysis for a load-bearing structure such as the strength frames of the agricultural machinery is still, a debatable problem, but it can be clarified then when there will be enough examples of analyzes at different levels of complexity for the same structure and the they results will be compared. Until then, there will be enough learned or brave people who will "shoot with the cannon at the sparrows". Generating 3D drawings to obtain the execution documentation is an operation that has transformed in these times, many engineers into simple designers. Many know how to generate complex curves and surfaces, which extruded give bodies with a very complex geometry. Then, such bodies are joined into assemblies by operations of translation and rotation in space. How many of these operators still know how to generate numerically, if not the curves and complex surfaces, at least the line, the circle, the spherical surface, the ellipsoid, etc. ? The black boxes, which are CAD programs, thus generate structural models that have small defects, called gaps or interferences, just as parts or subassemblies which are sets of disjoint points, or have non-empty intersections of non-zero 3D measurement. Such drawings or flat drawings, 2D or classic projections can be used successfully for the execution of physical models, because the human mind solves small errors such as gaps or interferences, obviously, when they are small enough and do not generate malfunctions. Solving such problems in execution practice is a matter of compromise between designer and executor, more difficult when the executor is a machine and needs the exact numerical model for an exact execution, with standardized tolerances.

The structural model prepared for the structural simulation must be carefully prepared so that the gaps and interferences are sufficiently reduced so that the calculation can tolerate them, possibly by introducing additional hypotheses. CAD-CAM models prepared by these 3D drawing programs (which generally have gaps and interference) have some errors generated by the set errors of the program when coupling different parts or assemblies, but at the same time they respect the tolerances of the standards, generating games without which the coupling of parts or assemblies becomes impossible. Structural calculation programs cannot tolerate large gaps and interferences. Small interferences are also difficult to tolerate and give rise to difficulties of mesh generation. However, small gaps are not only accidental but can often be necessary for installation. Even being small and regulatory from the point of view of execution, the existence of the necessary gaps requires, for numerical simulation, very large complications with contact problems. For this in the first stage of study, such gaps are tried to be tolerated by various procedures. However, these procedures, in general, do not correspond to reality. After eliminating gaps and interferences through various procedures (couplings, for example), the CAD-CAE model is obtained which will be subjected to structural analysis.

It is found that the analysis activity of a 3D structural model, meshed with solid elements, is long-lasting, expensive. It is not recommended to use such a method in the early stages of design, in which many solutions change (profiles, materials, shapes, variants). Changes in the structural geometry of these solutions will require redesign as well as all other operations described above. It is much more convenient to use, especially for load-bearing structures (resistance frames), 3D models, simple meshed with 1D finite elements, possibly mixed (hybrid). Such a model greatly reduces or even cancels the number of gaps and interferences. Complex 3D models are useful in the final design stage for determining stress concentrators, structure optimization or improvement activity based on information about the operation of the experimental model (elimination of intense vibrations, premature wear, etc.). In order for both types of models (meshed with 1D and 3D elements, respectively) to be fully accepted, in addition to experimental validation, a convergence study would also be needed, which is not simple, [18].

5 Conclusions

The results of the analysis show that the structural analysis of a resistance frame for agricultural machinery is useful primarily for checking the resistance, in the case of an experimentally determined work regime (demands, conditions on the border, etc.). This application is classic. The Cauchy tensor field is used and for global evaluation the equivalent stress or main stresses.

The 3D structural model allows repeated changes to the loading assumptions or boundary conditions, even the materials used. However, it is difficult to probe different solutions by changing the geometry. This operation requires the regeneration of the geometry and the resumption of the problem of eliminating gaps and interferences. In addition, a complete analysis would necessitate an operation that can usually be considered a luxury in solving the problem, namely the study of the convergence of the solution.

It is recommended that 3D structural models be used only in the final stage of calculation and simulation when there are no frequent changes in geometry or changes in the geometry of the loading and border conditions areas. Such models are mainly used to remedy the deficiencies noted in the experimental stage or in operation (as in the example presented in the article).

In order to optimize the structural analysis activity, it is recommended to seriously examine the efficiency of the structural analysis on complex models in approaching the problems of a structure from the category of those approached in this paper. There are a large number of simple structures for complex studies on 3D models are justified only at most for demonstration or learning purposes.

It must be taken into account that the CAD-CAE structures used in the structural analysis are through the hypotheses used, different from reality (couplings made through false hypotheses, the boundary conditions hypotheses and even the loading hypotheses). As a result, the validation of the structural models must be done experimentally, an activity in which it will be possible to motivate the differences between the theoretical and the experimental results.

Acknowledgement

This work was supported by a grant of the Romanian Research and Innovation Ministry (RRIM), through Programme 1 – Development of the national research-development system, subprogram 1.2 – Institutional performance – Projects for financing excellence in RDI, contract no. 16PFE, and was done by "NUCLEU" Programme, developed with the support of the RRIM, project PN 19 10 01 02.

References

1. A.W. Hendry, B.P. Sinha, S.R. Davies, *Design of Masonry Structure*, Taylor & Francis, 7 (2004)
2. L.F.A. Bernardo (Editor), *Architectural Structure*, Applied Science, MDPI, (2020)
3. P. Cardei & others, , *Structural analysis and new materials focused on mechanics, mechatronics, maintenance and operation of technical equipment for agriculture and food industry*, Terra Nova Publishing Iasi, (2012)
4. R. A. Tabile, E.P. Godoy, R.R.D. Pereira, G.T. Tangerino, A.J.V. Porto, R.Y. Inamasu, Design and development of the architecture of an agricultural mobile robot, Eng. Agric., 31 (1), 130-142, (2011)
5. J. Selech, D. Ulbrich, W. Kęska, Ż. Staszak, J. Marcinkiewicz, D. Romek, P. Rogoziński, Design of a cultivator mounted on a tractor with a power of up to 20 kW, MATEC Web Conferences **254**, 1-7, (2019)
6. R.Y. Garud, S.C. Tamboli, Structural Analysis of Automotive Chassis, Design Modificationnand Optimization, International Journal of Applied Engineerin Research, 13 (11), 9887-9892, (2018)
7. S.Şt. Biriş, E. Maican, V. Vladuţ, S. Bungescu, *Stress and strain distribution in the frame of agricultural cultivators using the finite element method*, Proceedings of The 44 International Symposium on Agricultural Engineering "Actual Tasks on Agricultural Engineering" Opatija, 111-117, (2016)
8. N. Makange, R. Parmar, V. K. Tiwari, Stress Analysis on Tyne of Cultivator Using Finite Element Method, Trends in Biosciences **8**(15), 3919-3923, (2019)
9. V. Vladut, E. Maican, L. Apostol, N. Ungureanu, I. Dumitru, R. Oprescu, Verification of Stress by FEM Analysis Mechanical Testing of Agricultural Mobile Aggregates Coupling Devices, INMATEH Agricultural Engineering, **54**(1),39-46, (2018)
10. S. Şt. Biriş, Ungureanu N., Cujbescu D., Modelling of soil compaction under heavy-duty tractors, INMATEH Agricultural Engineering, **57**(1), 261-270, (2019)
11. G.A. Constantin, Gh. Voicu, B. Olac, F. Ilie, G. Paraschiv, V. Ştefan, G. Musuroi, Structural analysis with finite elements of a subsoiler working part, INMATEH Agricultural Engineering, **61**(2), 61-26, (2020)
12. G. Ipate, G. A. Constantin, S. Şt. Biriş, Gh. Voicu, C. O. Rusănescu, V. Ştefan, Numerical 3D analysis a mini wind turbine with horizontal axis, for implementation in agricultural farms, INMATEH Agricultural Engineering, **61**(3), 325-333, (2020)
13. V. Niţescu, P. Cardei, V. Muraru, I. Cojocaru, R. Sfîru, Modalităţi de dimensionare a cilindrului hidraulic pentru mecanismul de reversare al unui plug reversibil, folosind cinematica şi dinamica procesului de reversare, INMATEH Agricultural Engineering, **16**(1), 163-168, (2006)
14. V. Muraru, N. Constantin, P. Cardei, R. Sfîru, Structural Analysis of MATINA soil working machine, INMATEH Agricultural Engineering, **31**(2), 17-26, (2010)
15. M. Boryga, P. Kolodziej, Stress analysis of the sugar beet lifter with the finite element method (FEM), Agricultural Engineering, **23**(3), 29-39,(2019)
16. Dassault Systemes, SolidWorks Help, https://help.solidworks.com/, (2021)
17. I. I. Ştefănescu, C. Spanu, *Organe de maşini*, vol. **I**, Editura Europlus, Galaţi, (2009)
18. P. Cardei, R. Sfîru, S. Muraru, V. Muraru, *The convergence of structural models used in the geotechnical software*, Proceedings of 17th International Multidisciplinary Scientific GeoConference SGEM, (2017)

19. S.Şt. Biriş, V. Vlăduţ, N. Faur, A. Cernescu, O. Kabaş, M. Matache, I. Voicea, S. Bungescu, C. Popescu, *FEM analysis / testing resistance of a tractor seat*, Proceedings of the 43 International Symposium On Agricultural Engineering "Actual Tasks on Agricultural Engineering", Opatija, 189-200, (2015)
20. S.Şt. Biriş, N. Ungureanu, N.E. Gheorghiţă, E. Maican, V. Vlăduţ, *Theoretical research on enhancing the performance of cultivators working bodies,* Proceedings of the 45 International Symposium On Agricultural Engineering "Actual Tasks on Agricultural Engineering", Opatija, 121-130, (2017)
21. D. Cujbescu, S.Şt. Biriş, Gh. Voicu, M. Matache, G. Paraschiv, V.Vlăduţ, M. Bularda, N. Ungureanu, *Determination of sowing precision in simulated laboratory conditions*, INMATEH – Agricultural Engineering, **61**(2), 209-216, (2020)
22. D. Cujbescu, N. Ungureanu, V. Vlăduţ, C. Persu, R. Oprescu, E. Gheorghiţă, *Field testing of compaction characteristics for farm tractor universal 445*, INMATEH – Agricultural Engineering, **59**(3), 245-252, (2019)
23. D.I. Vlăduţ, S. Biriş, V. Vlăduţ, D. Cujbescu, N. Ungureanu, I. Găgeanu, *Verification of stress by FEM analysis / mechanical testing of agricultural mobile aggregates couplig device*, INMATEH – Agricultural Engineering, **54**(1), 37-462, (2018)
24. D.I. Vlăduţ, V. Vlăduţ, I. Grigore, S. Biriş, *Experimental research on qualitative indices of work for equipment for seedbed preparation in conservative system*, 16th International Scientific Conference "ENGINEERING FOR RURAL DEVELOPMENT", 1174-1179, Jelgava, Latvia, (2017)
25. V. Vlăduţ, V. Gângu, I. Pirnă, S. Băjenaru, S. Biriş, S. Bungescu, *Complex tests of the resistance structures in simulated and accelerated regime on hydropulse installation*, Proceedings of the 35 International Symposium On Agricultural Engineering "Actual Tasks on Agricultural Engineering", 393-404, Opatija – Croaţia, (2007)
26. V. Vlăduţ, Matache M., Faur N., Biriş S., Paraschiv G., Bungescu S., *Traction bar testing of a 161 kW tractor in simulated and accelerated regime and its analysis by finite elements method for determining its rupture resistance*, Proceedings of the 37 International Symposium On Agricultural Engineering "Actual Tasks on Agricultural Engineering", 77-84, Opatija – Croaţia, (2009)
27. V. Vlăduţ, S. Biriş, S. Bungescu, N. Faur, A. Cernescu, M. Matache, O. Kabaş, G. Paraschiv, At. Atanasov, Gh. Ivan, *The verification of stress by MEF analysis/ mechanical testing of a traction bar*, Proceedings of the 43 International Symposium On Agricultural Engineering "Actual Tasks on Agricultural Engineering", 141-152, Opatija – Croaţia, (2015)
28. V. Vlăduţ, E. Marin, I. Grigore, S.Şt. Biriş, N. Ungureanu, G. Gheorghe, M. Matache, C. Persu, I. Voicea, D. Cujbescu, S. Bungescu, I. Kiss, C. Vlad, *Field and laboratory wear testing of integral seedbed implement's chisel tines point*, Proceedings of the 45 International Symposium On Agricultural Engineering "Actual Tasks on Agricultural Engineering", 131-140, Opatija – Croaţia, (2017)
29. L. Vlăduţoiu, P. Cârdei, V. Vlăduţ, L. Fechete - *Modern trends in designing and selecting the machine / equipment for soil deep tillage*, 16th International Scientific Conference "ENGINEERING FOR RURAL DEVELOPMENT", 1415-1420, Jelgava, Latvia, (2017).

Storage of grains and technical plants through active ventilation for the purpose of maintaining the quality of stored products

Anişoara Păun[1,*], *Gheorghe* Stroescu[1], *Alexandru* Zaica[1], *Samir* Yasbeck Khozamy [2], *Ana* Zaica[1] , *Oana* Cristea[1], *Vasilica* Ştefan[1], and *Carmen* Bălţatu[1],

[1]National Institute of Research - Development for Machines and Installations designed to Agriculture and Food Industry - INMA, Bucharest 013813, Romania
[2] Schneider Electric / Romania

Abstract. The problem of grain storage and storage can be approached from two perspectives: storage and storage of cereals and technical plants within individual agricultural holdings; storage and storage of cereals and technical plants within agricultural associations and companies. At the level of Romania, the storage and storage of cereals at small agricultural producers are deficient in terms of ensuring optimal storage conditions according to standards. These variants include: preservation at low temperatures; preservation by active aeration; self-preservation; chemical preservation; conservation with the help of radiation. In the paper we will approach the conservation by active aeration because it folds better for small agricultural producers. Proper aeration is one of the most important processes in a grain storage system and is essential for maintaining the quality of stored products. Aeration means the movement of a relatively small volume of air through the grain mass in order to control the temperature of the grain and to reduce the risk of product degradation. The two main objectives of aeration are to maintain a uniform temperature in the grain mass and at the same time the temperature must be as low as possible from a practical point of view. The paper will present the results of experiments for storing bulk grain in a metal cell with central aeration system, which is based on a patent. This type of storage system can be used successfully in small agricultural producers.

1 Introduction

Prior to processing, agricultural products must be stored in optimal storage conditions to ensure maximum preservation of long-term nutritional qualities and to eliminate depreciation of biological properties. Cereals are the most important staple food for humans. Of the amount of grain harvested worldwide, more than 20% is lost annually, mostly due to insect activity and the development of fungi and molds, due to inadequate storage. The technical storage conditions are closely correlated with the technologies for

* Corresponding author: ani_paun@yahoo.com

storage and conservation of bulk cereal seeds, with the crop species, with their subsequent destination and with the geographical area in which they are located (meteorological and vegetation conditions specific to the products reaching humidity and the optimum storage temperature) [1,2]. Keeping the seeds after harvest even just a few days with a high moisture content (over 18%) reduces the germination below the standardized limit (90%), thus eliminating the possibility of using these seeds in the process of establishing new crops. The seeds must be cleaned before being stored [3,4]. For seeds intended for storage, the conditions for preserving the nutritional and organoleptic qualities (taste, smell, aroma, etc.) must be ensured. The presence of weed seeds can lead to degradation of the main crop seeds during storage, as they have a higher humidity and create foci of heat, causing mold and spoilage, they are also favorable environments for the development of microorganisms and pests [5]. Cracks, undeveloped or damaged seeds are less resistant to storage, being easily attacked by molds and pests, etc. All these considerations impose the necessity and the obligation of the pre-cleaning and cleaning of the seeds in order to ensure optimal conditions for storage, storage, marketing, consumption, sowing material, etc. Regarding the storage and conservation systems of agricultural products in the form of seeds, the most rational and economical system for Romania is that of dry storage (for a long time). According to this system, cereal and legume seeds are harvested when they have reached biological maturity and have a humidity below 14% in the case of straw cereals and seed corn, less than 11% in soybeans and less than 8% in sunflower In recent years, there have been more efficient ways of preserving cereals with greater or lesser applicability in agricultural practice. These variants include: preservation at low temperatures; preservation by active aeration; self-preservation; chemical preservation; radiation preservation. Aeration preservation is a method that can be applied both to large agricultural farmers but also to small agricultural producers, both for cereal seeds and technical plants and for feed (hay) [6-10]. Proper aeration is one of the most important processes in a grain storage system and is essential for maintaining the quality of stored products. A properly used aeration system helps control insect infestations and moisture migration, reducing grain damage and saving money. Normally, cereals stored for more than a few weeks should be aerated. Aeration means the movement of a relatively small volume of air through the mass of grain in order to control the temperature of the grain and reduce the risk of product degradation.

In commercial practice, the method of transferring cereals (moving them from one silo cell to another) for temperature control and eliminating hot spots has largely been eliminated. Within INMA, a series of technologies, technical equipment for cleaning and storage have been developed and tested [11-13]. The paper presents some of the results obtained and the conclusions regarding the optimal aeration system based on a patent [14].

2 Materials and methods

The humidity in the seed, the high air temperature, the thickness too large of the stored seed layer, can determine a fast and intense evolution of the infections, leading to the decrease of germination and to the depreciation of the biological and agronomic value of the seeds. Damage that occurs to seeds stored for a long time is reflected by [5]:

- loss of germination capacity;
- increasing the number of abnormal germs;
- increasing sensitivity to unfavorable environmental conditions;
- decrease in metabolic activity.

Cereals should be aerated as soon as possible after being stored. This is particularly important if there are large variations in temperature or humidity or if the whole cereal mass is initially hot. In autumn, the temperature of the grain must be reduced to approach the average outside temperature. INMA developed and tested storage equipment for cereal seeds and industrial plants, figure 1. Installation for storage and preservation of cereals is a complex, modern plant for straw cereals (wheat, barley, rye), corn, legumes and oilseeds, species with a large share in all agricultural production on small and medium farms, which ensures the cleaning of cereals and their aeration during storage.

Fig. 1. Installation for storage and preservation of cereals ISC10

The ISC 10 storage and preservation installation has the following components: 1-screw conveyor; 2-group cleaning; 3- storage cell; 4- mobile pneumatic transport installation; 5- mobile aeration installation; 6-power and control electrical installation. The constructive solution for the storage cell of figure 1, was the basis for the realization of the patent no. 126266 / 30.05.2016, figure 2,

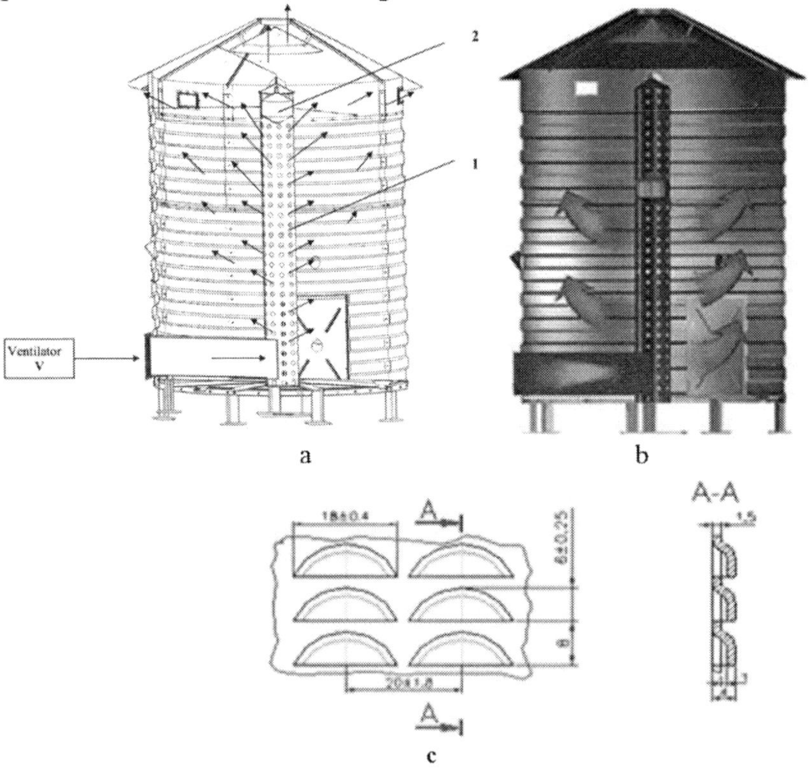

Fig. 2. Storage cell according to the patent [14]

a: 1- central aeration column; 2- adjusting device; b-storage cell 3D; c-storage cell wall detail

The storage cell is a metal construction, cylindrical in shape, removable, with corrugated galvanized sheet walls that alternate with special walls, also made of galvanized sheet, which allow natural and artificial aeration. At the top inside the storage cell is mounted a cone that achieves a uniform distribution of the product in the cell, around the central aeration column. According to the patent, figure 2, the central aeration column is a metallic, cylindrical construction, made of perforated sheet. Aeration is done both naturally, through the storage cell walls (cell wall detail) and controlled, through the central aeration column, made of perforated sheet. The air flow required for this process is ensured by the mobile aeration installation connected through the special piping, at the base of the central aeration column of the storage cell. Given the need for the correct development of the process and its special importance, the air flow required for aeration was calculated. Insufficient amount of blown air causes water vapor to condense in the upper layers of the seed mass. Proper aeration occurs when the air leaving the seed mass has a relative humidity below 80%. The cooling and drying process during active aeration proceeds as follows: during the insufflation of cold and dry air, it first meets the lower layer, where it forms a transitional front, in which there is a decrease in seed temperature and humidity, while in the upper area the temperature and humidity of the seeds increase slightly. As the operation continues, this front or transition zone moves in the direction of the blown air stream until it reaches the surface of the bulk. The area where the changes in temperature and humidity between the blown air and the seeds take place, limited by both the properties of the air current and those of the grain mass, ensures a relatively large exchange area, depending on the species.

For a more efficient distribution of the blown air in the areas that require it, the aeration column is provided inside, with an adjusting device that can move vertically by closing / releasing various areas of the side surface of the column. The purpose of aeration consists in: controlling the temperature of cereals; low temperature storage; moisture migration, figure 3; elimination of foreign odors; modification or equalization of humidity in the product mass; fumigation; keeping cereals with higher humidity.

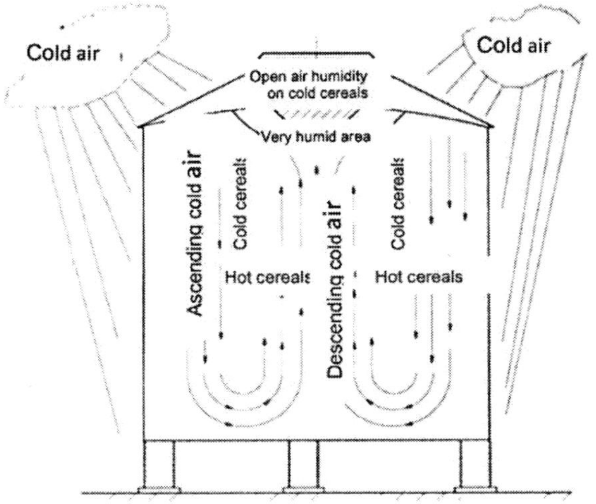

Fig. 3. Moisture migration in autumn and winter

Temperature deviations between seed mass and ambient outside air cause moisture migration problems. Figure 3 illustrates a typical migration phenomenon in cold weather (autumn and winter), caused by lower ambient temperatures. Because the dry product mass has a thermal insulating effect, the product on the sides of the cell changes its temperature faster than the product mass located in the center of the storage cell.

Temperature differences in the mass of the product cause air currents from the hot mass to the cold mass of the product. The direction of these currents depends on how the product cools if the temperature outside the cell drops or heats up if the temperature outside rises during spring and summer. In autumn and winter, the product mass, figure 3, next to the cell walls and under the roof is colder, the product mass in the middle of the cell having a higher temperature. Moisture-laden air rises, passes through the colder product layer where it leaves moisture. The airflow then moves down the cell wall to replace the volume of warm air in the center of the cell, thus closing the convection cycle.

The balance of moisture content for a given species depends on the relative humidity of the air and the storage temperature. Relative humidity is given by the relation:

$$\varphi = \frac{g_v}{g_s} x100 \ \% \tag{1}$$

where:

g_v - weight of water vapor in a m^3 of humid air, kg / m^3;

g_s - the maximum weight of vapors at the same volume, kg/m^3.

Amount of air to be blown (Q) is expressed in m^3 air / hour / ton of seed (sometimes in m^3 air / hour / m^3 of seed) and is calculated using the relation:

$$Q=\frac{D}{G} \tag{2}$$

in which:

D- air flow supplied by the fan, m^3 / h;

G- weight of the batch subjected to aeration, t.

The number of air exchanges must also be taken into account when calculating the amount of air required for active aeration in the storage cells. The required number of hourly air changes in the intergranular space varies depending on the seed moisture and is calculated using the relation:

$$N=\frac{D}{P} \tag{3}$$

in which:

D- the air flow supplied by the fan, m^3/h;

P- the porosity of the seed mass.

For the seeds of the main species the porosity of the mass is the following: wheat 35-45%, barley 45-55%, oats 50-79%, corn 30-50%, sunflower 60-80%, linseed oil 30-50%.

3 Results and discussions

The experiments were performed with the installation shown in figure 1. After the pre-cleaning process, experiments were performed following the equilibrium humidity of the cereal seeds, at different temperatures. The experiments were performed on seed lots of: wheat, corn, barley and rice, (tables 1, 2 and 3). For the correct evaluation of the results were determined in the laboratory the characteristics of the seeds subjected to the pre-cleaning operation. The results of the experiments are presented in the graphs in figures 4, 5 and 6.

Table 1 Operating indices - the equilibrium humidity of cereal seeds at different temperatures

Seed product / temperature, °C	Air Relative Humidity φ, (%)								
	10	20	30	40	50	60	70	80	90
Wheat/20⁰C	0	3.9	7.2	10.7	11.8	13.1	14.8	16 .5	20.1
Barley/20⁰C	0	8.3	9.5	10.9	12.0	13.4	15.2	17.5	20.9
Barley/25⁰C	0	6.4	8.8	10.7	11.8	13.3	15.9	19.1	24.9
Corn/20⁰C	0	8.2	9.4	10.7	11.9	13.2	14.9	16.9	19.2

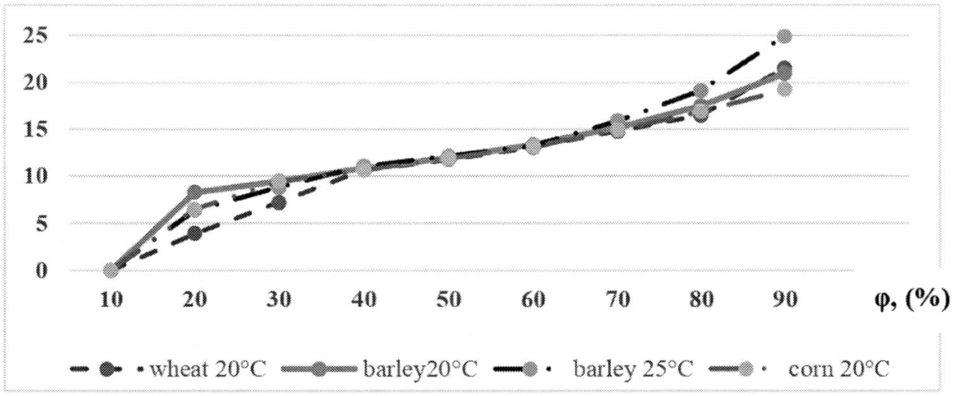

Fig.4 Balance curves for different grains

Analyzing figure 4 we find the balance of moisture content for a particular species depends on the relative humidity of the air and the storage temperature.

Table 2 Operating indices when drying by active aeration of cereals.
The difference between the temperature of the seed mass and the air outside the cell. (0C)

Grain moisture, %	Air Relative Humidity φ, (%)								
	60	65	70	75	80	85	90	95	100
17	-	-	-	0	0.4	1	2	2.9	3.4
16	-	-	-	0	1	1.9	2.7	3.6	4.3
15	-	-	0	1	1.9	2.9	3.9	4.9	5.5
14	-	0	1.3	2.2	3.1	4.1	5	6	7

Mobile aeration installation position 5, within the ISC10 grain storage and preservation installation, figure 1 can ensure through the provided fan an air flow of 4500 m^3/h.

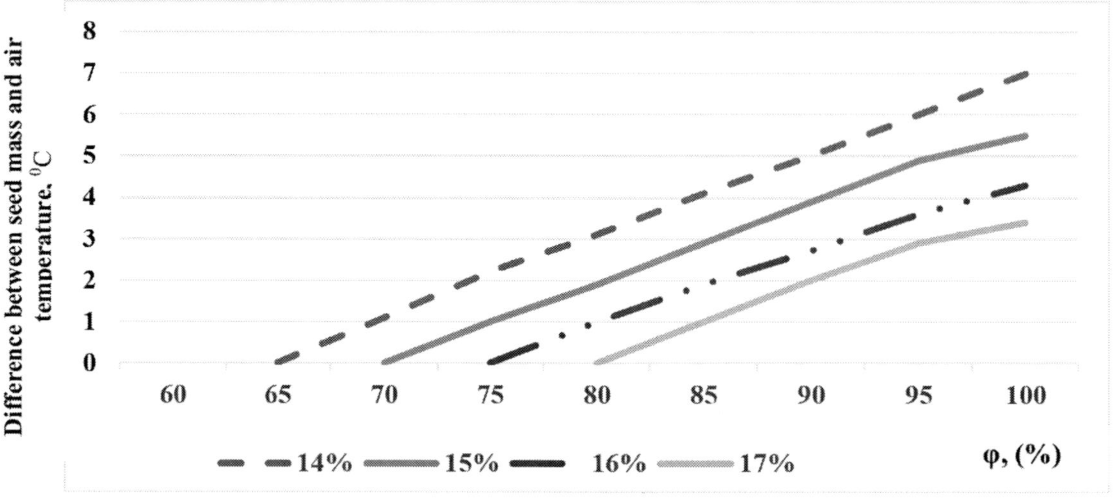

Fig. 5 Drying cereals by active aeration

Analyzing the graph in figure 5, it is found that if there is no temperature difference between the seed mass and the air, the drying can take place only at a relative humidity φ, of the air below 65%. If a relative humidity φ of the air is 95% the temperature difference must be 60 ^{0}C, to bring the seeds to a humidity of 14% or 2.9 0 ^{0}C to be brought to a humidity of 17%.

Table 3 Operating indices when cooling cereals by active aeration

ΔT ^{0}C	Required air flow, m^3/h/t			
	50	100	150	200
1	0.03	0.04	0.06	0.09
2	0.06	0.07	0.11	0.15
3	0.08	0.1	0.15	0.23
4	0.09	0.15	0.23	0.29
5	0.12	0.2	0.3	0.35
6	0.13	0.25	0.35	0.45
7	0.15	0.29	0.42	0.52
8	0.18	0.33	0.51	0.58
9	0.18	0.37	0.6	0.72
10	0.2	0.4	0.68	0.8

Note: It was noted with **ΔT** temperature reduction, in an aeration hour ^{0}C.

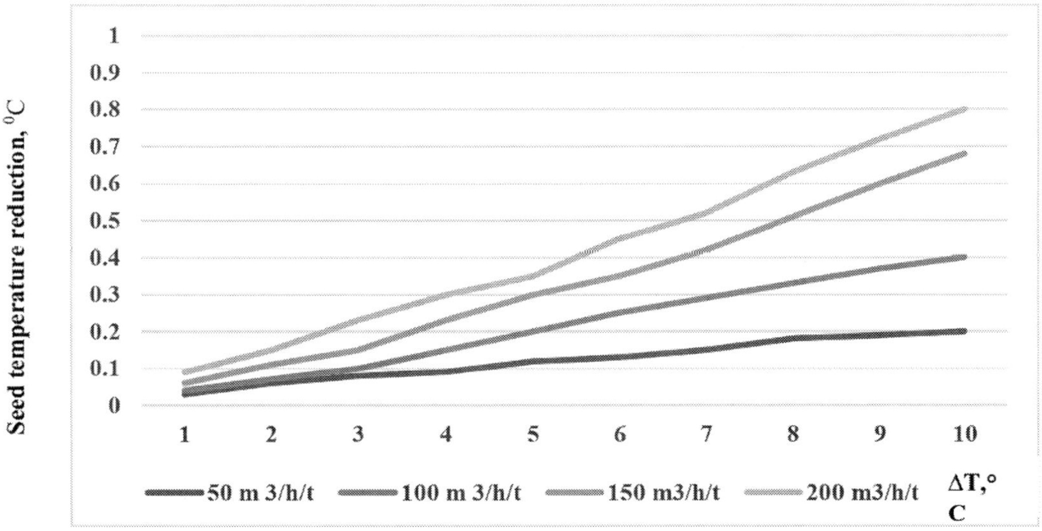

Fig. 6 Cooling of cereals by active aeration

Aeration should take place as soon as the average outside air temperature is at least 5 ° C colder than the hottest temperature in the silo. Cooling takes place in stages until whole grains reach 0 to -5°C. In the spring, the heating of the cereals should be done if it is planned to be kept for an extended period of time (after June) or if the temperature of the cereal mass is below 0 ° C. If necessary, the fans are activated as soon as the average outside temperature reaches 5-7°C above grain temperature.

4 Conclusions

- For short-term storage, cereal seeds must be pre-cleaned (when many foreign bodies have resulted from the threshing operation), not infested and have a maximum moisture content of 16% in the case of straw cereals or maize, 13% in soybeans and a maximum of 11 % for sunflower;
- For long-term storage, the seeds must be cleaned and have the qualitative parameters of the finished product standard, in correlation with the stored species and its destination (for sowing, consumption or industrialization);
- Cereals with more than 17% humidity can be stored, until drying, for short periods of time, in spaces provided with active aeration installations;
- It should be noted that the humidity increases with decreasing temperature and decreases with increasing temperature;
- Cereals, legumes and oilseeds with humidity above the limits provided in the delivery standards, will be dried in a shorter period of time the higher the humidity;
- The storage life of stored cereal seeds depends on a series of technical, economic and strategic factors, all of which are correlated with atmospheric conditions, the quality of thermal insulation, etc.

This paper was financed with:
- the support of National Agency for Scientific Research and Innovation, Programme FARMING TOMORROW - ENSURING ENVIRONMENT SUSTAINABILITY BY INCREASING COMPETITIVENESS OF IMPROVEMENT FOR ECOLOGICAL AGRICULTURE (UEFISCDI PN-III-P1-1.2-PCCDI-2017- 0850, Contract: 14PCCDI/06.03.2018), project component 4 "Making efficient the technology of producing ecological seeds for legume species by achieving a module of seed conditioning".

References

1. N. Bucurescu, D.Roman, Seed and its preparation for sowing, Ceres Publishing House, Bucharest (1992)
2. A. Paun, P.Gageanu, I. Chih Li-Hua, Al. Zaica, A.Canpeanu , Symposium - INMATEH Agricultural Engineering, 46-50, Vol. 29, No.3 (2009)
3. Gh. Stroescu, A. Paun, I.Voicea, C.Persu, A. Matache, G. Bunduchi, , Proceedings of 19th International Scientific Conference Engineering for Rural Development Jelgava, 1277 – 1283, DOI:10.22616/ERDev.2020.19.TF315 (2020)
4. Gh. Stroescu, A. Paun, C. Bracacescu, M. Olan, Proceedings of 18th International Scientific Conference Engineering for Rural Development Jelgava, 34-38 (2019)
5. D.D.Plăcintă Micromycetes that are transmitted through the seeds of straw cereals prevention and control, Editura Universităţii din Suceava, (2007)
6. A. Zaica, A. Nedelcu , R. Ciuperca , L. Popa, Al. Zaica, C. Sorica, ISB, 723-730, (2016)
7. A. Zaica, A. Nedelcu , R. Ciuperca , L. Popa, Al. Zaica, C. Sorica, 45th International Symposium "Actual Tasks on Agricultural Engineering", Croatia, 546-556 (2017).
8. A. Zaica, A. Nedelcu , Al. Zaica, R. Ciuperca , L. Popa, C. Mircea, TE-RE-RD, 495-500, (2017).
9. A. Zaica, A. Nedelcu , R. Ciuperca , Al. Zaica, 17th International Scientific Conference "Engineering for Rural Development", Jelgava, LATVIA, 115-123 (2018).
   10.A. Zaica, A. Nedelcu, R. Ciuperca, Al. Zaica, A.-L.Visan, TE-RE-RD, 2457-3302, (2019).
10. A. Zaica, R. Ciupercă , Al. Zaica, Diana Popa, 19th International Scientific Conference "ENGINEERING FOR RURAL DEVELOPMENT", Jelgava, LATVIA, 1641-1647,(2020)
11. Ciobanu V.G., A.L. Visan, Paun A, Nedelcu A. JoKULL Journal, Vol.64, No.9, 91-100 (2015).
12. A.Paun, Ghe. Stroescu, Al.Zaica ,G. Bunduchi, A. Zaica, C-tin. Bogdanof, D-tru. Milea, S.Yasbeck Khozamy, O.D.Cristea, L.C.Cristea, ISB-INMA TEH' ,814-821, (2020)
13. Experimentation report, ISC10 grain storage and preservation plant (2014)
14. A. Paun, I. Pirna, I. Cojocaru, I. Ganea-Christu, Patent no. 126266/30.05.2016

Experimental research of equipment for burying in the ground the drip irrigation lines

Dragoş Manea[1,*], *Eugen* Marin[1], *Gabriel* Gheorghe[1], *Cătălin* Persu[1], and *Roxana* Mitroi[2]

[1]National Institute of Research - Development for Machines and Installations designed to Agriculture and Food Industry - INMA, Bucharest 013813, Romania
[2]University Politehnica of Bucharest, Faculty of Biotechnical Systems Engineering, 313 Splaiul Independenţei, 060042, sector 6, Bucharest, Romania

Abstract. The equipment for burying in the ground the drip irrigation lines was designed and built at INMA Bucharest, within a complex research project carried out in partnership with other institutes, universities and research stations in Romania. The equipment is intended for drip irrigation technology for field crops. Subsurface crop irrigation technology is a variant of the classic drip irrigation technology, in which the drip lines are buried below the soil surface, providing water directly to the root zone of the plants. The depth of burial and the distance between the drip lines depends on the type of soil and the structure of the roots of the crop. The activities prior to the experimental research consisted in determining the characteristics of the test field (soil compactness, soil moisture, geographical coordinates of location). Experimental research of the equipment aimed at determining the energy indices (working speed, traction force, traction power, working capacity and fuel consumption).

1 Introduction

Subsurface crop irrigation technology is a variant of classical drip irrigation technology, in which drip lines are buried below the soil surface, providing water and / or fertilizer directly to the plant root zone. Compared to surface drip irrigation, subsurface irrigation technology has the following advantages: it eliminates surface evaporation; reduces water stress of the crop; prevents weed germination and reduces the use of herbicides; reduces the need for labour; reduces maintenance costs and ensures the safe and efficient delivery of fertilizers.

The equipment for burying pipes for subsurface drip irrigation is composed of a towed frame on which are mounted drums with tubes in the form of a coil which run through a section for burying in the soil at different depths.

There are relatively little research in the literature on equipments for burying drip irrigation lines in the ground. It is known from the prior art document [1], where the gutter opening organs are simultaneously assisted by other organs, which, as the irrigation tube is unrolled, cover the ditch with the soil resulting from the excavation.

Internationally, various equipments are being built for the placement of irrigation hoses in the ground, manufactured by prestigious companies [2, 3, 4]. Compared, these equipments differ from each other by the width and depth of work, the number and construction of the working organs, the power of the tractor with which they are intended to work in the aggregate, etc. Instead, the technology of subsurface drip irrigation has been studied over time by several researchers [5 - 8].

* Corresponding author: manea_dragos_05@yahoo.com

Subsurface drip irrigation gives higher crop yield compared to surface drip irrigation [9], it can reduce the environmental risks and decrease nitrate leaching rates [10], but it has the disadvantage that it is very dificult to observe when a transmitter gets clogged [11].

Within a complex project developed in partnership with other research institutes, public universities and research resorts in Romania, INMA Bucharest designed and built equipment for burying drip irrigation hoses in the ground (fig. 1), intended for agricultural farms who want to implement the technology of subsurface irrigation of hoeing plants or vegetable crops, in the open field. This equipment works in aggregate with wheeled agricultural tractors, provided with suspension mechanisms in three points category III according to ISO 730: 2012. The equipment is of the type carried on the hydraulic lift of the tractor, it has two working elements of vertical blade type and chisel knife, the distance between the organs being adjustable in the interval 70-140 cm and the working depth being between 15 and 45 cm.

Fig. 1. The research object: equipment for burying drip irrigation lines in the ground

This paper presents the results of experimental research on the equipment for burying drip irrigation lines in the ground, research that aimed to determine energy indices (working speed, traction force, traction power, working capacity, fuel consumption), in the conditions of an arid area in Romania and a sandy soil.

2 Material and method

The experiments of the technical equipment for burying drip irrigation lines in the ground were carried out within the Testing Department of INMA Bucharest and in the experimental plot of SCDCPN Dăbuleni, during October - November 2020. The experimental research was carried out in several stages.

In the first stage, in the most critical points on the equipment were glued with special adhesives tensometric marks KYOWA model KFSG-6-120-C1-11N15C2, 120 Ω, temperature coefficient 11 ppm/°C, which measures the components force parallel to the direction of movement of the equipment and perpendicular to this direction (Fig. 2).

Fig. 2. Tensometric marks glued to critical points on the equipment frame

The next step was to assemble the equipment on the experimental laboratory stand and calibrate the tensometric marks (Fig. 3). The equipment used consisted of: Hydropuls installation (hydraulic actuator; load cell 10 - 100 kN, Type FC-10, series 8608, with reading on the deck SERIES 366.90 / 87/13/1987, measurement uncertainty 0.89%), laptop with specialized software package; QuantumX 1615 data acquisition system, 2 modules with 32 measurement channels each module.

Fig. 3. Experimental stand for calibration of tensometric marks in the laboratory: *1 - fixing elements on the stand of the equipment; 2 - drawbar; 3 - tensometric marks glued in the critical points; 4 - data acquisition system; 5 - laptop; 6 - hydraulic actuator; 7 - load cell*

The third step was to determine the characteristics of the test plot (Fig. 4). The soil of the experimental polygon of SCDCPN Dabuleni is a sandy soil, poorly developed (psamosol), with a coarse sandy texture (74-84% coarse sand), contains little clay (0.7-3%) and dust (2-4%).

location of the experimental polygon (geographical coordinates:N 43°48'30.24", E 24°0.3'5.76") | *measuring soil compactness* | *soil moisture measurement*

Fig. 4. Aspects during the determination of the characteristics of the test plot

The FIELDSCOUT SC 900 digital cone penetrometer was used to measure the compactness of the soil. The measurements were performed at five points located on the diagonal of the experimental polygon. The average value of the resistance forces to the penetration of the cone in the soil layers in the depth range 0÷40 cm is 4694 kPa, which means that the soil falls in the middle class of compaction. Soil moisture was measured with the Delta - T Devices portable humid - meter, model HH2 with Theta Probe ML2x probe, at the same points where the compactness was measured and on the same depth range. The average value of soil moisture at a depth of about 20 cm was 6.5%, and at a depth of about 40 cm it was 16.8%.

Subsequently, the traction force was measured in the experimental plot from SCDCPN Dăbuleni (Fig. 5). The traction force was determined by the tensometric method, which consists in the direct measurement of the forces that appear in the structure of the equipment frame. The traction force of the equipment is the algebraic sum of the components of the force parallel to the direction of travel. The tractor used was Lamborghini with a power of 150 HP. The measurements were performed at two working depths, 20 cm and 40 cm, for three working speeds (2.46 km / h; 3.30 km / h and 4.50 km / h).

traction force measurement

measuring the amount of fuel consumed during a test

Fig. 5. Aspects during the determination of the energy indices of the technical equipment for burying drip irrigation lines in the ground , in the experimental plot from SCDCPN Dăbuleni

The working speed was determined by timing the time required to cover the length of the test plot and reporting the space to time, using for measurement roulette, stopwatch and milestones. The slippage of the tractor was determined by recording the number of rotations of the drive wheels when idling and when driving, by traversing the length of the polygon. Hourly fuel consumption was determined by measuring with a graduated cylinder the amount of fuel consumed over the length of the test plot and reporting the amount of fuel consumed in time.

For the calculation of the values of the energy indices, the formulas presented in table 1 were used.

Table 1. Calculation formulas for energy indices.

Index name, symbol	**Calculation formulas**	**Significance of variables**
Working speed, v_l	$v_l = \frac{3.6 \cdot s}{t}$, km/h	s - linear space travelled, in m; t - space travel time, in s.
Skating of tractor wheels, δ	$\delta = \frac{n_s - n_g}{n_s} 100, \%$	n_s - the number of revolutions of the drive wheel under load; n_g - the number of idle wheel rotations.
Traction power, P_{tr}	$P_{tr} = \frac{F_{tr} \cdot v_l}{270}$, CP	F_{tr} - traction force measured by the tensometric method, in daN; v_l - working speed, in km/h.
Hourly work capacity, W_{ef}	$\mathbf{W_{ef} = 0.1 \cdot B_l \cdot v_l}$, ha/h	B_l - working width of the technical equipment, in m; v_l – working speed, in km/h.
Hourly fuel consumption, q	$q = \frac{3.6 \cdot V}{t}$, l/h	V - the volume of fuel measured with the graduated cylinder, in cm^3.
Fuel consumption per hectare, Q	$Q = \frac{q}{W_{ef}}$, l/ha	q - hourly fuel consumption; W_{ef}- hourly work capacity.

The precise location of the subsurface irrigation lines was made with the help of the Matrix 430 agricultural GPS made by the TeeJet company, located in the tractor cab. The major advantage of using GPS technology is that the distance between the subsurface drip lines, at two successive passes, is kept constant.

3 Results

The results obtained for the traction force, for the working depths 20 cm and 40 cm and the travel speeds 2.46 km/h, 3.30 km/h and 4.50 km/h, following the processing of the experimental data are presented graphically in Figure 6.

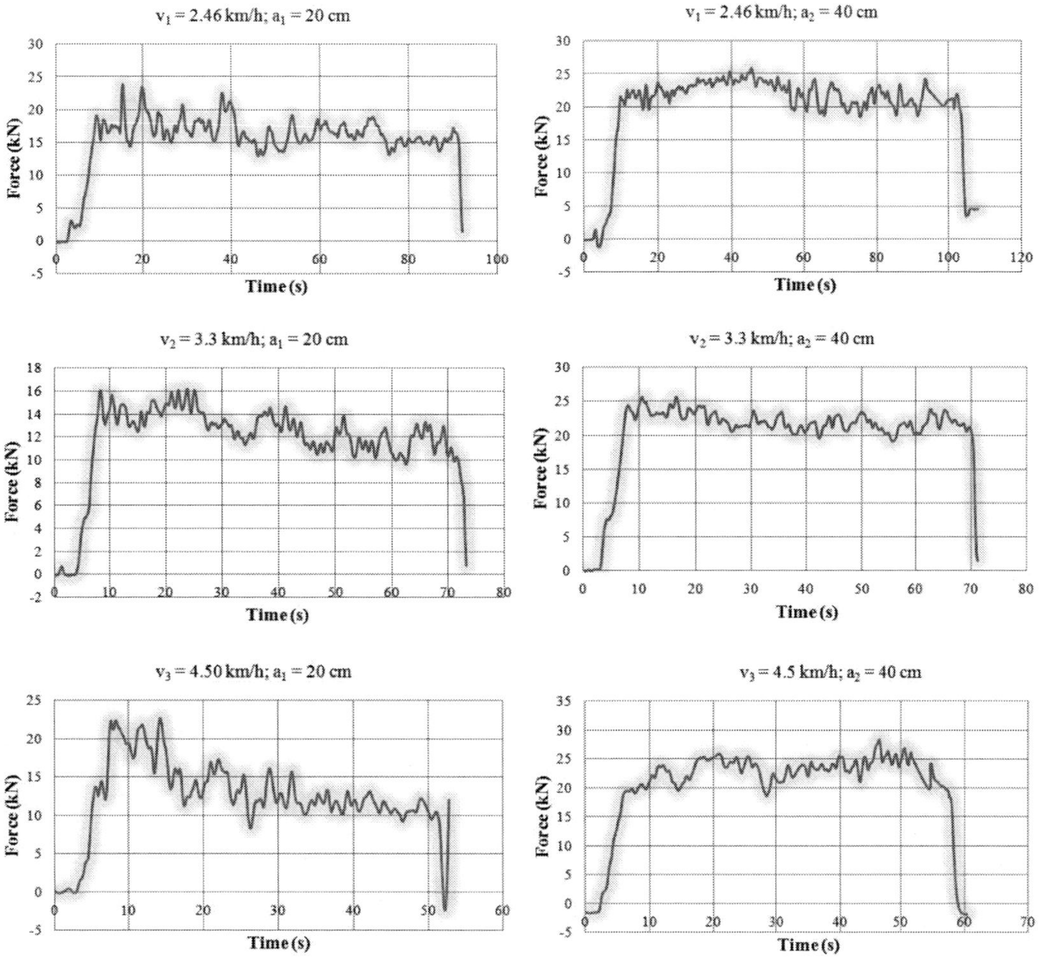

Fig. 6. Graphs of traction force variation

Analyzing the graphs in Figure 6, the average values of the traction forces were determined, which are presented centrally in table 2.

Table 2. Average values of traction force.

Working depth (cm)	Working speed (km/h)		
	2.46	**3.30**	**4.50**
	Traction force (kN)		
20	16.66	12.92	13.53
40	22.15	22.21	23.29

The results obtained for energy indices, by applying the calculation formulas from chapter 2, are presented in tables 3 and 4.

Table 3. The values of the energy indices, for the working depth 20 cm.

Working speed (km/h)	Skating of tractor wheels (%)	Traction force (daN)	Traction power (CP)	Hourly work capacity (ha/h)	Hourly fuel consumption (l/h)	Fuel consumption per hectare (l/ha)
2.46	5.0	1666	15.17	0.24	9.59	39.95
3.30	6.4	1292	15.79	0.33	7.90	23.93
4.50	7.1	1353	22.55	0.45	6,00	13.33

Table 4. The values of the energy indices, for the working depth 40 cm.

Working speed (km/h)	Skating of tractor wheels (%)	Traction force (daN)	Traction power (CP)	Hourly work capacity (ha/h)	Hourly fuel consumption (l/h)	Fuel consumption per hectare (l/ha)
2.46	9,5	2215	20.18	0.24	9.73	40.54
3.30	9,7	2221	27.14	0.33	9.00	27.20
4.50	10,2	2329	38.81	0.45	8.36	18.57

Analyzing the values in tables 3 and 4 for traction power, it is found that the required power is clearly lower than the power that the used tractor can develop (150 HP). For example, for a working depth of 40 cm and a working speed of 4.5 km / h, a traction power of 38.81 hp was obtained. The evolution of the hourly fuel consumption is inscribed on the tractor consumption curve, but without presenting significant differences depending on the working speed.

4 Conclusions

The experiments with the technical equipment for burying drip irrigation lines in the ground have led to the conclusion that it can be used successfully in the technology of subsurface crop irrigation.

Considering the necessary power determined during the experiments, we can strongly state that the equipment can also be used in aggregate with a tractor with a power of less than 150 HP, 100 HP being recommended.

Future research directions will focus on determining the degree of wear of working organs in the conditions of sandy soils in the Dabuleni area and on determining the coefficients of use of working time. Also, higher working speeds must be taken into account, given the very good behaviour of the equipment during the tests performed, in order to increase productivity.

This work was supported by a grant of the Romanian Ministry of Research and Innovation CCDI - UEFISCDI, Project INNOVATIVE TECHNOLOGIES FOR IRRIGATION OF AGRICULTURAL CROPS IN ARID, SEMIARID AND SUBHUMID-DRY CLIMATE, project number PN-III-P1-1.2-PCCDI-2017-0254, contract no. 27PCCDI / 2018.

References

1. C. R. Schultz, M. S. Tollefson, H. A.Wuertz, US Patent 4447173 (1984)
2. https://www.rainfloirrigation.com/equipment/drip-applicators
3. http://www.buctraco.com/3%20Online%20Catalog/DripIrrigation.htm

4. https://andros-eng.com/irrigation/boss-installation-implements/
5. E.D.Vories, P.L. Tacker, S.W. Lancaster, R.E. Glover, Agricultural Water Management **96**:6, 912-916 (2009)
6. S. J. van Donk, J. L. Petersen, D. R. Davison, Irrig Sci **31**:599–609 (2013)
7. C. Murley, S. Sharma, J. Warren, B. Arnall, W. Raun, Agricultural Water Management **208**, 357-362 (2018)
8. S. Evett, G. Marek, P. Colaizzi, D. Brauer, S. O'Shaughnessy, *Transactions of the ASABE 62*, 1377-1393 (2019)
9. N. Patel, T Rajput, Agric. Water Manag. **88**, 209–223 (2007)
10. V.K. Tripathi, T.B.S. Rajput, N. Patel, J. Clean. Prod. **139**, 396–406 (2016)
11. M. Dorta-Santos, M. Tejedor, C. Jiménez, J.M. Hernández-Moreno, M. Palacios-Díaz, F.J. Diaz, Ecol. Eng. **79**, 60–68 (2015)

Kinematics of the seedling plant with vertical distributor and buckets

Paul-Marius Mitrache[1], *Radu* Ciupercă[2], and *Ioan* Sărăcin[3]

[1,3] University of Craiova, Romania
[2]National Institute of Research Development for Machines and Installations Designed to Agriculture and Food Industry – INMA, Bucharest, Romania

Abstract. In this paper are presented some theoretical and experimental research on the kinematics of planting seedlings of vertical disc type and buckets, the variant of planting directly in the cultivated soil, the kinematics that must ensure the correlation between the working speed of the planting equipment and the speed periphery of the bucket, so that the seedling is planted in an upright position. After a brief presentation of the research carried out worldwide in the field, the kinematic scheme of the planting apparatus and the condition that must be fulfilled by the construction and operation of the apparatus for the correct planting of the seedling are presented. The relations for the projections of the studied point, called the tracer (root of the seedling) and of the afferent angular velocity, are written in a system of xOy coordinates. Using the real values of a studied planting machine, the real and corrected trajectories of the studied tracer point were drawn, using the MatLab program. Following the research, the related conclusions and recommendations for improvement were formulated.

1 Introduction

The realization of a mechanized work for planting vegetable seedlings corresponding to quality, with final implications in obtaining high yields per unit area, involves compliance with agrotechnical requirements the work of planting seedlings that can be divided into three categories, as follows [1,2]:
- Agrotechnical requirements regarding land preparation;
- Agrotechnical requirements regarding the planting work;
- Requirements imposed by regulations.

The cultivation of vegetables through seedlings has an important place, one of the important problems in the technological process is planting, an operation that is overwhelmingly mechanized.

Seedling planting equipment has experienced a spectacular development over time, the technical solutions developed, especially for the planting apparatus, have been very diversified [3], and following multiple comparative research, it has been concluded that those equipped with vertical discs and buckets achieve working parameters in accordance with the agrotechnical requirements at the highest level [4,5].

The improvement of the planting devices took place in the direction of their automation, a situation in which it was necessary to find technical solutions that would especially take over the seedling from the support and place the cold in the gutter or directly in the soil.

Worldwide, the issue of solving the requirement of planting seedlings in an upright position has been addressed by countless specialists, who have developed several types of pick-up and planting mechanisms for automatic planting machines, the results being published in well-known journals published on world plan.

We mention thus a mechanism for taking over the seedlings with a nutritious bale, of the type with planetary gear train, mechanism that equips an automatic planting system that takes the seedling from the pot and positions it so that, when it is released, the seedling is released in the gutter. sol fig. 1 [6].

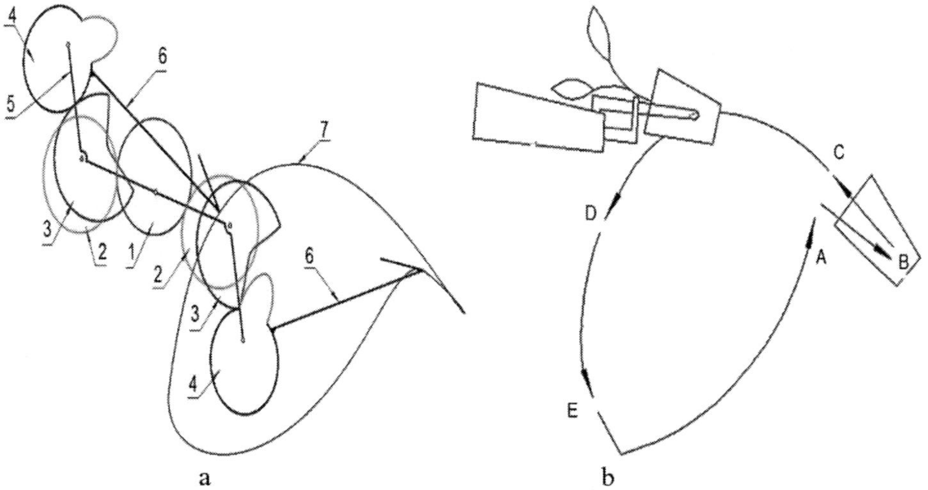

Fig. 1. Seeding bale picking mechanism with gear train planetary and take-up arms
a)1. Two-jaw pattern of the seedling pick-up mechanism; 2. Sun gear, 3. First middle gear, Second middle gear, 4. Planet gear, 5. Planet frame, 6. Seedling arm, and 7.Seedling pick-up trajectory. b) trajectory characteristics of the seedling mechanism

An analysis of the trajectory characteristics of the seedling mechanism indicated that the ideal seedling trajectory requires an arm of a certain length that makes a path of the seedling A - B - C - D - E.

Another achievement is a device for taking seedlings from pots, with reciprocating movement and control system of the seedling planter, fully automatic [7].

On this system it was found that the positioning accuracy of the linear transmission device gradually decreases with increasing speed of seedlings, especially when the pulse frequency of the electric control motor exceeds 20,000Hz. The engine goes out of gear and the accuracy will decrease rapidly. The motor parameters can be optimized by setting the different pulse frequencies and setting the range of error values between 0.2 mm and 0.8 mm, which depends on the requirement of seedling positioning.

Another achievement is an automatic mechanism for taking over and planting seedlings with a high-precision asymmetric transmission train, fig. 2.

The pick-up mechanism removes the seedlings from the seedling tray and transports them to the floppy planter that plants them in the soil [8].

As shown in fig. 2, the angle between the seedling tray and the horizontal plane is 60 °, ie relative to the suppling position of the seedling, the rotation angle of the execution component (seedling gripping component) being approx. 150 °. When placed on the floppy disk, the seedling has a horizontal position.

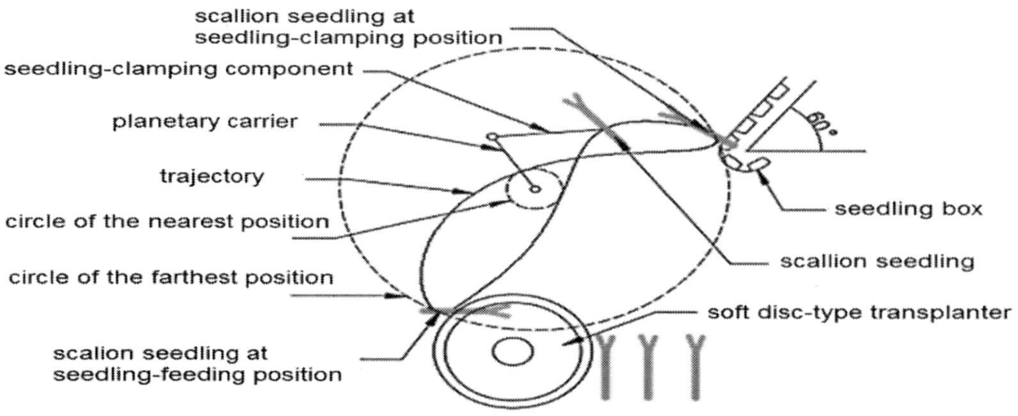

Fig. 2 Automatic mechanism for taking and planting seedlings with asymmetric transmission train, high precision [8]

A bucket-type planting machine, used for planting seedlings with a nutrient bowl, consisting of conical buckets, a guide tube and a soil compaction wheel, is another notable achievement [9].

Another model has conical cups attached to the drum which is driven by a soil compaction wheel [10], fig. 3. It is a high speed planter that extracts the seedlings directly from a tray with seedlings, the functions of making holes in the soil, withdrawing and depositing the plants, being performed at absolute zero speed of the seedling, at the time of planting.

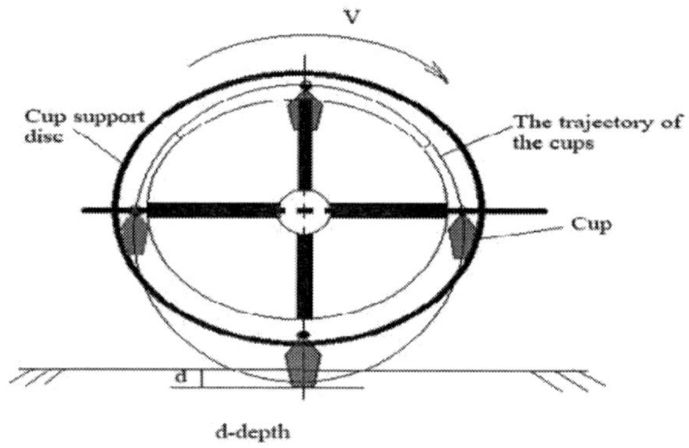

Fig. 3. Mechanism equipped with disc-type devices with cups for planting vegetable seedlings [10]

2 Materials and methods

To carry out the experiments, a planting machine was used, equipped with a distributor with articulated cups, in aggregate with a 45 hp vegetable tractor fig. 4.

Fig.4 Machine for planting vegetable seedlings and herbs in a row, MPA symbol (**** INMA Bucharest, 2018)
1-Frame, 2- planter, 3- transmission, 4- rear support wheels, 5 –compaction wheels, 6- rack, 7- track marker

The condition for the planted material to be placed in a vertical position is that the absolute velocity *Va* of the material, at the time of planting, be approximately zero (*Va = 0*). This becomes possible when the ratio λ_T is defined according to relation (1), in this case, the considered point moving after a normal cycloid, fig. 5, [11].

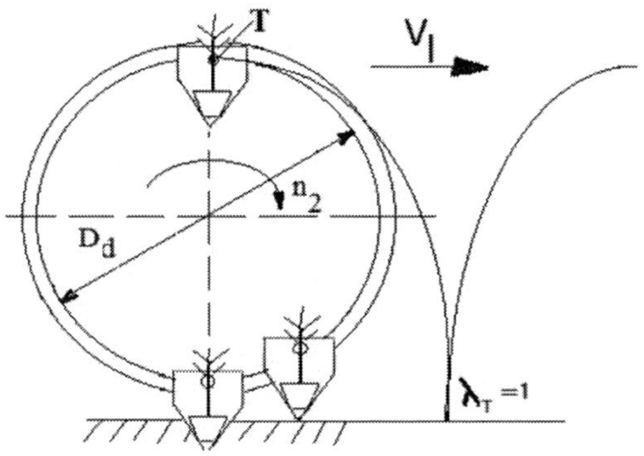

Fig. 5. Diagram of the working process of the disc planting machine with cups, with the placement of the seedling directly in the soil

$$\lambda_T = \frac{V_P}{V_l} = 1 \qquad (1)$$

in which:

Vp - peripheral speed of the planting device, m/s;

V_l - working speed of the machine, m/s.

The condition that the seedlings be planted at equal distances between rows can be achieved if there are no losses in the transmission chain, or the losses are minimal, so that the deviations from the distance between plants in a row fall within the values accepted by users.

The mentioned deviations can be caused by the losses in the kinematic chain of the transmission or by the non-uniformity of the drive wheel movement at the contact with the soil, which can be manifested by the sliding of the wheel, this cause being most frequently encountered when operating planters.

According to the kinematic diagram of the machine fig. 6 [12], the transmission of the movement from the drive wheel to the bucket distributor is very simple, this being made up of only two chain transmissions, thus achieving minimal losses in the powertrain.

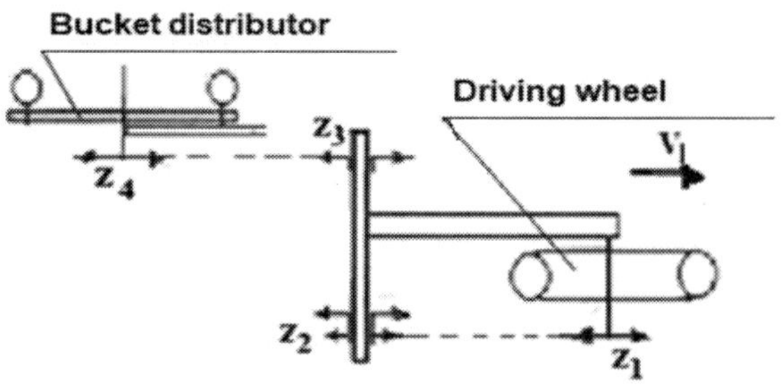

Fig. 6. Kinematic diagram of the transmission of the planter with vertical distributor and cups [12]

The rotational frequency of the rotary distributor is determined by the relation (5.2).

$$n_2 = n_1 \cdot i = n_1 \cdot i_1 \cdot i_2 = n_1 \cdot \frac{z_1}{z_2} \cdot \frac{z_3}{z_4} = \frac{30 \cdot v_l}{\pi \cdot R_a} \cdot \frac{z_1}{z_2} \cdot \frac{z_3}{z_4} \quad (2)$$

in which:
Ra - radius of the drive wheel, m;
v_l– speed of movement of the aggregate, m/s;
n_1 - drive wheel speed, s^{-1};
n_2 – speed of the vertical distributor with buckets, s^{-1}.

Kinematics of the planter with vertical disc and cups

To meet the requirement to place the seedling in the ground in a vertical position, expressed by relation (1), the equations for defining the coordinates on the *X* and *Y* axes of the reference point for the cycloid root of the seedling and the projections of the velocity of the drawer point *T*, which has a trajectory similar to point *A* defined above, on the axes of the orthogonal system *XOY*.

The expression of the motion relations for the tracer point *T* was adopted, fig. 6, for reasons of accuracy of defining its actual position on the distributor disk, this point representing the point of fixing the cups on the disk.

It know:

V_l - working speed of the planting machine, m/s;
ω_l– angular speed of the drive wheel, rad/s ;
Z_1, Z_2, Z_3, Z_4 - the number of teeth of the chain wheels ;
Ra– radius of the drive wheel, m;
Rd - the radius of arrangement of the cups on the distributor, m.

It is considered a kinematic cycle of the drive wheel. The angular velocity of the distributor is higher than the angular velocity of the drive wheel to compensate for the difference between the radius of the drive wheel and the radius of arrangement of the cups on the distributor disc. This was achieved by the transmission ratio between the chain wheel Z_1 and Z_2.

It is determined:

ω_2– angular velocity of wheels 2 and 3, rad/s;
ω_4– angular speed of wheel 4 (angular speed of distributor), rad/s;
X_T, Y_T - the coordinates of the center of the cups (tracer point *T*), m;
V_{TX}, V_{TY} - the projections of the speed of the tracer point *T*, on the axes of the orthogonal system, m/s.

The transmission ratios expressed by angular velocities are according to the relations (3).

$$i_{12} = \frac{\omega_1}{\omega_2} = \frac{z_2}{z_1} \Rightarrow \omega_2 = \omega_1 \cdot \frac{z_1}{z_2}; \omega_1 = \frac{V_l}{R_a} \tag{3}$$

$$i_{14} = \frac{\omega_1}{\omega_4} = i_{12} \cdot i_{34} = \frac{z_2}{z_1} \cdot \frac{z_4}{z_3} \Rightarrow \omega_4 = \omega_1 \frac{z_1 \cdot z_3}{z_2 \cdot z_4}$$

The position of the tracer point *T* on the bucket distributor, relative to the reference system *XOY* is represented in fig. 7 and is expressed by the relations (4). T_0 is the starting point from which the movement of the cup ditributor is considered.

$$\begin{aligned} X_T &= X_{C4} + V \cdot t + R_d \cdot \sin \varphi_4 \\ Y_T &= Y_{C4} + R_d \cdot \cos \varphi_4 \end{aligned} \tag{4}$$

Or relations (5).

$$\begin{aligned} X_T &= X_{C4} + V \cdot t + R_d \cdot \sin \omega_4 \cdot t \\ Y_T &= Y_{C4} + R_d \cdot \cos \omega_4 \cdot t \end{aligned} \tag{5}$$

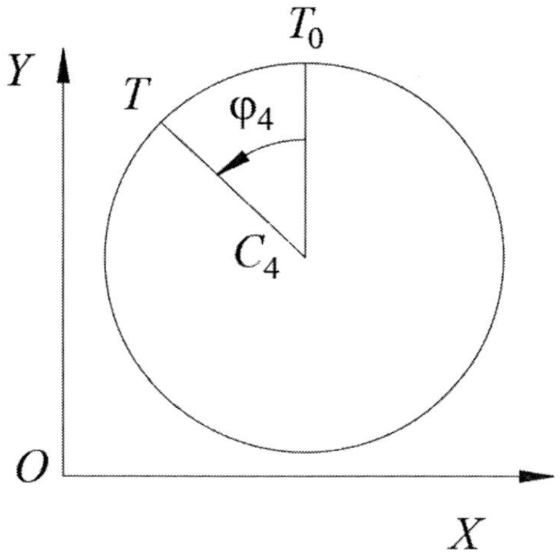

Fig. 7. Tracer point position *T*

The projections of the speed of the tracer point *T*, with respect to the reference system *XOY*, are expressed according to the relations (6).

$$\begin{aligned} V_{TX} &= V_l - R_d \cdot \omega_4 \cdot \sin \omega_4 \cdot t \\ V_{TY} &= R_d \cdot \omega_4 \cdot \cos \omega_4 \cdot t \end{aligned} \tag{6}$$

3 Results and discussion

For plotting the kinematic curves of the planting machine of the tested planting machine [12], the real values of the dimensions of the components of a seedling planting equipment used, according to fig. 8.

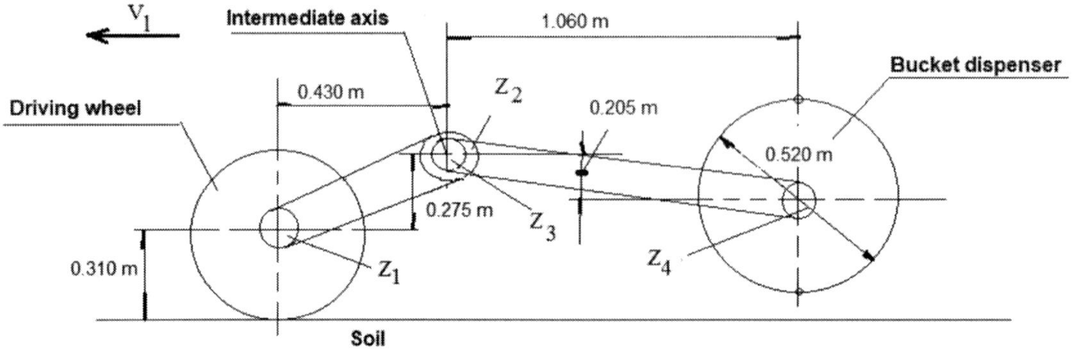

Fig. 8. Schematic of the planting machine [12]

$z_1 = 25$ teeth;
$z_2 = 23$ teeth;
$z_3 = 15$ teeth;
$z_4 = 15$ teeth;
$R_a = 0.31$ m;
$R_d = 0.26$ m.
V_1 = 1.15; 1.55; 2.15 km/h or 0.319; 0.431; 0.597 m/s - speeds used for experiments.

The real kinematics of the studied planting apparatus is shown in fig. 9

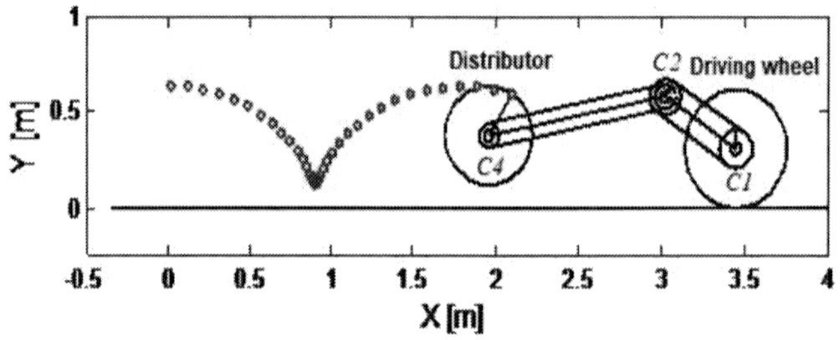

Fig. 9. The real kinematics of the studied planting apparatus

Using the real values of the planting machine used in the testers, the variation diagrams of the projection of the speed of the tracer point T, were drawn. fig. 10-12.

The real values of the constructive parameters of the machine used in the experiments are presented by quotation on the diagram in Fig. 8 and were measured stationary on a concrete platform, and the values for Ra; Rd, and the number of teeth of the chain wheels are as follows: $z_1 = 25$ teeth; $z_2 = 23$ teeth; $z_3 = 15$ teeth; $z_4 = 15$ teeth;$R_a = 0.31$ m;$R_d = 0.26$ m; $V_1 = 0.319$; m/s.

The diagrams in fig. 10-12 were drawn using relation (6) the variable is time and ω4 was calculated with relation (3), using the values mentioned above.

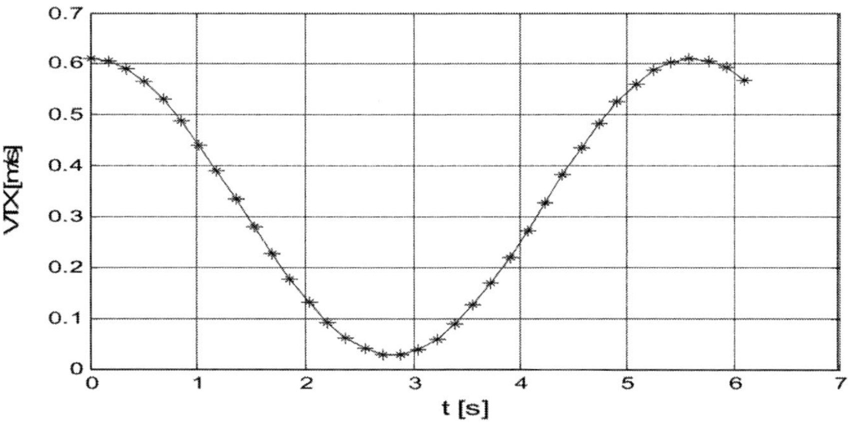

Fig. 10. Diagram of the variation of the speed projection of the tracer point T (for Z_2 = 23 teeth)

To satisfy the condition expressed by relation (1), the curve should have been tangent to the *0* line of the diagram. The difference is caused by the drive wheel slipping on contact with the soil, a phenomenon often encountered in the operation of such equipment.

Taking into account the above, the diagram was corrected by reducing the number of teeth on the wheel Z_2 to 21 teeth, thus resulting in the ideal curve according to fig. 11.

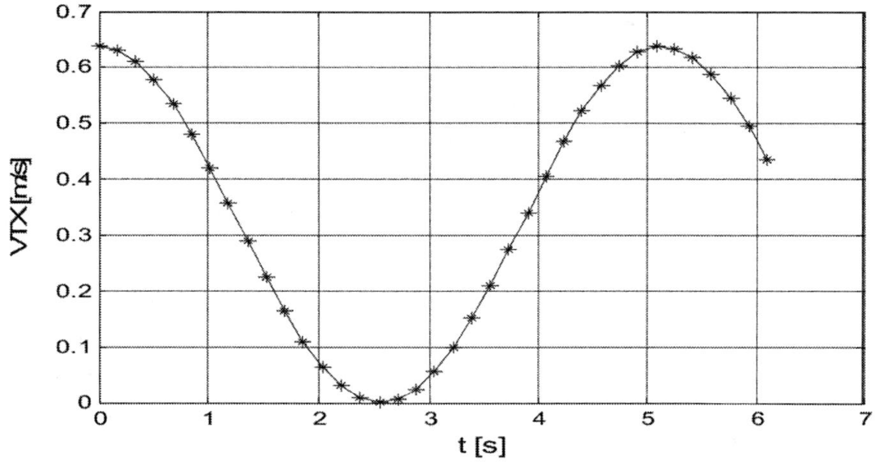

Fig. 11. Corrected variation diagram of the speed projection of the tracer point T, (for Z2 = 21 teeth)

To draw the variation diagram of the projection of the velocity of the point representing the root of the seedling, point T, fig. 5, the radius of the distributor R_d was corrected, thus resulting in the diagram according to fig. 12.

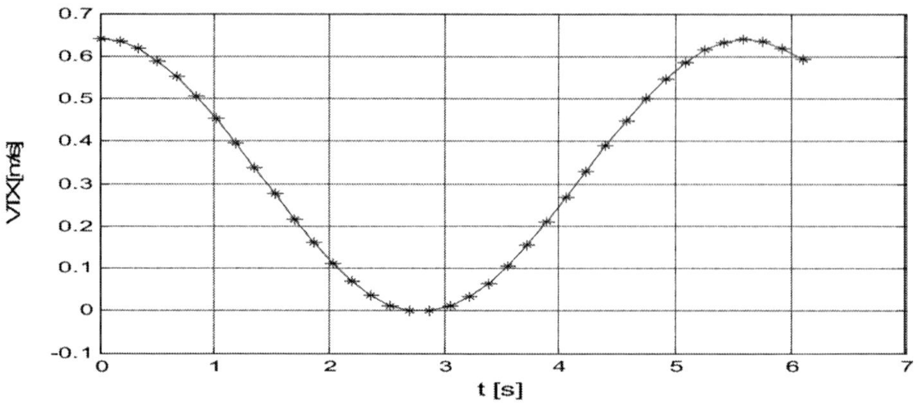

Fig. 12 Diagram of variation of seedling root velocity projection (R_d+0.02627 m and z_2=23)

4 Conclusions

- the condition that the planted material be placed in a vertical position, is that the absolute speed *Va* of the seedling, at the root level, at the time of planting, be approximately zero (*Va = 0)*. This becomes possible when the ratio λ_T defined as the ratio *V1/Vp* is as close as possible to the value *1*, in this case, the point considered moving after a normal cycloid;
- for the analysis of the way of fulfilling the condition *Vl/Vp = 1*, the kinematics of a planting device with vertical disc and buckets were made, the equations of motion of the tracer point (distributor bucket joint) and of its angular velocity an xOy coordinate system and the diagrams of variation of the projection of the speed of the tracer point were drawn, according to its constructive and functional parameters;
- from the determined diagrams, using the real parameters of the planting equipment, resulted minor deviations from the ideal trajectory (normal cycloid), reason for which it was necessary to correct it by varying the number of teeth of the chain wheel $z_{2,}$ from 23 to 21 of teeth, fig. 10-11;
- for plotting the variation curve of the projection of the seedling root speed, the radius of the distributor was corrected, at a value *Rd + 0.02627m*, following which the ideal curve resulted, fig. 12.

5. References

1. V. Popescu, A. Popescu, *Cultura legumelor din camp şi solarii* (Vegetable crop in field and solariums), Editura MAST (MAST Publishing House), Bucharest, (2006).
2. T. Stan, *Tehnologia cultivării legumelor*, Editura Alfa, (2005).
3. C. Muraru-Ionel, *Maşini cu discuri flexibile pentru plantat răsaduri legumicole (Machines with flexible disks for planting vegetable seedlings),* Editura Terra Nostra (Terra Nostra Publishing House), Iaşi, Romania, (2009).
4. P.M. Mitrache, I. Saracin, R. Ciuperca, *Comparative study on seedlings planting machines with cups dispensers ISB-INMA-TEH*, International Symposium, Bucharest, Romania, pp. 796-803, (2019).
5. SR 13215:1994 - *Maşini de plantat răsaduri. Condiţii tehnice de calitate (Seedlings planting machines. Quality technical conditions*), ASRO, Romania, (1994)..

6. T. Zhipeng, Y. Gaohong, Z. Xiong, L. Pengfei, Y. Bingliang, Design of Vegetable Pot Seedling Pick-up Mechanism with Planetary Gear Train, Chinese Journal of Mechanical Engineering, Volume **33,** Article number: 63, (2020).
7. Q Z Yang, L Xu, X Y Shi, et al. *Design of seedlings separation device with reciprocating movement seedling cups and its controlling system of the full-automatic plug seedling transplanter. Computers & Electronics in Agriculture*, 147: pp. 131-145. https://doi.org/10.1016/j.compag.2018.02.004, (2018).
8. Z. Xiong, Y. Jun, C. Mengyan, C. Jianneng, Automatic Scallion Seedling Feeding Mechanism with an Asymmetrical High-order Transmission Gear Train, Chinese Journal of Mechanical Engineering, Volume **33**, Article number 10, (2020).
9. A. Khadatkar, S. M. Mathur, B. B. Gaikwad, 2018, *Automation in transplanting: a smart way of vegetable cultivation*, CURRENT SCIENCE, Volume. 115, NO. 10, 25 NOVEMBER, pp.1884-1892. (abhijitnu2@gmail.com), (2018).
10. R. D. Munilla. L. N. Shaw, *A high speed dibbling transplanter*, Trans. ASAE, 30(4), pp. 904–908, (1981).
11. V. Scripnic, P. Babiciu, *Masini Agricole*, Editura Ceres, pp. 206, Bucuresti, (1979).
12. *Tehnologie de înființare și valorificare superioară a materiei prime vegetale obținută din plante medicinal (Technology for the establishment and higher recovery of plant raw material obtained from medicinal plants)* , INMA Bucharest, Romania. Availabile at: www.inma.ro/Proiecte/Proiecte CDI Program NUCLEU/Project PN 18 30 02 03, (2018).

Biological water treatment in a recirculating aquaculture system

Evelin-Anda Laza[1*], *Ioan Ladislau* Caba[1], *Mihai* Olan[1], and *Valentin* Vladut[1]
[1]National Institute of Research - Development for Machines and Installations designed to Agriculture and Food Industry - INMA, Bucharest 013813, Romania

ABSTRACT Biological water filtration is a process by which toxic compounds are removed from water using organisms. A well-sized biological filter is extremely important as part of a recirculating aquaculture system for fish farming. Biological water filtration equipment in aquaculture recycling systems is a technological set that restores the vital qualities of wastewater from fish ponds, thus allowing its reuse

1. Introduction

Biological water filtration equipment in recirculating aquaculture systems is a technological set that restores the vital qualities of wastewater from fish ponds and allowing its reuse. Biological filters include a filtration technique that uses living organisms to remove a number of toxic compounds from the water and is intended to neutralize ammonia and nitrites in recirculating aquaculture systems (RAS) for superintensive fish farming[1,6,7].

The main technological desideratum to be achieved in a SAR is to ensure medial conditions that correspond, to the greatest extent possible, to the ecophysiological particularity of the crop species [2,4,5].The decomposition of nitrogen compounds is of particular importance in the field of aquaculture. Ammonia NH_3 and Nitrites NO^-_2, are toxic substances for fish. Nitrates NO^-_3, are also toxic when they accumulate in high concentrations. With the help of *Nitrosomonas* bacteria, ammonium is converted into nitrites NO^-_2 [2,3,6]

The next phase under the action of other types of *Nitrobacter* bacteria, the nitrites resulting in the first phase are transformed into nitrates NO^-_3 , because nitrates can also become toxic when they accumulate in a large amount, it is recommended to limit the bacterial conversion $NH^+_4 \rightarrow NO^-_3$ [2,3,6].

Water from recirculating aquaculture system can come *from cattle farm if are advanced electrochemical treatment* [8] or various other sources of recovered water [9], using smart system to monitor wastewater treatment [10].

Figure 1 describes the characterization of nitrification and denitrification processes that take place during the process of biological filtration of water

* Corresponding author: eveline_anda@yahoo.com

 https://doi.org/10.1051/e3sconf/202128603013

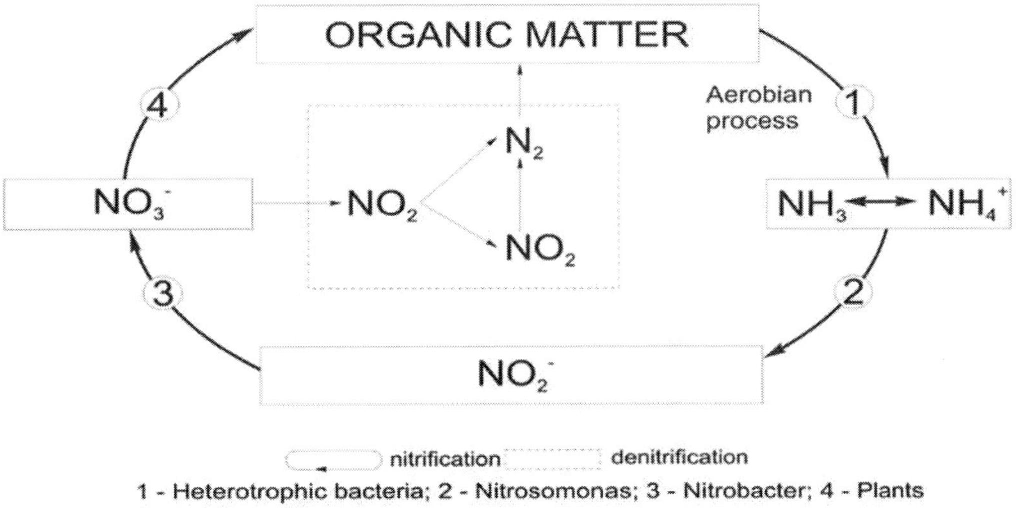

Fig.1. Characterization of nitrification and denitrification processes that take place during the process of biological filtration of water

In figure 2 a biological filter can be seen and figure 3 shows the aeration system of the biological filter.

Fig.2 Biological filter with ball

Fig.3 Aeration system of the biological filter

2. Methodology

Figure 4 represents the technological scheme of a recirculating aquaculture system, in which the position of the biological filter can be observed.

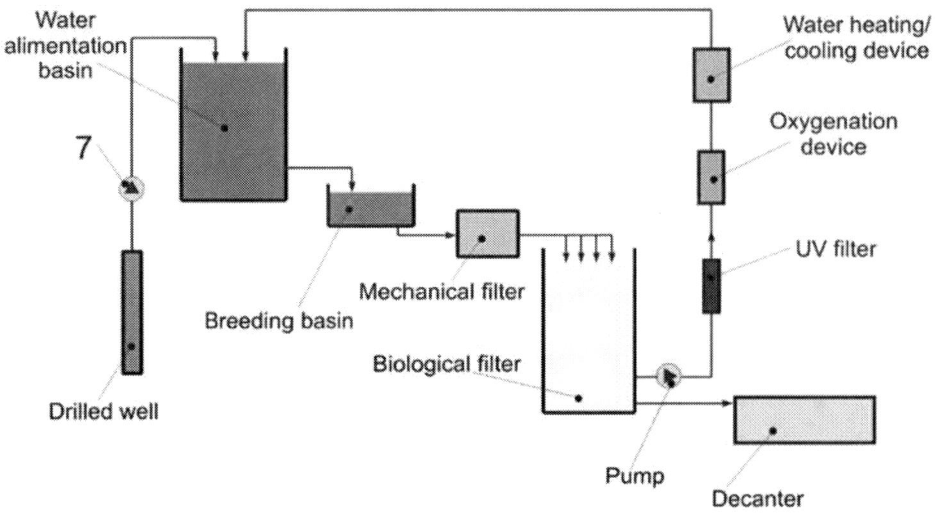

Fig. 4 Water circulation scheme in a recirculating aquaculture system

The biological filter on which the experimental determinations were made was supplied with wastewater from fishs basins through the exhaust connection on the top of an outer wall. Wastewater drains gravitationally through the upper grate, the filter element layer and the lower grate in the first cavity of the filter, passes through the slot provided for this purpose in the second cavity, after which, due to the principle of communicating vessels, passes up, in turn through the grate bottom, the filter element layer and the upper grill, reaching the discharge opening.

Biological filtration of water takes place during its passage through the layers of filter element, which are made of round plastic supports populated with aerobic bacteria.

For a good functioning of the biological filter, it is necessary to produce a strong aeration of the water, and in some situations also its additional oxygenation.These operations are performed by insufflating air and oxygen, respectively, under pressure, in the exhaust connection provided for it, which reaches the water in dispersed form, with the help of diffusers. Sediments deposited on the bottom of the biological filter cavities are evacuated by periodic purging.

The measurement of water temperature, pH level and dissolved oxygen was performed for each sample taken in order to perform physico-chemical analyzes of water, in order to individualize each sample and to monitor the influence of these factors on the quality of biological filtration. The quality parameters of water after the biological treatment will be by taking water samples before and after the biological filtration process.

Figure 5 shows the flow of water in the biological filter

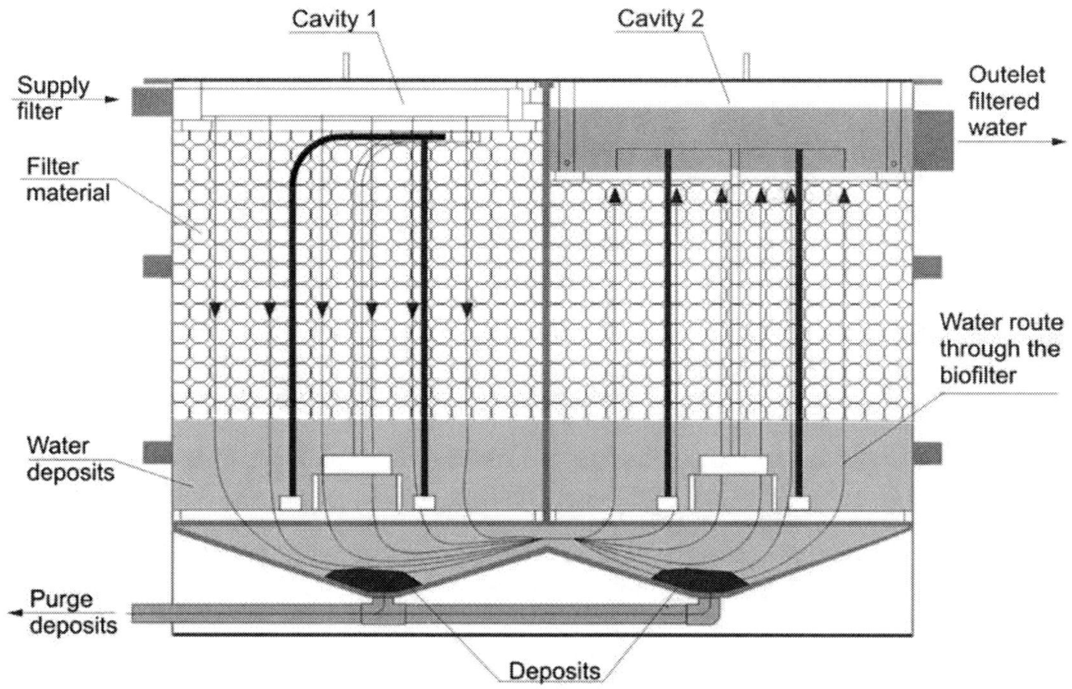

Fig.5 The water flow in the biological filter

The flow rates of filtered water with which each basin was supplied separately were determined by timing the flow (in seconds) of ten liters of water and making the ratio:

$$Q_n=600/t_n \text{ [l/min]} \qquad (1)$$

In this way the following results were obtained:

t_1=5 s	Q_1=150 l/min
t_2=5 s	Q_2=150 l/min
t_3=11 s	Q_3=60 l/min
t_4=11 s	Q_4=60 l/min
t_5=11 s	Q_5=60 l/min
t_6=40 s	Q_6=15 l/min

3. Results and discussions

As it was found, the flow thus determined falls within the range of 30 ... 50 m3 / h of the maximum flow provided for the biological filter.

$$Q=570 \text{ l/min} = 34{,}2 \text{ m}^3\text{/h} \qquad (2)$$

The useful volume of the biological filter, defined as the volume occupied by the filter element, was determined by measuring the thickness of the filter element layer in the two filter cavities, which was multiplied by the section of the cavities.

$$V_{usfl}=S_1xh_1+S_2xh_2 \tag{3}$$

Where they were measured:

$$S_1=S_2=1,47x0,9775=1,437m^2 \tag{4}$$

$h_1=0,87$ m

$h_2=0,80$ m

$$\text{Result: } V_{usfl}=2,4 \text{ m}^3 \tag{5}$$

The total volume of the two cavities of the biological filter is 3.75 m^3, which means that the useful volume occupies 64% of the total volume. The remaining 36% is occupied by water, air, deposits and grills.

The air flow blown into the biological filter was determined by measuring with a thermo-anemo-manometer using a Pitot tube.

The flow determined by the air blown into the biological filter through the two diffusers provided for this purpose was 5.5 m^3 / h

The flow of blown oxygen was determined using the flow meter belonging to the oxygen distribution system of the recirculating aquaculture system, provided for this purpose at the entrance to the biological filter. This flow depends on the oxygen pressure and can be adjusted with the help of a valve when the maximum opening of the oxygen reached 8 l / min. Due to the high costs required by the oxygen supply, the use of oxygen is only used if the aeration does not meet the requirements, especially in the hot summer months.

The determination of the purged water flow was performed by opening the purge valve to the maximum and timing the flow of 10 liters of water with deposits, the following results were obtained: tp = 25 s, and we got the result:

$$Q_p = 24 \text{ l/min} = 1,44 \text{ m}^3/\text{h} \tag{6}$$

The determination of the water temperature, the pH level and the dissolved oxygen was performed for each sample taken in order to perform the physico-chemical analyzes of the water, in order to individualize the sample and to be able to follow the influence of these factors on the quality of biological filtration. The determination of water temperature, pH level and dissolved oxygen was performed with C863 Electrochemical Multimeter, using the probes provided.

The quality parameters of water after the biological treatment were measured before and after the biological filtration process.

The results in table 2 show the values of the parameters of water before the biological filtration.

Table 1 Parameters of water before the biological filtration

Crt. no	Physico-chemical parameter	M.u.	Measured values
1.	pH	upH	7,8
2.	Oxygen	mg/l	5
3.	Ammonia (NH_3)	mg/l	0,06
4.	Nitrates (NO^-_3)	mg/l	0,6
5.	Nitrites (NO^-_2)	mg/l	0,4
6.	Ammonium (NH^+_4)	mg/l	0,03

The water parameters have normal values for the water in fish basins and nitrogen compounds are approaching the limit. By analyzing the water after the biological filtration and determining the physico-chemical parameters, the following values resulted:

Table2 Parameters of water after the biological filtration

Crt. no	Physico-chemical parameter	M.u.	Measured values
1.	pH	upH	7,5
2.	Oxygen	mg/l	4,5
3.	Ammonia (NH_3)	mg/l	0,01
4.	Nitrates (NO^-_3)	mg/l	0,2
5.	Nitrites (NO^-_2)	mg/l	0,1
6.	Ammonium (NH^+_4)	mg/l	0,0

After the water analysis it was found that the biological water filtration process was efficient and that the water parameters reached optimal values for the good development of the fish.

4. Conclusion

Following the biological filtration of water the following conclusions were found:

- the working regime of the biological filter can vary greatly, depending on a number of factors, such as: the species of fish raised in the ponds, its age and density in the ponds, the type and amount of feed, the water temperature, its pH, quality of aeration and /or oxygenation;
- the nitrification process is a process with high oxygen consumption, it is important to ensure additional aeration of the water;
- due to the low passage of water, fine solid impurities in the water were deposited on the bottom of the basins. By periodically opening the purge valve (once every 24 hours, or as many times as necessary), these impurities have been removed to the channel;
- because atypical situations can occur occur, especially on hot summer days, when the biological activity of fish is more intense, given the importance of reconditioning the water that feeds the fish basins, it is recommended that the biological filter be slightly oversized.

Acknowledgement: The work has been funded by the Sectorial Programme MADR – ADER 2019-2022, ADER 25.3.1. Recirculating aquaculture systems used in the stage preceding the repopulation of natural waters with fish material, Contract no. **ADER 25.3.1 / 24.09.2020/ AA 2 / 11.05.2021.**

References

[1]. Avnimelech Y. Bio-filters The need for an new comprehensive approach, Aquacultural Engineering, Elsevier, vol. 34 (3), (2005)

[2]. Andrei S.G., Pop A., David E., Ştefanov C. - Contributions regarding the technical solutions for biological filtration within recirculating aquaculture systems, ISB INMA TEH' 2014, International Symposium, 30 octombrie – 1 noiembrie Bucureşti, ISSN 2344-4118(2014)

[3]. Crab R., Avnimelech Y., Defoirdt T., Bossier P., Verstraete W. - Nitrogen removal techniques in aquaculture for a sustainable production, Aquaculture, Elsevier, vol. 270 (1-4), p. 1-14, (2007)
[4]. Cristea V., Grecu I., Ceapa C. Recirculating Aquaculture Systems Engineering, Ed. Didactical and Pedagogical R.A. Publishing, Bucharest, (2002) .
[5]. Malone R.F., Pfeiffer T.J. Rating fixed film nitrifying biofilters used inrecirculating aquaculture systems, Aquacultural Engineering, Elsevier, vol. 34 (3), (2006)
[6]. Pop A., Ştefanov C., Andrei S.G. Modular biofilter for recirculating aquaculture systems, Patent Application A01435 / 22.12.2011, INMA Bucharest, (2011)
[7]. Ştefanov C., Technical equipment for biological water filtration for superintensive fish growth in recirculating aquaculture systems (Biological Filter 2.4), (2011)
[8]. Ungureanu N., Vlăduţ V., Istrate I., Zăbavă B., Tociu C., Ferdeş M., Dincă M. - Advanced electrochemical treatment of the wastewater from cattle farm, PROCEEDINGS OF THE 47 INTERNATIONAL SYMPOSIUM ON AGRICULTURAL ENGINEERING "Actual Tasks on Agricultural Engineering", pag. 147-158, Opatija – Croaţia, http://atae.agr.hr/, (2019);
[9]. Ungureanu N., Vlăduţ V.*, Voicu gh. - Water Scarcity and Wastewater Reuse in Crop Irrigation, SUSTAINABILITY 12(21), 9055; https://doi.org/10.3390/su12219055, (2020);
[10]. Zăbavă B., Ipate G., Voicu Gh., Dincă M., Ungureanu N., Ferdeş M., Vlăduţ V. - Smart system to monitor wastewater treatment based on Raspberry Pi computer, PROCEEDINGS OF THE 47 INTERNATIONAL SYMPOSIUM ON AGRICULTURAL ENGINEERING "Actual Tasks on Agricultural Engineering", pag. 159-166, 2019, Opatija – Croaţia, http://atae.agr.hr/, (2019).

Technology for biohumus production, an alternative to conventional fertilizers for bio agriculture

Evelin-Anda Laza[1*], *Oana* Cristea,[1] *Nicoleta* Ungureanu [2]

[1]National Institute of Research - Development for Machines and Installations designed to Agriculture and Food Industry - INMA, Bucharest 013813, Romania
[2] Polytechnic University of Bucharest, Faculty of Biotechnical Systems Engineering, Bucharest

Abstract. Composting is the process of decomposition and transformation of solid organic substances by microorganisms (mainly bacteria and fungi) into a stable material, which can be used (depending on characteristics) in agriculture, instead of chemical fertilizers or land improvement works (soil improvement. The composting process consists of two stages: mechanical treatment and decomposition (fermentation). In the case of mechanical treatment, the waste is crushed, homogenized and prepared for biological treatment, fermentation. The fermentation stage consists in three main stages, namely the mesophilic fermentation stage, the thermophilic fermentation stage and the maturation stage of the compost.

1. Introduction

This paper aims to bring to the attention of farmers the benefits of reusing animal and agricultural waste through composting. Composting is the process of decomposition and transformation of solid organic substances by microorganisms (mainly bacteria and fungi) into a stable material, which can be used (depending on characteristics) in agriculture, instead of chemical fertilizers or land improvement works (soil improvement) [1,2].

Compost is the best mulch and natural soil amendment and it can be used instead of commercial fertilizers. But the most important thing is that it is a cheap product. The use of compost leads to improved soil structure, improved excessive textures, improved aeration and increased water storage capacity, increased soil fertility and stimulated the development of a healthy root system of plants. Organic matter applied through the compost is like food for microorganisms in the soil[1,2] .

Composting can be done in households, in open piles, in special plastic compost bins or in other containers. In the case of farms and composting stations, composting can be done on specially designed platforms, the material for composting being arranged in long rows (furrows) and turned periodically (using special furrow reshaping equipment), in systems of aerated static piles using perforated pipes or in special containers. Good composting consists in ensuring adequate environmental conditions for microbial life. Organisms responsible for composting need three elements to live [2,3,4]:

* Corresponding author: eveline_anda@yahoo.com

• *Food for microorganisms*, composed of a mixture of carbon-rich ingredients (brown-hard-dry, such as straw, tree branches, sawdust) and nitrogen-rich ingredients (soft-green-moist such as weeds, leaves, residues of fruits and vegetables). It is important that both dry and green waste be used in the composting process to obtain a high quality compost [4,5,6].

• *Humidity*, especially present in nitrogenous ingredients, or from water added separately; the humidity ensures an environment favorable for the multiplication of aerobic bacteria and accelerates the composting process [4,5,6].

• *Air*, whose circulation is favored by hard carbonate ingredients. The microbes involved in the composting process are aerobic, they need air to grow and decompose waste. In the absence of proper aeration of the mixture, rotten compost is formed (anaerobic), a low quality organic fertilizer and dangerous for the environment. It generates unpleasant odors and methane gas, while the leachate resulting from rot drains into the soil and later reaches the groundwater [4,5,6] .

2. Methodology

A suitable location was found for the production of compost (platform for its production. The organic materials were stored on the composting platform so that they could be transformed into compost. They were carried out regular inspections of temperature and humidity and to prevent dispersal of the waste and the penetration of rodents and insects. A compost aerator was used to aerate the compost. In order to speed up the composting process of the plant residues that were to be used, they were chopped as small as possible. On the platform for composting was added a first layer of resulting waste from trimming trees or other plant debris will be obtained until the film thickness of residues 10-15 cm. This is actually a base layer for the rest of the waste added later in the composting process.

To activate the composting process, soil and manure was added. Once the "bed" of garden waste was finished, the addition of vegetable waste and and other garden waste began.

As main operational stages in the production of compost we list: humidity regulation in hot periods twice a week, aeration at least once a week, the compost that went through the thermophilic phase was covered with foil. The prisms were covered again when lower temperatures were recorded in order to better maintain the temperature overnight (higher thermal inertia), the parameters were monitored once every 2 weeks (internal) and general outsourced analyzes 1 time per month and the activity of bacteria / fungi / insects present in the compost was monitored. A cylindrical sieve was used to sort the resulting compost.

The determination of the macroelements, microelements and other elements necessary for the good development of the plants in the obtained compost was performed with the help of the Elemental Analysis Kits. Compost samples subjected to chemical analysis were taken from the composting platform.

Figure 1 shows a mobile compost kit for on-site analysis and figure 2 shows a thermometer for measuring the temperature of the compost

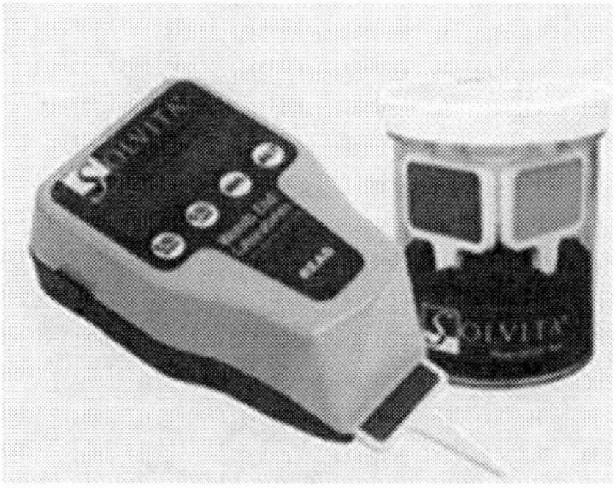

Fig.1. Mobile compost test kit

Fig. 2 Thermometer

When making the composting recipe, the specialized information on the carbon / nitrogen ratio in table 1 was taken into account.

Table 1 Materials rich in Nitrogen and carbon

Type of waste	**C / N ratio**	**Type of waste**	**C / N ratio**
Waste rich in Nitrogen		**Waste rich in Carbon**	
Liquid manure	2 – 3	Fruit leftovers	34
Solid manure	10	Leaves	50-65
Cut grass	12-15	Straw, oats	50-63
Vegetable waste	13	Tree bark	110-140
Kitchen waste	23	Remains from bush cleaning	110-160
Potato plant	25	Sawdust	120-510
Horse manure	25	Paper / cardboard	220-520

With the help of this information we adjusted the C / N ratio of the material introduced in to the pile.

The process of decomposition depends on several factors. These factors and the relationships between them, influence the speed of the decomposition process, the decomposition phase and the activity of microorganisms.

The type of substrate , by using a certain material or a mixture of different materials, many properties of the process can be influenced, such as: the volume of interstices, humidity, or particle size of the material introduced in the pile, but also the quality of the compost. Therefore, the mixture of introduced materials (the substrate used) is the most important step in producing a quality compost.

Temperature is a value for determining the degree of putrefaction.

Humidity, the microorganisms need water to survive and a constant humidity is required to ensure the continuation of the decomposition process.

pH level: The pH values must be betwen 7 and 11.

C / N ratio:. The optimum carbon : nitrogen ratio should be between 20: 1 and 35: 1

Figure 3 shows the materials used to prepare the compost, and figure 4 shows Images during the experimentation of the recipe for composting

Fig.3 Materials used to prepare the recipe for composting

Fig. 4 Images during the experimentation of the recipe for composting

3. Results and discussions

When the temperature is below 40 ° C, the composting process it's over.
In table no 2 are listed values of the quality of the resulting compost

Table 2 The quality of the resulting compost

Characteristic	U/M	Characteristic value
Hygiene		A hygienic product that ensures the exclusion of germs
Impurities	%	Impurities greater than 2 mm is not more than 0,6%
Glass, plastic, metal		0,73
Vegetable residues (2 mm)		0,6
Stones		0,5
Type of compost and material introduced in the pile		Green plant residues (plant removed from the vegetable crops) Chopped apples Chopped vegetables (waste tomatoes and peppers) Waste from a poultry farm Deciduous leaves and vegetable scraps from harvesting onions that limit dehydration
The compost maximum particle size	mm	15
Material density	Kg/m^3	410
Salt content		0,09
pH value		9
The decomposition degree		Mature compost: phase IV (degree of rot)
The compost water content	%	43 %
Zinc (Zn)		400 mg/kg DS
Lead (Pb)		150 mg/kg DS
Copper (Cu)		100 mg/kg DS
Chrome (Cr)		100 mg/kg DS
Nickel (Ni)		50 mg/kg DS
Cadmium (Cd)		1,5mg/kg DS
Mercury (Hg)		1,0mg/kg DS

4. Conclusions

So, in conclusion we can say that the temperature is the most important value that must be monitored during the process of transformation into compost, because it can be very easily be measured and shows the progress of the process. The decomposition of organic substances as a result of the activity of microorganisms due to their self-heating capacity is the reason for the temperature differences in the center of the pile and the surrounding temperature. Compost is the result of microbial decomposition of dead matter by oxygen.

With the use of compost, there is a wide range of benefits. Factors, such as the input material or raw material, the method of composting, storage and application of compost, all these things have an influence on the characteristics of the compost.

References

1. Caba I.L PN 19 10 02Contract no.: 5N / 7.02.2019 Research regarding the realization of a technology for the production of biohumus (vermicompost) "Prospective studies on innovative technologies for the production of biohumus existing worldwide", (2019)
2. Glenn, M., Manual of On-Farm Vermicomposting and Vermiculture. Organic Agriculture Centre of Canada, (2009).
3. G.E.O.R.G., Feasibility of developing the market of organic and transitional farms for the processing of organic farm waste and municipal waste using large-scale vermicomposting. Ed. Good Land Organic Resources Group , Halifax, Nova Scotia, (htpp://alternativeorganic.com) , (2004).
4. Rink, R. (Editor), Farm Composting Guide. Natural Resources, Agriculture and Engineering Service (NRAES-54), (1992).
5. http://asociatiacompostului.ro
6. http://www.ecoteca.ro/compostarea-deseurilor-aspecte-esentiale.html

Precision sowing of vegetable seeds using electrically operated distribution devices

Adriana Muscalu[1,*], *Cătălina* Tudora[1], *Laurențiu* Vlăduțoiu[1], *Constantin* Vlad[2], and *Angela* Dorogan[3]

[1]INMA, 6 Ion Ionescu de la Brad Blvd., Bucharest, 013811, Romania
[2]SCDL, 23 Mesteacanului Street, Buzău, 120024, Romania
[3]INCDPT, 6 Ion Ionescu de la Brad Blvd., Bucharest, 013811, Romania

Abstract. The high efficiency of vegetable crops is influenced, in addition to soil preparation works, weed, diseases and pests control, irrigation, etc. by precision sowing, which involves placing a certain number of seeds at an exact depth and distance. Thus, cost savings, a high reliability of production and especially a greater uniformity of crops are achieved, most often uniform and quality productions being obtained. For vegetable species in the case of which harvesting is practiced only once, uniformity of seed distribution is particularly important. In most cases, vegetable seeds have small size, irregular shape and small mass. These characteristics make the development of precision seeders for this type of seed quite difficult. Precision sowing of vegetable seeds ensures a correct sowing rate, which implies an optimal space for plant growth and development, thus improving crop yields. The paper presents a summary of the current state of precision seeders equipped with electrically operated distributors for vegetable seeds, as well as the preliminary results of research conducted by authors in this field.

1 Introduction

Precision agriculture aims to optimize agricultural production and environmental quality, being defined as a complex system that applies advanced technologies combined with agricultural management practices and crop-specific information, taking into account aspects of their temporal and spatial variability [1, 2, 3, 4, 5]. In the current context, it can contribute substantially to meeting the market demand for agricultural products, increasingly oriented towards healthy and high quality products [6, 7, 8].

Precision sowing is a way to apply the general and current concept described above. The improvement of specialized equipment, which can perform this operation with precision for various types of crops, is a continuous concern for specialists in the field [9, 10, 11].

For vegetable crops, the placement of the correct amount of seeds per unit area, respecting the prescribed sowing depth, together with achieving uniformity of distance between seeds on the row and between rows, significantly influences the yield. A correct

* Corresponding author: amuscalis@yahoo.com

distribution of vegetable seeds is the premise of crops with uniform and quality production, achieving cost savings [3, 10, 12].

The efficiency of vegetable crops, especially of those with only one harvest per year, is conditioned by the uniform distribution of seeds in the soil, in addition to land preparation, irrigation, crop maintenance, etc. The correct sowing norm ensures a better germination and an optimal space for plant development, with weed suppression effects [13, 14]. Also, weed control works in vegetable crops can be done efficiently, both by using classical and intelligent systems, knowing that these crops, compared to others, have a high sensitivity to attack by harmful plants. [15, 16, 17, 18]

For most vegetable species, the seeds have irregular shapes, small mass and small size (average diameter <3mm), so the development of dedicated precision seeders is quite difficult [19, 20].

Numerous research units and companies producing agricultural machines have dealt with the study and development of electrically operated distribution devices, intended for precision sowing. Of these, some have interesting technical features.

The research and results presented in the paper are a premise for optimizing precision seeders for vegetable seeds.

2. Electrically operated distribution devices

2.1 The current state of electrically operated distribution devices

Large companies producing precision seeders have developed and implemented electrically operated distribution devices intended mainly for maize, soybean, sunflower, sugar beet seeds.

John Deer implemented the type of row unit ExactEmerge ™ (fig.1), electrically driven, which uses the "Brush Belt Delivery System", patented by John Deere company. This system refers to the replacement of the seed driving tube with a brush belt that places the seeds, in the furrow according to the agrotechnical norms of corn and soybean crops, especially at high working speeds (over 12 km/h). The ExactEmerge™ electric drive system is powered by two brushless motors, on each unit, that work independently to control the meter and the BrushBelt. During the distribution process, the seeds pass in front of a sensor, each row unit is equipped with this. It has the role of monitoring the singularity, the precision of the distance and the seed norm [21].

Fig.1 Row unit ExactEmerge™ (1 – brush system for seed transport; 2 – electric drive motor of meter; 3 – BrushBelt electric drive motor) [21]

MaxEmerge ™ 5 is a new type of row units, electrically driven, used for 1775 NT pneumatic planter (produced by John Deer), so that maize and soybean can be sown with precision at high working speeds of 16 km/h [21].

Kinze company produces seed meters (4000 series), vacuum driven distributors with electric drive (fig.2), which achieve a seed distribution precision of 99%, at sowing speeds of over 10 km/h. The seeder discs have a configuration that ensures the filling of all the holes, in order to prevent the appearance of gaps when sowing. For this, some channels were made on the surface of the discs with the role of directing the seeds [11, 21].

Fig.2 Seed meter Kinze 4000 series (1 – scraper; 2 – electric motor; 3 – seeder disc; 4 – seed driving tube) [21]

For precision sowing of beet seeds at the Department of Environment, Resources and Technology at the Royal Veterinary and Agricultural University KVL, Denmark, a precision planter with electric drive was developed. For driving, stepper electric motors were used, providing high-precision control for the seed distribution discs. In addition, the seeds are monitored electronically, their position being permanently known. Thus, it was possible to determine the points of fall, respectively their placement points. Have been used 4 row units mounted on a Kverneland Accord seeder, Germany, Kleine Unicorn 3, attached to an autonomous tractor. A GPS-controlled lateral displacement system (own conception) was also used to synchronize seed distribution and transverse overlap, which guides the aggregate to predefined virtual field passes [22].

The machine was tested in the laboratory to evaluate the uniformity of the distance between the seeds and the synchronization of the seeder units on the working width. Thus, it was found that the standard deviation of the distances between seeds on the row is very small, the electric motors being able to operate the distribution disks with precise and constant speed [22].

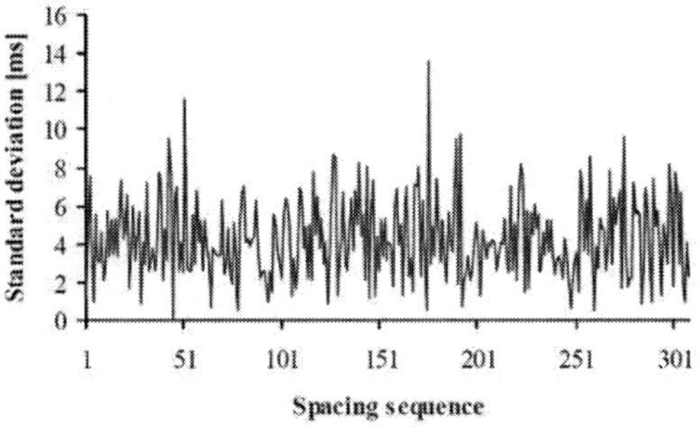

Fig. 3. Synchronization of the four seeder units expressed in standard deviation versus spacing sequence (for 6000 Hz motor speed) [22]

To evaluate the synchronization, the standard deviation of the accumulated time per seed detection was determined, for the same spacing sequence, between the four row units. The values of this parameter shown in Figure 3 were in a constant range during the test, which shows that the seed distribution is synchronous across the working width. The 4 row units place the beet seeds in the soil, at about the same time. This synchronization is important for the efficiency of the maintenance works, which can be performed longitudinally and transversely to the rows, in the crop [22].

Wendte K.W. is the author of a patent that refers to a seed distributor with electric drive (fig. 4). It consists of a seed plate (4) rotatably mounted concentrically in the meter housing (7) for singulating the seeds. The seed plate is driven by a gear transmission. The gear 2 is mounted on the electric motor shaft. The intermediate gear 3 has the same number of teeth, having the role of ensuring the correct direction of the distribution disc. The distribution disc provided with seed holes (6), is in fact a gear, made so that the transmission ratio is 1:5, so that its rotation speed will not exceed 600 rpm. The electric motor is connected via connectors (4) to a controller and a power supply, connected to the tractor's 12V DC electrical system. The roe unit is equipped with sensors connected to the controller to determine the working speed during the sowing operation [23].

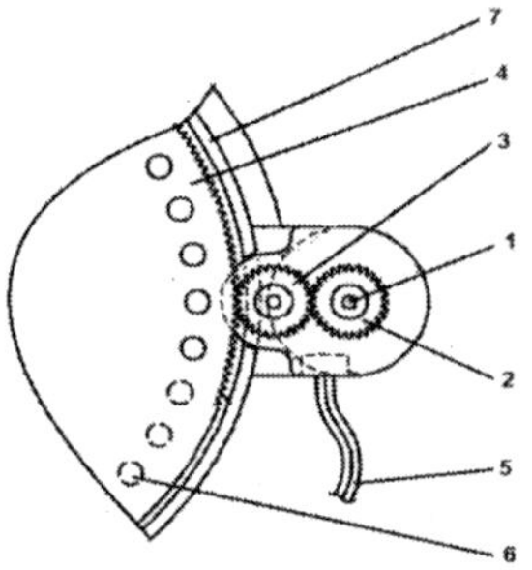

Fig. 4 Side view of the electric drive system of the seed plate [23]

For small vegetable seeds, with diameters between 2-10 mm, an electric precision seeder prototype was developed and tested. During the sowing operation, the seeder is electrically operated. It is also equipped with a sowing quality monitoring system consisting of the following main elements: Holzer speed measuring sensor, sowing monitoring fibre sensor, controller, computer, display etc., plus the host computer software. The seeder (fig. 5) is intended for use on small areas. The equipment uses a powerful brushless DC motor installed in the rear wheel, which can adjust the speed of travel, thus providing the necessary running power to save labour and improve efficiency. The seeder is equipped with 4 seeding units, each having distribution discs, interchangeable, depending on the type of seeds [19].

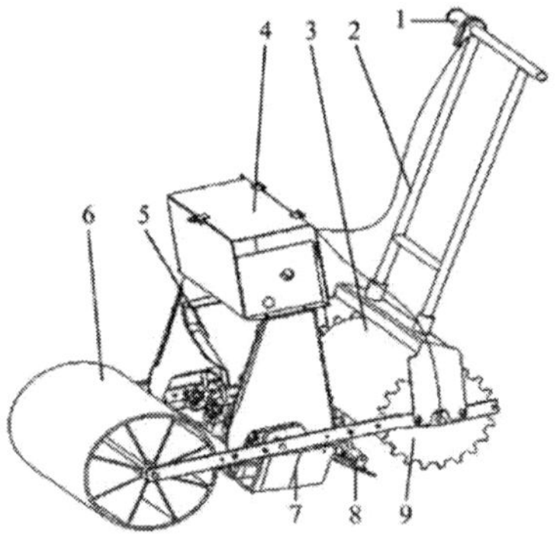

Fig. 5 Overall structure sketch and overview of prototype [19]

According to overall structure sketch, the seeder consists of: 1. Speed governing knob, 2. Rack, 3. Rear repression wheel, 4. Power supply box, 5. Seeding unit, 6. Front pressing wheel, 7. Ditching device, 8. Earthing device, 9. Built-in drive motor. The construction of the seeder allows the quick change of the sowing disc, as well as the easy adjustment of the distance between the rows and of the sowing depth, according to each crop's cultivation technology.

The monitoring system can perform control functions of the sowing operation during the operation of the equipment, of its starting or stopping. It can check the correct selection of the distributor disc by correlating it with the seed diameter. When the distributor disc works, the type of seeds, the sowing conditions, the number of seeds, etc. are displayed. The system can detect the lack of sowing, being equipped with an alarm function. The structure of the system is designed for easy use [19].

Field experiments were performed for coriander seeds, Chinese cabbage and radishes. Thus, using the distribution discs suitable for each type of seed, a sowing precision of over 95% was obtained for each species. At the working speeds of the seeder of 3 km/h and 4 km/h, the relative error of the monitoring precision of the system was less than 6% [19].

2.2 Research on the realization and experimentation of a vegetable seed distributor with electric drive

2.2.1 Material and Methode

INMA Bucharest designed and made an Experimental Model of an *Electrically operated vegetable seed distributor* (fig.6) consisting of the following main subassemblies: 1. seed box, 2. box lid, 3. assembled depression chamber, 4. gear motor, electrically operated, 5. supply chamber, 6. distribution disc. The distribution disc has a ring shape, having 2 rows of large and small holes, respectively, on the outer circumference, for bean and onion seeds.

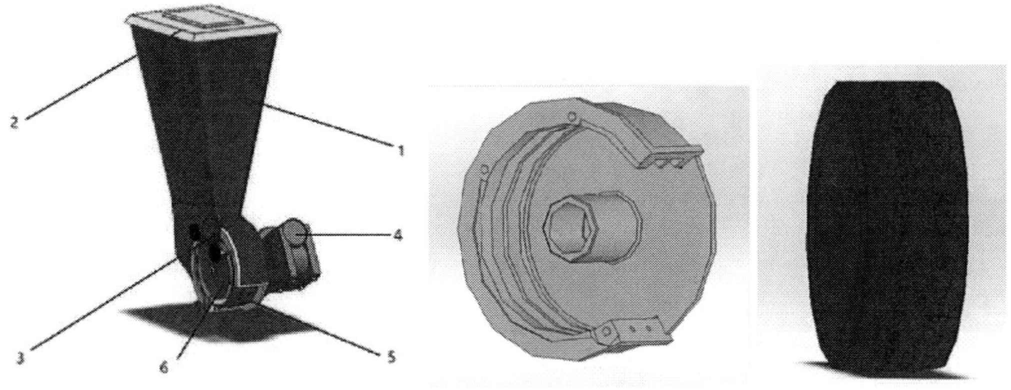

Fig. 6 Vegetable seed distributor, Supply chamber, Seed distribution disc 3D modeled with *SolidWorks* program

There are 32 small holes and 14 large holes in order to ensure a distance between seeds, in a row, of 30 mm for onion and 200 mm for bean, respectively, according to cultivation technologies. Depending on the type of seed, the distribution disc slides on the drive shaft bringing one of the rows of holes into the supply chamber. Both the supply chamber and the distribution disc were made of composite materials, through 3D printing technology.

For the experiments carried out in the laboratory, 2 species of vegetable seeds were used: for large seeds - garden bean *Doina* variety and for small seeds - onion *Aurie de Buzau* variety, (fig.7).

Fig. 7 Types of seeds used in experiments (left - onion, right - bean)

The seed distributor was tested in the laboratory, at INMA. For experimentation, the aim was to verify the distributor's mechanical principle of operation, emphasizing the determination of the distribution norm coefficient of variation, without pneumatic input. For this, the assembly consisting of the seed supply box and the seed distributor was mounted on a support, after which the scraper was adjusted according to the dimensions of the seeds. The seed supply box was loaded at a filling level considered so that, after the tests, the seed level is above the slots in the bottom of the box and on the seed distributor, more precisely in the supply chamber.

For a working speed of 3 km/h, the disc rotation frequency was 52 rpm for onions and 18 rpm for beans. A stabilized voltage source was used to power the electric motor, for which the value of the supply voltage of 18 Vdc for onion and 7 Vdc for bean was set.

Tests were performed in three repetitions, over a period of 60 s, for each of them and for each type of seed, respectively. The seeds were collected in the container and weighed after each test. The following indicators were determined: average mass, empirical standard deviation and variation coefficient of the distribution norm.

Average seed mass:

$$M_m = \frac{\sum Mi}{n}, \text{ [g]} \quad (1)$$

where: M_i is the mass obtained in the test "i";

n= number of repetitions.

Empirical standard deviation:

$$\sigma_N = \sqrt{\frac{\sum(M_i - M_m)^2}{n-1}} \quad (2)$$

Variation coefficient of the distribution norm:

$$C_{VN} = 100 \cdot \frac{\sigma_N}{M_m}, \text{ [\%]} \quad (3)$$

2.2.2 Results

The results obtained are presented in table 1.

Table 1. Experimental results.

Seed type	Adjustment parameters T, U, n	Repetition	Mass M [g]	Average mass M_m [g]	Empirical standard deviation	Variation coefficient of the distribution norm [%]
Onion	Time T=60 s Voltage U=18 Vdc Disc speed n=52 rpm	R1	28.50	27.87	0.79	2.84
		R2	26.98			
		R3	28.13			

Bean	Time T=60 s Voltage U=7 Vdc Disc speed n=18 rpm	R1	190.16	192.69	6.24	3.24
		R2	188.11			
		R3	199.80			

Aspects during the experiments performed within INMA are presented in figure 8.

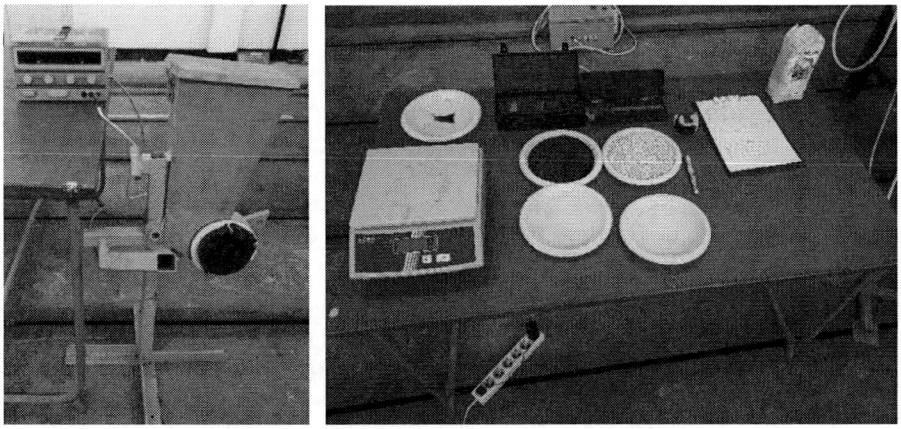

Fig. 8 Experimenting the electrically operated vegetable seed distributor

For INMA-type electrically operated vegetable seed distributor, intended for large and small seeds, at the recommended working speed was obtained a variation coefficient of the distribution norm of 2.84% for onions and 3.24% for beans, respectively, values that fall below the maximum value accepted by the specific norms (max. 5%).

3 Conclusions

The experiments showed that for INMA-type electrically operated vegetable seed distributor, the values obtained for the variation coefficient of the distribution norm were lower for the small seeds than those for larger seeds, however both values were found within the allowed limits.

Given that the distributor will operate exhaustively, it is expected to obtain lower variation coefficients of the norm, increasing the sowing precision.

By using electric motors, characterized by high reliability, to drive the seed distributors, the mechanical transmission elements that usually require additional adjustments and high maintenance are eliminated. Thus, the power supply of the seed distributors represents an efficient way to maintain the uniformity of the seed distribution and implicitly to achieve a precision sowing, according to the current trends.

"The work was supported by a grant of the Romanian Ministry of Research and Innovation, CCCDI - UEFISCDI project number PN-III-P1-1.2-PCCDI-2017-0659/ No.11 PCCDI/2018, within PNCDI III".

References

1. A. Nabi, S, Narayan, B. Afroza, F. Mushtaq, S. Mufti, H.M. Ummyiah, A. Malik, Journal of Pharmacognosy and Phytochemistry, 6, (6), 6, (2017)

2. F. J. Pierce, P. Nowak, Aspects of Precision Agriculture, Advances in Agronomy, 67, 86, (1999)
3. E. M. Brandelero, P. F. Adami, A. J. Modolo, M. M. Baesso, A.J. Fabian, Journal of Agronomy 14 (3), 7, (2015);
4. A. Balafoutis, B. Beck, S. Fountas, J. Vangeyte, T. van der Wal, I. Soto, M. Gómez-Barbero, A. Barnes, V. Eory, Sustainability 9, 1339; 28, (2017)
5. D. Cujbescu, Gh. Voicu, Gh. Bolintineanu, V. Vlăduţ, D. Manea, C. Persu, S. Bungescu *Comparative study regarding the sowing precision of precision sowing machines distribution devices*, Proceedings of The 43 International Symposium on Agricultural engineering "Actual Tasks on Agricultural Engineering", Opatija, Croatia, 24th - 27th February 2015, 12, (2015);
6. L. Trivelli, A. Apicella, F. Chiarello, R. Rana, G. Fantoni, A. Tarabella, British Food Journal, 121, 8, 14, (2019)
7. D. Cujbescu, V. Vlăduţ, Gh. Voicu, N. Ungureanu *Researches on obtaining higher qualitative indices for the sowing work*, 15th International Scientific Conference "Engineering for Rural Development", Proceedings, (15), Jelgava, Latvia, 25th – 27th May 2016, 6, (2016);
8. S. P. Singh, Mukesh K Singh, U. Ekka, M. K. Singh, Indian Journal of Agricultural Sciences 89 (12), 5, (2019)
9. D. Cujbescu, S.St. Biriş, Gh. Voicu, M. Matache, G. Paraschiv, V. Vlăduţ, M. Bularda, INMATEH Agricultural Engineering, 61, 2, 8, (2020)
10. V.P. Khambalkar, D.S.Karale, U.S. Kankal, Internat. J. Agric. Eng., 7 (1), 4, (2014)
11. N. Sharaby, A. Doroshenko, A. Butovchenko, A. Legkonogih, E3S Web of Conferences 135, 01080 (2019)
12. T. Kaur, D. Kumar, International Journal of Computer Science, Engineering and Applications (IJCSEA) 3, (3) 8, 2013
13. E.A. Alhassan, D.A. Adewumi, B. Okpodjah, International Journal of Mechanical Engineering and Technology (IJMET), 9 (10) 8, (2018)
14. T. V. Maxsudovich, M. G. Usmonjonovich, K. S. Rasuljonovich, R. A. Yakubjonovich IJSTR 9 (9), 7, (2020)
15. E. Pannaci, B. Lattanzi, F. Tei, J. Crop Protection 96, 15, (2017)
16. S. Korav, A. K. Dhaka, R. Singh, N. Premaradhya, G. Chandramohan Reddy, Journal of Pharmacognosy and Phytochemistry 7 (4), 6, (2018)
17. S. A. Fennimore, D. C. Slaughter, M. C. Siemens, R. G. Leon, M. N. Saber, Weed Technology 30 (4), 15, (2016)
18. B. Melander, B. Lattanzi, E. Pannacci, Crop Protection 72, 8, (2015)
19. X. Jin, Q. W. Li, K. X. Zhao, B. Zhao, Z.T. He, Z. M. Qiu, Int J Agric & Biol Eng, 12 (2), 7, (2019)
20. .Z. Gobor, H. Heuberger, R. Rinder, *Precision plot seeder for medicinal and aromatic plants – concept, development and optimisation*, Hungarian Agricultural Engineering (HAG) 27, 5, (2015)
21. C. Persu, D. Cujbescu, E. Marin, G. Gheorghe, C. Vlad, F. Burnichi, O. Kabas, *Seed meters with electric drive at precision seeders*, Proceeding International Symposium, ISBINMATEH 2019, 4, (2019)
22. H.W. Griepentrog, P.T. Skou, J.F. Soriano, B.S. Blackmore, *Design of a seeder to achieve highly uniform sowing patterns*, 5th European Conference on Precision Agriculture, Uppsala, Sweden, 9th – 12th June 2005, 8, (2005);
23. Keith W. Wendte, *Direct drive electric seed metering system*, Patent no. U.S. 7,617,785 B2/ 17 Nov. 2009.

Advanced method of managing soil conservation works in Smart Farms

Eugen Marin[1,*], *Marinela* Mateescu[1], and *Carmen* Bălțatu[1]

[1]National Institute of Research-Development for Machines and Installations designed to Agriculture and Food Industry-INMA Bucharest 013813, Romania

Abstract. The paper presents the results obtained in the SMART farm by using an advanced method of managing soil conservation works. These works involve, first of all, the collection and storage of data on the spot regarding the following parameters in which the plants will grow: disease climate (air temperature, dew point), growth climate (air temperature, solar radiation, deficiency vapours pressure, relative air humidity, precipitation, wind speed, maximum wind speed, daily evaporation) and soil monitoring (soil moisture, precipitation, soil salinity, soil temperature). Data collection is done through intelligent sensors from a wireless weather station and the reception of this information in real-time on a computer/smartphone by the farmer. Therefore, the farmer will be able to make instant decisions on soil conservation work, thusly saving time and workforce for additional on farm inspections.

1 Introduction

International agricultural research could make important contributions to increasing production and making the most rational use of existing resources [1].

The declining availability of agricultural land, ecological losses and environmental degradation are leading to the adoption and diffusion of digital agricultural technologies that contribute to the transformation of current agricultural systems towards sustainability [2], [3], [4], [5].

Farmers are increasingly determined to implement smart solutions where agricultural technologies are digitally monitored and each agricultural work is managed with the help of IT systems, applications specially developed for agricultural activity [6].

Smart farming involves, among other things, real-time management of meteorological information on a farm to maximize agricultural production. Wireless weather stations are important tools for monitoring the soil (soil moisture, precipitation, soil salinity, soil temperature) within a farm, as many of the problems related to agricultural activity can be solved by a better understanding of the surrounding weather conditions [7].

The iMETOS 3.3 station is designed for all types of environmental conditions, has non-volatile memory, communicates and transmits real-time weather information via GSM [8].

* Corresponding author: marin_43eu@yahoo.com

2 Material and Methodology

An experimental model of an innovative soil conservation system for SMART farm was developed within INMA Bucharest. This system uses an advanced management method that involves collecting and storing on-site data on soil monitoring (soil moisture, precipitation, soil salinity, soil temperature) via an iMETOS 3.3 wireless weather station. The weather station is equipped with high quality sensors, a datalogger and a GPRS system for uploading data regularly on the FieldClimate by Pessl Instruments platform. The agrometeorological information transmitted on the internet platform and communications of Pessl Instruments were downloaded, interpreted and analysed from any place with internet access at any time, in real-time on desktop or mobile phone, by researchers for instant decision making, saving time and manpower for the necessary field checks [7].

Aspects of the configuration of the iMETOS 3.3 wireless weather station, located in the experimental field of INMA Bucharest and viewed on the FieldClimate by Pessl Instruments platform on the desktop or on the mobile phone are presented in figure 1.

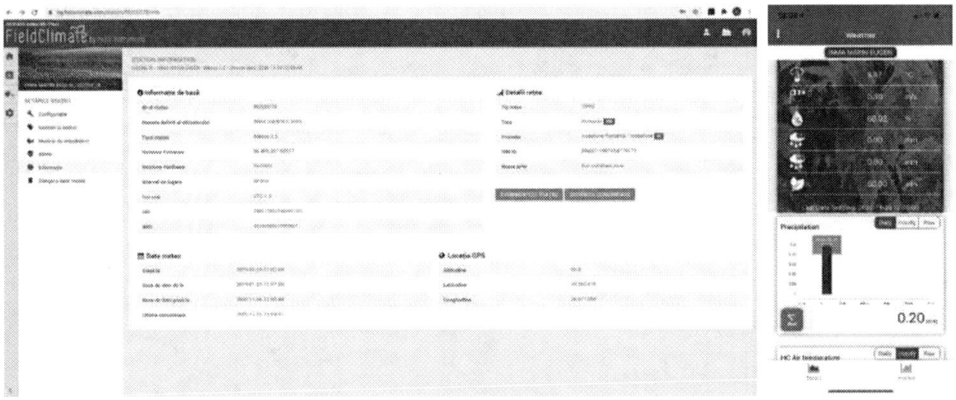

Fig. 1. Aspects of the configuration of the iMETOS 3.3 wireless weather station viewed on the FieldClimate by Pessl Instruments platform on the desktop or on the mobile phone

The access to the weather station data was made from the platform www.fieldclimate.com, from where it was possible to configure and SMS alerts for different parameters measured by the iMETOS 3.3. The FieldClimate platform is designed by Pessl Instruments for collecting, analysing, displaying agronomic, meteorological, soil, insect and tracking data (Fig. 2) by integrating field climatic data that are collected from all Metos-type climate stations [9].

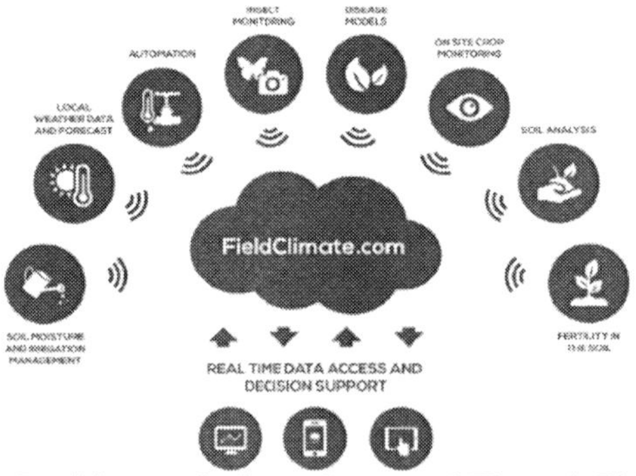

Fig. 2. Holistic Solutions for Decision Agriculture, Supported Through FieldClimate

Using the remote data exchange (field office) the researchers were able to inform the operator (mechanizer) when to prepare the tillage equipment in the arable substrate adapted to the conservative technology.

An Agrotron X720 tractor was used to carry out experimental research in field conditions, in order to determine the qualitative work and energy indices of the conservation tillage technology. The tractor has the following main technical characteristics: overall dimensions (length × width × height), mm: 5399 × 2749 × 3270; tractor mass, kg: 9995; wheelbase, mm: 3095; front / rear axle track, mm: 2104/2014; engine power: 198 kW (269 hp) [11].

The application of the conservative tillage technology was made with the tillage equipment in the arable substrate adapted to the conservative technology (fig. 3), which is of the type carried on the three-point suspension mechanism mounted on the back of the 220- wheel tractors. 240 CP, category 3, SR ISO 730: 2012 [12].

Fig. 3. Tillage equipment in arable substrate adapted to conservative technology

The main technical characteristics of tillage equipment in the arable substrate are presented in Table 1.

Table 1. The main technical characteristics of tillage equipment in the arable substrate

Characteristic	**U.M**	**Value**
Operation source	kW (CP)	161…176 (220…240)
Number of loosening organs	pieces	5
Maximum working depth	cm	60
Working width	m	2,3
Mass	kg	1800

The tests were performed on the experimental fields of INMA Bucharest in accordance with a specific test procedure. The additional equipment for determining the soil characteristics in the experimental field were the following:
- Electronic penetrometer with digital cone FIELDSCOUT SC 900;
- Portable soil humidometer type HH2 + ThetaProbe ML2x precision sensor.

During work, to avoid overlaps and replace the landmark, the operator (mechanic) used a guidance system (fig. 4), which consists of a MATRIX® 570GS console, a 26 dB RXA-30 antenna and a camera RealView with night vision.

Fig. 4. MATRIX® 570GS + RXA-30 + RealView manual guidance system [13].

One of the main indicators used to determine a wide variety of physical and mechanical properties of the soil is the resistance to penetration. Measurements of penetration resistance were performed with the digital penetrometer, working in 3 repetitions for the experimental group.

Figure 5 shows graphically the average values of the distribution of resistance forces to the penetration of the cone in the soil layers, in kPa, determined in 12 test points up to a maximum depth of 30 cm.

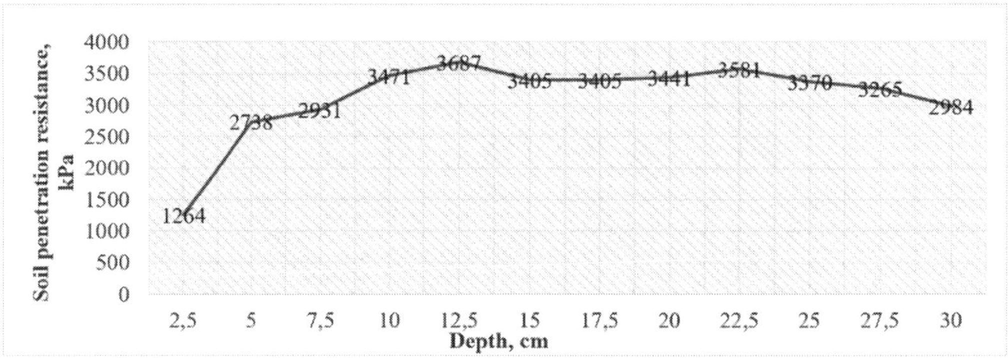

Fig. 5. Graphical representation of the average values of the distribution of resistance forces to the penetration of the cone in the soil layers of the experimental plot

On the experimental group, the water content expressed as a percentage of the soil volume was measured in 3 repetitions using the HH2 portable soil humidometer + ThetaProbe ML2x precision sensor.

Figure 6 shows graphically the average values of soil moisture determined as a percentage of soil volume.

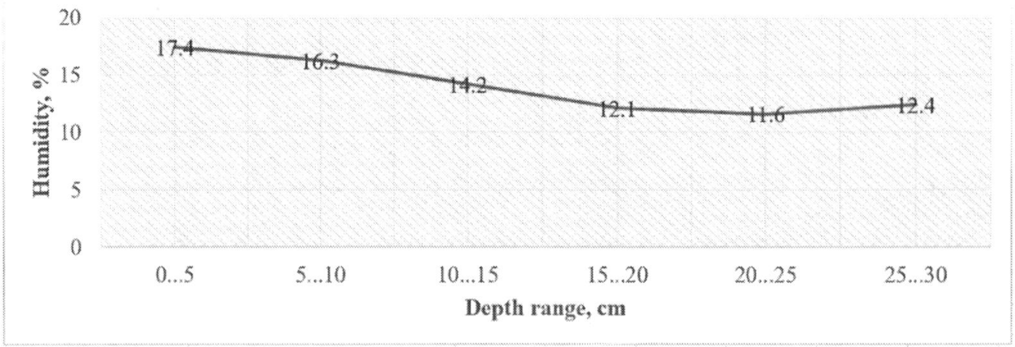

Fig. 6. Graphical representation of the average values of soil moisture on the experimental group

Given that it has led to a shortage of water in the soil, the conservation of still existing water is a very important thing. Conservative tillage leads to soil consolidation in order to keep the water even at shallow depths.

The iMetos 3.3 weather station allowed to monitor in real time the soil characteristics, by transmitting data on computer / smartphone on: soil moisture [%], precipitation [mm], soil salinity [VIC], soil temperature 1 [°C].

Figure 7 shows graphically the soil monitoring (soil moisture, precipitation, soil salinity, soil temperature) on the experimental group INMA Bucharest.

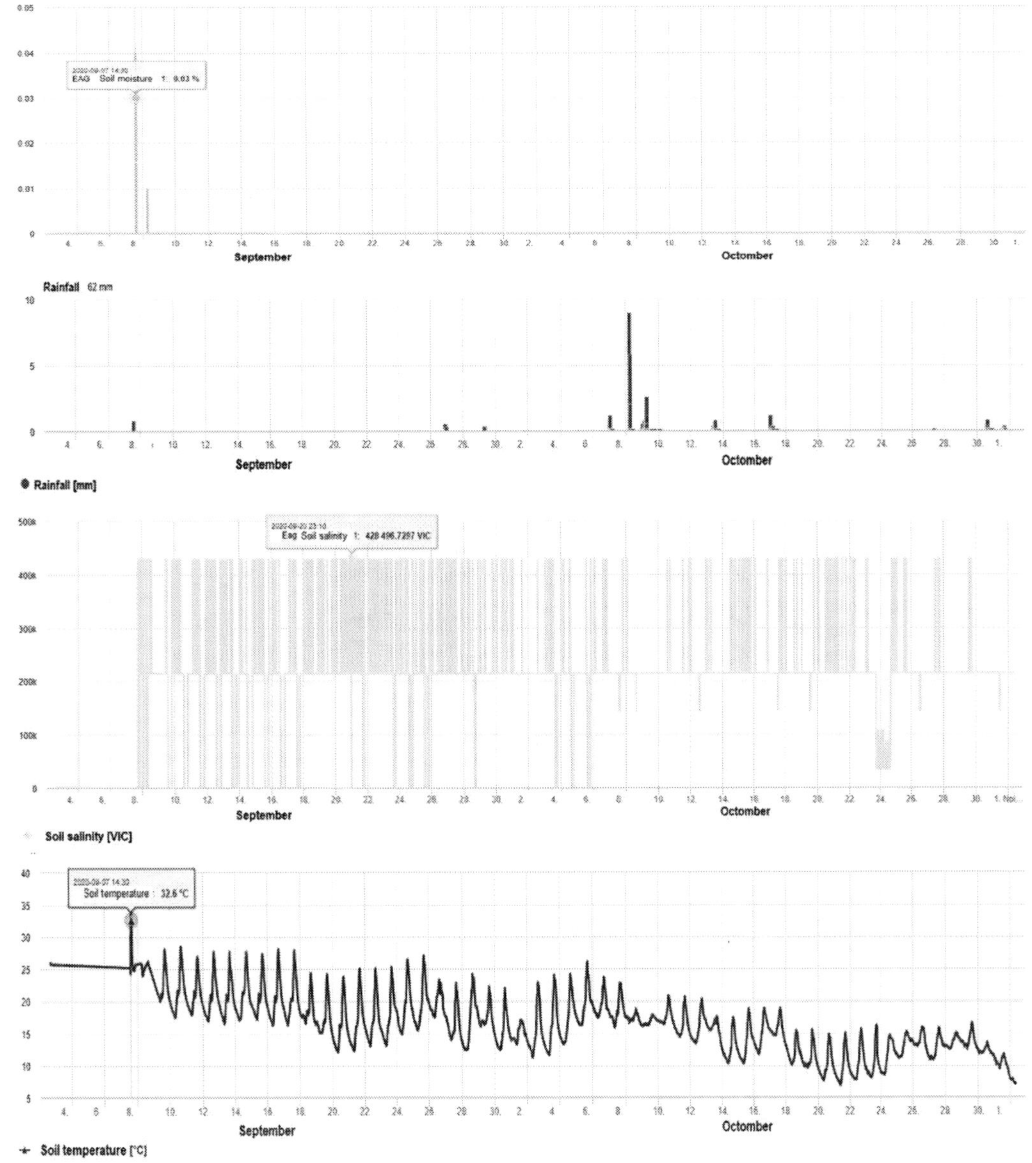

Fig. 7. Diagram regarding soil monitoring on the INMA Bucharest experimental lot

The following quality indicators were determined with the technical equipment in the Agrotron X 720 197 kW (269 hp) tractor unit:

- the variation index of working depth, V_a;
- the variation index of working width, V_b;
- loosening degree, G_{as}.

Equations (4), (5), (6), (7), (8), (9) and (10) presented in [14] were used for the calculations:
- *Loosening degree* was calculated according to the following equation:

$$G_{as} = \frac{\sum_{1=1}^{n} \frac{h_1 - h_2}{h1}}{n}, [\%] \tag{4}$$

where: h_1 is the ordinate at a certain point on the ruler at the soil surface before the passage of technical equipment [cm];
h_2 – the ordinate at the same point on the ruler at the soil surface after the passage of technical equipment [cm];
n – number of measurements.
- *Average working depth* was determined by measuring the distance between soil surface and the bottom of the furrow left by the active body and was calculated according to the following equation:

$$a_m = \frac{\sum_{1=1}^{n} a_i}{n}, [\text{cm}] \tag{5}$$

where a_i is the measured working depth, cm and n – number of measurements.

The standard deviation of working depth was calculated according to the following equation:

$$\sigma = \pm\sqrt{\frac{\sum_{1=1}^{n}(a_i - a_m)^2}{n-1}}, [cm] \tag{6}$$

The variation index of working depth was calculated according to the following equation:

$$V_a = \frac{\sigma_a}{a_m} \times 100, [\%] \tag{7}$$

Average working width was calculated according to the following equation:

$$B_m = \frac{\sum_{i=1}^{n} B_i}{n}, [\text{m}] \tag{8}$$

where: B_i is the measured working width, m and n – number of measurements.

The standard deviation of working width was calculated according to the following equation:

$$\sigma = \pm\sqrt{\frac{\sum_{1=1}^{n}(B_i - B_m)^2}{n-1}}, [cm] \tag{9}$$

The variation index of working width was calculated according to the following equation:

$$V_B = \frac{\sigma_B}{B_m} \times 100, [\%] \tag{10}$$

Table 2 shows the average values of the qualitative working indices determined following the tests carried out with the tractor unit and the tillage equipment in the arable substrate adapted to the conservative technology.

Table 2. Qualitative work indices

Qualitative work indices	U.M	Value
The variation index of working depth, V_a	%	1,68
The variation index of working width, V_b	%	1,44
Loosening degree, G_{as}	%	18,1

Figure 8 shows graphically the variation of the energy indices determined for three speed stages of the unit with the Agrotron X 720 tractor.

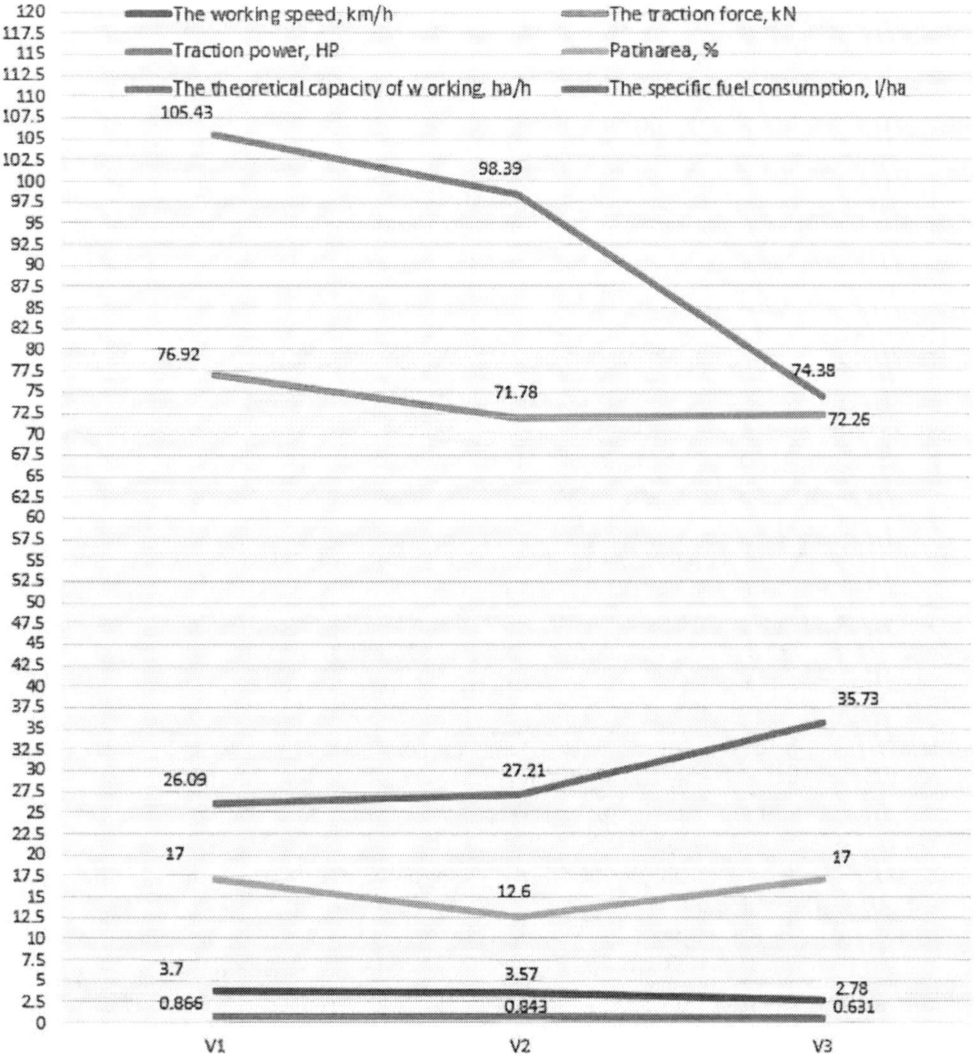

Fig. 8 Graphical representation of the variation of indexes determined for three speed stages

After the experimental tests performed with the technical equipment for soil conservation work, the following obtained results can be presented:
- in terms of operation, the equipment has obtained qualitative work indices that are in accordance with the agro-technical requirements provided for working the soil in the arable substrate, in accordance with the standard in force. This was proved in the variation index of working depth - 1.68%, the variation index of working width - 1.44% and loosening degree - 18.1%;
- at the tests performed the equipment had a good performance, achieving an average working capacity of 0.778 ha / h at the real working time in the average working speed of 3.35 km / h.

3 Conclusions

- experimental research has allowed the validation of technical and technological solutions adopted in the design and implementation of the experimental model of innovative system of conservative tillage for SMART farm;
- the results of experimental research allow useful recommendations for farmers who apply innovative system of conservative tillage in order to increase efficiency and ensure efficient soil management for sustainable production.

This work was supported by a grant of the Romanian Ministry of Education and Research, through Programme NUCLEU, ctr. nr. 5N / 07.02.2019, project PN 19 10 02 01-Development of innovative technologies inside of Smart Farms.

References

1. W. Martin, A research agenda for international agricultural trade. Applied Economic Perspectives and Policy, 40(1), 155-173, (2018)
2. L. Shang, T. Heckelei, M.K. Gerullis, J. Börner, S. Rasch, Adoption and diffusion of digital farming technologies - integrating farm-level evidence and system interaction, Agricultural Systems, Volume 190, (2021)
3. N. Ungureanu, V. Vlăduţ, Biriş, FEM modelling soil behaviour under compressive loads, the International Conference of Applied Sciences (ICAS2016), Materials science and engineering, Vol 163 (2017), 012001, pg. 1-9 (2017);
4. L.Vlăduţoiu, P.Cârdei, V.Vlăduţ, L. Fechete, Modern trends in designing and selecting the machine/equipment for deep soil tillage, 16th International Scientific Conference Engineering for Rural Development", pg. 1415-1420, Jelgava, Letonia (2017);
5. N. Ungureanu, V. Vlăduţ, S.-Ş. Biriş, G. Paraschiv, M. Dincă, B. Ş.Zăbavă, V. Ştefan, NE Gheorghiţă, FEM modelling of machinery induced compaction for the sustainable use of agricultural sandy soils, Proceeding of the 46th International Symposium „Actual Tasks on Agricultural Engineering", 2018, pg. 201-211 (2018)
6. R. Finger, S.M. Swinton, N. El Benni, A. Walter, Precision farming at the nexus of agricultural production and the environment, Ann. Rev. Resour. Econ., 11 (1) pp. 313-335 (2019)
7. Tenzin, S. Siyang, T. Pobkrut and T. Kerdcharoen, Low cost weather station for climate-smart agriculture, 9th International Conference on Knowledge and Smart Technology (KST), Chonburi, Thailand, 2017, pp. 172-177, (2017)
8. D. Strigaro, M. Cannata, M. Antonovic, Boosting a Weather Monitoring System in Low Income Economies Using Open and Non-Conventional Systems: Data Quality Analysis. Sensor,s 19(5):1185, (2019)
9. V. Neagoe, M. Neagoe, E. Marin, I. Voicea, Contributions to the improvement of technical equipment for soil substratum tillage, adapted to conservation technology, International Symposium, ISB-INMA-TEH, Agricultural and Mechanical Engineering, 605-608, INMA Bucharest (2020)
10. iMetos 3.3 weather station and Field Climate platform – Website https://metos.at/imetos33/; https://metos.at/fieldclimate/; Austria.
11. Tractor Agrotron X 720, website https://www.deutz-fahr.com/en-eu/products/tractors/6404-agrotron-x;
12. SR ISO 730:2012, – website https://magazin.asro.ro/ro/standard/200754
13. Agricultural GPS Matrix pro 570GS for tractor, website https://agri-inovatie.ro/ghidaj-gps-agricol-tractoare/278-matrix-pro-570gs.html
14. Şt. Croitoru, E. Marin, M. Bădescu, V. Vlăduţ, Agrotechnical and energetic characteristics of new designed subsoiler, in Proc. 43rd International Symposium on Agricultural Engineering. Actual Tasks on Agricultural Engineering, pp. 165-176, Opatija, Croatia (2015)

Photodegradable organic pollutants from wastewater

Oana Stoian[1], *Cristina Ileana* Covaliu[1,*], *Gigel* Paraschiv[1], *Mihai* Nita-Lazar[2], and *Ioana- Corina* Moga[3]

[1]Politehnica University of Bucharest, 313 Splaiul Independentei, 060042, Bucharest, Romania
[2]Research and Development Institute for Industrial Ecology – ECOIND, 71-73 Drumul Podu Dambovitei Str., 060652, Bucharest, Romania
[3] Dfr Systems SRL, Drumul Taberei 46, Bucharest, Romania

Abstract. Nowadays, environmental pollution is constantly increasing, and depollution is the most important challenge of the 21st century. This paper is a review of organic pollutants in wastewater that can be removed by the photocatalysis process. Also are presented specialized studies that demonstrate the removal of organic pollutants from wastewater and the experimental conditions necessary to obtain high treatment efficiencies. The important classes of pollutants in wastewater are hydrocarbons, halo compounds, carbonyl derivatives, alcohols, phenols, pesticides, surfactants, nitrogen containing compounds and dyes, and represent a major source of water pollution. These organic pollutants are photodegradable and can be easily removed by the photocatalysis process.

1 Introduction

Toxic organic pollutants from domestic use or from the activity of various industries are a major environmental problem through water pollution, and their remediation is a considerable concern [1]. A large number of organic pollutants such as dyes, pesticides or pharmaceuticals are released daily into many wastewaters, which are then discharged into natural waters and accumulate in the aquatic environment [2-6]. Pollution comes from domestic, industrial and agricultural activities and cannot be stopped by conventional water treatment methods. Thus, synthetic organic substances have been detected in small quantities in lakes, rivers, oceans or drinking water. In order to remove organic pollutants from wastewater, safe, efficient, economical and simple technologies are developed to avoid toxic consequences on human health [7]. Organic pollutants removed / degraded from wastewater by the photocatalysis process are described in the following subchapters.

Factors that may influence the photocatalysis process used to degrade wastewater pollutants are pollutant nature, concentration and the existence of other compounds in wastewater, the amount of photocatalysts added, their size and structure, water pH and temperature, light intensity and irradiation time [8]. Also, the reactor design, the number of UV lamps and their characteristics influence the photocatalysis process. High concentration

* Corresponding author: cristina_covaliu@yahoo.com

of pollutants in wastewater can saturate the surface of the photocatalyst and thus, at some point, can reduce photonic efficiency or deactivate the photocatalyst [9]. As mentioned above, the photocatalysis process also depends on the nature of the pollutants. If the pollutants are compatible with the photocatalyst used to adhere efficiently to its surface, the process will be more efficient and faster in order to remove these pollutants from the wastewater [8]. The degradation of the pollutant increases with the increase of the amount of catalyst added in the wastewater. This phenomenon is explained by the fact that the addition of a larger amount of photocatalyst leads to an increase in the number of active sites that are needed to remove pollutants from wastewater [10]. The surface morphology is a major factor that influences the photocatalysis process due to the existence of a direct relationship between the pollutant and the photocatalyst surface [11]. Changing the pH of the wastewater will change the potential of photocatalytic reactions and will influence the reaction rate. Depending on the acidic or alkaline aqueous medium, the surface of the photocatalyst can be protonated or deprotonated [12]. The increase of the reaction temperature will lead to an increased photocatalytic activity. However, a temperature higher than 80°C disadvantages the removal of pollutants using photocatalysts due to the recombination of charge carriers. The temperature range considered adequate in the photocatalysis process is 20-80°C [13].

1.1 Hydrocarbons

Water pollution with hydrocarbons has negative effects on all life forms, forming a toxic film at the water surface, thus reducing the power of the sun to penetrate. In this case, the aquatic plants photosynthetic activity will decrease. The main sources of water pollution with hydrocarbons are represented by the oil platforms as well as the oil transport ships, by accidental pollution or by the discharge into the sea of the washing water of the ship tanks. By forming water-impermeable film due to crude oil and petroleum products, the exchange of oxygen between the environment and water can no longer be achieved, causing damage to animals, plants and humans [14].

1.2 Halo compounds

Halo-organic compounds are pollutants that are used as dyes, agrochemicals, solvents, drugs, etc. These compounds are consumed by humans and animals by accumulating in aquatic plants and animals [15].

In order to transform halo-organics into other non-polluting compounds, treatment methods such as incineration, extraction, electrochemical treatment, chemical degradation, sonochemical destruction and photochemical process are applied. The latter was found to be the most useful method used in the degradation process of halo compounds in wastewater [16-18].

1.3 Carbonyl derivatives

Carbonyl compounds are recognized as carboxylic acids, ketones, aldehydes and quinones [15]. Table 1 shows carbonyl derivatives organic pollutants removed from wastewater by photocatalysis using photocatalysts containing TiO_2. Also, the experimental conditions and the treatment efficiencies are presented in this table.

Table 1. Examples of carbonyl derivatives removal from wastewater, conditions and efficiencies

Photocatalytic material	*Pollutant*	*Initial concentration [mg/L]*	*Time [min]*	*Efficiency [%]*	*Ref.*
Pt- TiO_2	Oxalic acid	1,98	60	100,00%	[19]
TiO_2	Salicylic acid	-	90	17,00%	[20]
Polyacrylanitrile – TiO_2 + H_2O_2	Formaldehyde	90,00	250	≈86,00%	[21]

1.4 Alcohols

The alcohol industry is considered a major source of environmental pollution. After proper treatment of these waters polluted with alcohol, they can be used for fertilized irrigation [22]. Table 2 shows alcohols as organic pollutants in wastewater that are removed by the photocatalysis process using photocatalysts containing TiO_2. Also, the initial pollutant concentrations, the required treatment time and the efficiencies are presented in this table.

Table 2. Removal of alcohols from wastewater, conditions and efficiencies

Photocatalytic material	Pollutant	Initial concentration [mg/L]	Time [min]	Efficiency [%]	Ref.
TiO_2 Degussa 25	n-butanol	580,00	30	99,00%	[23]
Photocatalytic media containing TiO_2	Isopropyl alcohol	1,00	120	97,20%	[24]
	Ethanol			99,00%	

1.5 Phenols

Phenolic compounds from wastewater come from natural, domestic, industrial and agricultural activities. Following the decomposition or degradation of natural organic matter in water, the disposal of household and industrial waste in water or runoff from agricultural land, phenolic compounds may occur in wastewater [25]. They are aromatic compounds, having attached on the aromatic ring one or more hydroxyl groups. The discharge of toxic phenolic compounds into lakes, rivers or the sea comes from the oil industry, the manufacture of phenolic resins, plastics, wood products, textiles, pharmaceuticals, detergents, cellulose, pesticide synthesis and coking operations [26, 27]. Also, in aquatic environments can be generated phenolic pollutants, from the secretion of algae, flavonoids, hydrolysable tannins and lignin transformation [28]. Due to their corrosive nature, phenols can cause irritation to the eyes, skin and airways if they are absorbed by the lungs [29]. Table 3 shows the removal of phenols from wastewater by using photocatalysts containing TiO_2.

Table 3. Examples of phenols removal from wastewater

Photocatalytic material	*Pollutant*	*Initial concentration [mg/L]*	*Time [min]*	*Efficiency [%]*	*Ref.*
TiO_2	Phenol	30,00	300	50,00%	[30]
TiO_2 supported on ZSM-5	Phenol	90,00	30	>90,00%	[31]
TiO_2 P25 Degussa	3 - aminophenol	10,91	160	52,80%	[32]

1.6 Pesticides

Pesticides are an important class of water pollutants. Pesticides such as beta-hexachlorocyclohexane organochemicals, organophosphorus (representing a group of insecticides) cypermethrin, dicofol, etc. are commonly used [15].

The efficient methods used to degrade pesticides are mechanical-chemical destruction, precipitation, microbial degradation, thermal degradation, adsorption on activated carbon and photocatalytic degradation. Due to the long contact time required and the high cost, some methods are not so suitable for pesticide degradation. An easy-to-operate, fast and low-cost process is required. Specialized studies show that the technology successfully used to remove or degrade pesticides is photocatalytic oxidation [15]. Table 4 shows that pesticides are efficiently removed from wastewater through the photocatalysis process by using titanium dioxide as a catalyst.

Table 4. Degradation of pesticides from wastewater

<table>
<tr><th>Photocatalytic material</th><th>Pollutant</th><th>Initial concentration [mg/L]</th><th>Time [min]</th><th>Efficiency [%]</th><th>Ref.</th></tr>
<tr><td>TiO_2 Degussa P-25</td><td>Pirimiphos – methyl</td><td>9,70</td><td>40</td><td>100,00%</td><td>[33]</td></tr>
<tr><td rowspan="3">Anatase TiO_2</td><td>Lindane</td><td rowspan="4">40,00</td><td rowspan="2">150</td><td>77,00%</td><td rowspan="4">[34]</td></tr>
<tr><td>Methoxychlor</td><td>>99,00%</td></tr>
<tr><td rowspan="2">p, p´ - DDT</td><td>120</td><td>95,00%</td></tr>
<tr><td>Rutile TiO_2</td><td>150</td><td>85,00%</td></tr>
</table>

DDT - dichlorodiphenyltrichloroethane

1.7 Surfactants

Surfactants are composed of a water-soluble hydrophilic group and an oil-soluble hydrophobic group, two groups having the opposite nature. Surfactants are used in the fields of research, technology, cosmetics, pharmacy, textiles, biotechnology or agriculture on a large scale due to the fact that they possess favourable physico-chemical properties.

Due to the foaming tendency and the low oxygenation potential, surfactants pose a great threat to the aquatic environment. Types of surfactants can be ionic and nonionic. Ionic surfactants are cationic, anionic and amphoteric surfactants [15]. Table 5 shows the water treatment efficiency with surfactant content through the photocatalysis process.

Table 5. Surfactants degradation from wastewater

Photocatalytic material	*Pollutant*	*Initial concentration [mg/L]*	*Time [min]*	*Efficiency [%]*	*Ref.*
TiO_2	SLS, SDDBS	50,00	60	80,00%	[35]
	DBS	20,00	120	>80,00%	[36]
	SDS			>90,00%	
$TiO_2 - Cu_2O$	DBS	100,00	-	97,30%	[37]

SLS – sodium lauryl sulphate,
SDDBS – sodium dodecylbenzenesulphonate,
SDS – sodium dodecyl sulphate,
DBS – dodecylbenzenesulfonate.

1.8 Nitrogen containing compounds

Nitrogen-containing compounds include proteins, amino acids, pesticides, herbicides, drugs, dyes and more. Due to their toxicity and stability in the environment, they can pose a serious threat [15]. Nitrobenzene, as an example of a nitrogen containing compound, is presented in the Table 6 as an organic pollutant that can be degraded by photocatalysis, obtaining high water treatment efficiencies.

Table 6. Nitrobenzene degradation from wastewater

Photocatalytic material	*Pollutant*	*Initial concentration [mg/L]*	*Time [min]*	*Efficiency [%]*	*Ref.*
Nanocrystalline TiO_2	Nitrobenzene	50,00	180	100,00	[38]
P-25 Degussa TiO_2				100,00	
TiO_2 nanotubes		40,00	240	100,00	[39]
TiO_2 powder				85,00	

1.9 Dyes

Dyes represent the class of pollutants that is visually observable due to the coloration of the water. Because of this, but also because of their toxicity, dyes have received a lot of attention [40]. 10,000 dyes are commercially available and are used in the production of textiles, paints, plastics, paper, cosmetics, rubber or tanneries [41]. Dyes are a major danger to living organisms in water and negative impacts on humans are represented by symptoms such as nausea, abdominal discomfort, vomiting and irritation [40]. Table 7 shows the removal of dyes from wastewater by using photocatalysts containing TiO_2. Also, the initial pollutant concentrations, the required treatment time and the efficiencies are presented in this table.

Table 7. Dyes degradation from wastewater

Photocatalytic material	*Pollutant*	*Initial concentration [mg/L]*	*Time [min]*	*Efficiency [%]*	*Ref.*
Nano – $SrTiO_3$	Direct Green	20,00	180	≈84,00	[42]
	Reactive Orange			≈88,00	
CA - $CNT/TiO_2 - NH_2$	Methylene blue	10	300	100,00	[43]
		30		80,00	
		50		70,00	
CA - $CNT/TiO_2 - NH_2$	Indigo carmin	10	180	100,00	[43]
		30		90,00	
		50		80,00	
Fe_2O_3/TiO_2	Methyl orange	30	210	61,50	[44]
	Congo red			46,80	

CA – cellulose acetate

2 Conclusions

The heterogeneous photocatalysis process is found in Advanced Oxidation Processes (AOP) and has great potential for degradation of wastewater pollutants. In this review were presented the organic pollutants that can be degraded by the photocatalysis process. Applying this process were obtained high treatment efficiencies for the removal / degradation of organic pollutants from wastewater. From all the catalysts titanium dioxide plays an important role. The application of the photocatalysis process has advantages such as the degradation of organic pollutants or their transformation into non-toxic chemicals by using TiO_2-based catalysts. Using this process does not generate other pollutants or the necessary energy resource comes largely from natural solar energy which is unlimited.

The photocatalysis process could have some limitations represented by the possibility of not reaching a maximum treatment efficiency, the difficulty of distributing the photocatalyst particles uniformly, the difficulty of recovering nanoparticles from the system, agglomeration of nanoparticles or inefficient use of visible light. However, these limitations can be remedied by improving the process or photocatalysts used.

This work was supported by a grant of the Romanian Ministry of Education and Research, CCDI - UEFISCDI, project number PN-III-P2-2.1-PTE-2019-0628, within PNCDI III.

References

1. J. Zhao, C. Chen, W. Ma, Catalysis, **35**, 3–4 (2005)
2. J. P. Bound, N. Voulvoulis, Chemosphere **56**, 1143 (2004)
3. S. K. Khetan, T. J. Collins, Chem. Rev. **107**, 2319 (2007)
4. K. Kümmerer, Chemosphere **75**, 417 (2009)
5. C. A. Martínez-Huitle, E. Brillas, Appl. Catal., **87**, 105 (2009)
6. Y. Luo, L. Xu, M. Rysz, Y. Wang, H. Zhang, P. J. J. Alvarez, Environ. Sci. Technol. **45**, 1827 (2011)
7. E. Brillas, J. Braz, Chem. Soc., **25**, 3, 393-417 (2014)
8. A. Kumar, G. Pandey, Material Sci. & Eng. Int. J., **1**, 3, 106-114 (2017)
9. A. Sobczynski, A. Dobosz, Pol. J. Environ. Stud., **10**, 4, 195–205 (2001)
10. H. M. Coleman, V. Vimonses, G. Leslie, et al., J. Hazard. Mater., **146**, 3, 496–501 (2007)
11. C. Kormann, D. W. Bahnemann, M. R. Hoffman, Environ. Sci. Technol., **22**, 7, 798–806 (1988)
12. R. J. Davis, J. L. Gainer, G. O. Neal, et al. Water Environ. Res., **66**, 1, 50–53 (1994)
13. G. Mamba, M. A. Mamo, X. Y. Mbianda, et al., Ind. Eng. Chem. Res., **53**, 37, 14329–14338 (2014)
14. M. Srivastava, A. Srivastava, A. Yadav, V. Rawat, *Hydrocarbon Pollution and Its Effect on the Environment* (2019)
15. R. Ameta, S. Benjamin, A. Ameta, S. C. Ameta, Mat. Sci. Forum, **734**, 247-272 (2013)
16. D. F. Ollis, Environ. Sci. Technol., **19**, 480-484 (1985)
17. D. Cesareo, D. A. Di, S. Marchini, L. Passerini, M. L. Tosato, Homo.-Hetero. Photocatal., **174**, 593-627 (1986)
18. D. F. Ollis, E. Pelizzetti, N. Serpone, Photocatalyzed destruction of water contaminants, Environ. Sci. Technol., **25**, 1522-1529 (1991)
19. Y. Li, G. Lu, S. Li, Applied Catalysis A: General, **214**, 179–185 (2001)
20. V. Sukharev, A. Wold, Y-M. Gao, K. Dwight, J. Solid State Chem., **119**, 339-343 (1995)
21. R. Bonora, C. Boaretti, L. Campea, M. Roso, A. Martucci, M. Modesti, A. Lorenzetti, Nanomat., **10**, 148 (2020)
22. P. Chowdhary, R. N. Bharagava, *Emerging and Eco-Friendly Approaches for Waste Management* (2018)
23. J. Kirchnerova, M.-L. Herrera Cohen, C. Guy, D. Klvana, Appl. Cat. A: General, **282**, 321–332 (2005)
24. J. Chung, S. Chung, G. Lee, Y-W. Lee, Int. J. Environ. Res. Public Health, **17**, 4203 (2020)
25. W. Anku, M. Mamo, P. Govender, *Phenolic Compounds in Water: Sources, Reactivity, Toxicity and Treatment Methods* (2017)
26. S. Das, N. Kuppanan, V.A. Channashettar, B. Lal,. Springer, 165-177 (2018)

27. L.G.C. Villegas, N. Mashhadi, M. Chen, D. Mukherjee, K.E. Taylor, N. Biswas, J. Current Poll. Rep., **2**, 3, 157-167 (2016)
28. N.F. Zainudin, A.Z. Abdullah, A.R. Mohamed, J. Hazard Mater., **174**, 1-3, 299-306 (2010)
29. R. Rathna, E. Nakkeeran, Springer, 253-276 (2018)
30. A. Nickheslat, M. M. Amin, H. Izanloo, A. Fatehizadeh, S. M. Mousavi, J. Environ. Public Health, **2013** (2013)
31. J. Chen, L. Eberlein, C. H. Langford, J. Photochem. Photobiol. A: Chem., **148**, 183–189 (2002)
32. N. San, A. Hatipoglu, G. Koçtürk, Z. Çınar, J. Photochem. Photobiol. A: Chem., **139**, 225–232 (2001)
33. J. M. Herrmann, C. Guillard, M. Arguello, A. Agüera, A. Tejedor, L. Piedra, A. Fernández-Alba, Cat. Today, **54**, 353–367 (1999)
34. A. Zaleska, J. Hupka, M. Wiergowski, M. Biziuk, J. Photochem. Photobiol. A: Chem., **135**, 213–220 (2000)
35. C. Lizama; C. Bravo; C. Caneo, M. Ollino, Environ. Tech., **26**, 909-914 (2010)
36. A. M. Amat, A. Arques, M. A. Miranda, S. Segu, Solar Energy, **77**, 559–566 (2004)
37. C. Han, Z. Li, J. Shen, J. Haz. Mat., **168**, 215–219 (2009)
38. R. J. Tayade, R. G. Kulkarni, R. V. Jasra, Ind. Eng. Chem. Res., **45**, 922-927 (2006)
39. R. J. Tayade, D. L. Key, Mat. Sci. Forum, **657**, 62-74 (2010)
40. S. Shakoor, A. Nasar,, J. Taiwan Inst. Chem. Eng., **66**, 154-163 (2016)
41. N., Peng, D., Hu, J., Zeng, Y., Li, L., Liang, C., Chang, J ACS Sustainable Chem. Eng., **4**, 12, 7217-7224 (2016)
42. L. Karimi, S. Zohoori, M. E. Yazdanshenas, J. Saudi Chem. Society, **18**, 581–588 (2014)
43. A. Salama, A. Mohamed, N. M. Aboamera, T. A. Osman, A. Khattab, App. Nanosci., **8**, 155–161(2018)
44. M. Nasirian, C. F. Bustillo-Lecompte, M. Mehrvar, J. Environ. Management, **196**, 487-498 (2017)

E3S Web of Conferences **286**, 03018 (2021) https://doi.org/10.1051/e3sconf/202128603018
TE-RE-RD 2021

Pollutants resulting from intensive poultry farming activities and their impact on the environment

Roxana Mitroi[1,*], *Oana* Stoian[1], *Cristina Ileana* Covaliu[1], and *Dragoş* Manea[2]

[1]University Politehnica of Bucharest, Faculty of Biotechnical Systems Engineering, 313 Splaiul Independenței, 060042, sector 6, Bucharest, Romania
[2]INMA Bucharest, 6 Ion Ionescu de la Brad Bld, sect. 1, Bucharest, Romania

Abstract. Poultry farming activities can have a negative impact on the environment, namely: acidification (with NH_3 mainly, H_2S, NO_x etc.); pollution of surface water and groundwater (with NO_3^- and NH_4^+); eutrophication (N, P); air pollution with NH_3, N_2O, NO, dust (PM_{10} and $PM_{2.5}$), bioaerosols; increasing the greenhouse effect (CO_2, CH_4, N_2O etc); drying (use of groundwater); smells, noise; pollution with heavy metals, pesticides and toxic substances; the spread of pathogens that are resistant to antibiotics; pollution of waters with residues of pharmaceutical products. As a result, in order to prevent or reduce the negative impact on the environment, in the production process it is necessary to use BREF/BAT techniques (Best Available Techniques). The activity of poultry farming must be done according to the best available techniques: the farming system, production halls and related facilities are designed and built according to the latest standards in the field; implicitly the consumption of raw materials, waste emissions, wastewater, air pollutants values being according to the legislation [1].

1 Introduction

Bird shelters are zootechnical constructions intended for poultry or laying hens raising and exploiting. These constructions are designed according to the intended production (meat or eggs) and some aspects must be considered such as the technological solution related to the maintenance systems, the feeding system, watering, manure evacuation, the mechanization and automation degree of the production processes [2].

Projects aimed at intensive poultry farming activities are subject to an environmental impact assessment and this is due to the fact that poultry farming facilities involve a very large number of environmental impacts [3].

Regarding to the poultry farming activity, the potential impact on environmental factors refers to ammonia emissions to air, nitrogen and phosphorus leaks from manure into the soil, groundwater and surface water.

Poultry manure can be used as an organic fertilizer for plants, it has a high content of essential nutrients [4, 5], but improper management and excessive application of poultry manure lead to ammonia volatilization, nutrients leaching, higher nitrogen mineralization, damage of sensitive crops or surface water contamination by excessive phosphorus [5, 6, 7].

2 Impact on environmental factors

The environmental impact assessment process aims to ensure the proper assessment of potential negative effects on the environment, the identification of alternatives, the establishment of measures to reduce and / or prevent negative effects on the environment.

The environmental impact evaluation process assesses feasible alternatives to the same project and makes recommendations for preventing, minimizing or mitigating adverse effects.

Assessing the impact of activities related to the intensive rearing of birds on environmental factors aims to determine the correct forms of impact.

In order to be able to establish measures to reduce or eliminate the negative impact of activities aimed at the intensive farming of birds on the environment, it is very important to know all the risk factors from the outset [3].

The impact forms involve the analysis of some key aspects that are presented in Table 1.

Table 1. Key aspects necessary to establish the impact forms on the environment [3]

The characteristics and duration impact	Determining the environmental aspects that may be affected. Positive, neutral or negative nature of the impact. Highlighting the forms of significant impact (positive and negative). Description of impact (for example cumulative, continuous, intermittent, occasional, temporary, short, medium or long term, direct/indirect, reversible/irreversible).
Duration, scope and complexity	Quantification of the quantity or intensity with which the character / quality of any aspect of the environment will change. Indication of the geographical extent of the effects (if some, a large part or all of the areas will be affected). Description of the change degree (imperceptible, slight, observable or significant).
Consequences	Indication of whether the impact can be avoided, mitigated or remedied. Highlighting the forms of reversible impact. Indication of whether a form of compensation is available, possible or acceptable.

The impact on the environment will be lower as the poultry farming plant is more modern and efficient. For this reason, it is recommended that, when assessing the impact on the environment, to take into account the age of the farm, in order to adopt appropriate measures, such as air filtration techniques, equipping the farm with technologies that allow reducing emissions as much as possible, changing manure transfer methods, imposing specific conditions regarding the manure storage [8].

2.1 Impact on environmental factors

In the case of an intensive poultry farming activity, potential cumulative effects should be considered, as presented in the following subchapters.

2.1.1 Surface water quality

By presenting measures to prevent/reduce/compensate the impact of the zootechnical farm and the available evidence it must demonstrate that the zootechnical activity will not significantly affect the water quality [9, 10].

2.1.2 Groundwater quality

The groundwater can be affected by rainwater infiltration or manure that changes the quality of the groundwater, especially with regard to the nitrogen content [9, 10].

This potential effect should be considered in the context of the existence of other potential sources of groundwater damage [9].

2.1.3 Metabolic gas emissions

Poultry farms are a source of metabolic gas emissions (ammonia, methane). The analysis of the effects on air quality must be done taking into account the existence of other emission sources in the vicinity [9, 10].

2.1.4 Potential impact on soil quality

Animal manure contains nutrients and trace elements, which can be accumulated in the soil. Indirect impacts on the soil can also be represented by the destruction of ecosystems, the erosion of biodiversity by expanding the production of fodder crops in natural habitats or the overexploitation of non-renewable resources for fodder production [11].

3 The main types of pollution generated by activities aimed at intensive poultry farming

3.1 Air pollution

Air emissions from poultry farms are shown in Table 2.

Table 2. Air emissions associated with intensive poultry farms [12]

Air	Production system
Ammonia (NH_3)	Shelter for poultry, storage of manure
Methane (CH_4)	Shelter for poultry, storage of manure
Nitrogen oxide (N_2O)	Shelter for poultry, storage of manure
NO_x	Indoor heating installations
Carbon dioxide (CO_2)	Shelter for poultry, fuel used for heating and transport, burning of vegetable waste
Scents (H_2S)	Shelter for poultry, storage and manure spreading
Dust	Food preparation, food storage, production halls, storage and spreading of manure
Smoke/CO	Burning of vegetable waste

From an environmental point of view, manure is the most important residue from farms activity. The measures applied to reduce emissions, associated with the collection, storage and manure treatment affect the structure and composition of manure and, finally, influence the emissions associated with the application of manure on agricultural lands. Associated with the storage of manure in the halls, N_2O, CH_4 and, in small quantities, H_2S are released. The ammonium concentration can have high values, up to 40 ppm in the halls where poultry are farmed and the concentrations of N_2O and CH_4 are slightly higher than in ambient air. Dust concentrations reach values of 10 mg/m^3 in these intensive poultry farms (from 210 mg/m^3 for non-respirable particles and between 0.3-1, 2 mg/m^3 for respirable particles) [12].

3.2 Water pollution

Domestic wastewater from the sanitary filter drains into the internal sewer then will be collected in a concrete basin, from where it will be sent to the wastewater treatment plant.

The technological wastewater from the washing and rinsing of the poultry farming halls will be collected and directed to the concrete basin, then it will be sent to the wastewater treatment plant.

Given the variation of wastewater flow over time, their composition during the process of poultry farming on permanent bedding and the nature of pollutants, wastewater treatment can be achieved biologically.

Technological wastewater is not an important source of emissions from poultry farming activity, as it has a discontinuous regime and can be collected and treated. This wastewater will be treated in the treatment plant.

The disinfection of the poultry halls is done when they are empty, by spraying disinfectant solutions. Pollutants emitted into surface water associated with intensive poultry farms are presented in Table 3.

Table 3. Pollutants emitted into surface water associated with intensive poultry farms

Wastewater source	Emitted pollutants
Washing the halls	organic substances suspended matter detergents
Hygienic - sanitary needs of the staff	organic substances suspended matter detergents
Rainwater collected from the platforms	suspended matter

Wastewater discharged directly into surface water can come from a variety of sources, but only emissions from manure treatment systems are normally allowed. Emissions from these sources contain N and P, but an increase in CBO levels may also occur; especially in wastewater collected from farmyards and manure collection areas. However, of all the sources, field scattering is the activity responsible for the emissions of many compounds into air, soil, groundwater and surface water. Although manure treatment techniques are available, their application in the field is still the most widely used technique. Manure can be a good fertilizer, but where it is applied in excess of the capacity of the soil and the need for crops it becomes a major source of emissions [13].

Water pollution with nitrates, pathogenic phosphates (usually fecal germs and Salmonella) or heavy metals is the main cause for concern in the case of poultry farming.

3.3 Soil pollution

After the commissioning of the objective, a source of soil pollution could be the storage of manure, following the activity resulting in a relatively large amount of nitrogen and phosphorus. Pollution sources of nitrogen and phosphorus are presented in Table 4. Manure contains variable amounts of these nutrients as well as a number of minerals and essential elements: sulfur, magnesium, etc.

Table 4. Sources of nitrogen and phosphorus pollution [12]

Pollutant	Pollution sources
Nitrogen	accidental leaks of washing water processes for denitrification of compounds with nitrogen (NO_2, NO, N_2) washing with rainwater of the soil on which the manure was spread nitrogen and phosphorus accumulation in the organic component of the soil
Phosphorus	accidental leaks of washing water washing with rainwater of the soil on which the manure was spread nitrogen and phosphorus accumulation in the organic component of the soil

3.4 Noise pollution

Zootechnical activities can generate noise pollution due to:
- Fans in the production halls;
- Vehicles, pumping stations, etc.

The maximum admissible limits, based on which the state of the environment from an acoustic point of view in the area of such an objective is assessed, outdoors, are specified in the STAS 10009-88 Norm. In the case of buildings, the noise level limit must not exceed the maximum value of 50 dB (A) for the noise level outside the building, measured at a distance of 3 meters from it according to STAS 6161/189 [14].

4 Conclusions

This article presents and describes the impact of pollutants resulting from poultry farming on the environment. This study also shows the need to prevent, minimize or mitigate the negative effects produced by poultry farms. Pollutants from this industry can be mainly ammonia and methane, causing pollution of air, nitrogen and phosphorus from manure, causing pollution of soil, groundwater and surface water.

In conclusion, pollutants from poultry farming must be identified, and their concentration must be minimized by modernizing and streamlining the poultry plant. The endowment of the farm with technologies should be considered both for reducing emissions and for the correct transfer and storage of manure.

Potential benefits for farmers and environment can be obtained by using manure as fertilizer. Instead of facing an environmental problem, poultry manure can be a sustainable source of nutrients and energy.

In order to reduce the impact of pollutants on the environment, technologies have been developed, but one of the problems of their application in farms is the cost.

References

1. Directive 2010/75 / EU of the European Parliament and of the Council for the intensive rearing of poultry and pigs.
2. L. C. Stafie, R. Antohe, Laying hens in free range system PRACTICAL GROWING GUIDE, Waldpress Publishing House, Timisoara (2018)
3. I. Punga; coord .: I. R. Iordanov Environmental impact assessment: Case study /; AO "EcoContac". - Chisinau: S. n. (2016)
4. Y. Hu, H. Cheng, S. Tao, Environ. International, **107**, 111-130 (2017)

5. "Guide on facilities for the intensive rearing of farm animals, including poultry, laying birds, pigs and sows", developed within the Project "Development of guidelines for improving the administrative capacity of environmental protection authorities for the purpose of the unitary assessment procedure of environmental impact (EGEIA) ", SIPOCA code 19 - contracting authority: Ministry of Environment
6. Decision no. 964/2000 on the approval of the Action Plan for the protection of waters against nitrate pollution from agricultural sources with subsequent amendments and completions.
7. Government Decision no. 321/2005 on the assessment and management of ambient noise, republished, with subsequent amendments and completions.
8. O. Dikinya, N. Mufwanzala, J. Soil Sci. Environ. Manage, **1**(3), 46-54 (2010)
9. G.. Ren, T. Lin, Y. Ying, G.. Chowdhary, K. C.Ting, Computers and Electronics in Agriculture, **169**, 105216 (2020)
10. Law no. 292/2018 on assessing the impact of certain public and private projects on the environment.
11. P. Gerber, C. Opio, H. Steinfeld, *Animal Production and Health Division, Food and Agriculture Organization of the United Nations,* (2008)
12. J. Muhammad, S. Khan, M. Lei et al., Environ.Tech. & Innovation, **19**,100909 (2020)
13. K.Y. Chan, C. G. Dorahy, S. Tyler, A. T. Wells, P. P. Milham, I. Barchia, Soil Res. **45**(2), 139-146 (2007)
14. K. C. Reddy, S. S. Reddy, R. K. Malik, J. L. Lemunyon, D. W. Reeves, Agro. J. **100**(4), 1047-1055 (2008)

Considerations on the importance of the main chemical properties of the soil in agrochemical studies

Sebastian Muraru[1, *], *Vergil* Muraru[1], *Paula* Condruz[1], *Cornelia* Muraru Ionel[1], and *Raluca* Sfiru[1]

[1]National Institute of Research – Development for Machines and Installations Designed for Agriculture and Food Industry – INMA Bucharest,Computer Engineering Department, no 6, Blv. Ion Ionescu de la Brad , Romania

Abstract. Information on soil nutrition level or soil pollution level can lead to the establishment of some recommendations or a strategic decision-making program for the reference area development. The specific soil analysis is essential both from an economic point of view and from an environmental point of view. To achieve an adequate management of the nutrients into an agricultural field, it is necessary to conduct an agrochemical study at an interval of 3–5 years, in which, the relevant agrochemical indicators for the evaluation of the nutrient supply state must be analysed, to determine fertilizer doses and/or recommended amendments for the expected crops. The agrochemical study must contain the analysis bulletin that provides the following information: pH value, humus and total nitrogen content, phosphorus (P) and potassium (K) content, C/N ratio (carbon / nitrogen), nitrogen index (NI), as well as the main microelements.

1 Introduction

Soil, the natural environment for plant growth and ecosystem survival, is a heterogeneous polydispersion system consisting of three states of aggregation of matter: solid, liquid and gaseous.

Most of the nutrients contained in the soil in larger quantities are: Si, Al, Ca, Mg, K, Fe, Na, P and are fixed in the primary and secondary minerals from which the soil comes from. They cannot be absorbed directly by plants, but only after their passage into the liquid phase, as a result of the disaggregation and alteration processes.

The ions released by alteration can be retained by adsorption processes on the colloidal side of the soil, or they can recombine with other ions and precipitate. The intensity of alteration of primary and secondary minerals primarily depends on temperature, humidity and pH. [1, 2].

Agricultural lands cannot maintain their production capacity because impoverish in nutrients. Knowing the relationships between plant, soil and environment must reach a balance.

To do this, we must first determine the soil fertility. Soil properties of composition, dispersion, indoor activity as well as physical, chemical, biochemical processes, which determine the soil fertility.

An agrochemical study can provide the necessary data regarding the pedoclimatic situation from the agricultural field, and must contain the analysis bulletin that provides the following information: soil reaction by pH value, the content of organic matter (humus); total nitrogen; phosphorus (P) and potassium (K) content, as well as the main microelements, [3].

Depending on the soil reaction, other agrochemical fertility indices can also be determined: alkaline earth carbonates and soluble salt content and cation exchange capacity, as well as the calculation of synthetic indices such as: C/N ratio (carbon / nitrogen) and nitrogen (IN).

An accurate knowledge of the soil characteristics gives us exact information about what amounts of fertilizer, we will have to use in the coming years, [4]. Thus, we will be able to save money because we will not waste money on chemicals that exist in the soil and which, if they are in excess, can damage crops.

pH- It is a characterization indicator of **the soil reaction** and expresses the property of the soil to dissociate in its own hydrogen ions (H +) solution or hydroxyl ions (OH-).

Soil acidity represents both hydrogen ions (H +) absorbed at the surface of colloidal particles and those in the free state in solution. It is characterized by two parameters: one of intensity and another of quantity, [5].

* Corresponding author: sebi_muraru@yahool.com

The intensity parameter represents the activity of hydrogen ions absorbed at the particles surface, which can pass from the solid part of the soil in the form of ions, in water or in saline solutions with which the soil comes into contact. It is called **current acidity** and is **practically materialized through the pH values**.

In fact, when determining the pH of the soil in aqueous or alkaline solution, only the current acidity is determined. Theoretically, the pH values oscillate between 0 and 14. The average value of the pH scale (7), corresponds to a neutral reaction, and the values lower and higher than 7 show an acidic reaction, respectively, basic. Basically, for the soil, the pH scale is narrower, between the values of 4 and 11.

The actual acidity of the soil depends on the environment in which it is determined. The pH indicator of acid soils is numerically lower when is measured in the suspension than if it were measured in the filtered solution, that was in equilibrium with the soil. It is necessary to standardize the conditions for measuring the pH in the laboratory. Among the methodological effects we mention the dilution effect, the CO_2 effect in the air and the saline effect.

Decreasing the soil: water ratio (increasing of the soil suspension dilution) increases the pH value in the case of both acidic and alkaline soils. Hence the need to standardize the soil: water ratio (mass / volume): for arable soils, meadows and orchards and vineyards is 1: 2.5, for greenhouse soils intended for vegetables is 1: 5; in greenhouse soils intended for flower crops, composts, nutrient mixtures, peat with a content of organic matter determined by loss during calcination (600^0C), as well as soil horizons with a content of over 30% organic matter (humus) is used a ratio of 1:10.

For all soil types, the saline suspension pH (pH_{KCl}) is determined in a saline suspension of potassium chloride, 0.1n KCl, at a soil: solution ratio of 1: 2.5 (mass / volume).

The influence of CO_2 from atmospheric air dissolved in the suspension on pH differs depending on the reaction of the soil. In the case of soils with pH> 6.0 CO_2 the pH values decrease, and in soils with pH <6.0 CO_2 does not significantly influence the pH value (dissociation H_2CO_3 is insignificant).

In the field, the soil pH changes depending on the variation in water content and soluble salt content: in the hot and dry season increase the content of soluble salts-nitrates, sulfates, chlorides-accumulated in the soil , which leads to a lower pH value; in the rainy season the salts are dissolved and transported in the groundwater (leaching), the pH increasing up to its maximum value in the respective soils.

Fertilization in soil with soluble salts (NH_4NO_3, K_2SO_4) leads to an immediate significant decrease in soil pH. This effect decreases rapidly through consumption by plants, through ion exchange reactions, by fixation in complexes and by leaching.

The pH value defines the soil reaction class and the representation colors on the agrochemical cartograms:

Strongly acidic	Moderately acidic	Weakly acidic	Neutral	Weakly alkaline	Moderately and strongly alkaline
pH ≤3,55	5,8	6,8	7,2	8,4	>8,41

The quantity parameter is given by the hydrogen ions adsorbed on the colloidal particles surface of clay minerals and humic substances, determined as the titratable acidity of the soil to a conventionally established end point and represents the **potential acidity**, [5].

Perfectly dry soil contains only potential acidity; as the hydrogen ions pass into solution it becomes current acidity. A dynamic balance is maintained between the potential acidity and current acidity, the decrease of the potential acidity decreases the current acidity. The potential acidity of the soil has 3 components: exchange acidity (Ea); hydrolytic acidity (Ha); neutralizing acidification.

2. Materials and methods

2.1 The importance of determining the pH value

The pH parameter of the soil is a very important element, it needs to be known and monitored because it intervenes in many physico-chemical and biological mechanisms in the soil. About the importance of the value of the pH index in the general chemistry of the soil we can say the following:

1) It sufficiently characterizes the soil, offering a global image of its trophicity. Because the activity of hydrogen ions also determines the activity of other ions in the soil solution, the pH value is important in assessing soil fertility, [4].

2) Appreciates the state of opportunity for the amendment of acid soils and the improvement of alkaline soils. Most crop plants do not tolerate a pH below 4.5 or 8.3 pH units, [6]. Most plants grow best on neutral soils, meaning pH = 6.8–7.2 or slightly acidic-pH = 6.3–6.8. The pH of the soil is adjusted according to the plants that will be grown.

To the acidic soils, the colloidal complex is saturated with hydrogen ions, leading to unfavorable physical properties (the structure is poorly formed or degraded, the porosity is low, small permeability for water and air). Therefore, the acid reaction imposes unfavorable physical, chemical and biological properties in the soils, which have a low productive capacity. Therefore, the pH value is an index of appreciation of the application of calcareous amendments to acid soils, along with the degree of saturation in basic cations and the presence of exchangeable aluminum.

In addition, the strongly alkaline reaction has an unfavorable influence on the properties and productive capacity of soils. Basically, plants start to suffer when the pH value rises above 8.5.

If the soils contain salts that hydrolyze alkaline (Na2CO3 has the strongest basic character), they release OH- ions in solution and give to the soil solution an alkaline reaction. By the hydrolysis of carbonates in the soil results bases and carbonic acid, an acid that decomposes into carbon dioxide (CO2) and water (H2O). The bases strongly dissociate in aqueous solution, releasing large amounts of OH-.

The high concentration of soluble salts in the soil solution causes to acquire a high osmotic pressure, which exceeds that of sucking roots, in these conditions, the plants can no longer absorb water from the soil with the nutrients, manifesting strongly the physiological drought phenomenon

The strongly alkaline reaction determines the blockage of some microelements (Cu, Zn, Bo, Mo, etc.), and the presence of sodium in large quantities imprints to the soil completely unfavorable physical properties. In strongly alkaline soils, the physical, chemical and biological properties are more unfavorable than the acid soils, they being practically infertile. The improvement of the strongly alkaline reaction is much more complicated, being achieved by administering amendments based on gypsum or phosphogypsum at the same time with the application of washing operations and drainage works, [3].

3) Affects the mobility and accessibility of soil nutrients. The soil pH can determine whether a plant can access or not the nutrients it needs. The solubility of soil constituents, both in soils that contain ions needed in plant nutrition and as well as those that can release ions that can become toxic to plants beyond certain limits, depends on pH, [5].

If the soil reaction is acidic, with pH <5, the amount of exchangeable calcium (Ca^{2+}) and magnesium (Mg^{2+}) ions decreases and the solubility of zinc (Zn^{2+}), aluminum (Al^{3+}), manganese (Mn^{2+}), copper (Cu^{2+}) ions increases. It also decreases the solubility of elements such as phosphorus and molybdenum, which are immobilized in insoluble complexes of iron, aluminum in the soil.

The decrease in pH decreases the cation exchange capacity and an increase in the anion exchange capacity by adding positively charged particles, H + ions.

If the soil reaction is a basic (pH > 7.8), in the presence of exchangeable calcium and magnesium ions, hardly soluble salts are formed, such as monocalcium phosphate ($CaHPO_4$) and tricalcium phosphate [Ca_3 (PO_4)2]. Also, the compounds of iron, zinc and manganese pass into hardly soluble forms, inaccessible for plants.

The basic elements for a balanced soil, which decisively influences the development of N, P, K plants (nitrogen, phosphorus, potassium), are assimilated by them depending on pH values as follows: nitrogen is assimilated at pH 6–8.5, phosphorus at a pH 6.5–8 and potassium at a pH> 6.

The optimal pH range in which nutrients become accessible to the plants is between 6.0 and 7.0, [7]. The mobility of nutrients under acidic condition changes. The activity of microorganisms decreases, the N content decreases, implicitly the humus content decreases. Phosphorus retrograde by increasing the content of Fe and Al, Al and Fe ions have a toxic effect on plants if they exceed certain limits. The soluble forms of K decreases with increasing acidity due to the immobilization of K in the structure of clay minerals. In the graphical representation below shows the quantities of macroelements (N, P, K) adsorbed by plants at different pH values of the soil solution (Fig. 1):

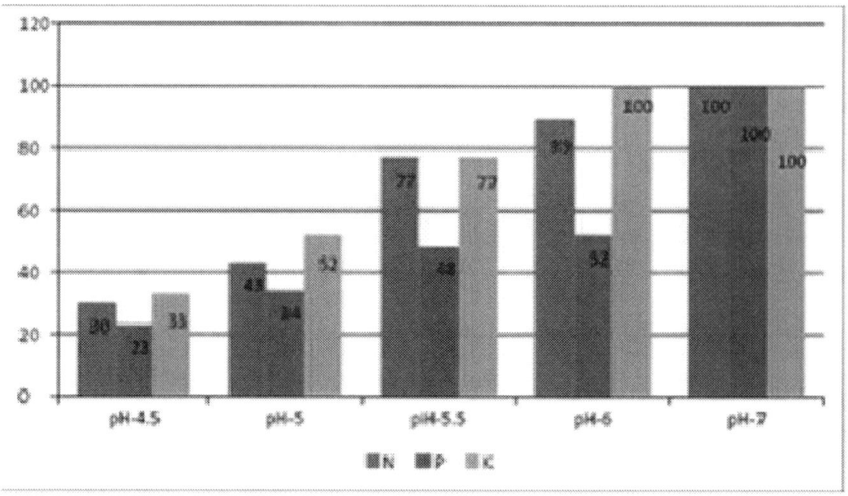

Fig. 1. Adsorption N, P, K by plants depending on the soil pH

4) Influences the activity of microorganisms. Soil microbial diversity is responsible for the greatest transformation of nutrients in the soil, the regeneration of minerals that limit crop productivity. The reaction of the soil has a strong influence on the development of microorganisms in the soil. Thus, bacteria prefer a reaction from weakly acidic to alkaline (pH 6–8), actinomycetes prefer a neutral or alkaline reaction (pH 7–7.5), and fungi grow under acidic reaction conditions (pH 4–5).

At certain pH values, the activity of microorganisms shows an optimal activity [such as bacteria that transform atmospheric nitrogen (N2) into nitrogen from soil available in plants (NH4), [8]. In strongly acidic soils, bacteria living in symbiosis with leguminous plants, nitrogen-fixing bacteria, and nitrifying bacteria are almost completely absent. A low soil pH tends to favor the appearance of fungi, while a higher pH tends to favor bacteria, so that the importance of balancing the two becomes obvious: too low a pH can lead to the proliferation of fungal diseases. Fungi and bacteria are needed, but too much can lead to problems.

5) Influences the application of fertilizers. When fertilizing soils with mineral fertilizers, the soil reaction is taken into account, [3]. Usually, basic reaction fertilizers are applied on acidic soils, and acidic fertilizers are applied on alkaline ones. Particular attention should be paid to phosphorus fertilizers. On acid soils, due to their high solubilizing power, fertilizers based on tricalcium phosphate and even crude phosphorites are applied. The gradual release of phosphorus ions on these soils also has the advantage that it cannot be blocked in the form of iron or aluminum phosphates. On alkaline soils, it is recommended to apply slightly soluble superphosphate (monocalcic and dicalcic phosphate), [9].

6) The pH value is used to calculate synthetic agrochemical indices: *Zinc deficiency index* - ICZn; *Mobile phosphate reaction index* - IRPM; *Molybdenum index* – IMo; *Reaction factor calculation of PAL values.*

Humus. Soil humus is advanced organic matter in an advanced stage or various stages of transformation. It is a complex of organic compounds in a close correlation with the mineral part of the soil. The essential part of humus is humic substances (85%-90%) consisting mainly of humic acids, fulvic acids and their organo-mineral products.

Humic acids (black colour) are macromolecular compounds with a cyclic structure that contains carbon (50%-65%), hydrogen, oxygen, nitrogen (2%-6%). They are insoluble in water and acids, but slightly soluble in alkaline solutions. The presence in the molecular structure of the carboxylic and phenolic functional groups leads to exchange reactions between humic acids and the mineral part of the soil. Humic acids fixed to the mineral part of the soil can no longer pass into the solution when treating the soil with alkaline solutions. They have a high adsorption capacity and form a series of water-insoluble organo-mineral compounds, favoring the accumulation of nutrients in the upper part of the soil profile. Freshly formed humic acids, saturated with bases, especially with calcium, coagulate irreversibly, thus contributing to the formation of hydrostable structural aggregates

Humic acids are formed mainly from organic waste derived from grassy vegetation. When interacting with the mineral part of the soil, humic acid form compounds, called humites, which improve the structure and properties of the soil.

Fulvic acids of pale yellow colour (in low concentration in solution) up to an orange colour, are soluble in water, acids and alkaline solutions. They contain carbon (40%-45%), hydrogen, oxygen and nitrogen (3%-4%). It exerts strong alteration action due to the strong acid reaction (they have pH = 2.6–2.8), they allow the mobilization and deep displacement of different soil components. When interacting with the mineral part of the soil, fulvic acids form compounds, called fulvates, which also favor the alteration of the mineral part of the soil and the movement of nutritive elements.

1.2 The importance of knowing the total humus content of the soil

The humus content indicates the degree of soil fertility.

The soil with 3%-4% humus can provide during a vegetation period of 10–30 kg of N/ha in dry years and 80–100 kg of N/ha in years with abundant rainfall. The soils have certain C/N ratios, which, to a certain extent, reflect the nature of humus and as a result serves to it's characterization in relation to soil types. Along with other data, it gives indications of the richness of the soil in nitrogen, the main element of plant nutrition. Thus, the low C/N ratio for chernozems (10–12) shows their nitrogen richness compared to podzolic soils, where the ratio is higher (13–14).

The C/N ratio indicates the intensity of the decomposition of organic matter in the soil by microorganisms.

C/N = (Humus% x 0.6767) / Nt.

C/N <15-very intense decomposition (humus). For soils in Romania, Cernescu (1941) established the following classifications: chernozems where C/N is 12.5–13.5; leachate chernozems where C/N is 14–15, brown forest soils where C/N is 12.5–14; secondary podzols where C/N is 15.5–16

C/N = 15–30-average decomposition

C/N> 30-reduced decomposition.

H% = organic C x 1.7241 where, 1.7241 - represents the amount of humus, in g, corresponding to 1 g of organic carbon (corresponding to the average percentage of carbon content in humus is 58%);

Due to their cation exchange capacity, humic substances can retain a series of cations (Ca^{2+}, Mg^{2+}, Na^{+}, K^{+}), which, also through cation exchange, can be released and made available to plants. Together with clay, which also has this property, humus forms the adsorbent complex of the soil.

In the presence of clay, humic acids saturated with calcium cations cause the formation of the grainy structure of the soil. The presence of aggregates with high water stability creates in the soil an aerohydric regime favorable to the normal development of plants. Soil fertility depends not only on the humus content in the soil, but also on the degree of saturation of the adsorbent complex with Ca^{2+}, which imprints the soil with the best physical and chemical properties. Due to its ability to stably cement mineral particles into structural aggregates, humus improves the physical properties of clay soils, reducing their cohesion and at the same time increasing their permeability to air and water. Humus increases the cohesion of sandy soils, reduces their permeability to water and increases their retention capacity. Also, humus influences the thermal regime of the soil by the fact that imprinting dark colour contributes to the increase in caloric capacity. In the humus of the soil is concentrated a large amount of energy from solar energy. The energy, with the nutrients in the humus, releases the necessary amounts of nitrogen, phosphorus, potassium and carbon dioxide of plants.

In terms of plant nutrition, humus is the main source of nitrogen and phosphorus (30%-50% phosphorus in the soil from humus and partially decomposed or undecomposed organic matter) from the soil as well as other nutrients. Humic substances in the soil also exert a direct influence on plants, stimulating their growth and development

Total nitrogen Nt. Over 90% of the nitrogen in the soil is organic in nature. It does not come like other nutrients from the rock on which it was formed, resulting from the biological activity that occurs in the organic matter in the soil and in which it is about 5%. In the soil, the amount of total nitrogen varies on average in depth from 0 to 20 cm, depending on the humus content, from 0.03% in soils poor in organic matter, to about 0.5% in chernozems (2, 5–10 t / ha), and in the deeper horizons, it decreases, so that at a depth of 2–3 m it is almost missing. It was determined that at a depth of 1 m - in relation to the type of soil - the amount of nitrogen is 9 - 30 t/ha. The soil provides a good part of the nitrogen necessary for plant growth. It is present in the soil in the form of organic and mineral compounds, which, together, give the total available nitrogen. Nitrogen is taken from the soil mostly in the form of NH_4^+ or NO_3^- ions and to a lesser extent in the form of NO_2^- (because it oxidizes rapidly to nitrates), forms that come from the organic matter in the soil.

1.3 The importance of knowing the total nitrogen content of the soil

The regulation of the N regime in the soil, in relation to the requirements of the plants is one of the agrochemical measures. The values of the total N content are used to calculate the C/N ratio, for assessing the state of nitrogen supply of soils, for characterizing organic matter.

C/N = Humus,% x 0.6767 / Nt,%

The appreciation of the level of total N%: very small <0.100; small 0.100 - 0.140; medium 0.141 - 0.270; high 0.271 - 0.600 and very high> 0.600, [8].

Indirectly, the assessment of the nitrogen supply level of the soil can be done with the help of the nitrogen index (IN). It is a synthetic index that shows the ability of the soil to make available to the crop the amount of N

from the humus mineralization. To determine the IN, the following must be determined: the sum of the exchangeable bases (SB) by the Kappen method, the hydrolytic acidity (Ah), by means of the Kappen-Daikuhara method and humus by the oxidimetric method, in the Walkley - Black variant.
IN = (Humus,% x VAh) / 100 where VAh represents the degree of saturation in bases, [4]. The IN value represents the appreciation of the state of soil supply with nitrogen as follows: ≤2.0 - poor; 2.1–4.0 - medium; 4.1–6.0 - good; ≥ 6-very good.

Phosphorus (P). The total phosphorus in the soil is given by the phosphorus of organic nature, 30%-50% (from humus and from the partially decomposed or undecomposed organic matter) and by the inorganic phosphorus, 50-70% (simple phosphates of calcium, magnesium, iron, aluminum, of complex phosphates and free ions in the soil solution or absorbed in the colloidal micelle). In acid soils, iron and aluminum phosphates predominate, in neutral or alkaline soils, calcium phosphates. The decomposition of organic matter results in orthophosphate anions soluble in the soil solution. The C: N: P ratio of soil organic matter is approx. 100: 10: 1. If the C: P ratio is less than or equal to 200: 1, phosphate ion mineralization occurs, if it is greater than or equal to 300: 1 phosphorus immobilization occurs, [9].

Knowing the total phosphorus content in the soil (% P), we can assess the fertility status of the soil as follows: <0.044 low fertility; 0.044–0.065 average fertility; > 0.065 high fertility. Of the total phosphorus content in the soil, only 0.5 - 1% is accessible to plants, its mobility in the soil being low.

Phosphorus is absorbed from the soil by plants either as H2PO4 monophosphate or as HPO42 diphosphate. The relative abundance of the two anions in the soil solution depends on the soil pH: if the soil pH increases, the proportion in the H2PO4 form - decreases and the HPO42 form - increases. At pH = 7, the concentration of the two anions is practically equal.

1.4 The importance of knowing the content of soil PAL

Based on the analytical data on the reaction and soil content in soluble phosphates extracted in ammonium acetate-lactate reagents or calcium, we can assess the state of supply with plant accessible phosphates (PAL).

In order for the PAL values to be interpreted using the same limit values for the different insurance classes, the analytical values obtained are corrected with a factor (FC), a correction factor that depends on the soil pH.

Potassium (K). Potassium is the second most abundant nutrient after N, it is 4–6 times more abundant than macronutrients such as P, Ca, Mg, and S, and the most important cation for living organisms. Most of the potassium in the soil (90%-98%) is in the crystal lattice of minerals (mica, feldspar, illite, chlorite, vermiculite, smectite and is inaccessible to plants except after alteration processes). Fixed potassium - retained between the sheets of the clay mineral network and can be assimilated by plants under certain conditions. Exchangeable potassium - is found in the form of ions retained in the colloidal complex of the soil and represents less than 1% of the total K. It is the main source of accessible potassium for plants, [10]. Soluble potassium - is found in the form of ions in the soil solution and represents only 1% of exchangeable potassium. Therefore, usually by exchangeable potassium in the soil, the sum between exchangeable potassium and soluble potassium is understood. There is a balance between all these forms of potassium in the soil.

Total potassium in the soil has no significance for its accessibility to plants but only rarely for the study of accumulation in soils in case of excessive application of potassium fertilizers.

Soluble and exchangeable potassium (K +) is the immediate source of plant nutrition with this element. As it is consumed by plants in the soil solution, the balance is restored by switching to solution (by cation exchange reactions) of exchangeable K. To assess the soil potassium reserve, the value of the total potassium content in the soil is determined. The determined value of the mobile potassium content, extractable in ammonium acetate-lactate solution (KAL), directly significant value for plant nutrition, establishes the soil content classes: ≤40 mgK/Kg soil - extremely small; 41–65 mgK/Kg soil - very small; 66–130 mgK/Kg soil - small; 131–200 mgK/Kg soil-medium; 201-300 mgK / Kg soil- high; ≥300 mgK / Kg soil - very high.

The main chemical properties that characterize soil well enough are: soil reaction (pH), amount of humus (H %), total nitrogen in the soil (Nt), phosphorus (P), potassium (K).

3 Results and discussion

Laboratory analyses were performed on 12 average soil samples collected from lot 1 and on 13 average soil samples from lot 2, presented in table 1. The lots are at a distance of 10 km from each other. The supply status with N, P, and K of the respective lots was taken into account.

Table 1

no	lot	sample no.	harvest depth cm	Analyses performed								
				pH	H	Nt	VAh	IN	PAL	KAL	Pt	Kt
				Unit pH	%	%	%	IN= (H,% xVAh)/100	mg/Kg	mg/Kg	%	%
1	1	01	0–25	6,44	2,96	0.144	89,5	2,65	75	193	0,072	1,71
2		02	0–25	6,25	3,20	0.151	87,7	2,80	45	205	0,058	1,08
3		03	0–25	6,26	3,14	0.149	87,5	2,75	28	209	0,063	1,23
4		04	0–25	6,24	3,61	0.174	87,7	3,16	43	201	0,065	1,28
5		05	0–25	5,93	3,67	0.178	85,7	3,15	26	211	0,054	1,27
6		06	0–25	5,97	3,55	0.176	87,5	3,11	63	217	0,071	1,31
7		07	0–25	6,37	3,55	0.180	89,1	3,16	45	213	0,066	0,93
8		08	0–25	6,27	3,43	0.172	86,9	2,98	27	184	0,070	0,90
9		09	0–25	6,25	3,73	0.174	86,8	3,24	26	203	0,067	0,95
10		010	0–25	6,17	3,37	0.151	88,4	2,98	37	203	0,070	1,00
11		011	0-25	6,26	3,20	0.163	89,0	2,85	41	201	0,062	0,96
12		012	0–25	6,76	3,26	0,169	92,7	3,02	39	203	0,064	0,74
13	2	52	0–25	8,02	3,55	0,207			21	182	0,080	0,97
14		53	0-25	8,21	2,78	0,152			172	254	0,142	1,00
15		54	0-25	8,27	2,49	0,136			115	303	0,118	0,84
16		55	0-25	8,09	3,14	0,168			64	166	0,091	0,94
17		56	0-25	8,21	2,55	0,153			171	297	0,131	0,67
18		57	0-25	8,07	3,26	0,197			84	189	0,093	0,72
19		58	0-25	8,06	3,91	0,262			58	231	0,086	0,73
20		59	0-25	8,08	2,84	0,170			116	244	0,107	0,88
21		60	0-25	7,90	3,73	0,205			78	189	0,091	1,01
22		61	0-25	8,11	3,67	0,219			37	186	0,072	0,73
23		62	0-25	8,06	3,67	0,234			47	272	0,077	0,52
24		63	0-25	8,16	3,67	0,162			39	221	0,076	0,81
25		64	0-25	8,11	3,67	0,259			53	290	0,085	0,82

3.1 Interpretation of the pH analysis performed

From the analysis of the pH of the soil samples from the 2 lots, it is found that:

The pH values in the samples of lot 1 fall into the reaction class of weakly acidic soil, so it is in the pH range between 5.8 and 6.8. At pH 5.6 - 6.2 the need for amendment is medium; at pH > 6.3, it is not necessary to correct the acid reaction. For this lot, we recommend the cultivation of sunflower, wheat, corn, soybeans, barley, barley, hemp, peas, spinach, celery, potatoes, zucchini, tomatoes, cucumbers, watermelon, basil, rosemary, thyme, sage, parsley, iris, marigold, watercress, plum, and currant. The pH values in Lot 2 samples fall into the reaction class of weakly alkaline soil, so it is in the pH range between 7.2 and 8.4. For this lot, we recommend the cultivation of sugar beet, fodder beet, lucerne, rapeseed, cauliflower, cabbage, salad, beans, mint, quince, sour cherry.

3.2 Interpretation of humus analyses performed

From the analysis of the soil samples collected from the two lots, we observed that we have a humus content in the range of 3-4% (except for 4 samples), which corresponds to a soil with medium fertility. Calculating and analyzing the C/N ratios C/N = (Humus% x 0.6767) / Nt) it is found that they are less than 15, so the process of decomposition of organic matter in the soil by microorganisms is very intense which shows us that along with humification occurs mineralization humus, thus returning to the soil variable amounts of nutrients. With the help of humus content and the degree of saturation in the bases, we can calculate the nitrogen index, a synthetic index that shows the ability of the soil to make available to the crop during the vegetation period, certain amounts of nitrogen by humus mineralization. IN = (Humus,% xVAh) / 100

3.3 Interpretation of total N analyses performed

The content of the samples collected from lot 1 and lot 2 indicates a total nitrogen level in the middle soil (it is in the range 0.140% - 0.270%). Both batches have a pH in which nitrogen is assimilated very well (pH 6 - 8.5). Calculating the nitrogen index for the soil samples from lot 1, we can appreciate that the soil has an ability to make available to the crop during the vegetation period, certain amounts of nitrogen that fit it into the soil with medium insurance. Depending on the type of crop, the expected production, the degree of soil supply, recommendations can be made for fertilization with nitrogen fertilizers.

3.4 Interpretation of phosphorus analyses performed

Lot 1 of soil has a good supply of phosphorus accessible to plants for field crops, natural and cultivated meadows, extensive plantations of trees and vines. For vegetables grown in the field, intensive plantations of fruit trees, vines, hops, nursery vineyards and orchards, insurance with phosphorus accessible to plants is poor. Lot 2 of soil has an excellent supply of accessible phosphorus for vegetables grown in the field, intensive plantations of fruit trees, vines, hops, vine and fruit nurseries, but the supply is excessive for other plants (even toxic). Both batches fall within the pH range in which phosphorus is well assimilated (pH 6.5 and 8).

3.5 Interpretation of potassium analyses performed

Both batches have good availability for potassium, pH> 6 favoring the assimilation of potassium in the proportion of 100%.

4. Conclusions

Along with the analysis of pH, humus, degree of saturation in the bases, the analysis of macronutrients N, P, K is very important in establishing the degree of fertilization of a soil in making fertilizer recommendations with these nutrients and not only. Depending on the result of the analysis, the type and facility of fertilizer application differ from case to case. To better exemplify the link between soil qualities and fertilization, the main problems should be shown, both on alkaline and acid soils. Plants prefer slightly acidic solutions, which are easily metabolized by plants. Where the pH is alkaline, it is good to apply foliar fertilizers. By administering phosphorus-based fertilizers in alkaline soils, we make the situation even more difficult and the action of nitrogen is blocked. Alkaline pH also confuses the situation with seed treatment. Due to the alkalinity, the effect of insecticides is neutralized. In addition, iron, phosphorus and manganese are blocked in alkaline soils. Fertilization recommendations must consider the cultivated plant, the production, we want to obtain, the degree of soil supply of nutrients and last but not least, must respect the principle of soil fertility conservation. In the elaboration of the fertilization plans, the specific nutritional consumption of the crop per ton of product, the expected quantity of product obtained per hectare, the reserve of nutrients in the soil is taken into account. The methods of fertilization, the technologies used, involve an optimal recovery of nutrients in the soil, a fertilization with minimal risks of soil pollution and groundwater.

This work was supported by a grant of the Ministry of Research, Innovation and Digitization (MRID), CCCDI – UEFISCDI, project number PN-III-P1-1.2-PCCDI-2017-0560 "Eco-nanotechnologies and intelligent equipment for soil properties mapping and evaluating the dynamics of the plant in order to improve agricultural production and environmental protection, within PNCDI III

References

1. P. Cardei, S.L. Muraru, R. Sfiru, V. Muraru.*General structure of tillage draft force. consequences in experimental and applicative researches*, INMATEH - Agricultural Engineering,Vol. 59, No. 3, 256-260,(2019)
2. Z. Borlan, C. Rauta, *Methodology for agrochemical analysis of soils to establish the need for amendments and fertilizers*, vol 2, Methods, guidance reports Series, ICPA no. 3, (1981)
3. D. Davidescu, L. Calancea, V. Davidescu, Gh. Lixandru, *Agrochemistry,* (Didactic and Pedagogical Publishing House Bucharest, 1981)
4. D. Davidescu, Z. Borlan, V. Davidescu, C. Hera, *Phosphorus in agriculture*, (Publishing House of the Academy of the Socialist Republic of Romania, Bucharest, 1974)

5. D. Davidescu, V. Davidescu, L. Calancea, M. Handra, *Nitrogen in agriculture,* (Publishing House of the Academy of the Socialist Republic of Romania, Bucharest, 1976)
6. D. Davidescu, V. Davidescu, *Potassium in agriculture* (Publishing House of the Academy of the Socialist Republic of Romania, Bucharest, 1979)
7. R. Lacatusu, *Agrochemistry,* (Helicon Publishing House, Timisoara, Romania, 2000)
8. R. Lacatusu, *Mineralogy and soil chemistry*, (Publishing House of „Alexandru Ioan Cuza" University of Iaşi, Romania, 2000)
9. R. Lacatusu, M. Lungu, N. Rizea, *Global soil chemistry,* Terra Nostra Publishing House, 2017)
10. R. Madjaru, V. Davidescu, *Agrochemistry*, Distance learning course, Faculty of Horticulture, Bucharest, 2009.

Developing efficient practices for producing bioethanol from sweet sorghum, evaluating several varieties of the plant and growing environments

Florin Nenciu[1,*], *Valentin* Vlăduţ[1], *Gabriel* Nae[1], *Lorena-Diana* Popa[2], and *Oana Emilia* Constantin[3]

[1]INMA Bucharest, Testing Department, Ion Ionescu de la Brad Blv. No. 6, Bucharest, Romania
[2]University of Craiova, Faculty of Horticulture, Craiova, Dolj, Romania
[3]University Dunărea de Jos, Galaţi, Faculty of Food Science and Engineering, Galaţi, Romania

Abstract. The challenges we face today envisaging environmental protection, starting from global warming and climate change, up to natural resources depletion, creates opportunities for developing new technologies for biofuel production. Considering that most biofuels used at the moment are based on energy-intensive crops, the question arises whether occupying large areas of land to the detriment of food is justified or not, especially as these crops generally deplete the soil of nutrients. In the present paper we propose a different approach for assessing an easily adaptable energy plant named Sweet Sorghum, that can grow with low input requirements, in difficult growing conditions, on contaminated or poor-nutrient and dry soils, and yet having a high productivity potential. Our research directions are targeted towards identifying the best opportunities for producing bioethanol from Sweet sorghum testing different varieties and sustainable planting conditions. We have been established several experimental cultures and evaluated the potential for producing biofuels using different techniques, such as juice extraction and fermentation and lignocellulose processing.

1 Introduction

There have been made important advancements towards reducing dependence on fossil fuels, both through several policies that encourage electric vehicles sale, but also by increasing the production and consumption of biofuels. A new challenge now is to reduce the fertile land area that is being cultivated with energy-intensive crops to the detriment of food and to increase the efficiency of biofuel production technologies.

A plant that is not yet used to its full potential is Sweet sorghum, which has excellent characteristics of adaptation and growth and high potential for producing bioethanol, using both first and second-generation technologies. In addition, this technical plant outperforms other similar cultures as an attractive climate-resilient crop in terms of reducing greenhouse

[*] Corresponding author: florin_nenciu2000@yahoo.com

gas emissions, being less demanding in agricultural work and having a high resistance to drought [1]. Various management practices may be used to reduce GHG emissions from the agricultural sector, the most promising being cultivation of crops with a higher potential for carbon sequestration (C4 photosynthesis cycle crops), improve fertilization management, implementation of organic fertilizers and alternative soil amendments to replace synthetic fertilizers, precision farming, etc. [2].

Although many farmers consider sugar cane a much more effective plant for ethanol production, a large number of recent studies are demonstrating than Sweet sorghum has many additional strengths, in addition to the productivity factor [1]. Sweet sorghum is an extremely versatile plant that can be used successfully in the food industry for extracting natural sweeteners, or as animal feed, given the increased nutritional value of the vegetal part of the plant.

In terms of energy production, first generation bioethanol can be produced by direct fermentation using its juice, and second-generation bioethanol made from bagasse and seeds. From the vegetable mass can obtain biogas by anaerobic digestion, cogeneration of power by incineration of dried plants, or by applying various composting techniques may produce high quality fertilizers [2].

Other financially advantageous uses may also include roofing, paper production, or obtaining several chemical products. The multitude of possible uses gives farmers flexibility and the opportunity to comply with some stringent legislative provisions related to food versus fuel conflict, since they can produce simultaneously biofuels, edibles, or animal feed.

The sugars found in Sweet sorghum juice are a combination of sucrose, glucose, and fructose, with proportions varying by each plant genotype [3], this feature being an advantage over other possible sources that use carbohydrates to produce alcohol.

Sweet sorghum has a highly efficient photosynthetic system (C4), it is very efficient in the utilization of soil nutrients, even from lands depleted in nutrients and is resistant to pests without treatments.

Sweet sorghum offers the opportunity to establish two crops per year. Two annual crops involve an extended vegetation interval, therefore plants have to benefit of very good adaptation qualities to various environmental conditions, such as temperature, precipitations, seasonal diseases, or other soil properties variations that may occur over the year. These qualities have been studied in several papers that validate the high level of adaptability, tolerance to various abiotic stresses [4], or higher water, nitrogen, and radiation use efficiencies. The low need for irrigation is an important advantage compared to sugar cane, given that we expect in the coming years to be imposed several restrictions on the use of potable water in agriculture.

Second generation lingo-cellulosic ethanol production have developed rapidly in recent years due to the evolution of biocatalysts, such as genetically engineered enzymes, yeasts and bacteria. The processes allow producing ethanol from ligno-cellulose biomass including from cereal crop residues.

Some research studies show that producing biofuels from plants can bring harm to the ecosystem, since the leaves, stems and roots might be used for land conditioning [5]. The process involves letting the dried stalks and leaves of a cereal crop after the grain has been harvested, improving soil quality and sequestering carbon. Therefore, a strategy must be developed to produce as much bioethanol from lignocellulose, but without adversely affecting the soil properties.

2 Materials and methods

Replacement of fossil fuels with biofuels brings improvements not only in terms of absorption of greenhouse gases by the plants during growth, but also positive phenomena is associated to combustion processes. Biofuels develop chemical reactions during combustion that cause polluting gases to be generated in much smaller quantities. Koppen et al., [6] evaluated the equivalent of GHG emissions saved from sweet sorghum cultivation, resulting between 1.4 and 22 kg per hectare emission reduction depending on yields, production methods, and the land cover. The assessment is important to highlight the better impact of reducing greenhouse gases by planting sweet sorghum compared to other similar crops or energy crops.

In addition to the environmental benefits, another considered aspect is the total energy gained from Sweet sorghum, since it is very important that the total energy consumed to produce the biofuel must be lower than the energy gained by bioethanol use.
This is not an easy task, because the agricultural works may have many variables related to the area, climate and soil. We have varied for each of the four pilot crops established, several agricultural works as sowing, pest control, irrigation, harvesting and processing.

According to United States Department of Agriculture (USDA) Sweet sorghum ethanol could produce between 12-16 Btu (British thermal units), for every Btu of fossil energy used in production.
Ethanol is produced from Sweet sorghum juice through direct fermentation technologies, from lignocellulose mass by processes of decomposition of vegetable mass into sugars, or by processing seeds, similar to other grains.
Productivity of Sweet sorghum crops may vary widely depending on the region, climate, crop management systems, yield, genotypes, efficiency of juice extraction and the manner fermentation is performed. Therefore, we can make some evaluations comparing to other similar energy crops, analyzing main benefits and weakness.

Our theoretical studies have shown that in order to obtain an increased efficiency, there must be an extraction rate of at least 35% from the clean stalks, while the bagasse can be used as animal feed, to be pelletized and burned for heat, or to produce electric power. However, the source of energy production and cattle for feeding should be in the vicinity of the crop, otherwise transportation costs can reduce efficiency to zero.
Sweet sorghum reaches maturity much faster than other plants such as sugarcane, requiring less water for development and has a higher fermentable sugar, therefore is better suited for ethanol production.

Sweet sorghum is a multi-functional culture, managing to generate products associated with the food industry (syrup, sugars), animal feed, bulk organic chemicals, industrial alcohol, restoration of soil quality and energy.
The main advantage Sweet sorghum has compared to other energy plants consist in maintaining high productivity without negatively affecting food production, since it can be cultivated on less fertile soils, in marginal lands or semi-arid regions, being a drought tolerant crop. The characteristic of having shorter growth duration allows establishing up to two crops per year, requiring low water and fertilizer input for development.

Hallam et. all. [7] evaluated the biomass production on both fertile and marginal lands and the annual expenses incurred with several annual Sorghum species in terms of biomass production compared to other crops. Have been showed that Sorghum had superior qualities compared to other annual and perennial plants that have been analyzed.
Sweet sorghum is considered an efficient method for nitrogen fertilizer of depleted soils [8]. Stevens et. all. [9] registered improvements in soil fertilization following planting of Sweet sorghum, when applying some methods of crop management, involving letting some of the bagasse or leaves on the field.

Another aspect that have been studied was carbon sequestration in the soil layers using several sustainable land management practices, that allow to simultaneously improve soil quality and capture carbon dioxide. Sequestration of atmospheric carbon dioxide as organic carbon in the biosphere attracts attention as an alternate way to reduce the rates of greenhouse gases growth. Soil carbon sequestration is the process by which some plants with abundant development of the vegetal mass take the atmospheric carbon dioxide and store it into their biomass and land [10, 11].

Technical and economic analyses conducted by Farrell et al. [12] have shown that producing biofuels from lignocelluloses is projected to have the smallest contribution to the emission of CO2 and the largest net energy production, reaching up to quantities of 19400 l per hectare, when is used the entirely biomass of the plants and when applying several efficient pre-treatment techniques. Considering the negative effects when using the entire mass of plants, a compromise solution could be partial exploitation of the plant mass for producing energy [13, 14]. If only the stems and tips are used to produce bioethanol, while the leaves are left on the soil as a mulching surface, this practice could generate positive effects on the quality of agricultural land. The remaining leaves and roots are contributing to the sequestration of carbon in the soil, will transform into nutrients by decomposition, and help in fighting weeds. The constraints for large scale cultivation are the limited availability of genotypes suited for different agroclimatic conditions with all built-in resistances for biotic and abiotic stresses, photoperiod sensitivity and non-availability of required quantity of feedstock suited to off-season crushing in sugar industries [15, 16].

There are several ways of sequestrating carbon into the soil layers, either by rotating cultures, mulching or incorporating residues into the soil layers.

The study will further analyze aspects related to the production of bioethanol from 4 varieties of Sweet Sorghum crops, considering all the positive effects the crop may have on the environment. The experimental research approached the following main phases:

- Establish 4 Sweet Sorghum crops, on different soil types and in different climatic areas. Analyze growth and development for the four plant varieties
- Evaluate the potential of producing vegetal mass and quantify the percentage of each plant part: Leaves, plant tips with seeds, cleaned stems.
- Evaluate the techniques and equipment for processing sweet sorghum strains, in order to obtain juice.
- Fermentation and distillation of the juice, obtained by crushing sweet sorghum strains, highlight the maximum amounts of bioethanol that can be obtained from juice.
- Theoretical evaluation of the quantity of alcohol obtained by processing the vegetal mass in order to produce alcohol (by processing seeds and bagasse).
- Evaluating the levels of carbon sequestration in the land, by integrating sorghum leaves in soil layers.

3 Results and discussion

The experimental study was conducted during the period 2018-2020, and aimed on testing several Sugar sorghum species, that were established on soils having different characteristics, and in various geographical areas. Depending on the specificities of each area, have been applied different agricultural works, in order to optimize the production of vegetal mass. In case of the fourth crop, only minimal works have been performed, for testing sweet sorghum yields when adopting a management strategy that aims at minimum costs.

Four sweet sorghum varieties have been planted in experimental crops and analyzed over long periods of time. The agricultural lands were previously planted with wheat and corn crops (the soil had an efficient crop management in previous years), while for the establishment of culture, the soil have been ploughed and harrowed, nutrients have been added to bring it as close as possible to the optimal growth of sorghum. The planting operations were performed keeping a distance of 15 cm between plants and 60 cm between rows. Weed growth was controlled by mechanical weeding and an insecticide treatment was applied for pest control. Harvesting of sweet sorghum have been performed using an automatic threshing equipment, which cuts the stem from 2 cm above the ground, when the plant has reached maturity, and when the Brix index has exceeded the value of 19 Brix degree.

The equipment has a juice storage tank that can either be attached to a pump for driving the juice into a larger tank, or can accumulate the juice until filling.

The equipment is designed with three adjustable rollers, for a more efficient pressing of the stems. The mechanical part is driven by a motor that has adjustable speed, to customize the working advancement, depending on the operator number.

Fig. 1. Equipment for pressing sorghum stems to obtain bioethanol

The stems were cleaned of leaves and tips and transported to the laboratory, for further processing. In order to sequester carbon in the soil, the leaves were placed on the ground, forming a mulching layer. In this way, the decomposition of leaf mass will provide nutrients for next year's crop, and weeds will be more easily controlled. The plant tops and seeds were not left on the ground, in order to avoid germination in the following year.

The moisture level of the stems was determined using a hood and the Brix index was determined with an electronic refractometer.

Evaluations of the plant mass are also effective indicators for estimating the potential of sorghum plants, to be used as animal feed, the potential for sequestering carbon in the soil by integrating leaves into the soil layers, seed productivity for producing sorghum flour, or for assessing plants growth on different soils types.

The total amount of sugars and the mass of the plants may vary depending on the variety and on the harvest period, therefore it is essential to choose for planting the Sweet sorghum types that are able to develop vigorous stems and high amounts of sugars, but are also suitable to adapt to various soils and climatic conditions.

Table 1. Centralization of the main characteristics of four varieties of Sweet sorghum, planted in different climatic conditions

		Sweet Sorghum plantations							
		Variety 1		*Variety 2*		*Variety 3*		*Variety 4*	
Plant characteristics	Number of sugar sorghum plants, per square meter	17-20	*plants per square meter*	18-21	*plants per square meter*	19-21	*plants per square meter*	18-20	*plants per square meter*
	Average diameter of plants	17.76	*mm*	17.75	*mm*	18.11	*mm*	16.72	*mm*
	Average plant height	2.6	*m*	2.9	*m*	3.2	*m*	1.7	*m*
Quantities obtained per hectare	**Total plants mass harvested per hectare**	72.26	*t/ha*	78.1	*t/ha*	86.2	*t/ha*	51.2	*t/ha*
	Leaves	13.12	*t/ha*	17.42	*t/ha*	20.68	*t/ha*	9.6	*t/ha*
	Plant top with seeds	4.46	*t/ha*	4.44	*t/ha*	5.72	*t/ha*	5.8	*t/ha*
	Stem mass after removal of leaves and tips	54.68	*t/ha*	56.24	*t/ha*	59.8	*t/ha*	35.8	*t/ha*
	Bagasse	38.28	*t/ha*	38.6	*t/ha*	38.87	*t/ha*	29.46	*t/ha*
	Sorghum Juice produced	16.4	*kl/ha*	17.64	*kl/ha*	20.93	*kl/ha*	6.146	*kl/ha*
	Seed mass	1.09	*t/ha*	1.13	*t/ha*	1.21	*t/ha*	0.71	*t/ha*
Juice and bagasse analysis	Averege Brix	19	*Brix degree*	18	*Brix degree*	19	*Brix degree*	19	*Brix degree*
	Moisture bagasse	47%	*%*	46%	*%*	48%	*%*	47%	*%*
Percentages calculation	Sorghum juice	23%	*%*	21%	*%*	25%	*%*	12%	*%*
	Bagasse	53%	*%*	49%	*%*	45%	*%*	58%	*%*
	Plant top with seeds	6%	*%*	6%	*%*	7%	*%*	11%	*%*
	Leaves	18%	*%*	24%	*%*	23%	*%*	19%	*%*
	Percentage of juice raported to the processed stems (without leaves, tips and signs)	30%	*%*	31%	*%*	35%	*%*	17%	*%*
	Percentage of bagasse, raported to the processed stems (without leaves, tips and signs)	70%	*%*	69%	*%*	65%	*%*	82%	*%*
Alcohol production	Fist generation biofuels potential (Juice fermentation)	1.31	*t/ha*	1.41	*t/ha*	1.67	*t/ha*	0.49	*t/ha*
	Second generation biofuels potential (from bagasse)	0.96	*t/ha*	0.97	*t/ha*	0.97	*t/ha*	0.74	*t/ha*
	Biofuel generation by seeds processing	0.43	*t/ha*	0.45	*t/ha*	0.48	*t/ha*	0.28	*t/ha*
Quantifying carbon sechestration impact	Carbon sechestration by burying the leaves in the ground and optimizing agricultural works	0.42	*tones CO2 abatement rates estimated per ha per year*	0.48	*tones CO2 abatement rates estimated per ha per year*	0.47	*tones CO2 abatement rates estimated per ha per year*	0.43	*tones CO2 abatement rates estimated per ha per year*

Figure 2 shows the percentages of each part of the plant, after processing (juice, top with seeds, bagasse and leaves), for each variety of Sweet sorghum.

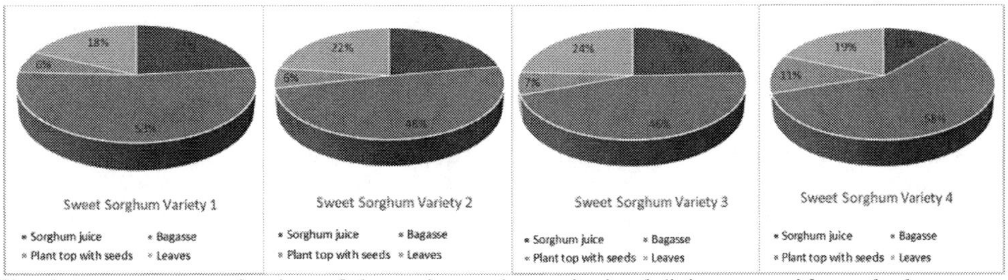

Fig. 2. Percentage evaluation of the main products obtained (juice, top with seeds, bagasse and leaves) after processing Sweet sorghum plants

If biofuel is obtained only by processing the juice, then a superior performance is observed for the crops that are benefiting from the best agricultural works, including irrigation. Variety 4, which had the least works performed, produced only 12% of the total plant mass (meaning 17% of the mass of cleaned stems), while variety 3 produced 25% of the total plant mass (meaning 35% of the mass of cleaned stems). The other varieties did not have significant variations of the 4 analyzed parameters.

If the biofuel is produced by processing the lignocellulosic mass, then we can evaluate the production per hectare for the 4 varieties, as can be seen in figure 3.

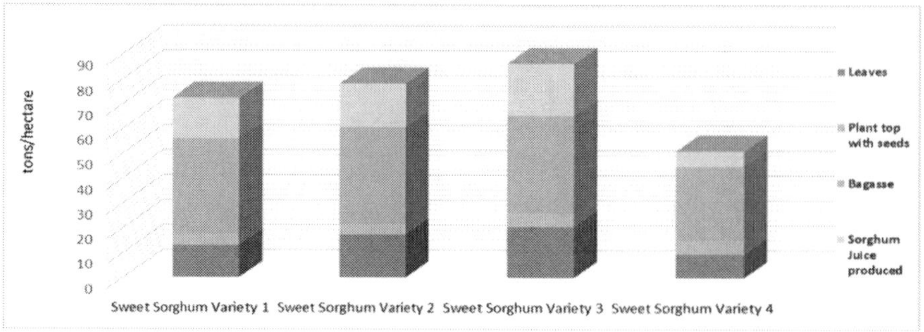

Fig. 3. Biomass expressed according to the main categories of products after processing: leaves, tops with seeds, bagasse and juice

Analyzing the amount of biomass obtained per hectare for several varieties of sugar sorghum in different cultivation areas and considering different agricultural activities applied to crops, can be seen that the quantities may vary significantly. It is therefore necessary to identify the best varieties adapted to local soil and climate conditions, in order to obtain maximum bioethanol production. Figure 4 shows the amount of juice and seeds obtained annually per hectare, in relation to the total biomass. The considerable difference between the total value of the plant mass and the juice and seeds shows the high opportunity to produce alcohol from the vegetable mass.

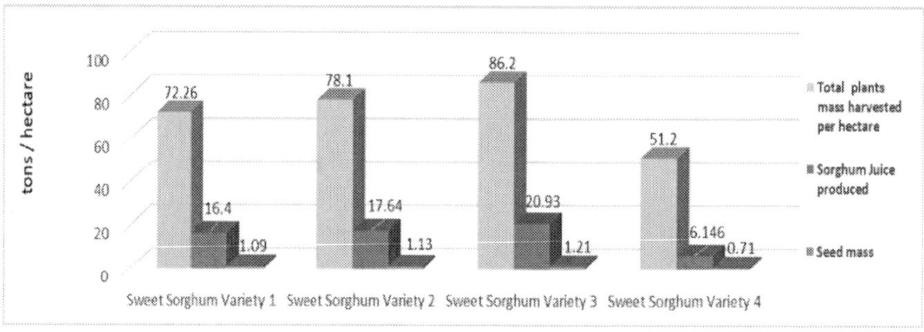

Fig. 4. Amount of juice and seeds obtained annually per hectare, in relation to the total biomass for Sweet sorghum crops

Although the vegetal mass is found in a much higher extent, due to the technological limitations, the maximum amount of bioethanol that can be obtained from bagasse remains relatively low, as can be seen in figure 5.

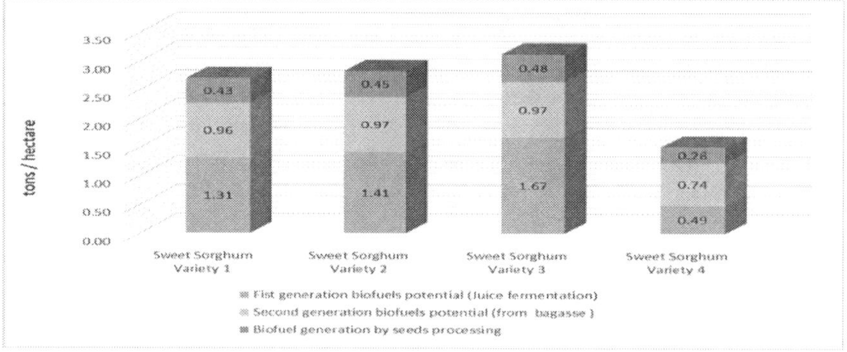

Fig. 5. Evaluation of the maximum amounts of bioethanol that can be obtained from each part of the Sweet sorghum plant

Soil carbon sequestration is the process that allow to atmospheric carbon dioxide to be taken up by plants through photosynthesis and to be stored as stable carbon in biomass and soils. In addition to reducing air pollution, the process also helps in supplementing the carbon lost from the soil layers by adding a new amount of organic carbon. We tried to adopt a crop management that balances the efficient production of biofuels with protecting the health of the biosystem, by introducing some of the biomass in the soil. In this way, we were able to contribute to the increase of the nutrient content and to sequester carbon in the soil.

Sustainable land management practices lead to a decrease in the maximum potential for the use of biomass in the production of bioethanol, in favor of maintaining or improving soil quality. Eliminating only the leaves on the field, they will behave as mulch for next year controlling weeds, and as fertilizer when decomposing. In Figure 8 can be seen the estimated amounts of carbon that will be sequestered in the soil due to sustainable land management practices that have been applied, involving incorporating leaf residues and optimizing agricultural works.

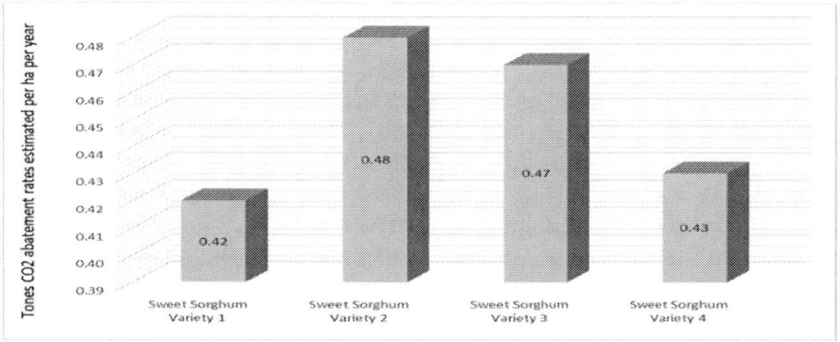

Fig. 6. Annual estimations of carbon sequestration in the soil when incorporating leaf residues and optimizing agricultural works for the Sweet Sorghum varieties

The application of the bagasse on the soil after the juice extraction was not considered, even if this option would have improved carbon sequestration values, considering the very high transportation costs.

3 Conclusions

In order to further reduce fossil fuels global use, we must simultaneously develop new processing technologies that will allow producing biofuels from juice extracts, lignocellulosic waste and seeds. In addition, we have to grow new varieties of plants adapted to severe climatic conditions and various soil characteristics that can be planted on marginal or depleted soils.

Have been showed that Sweet sorghum is a versatile crop, that has a high potential to produce ethanol from different parts of the plants, using different technologies, and due to its adaptability and resistance to extreme drought conditions, can be planted on marginal, polluted, nutrient-free or floodable soils.

Sweet sorghum is a promising solution to the food-versus-fuel debate, maintaining high ethanol productivity without negatively affecting food production. This approach leads to a sustainable crop management that protects and improves soil quality.

The characteristic of having shorter development duration allows establishing up to two crops per year, requiring relative low water and fertilizer input for development. However, has been shown that in the case of bioethanol production from fermentable juice, irrigation has a major impact in improving the efficiency of alcohol production.

We consider that future research should focus on new techniques and new genetic resources that could enhance productivity, improving stress tolerance characteristics, for crops that have a less impact on food supplies and on natural environment.

When using the bagasse for animal feeding or in direct combustion for producing energy, the farm or the power plant should be located in the vicinity of the plantation, or else there is a high possibility these activities generate higher costs than benefits.

The major challenges to biofuel technology from Sweet sorghum are the rapid degradation of the juice, high labor requirements, seasonality of the crop and the difficult process of turning lignocellulose into sugars.

This work was supported by a grant of the Romanian Ministery of Research and Innovation CCDI - UEFISCDI, project number PN-III-P1-1.2-PCCDI-2017-0566, Contract no. 9PCCDI/2018, and PNCDI III, contract no. 16PFE

References

1. C.S. Wortmann, A. Liska, R.B. Ferguson, D.J. Lyon, R. Klein, I. Dweikat,. Agron. J., **102**, 319-326 (2010)
2. M. Molaverdi, K. Karimi, M. Khanahmadi, A. Goshadrou, Ind. Crops Prod., **49**, 580-585. (2013)
3. A.V., Umakanth, A. Kumar, W., Vermerris, V.A., Tonapi, Breeding Sorghum for Diverse End Uses. Sweet Sorghum for Biofuel Industry, Chapter 16, Elsevier, 255-269. (2019)
4. T.T., Tesso, L.E., Claflin, M.R., Tuinstra, (2005). Analysis of stalk rot resistance and genetic diversity among drought tolerant sorghum genotypes. Crop Sci., **45**, 645-652
5. B.V.S., Reddy, S., Ramesh, A.A., Kumar, S.P., Wani, O. H., Ceballos, T.K., Sreedevi,. Bioenergy Res. **1**, 248-258, (2008)
6. S. Koppen, G. Reinhardt, S. Gartner,. Assessment of Energy and Greenhouse Gas Inventories of Sweet Sorghum for First and Second Generation Bioethanol, **77**, (2009)
7. A. Hallam, I.C. Anderson, D.R. Buxton, Biomass and Bioenergy, **21**, 404-424, (2001)
8. M. Bitzer,. Research report: Early deheading of sweet sorghum. National Sweet Sorghum Producers and Processors Association. http://nssppa.org/, (2009)
9. G. Stevens, R. Holou, D. Dunn, A. Wrather. Switchgrass and sweet sorghum fertilization for bioenergy feedstocks. Proc. Southern Plant Nutrition Management Conf., MS, 38-45, (2009)
10. A.C. Abdullahi, C. Siwar, M.I. Shaharudin, I. Anizan. Intechopen, Carbon Capture, Utilization and Sequestration, (2018)
11. A., Almodares, M.R., Hadi, Afr. J. Agric. Res. 4, **9**, 772-780; (2009)
12. A.E. Farrell, R.J. Plevin, A.D. Turner, M. Jones, D.M. O'Hare Kammen. Science **311** (5760), 506–508. (2006)
13. M.O. Oyier, J.O. Owuoche, M.E. Oyoo, E. Cheruiyot, B. Mulianga, J. Rono. The Scientific World Journal., 1-10, (2017)
14. S.R. Pinnamaneni, U. Akula, N. Seetharama, S.S. Rao, The Internat. Crops Research Inst. for the Semi-Arid Tropics (ICRISAT), 1-71. (2009)
15. R. Silva e Souza, R.A. da Costa Parrella Vander, F. de Souza. N.N.L.D. Parrella Cienc. E Agrotecnol. **40**, 46-56, (2016)
16. K.S. Vinutha, L. Rayaprolu, K. Yadagiri, A.V. Umakanth, P. Srinivasa Rao, Sugar Tech. **16** (2), 133-143; (2014).

Studies and research on the tribological behavior of the braking systems of vehicles. Review

Andreea-Catalina *Cristescu*[1], Ilie *Filip*[1], George *Ipate*[1], Gheorghe *Voicu*[1], and Vasilica *Stefan*[2]

[1] Faculty of Biotechnical Systems Engineering, University Politehnica of Bucharest, Romania
[2] National Institute of Research – Development for Machines and Installations Designed for Agriculture and Food Industry – INMA Bucharest, Romania

Abstract. Given the current context and the advanced stage of technology, which the braking systems of vehicles have reached, the main purpose of the studies and research is focused on presenting the main tribological aspects that contribute at improving the performance of braking systems, in order to ensure the vehicles' safety and stability at braking, in any conditions. The performance of the braking system is a key factor for both producers and vehicle passengers, due to safety requirements, ever-increasing. Thus, over time, numerous studies and research have been carried on in order to improve the performance of the braking system. In this paper are studied tribological phenomena through those, which contribute to the improvement of the braking system as performance, safety and stability.

1 Introduction

Given the economic importance of the automotive industry, the impact it has on modern society, many studies have focused on improving the performance of vehicles [1].

Braking systems are essential for operating a vehicle in safe condition for both the driver, and for the other traffic participants. Therefore, the braking system is submitted to strict legal conditions. The efficiency and reliability required for braking systems over time have implied their continuous improvement [2].

Considering the fact that braking is a vital process, the elements of the system must be thoroughly designed, produced, maintained, and controlled. Research and development of new braking systems should be based on previous systems by testing their characteristics which, to a certain extent, have not met the needs, as well as the reliability of the system itself. The development of the system should be based on this experience, in order to not repeat the same disadvantages in the structure of the system, and to make efforts in order to obtain total quality and reliability [3].

Continuous improvement is needed to achieve a high-performance braking system that ensures a safe and stable braking at any speed. One such improvement to the braking system, is the addition of ferrite magnets to the disc brake system, so that the stopping power of the vehicle increases. Thus, both the distance and the braking time decrease considerably, so that the braking is smooth and stable at high speeds [4].

The occurrence of squeaking and other phenomena in the braking system can be reduced by decreasing the coefficient of friction, but with its decrease, the performance of the braking system decreases [5]. A good comprehension and prognosis of the tendency of squeaking in car braking systems is an essential issue encountered by automotive engineers. As an example, a significant number of studies on finite element models (FEM) for car brake squeak have been carried on in the last two decades [6-13]. Many techniques are accessible in order to predict this behavior. Two main categories are distinguished: dynamic transient analysis and complex eigenvalue analysis (CEA) of car braking system. Researchers proposed complicated models with several degrees of freedom, because the model with movable caliper neglects the interaction between the disc and the plate. Shin [14-15] simplified the physical model into a system model with two degrees of freedom, coupled with a pair of disc pads and a disc; this study found the periodic and chaotic vibration of the system [3].

In order to avoid the squeaking as much as possible, various uncertain parameters were taken into account, namely the coefficient of friction and two insignificant wights, added to the brake caliper, which correlate to the structural changes of the braking system. Thus, it was possible to estimate the impact of structural changes on the stability of the braking system and to obtain information to select the best design of the braking system [4].

Thus, over time, numerous studies and research have been carried out in order to improve the performance of the braking system, such as: the design of the components and the whole assembly, of the braking system; materials used for contact surfaces; type of braking system: mechanical system, hydraulic system, pneumatic system, vacuum system, magnetic system, electrical system, regenerative electro-mechanical system.

Research and development of the new braking systems should be based on previous systems, to a certain extent, until when they will no longer meet the requirements, needs and reliability of the system itself, so as not to repeat the same disadvantages in the structure of the system and for making efforts, in order to obtain the total quality and reliability.

The reliability of the braking system, or any other technical system, is the ability to perform the required function without interruptions or defects. Nowadays, many auxiliary control and assistance systems have been incorporated into the braking system to help the driver in critical and challenging situations, as well as to provide it with more comfort in traffic conditions. Therefore, the paper highlights those tribological phenomena, through different parameters, which must be considered in choosing the braking couple that will be performing efficient, stable and safe.

2 Theoretical aspects

The disc brake incorporates a rotating disc with a circular plate. The disc is attached and rotated by the wheel hub. A caliper, which moves on the disc, is assembled on the suspension bracket, the hub shaft or shaft housing (Figure 1).

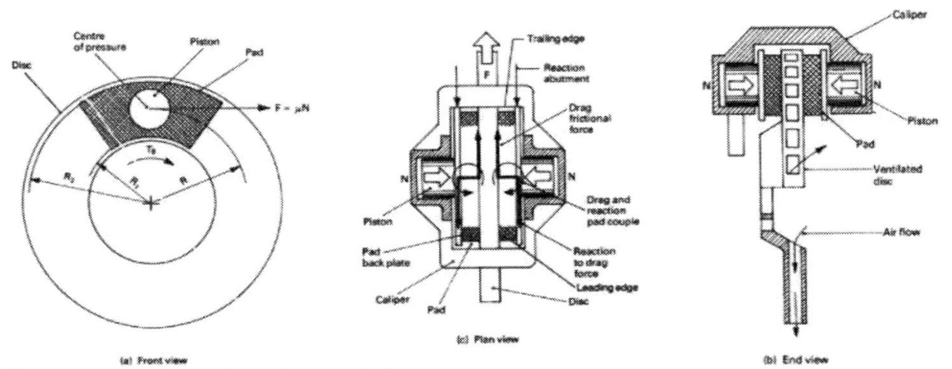

Fig. 1(a–c). Disc and pad layout [16]

In fig.1 the caliper is assembled with a pair of pistons and brake pads. When the brake pads are applied on the disc, they are locking it, causing the speed to be reduced, in accordance with the hydraulic pressure present in each piston. The hydraulic pressure is generated by the effort made by operating brake pedal [16].

Having regards to the fundamental principles of the braking system during movement, a vehicle has a kinetic energy, whose value depends on the mass and speed of the vehicle. The engine produces the necessary energy that accelerates the vehicle from starting to the desired speed, but this energy must be dissipated in part or in full, when the vehicle speed is slowed down or completely stopped. Tus, the function of the brake system is to convert the kinetic energy held by the vehicle at any given time, into thermal energy by friction (Figure 2) [17].

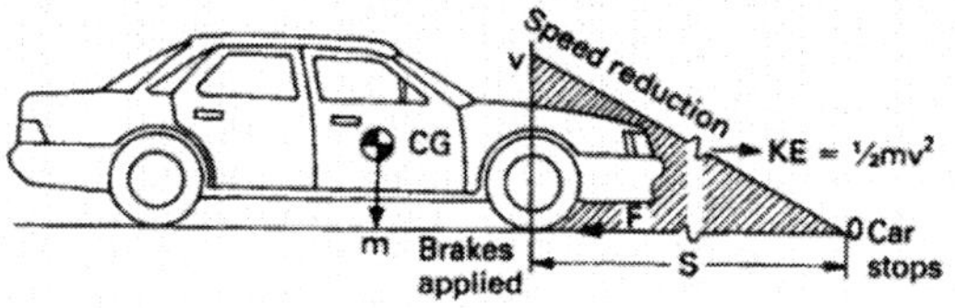

Fig. 2. Braking conditions of the vehicle [17]

The equation of kinetic energy, or the energy of motion, can be given by

$$E_c = \frac{mv^2}{2} \tag{1}$$

where:

E_c - vehicle kinetic energy (J); m - mass of the vehicle (Kg); V = vehicle travel speed (m/s).

The effort required to stop the vehicle is given by the relation:

$$E_w = F \times s \tag{2}$$

where: Ew - effort required to stop (J); F - average braking force (N); s - distance traveled (m) [16].

If a vehicle brakes until it stops completely, the effort exerted by the braking system must be equal to the initial kinetic energy of the vehicle, meaning:

$$E_w = E_c \tag{3}$$

$$F \times s = \frac{mV^2}{2} \tag{4}$$

So:
$$F = \frac{mV^2}{2s} \tag{5}$$

Braking distance and efficiency

Braking involves the process of generating a force that opposes the movement of the vehicle's wheels, thus minimizing the speed of the vehicle or stopping it. The force or resistance applied to stop a vehicle or reduce its speed is known as the braking force.

The braking efficiency of a vehicle is defined as the braking force produced as a percentage of the total weight of the vehicle, i.e. [17]:

$$Braking\ efficiency = \frac{Braking\ force}{Vehicle\ weight} \times 100\% \tag{6}$$

When the braking force is equal to the total weight of the vehicle that is being stopped, the braking efficiency is noted as 100%. Braking efficiency is usually less than 100% due to roads, the vehicle travels on a downhill slope or the braking system is inefficient.

The braking efficiency can be estimated by value the coefficient of friction, which is the ratio between the friction force and the normal load between the friction surfaces [17].

$$Friction\ coefficient = \frac{Friction\ force}{Normal\ load}, \tag{7}$$

$$\mu = \frac{F_f}{N} \tag{8}$$

$$\mu = \frac{F_f}{N} = \eta\ \% \tag{9}$$

where: μ - coefficient of friction; η - braking efficiency (%); F_f - friction force; N - normal load.

Therefore, a 100% braking efficiency is equal to a coefficient of friction, so equal to 1:

$$\eta(100\%) = \frac{F}{N} = 1 \tag{10}$$

Determination of braking distance

An approximate estimate of a vehicle's brake performance can be made by applying one of the equations of motion, assuming that the brakes have a 100% efficiency [17], namely.

$$V^2 = U^2 + 2gs \tag{11}$$

where: U - initial braking speed (m/s); V - final speed (m/s); g - deceleration caused by gravity ($\approx$ 10 m/s^2); s - stopping distance (m)

If the final speed of the vehicle is zero ($V = 0$), then:

$$0 = U^2 + 2gs$$

$$s = \frac{U^2}{2g} = \frac{U^2}{2 \times 9.81} \approx \frac{U^2}{20} \tag{12}$$

It is necessary to transform from km/h to m/s:

$$U\left(\frac{m}{s}\right) = \frac{1000}{60 \times 60} U = 0.28\, U\, (\frac{km}{h}) \tag{13}$$

$$s = \frac{(0.28U)^2}{2g} = \frac{(0.28U)^2}{2 \times 9.81} = 0.004U^2(m) \tag{14}$$

A table of vehicle breaking distances for distinct vehicle speeds and braking efficiency is presented in Table 1 [17].

Table 1. Braking distances for different vehicle speeds and braking efficiency [17]

Travel speed	Braking distance for different braking system efficiencies					
Km/h	**100%**	**90%**	**80%**	**70%**	**60%**	**50%**
10	0.4	0.4	0.5	0.6	0.7	0.8
20	1.6	1.8	2.0	2.3	2.7	3.2
30	3.6	4.0	4.5	5.1	6.0	7.2
40	6.4	7.1	8.0	9.1	10.7	12.8
50	10.0	11.1	12.5	14.3	16.7	20.0
60	14.4	16.0	18.0	80.6	24.0	28.8
70	19.6	21.8	24.5	28.0	32.7	39.2
80	25.6	28.4	32.0	36.6	42.7	51.2
90	32.4	36.0	40.5	46.3	54.0	64.8
100	40.0	44.4	50.0	57.1	66.7	80.0

3 Materials and methods

Friction materials play only a part in the friction. The vehicle's brakes consist of discs or drums, usually fabricated with cast iron as contact surface. Moreover, the category of cast iron (graphite flakes, spheroidal graphite, etc.), but also the precise metallurgical components, hold a highly important role. Some elements, such as vanadium, even in small amounts, have a great impact over the friction and wear performance of certain types of materials, while other elements, such as titanium, in large amounts, will affect friction and wear performance, but it can also make it indeterminate [18-19].

The evolution of friction materials is presented in Figure 3. All the main issues that were encountered in the past 20 years are described with their solutions. Many studies and researches lead to materials that are used today, with different metals and sulfides, although the current issue represents reducing the percentage of heavy metals used in friction materials.

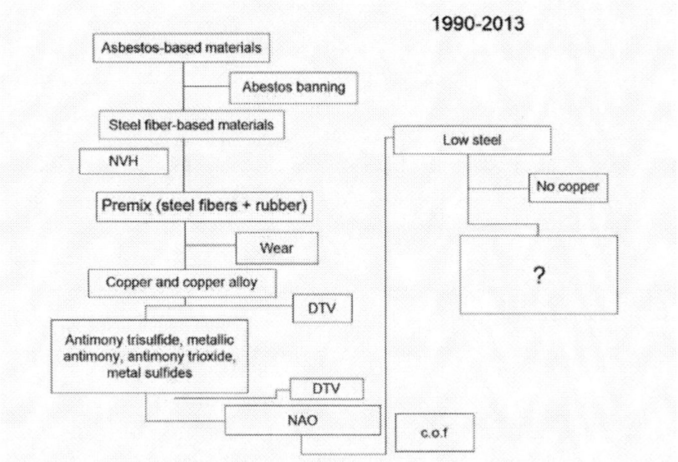

Fig. 3. Evolution of friction material formulas in the past two decades
NVH = noise, vibrations, hardness; DTV= Disc thickness variation;
NAO= Non asbestos organic materials; c.o.f= coefficient of friction [20]

Table 2 shows that the selection of the friction material is of paramount importance. For cars and trucks, disc brakes are mainly designed with cast iron as friction material, while for motorcycles disc brakes are made of stainless steel. The wet performance of stainless steel is significantly lower than the dry performance, in some case even 3 times lower. [22].

Table 2. Friction coefficients measured for different materials in the disc - brake pads assembly [20]

Material	**Coefficient of friction measured on diameter of 1 cm and specimen material at equilibrium, with slip speed 1 m/s, interface pressure 0.689 N/m^2, Initial temperature 80 °C**
Graphite cast iron flakes	0.40
Mild steel	Very irregular frictional forces measured
Stainless steel	0.60

Organic friction materials

a) Semi metallic materials

The so-called semi-metallic formulas are defined by a high metal percentage, usually exceeding 50% of the total mass. Metals components are abundant in steel fibers, copper, copper alloys (bronze and brass), tin and zinc in some cases. Other components are represented by a mixture of strong abrasives and lubricants. In Table 3, the components of non-asbestos friction materials are presented [20].

Table 3. Percentage of lubricants and metals in NAO materials, low steel, low metal [21]

Raw materials	**Ceramic materials (%)**	**Materials with low percentage of steel (%)**	**Materials with low percentage of metal (%)**	**Semi-metallic Materials**
Lubricant				
Fine graphite	4-9	2-8	4-10	4-10
Coarse graphite	2-5	0-7	4-8	5-10
Fine graphite				
Fine graphite	2-5	2-5	2-5	2-7
Cu sulfide	2-5	2-5	2-5	2-10
Other sulfides	0-4	0-4	0-4	2-5
Ferrous metals				
Iron fibers	0	10-18	5-10	20-35
Steel powder	0	2-5	1-2	2-7
Non-ferrous metals				
Copper	10-16	6-15	0-6	2-10
Copper alloys	5-10	5-10	0-5	2-10
Zinc / Sn	0-4	0-4	0-2	0-4

Figure 3 shows some classic proportions between the main component classification: abrasives, lubricants, total metals, including steel fibers, non-ferrous metals, graphite and sulphides for ceramic formulations, alloy steel and semi-metallic formulas. Abrasives in semi-metallic formulas are typically strong abrasives (alumina, and rarely corundum) that have the role of removing oxides from the contact surface of the disc. Ceramic materials require only weak abrasives (silicate of zircon).

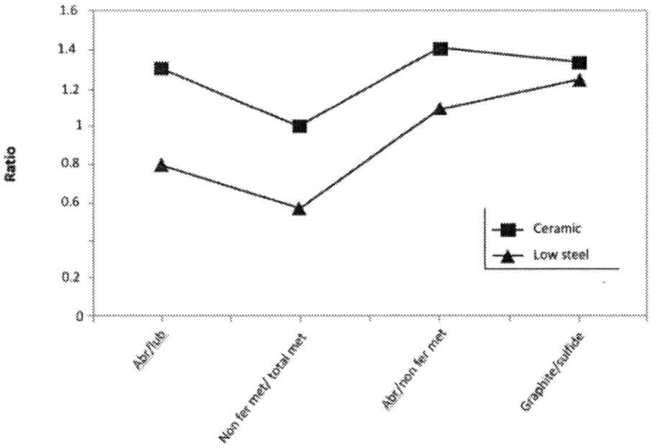

Fig. 3. Typical ratios between the main categories of materials [21]

b) Materials with a small amount of metals

Low metal materials are defined by a low percentage of metals in their formulas. Since squeaking and noise are caused by a higher quantity of metals, the main role of this type of material, is to increase the comfort. Still, this type of materials, are sully used for the rear axle, as their performance and coefficient of friction (0.35-0.38) is lower than the coefficient of friction of semi-metallic materials, which is 0.4. An application with a copper composite that replaces the classic cast iron for the brake disc is shown in Figure 4. [21]

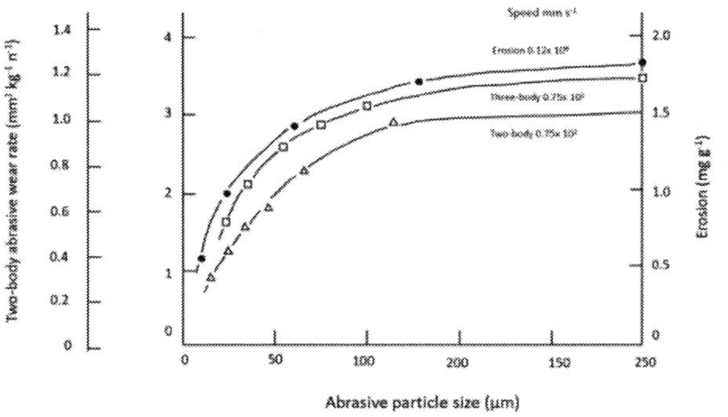

Fig. 4. Rate of wear of copper under conditions of abrasion (with two-body, three-body and the erosion) caused by silicon carbide particles of different sizes [21]

c) Materials with a small amount of steel

Formulas with a small amount of steel, are similar to formulas with a low metal content. The higher content of steel fibers provides a very high coefficient of friction, which makes these materials appropriate for the front axle. As fewer abrasives are required for this type of materials, the negative effects of noise and vibrations can be minimized. This damping system stands in need of the use of rubber and lubricant premix, as in the case of semi-metallic formulas. In order to stabilize the coefficient of friction, MgO is used as a gentle abrasive. The proper balance between the oxide tribofilm and the growth can achieve the stabilization of the coefficient of friction, as there is essential to maintain a constant coefficient of friction, when there is only one source of iron. [21]. An important study conducted by Kumar and Bijwe [21] describes, the importance of metals in friction materials. The study is concentrated on steel fibers, brass fibers and copper powders introduced in an asbestos-free formula, equilibrating with barite, the whole formula, and preserving continuously the other elements of the formula. The metal percentage was studied in two cases, 10% and 20%. The results highlighted that the main impact of these metals, represents a decrease in the coefficient of friction cause by either the increase in speed or the load. Figure 5 shows the impact of speed over the coefficient of friction [21].

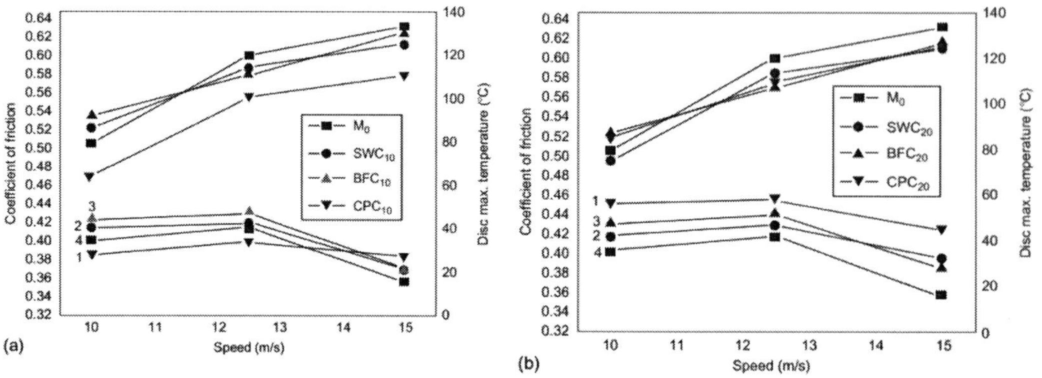

Fig. 5. The effect of speed on different friction materials with metals, with a percentage mass of 20% , determined with the disc tribometer. M0 - reference material without metals; SWC- steel wool; BFC - brass fiber; CPC - copper powder (sub-index 20 indicates an amount of metal is 20%) [21]

d) Ceramic materials

These materials are in need of copper to promote heat dissipation, as they do not have steel fibers in their formulas. Steel is commonly used in other types of material and has a great impact in thermal conductivity. Metals, in addition to their influence on the coefficient of friction, have an essential role in heat transfer between the brake pads and friction surface. Materials with a low quantity of metals, are an important replacement to this laborious issue. [21]. Thus, as it is specific in friction materials, in order obtain a proper properties balance, combining ceramic materials, which have a low thermal conductivity (zirconium), with others that have a good thermal conductivity (magnesium oxide and silicon carbide), is essential (Figure 6) [21-22].

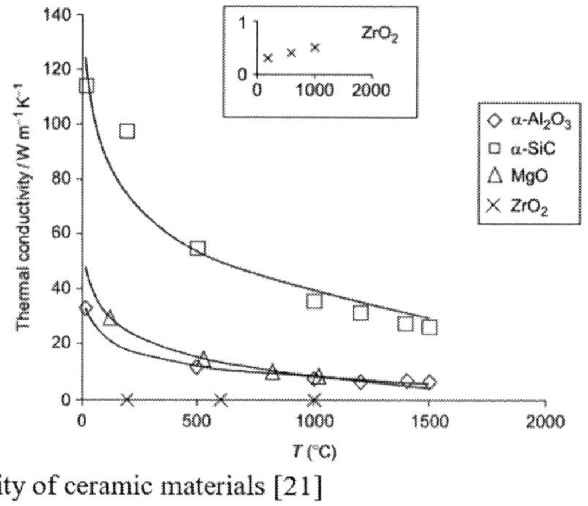

Fig. 6. Thermal conductivity of ceramic materials [21]

e) Carbon-ceramic discs and friction materials

In the last 10 years, a new type of material has been designed: carbon-ceramic materials, especially carbon / carbon-SiC (C/C - SiC) (figure 6). This type of material has a low density, such as 2 gcm^{-3}. Carbon-ceramic discs are currently used for the high-end cars, as they are a lot more expensive than other materials, but still more affordable than carbon-carbon materials [21]. Figure 7 presents the aspect of carbon / carbon-SiC during different stages of fabrication.

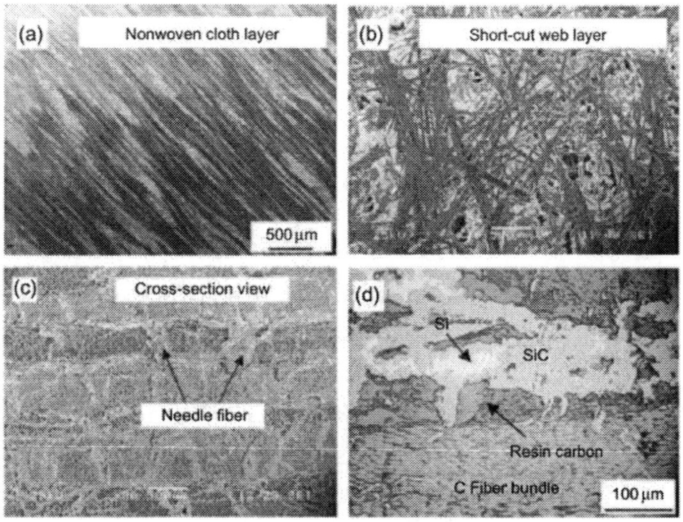

Fig. 7. Microstructure of composite C / C - SiC: (a) Nonwoven layer, (b) Cut fabric layer, (c) needle fibers and (d) the various components of the final material [21].

In the compression test, for the perpendicular direction, both the matrix and the fibers support the load, and the cracks cross both frames. The behavior is elastic until the material suddenly breaks. In Table 4, mechanical properties of a C / C - SiC composite are presented. It must be noted that the strength in the parallel direction is significantly lower that the strength in perpendicular direction, due to the contribution to the mechanical strength of the matrices and fibers. [21-22].

Table 4. Mechanical properties of C / C - SiC composites [21]

Bending strength	⊥	174±22 kN/m
	∥	134±31 kN/m
Compression strength	⊥	241±45 kN/m
	∥	188±36 kN/m

4 Results and discussions

In Figure 8, the wear in different conditions, such as load and speed tests, is discussed. The results show that materials with 20% wt of copper present a significant wear in comparation with the reference material, while the material with 10% wt copper has the least wear. This concludes that the wear is caused by copper and its alloys, while steel fiber impacts positively the strength of the formula. Copper also has a good thermal conductivity (401Wm − 1K − 1), which implies and excessive heat transfer to the brake pads, resulting in the destruction of the binder near the friction material, causing an increase in wear. [21-22].

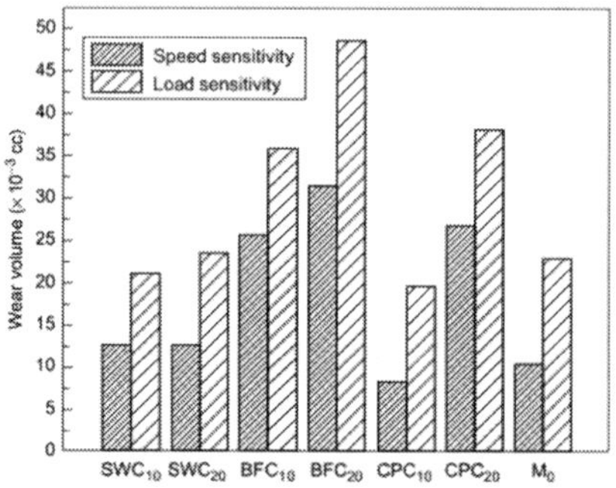

Fig. 8. Cumulative wear behavior of the selected composites in two test modes [21]

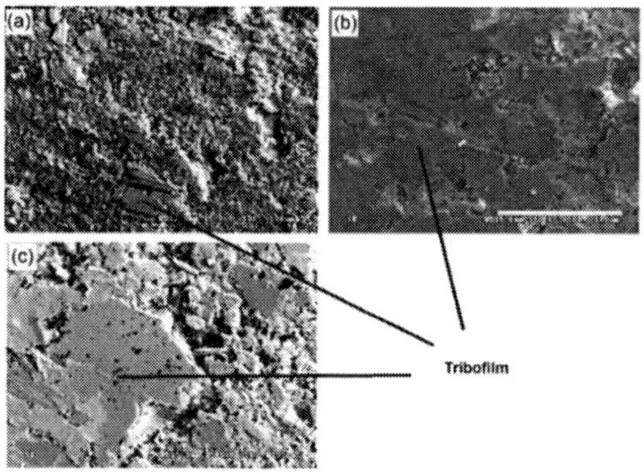

Fig. 9. SEM images: (a) CPC_{10} (low wear), (b) M_0 (reference) and (c) CPC_{20} (high wear) [21]

In principle, Amontons' friction laws apply to friction materials. Still, for resin/cast iron composites, the friction toque and coefficient of friction do not remain constant, due to the variation of the temperature. When the bakes are applied, the temperature at the friction interface increases, resulting in a high temperature of the friction material, even in low operating condition, due to low thermal diffusivity of the friction material. As a result, the friction material is thermally destroyed. Thus, the coefficient of friction, μ, changes with temperature; typically, μ increases slightly until the temperature of the disc or drum is approximately 200-250 °C and then decreases as shown in Figure 9. The exact temperature variation depends on the characteristics of the friction material (Table 5) [20].

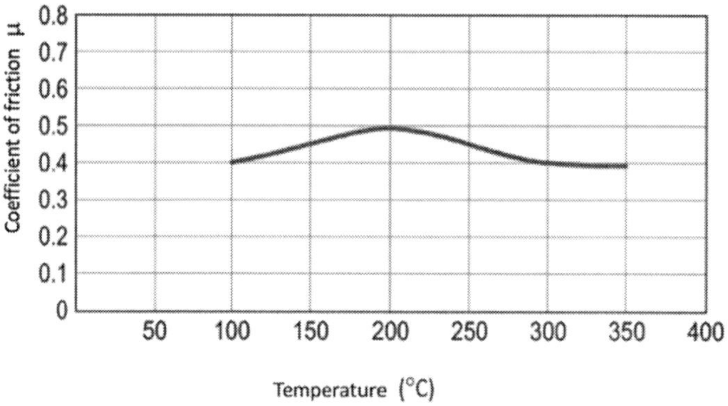

Fig. 10. Variation of the friction coefficient with temperature for resin-bonded composite material [20]

Table 5. Example of specifications of the characteristics of friction materials [20]

Example of specifications of the characteristics of friction materials	
Maximum tensile strength	15 MPa/m^2
Maximum tensile strength	25 MPa/m^2
Rockwell hardness	25 MPa/m^2
Density	1950 kg/m^3
Recommended maximum working temperatures	
Ongoing	250°C
Flashing	350°C

For extended operation at temperature above 300 °, resin-bonded materials associated with conventional cast iron disc, will develop changes in the friction material and probably in the thickness of the plate or friction material. The organic elements that have the role of controlling wear and friction properties, begin to deteriorate thermally. Therefore, the mechanical strength of the material is decreasing, and the friction material is severely impacted. As a result, the surface of the friction material will be "denatured", because all organic elements are burned and only the temperature-resistant elements remain unaffected. Friction and wear resistance are irreversibly damaged [20].

5 Conclusions

Friction materials represent a vital part of car and vehicle brakes. Design engineers must recognize the behavior of wear and friction and the origins of the discrepancy in their properties [15].

Requirements for commercially friction materials are represented by mechanical, friction, and wear properties. These properties indicate recommendation for operating condition and the impact of the temperature over the friction material. Their coefficient of friction is in a strong dependence with their composition, thus automakers will choose friction materials that meet their requirements; for example, for niche cars, friction materials that have a nominal value of the coefficient of friction between $0.38 < \mu < 0.45$, while trucks require durability, thus friction material for trucks require a coefficient friction between $0.35 < \mu < 0.40$. These values represent the starting point for designing the braking system [20].

A friction torque practically follows the friction laws of Amontons, but when the braking load increases and the temperatures generated at the friction interface increase significantly, the coefficient of friction will vary. Often, variations in the coefficient of friction are in accordance with the temperature. These changes are in close dependence with the properties of the friction material. The coefficient of friction of any friction material can only be accurately determined by testing.

References

1. B. Novaković, M. Kavalić, S. Stanisavljev, M. Vlahović, S. Borić, *VI International Conference Industrial Engineering and Environmental Protection 2016, Zrenjanin, Serbia,* (2016)
2. W. Post, *Car braking systems*, Brakes, Brake Control and Driver Assistance Systems, (Springer Vieweg, 2014)

3. S. Vinodkumar, S. Jairaman, P. Anvesh, G. Viswanath, International Research Journal of Engineering and Technology (IRJET), **04**, 08 (2017)
4. S. N. N. Baba, M. N. A. Hamid, S. N. M. Soid, M. N. Zahelem, M. S. Omar, Engineering Applications for New Materials and Technologies, (2018)
5. O. Fazio, S. Nacivet, Applied Acoustics, **91**, (2015)
6. T. Tison, A. Heussaff, F. Massa, I. Turpin, R.F. Nunes, Jornal of Sounds and Vibrations, **333** (2014).
    A. Nobari, H. Ouyang, P. Bannister, Mechanical Systems and Signal Processing, **60-61** (2015)
7. Y. Dai, T. Lim, Applied Acoustics, **69** (2008)
8. Hui Lu, Wen-Bin Shangguan, DejieYu, Mechanical Systems and Signal Processing (MSSP), **93** (2017)
9. S. Oberst, J. Lai, Journal of Sound and Vibration, **342** (2015)
10. H. Festjens, C. Gaël, R. Franck, D. Jean-Luc, L. Remy, Applied Acoustics, **73**, (2012)
11. K. Shin, M. J. Brennan, J.-E. Oh, C. J. Harris, **5**, 254 (2002)
12. K. Shin, J.-E. Oh, M. J. Brennan, JSME Int. J., **2**, 45 (2002)
13. M. Paliwal, A. Mahajan, J. Don, T. Chu, P. Filip, J. Sound Vib., **282** (2005)
14. E. Denimal, J-J. Sinou, S. Nacivet, Journal of Sound and Vibration, **463** (2019)
15. E. Denimal, J-J. Sinou, S. Nacivet, Journal of Sound and Vibration, **463** (2019)
16. S. Vinodkumar, S. Jairaman, P. Anvesh, G. Viswanath, International Research Journal of Engineering and Technology (IRJET), **04**, 08 (2017)
17. P. R. Childs, *Mechanical Design Engineering Handbook. Chapter 13-Clutches and brakes,* (Butterworth-Heinemann, 2014)
18. I. Hutchings, P. Shipway, *Tribology (Second Edition). 9-Applications and case studies*, (Butterworth-Heineman, 2017)
    A. Day, *Braking of Road Vehicles, Chapter 2 - friction and friction materials*, (Butterworth -Heinemann, 2014)
19. R. C. Dante, *Handbook of Friction Materials and their Applications. 3 - Types of friction material formulas*, (Woodhead Publishing, 2016)
20. M. Kumar, J. Bijwe, Tribol. Int., **43(5-6),** 965-974, (2010)

Identifying new criteria for assessing the impact on the ecosystem generated by the establishment of Jerusalem artichoke crops on difficult and floodable soils

Florin Nenciu[1,*], *Gabriel Nae*[1], *Gabriela* Milian[1], *Iulian* Dumitru[1], *Gheorghe* Matei[2], *Simona* Isticioaia[3]

[1]INMA Bucharest, Testing Department, Ion Ionescu de la Brad Blvd. No. 6, Bucharest, Romania
[2]University of Craiova, Faculty of Horticulture, Craiova, Dolj, Romania
[3]Agricultural Research-Development Station Secuieni, 377 Principală, Secuieni, Romania

Abstract. Jerusalem Artichoke (Helianthus tuberosus L.) is a technical plant that manages to adapt very well to unfriendly environments, even in polluted or poor in nutrients soils and may serve in the near future as an important raw material for the food, chemical, and pharmaceutical industries. Furthermore, the plant is growing rapidly, producing large amounts of biomass and offers the possibility to be harvested up to 2 times a year, therefore presents a high potential to be used in the field of biofuel production. The plant's capacity to multiply easily by developing small tubers in the soil, allowing the production of economically advantageous crops, can be considered an advantage for biofuel producers, however this feature might be a drawback for agricultural land owners considering the very invasive behavior. The present paper aim to assess two Jerusalem Artichokes crops established on marginal soils, assessing productivity and the extensive impact on the ecosystem, paying a special attention to plant invasiveness tendencies.

1 Introduction

The biofuels market has grown rapidly in recent years, due to the global goal of limiting the negative effects produced by the widespread use of fossil fuels. In order to meet the increased demand for raw material needed to produce biofuels, very large areas of fertile land have been cultivated with energy plants, creating pressure on the land requirements. Commonly cultivated energy crops are also generating problems associated to the competition with food production, soil depreciation or intensive treatment using various chemicals.

Finding other sources for producing high quality biofuels from new crops, that might grow on marginal, low-grade or contaminated soils, and can develop properly without

* Corresponding author: florin_nenciu2000@yahoo.com

intensive agricultural treatments, led us to consider cultivation of Jerusalem artichoke plant. Jerusalem artichoke is a technical plant, that can be used to produce ethanol either from tubers by fermentation and distillation, either from its rich vegetal mass by complex biochemical processes [1-3].

The need to produce large quantities of high-quality biofuels for partially or completely replacement of fossil fuels in global energy production, puts considerable pressure on the use of several plant species with a high potential to be invasive, that can harm either the native species or the ecosystem as a whole [4].

In order to reduce the potential damage, the best solution would be to design easy to use risk assessment systems, capable to evaluate both the positive and negative effects of introducing high-risk energy crops. Although such assessments have been made in the past, their accuracy remains insufficient to produce positive net economic benefits [5]. In addition, risk assessments must be able to assess some non-invasive species that may have negative effects on soils through the economic problems caused by these plants after harvest. The field of risk assessment for potentially dangerous plant species is now in an early stage of development [6], there is a gap of assessment principles and reliable procedures for identifying the invasive potential of plants in various geographic areas. Identifying the potential hazards and opportunities is the first step to build a valuable risk assessment, being needed the design extensive research studies that could assess aspects such as carbon sequestration [7,8], or certain advanced biofuel production techniques [9-11].

Pest Risk Analysis is commonly defined as a three-stage process (identification, assessing and management of the potential risk) of analyzing biological and other scientific and economic evidence to determine whether a plant should be regulated and the level appropriate measures to be taken [12].

The effort to evaluate and then to manage risks associated with potentially dangerous plant species have to be implemented in a such way to protect human, animal, local plants and soil, without creating unjustifiable obstructions to trading or agriculture. This subject is extremely sensitive, especially if the conclusions of a risk assessment study are taken over by legislative norms that implement various prohibitions.

Plant species that cause serious problems in agriculture are rarely eradicated once they were introduced as crops, and their damages occurs for long periods of time, sometimes imposing aggressive chemical treatment of soils. In order to avoid the costs generated by these management errors and to protect ecosystems in the long term, there is an urgent need to formulate scientific evaluations based on reasonable assumptions.

Most risk studies are performed on whole plant families, either by assuming some probabilities of spread, or by identifying some common hazardous characteristics concentrated within a country or a continent [13, 14].

We aim to analyze the characteristics of a single plant, namely Jerusalem artichoke, establishing the crop on a difficult soil (low nutrient level, floodable land, with a high level of compaction, covered with a large number of other persistent weeds). Although the solution to establish a Jerusalem artichoke crop, where other plants have failed to grow in a proper manner (taking into account the plant's ability to adapt to difficult soils), can be beneficial, a number of characteristics regarding the impact on the ecosystem must still be considered.

2 Materials and methods

The presented methodology examines the main elements that can be used to determine how a Jerusalem artichoke crop established on marginal land might pose risks on environment, providing information and guidance to assist farmers in managing the possible hazards generated by this highly adaptive technical plant, based on a range of reasonable assumptions and indicators [16, 17].

In the first phase we established what were the main pathways that will allow the plant to spread, and to become a negative issue to the soil, since the identification of high-risk pathways is an important part of an overall weed-risk assessment process.

Historical data on the presence of the plant in the area: We established that Jerusalem artichoke was identified in the history of the area, being eradicated twenty years ago, once the agriculture grow in the area. The plant was not quarantined, however suffered from natural eradication due to the way it multiply, by spreading the tubers.

Pathways to spread: Given that this plant existed in the area and has been eradicated naturally, and additionally it needs agricultural activities (for example plowing), to spread to neighboring lands, for the indicator "Pathways to spread ", Jerusalem artichoke received the low grade for rapid spread on arable land. The fact that it has a well-known dispersal mode that is influenced by anthropic activities and can be easily anticipated, there was another advantage in reducing the level of risk regarding this indicator.

Adaptability on new territories: Another indicator that have been analyzed was the level of adaptability of the plant to new territories. Once entering to new lands, plant species must overcome a range of abiotic and biotic barriers in order to establish. Human activities have an important role in assisting these species to make adaptation easier, because the plants that arrive in small number cannot overcome these barriers.

Accessibility and the cost of intervention on culture: An important role in this case is the accessibility and the cost of intervention on the culture. If in the case of some species of trees introduced for example to stop soil erosion, in the case of turnip the situation is easier.

Monitoring in time: The proper evaluation of plant species becomes relevant when analyzing their behavior through time, in this way the reliability of the prediction will increase.

Ease of eradication: The weed control intensity is actually the difficulty of eliminating an individual plant according to some specific attributes of the plant that makes it persistent. It is calculated as the product between the difficulties of an elimination attempt multiplied by the number of attempts, for a single plant. If in the case of normal soils, this indicator would not have been so important because large tubers can be harvested relatively easily, in the case of difficult soils the tubers usually remain very small and are difficult to harvest.

Reproductive capacity: Species with more than one reproductive system are more invasive, than those with only one system (especially plants that multiply through seeds) Jerusalem artichoke has only one way of propagation, through tubers. Given the particularly high resistance of tubers to drought and unfavorable conditions, we considered this indicator as having a relatively high importance for our evaluation.

Controlling local weeds: The Jerusalem Artichoke plants have proven to be extremely effective in controlling local weeds (2-5% of other plants have been identified in the crop), and for marginal agricultural land, this is a key factor.

Invasive behavior of the plant: Establishment of other crops after Jerusalem Artichoke culture is difficult especially on difficult/marginal soils, because many tubers will be too small to be harvested, and will germinate next year, affecting future crops.
Carbon sequestration: The crop has a rich plant mass even in the case of wild propagation, therefore in the case of integration into the soil it has superior advantages regarding carbon sequestration.
Connections with local animals and birds: Species of birds (pheasants) and animals (rabbits) have been identified in the Jerusalem artichoke culture, the high and dense vegetation representing shelter for them.

3 Results and discussion

Establishing and evaluating Jerusalem artichoke crops

Jerusalem artichoke (*Helianthus tuberosus*), is an herbaceous perennial plant that has a high productivity of tubers rich in inulin, which is a fructan polymer [15]. These inulins can be broken down into fructose and glucose for conversion into ethanol by fermentation.

We set ourselves the goal of identifying the differences between the characteristics of a Jerusalem artichoke crop established on difficult soil (difficult to process, floodable and deprived of nutrients) and a crop established on a productive land, having optimal planting conditions and works. In the two established plant cultures we evaluated some indicators envisaging mainly the health of the ecosystem. In order to observe the real invasive effect of the Jerusalem artichoke on the difficult terrain, we monitored the crop for three years (in the first year the tubers were planted and harvested, while in in the next two years we let the crop grow in the wild environment, the single work done being plowing the land in autumn).

A planting installation designed and executed by INMA Bucharest (Figure 1) was used for planting the tubers, towed by an agricultural tractor TD 80D New Holland 65 CP. Have been planted tubers without mechanical damages (having length between 3-10 cm), the machine being adjusted to plant at 100 cm distance per row, at 9 cm depth, the distance between the rows being 80 cm The working speed of the equipment was set to 1.5 ÷ 1.9 km / h, being equipped with a water spraying installation incorporated in the system.

Fig.1. Jerusalem artichoke crop planting using an equipment designed by INMA.

Measurement and calculation of the main indicators in order to assess the impact on the ecosystem

In addition to the potential for ethanol production on the two crops, we aimed to analyze the effect that wild-grown culture has on the ecosystem. Unlike classical invasiveness studies, we also considered the positive elements that this plant can have on the ecosystem when it is planted on difficult agricultural soils. Figure 2 is suggestive for the plant grown wild from the bulbs remained in the soil, in the 3rd year after planting. Can be seen that Jerusalem artichoke plant growth is not much affected by the e soil type even if it is low in nutrients, managing to adapt and thrive in difficult soils.

Fig. 2. Jerusalem artichoke crop grown wild from the tubers that remained in the soil, in the third year after planting

Although the vegetal mass was abundant, have been noticed that productivity on difficult soil was much lower, the tubers remaining very small and undeveloped.

The main elements that contributed to the underdevelopment of the tubers were the lack of precipitation (the crop is not irrigated), the high level of soil compaction and the high density of plants grown per square meter.

Figure 3 exemplifies the measurements made to evaluate the culture parameters, such as the number of plants grown per square meter, plant sizes, the number of tubers harvested per square meter, the number of inflorescences and plant vegetal mass.

Fig. 3. Evaluating the development parameters of a Jerusalem artichoke crop grown wild on difficult soils

The benefits of the crop such as high nectar production, rich plant mass that allows carbon sequestration in the soil, the potential to represent a habitat for birds and animals [18], were all assessed using techniques based on indicators and scoring. (Fig. 4).

Fig. 4. Highlighting the honey potential of Jerusalem artichoke culture in different time periods

We also evaluated the potential of Jerusalem artichoke crop to control local weeds on difficult soils, considering that this plant might impose when competing with very persistent plants. In the case of wild-grown crops, Jerusalem artichoke have proven to be very effective in controlling local weeds, leaving only a 1-4% percentage of other plants to grow, and even these have been found underdeveloped.

The weeds that could still be identified in the Jerusalem artichoke culture were Amaranthus retroflexus, Setaria viridis, Carduus nutans, Senecio Vulgaris, but in a very small percentage (Figure 5).

Fig. 5. Jerusalem artichoke control of local weeds on difficult soils

Productivity for the crops established on soils with different characteristics (fertile soils and optimal conditions versus difficult soils and unfavorable conditions), showed a much lower productivity of the total quantity of tubers produced on wild-grown crops. Analyzing 1 square meter from twenty plots, collected from both cultures, were found differences up to 40% productivity in favor of fertile lands (Figure 6).

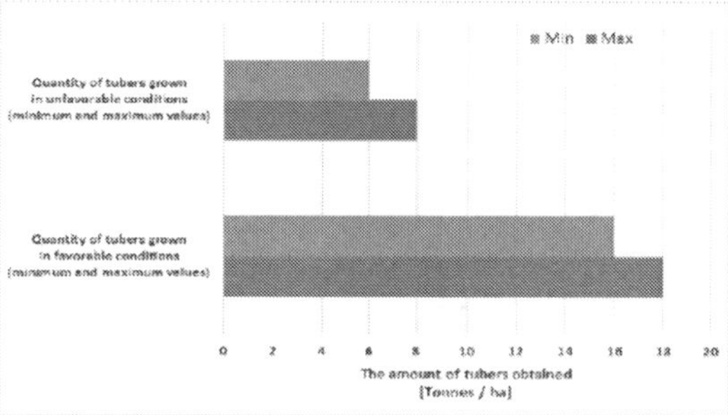

Fig. 6. Comparative analysis of the mass of tubers produced both in favorable planting conditions and in unfavorable planting conditions (minimums and maximums recorded for each category)

Fig. 7. Differences between the dimensions of the tubers obtained in the two cultures favorable planting conditions first picture) and in unfavorable planting conditions (second picture)

Not only the number but also the size of the tubers was affected, being smaller by up to 10 times, as can be seen in figure 7. The size of the tubers was actually the element that made the difference. The size of the tubers was actually the element that made the difference in the productivity for the two cultures. A determining factor here is the amount of rainfall, which influenced decisively the wild-grown crop, that has not been irrigated.

The biomass production of the two crops have been comparatively assessed in order to associate with carbon sequestration and with the potential to produce ethanol from conversion of the stems to alcohols, using thermochemical process. By incorporating the vegetal mass in the furrow, can also obtain an improvement of the soil quality [18]. In the thicket formed by the Jerusalem artichoke stems, different types of birds and animals can find shelter against predators and can nest, this being a very valuable element for the ecosystem. There have been determined low variations in the amount of stems mass per hectare (Figure 8), the wild-grown crop developing larger number of plants while the fertile soils allowed the development of taller and thicker stems, but fewer in number.

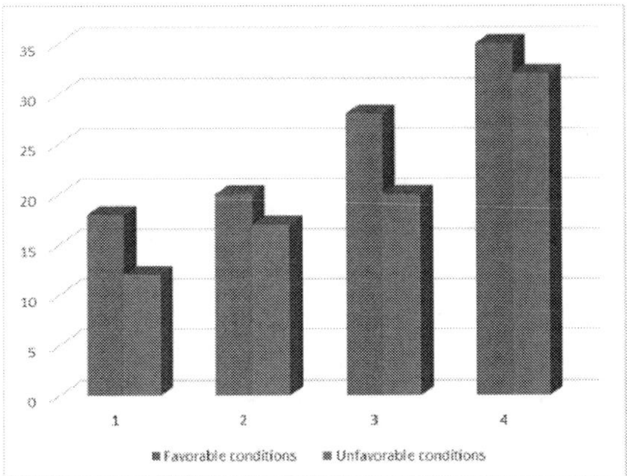

Fig. 8. Comparative analysis between plant mass of stems for favorable conditions versus unfavorable conditions

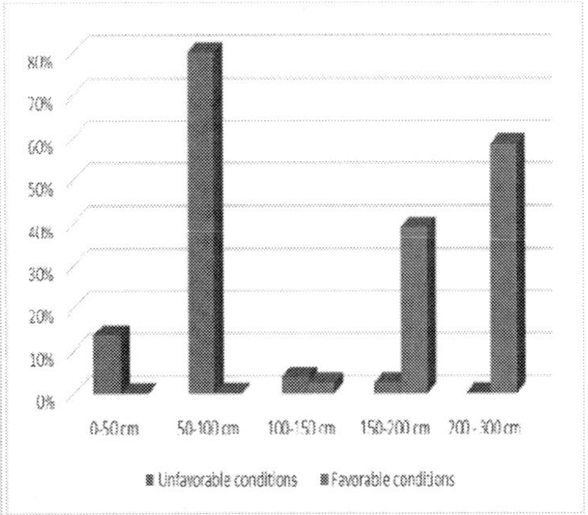

Fig. 9. Plant height analysis for Jerusalem artichoke crops

Numerous differences have been identified regarding the plants size assessment on the two cultures. Plants grown on fertile land were much taller, a large share being in the height ranges between 150-200 cm and 200-300 cm. In the case of crops established on difficult land, the vast majority of plants were in the range of 50-100 cm, but they were in a much larger number. Among the causes that generated the difference, were on the one hand the drought (the crop established on the difficult soil has not being irrigated) and the fact that on the fertile land the planting technology was established with the standardized distances between rows, while on the wild culture, the plants grew chaotically (Figure 9).

In terms of alcohol production, Jerusalem artichoke has a lower potential than other similar plants (such as potatoes), due to the difficulties in converting inulin to sugars. From a qualitative point of view, the differences between the tubers harvested from the two soil types was not different, although from a quantitative point of view, the differences in tonnage of tubers per hectare were large.

Although the main indicators are presented in the following table according to the impact on the ecosystem, the results have been assessed chronologically on three stages: **Planting stage** (ease of planting, planting mechanization, planting efficiency), **Plant development and monitoring** (maintenance works, effect on weed eradication, plant adaptability, easy monitoring, pest resistant, etc.) and **Harvesting stage** (tuber productivity, vegetal mass productivity, easy eradication etc.).

Table 1 Centralizing table of the main indicators for assessing the impact on the ecosystem, following the establishment of a Jerusalem artichoke crop on difficult soils.

Jerusalem artichokes evaluation indicators for impact on ecosystem when planting on difficult soils												
Indicator		Observation	Scoring impact									
			1	2	3	4	5	6	7	8	9	10
1	Historical data on the presence of the plant in the area	Historical data were found on the existence of Jerusalem artichoke crops in the area, but this plant is eradicated through intensive agriculture.	1	2	3	4	5	6	7	8	9	10
2	Pathways to spread	The plant multiplies by tubers, helped by farmers' activities (especially plowing). The spreading of the tubers to the neighboring lands was not significant. Unlike other plants that multiply by seeds, the culture has been reduced to this indicator.	1	2	3	4	5	6	7	8	9	10
3	Adaptability on new teritories	Although Jerusalem artichoke has a good high adaptability to new and fertile soils, planting on these lands is not recommended, unless the aim is to obtain food or medicine (our analysis aims to obtain biofuels).	1	2	3	4	5	6	7	8	9	10
4	Adaptability on difficult terrain	Although lower productivity is generally obtained, Jerusalem artichoke adapts very well to difficult terrain.	1	2	3	4	5	6	7	8	9	10
5	Accessibility and the cost of intervention on culture	A negative factor is the cost of intervention on the culture for eradication. It is made through expensive processes, and accessibility to small tubers in the soil is low.	1	2	3	4	5	6	7	8	9	10
6	Monitoring in time	The crop was monitored for 3 years, two years it was worked only in autumn by plowing. In this way the management of a marginal agricultural land was simulated.	1	2	3	4	5	6	7	8	9	10
7	Ease of eradication	The plant is difficult to eradicate	1	2	3	4	5	6	7	8	9	10
8	Reproductive capacity	High reproducibility through tubers, especially in rainy years	1	2	3	4	5	6	7	8	9	10
9	Controlling local weeds	A very high level of weed control, the crop almost completely suffocated the other plants	1	2	3	4	5	6	7	8	9	10
10	Dominant behavior of the plant	A dominant behavior in relation to other plants	1	2	3	4	5	6	7	8	9	10
11	Drought resistant	High resistance to drough, representing one of the advantages of the plant in cultivation on marginal lands. However, there is the disadvantage of low productivity in dry periods of time on unirrigated land	1	2	3	4	5	6	7	8	9	10
12	Pest resistance	Highly resistant to pests of all types, the crop, even if untreated, shows very good results	1	2	3	4	5	6	7	8	9	10
13	Vegetable mass production on difficult lands	Produces large amounts of vegetable mass even if the plant is not sown, but it grows from the tubers left in the soil from previous years	1	2	3	4	5	6	7	8	9	10
14	Carbon sequestration	If the Jerusalem artichoke crop has a large amount of moisture and produces a large amount of plant mass, then by integrating it into the soil it has a major impact on carbon sequestration.	1	2	3	4	5	6	7	8	9	10
15	Potential invazivitaty on difficult soils	Difficult soils favor the invasive nature of the plant, due to the small tubers that remain in the soil after harvest	1	2	3	4	5	6	7	8	9	10
16	Potential controlling local weeds	Given that it is considered to plant the crop on marginal / difficult **agricultural land** (and not in protected areas), weed control is a positive indicator.	1	2	3	4	5	6	7	8	9	10
17	High nectar production	The crop has a particularly good effect on insect species in the area (especially bees), by producing large amounts of nectar. Jerusalem artichoke grown in wild areas blooms in the fall when there are not many sources of nectar in nature.	1	2	3	4	5	6	7	8	9	10
18	Connections with local animals and birds	Animals and birds tend to use wild-raised Jerusalem artichokes to make their burrows / nests. Rabbits and pheasants were found in the Jerusalem artichoke crop.	1	2	3	4	5	6	7	8	9	10
19	Problems in establishing new crops after Jerusalem artichoke crop	Small amount of tubers obtained per hectare	1	2	3	4	5	6	7	8	9	10

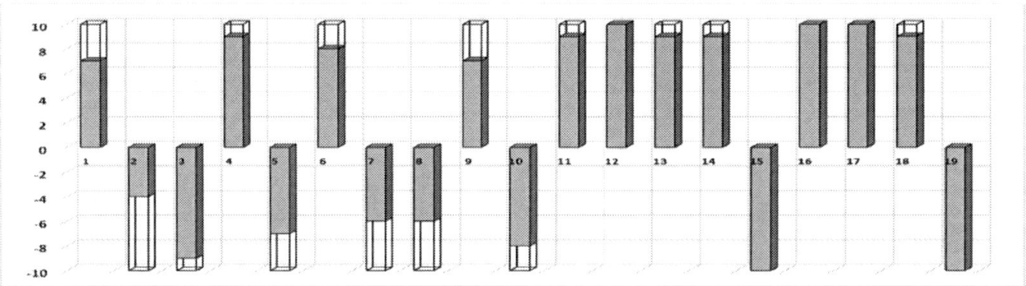

1	Historical data on the presence of the plant in the area	8	Reproductive capacity	15	Potential invasiveness on difficult soils
2	Pathways to spread	9	Controlling local weeds	16	Potential controlling local weeds
3	Adaptability on new territories	10	Dominant behavior of the plant	17	High nectar production
4	Adaptability on difficult terrain	11	Drought resistant	18	Connections with local animals and birds
5	Accessibility and the cost of intervention on culture	12	Pest resistance	19	Problems in establishing new crops after Jerusalem artichoke crop
6	Monitoring in time	13	Vegetable mass production on difficult lands		
7	Ease of eradication	14	Carbon sequestration		

Fig. 10. Graphic representation of the influence of a Jerusalem artichoke crop established on difficult soils on the ecosystem, where the indicators 1-19 are listed above

3 Conclusions

In order to ensure the reduction of our dependence on fossil fuels, it is mandatory to develop the domain of biofuels, both for those obtained from planting technical crops marginal or depleted soils and biofuels obtained from biomass waste.

We have highlighted in this article that Jerusalem artichokes crop have a high potential for ethanol production, and due its special properties of adaptability and resistance to extreme drought conditions, can be planted on marginal, polluted, nutrient-free or floodable areas.

By analyzing the problems generated by this crop in terms of invasiveness and the effects generated on the ecosystem, farmers can decide consciously whether Jerusalem artichoke present a high risk or is a good opportunity.

Jerusalem artichokes crops showed a good growth in difficult conditions and in less fertile soils, even though the humidity have been regulated only by rain. Without irrigation and agricultural work, the culture succeeded to grow a vegetable mass in the less fertile land was maintained at values close to the values obtained on fertile lands.

However, the quantity of tubers produced, suitable for producing alcohol was very small in the case of crops on difficult lands, in the best cases being 40% smaller than those in ordinary lands.

Jerusalem artichokes shoed a very good control of local weeds, only 1-4% of other plants survived in a culture dominated by Jerusalem artichoke for a period of 3 years.

Planting a crop on marginal lands does not necessarily affect other planned crops. However, in the case of fertile lands, if the farmer desired to establish another crop after Jerusalem artichoke, he must consider planting a very resistant variety to compete with the tubers remained in the soil.

In terms of impact on the ecosystem, Jerusalem artichokes crops offers important benefits, especially for wild-grown plants. It is a shelter for some species of animals and plants, and is a honey producing plant very appreciated by insects because it produces nectar in late autumn periods.

Physico-chemical analyzes showed similar potential for alcohol production in the two crops, however on the culture established on difficult soils a much smaller quantity of tubers was obtained. Because there are many variables in the analysis of production from a quantitative point of view, such as the presence of precipitation (since in the analyzed year the average annual amount of precipitation in the area was very low), for a better estimation a multiannual analysis is needed.

This work was supported by a grant of the Romanian Ministery of Research and Innovation CCDI - UEFISCDI, project number PN-III-P1-1.2-PCCDI-2017-0566, Contract no. 9PCCDI/2018, and PNCDI III, contract no. 16PFE

References

1. A.Eliasson, Master thesis, Lund University, Department of chemical Engineering, Sweden, (2017)
2. S.J.Kays, S.F., CRC Press Taylor & Francis Group, New York, United States, (2008).

3. Y. Qiu, P. Lei, Y. Zhang, Y. Sha, Y. Zhan, Z. Xu, S. Li, H. Xu, P. Ouayang, Biotechnology for biofuels, **11**, 151, (2018)
4. R. Meyer, A. Grunwald, C. Rösch, A. Sauter, Working report, **121**, Germany, (2007).
5. Y. Wu, F. Zhao, L. Wang, L. Qiu, G. Alexandrov, Geoscience Letters, **5** (14), (2018)
6. A. Bielecki, S. Ernst, W. Skrodzka, I.Wojnicki, Environmental Science and Pollution Research, **27**, 11506-11530, (2020)
7. E.C. Toochi, Forestry Research and Engineering : International Journal, **2** (3), Nigeria, (2018)
8. B. Zheng, D. TongLi, M. Liu, F. Hong, C. Geng, G. Li, H. Li X, L. Peng, J. Qi, L. Yan, Y. Zhang, H. Zhao, Y. Zheng, K. He, Q. Zhang, Atmos. Chem. Phys., Trends in China's anthropogenic emissions, **18**, 14095–14111, (2018)
9. S.S. Singh, R. Dhaliwal, M. Puri, Society for Industrial Microbiology, **35**, 777-782, (2008)
10. P. Thanokeo, S. Thanonkeo, K.Charoensuk, M. Yamada, African Journal of Biotechnology, Research Gate, **10**(52), 10612-10617, (2011)
11. V. Starovoytov, O. Starovoytova, N. Aldoshin, A. Manohina, Acta Technologica Agriculturae, **1**, 7-10, Slovacia, (2017)
12. European Environment Agency, EEA Technical report No 4/2008, **4**, ISSN 1715-2237, (2008)
13. J. Popp, Z. Lakner, M. Harangi-Rákos, M. Fári, Renewable and Sustainable Energy Reviews, Elsevier, 559-578, (2014)
14. A. Hina Fathima, K. Priya, T. Sudakar Babu, K.R. Devabalaji, M. Rekha, K. Rajalakshmi, C. Shilaja, Ijiret, **3** (1), (2014)
15. S. Bhagia, H. Akinosho, J.F.S. Ferreira, Biofuel Research Journal, **14**, 587-599, (2017)
16. E. Aguaron, E.G. McPherson, Urban Ecosystems and Social Dynamics Program, 43-71, USA, (2012)
17. U.S. Environmenal Protection Agency, Combined Heat and Power Partnership, USA, (2015)
18. F. Nenciu, V. Vlăduţ, IOP Conf. Ser.: Earth Environ. Sci., **635**, 1-13, (2021)

Estimation of the stress state of agricultural harrows with vertical axis rotor, using mathematical models with finite elements

Valentin Cornel Iordache[1,*], *Ion* Saracin[1]

[1] University of Craiova,Romania

Abstract. In this paper we simulated through mathematical models the interaction between the agricultural harrow with vertical rotor (equipped with 4 boxes per rotor) and soil. Due to the large or even very large number of mathematical models that can be developed and can receive a solution using the finite element method, there are also many new possibilities to approach and improve some already developed models. The most important benefits of using finite element modeling are: increasing the quality of projects by checking the resistance, noticing dangerous vibration problems, remedying deficiencies reported in the testing of experimental models or even in operation. Normal optimization aims to reduce material consumption, its own vibration spectrum or other mechanical qualities.By using mathematical methods, a resistance check of the subassemblies of the working member (knife), connected to the action device, is obtained.

1 Introduction

In this paper we will approach a slightly more complicated model, which contains in addition to four knives and a ring for fixing the knives to the drive system. In addition, the knives do not have the simplest shape, but have the shape of a curved bar that shapes (approximately) the shape of the knives of the agricultural harrow with vertical rotor[7], which can be observed in detail in fig.1. The simplest possible model could be the knife (working organ), simple organ, most often in the form of a bar fixed rigidly at one end and actuated laterally and frontally in the cycloidal direction of the knife.

The geometry of the model, the orientation and the elements of discretization (nodes and finite elements), are presented in fig. 2. The finite elements used are all Tymoshenko type bars, with the characteristics given in the documentation of the COSMOS / M 2.8 program, used for this modeling, [4.8].

In fig. 3 is show the structural model including boundary conditions (connection to the environment or fixation) and loading (forces)[5]. The fixing is done by canceling all the six components of the displacement (three translations and three rotations) in four nodes located equidistant in relation to the nodes in which the working organs (knives) are

* Corresponding author: valentin_jordache@yahoo.com

connected on the ring[2.3]. The loads of 100 N on each knife are evenly distributed along the length of the lower segment of the knife, a length which, in this example, represents the working depth[1.6]. The forces are oriented in the tangent direction to the circles on which the knife points rotate, simulating the reaction of the ground on the knives when the difference between the mechanical characteristics of the ground through which the knives did not pass and those of the ground through which the knives passed is not taken into account. There are also alternatives closer to reality for loading knives. Given the low speed of movement of the aggregate, it can be hypothesized that on a portion of a quarter circle from the tractor, the knives process hard soil, on the sides work in partially processed soil, and behind the equipment work in already processed soil. With the help of this hypothesis, results closer to reality are obtained. However, the hypothesis used in this article is more covering, taking into account the harsh regime of knife operation (shocks and intense friction).

The structural models presented in this paper are three-dimensional, combined with finite elements of bar type and solid type. The main purpose of these mathematical models is to estimate as close as possible to reality, the stress state of the knifes material, in the context in which experimental validation on knives is difficult to achieve.

Fig. 1. Detail of the four rotors that equip the agricultural harrow with vertical rotors.

2 Materials and methods

The material analyzed in the article consists of the agricultural harrow with vertical rotor (fig.1), formed by working organs (knives), connected to their rotation system.
The analyzed harrow has four rotors that can be equipped with two, three or 4 knives per rotor. The dimensioning of the main dimensions used for the abstraction of the puree model with 1D bar finite elements and for the geometric construction of the puree model with solid finite elements (3D) is observed in fig. 4. A detail of the working element is presented in fig. 3.

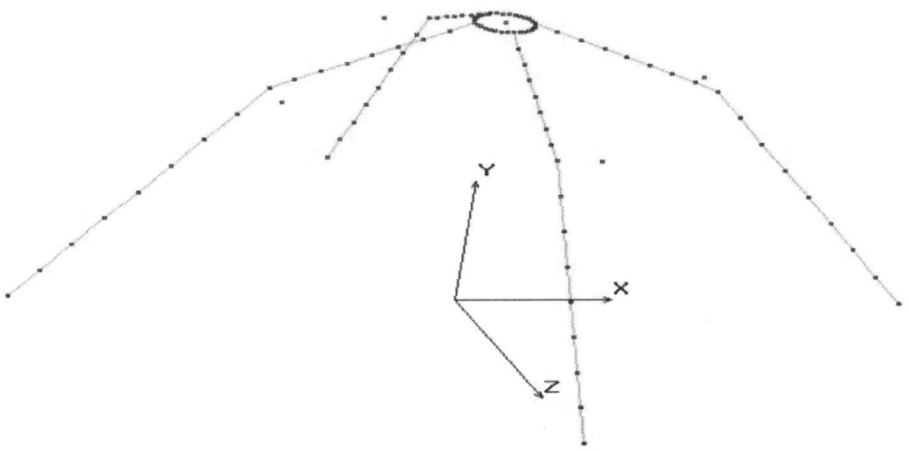

Fig. 2. The structural model of the rotor subassembly formed by knives and a connection ring with the motion transmission system.

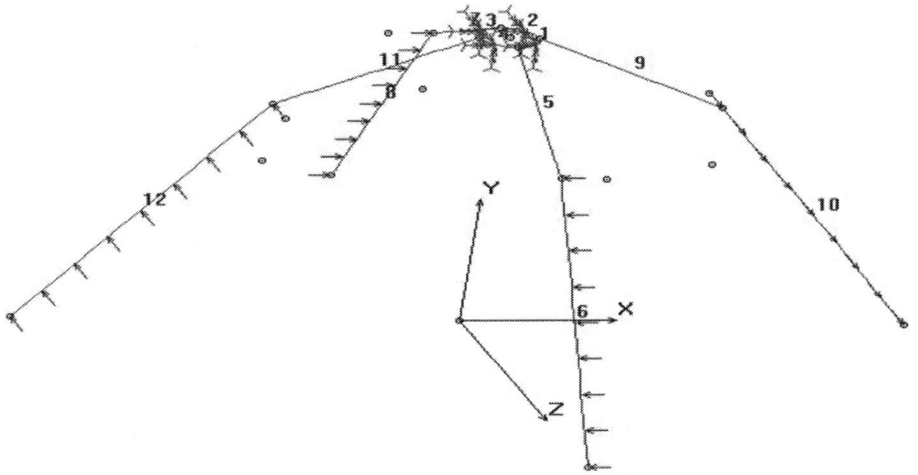

Fig. 3 The structural model of the rotor subassembly formed by knives and a connection ring with the motion transmission system.

The geometry of the mashed structural model with bar type elements, 1D, is given in fig. 4.
For the structural model with 3D dimensional finite elements, the geometry is represented graphically in fig. 5. For the construction of the geometry, in the case of both models, we used values of the main dimensions defined in fig. 2 and 3: a_1 = 0.15 m, a_2 = 0.2 m, b = 0.02 m, c = 0.05 m.

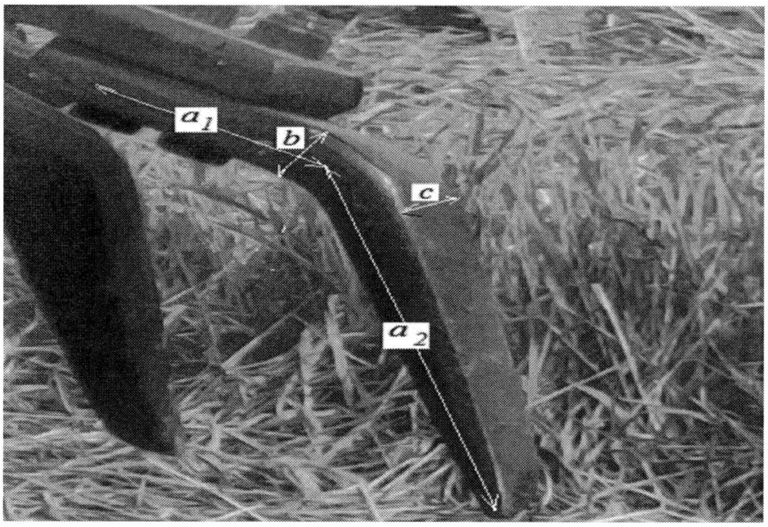

Fig. 4 Average dimensions for basic models

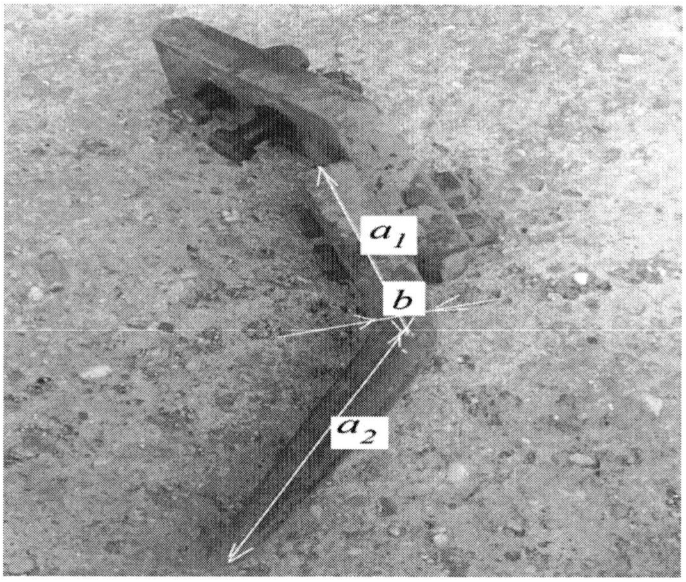

Fig. 5 Average dimensions for basic knife

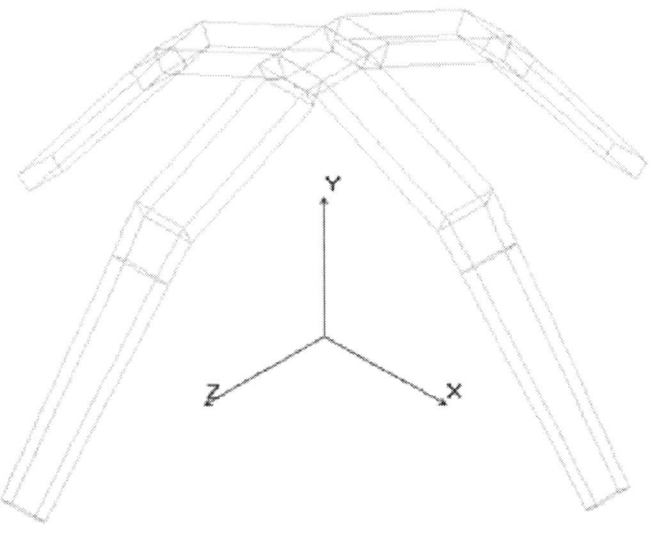

Fig. 5 Structural model of the subassembly of workpieces connected to the actuator, 3d model

The material of the bars of the structural model of the garv rotor subassembly works (except for some accidents) only in the linear elastic field. Consequently, the material is characterized by the following three constants: the modulus of linear elasticity, E=2.1•10[11] N/m2 the transverse contraction coefficient, ν = 0.29 and the mass density, ρ = 7850 kg / m[3]. The inertial constants of the sections of the three types of bars used (ring bar, knife top bar and knife bottom bar) are given in table 1.

Table 1 the inertial characteristics of the bar sections of the structural model in fig.2-3.

Characteristic	Ring bar, circular crown profile with a diameter of 40 mm and a thickness of 4 mm	The upper part of the knife, rectangular profile 50x20 mm	Bar of the lower part of the knife, rectangular profile 35x15 mm
Area of the section, m^2	$4.521599 \cdot 10^{-4}$	0.001	$5.25 \cdot 10^{-4}$
Moment of inertia Iy, m^4	$7.415422 \cdot 10^{-8}$	$2.083333 \cdot 10^{-8}$	$9.84375 \cdot 10^{-9}$
Moment of inertia Iz, m^4	$7.415422 \cdot 10^{-8}$	$3.333333 \cdot 10^{-8}$	$5.359375 \cdot 10^{-8}$
Depth of section, m	0.04	0.05	0.035
Section width, m	0.04	0.02	0.015

For the 1D and 3D structural models shown in fig (3-5), the forces are applied on the faces of the working organs that interact frontally with the ground, and on the support of the structures it is done by cancelling all degrees of freedom (three translations and three rotations), in the upper part of the structure. The forces are applied to the faces of the working body that interact frontally with the ground. The tensile strength is intensely variable over time, so it, for static analysis, we applied an average on each organ, taking into account the fact that the four working organs move in uncultivated soil on a single already fully processed. The theoretically calculated forces have been corrected to achieve the total traction force and the total power consumed in experimental tests of the rotary agricultural harrow tractor with vertical axis rotors. In this valuable way for the 1D model, they are distributed in 9 nodes on each knife, and for the 3d model on all the nodes of the faces exposed to the frontal presence of the ground.

In fig.6, is represented the distribution of the reaction forces in the fixing nodes of the structures, located on the ring. The maximum value of 1109.2 N is reached in the bearing points. As it is observed the application of some pairs of strong structures as a whole, it leads to higher forces in the bearing points. This fact draws attention to the need for projects and the realization of safe mobile couplings in order not to wear them prematurely and / or to produce work accidents with possible very serious consequences.

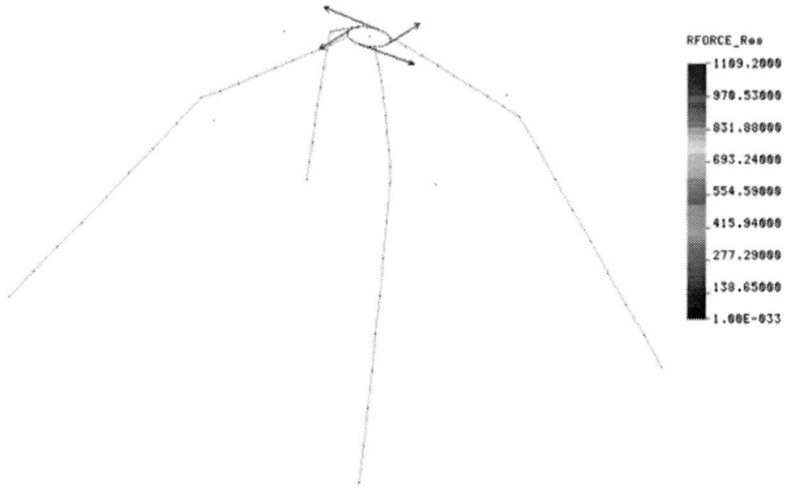

Fig.6 Reactions at the fixing points of the knives on the ring, in N

3 Results and discussion

The results of loading the structural model of the vertical subassembly are: the relative displacement resulting in the structure, having the maximum value 0.00013 mm, located at the free ends of the knives, the maximum equivalent tension, with the value 18.831 MPa, located on the knives, in the vicinity of the connection area . support ring. The maximum value of 1109.2 N is reached in the bearing points. As can be seen from the application of pairs of strong structures as a whole, it leads to higher forces at the support points.
A result obtained using the three-dimensional model is the fundamental eigenfrequency with the level 121.716 Hz and the next eigenfrequency of the structures, with the value 137.628 Hz. These natural frequencies are useful in the design calculation, but especially in the calculation of improving the intensity of intense vibrations that can be generated by resonance phenomena. When making this statement, we take into account the values that can reach, for example, the speed of the rotor (higher than those used in the experiments we performed).

 https://doi.org/10.1051/e3sconf/202128603023

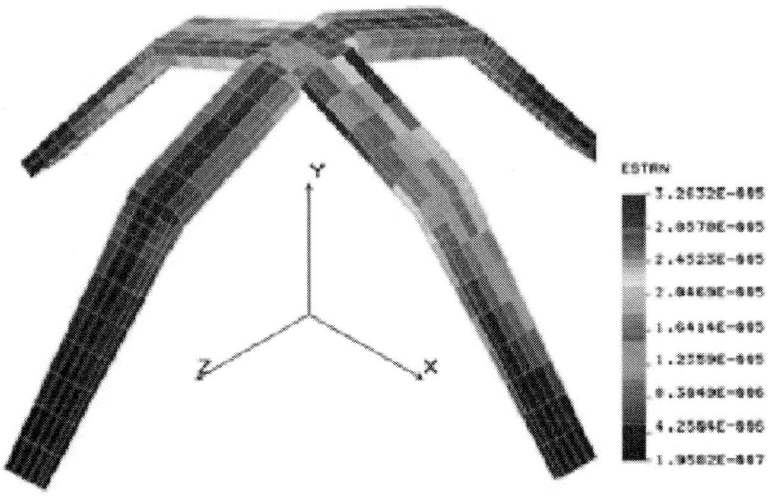

Fig. 7 Distribution of the total strain field on the boundary of the structure analyzed in 3D model

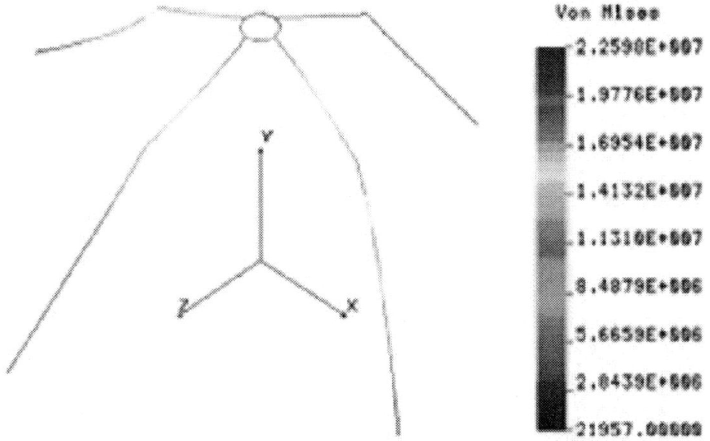

Fig.8 Equivalent stress state (Von Mises) in the structural model, in Pa in the 1D model.

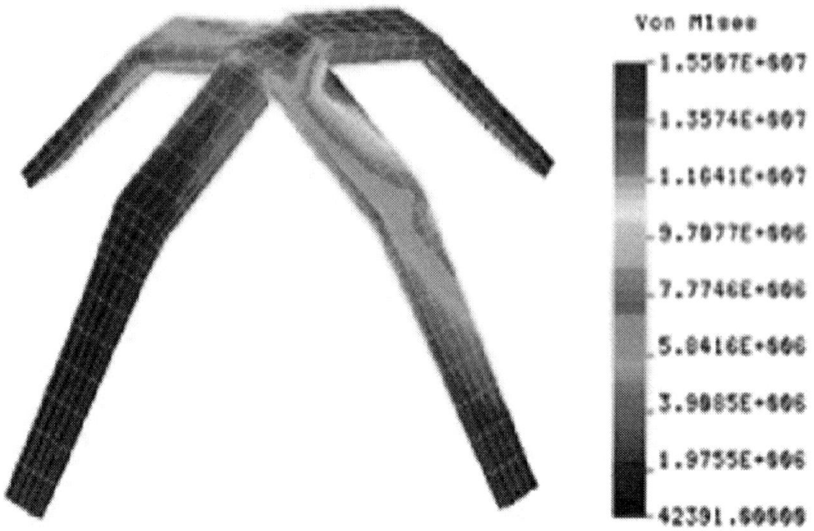

Fig.9 Diastribution of the equivalent stress field on the boundary of the structures, in Pa in 3D model

In order to compare the results obtained by the 1D and 3D methods, the distribution of the equivalent stress is represented in fig 8 and 9, obtaining the variant of the structural model 1D and respectively 3D.

4.Conclusions

- the structural models proposed and solved with the help of the finite element method, provide useful information for the activity of conception and design of the modeled elements and subassemblies. The stresses and deformations in the analyzed structure can be estimated and thus the safety factor can be estimated, so their resistance. From this point of view, the analyzed structure is very correctly dimensioned, apparently even oversized. In general, however, given the inhomogeneity of the soil, its anisotropy and high peripheral speeds, violent shocks can occur in operation, which means an increase of over 2.5 times the applied force, possibly in much smaller areas than in the examples.

-in the presented models you can change materials, geometry, loading to simulate various phenomena. The model with 1D element admits the fastest change of the geometry of the section, requiring only the change of the inertial characteristics of the cross section.

-the analyzed models can be used to diagnose in operation some vibration regimes with resonance. The eigenfrequencies of the analyzed structure are obtained by analyzing the spectrum of own frequencies of the elements and subassemblies of the modeled physical structure (in our case the subassembly of working organs). In this concrete case, the working frequencies of the rotors (17-22 Hz) are very low comparing with the first two natural frequencies of the examined subassembly (121.716 and 137.628 Hz).

-for the structural models built in this paper, it was considered that the material of the structures have an elastic linear behaviour. If the appearance of irreversible deformations is studied, then, materials with nonlinear properties must be considered - rigidity of the structure and the theoretically estimated values of the forces, used in mathematical models, exclude the occurrence of the irreversible deformations in the structure..

References

1. M. Voicu, Teoria sistemelor, Editura Academiei Romane, (2008)
2. *Foley, James D. (1996), "12.7 Constructive Solid Geometry", Computer Graphics: Principles and Practice, Addison-Wesley Professional, pp. 557–558, ISBN 9780201848403*
3. M. Li, Q. Zhang, H. Wang, H. Sun, J. Tong, A 3D FINITE ELEMENT SIMULATION ANALYSIS OF THE SOIL FORCES ACTING ON A ROTARY BLADE, Transactions of the ASABE (American Society of Agricultural and Biological Engineers) · January (2015)
4. Structural Research & Analysis corp., COSMOS/M 2.7 CAD Interface User Guide, 2007Foote R. L., The Volume Swept out by a Moving Planar Region, *Mathematical Magazine*, Vol. 79, No. 4, 2006, pp. 289-297
5. Rades M., Analiza cu elemente finite, (2006)
6. Tolvaly-Rosca F., Pasztor J., Work process analysis of the machines with working parts entrained, in seedbed preparation works, INMATEH Agricultural Engineering, vol. 58, 9:16, No. 2 / (2019)
7. Cardei P. &all, Structural analysis and materials focused on mechanics, mechatronics, maintenance and operation of machinery for agriculture and food industry, Terra Nostra Iasi publishing house, (2012)
8. ASABE (American Society of Agricultural and Biological Engineers, ASAE D497.7 MAR2011 Agricultural Machinery Management Data, Standard, (2011)

Research regarding the common problems of design of the bunkers – A review

Ancuta Alexandra Petre[1], *Mariana Mădălina* Stanciu, *Dumitru Bogdan* Mihalache[1] and *Dragoş* Dumitru[1]

[1] INMA National Institute of Research - Development for Machines and Installations designed to Agriculture and Food Industry, Bucharest

Abstract. Understanding the behavior of granular material in a bunker not only helps to design properly and achieve the correct granular flow, but also needs the engineer to ensure that the bunker is properly designed and conducted to withstand the tasks and demands exercised during its operation. Small and shallow bunkers are commonly used in industry when free space is limited. Shallow bunkers induce the flow of the funnel, which can cause problems such as segregation and agglomeration of the material inside it, which leads to considerable damage to the final product. The design and location of a suitable power supply and an appropriate outlet is still critical and is perceived rather as a science.
The paper presents some innovative solutions related to the design, modeling and construction of solid material storage and unloading bunkers.

Keywords: bunkers, seeds, design, problems

1 Introduction

Bunkers are used in the food and construction industry for the protection and storage of powdered materials or cereal seeds. Bunkers must be designed in such a way that they are easy to load so that they can be handled by operators. In particular, bunkers also be adequately designed so that granular or semifinal material is easy to unload [1].

The utility of adjustable bunkers is particularly important in the technologies that involve conditioning of wheat seeds or for various equipment used for separating impurities. Technologies that use rotating cylindrical sieves involve the need to create a continuous flow of material for processing, therefore the design of the bunker significantly influences the efficiency of the separation process [2].

The way the bunker is designed affects the flow rate of the powdery material, as granular particles can agglomerate inside it. Also, the way the bunker is designed affects the amount of stored material that can be discharged or the stationary space that reduces the actual storage capacity of the bunker [1].

The flow of gravity-driven granular material has fascinated human beings for millennia, either from hourglass devices for time-tracking or silos for storing grain and other granular products. Currently, industrial processing of granular materials is common and silos are regularly used as storage vessels. The physics of granular flow in general is underdeveloped compared to that of traditional Newtonian fluids, such as water, mainly due to the different states of aggregation (solid, liquid, gas type) in which the fluid can be found [3].

Many current studies on granular flow through an opening or slit are based on uniform grains. In almost all experimental or numerical relationships for calculating granular flow, the distribution of granule size or the effect of granule size is ignored. In nature, all phenomena related to granular discharge, such as seed injury due to the tilt of the bunker wall or due to faulty exhaust pipe, these problems occur frequently and the result is the uneven distribution of granules or seeds [4].

From the beginning of the silo storage to contemporary bulk storage of agricultural and industrial materials, bunkers have participated in a prominent function as some of the most common granular material storage structures in the world.

This has led to the need for complete bunker projects, which have also become imperative because of the disastrous malfunctions that have occurred recently [5].

Given every one of the issues identified with the construction and working of the bunkers, it is a fairly huge test to look at the conduct effect of granular material in ssiloss and to propose some creative specialized and factual arrangements on granular stream during filling, stockpiling and dumping measures.

The synthesis attempts to highlight certain clear distinctions between these different situations. The above-mentioned aspects presented in this paper are to be taken into account when designing storage and unloading bunkers.

2 Materials and methods

Silo is usually used for storing food seeds in most countries, through its prism cereals are protected from diseases and pests. Silos can be constructed of metal, mud, concrete, and plastic, although no construction has proved to be absolutely satisfactory. The storage time of grain in the silo varies from 6 months to a few years, and storage capacities also vary, depending on the size of the silo [6].

Historically, the development and design of silos has been entirely a practical problem. Their design was conducted by agronomists, farmers or engineers and adapted to their needs. Although a large number of researchers have suggested numerous ingenious geometric shapes, the classic design of storage silos for consumable cereals is based on vertical-cylinder multicell facilities. Within the vertical bunker, the seeds are forced to adapt to the shape of the silo, without considering the natural resting angle of the material.

Some researchers have shown that during the emptying process, large friction forces appear on the walls of the silo, as well as additional dynamic demands. In addition, these considerable bunker design problems lead to the enormous damage of the seed grains caused during the filling and emptying processes, being an additional impediment in their design [7, 8, 9].

The predominant importance of bunkers and silos for a wide variety of industries is evident by the intense efforts of many researchers to obtain models for the unloading behavior of the particular storage devices [10, 11, 12, 13]. The size of the mass flow itself and the difficulties in maintaining reliable and uniform flow behavior are of great interest, as these problems can have major impact consequences.

The sensitivity of metal silos to flame failure is the most important design criteria. The non-symmetric pressure regime that arises during the eccentric discharge of stored granular material frequently results in catastrophic buckling failures, and thin silos are especially vulnerable to this form of failure.(Figure 1).The most common way of damage in an eccentric thin-walled steel silo is the discharge of granular material that occurs through serious damage to the adjacent part [14].

Fig. 1. Metal bunker [28]

The most basic part of silos pressures is their impact on the planned design that should contain strong material. Since the properties of strong materials shift incredibly, pressing factors can likewise differ significantly, both as far as size, distribuiton and stability. A few conditions lead to eccentric pressing factor tops that can cause genuine harm [15].

The walls of a silo cell are sized for a specific load. It is critical to not use in a bunker, a certain type of granular material that has a specific mass larger than that for which the silo walls are designed. The risk of walls failing is high and this can cause the machine to overturn due to the large amount of material [16].

Initiating the gravitational discharge of granular solids from a storage silo is one of the riskiest operations because it is associated with powerful dynamic effects: rapid pressure pulsations and the transition from active to passive state of stress. Dynamic load pulse can be observed in all types of bunkers with high mass flow, with a filling height greater than the diameter of the bunker.

Pulsating tasks can lead to structural fatigue or even failure when the natural frequencies of silo filling and its structure are aroused by flow impulses [17]. The major differences between the various types of powder flow in bunkers are wall wear, feeding valve, outlet and variations in the discharge time of the granular material shown in Figure 2.

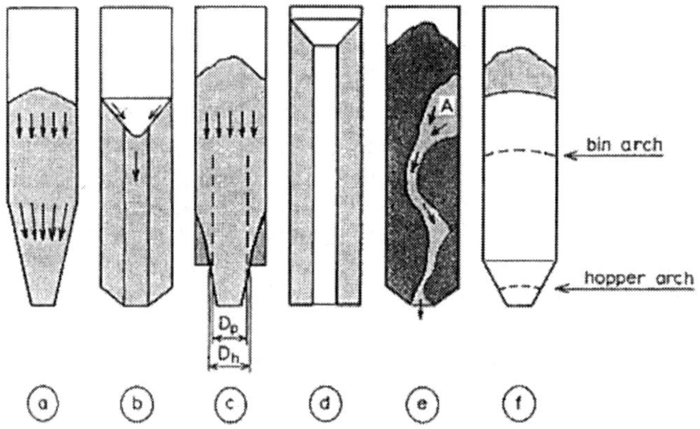

Fig. 2. Types of flow of powders in silos [18]

a. Mass flow b. Funnel flow c. Extended flow rate d. Mass flow of type "pipe" e. Ratholing f. Arching material at unloading

The discharge of the material through the outlet generates mass or funnel flow in the silo, depending on the design of the discharge system. During mass flow, the movement of the content material is slow enough and close to the equilibrium state for the inertial forces to be unobservable.

Consequently, the voltage state is similar to the active case, except the kinematic angle of friction of the wall to be used to explain the movement of the silo mass[19].

Different models are used for the design and construction of bunkers, namely: symmetrical cone, asymmetric cone, wedge-shaped bunker with converging walls, chisel-shaped bunker, the transition funnel-shaped bunker shown in Figure 3. The shape of the bunker geometry, the shape of the outlet and the supply tank (cylindrical, hemispheric, cone trunk, conical) influence its structure and stability [20].

The pressure of filling and unloading in the silo, which is predicted by the print flows (mass flow and funnel flow), determines the risk of rotation and displacements of the silo wall and other forms of structural deformation. In short, filling and unloading pressures are the most important secondary factors in the design of silos.

Primary cereal processing operations play a particularly important role in the preparation of the final product, as the percentage of seed impurities may be high if harvested improperly. The development cycles of the pollutants of agrarian items likewise vary as per their inclination and the objective they get after collect (preservation, utilization, industrialisation, marketing, planting, and so forth) These activities mean to take out pollutants of any sort, making better capacity conditions, and to decrease the volume of transport and storage.

If all these operations are properly treated, it influences the operation of the bunker at optimum parameters, without further modification [22].

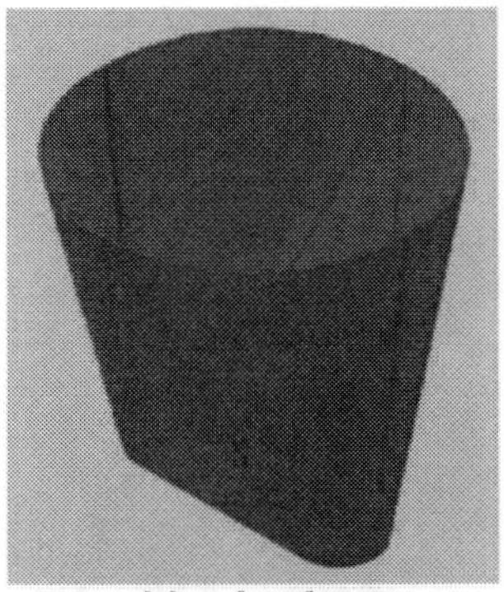

a) Transition bunker

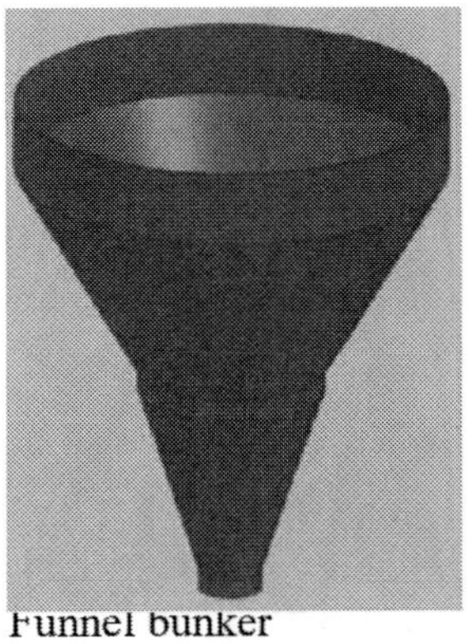

b) Funnel bunker

c) Chisel-type bunker

d) Flat-walled bunker with converging walls

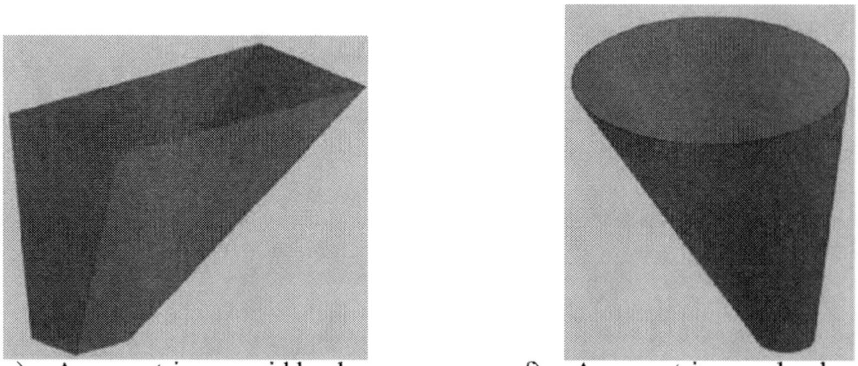

e) Asymmetric pyramid bunker f) Asymmetric cone bunker

Fig. 3. Most used geometric shapes for the design of a bunker [20]

3 Results and discussions

The specialists utilized some more established thoughts dependent on the rheology of granular materials and applied them and adjusted them to another sort of bunkers. The recommended geometry depends essentially on the regular resting point of the granular material, just as on the qualities of the stream and circulation of stress inside the storehouse. The new hexagonal geometry is represented in Figure 4 as a typical storage cell, and the cross-section of the cell is divided into two inner sections of the same area using two transverse plates. These plates are parallel to the lower walls of the silo at 35° in terms of horizontal. The angle has a value close to the natural resting angle of most commercial food cereals. At the bottom, the plates are tilted with the natural resting angle of the cereals, as well as the two internal transverse plates. The seed grains are fed by the feeding vat and the flow of the material is conducted through the outlet positioned at the bottom [23].

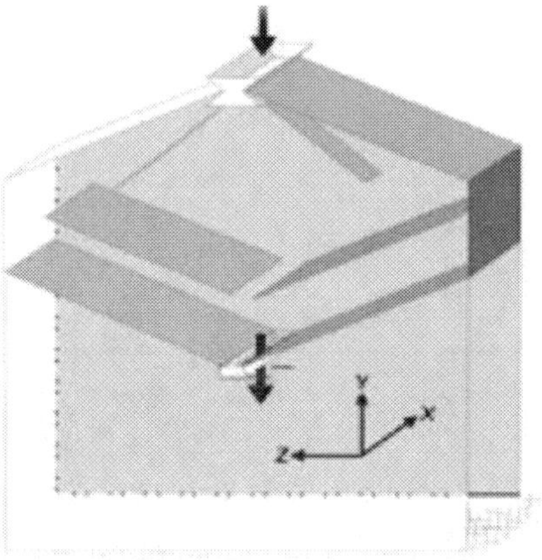

Fig. 4. Hexagonal bunker scheme [23]

Researchers have invented a new macroscopic bunker model that is used to discharge granular material from a lab-scale silo with aluminum walls and adjustable hole size. The silo shaft is 200 mm in height, 45 mm in width and 60 mm in depth. Depending on the size of the hole and the angle of inclination, the funnel has a maximum vertical height of about 100 mm.

In Figure 5, the angle tilt of the bunker is fixed at 15°, while the size of the hole varies between 5 mm, 10 mm and 15 mm. For the macroscopic model, areas without granular material have been added to the top and bottom of the silo to allow for better particle interaction, ensuring numerical stability [24].

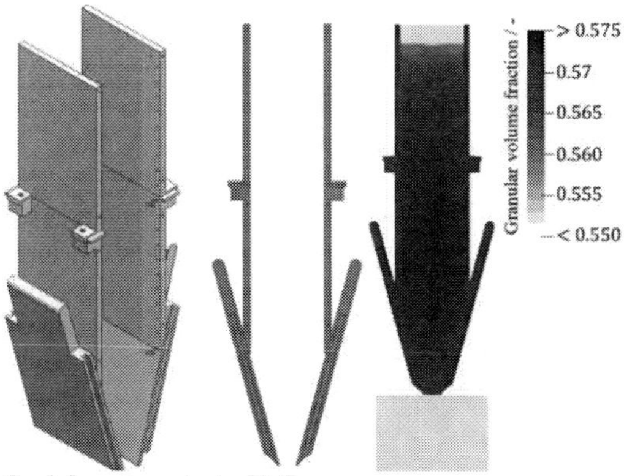

Fig. 5. Macroscopic bunker for laboratory tests [24]

Researchers have thought about the possibility of changing the geometric shape of the bunker that can influence the small mass flow when faced with mass or transition flow, as opposed to the funnel flow. This funnel-type bunker has been used in the past to determine whether a selectively perforated deflector could change the flow regime of the granular material from funnel flow to mass flow. (Figure 6).

While that was possible (a current approach used by the industry is to insert small plates or cones near the funnel outlet, turning it into mass flow), many of the perforated deflectors that produced mass flow had lower mass flow than at the original bunker [25].

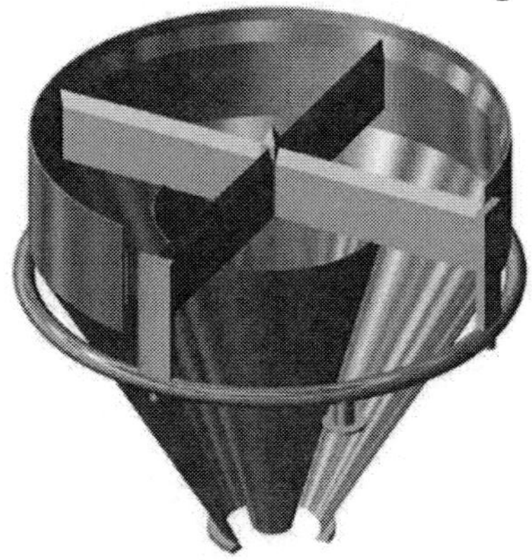

Fig. 6. Funnel-shaped bunker [25]

The researchers helped build the current bunker configuration, being presented as a funnel-shaped three-dimensional feather on the wall structure. (Figure 7). Funnel walls are considered rigid because they are treated as fixed limits, resulting in friction of granular materials (including at the bottom).

Friction is assumed to happen when the viscous damping coefficients of the walls are the same as those of the particles. The characteristic size of the socket, D, is assumed to be of the maximum diameter d. The size of the hole affects the density of the material, the propagation of the expansion wave during the flow can cause the material to lock [26].

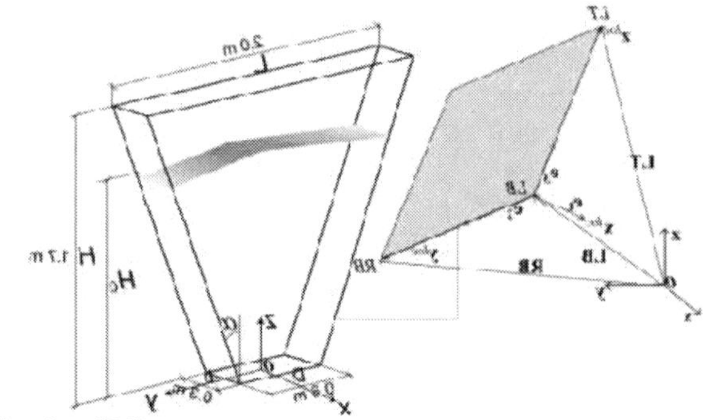

Fig.7. Three-dimensional bunker [26]

Researchers have faced some current concerns all about drum unloading bunkers due to the use of a large number of parts in the design, large amounts of horsepower and energy consumption, the free space (at rest), but especially its size and size. To overcome these concerns, the vertical discharger shown in Figure 8 has been developed and adapted to solve previous problems.

The powdery material entering the top, i.e. through the power supply valve passes through a rotating cone and subsequently through vertical discharger, resulting in a vertical flow. Adjusting the clearance between the cone and housing is used to allow the flow of material to pass (the intention is to obtain a consistent feed). This "thin cylinder" is gravitationally positioned toward the mixing chamber and the water sprayer [27].

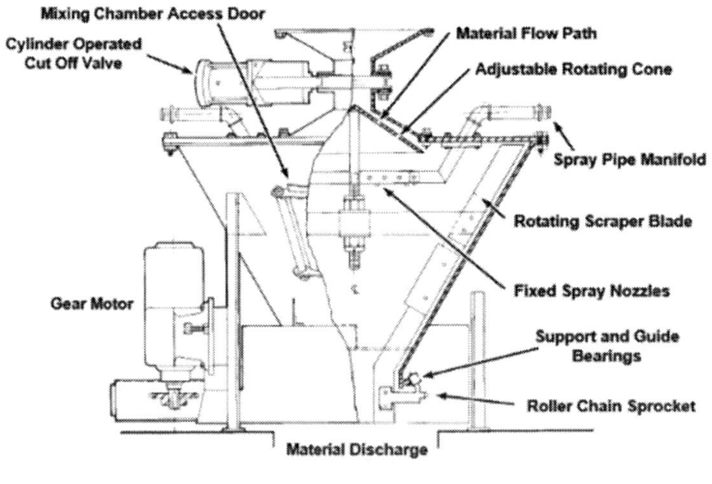

Fig.8. Vertical drum bunker [27]

Researchers have designed precise feeding devices to improve the controlled discharge of materials in the form of powder or granules from silos, bunkers, pneumatic transport systems, or cyclones, with maximum versatility in application.

The RV/RVR rotary valves with circular or square inlet shall consist of a tubular housing of cast iron or stainless steel, a rotor mounted horizontally with a certain number of V-shaped cross-section compartments, an drive unit and a housing cover opposite the end of the unit, presented in figure 9.

This ideal equipment solution has high efficiency of precise filling and measurement in the gravitational supply and pneumatic transport of products in the form of powder and granules [28].

Fig.9. RV/RVR rotary valves [28]

Researchers have implemented a new type of equipment called Hyde TDXZ Vibro which can be attached to any silo for unloading round materials with the help of vibration motion, presented in figure 10. This equipment ensures a uniform and gradual flow through which powdery or granular materials flow through the outlet without being blocked, ensuring a complete discharge without segregation under the vibration of the engine.

When vibrations are applied to a bunker as they discharge, the static bonds that accumulate between the product and the walls of the container, as well as the particle-particle friction that occurs between solid bulk materials such as seeds, dust and sand, are broken, even when moisture is present [29].

Fig.10. Hyde TDXZ Vibro Discharger [29]

Conclusions

Internal pressures are substantially higher toward the bottom of the silo, thus flaring becomes plastic and unavoidable. The spill channel or the lack of a flow channel in the actual bunker buildings makes the granular material solid and immobile, which is why this mode is not observed in practice.The presence, type, shape and position of the insertion influences the flow pattern and pressures exerted on the bunker wall.

For structural design, loads and forces occurring inside bunkers or silos must be assessed. Such loads depend heavily on the properties of bulk granular materials. Wall particle friction, stress, ratio and bulk density are important processes that must be determined in advance to avoid the occurrence of the stress interval in that silo.

The position of the stress peak acting on the silo wall is determined by the flow profile (mass flow or funnel flow). In a mass flow silo, the voltage tip is at the transition from the cylindrical section to the bunker, whereas in a funnel flow silo, the peak of stress cannot be predicted in advance.

Eccentric flow, which is generated by a poorly built feeder or eccentric discharge, is harmful because it produces in an asymmetrical voltage distribution along the silo wall, which raises the risk of metal wall flaring.

This work was supported by one founding source the NUCLEU Program, carried out with the support of ANCSI, Project PN 5N/07.02.2019 "Integrated management of works in agricultural, vitical and pomicole ferms"

References

1. G.G. Chase, Course Solids Notes 10, The University of Akron (2010);
2. C. Mircea, F. Nenciu, V. Vlăduț, G. Voicu, D. Cujbescu, I. Găgeanu, I. Voicea, Increasing the performance of cylindrical separators for cereal cleaning, by using an inner helical coil, *INMATEH Agric. Engineer.*, **62**, No.3, pp. 249-258, (2020) ;
3. L. A. Fullard, E. C. P. Breard, C. E. Davies, A. J. R. Godfrey, M. Fukuoka, A. Wade, J. Dufek and G. Lube, The dynamics of granular flow from a silo with two symmetric openings (2019);
4. M. Ostadi, Mechanics of Dry Granular Flow through an Opening, Master of Science in Geotechnical Engineering Department of Civil and Environmental Engineering, University of Alberta, (2019);
5. H.H.N.D. Haggalla, K.I.U. Nanayakkara and H.M.Y.C. Mallikarachchi, Simulating dynamic discharge of infill bulk material stored in cylindrical silos, Society of Structural Engineers, Sri Lanka - Annual sessions, (2019);
6. A. Jacobs, B. Moboladea, N. Bunindrob, D. Sahoob, Y. Rajashekarb, Traditional methods of food grains preservation and storage in Nigeria and India, *Annals of Agric. Scien.* **64**, pp. 196–205, (2019);
7. R. Jackson, *J. Rheol.* **30**, 907, (1986);
8. G. Gudejus, D. Kolymbas and J. Tejchman, *Powder Technology* **48**, 8, (1986);
9. A.V. Potapov, C.S. Campbell, *Physics of Fluids* **8**, 2884, (1996);
10. J. Tomas, Modelling of instationary discharge behaviour of cohesive particulate solids out of bunkers, *Chemie-Technik* 43(**8**), pp. 307–309, (1991);
11. D. Barletta, G. Donsì, G. Ferrari, M. Poletto, P. Russo, Solid flow rate prediction in silo discharge of aerated cohesive powders. *AIChE J.* 53(**9**), pp. 2240–2253, (2007);
12. R. Baserinia, I.C. Sinka, Mass flow rate of fine and cohesive powders under differential air pressure, *Powder Technology* **334**, 173–182, (2018);
13. W.A. Beverloo, H.A. Leniger, J. van de Velde, The flow of granular solids through orifices, *Chem. Engineer. Scien.*, 15(**3-4**), pp. 260–269, (1961);
14. A.J. Sadowski, J. M. Rotter, Study of Buckling in Steel Silos under Eccentric Discharge Flows of Stored Solids, *J. Eng. Mech.* 136:769-776, (2010);
15. J. M. Rotter, Silo and hopper design for strength, Bulk Solids Handling: Equipment Selection and Operation Edited by Don McGlinchey, Blackwell Publishing Ltd. ISBN: 978-1-405-15825-1, (2010);
16. H. Frederiksen, D. Dănuț, M. Mașinistru, A. Greculescu, Sisteme pentru depozitarea furajelor. Standarde de fermă, Danish Agricultural Advisory Service, (2010);

17. A.W. Roberts, Review of silo loadings associated with the storage of bulk granular materials, In: CIGR-AgEng International Conference of Agricultural Engineering, Valencia, Spain, pp. 8–12, (2012);
18. R. Zevenhoven, Powder mechanics & powder flow testing (flow / no flow), Åbo Akademi University, (2018);
19. A.W. Jenike, J.R. Johanson, Bin loads, Proc. Amer. Soc. Civil Eng., *J. Structural Div.* **94** (ST4):1011-1041, (1968);
20. J. Carsona, D. Craig, Silo design codes: Their limits and inconsistencies, The 7th World Congress on Particle Technology (WCPT7), Publishing Elsevier, (2015);
21. C. Maraveas, Concrete Silos: Failures, Design Issues and Repair/Strengthening Methods, *Applied Sciences*, MDPI, (2020);
22. C. Brăcăcescu, P. Găgeanu, G. Bunduchi, A. Zaica, Considerations on technical equipment used for cleaning and sorting seed mixtures based on aerodynamic principle, Engineering for rural development, Jelgava, (2018);
23. J. Hernández-Cordero, R. Zenit, E. Geffroy, B. Mena and R. R. Huilgol, Experiments on granular flow in a hexagonal silo: a design that minimizes dynamic stresses, *Korea-Australia Rheology Journal* **12**, No. 1, pp. 55-67, (2000);
24. R. Hessel, D. Gilberg , K. Steiner, S. Antonyuk, Modelling of granular flow on micro- and macroscopic scales with calibration using experimental and numerical setups, Proceedings of the 8th international conference on discrete element methods (DEM8), (2019);
25. M. J. Sandlin, An Experimental and Numerical Study of Granular Hopper Flows, Institute of Technology, Georgia, USA, (2013);
26. I. Sielamowicz, R. Balevičius, Experimental and computational analysis of granular material flow in model silos, IPPT Reports on Fundamental Technological Research, (2013);
27. K. P. Shah, Fundamentals, Troubleshooting & Maintenance of Ash Handling Plants and Pneumatic Conveying Systems for Bulk Materials, https://practicalmaintenance.net/, (2017);
28. www.wamgroup.com
29. https://maoshengmachinery.en.made-in china.com/product/YXvmroVjhkUa/China-Food-Processing-Flour-Mill-Vibro-Sieve-Discharger-Machine.html

https://doi.org/10.1051/e3sconf/202128603025

Integrating Green Tourism and Bioenergy for Sustainable Rural Development: the Example of the Hemp Cluster in Sumy Region in Ukraine

Hanna Shevchenko[1],*

[1]Institute of Market Problems and Economic-Ecological Research of the National Academy of Science of Ukraine, Department of Economic Management of Natural Resources, Frantsuz'ky Blvd, 29, 65044 Odesa, Ukraine

Abstract. Institutional support for the transition to sustainability, agricultural traditions, as well as land use regimes create favorable conditions for the development of bioenergy, in particular in the field of biomass related activities, in rural territories of Ukraine. The purpose of the study is to analyze the factors and substantiate the relationship between green tourism and bioenergy in the context of expanding areas of sustainable rural development, on the example of agro-tourism hemp cluster in Sumy region in Ukraine. The research methodology: the theoretical-game approaches to balance the interests of stakeholders in rural development are applied in the paper, namely in the context of renewable energy, waste management and recreational and tourism activities. SWOT analysis was also applied with rationale of strategic directions of agro-tourism clusters development in the conditions of transitive economy. The rationale of proposals for improving the institutional support of sustainable rural development at the regional level allows to form a basis for increasing the level of investment attractiveness of projects such as agro-tourism hemp cluster. Besides, recommendations for the integration of green tourism and bioenergy are focused on the advanced development of rural areas with the involvement of other sectors of the economy, in particular, waste management.

1 Introduction

For transition economies, in conditions of insufficient financial resources and at the same time efforts to accelerate market transformations, it is advisable to use all possible organizational innovations at both territorial and sectoral levels. It is not only about the economic vector of development, in particular such industries as agriculture or energy, but also about such complex areas of socio-economic and environmental activities as green tourism.

Institutional support for the transition to sustainability [1, 2], agricultural traditions, as well as land use regimes create favorable conditions for the development of bioenergy in

* Corresponding author: hannashevchenko@gmail.com

rural territories of Ukraine. A comprehensive approach to the use of agricultural land [3] in the context of their optimal diversification actualizes the search for intersectoral cooperation in the field of biomass related activities.

"The main sources of biomass in Ukraine are waste and agricultural residues.., and in the long run – energy crops grown specifically for this purpose"; as of today, 80% of renewable energy in Ukraine is biomass [4]. The potential for bioenergy development is presented in table 1.

Table 1. Bioenergy potential of Ukraine.

Type of biomass	Potential volume
Straw (cereals, rapeseed)	10.86 million tons
Waste (stems, rods, husks, pulp)	29.655 million tons
Wood biomass	3.78 million tons
Biogas (agro-industrial waste, landfills, wastewater)	1.234 billion m^3
Energy crops (willow, poplar, miscanthus, corn)	13.32 billion m^3

Source: expert assessment of the Ukraine Bioenergy Association [5]

Hemp as a biomass needs further study, not only in terms of energy, but also in terms of waste management and agritourism. Technological aspects of hemp production have been developed in the papers [6, 7]. The source [8] summarizes research on agritourism. The paper [9] analyzes renewable energy as a factor within the sustainable tourism development, the paper [10] identifies renewable energy and tourism as symbiotic socio-economic sectors. The source [11] develops the widespread use of bioenergy to grow the national economy, with the "local context, empowerment of local governments and encouragement of both public and private initiatives". The papers [12-14] explore sustainable, eco, and green tourism. Innovations in the field of sustainable tourism are considered in [15], in particular, which are aimed at reducing carbon emissions and ultimately contribute to the fight against climate change.

At the same time, the complex interrelationships between the above-mentioned industries require further research.

The purpose of the study is to analyze the factors and substantiate the relationship between green tourism and bioenergy in the context of expanding areas of sustainable rural development, on the example of agro-tourism hemp cluster in Sumy region in Ukraine.

2 Methodology

The theoretical-game approaches (games against nature [16, 17] and mechanism design [18-20]) to balance the interests of stakeholders in rural development are applied in the paper, namely in the context of renewable energy, waste management and recreational and tourism activities (Fig. 1).

The formalization of the balance of interests for actors interested in the hemp development on the basis of the theoretical-game approach application is as follows:

$$H_i(s^*) \geq H_i(s_i, s_{-i}^*) - \varepsilon \tag{1}$$

where $H_i(s^*)$ – the result of the i actor of hemp relevant activity (energy, tourism, waste management) in the equilibrium situation s^*, that belongs to a set of strategies $S=(s_1, s_2, \dots s_i, \dots, s_n)$; depends on the strategies of all actors in the situation s^*; ε – conditional equilibrium parameter.

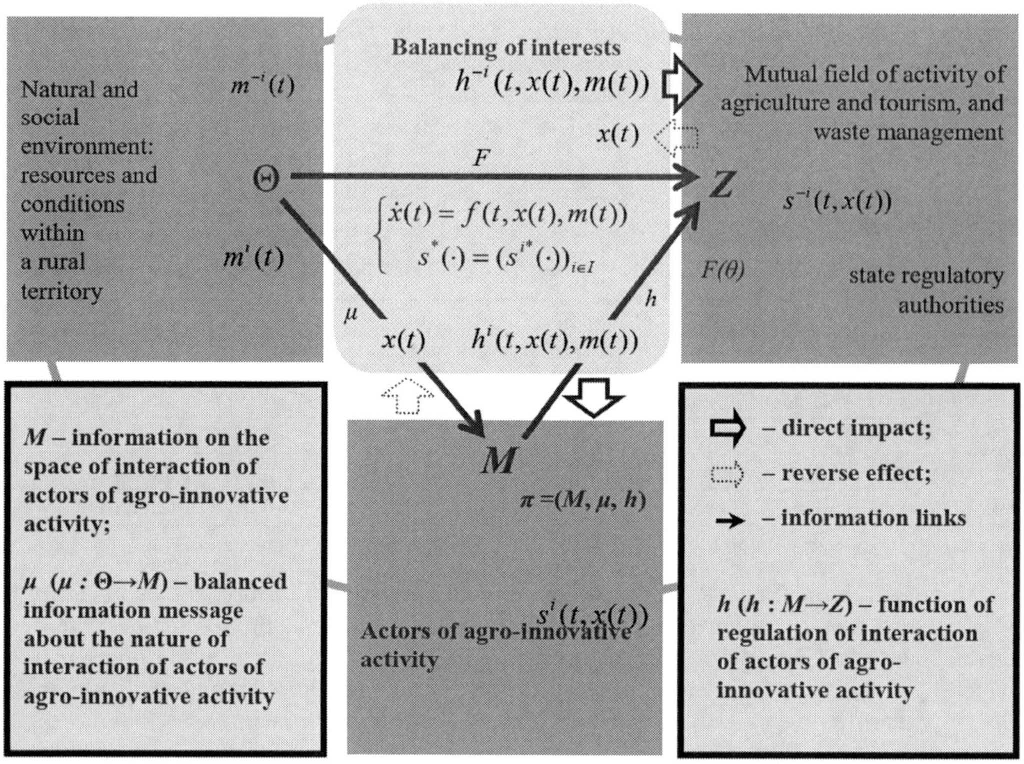

Fig. 1. A game-theoretical modeling of balancing the interests of actors within the agro-tourism cluster.

Source: Developed by the author based on [19, 20]

The data in the study cover regional and sectoral [21, 22], as well as national [1, 23] levels. These data have an economic, social and environmental context, as well as a socio-economic, economic-environmental and integrated context. The search for primary information, which is summarized in the secondary data, is also partially carried out during the study of relevant documentation of organizations.

SWOT analysis was also applied with rationale of strategic directions of agro-tourism clusters development in the conditions of transitive economy (the example of application is in [24]).

3 Results

For sustainable development of rural areas in transition economies, an innovative approach is needed, which provides a comprehensive combination of the advantages of different industries in their interaction and at the same time a return to the traditions of national agriculture, including hemp.

3.1 Innovative policies to support rural development

Starting in 2019, in Ukraine, despite the general socio-economic crisis in the country and as of 2020 the difficult situation associated with the COVID-19 pandemic, sustainable rural development continues.

Table 2. Innovative projects for sustainable rural development in Ukraine.

Project	Region	EU finance	Other finance
Restoration of irrigation of Kherson region*	Kherson	4700	2300
Incubator for starting a business in the field of aquaculture*	Mykolayiv	5887	2974.4
Agro-industrial park in Shepetivka*	Khmelnytskyi	21000	–
Preservation of forest plantations and improvement of ecological safety of rural areas*	Kherson	10308.4	1146
Production of eco-containers from biopolymers*	Lviv	7650	2150
Interregional Innovation Center for Rural Development**	Ukraine	12000.6	5182.56
Creation of agro-tourist cluster "Slobozhanske hemp"**	**Sumy**	**14123.031**	**6580**
Glampings in protected areas***	Ukraine	23986.04	4857.216
Human capital development in the agro-organic sector of the economy****	Rivne	3652.138	1218
Development of tourism in sparsely populated rural areas of Trostyanets UTC*****	Sumy	5107.036	2745.954
Slavutych – interregional cycling recreation for health and leisure*****	Kyiv	12322.66	2273.729
Measures for waste sorting and utilization in nature reserves of the Carpathians *****	Trans-carpathia	4870.4	1050.6
Creating a space for recreation in Zbarazh – overhaul of the coastal zone of the lake*****	Ternopil	9835.35	1735.98
Introduction of "smart city" elements in rural communities of Cherkasy region******	Cherkasy	1248.761	1274.161
Training of self-government representatives in sustainable development of communities ******	Kyiv	1600	400

Project groups: * Innovative economy, **Rural development, ***Tourism development, ****Human development, *****Support for the development of depressed areas, ******Effective management of regional development

Source: compiled by the author on the basis [21]

The main advantage of the researched project is first of all that it is complex and relevant to almost all programs (Table 2).

Thus, according to the results of the competition [21], 56 projects were selected for 7 programs (innovative economy and investments, rural development, tourism development, human development, support for the development of depressed areas, effective management of regional development, all-Ukrainian solidarity) in the amount of over UAH 535 million. The sources of funding are the state budget, formed with EU funds. The topics of most projects are tourism (219) and innovative economy (137), as well as the development of depressed areas. (106). Project selection criteria: their complexity, socio-economic effect for cities and regions, the applicant's ability to support the project in the future.

3.2 Development of agro-tourist clusters in Ukraine

The project revives traditions and diversifies the possibilities of agricultural production in Slobozhanshchyna, taking into account innovations in the cultivation of technical hemp, their processing, as well as the use of products of their processing in food and everyday life. To this end, the following is envisaged: comprehensive development of the hemp industry, with an increase in the scale of hemp cultivation and the provision of capacity for their processing; creation of a training center for the hemp development; development of rural and gastronomic tourism [22].

Extended project activities include the following main components:

– conducting trainings, "round tables", conferences;

– improvement of waste management processes according to the principles of circular economy;

– organization of festivals, museum and other tourist sites;

– informing the public about the project implementation.

An integrated approach to the formation of this cluster involves information and economic interaction of the three main activities relevant to innovative hemp, based on the design mechanism (Fig. 1). In turn, information interaction has been analyzed in the context of information asymmetry (1) and informational-innovation development (2); economic interaction is considered in terms of intersectoral processes (3) and the project approach (4), which requires the introduction of regulatory changes, namely:

1) promoting local openness and complete coverage of information on environmental and social activities, and on the other hand, significantly increasing control over the veracity of information data and the introduction of tougher fines for violators;

2) improving environmental monitoring with the use of innovations such as drone surveillance in the framework of public control, as well as strengthening the scientific component, including through citizen science;

3) strengthening coordination work in the field of industrial, science and technological parks, in the direction of strengthening intersectoral cooperation, in particular through its infrastructure;

4) a combination of participatory budgeting and public-private partnership practices in the field of innovative intersectoral activities.

In the energy context of cluster formation, the current issue in Ukraine is the "green" tariff, which purchases electricity produced on the basis of alternative energy sources, including bioenergy [25].

In Ukraine, the government guarantees timely current payment by the State Enterprise "Guaranteed Buyer" for electricity supplied to the United Energy System and repayment by the end of 2021 of debts to renewable energy producers, about 14 billion UAH. The government has also pledged to set and approve annual quotas to support green energy and ensure that auctions are held. The main condition of the memorandum is the refusal of the authorities from further attempts to change the "rules of the game" for the entire period of the "green" tariff – until 2030. The size of the "green" tariff for farm and home power plants will not change [26].

The basis for the introduction of special conditions of the "green" tariff for bioenergy is its environmental orientation and the promotion of rural development (more broadly – in table 3).

Despite the changes taking place in the socio-economic life of Ukraine and at the global level, investors are interested in renewable energy projects due to attractive conditions in the country. But high risks are an obstacle to investment. One of the risks is the choice of land for the location of the energy object [27]. To choose the right site for the construction of the power plant in Ukraine, it is necessary to analyze the following: natural and environment conditions, the possibility of connecting to the existing energy system (remoteness of the site from the grid, lack of energy infrastructure, especially transformer substations), land availability according to its purpose. In Ukraine, renewable energy facilities can be built on land for industry, transport, communications, and other purposes.

Table 3. SWOT-analysis of strategic directions of agro-tourist cluster development in the transitive economy conditions.

External environment / Internal environment		Opportunities (O)	Threats (T)
		Potential support from international organizations, in particular the UNWTO Growing interest and readiness of regional organizations Revival of interest and research activities of pro-environmental NGOs The presence of business interest	Lack of clear national standards Resource needs (finance, technology and human resources) Inertia of cannabis stereotypes Complexity of bureaucratic procedure (project documentation, permits, etc.)
Strengths (S)		Strategies (SO)	Strategies (ST)
	Agro-industrial traditions Land use mechanism (compared to solar or wind energy in rural areas in Ukraine) Complexity Existing regional development policy	Gradual introduction of the cluster with simultaneous broad development of energy and tourism Expanding the participation of local governments Special conditions of the "green" tariff Improving waste management practices	Mobilization of small and medium-sized businesses within the cluster through diversified financial support (both public and private investment) Introduction of training programs Strengthening institutional support
Weaknesses (W)		Strategies (WO)	Strategies (WT)
	Lack of experts Slight development experience Lack of business motivation due to distrust of the authorities	Search for new opportunities for circular development Consultations with international experts (e.g. UNWTO) Measures to increase trust in government	Comprehensive assessment of the feasibility and content of project proposals Ideologization and propaganda Involvement of all stakeholders in projects

Source: Developed by the author

Ukrainian legislation allows changing the purpose of land, but there are significant difficulties with agricultural land (which has a special status). Each case is investigated separately. The issue of land ownership is the most controversial, so it is advisable to alternatively lease state or communal land; taking into account such factors as the availability of the decision-makers' authority within the transfer of land for use, as well as the correctness of the contract.

Given this, bioenergy has the least barriers to development within rural areas in Ukraine, and, as described above, is an important component for the creation of agro-tourism clusters.

4 Conclusions

The organizational-economic provisions developed in the paper substantiate the following:

– preconditions for creating rules of the game, namely through the following measures: promoting openness and complete coverage of relevant information, improving environmental monitoring, strengthening coordination in the field of innovational parks, combining the practices of participatory budgeting and public-private partnership;

– strategic directions of agro-tourist cluster development, in particular such as: the gradual introduction of the cluster with simultaneous broad development of energy and tourism, the mobilization of small and medium-sized businesses within the cluster through diversified financial support, the comprehensive assessment of the feasibility and content of project proposals etc.;

– the possibility of compensating the negative consequences of the green tariff policy with additional benefits of sustainability of rural territories through the integration of bioenergy and green tourism in Ukraine.

Also, the rationale of proposals for improving the institutional support of sustainable rural development at the regional level allows to form a basis for increasing the level of investment attractiveness of projects such as agro-tourism hemp cluster. Besides, recommendations for the integration of green tourism and bioenergy are focused on the advanced development of rural areas with the involvement of other sectors of the economy, in particular, waste management.

The paper contains the results of research conducted under the National Academy of Science of Ukraine's grant "Formation and use of natural-resource assets of the recreational and tourism sphere" (No. 0120U100159) and project "Inclusion of the experience economy in the nature management" (No. 0119U000229).

References

1. https://sustainabledevelopment.un.org/content/documents/26295VNR_2020_Ukraine_Report.pdf [Accessed 24 May 2021]
2. https://ec.europa.eu/info/strategy/priorities-2019-2024/european-green-deal_en [Accessed 24 May 2021]
3. H. M. Shevchenko, V. M. Pakhomov, M. M. Petrushenko, Economic and legal issues of rural and recreational land use in Ukraine. Economic Annals-XXI. **156**(1-2), 54-58. https://doi.org/10.21003/ea.V156-0012 (2016)
4. https://www.ua.undp.org/content/ukraine/uk/home/presscenter/pressreleases/2019/study-tour--undp-shared-best-bioenergy-practices-with-ukrainian-.html [Accessed 24 May 2021]
5. https://saee.gov.ua/node/586 [Accessed 24 May 2021]
6. P. Găgeanu, M. Olan, I. Găgeanu, A. Zaica, Theoretical considerations on the factors influencing the processing of hemp stalks. E3S Web Conf. **180**, 03005. https://doi.org/10.1051/e3sconf/202018003005 (2020)
7. A. Păun, G. Stroescu, A. Zaica, R. Ciupercă, C. Bogdanof, E3S Web Conf. **180**, 03026. https://doi.org/10.1051/e3sconf/2020180030026 (2020)
8. S. Rauniyar, M. K. Awasthi, S. Kapoor, A. K. Mishra, Tourism Recreation Research. **46**, 1-19. https://doi.org/10.1080/02508281.2020.1753913 (2020)
9. M. Evanthie, Études Caribéennes. **11**. https://doi.org/10.4000/etudescaribeennes.3487 (2008)

10. H. Shevchenko, M. Petrushenko, A symbiosis between renewable energy and tourism: The case of Dnipro region in Ukraine. E3S Web Conf. **234**, 00087. https://doi.org/10.1051/e3sconf/202123400087 (2021)
11. C. S. Goh, T. Aikawa, A. Ahl, et al, Rethinking sustainable bioenergy development in Japan: decentralized system supported by local forestry biomass. Sustainability Science. **15**, 1461–1471. https://doi.org/10.1007/s11625-019-00734-4 (2020)
12. N. MacKenzie, M. J. Gannon, Exploring the antecedents of sustainable tourism development. Int. Journal of Contemporary Hospitality Management. **31**(6), 2411-2427. https://doi.org/10.1108/IJCHM-05-2018-0384 (2019)
13. S.-Y. Pan, M. Gao, H. Kim, et al., Advances and challenges in sustainable tourism toward a green economy. Science of the Total Environment. **635**, 452-469. https://doi.org/10.1016/j.scitotenv.2018.04.134 (2018)
14. N. D. Line, L. Hanks, L. Miao, Image matters: incentivizing green tourism behavior. Journal of Travel Research. **57**(3), 296-309. https://doi.org/10.1177/0047287517697848 (2018)
15. H. Buijtendijk, J. Blom, J. Vermeer, R. van der Duim, Journal of Sustainable Tourism. **26**(7), 1222-1240. https://doi.org/10.1080/09669582.2018.1433184 (2018)
16. M. Petrushenko, H. Shevchenko, B. Burkynskyi, N. Khumarova, A game-theoretical model for investment in inclusive recreation and wellness in Ukraine: the regional context. Investment Management and Financial Innovations. **16**(4), 382-394. http://dx.doi.org/10.21511/imfi.16(4).2019.32 (2019)
17. J. Szép, F. Forgó, Games against nature. In: *Introduction to the Theory of Games. Mathematics and Its Applications*. 17. https://doi.org/10.1007/978-94-009-5193-8_21 (Springer, Dordrecht, 1985)
18. O. V. Prokopenko, M. M. Petrushenko, Marketing and Management of Innovations. **1**, 254-266 (2013). https://mmi.fem.sumdu.edu.ua/journals/2013/1/254-266 [Accessed 24 May 2021]
19. L. Hurwicz, S. Reiter, *Designing economic mechanisms* (Cambridge University Press, New York, 2008)
20. M. M. Petrushenko, H. M. Shevchenko, Management of ecological-economical conflicts within the framework of the theory of optimal mechanisms for resource distribution. Actual Problems of Economics. **141**(3), 186-192 (2013). https://eco-science.net/downloads/ [Accessed 24 May 2021]
21. https://www.minregion.gov.ua/press/news/vyznacheno-peremozhcziv-konkursnogo-vidboru-proyektiv-regionalnogo-rozvytku/ [Accessed 24 May 2021]
22. https://sectoral.minregion.gov.ua/sectorals/128a0c70-1f94-11eb-ba4a-b356ee544eea [Accessed 24 May 2021]
23. http://www.ukrstat.gov.ua [Accessed 24 May 2021]
24. X. M. Zhang, Research on the development strategies of rural tourism in Suzhou based on SWOT analysis. Energy Procedia, **16**(B), 1295-1299. https://doi.org/10.1016/j.egypro.2012.01.207 (2012)
25. https://zakon.rada.gov.ua/laws/show/555-15#Text [Accessed 24 May 2021]
26. https://www.ukrinform.ua/rubric-economy/3044409-mirovu-ugodu-sodo-zelenogo-tarifu-ukladeno-geruskriza-trivae.html [Accessed 24 May 2021]
27. https://getmarket.com.ua/ua/news/yak-pravil-no-obirati-zemel-ni-dilyanki-dlya-budivnictva-proektiv-vidnovlyuvanoyi-energetiki [Accessed 24 May 2021]

Designing and dimensioning of a new type of pump that can be used in the field of land reclamation

Mariana Mirela Stoican (Prisecaru) [1,*] and *Nicolae* Băran[1]

[1]University Politehnica of Bucharest, Faculty of Mechanical Engineering and Mechatronics, Splaiul Independenței, no. 313, Bucharest, Romania

Abstract. The findings made during the operation of the pumping stations within the land improvement arrangements, energy balances, measurements performed and low efficiencies that are currently between 50% -77%, it is necessary to continue research in order to improve parameters and replacement of existing aggregates with energy efficient equipment. Thus, theoretical and experimental research was performed in order to improve the projected theoretical parameters of a rotating volumetric pump. For a certain water flow rate that should be transported to irrigate an area of agricultural land or discharged from a land area equipped with drainage works, a rotating pump is dimensioned and the flow rate according to the number of rotations diagram is plotted.

1 Introduction

A current research direction is to improve the performance of the machines that transports fluids. Optimizing the interior architecture of working machines is a very important issue, a problem studied both in our country and abroad.

Both motor and working machines evolve over time in the following direction:

- For motor machines, the aim is to produce the maximum mechanical work yielded outside.

- For working machines, the minimum consumption of mechanical work from the outside is sought.

The architectural problem of rotors, consists in finding solutions for optimizing their geometry and the choice of parameters that lead to a more efficient fluid transport.

For rotating working machines with profiled rotors, the problem regarding the architecture of the rotor is the optimization of its geometry and the choice of parameters that lead to a more efficient fluid transport.

Minimizing the driving power of volumetric pumps with profiled rotors and finding a new architecture of the profiled rotor are key elements in the conception and designing of a new rotating machine.

* Corresponding author: mirela.prisecaru@yahoo.com

Rotating pumps are included in the category of rotating working machines for liquid circulation.

In technics, there are two main criteria for pumps classification [1] [2]:

I. According to the operation principle;

II. According to the driving mode.

I. According to the operation principle, two categories are distinguished:

A. Volumetric pumps;

B. Non-volumetric pumps.

The main feature of volumetric pumps is the relative independence of the flow rate from the suction pressure values and especially at the pump discharge. For this type of hydraulic generators, the flow rate is given by the sum of the elementary volumes pumped in the unit of time.

II. ***There are three categories of pumps according to the driving mode*** [2]:

- Electro-pumps; pumps; turbopump.

Table 1 classifies the pumps according to the operating principle [3].

Table 1. Categories of pumps.

A) Volumetric pumps	**Piston pumps**	a) Single cylinder pumps
		b) Polycylindrical pumps
		c) Pumps with axial pistons
	Rotating pumps	d) Blade pumps
		e) Gear pumps
		f) Screw pumps
		g) Lobe pumps
		h) Pumps with profiled rotors
B) Non-volumetric pumps	Centrifugal pumps	
	Axial pumps	

Next, the paper will study the category of rotating volumetric pumps, namely, the one with profiled rotors (h).

2 The pump sketch and the operation principle

The main dimensions of the pump, which has two identical rotors, must first be determined:

- l - the rotor length [m];
- R_r - the machine rotor radius [m];
- z - the height of the rotating piston [m].

The case radius (R_c) results from the sum of the rotor radius (R_r) and the height of the piston (z) (figure 1).

$$R_c = R_r + z \; [m] \tag{1}$$

What is the connection between z and R_r? How big can z be in relation to R_r?

To solve this problem, we will consider a single piston (5) positioned on the lower rotor (1) [4] [5].

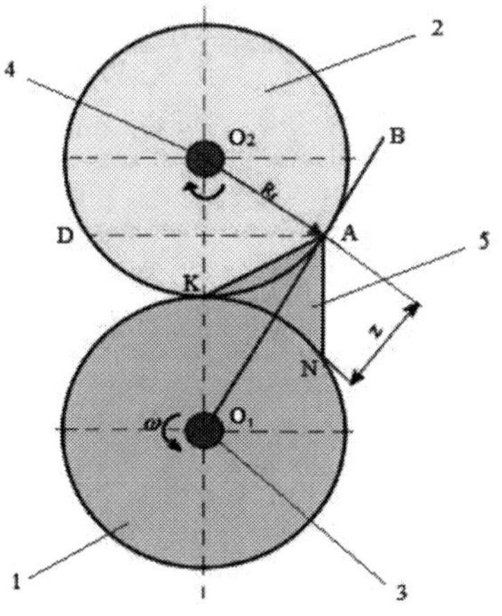

Fig. 1. Calculation notations
1 - rotor located at the bottom; 2 - rotor located at the top;
3 - the driving shaft of the machine; 4 - the driven shaft of the machine;
5 - rotating triangular piston.

The radius of the rotor (1) of the rotating machine is extended by a length (z) so that the line O_1B reaches at point A the rotor (2). In this way, the sealing between the two rotors is ensured by direct contact between the side surfaces of the rotors.

Analyzing the triangle O_1O_2A will result:

$$O_1O_2^2 = AO_2^2 + AO_1^2 \tag{2}$$

$$(2R_r.z)^2 = R_r^2 + (R_r + z)^2 \tag{3}$$

relations that become:

$$z^3 + 2R_r.z - 2R_r^2 = 0 \tag{4}$$

If in relation (4) the derivative is performed according to the height of the rotating piston (z) one can obtain:

$$z = -R_r \tag{5}$$

That is, z is maximum when it becomes equal to the rotor radius R_r; from the conditions of sealing and execution of the rotors, it is chosen $z < R_r$.

The rotating volumetric pump with profiled rotors consists of (figure 2) two profiled rotors (3, 7), which rotate at the same speed inside the cases (1, 4); the profiled rotors are engaged by two gears (8) mounted inside the side wall of the pump case, thus ensuring their synchronization. The shaft drives the lower rotor through a flange fixed with rotor screws; the flange rotates inside the side wall of the case (figure 2.b) [6] [7].

 https://doi.org/10.1051/e3sconf/202128604001

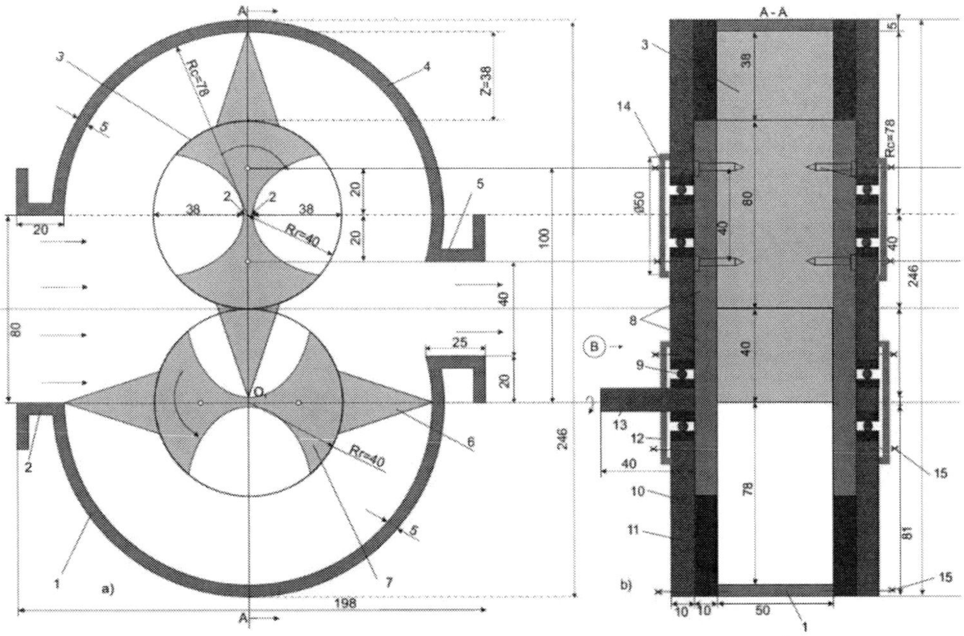

a b

Fig. 2. Cross section (a) and longitudinal section (b) through the rotating machine
a: 1 - lower case; 2 - fluid suction connection; 3 - upper rotor;
4 - upper case; 5 - fluid discharge connection; 6 - rotating piston; 7 - lower rotor;
b: 8 - gears; 9 - bearings; 10 - side wall (left) of the case;
11– intermediate wall; 12 - bearing cover; 13 - driving shaft;
14 - sealing caps; 15 - clamping screws.

The technical solution shown in figure 2 has the following characteristics:

a) The height of the piston is approximately equal to the radius of the rotor, thus increasing the useful volume of fluid conveyed;

b) The shaft for each rotor does not penetrate inside the rotor, so $z \rightarrow R_r$.

For the construction of the rotors, a mathematical calculation program was elaborated, which specifies the shape of the rotor contour, and their execution was performed with the help of a numerically controlled machine tool (C.N.C.).

Figure 3 shows the location of the rotors inside the case.

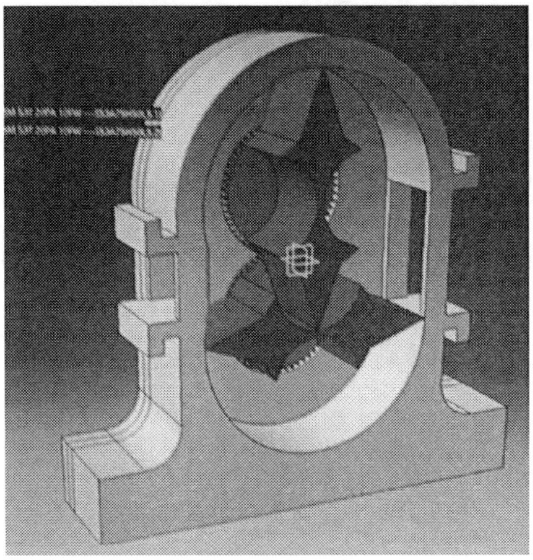

Fig. 3. Axonometric view of the case with the two rotors mounted inside it.

Figure 4 shows the operation of the rotating volumetric pump with profiled rotors. One can observe that after a 180° rotation the useful volume of the transported fluid (V_u) is discharged into the discharge chamber.

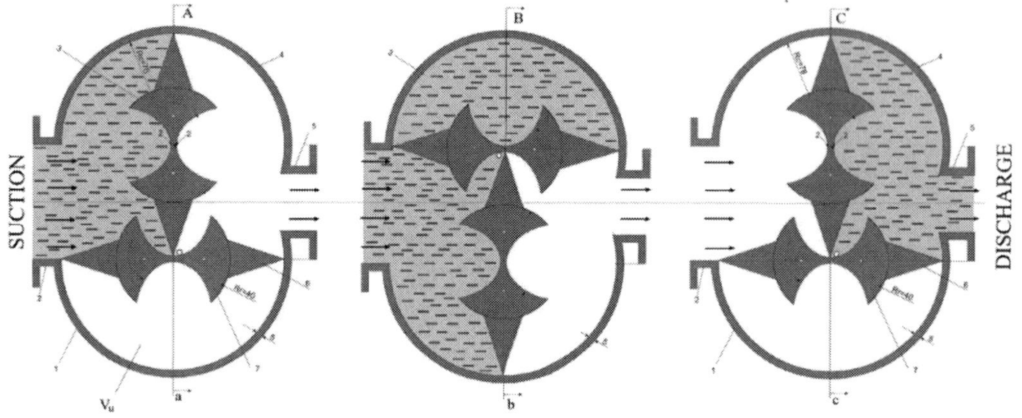

Fig. 4. Operating principle of the rotating volumetric machine
1 - lower case; 2 - fluid suction connection; 3 - upper rotor;
4 - upper case; 5 - fluid discharge connection; 6 - rotating piston; 7 - lower rotor.

The useful volume is the volume between two pistons and the case (figure 4.a).

When operating as a rotating volumetric pump with profiled rotors, two volumes (V_u) will be transported from the suction to the discharge at a complete rotation of the shaft.

$$V_u = 2\left(\pi R_c^2 + \pi R_r^2\right) l \quad \left[m^3 / rot\right] \tag{6}$$

where: l - the rotor length [m].

The case radius (R_c) is the sum of the rotor radius (R_r) and the height of the piston (z):

$$R_c = R_r + z \quad [m] \tag{7}$$

The volumetric flow rate of the fluid transported by the pump will be:

$$\dot{V} = \pi \cdot l \cdot z\left(z + 2R_r\right) \cdot \frac{n_r}{30} \quad \left[m^3 / s\right] \tag{8}$$

One can observe that the volumetric flow rate increases linearly with the length (l) and the rotor radius (R_r) and with the speed (n).

In the laboratory of the technical department of Faculty of Mechanics within the UPB, a prototype of this volumetric pump with the following characteristics was designed and built [8] [9]:

- the rotor length: l = 0.05 [m];
- the rotor radius: R_r = 0.04 [m];
- the height of the rotating piston: z = 0.038 [m].

So, the case radius will be:

$$R_c = R_r + z = 0.04 + 0.038 = 0.078 \; [m] \tag{9}$$

An overview of the pump mounted on the test stand is shown in Figure 4.

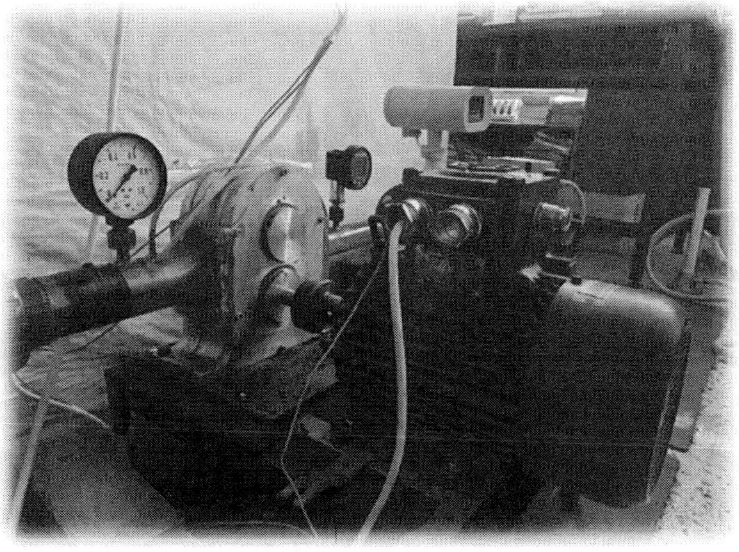

Fig. 5. Overview of the volumetric pump in the test stand.

Equation (8) is repeated, which specifies the flow rate of the pump, the rotating pistons are considered to have a triangular shape, the area of the section between the base of the prism and the rotor is neglected, and the volume of this prism will be:

$$V_p = A_{base} \cdot l = \frac{1}{2} \cdot b \cdot z \cdot l = \frac{1}{2} \cdot 0.03 \cdot 0.038 \cdot 0.05 \; ; \; V_p = 0.0285 \cdot 10^{-3} \quad [m^3 / rot] \quad (10)$$

The theoretical flow rate of the machine will be lower with V_p.

$$\dot{V}_u = \left[\pi l z \left(z + 2R_r \right) - V_p \right] \cdot \frac{n_r}{30} \; [m^3 / s] \quad (11)$$

$$\dot{V} = \left[3.14 \cdot 0.05 \cdot 0.038 \left(0.038 + 2 \cdot 0.04 \right) - 0.0285 \cdot 10^{-3} \right] \cdot \frac{500}{30} \quad (12)$$

$$\dot{V} = 0.011258 \; [m^3 / s] = 40.529 \; [m^3 / h] \quad (13)$$

For the pump data presented above, choosing the operating speed of the pump as 200 ... 500 [rpm] the values in Table 2 resulted.

Table 2. The values of $\dot{V} = f(n_r)$.

n_r [rpm]	200	300	400	500
$\dot{V}\left[dm^3 / min\right]$	270.18	405.3	540.42	675.48
$\dot{V}\left[m^3 / h\right]$	16.211	24.317	32.423	40.529

Based on the theoretical data in Table 2 and following the experimental researches, the graph of the function $\dot{V} = f(n_r)$ was drawn in Figure 6.

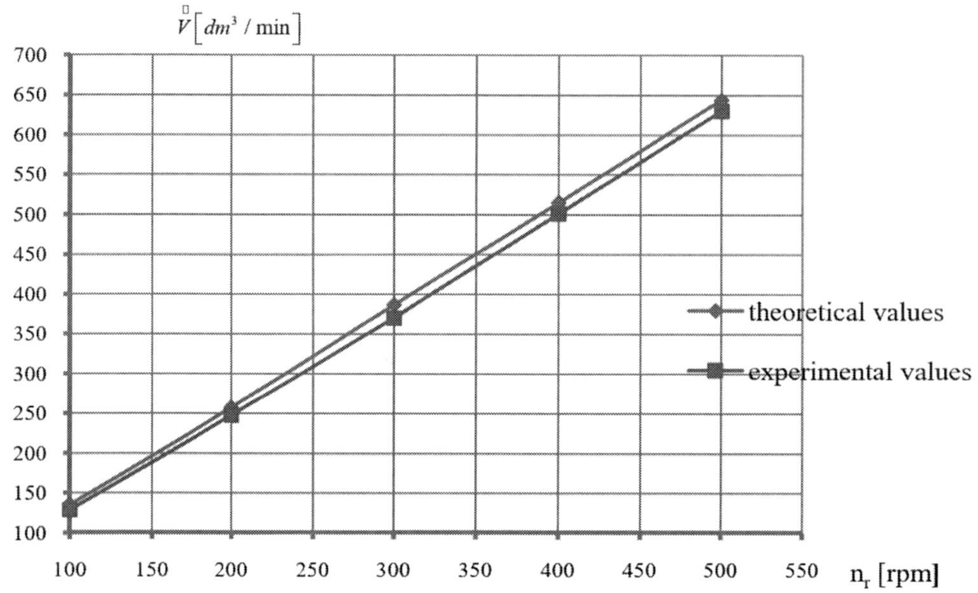

Fig. 6. The graphs of the function $\dot{V} = f(n_r)$
1 - theoretical values; 2 - experimental values.

Figure 6 shows that the two graphs are very close; as a result, the volumetric efficiency will be [10]:

$$\eta_v = \frac{\dot{V}_t}{\dot{V}_r} \tag{14}$$

Approximately, as the average value, it will be $\eta_v = 0.95$.

3 The use of the constructive solution of the rotating volumetric pump in the field of land reclamation

If the characteristic sizes of the laboratory prototype are multiplied by 10, one can obtain:

- the rotor length: l = 0.5 [m];
- the rotor radius: $R_r = 0.4$ [m];
- the height of the rotating piston: z = 0.38 [m].

The case radius will be:

$$R_c = R_r + z = 0{,}4 + 0{,}38 = 0{,}78 \; [m] \tag{15}$$

Substituting these values in relation (11), for different speeds 200 ... 500 [rpm] the data in table 3 are obtained.

Table 3. The values of $\dot{V} = f(n_r)$.

n_r [rpm]	200	300	400	500
$\dot{V}[m^3/h]$	*16211.712*	*24317.568*	*33790.056*	*40529.280*

 https://doi.org/10.1051/e3sconf/202128604001

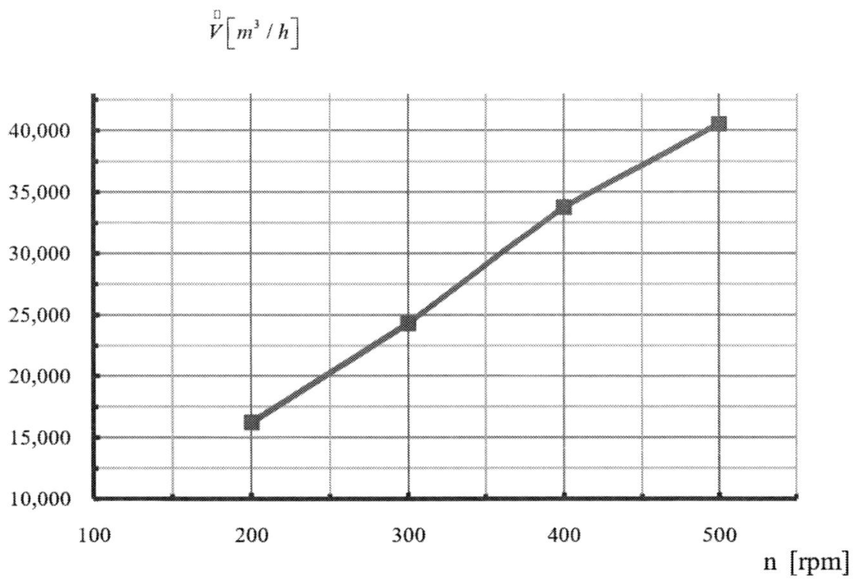

Fig. 7. The graphs of the function $\dot{V} = f(n_r)$

This flow rate for n_r = 400 [rpm] can ensure the irrigation of 28 hectares of agricultural land, once a day with 10 l /m^2.

4 Conclusion

a. Rotating working machines have the advantage of converting the motor torque received from the shaft into potential pressure energy with minimal losses.

b. The advantages of this original construction solution is that it can transport high flows rate water with lower energy consumption compared to existing pumps in irrigation stations.

c. As a result of the theoretical and experimental research carried out in the laboratory, the effective efficiency of the machine of 78% resulted. The actual efficiency of the machine is influenced by the nature of the fluid transported by the viscosity and of the machine speed.

d. Thus, we conclude that the driving power is influenced by the flow rate (i.e., by the parameters mentioned above) by the increase in pressure (Δp) achieved by the rotating machine between suction and discharge, the nature of the transported fluid.

e. Volumetric pump with profiled rotors can be used in the field of land reclamation, in wastewater treatment plants, in: mining, energy, petrochemical industry.

f. There is a good concordance between theoretical data and those experimentally determined.

References

1. N. Băran, P. Răducanu, a.o., *Bases of Technical Thermodynamics, Technical Thermodynamics* (in Romanian), POLITEHNICA PRESS Publishing House, Bucharest (2010).
2. C. Ţurcanu, N. Ganea, *Volumetric Pumps for Liquids*, (in Romanian), Technical Publishing House, Bucharest (1980).

3. M. Exarhu, *Pneumatic and hydraulic machine and installations*, (in Romanian), SC. ANDOR SRL, Bucharest, 2011.
4. N. Băran, A. Detzortzis, M. Hawas, D. Besnea, The correlation between the rotor shape and the energetic performance of a rotating machine with profile rotors, The Romanian Review Precision Mechanics, Optics & Mechatronics, no. 45,18-26 (2014).
5. M. Hawas, N. Băran, A. Detzortzis, Influence of the rotor architecture and of the speed on the volumetric efficiency of a new type of rotating volumetric machine, Advanced Materials Research, Trans Tech Publications, Switzerland, 905, 487-491, (2014).
6. M. Hawas, The influence of fluid viscosity on the performance of rotating machine with profiled rotors, PhD Thesis, Faculty of Mechanical Engineering and Mechatronics, University Politehnica of Bucharest (2015).
7. A. Motorga, Influence of constructive and functional parameters on the performances of rotating machines with profiled rotors, PhD Thesis, Faculty of Mechanical Engineering and Mechatronics, University Politehnica of Bucharest (2011).
8. N. Băran a.o., *Bases of Technical Thermodynamics,* (in Romanian), POLITEHNICA PRESS Publishing House, Bucharest, **3** (2010).
9. N. Băran, A. Motorga, D. Duminică, Research Regarding the Building of an Experimental Setup for the Testing of a New Type of Rotating Machine with Profiles Rotors, Mechatronics Magazine, no.1, 27-30, (2011).
10. N. Băran, D. Besnea, A. Motorga, Elements of computing the architecture and manufacturing technology for a new type of profiled rotor, Proceedings International Conference, 6th Workshop on European Scientific and Industrial Collaboration on promoting Advanced Technologies in Manufacturing, WESIC'08, Bucharest, 233-241 (2008).

Research on Exhaust Gas Cleaning System (EGCS) used in shipping industry for reducing SO_x emissions

Mihail-Vlad Vasilescu[1], *Dumitru* Dinu[1], *Mariana* Panaitescu[2,*] and *Fanel-Viorel* Panaitescu[2]

[1]Constanta Maritime University, Doctoral School of Mechanical Engineering, 104 Mircea cel Batran Street, 900663 Constanta, Romania

[2]Constanta Maritime University, Faculty of Electromechanics, 104 Mircea cel Batran Street, 900663 Constanta, Romania

Abstract. Nowadays maritime transportation develops environmental pollution issues. On 1st of January 2020 "International Maritime Organization (IMO)" give a new amendment which limits the sulphur content in the fuel oil used on board ships operating outside designated emissions control area (ECA) to 0.50% m/m – a significant reduction from previous limit of 3.5 %. In ECA the sulphur content in the fuel must be maximum 0.10% m/m. In order to comply with the new regulation maritime industry has to search for new methods of reducing the SO_x emissions. This article major points are: environmental pollution due to maritime transportation nowadays; methods for shipping industry to comply with IMO 2020 amendment; the use of an EGCS (Exhaust Gas Cleaning System) in order to reduce the output of SO_X; discharge washwater chemistry; titration curve and dilution requirement; use CFD modeling in order to determine the expected flow and mixing profiles for the discharge washwater of an acidic effluent waste stream from ship into the ocean; comparison between the manufacturer system specification and the CFD modeling results; conclusions.

1 Introduction

1.1 Environmental pollution due to maritime transportation nowadays

In the last 30 years the chemical structure of the atmosphere has change due to polluting emissions of different gases and PM which destroy the ozone layer. [1]

For this reason, IMO on 1st of January 2020 give a new amendment which limits the sulphur content in the fuel oil used on board ships operating outside designated emissions control area (ECA) to 0.50% m/m – a significant reduction from previous limit of 3.5 %. In ECA the sulphur content in the fuel must be maximum 0.10% m/m. In order to comply with

* Corresponding author: panaitescumariana1@gmail.com

the new regulation maritime industry has to search for new methods of reducing the SO_x emissions.

1.2 Methods for shipping industry to comply with IMO 2020 amendment

In order to comply with the new rule, the shipping industry has a few options:

- the first alternative could be to use a compliant fuel with lower sulphur content, an alternative fuel like LNG or methanol;
- a second alternative could be to install a green energy source like: Flettner Balloon with Helium, Vertical wind turbines, Solar panels, Kites; these alternative solutions are used for reducing the fuel oil consumption of the electric generators in the ship power plant (not for propulsion);
- the third alternative could be to clean its exhaust gases with a "scrubber" technology to reduce the output of SO_x.

1.3 The use of an Exhaust Gas Cleaning System in order to reduce the output of SO_x

1.3.1 System description

The Exhaust Gas Cleaning System (EGCS) is based on the principle of scrubbing. The EGCS is composed of one Exhaust Gas Cleaning (EGC) unit , appointed scrubber (Figure 1)[2], exhaust gas supplying and discharging system and seawater supplying and washwater discharging system. The EGC unit is designed to be continuously operated in a wet (scrubbing) condition. The EGC unit is designed to pre-treat the exhaust gas and intimately mix the gas with seawater to remove SO_x and particulate matter emissions generated from burned fuel.

Exhaust gas pipes are conducted to the EGC unit (Figure 2). Continuous Emissions Monitoring System (CEMS) is equipped in the gas outlet duct of the EGC unit to monitor CO_2 and SO_2 in exhaust gas.

Scrubber water monitoring sensor is also installed in the washwater discharging pipe for sensing polycyclic aromatic hydrocarbons (PAHs), pH, temperature and turbidity.

In the open loop mode, the scrubber water is supplied from the sea chest and pumped to the EGC unit, where the water is sprayed in two stages. The first stage is in the bottom spray at EGC body, in order to improve gas flow, reduce back pressure and catch particulate matter and sulphur. The second stage is in the top of the EGC body. Internal mist eliminator at the next of top spray effectively prevents water droplets to be carried away with the outlet gas flow. A bypass arrangement is needed and is permitting operation of the combustion units. In the open loop mode, the washwater is drained from the EGC unit and discharged overboard.

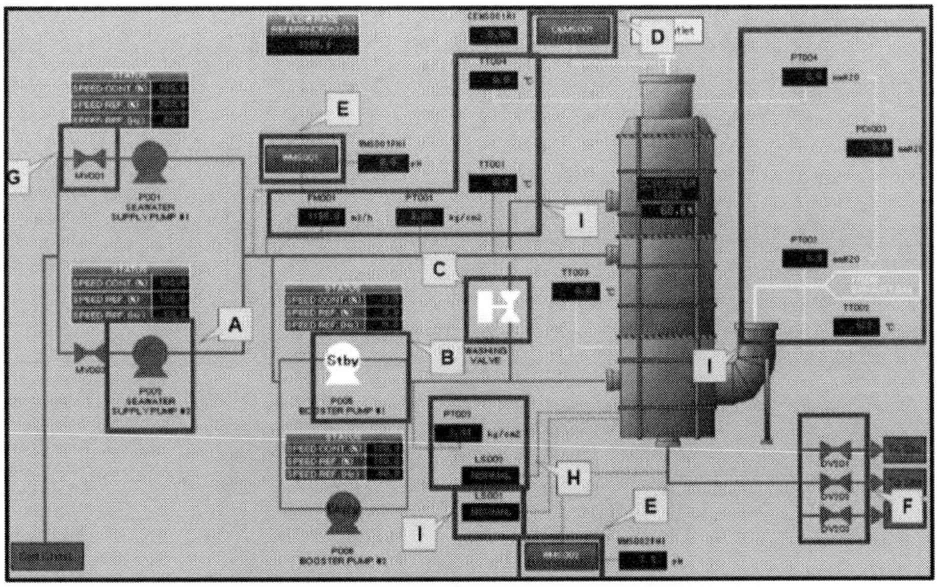

Fig. 1. Exhaust Gas Cleaning System –Scrubber

Fig. 2. Exhaust gas pipes

In the Figure 1:

- A is the seawater supply pump is to provide water from sea chest to scrubber.
- B are the booster pumps will serve as making the hot exhaust gas cool and washing the mist eliminator by the seawater sprayed on scrubber.
- C is the washing valve which provide seawater to wash the mist eliminator by washing water.
- D is the Control Environment Systems (CEMS) which can measure sulfur dioxide, carbon monoxide, ratio SO2 /CO2, H2O.
- E is the Water Measurement Systems (WMS) which can measure PAHs, turbidity, pH, temperature.
- F is the overboard valve which is used for discharging the seawater after scrubbing with the gas.
- G is the suction valve which is used to indicate inlet valve status, if this valve is closed, seawater supply pump must be stopped.

https://doi.org/10.1051/e3sconf/202128604002

- H is the level switch which is used to detect the water which is flowing into the scrubber body. When LS002 alarm occurs, the system should shut down to protect the water flowing into the M/E, G/E, Auxiliar Boiler.
- I is the measuring equipment:

FM001: Seawater supply flow (Figure 3);
TT001: Seawater supply temperature;
TT002: Inlet exhaust gas temperature;
TT003: Packing inlet temperature;
TT004: Outlet exhaust gas temperature;
PT001: Seawater supply pressure;
PT002: Booster pump pressure (Figure 4);
PT003 & PT004: Scrubber differential pressure.

Fig. 3. Seawater supply pumps

Fig. 4. Booster pumps

The ECGS [3] is shows in the Figure 3:

 https://doi.org/10.1051/e3sconf/202128604002

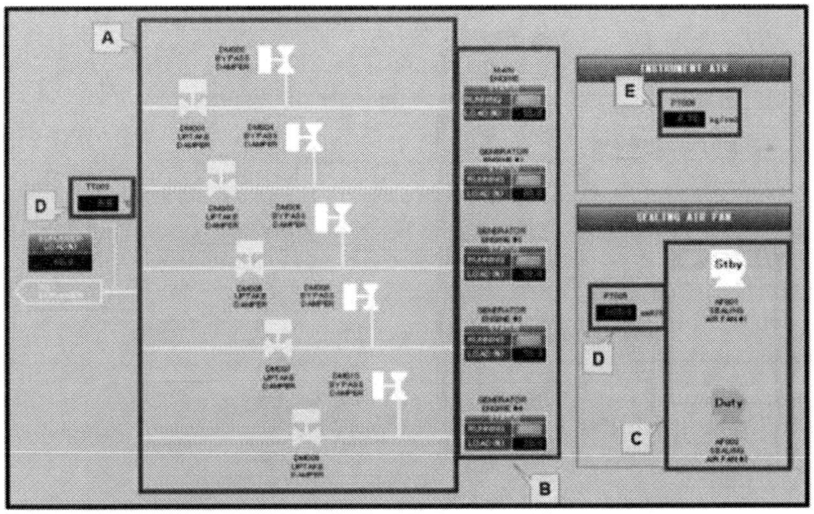

Fig. 5. Exhaust Gas Cleaning System

where:

- A is the uptake and bypass dampers which are used to lead the exhaust gas in the desired direction. When the scrubber is in operation, the exhaust gas is led from the engine or generator through the respective uptake damper to the scrubber.
- B indicates the information received from alarm monitoring system and affect the uptake & bypass damper control. Indicates the main engine, generators: 1, 2, 3, 4, Auxiliar Boiler run & load status which are received from alarm monitoring system.
- C is the sealing air fan which provides the sealing air necessary to close the valve of uptake and bypass dampers and prevent gas flowing into residential areas.
- D is the measuring equipment:

PT005: Sealing air fan pressure;

TT002: Inlet exhaust gas temperature;

- E is the instrument air

PT006: Instrument Air Pressure.

1.3.2 System performance

The EGCS is designed to reduce SO_x emissions to a level not exceeding an equivalent of 0.5% m/m sulphur content in the fuel (SO_2/CO_2 ratio of 21.7 or less) for global and/ or 0.10 % m/m sulphur content in the fuel (SO_2/CO_2 ratio of 4.3 or less), when the combustion units are operated on a fuel that contains a maximum of 3.5 % m/m sulphur.[4]

Table 1. Typical SO_2/CO_2 ratios based on fuel sulphur content

FO Sulphur content (% m/m)	Ratio emissions SO2 ppm/CO2 (% v/v)
0.50	21.7
0.10	4.3

 https://doi.org/10.1051/e3sconf/202128604002

1.3.3 System configuration

In the open loop mode, exhaust gas is fed the scrubber body via scrubber inlet manifold pipe. Manifold pipe is sized to optimize the flow uniformity in the scrubber. Exhaust gas passes through the packing material for ensures good contact between exhaust gas and liquid resulting in a highly efficient SO2 absorption. The treated flue gas from the scrubber is routed to the atmosphere through the stack. Fresh seawater is used for removal of SO_x. Seawater is routed from the sea chest to the seawater supply pumps and then divided into the scrubber and booster pumps. The water that goes to the scrubber body directly is used for the scrubbing water. For the scrubbing spray, HPS (Hyundai Power Systems) scrubbing system adopted nozzle-free spray system which lead lower electric power than the nozzle spray system (Figure 6) [5].

Fig.6. Nozzle spray system

The other water is pumped by booster pump, in order to have enough pressure for quenching spray and mist eliminator washing water. All used seawater is routed to the sea through the overboard valve.

4 Methods and researches

4.1 Discharge washwater chemistry

The SO_x in the exhaust gas is present mainly as SO_2, which in seawater is hydrated to H_2SO_3. The more SO_2 that is dissolved in the washwater, the lower pH value will be in the discharged water (Figure 7)[6]. The washwater in an EGCS is never exposed to atmospheric conditions after absorbing the SO_2. It is kept in closed tanks, scrubbers and pipes prior to release. It is thus important to include in the chemical model such conditions. After discharge of the washwater into the ocean, the mixing with pure seawater will cause the pH to increase. The pH increase is due to both a dilution effect and a further neutralization caused by the alkalinity in the seawater. The major contributor to the alkalinity in seawater is carbonate and bicarbonate, and the neutralization process leads to release of CO_2. The rate of stripping of CO_2 from the seawater is thereby a limiting factor for the pH-recovery, and less CO_2 stripping means that the dilution has to be increased.

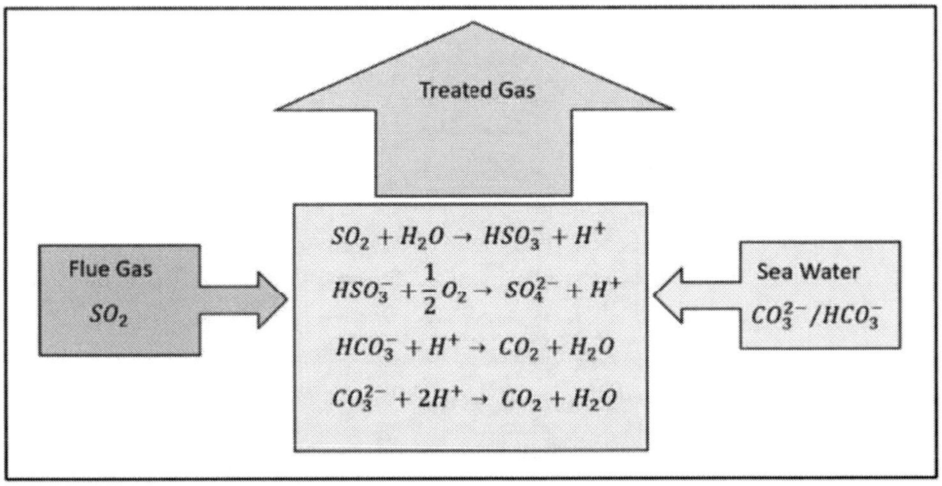

Fig. 7. Chemistry for SO_2 removal by seawater.

Figure 7 shows the chemical process in which SO^{-2}_2 generated after SO_2 is absorbed in seawater, is neutralized by carbonate, CO^{-2}_3, and bicarbonate,HCO_3^-, the base compounds in seawater.

4.2 Titration Curve and Dilution Requirement

The *Titration curve* is prepared by using a chemical equilibrium model for seawater [6]. The model includes inorganic carbon, boric acid, sulfate, fluoride and dissolved SO_2 equilibrums, the equilibrium constants are function of salinity (ionic strength) and temperature.
Basis for the calculation:

- released CO_2 retained in solution, i.e no forced stripping of CO_2 ;
- 10% of dissolved S (IV) oxidized to S (VI) inside scrubber system;
- seawater alkalinity 2.2 mml/l;
- seawater salinity 35 psu;
- seawater pH 8.2;
- seawater temperature 32°C (for the 0 class ships, we take always the worst condition for seawater temperature).

The fit equation is provided based on an empirical equation fit to the curve. The equation is:

$$pH = 3.84 - 0.2308\, x\, SO_2 + \frac{1.403}{(0.0403+\exp(2.966\, x(SO_2-0.189)))} + \frac{9.947}{(4.605+\exp(4.554\, x\,(SO_2-1.588)))} \quad (1)$$

Where the variable SO_2 absorbed in mmole/kg seawater. The "*fit equation*" is used for the determination of the dilution factor.

The seawater with dilution ratio, 1.6:8.4 (16%) is required to reach pH 6.5 (0.46 mmol SO_x /kg) when the pH of discharge washwater is 3.2. (3.9 mmol SO_x /kg). As can be seen from titration curve, lower dilution rate is required to reach the pH 6.5 before 4 m from discharge point if the pH of discharge washwater is more than 3.2.

$$Dilution\ factor = \frac{2.89}{0.46} = 6.23 \tag{2}$$

$$Dilution\ ratio = \frac{1}{6.23}x100 = 16\ \% \tag{3}$$

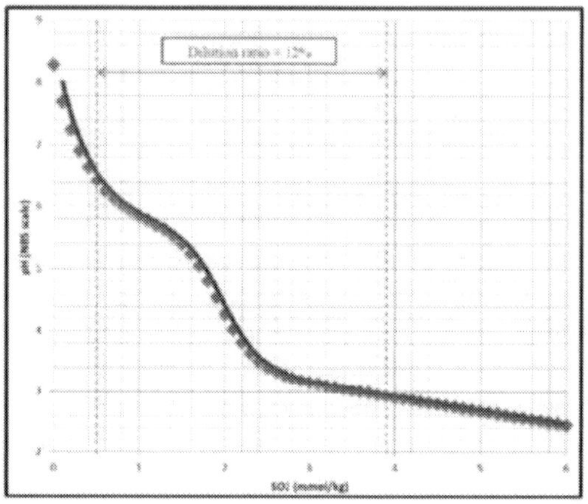

Fig. 8. Titration Curve.

4.3 Use CFD modeling in order to determine the expected flow and mixing profiles for the discharge washwater of an acidic effluent waste stream from ship into the ocean

4.3.1 CFD Model Geometry

The current CFD analysis was conducted on the basis of hull section information and considered discharge nozzle's location & light draft [6]. Especially, this analysis focused on 4 m distance from discharge nozzle.

Model geometry with 3 discharge nozzles was used with this vessel. Three 450 A discharge nozzles of which distance is 2.8 m were added to the vessel along the starboard side near the stern at a depth of ~ 2 m below the water level (Figure 9, Figure 10).

CFD analysis for the pH of discharge washwater has been performed to show that the pH of discharge washwater becomes 6.5 before 4.0 m from every discharge nozzle.

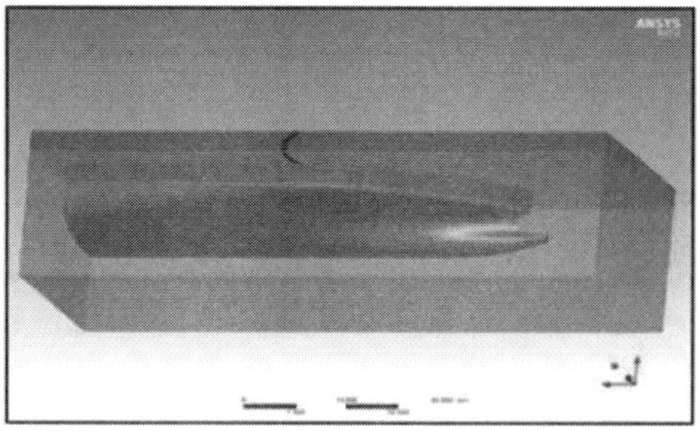

Fig. 9. Hull and discharge nozzles from CAD.

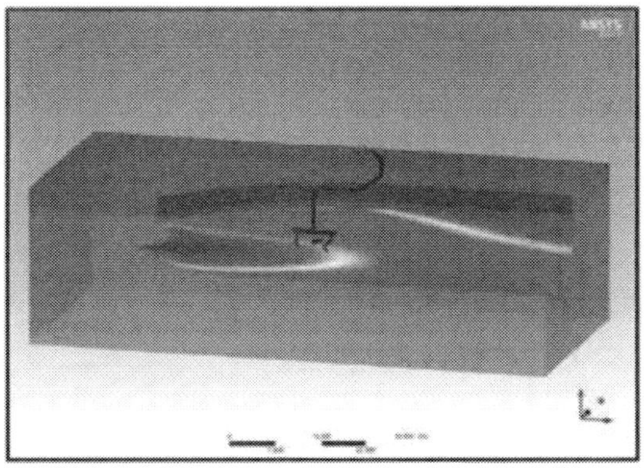

Fig. 10. Hull and discharge nozzles for CFD

4.3.2 CFD Methodology

The seawater flow induced by discharge washwater creates flow streamline in air phase which is in contact with the sea phase (Figure 11, Figure 12).

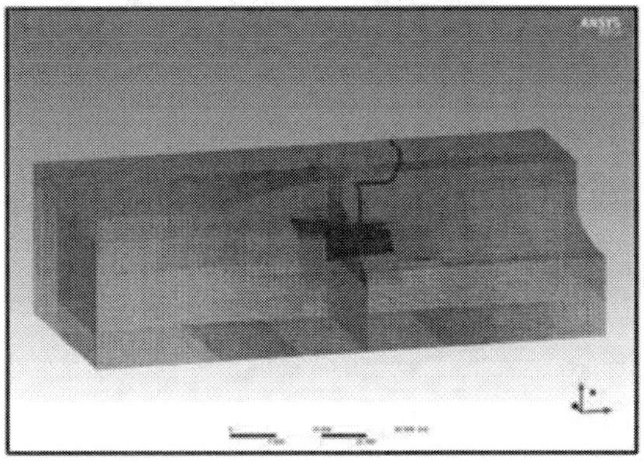

Fig. 11. Mesh structure for the CFD simulation.

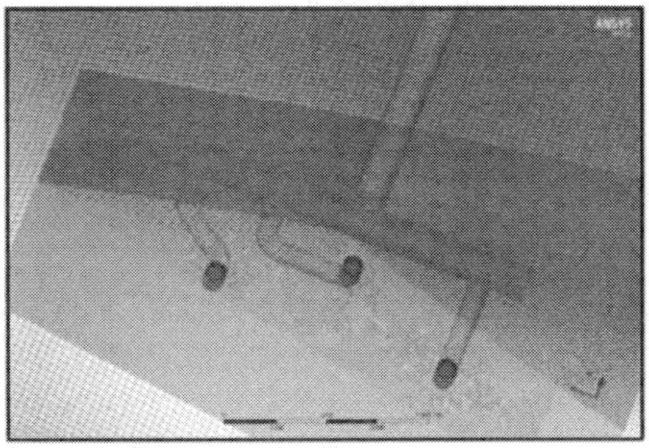

Fig. 12. Mesh structure for the pipes CFD simulation.

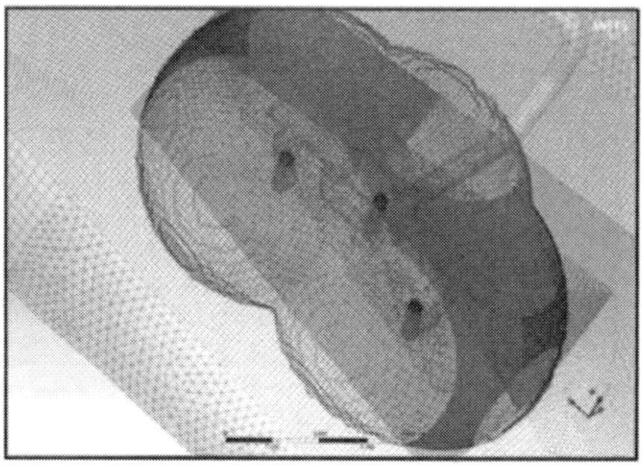

Fig. 13. Mesh structure for near the discharge nozzles.

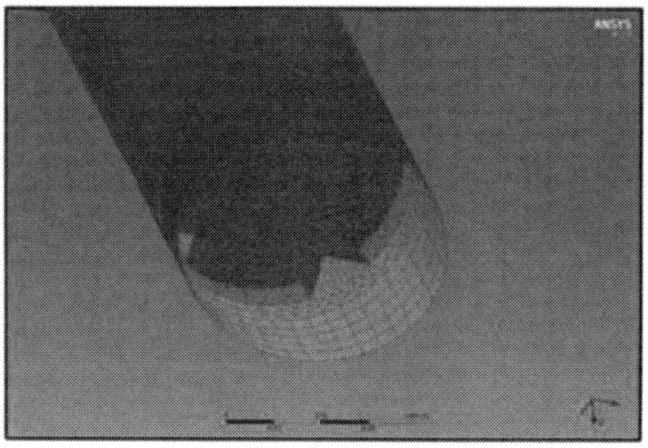

Fig. 14. Mesh structure forintire discharge nozzles.

For the current analysis, a mesh system was used which consists of 8,618,388 computational cells. The number of mesh for sectional area of discharge washwater was 100. The cell size at discharge outlet is 2 cm to be fine enough in discharge area and in the dilution area (at least 4 m away from the discharge point) and the cell size at 4 m distance from discharge points is 3~6 cm. The current analysis utilized the steady state, multiphase, 3D, turbulent, fluid solvers. The turbulence model used was the k-ε model. The size of computational domain was 49 m width, 25 m depth and 21 m height.

4.3.3 Boundary Conditions

For this CFD simulation, seawater phase and air phase are considered at the same time, multiphase flow, sea and air flow, analysis model was applied. The surface energy at the interface between air and seawater was given 0.017 N/m. The gravitational acceleration on seawater and air is 9.81 m/s^2. All the boundaries of sea except the face of discharge nozzle location is open sea. The lower part of computational domain, the ocean, is treated as a wall with a shear zero condition having 25 ^{0} temperature condition. Wall with two discharge nozzles installation by simplifying the hull were subjected to no slip wall condition and temperature of the wall was 25^0 C. The three sides of the computational domain were covered by opening boundary conditions in Ansys, with temperatures of 25^0 C and pH of

8.2. The upper surface and lateral surface of the air domain adjacent to the sea domain was treated as wall of 25^0 C. The gas entrained in the washwater was taken to be air at standard temperature. The air affects the variations in turbulence in the flow to the gas phase were incorporated in the flow model because of bubble break-up, bubble coalescence. All simulations were carried out with 10% air contained washwater, with homogenous bubble sizes of 2 mm diameter.

5 Results

EGCSA PPR2 (Exhaust Gas Cleaning System – Pollution Prevention and Response) was used to predict the change in pH of the emission water after it was discharged into the sea. Depending on the predicted dilution rate in EGCSA PPR2, the pH of the emission water can be checked and this analysis shows that the emission water rises to pH 6.5 before reaching 4.0 m point at the discharge point.
This method tracks the pH value of the fluid exiting the discharge nozzle and uses convection, turbulence, and diffusion to predict the pH value downstream within the entire flow domain based on the boundary conditions provided. The method is passive in that no chemical reactions are considered or modeled.
In this analysis, the pH is predicted by comparing the titration curve and seawater with the concentration result of the discharge washwater predicted by CFD analysis. The typical pH estimation results for discharge washwater stream according to the degree of dilution by seawater are shown in table below (Table 2)[5].

Table 2. pH correlation table according to dilution ratio.

Volume fraction		pH
Discharge water	Seawater	
1	0	6.5
0.65	0.35	6.13
0.39	0.61	5.74
0.28	0.72	5.13
0.21	0.79	3.93
0.19	0.81	3.52
0.17	0.83	3.31
0.16	0.84	3.23

5.1 Comparison between the manufacturer system specification and the CFD modeling results

Eulerian multiphase computer simulation of CFD was carried out for max load case, 2.200 m^3 /h. To check the boundary of pH 6.5, surface contour of volume fraction, 0.16 is confirmed and shown in the CFD results.

5.1.1 Max.Flow

The contour value is calculated automatically by CFX. It demonstrates that plume of discharged water dilutes to a pH 6.528 when pH at discharge point is 3.2 (Figure 15, figure 16, Figure 17). It satisfied the IMO regulations & MEPC 259 (68) 10.1.2.1.2 *(Marine Environment Protection Committee Resolution).*

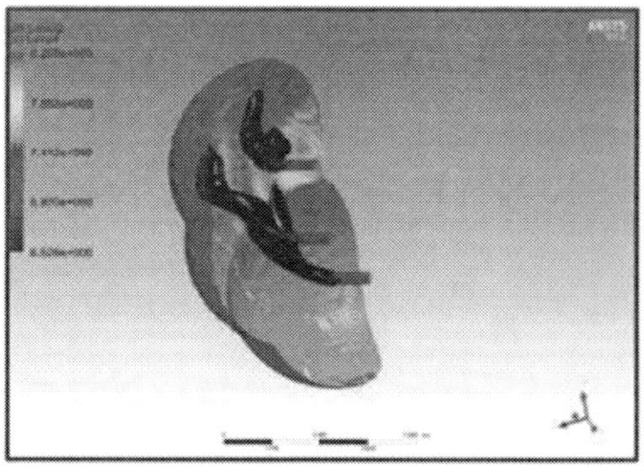

Fig. 15. pH 6.5 plume of discharge washwater and 4.0 m sphere with 2.200 m3 /h (Max. Flow).

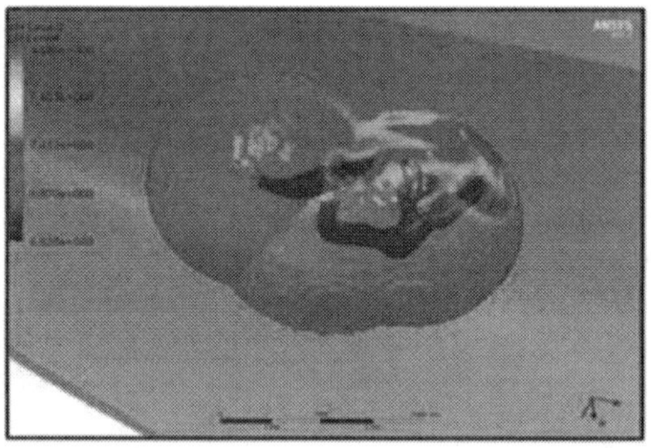

Fig. 16. pH 6.5 contour of discharge washwater stream and 4.0 m sphere with 2.200 m3 /h (Max. Flow).

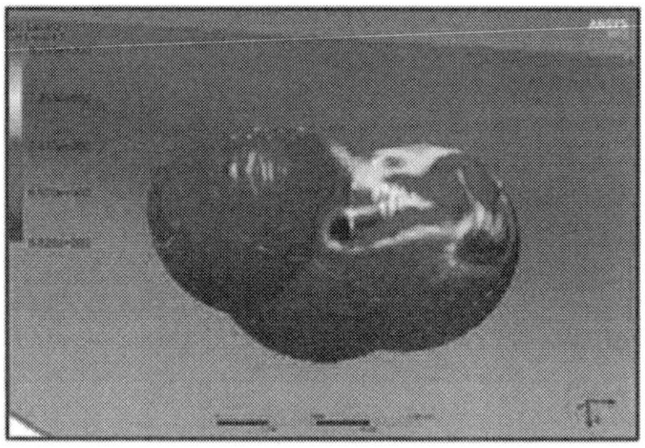

Fig. 17. Contour from front-view with 2.200 m^3/h (Max. Flow).

5.1.2 Min.Flow

The plume of discharged water dilutes to a pH 6.594 when pH at discharge point is 3.2 (Figure 18, Figure 19, Figure 20). It is satisfied with IMO regulations & MEPC 259(68) 10.1.2.1.2 (*Marine Environment Protection Committee Resolution*).

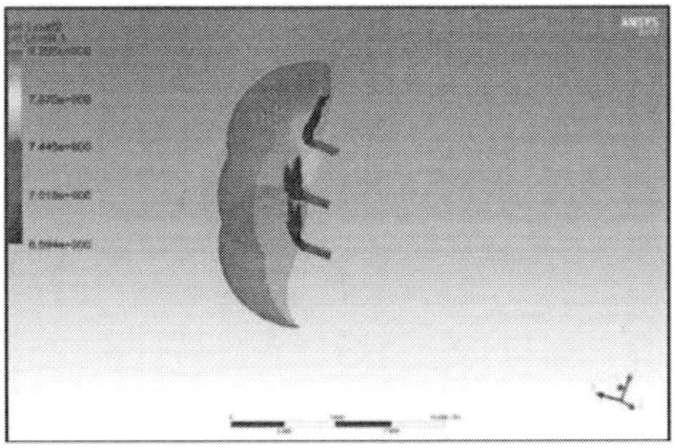

Fig. 18. pH 6.5 plume of discharge washwater and 4.0 m sphere with 990 m^3/h (Min. Flow).

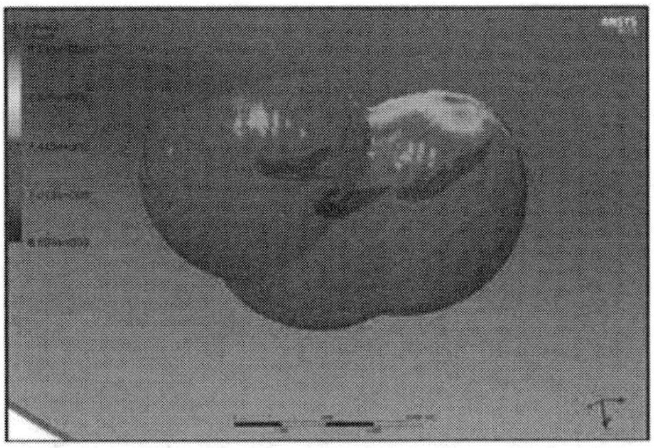

Fig. 19. Contour from front-view with 990 m3 /h (Min. Flow).

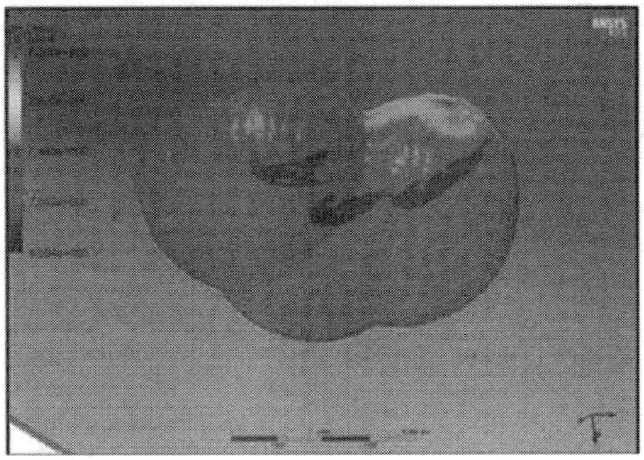

Fig. 20. Contour from side-view with 990 m^3/h (Min. Flow).

5.1.3 Comparison between IMO PPR 2/2/3 and our results

Area of 1 hole in discharge nozzle with diffuser is 0.0075 m^2. In this case, pH is calculated to 3.01 according to IMO PPR 2/2/3.

The results extracted from Ansys Fluent software are reliable are similar with IMO PPR 2/2/3 (Table 3).

Table 3. Comparison between IMO PPR 2/2/3 results and the results from Ansys Fluent software

Method by PPR2	Dilution Factor	SO_2 - Discharge	pH Limit to recover pH 6.5
Source 1	7.76	3.60	3.01
Source 2	9.31	4.32	2.84
Source 3	8.23	3.82	2.96
Source 3 Min	8.37	3.89	2.94
Source 3 Max 8	8.08	3.75	2.98
Our Result	6.23	2.89	3.20

6 Conclusions

From the pH prediction of discharged/diluted water and dilution modeling, we have concluded that:

- To reach the pH of 6.5 at 4 m from discharge, discharge washwater should be mixed to seawater with dilution ratio 1.6: 8.4 when the pH of discharge washwater is 3.2 on max load case of EGCS;
- In maximum load case, CFD simulation shows that the plume for 1.6: 8.4 dilution ratio is whining the range of 4 m from discharge;
- Discharge washwater with pH 3.2 or more confirms compliance to MARPOL regulations 14.1 and 14.4. on maximum load case;
- Discharge washwater is satisfied with a pH of no less than 6.5 at 4 m.

7 References

1. M. V. Vasilescu, Ph. D. Thesis, *Research on the design and operation of container vessels for improving energy efficiency*, **7** (2020)
2. Hyundai Power System, *Operation and maintenance manual. Project – Scrubber,* **12** (2019)
3. Hyundai Power System, *Operation and maintenance manual. Project – Scrubber*, **14** (2019)
4. Hyundai Power System, *SECP EGC System - SO_x Emissions Compliance Plan, Project – EGCS open for cntr.*, **4** (2019)
5. http://eng.hhi-power.com/business/scrubber/index.jsp.
6. Hyundai Power System, *EGC System Record Book/Log, Project – EGCS open for cntr.*, **7** (2019)

Acknowledgments

Authors gratefully acknowledge to this material support path received under Project PN-III-P1-1.2-PCCDI-2017-0404/ 31PCCDI / 2018, *Holistic on the Impact of Renewable Energy Sources on Environment and Climate*-HORESEC.

Models to estimate river temperature. Example for Danube, at Oltenița in the context of climate change and anthropic impact

Daniela-Elena Gogoaşe-Nistoran[1,*], *Cristina Sorana* Ionescu[1] and *Ioana* Opriş[2]

[1]University Politehnica of Bucharest, Faculty of Power Engineering, Dept. of Hydraulics, Hydraulic machines and Environmental engineering
[2]University Politehnica of Bucharest, Faculty of Power Engineering, Dept. of Energy Production and use

Abstract. Daily variation of Danube River temperature measured at Oltenitț gauging station over 9 years (2008-2016) was analysed in comparison with the air temperature measured by satellite in the same location between 1979-2020. Air temperature shows a nearly 2°C increase over the 40-years period, which can be attributed to both climate warming and anthropic impact. Water temperature was modeled with a sinusoidal function and variation with discharge was discussed. Long-term trend of hourly surface air temperature variation was obtained from Open Weather data. Air - water temperature dependency was fitted with a logistic function with good approximation. Resulting correlations help predict water temperature as a function of satellite - measured air temperature.

1 Introduction

Stream temperature is one of the most important parameters that indicates the health of the aquatic environment. River temperature is influenced both by natural (e.g., atmospheric conditions, precipitation, wind, river discharge, riparian vegetation) [1] and anthropic factors (effluent discharges temperature, high amounts of water intakes for multiple uses, such as agriculture and industry and increased urbanisation) [2, 3].

European Environment Agency assessed that the average annual temperature for the European land area for the last decade (2008–2017) was between 1.6 °C and 1.7 °C above the pre-industrial level, which makes it the warmest decade on record [4].

In Romania, during 1991-2019 air temperature has increased by about 1.5 °C [5]. As well, the frequency of extreme heat waves increased during the last 20 years [6, 7]. These phenomena also impact river temperature, particularly during summer periods, when precipitation and discharges are the lowest.

In the context of climate change, the study of extreme heat periods is of increasing interest in general and in particular in relation with their influence on river water temperature. Identifying possible patterns of river water temperature correlations with air temperatures is of utmost concern mainly in vulnerable regions, subjected to droughts and

* Corresponding author: daniela.nistoran@upb.ro

desertification. Romania is particularly prone to such climatic risks, due to its natural conditions such as its geographical location and relief [8], as well as to anthropogenic factors, such as deforestation and urbanization growth, that generates an increased stress on water resource availability and quality.

In Romania the area subjected to desertification, characterized by an arid, semiarid or subhumid-dry climate is about 30% of the total surface of Romania, being mostly situated in Dobrogea, Moldavia, the South of the Romanian Plain and the Western Plain. This area is mainly used in the sectors of agriculture (about 80% of the total, of which about 60% are arable land), forestry (about 8%) and waters, especially the Romanian Plain and the Danube Delta [9]. The location of the present case study was selected to be Oltenita City, within the Romanian Plane, on Danube River.

The objective of the paper is to provide quick estimates of Danube river temperature based on its long-term variability and correlations between air and water temperatures. These can be used as a first approximation for water temperature predictions in locations prone to excessive warming.

2 Site and data

Olteniţa is an industrial city located on the left bank of Danube river. During summers heat waves characteristic to Danube floodplain are very hot. Both water and air temperatures were available at the gauging station of Oltenita.

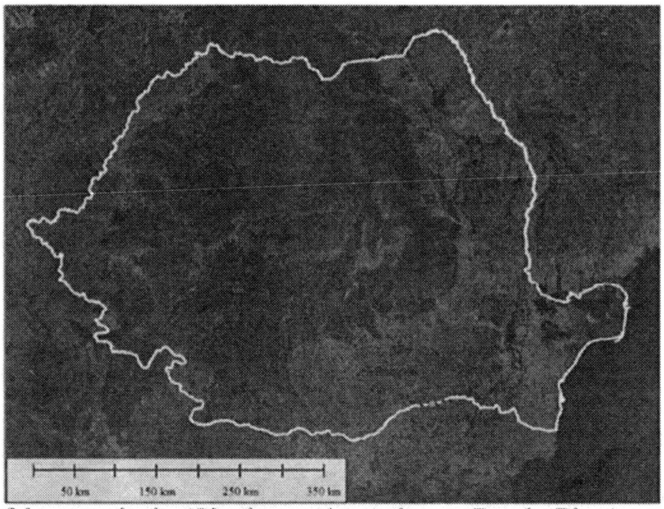

Fig. 1. Location of the example site (Oltenita gauging station, on Danube River).

Satellite hourly mean air temperature values near Oltenita gauging station (44.06°N and 26.63°E), over 40 years (1979-2020), were downloaded from openweather.com (Fig. 2). OpenWeather collects and processes weather data from a vast network of over 40,000 weather stations and from meteorological broadcast services worldwide, such as satellites, radars and uses global and local weather models (set of algorithms, to improve its quality and accuracy, which is reported to be of 99%) [10]. From the fitting line over this period, one may observe an increase of mean air temperature by nearly 2 degrees.

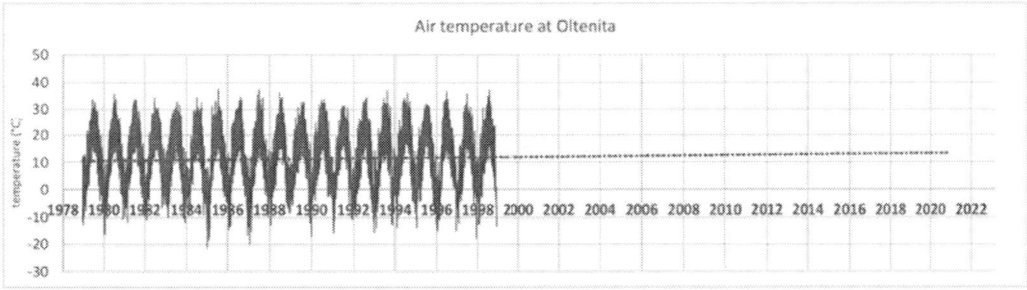

Fig. 2. Air temperature at Olteniţa gauging station - satellite data at 2m height from ground. Openweather.com, period 1979-2020

One may see that while until year 2000 air temperature during summer rarely overpassed 35 °C, after 2000 there are several years when air temperature during summer hot waves reached 40 °C: 2000, 2007, 2012, 2017 (with a quite constant period of 5 years).

Daily mean water temperature values registered at the Olteniţa Gauging station were available from edelta.ro website for the period 2008-2015, whereas for 2016 only some random weekly data were available (Fig.3). One may observe an increase of about 1.5 degrees over this 9-years period. Also, an approximate sinusoidal annual pattern of water temperature can be seen. The maximum water temperature was registered in 2012, when the hottest summer over the period 1961-2016 was recorded [8]. Accuracy of the measured water temperature is of 0.1 °C.

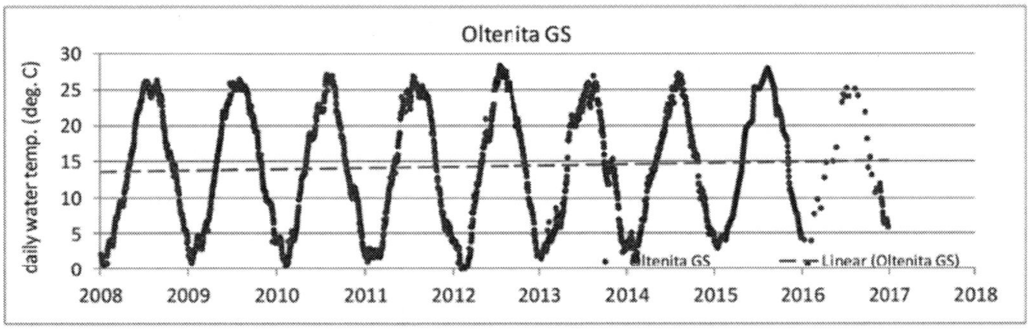

Fig. 3. Danube water temperature at Olteniţa gauging station

In order to compare air with water temperature variation during the overlapping period 2008-2016, the hourly air temperatures were averaged over each day using a HEC-DSS software, (Fig.4 and Fig. 5).

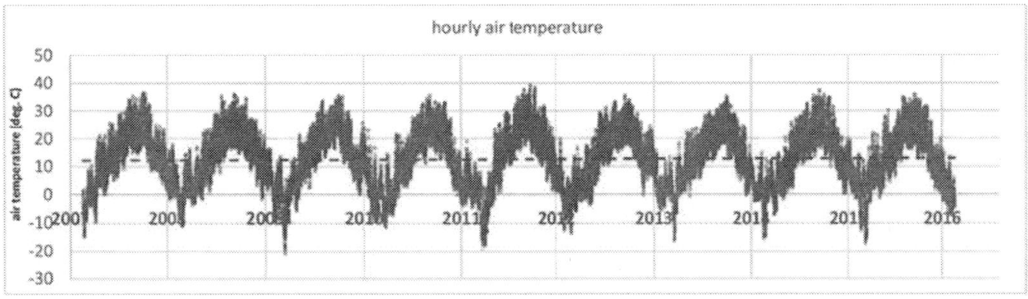

Fig. 4. Air temperature (hourly variation) at Olteniţa gauging station during the study period

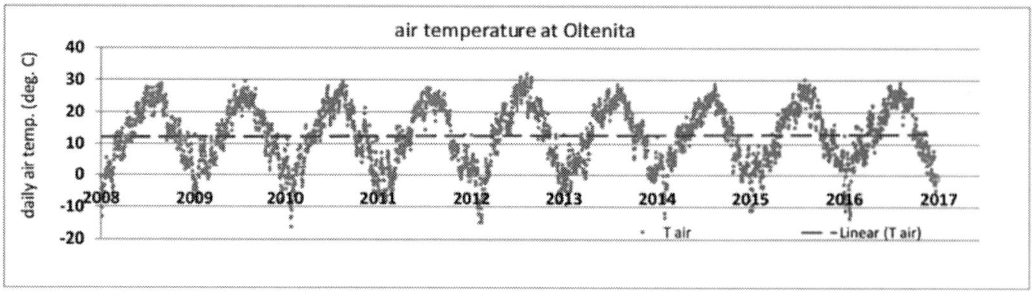

Fig. 5. Air temperature (daily average variation) at Oltenița gauging station during the study period.

Figure 6 depicts both daily water and air temperature variations at the Oltenita gauging station. Due to its higher thermal inertia, water temperature does not fall below 0 °C during winter periods.

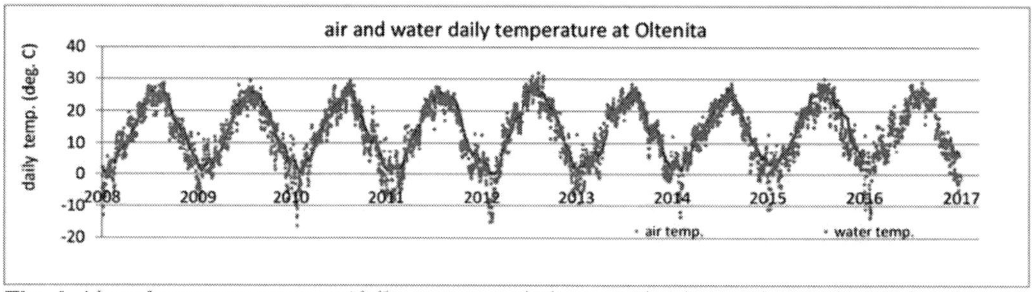

Fig. 6. Air and water temperature (daily average variation) at Oltenița gauging station during the study period.

3 Mathematical models

Two models were used to reproduce water temperatures. The first model depends on time and relies on a sinusoidal variation over a year of daily water temperature [11], $T_w = f(t)$, given by

$$T_w = T_{w0} + A\,sin\left(\pi \frac{t-t_C}{W}\right). \quad (1)$$

where the variables for a typical year whithin the 9-year period are :

t = current day (time), which may take values between 0 and 365/366 days, regardless of the year chosen between 2008 and 2016 (the small overall increase in temperature has been neglected for this approximation).

A = the amplitude over the 9-year period, i.e. the difference between maximum temperature (α) and minimum temperature (μ): A = 28.86 - 0.1 = 28.76°C (considered the same for each year)

W = half of the period of yearly variation = 365.25 days / 2 = 182.625 days

T_{w0} = mean water temperature over a year = 14.53°C

t_C = mean inflexion day of the year over the period, for the sinus function = 93 of 365.25 days.

The second mathematical model uses an S-shaped logistic function that relates water with air temperatures [12], as follows :

$$T_w = \mu + \frac{(\alpha-\mu)}{\left(1+e^{\gamma(\beta-T_a)}\right)}. \quad (2)$$

The parameters of the function are the following:
α = maximum water temperature over the study period = 28.86 °C
μ = minimum water temperature over the study period = 0.1 °C
β = air temperature of inflexion point of the S-curve
γ = a measure of the steepest slope of the function .

The model is calibrated and validated by determining the best-fitting β and γ parameters to the function of the paired daily data, T_w and T_a over the over a period of time, in this case a 9-year time period. Even though the parameters are determined for a certain past period of study (as long as possible), the logistic function may be used to extrapolate the relationship in the future with good approximation, under similar climate conditions [2].

4 Results

In Fig. 7 are sholw the results of the model calibration and validation performed with the sinusoidal function with the help of Sigma Plot mathematical software. On the same graph is plotted the daily water discharge at Oltenita gauging station, over a 8-years period. Besides the air temperature, the river discharge is another important factor that controls water temperature. One may see that when the discharge is very low during summers, the water temperature increases more than the average sinusoidal function (2012 and 2015). The reverse is also valid (for examples for the years 2008, 2009 and 2011). During the calibration period there have been both wet years (such as 2010) and dry years (such as 2011) to insure proper variability of the control factors.

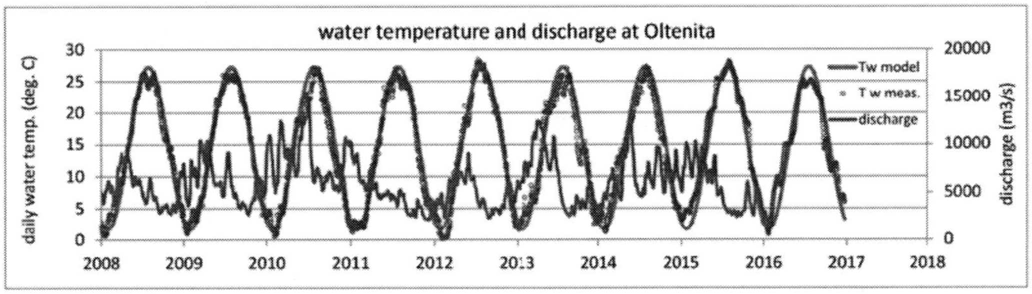

Fig.7. Air and water temperature at Oltenița gauging station over the period 2008-2017.

The absolute errors between measured and computed water temperatures defined by:

$$\Delta T = T_w^m - T_w^c \tag{2}$$

shows (Fig. 8) an overall good approximation, with values less than 4°C, besides the periods of lowest and highest temperatures from the end of winter and the end of summer (heatwaves), when these errors may even reach 6°C.

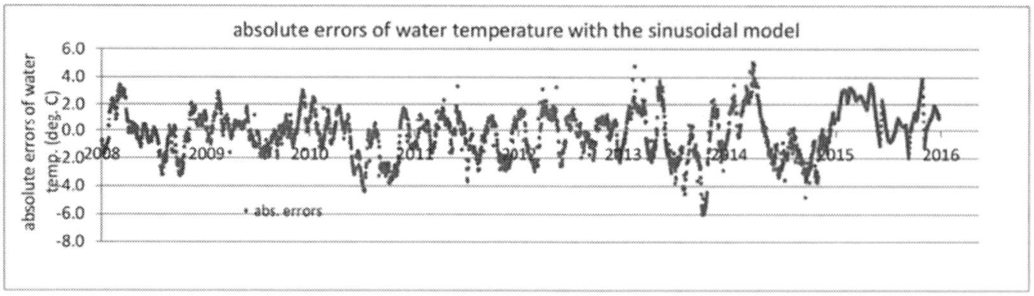

Fig. 8. Absolute errors between measured and computed water temperatures with the sinusoidal model.

Concerning the model based on a logistic function, its fitting parameters (Fig.9), were found to be: $\beta = 12$, $\gamma = 0.17416$. The adjusted R-square value for the calibration (a modified version of R-square) is 0.8353, which proves the model is a good predictor of the data.

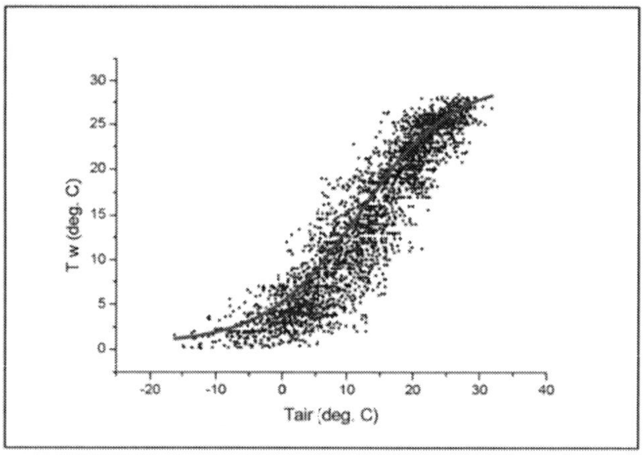

Fig. 9. Fitting of water-air temperature dependence with the logistic model.

In Fig. 10 are shown the measured and computed water temperature with the logistic model. One may see these computed values are more dispersed than the ones with the sinusoidal model, due to the sparse air temperature variation upon which depends.

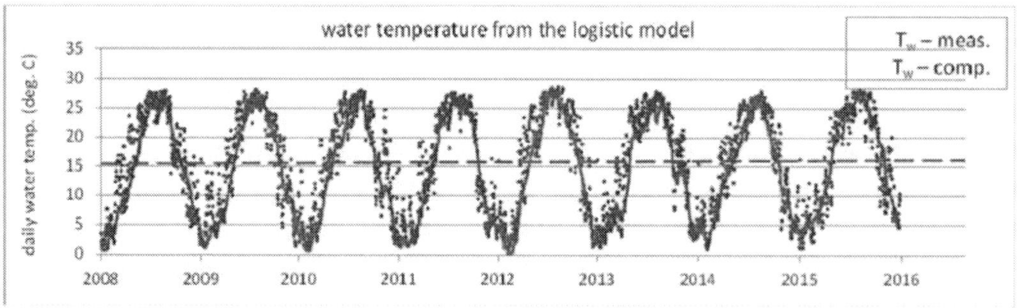

Fig. 10. Water temperature

5 Conclusions

Both temperature models reproduce water temperature values with good approximation, one as a function of time, and one as a function of air temperature. One may observe the second model reproduce water temperature with a greater dispersion, due to sparse air temperature data.

The 9 years calibration and validation period included both wet and dry years, heatwave and cold spell periods. Given the good approximations between measured and computed water temperature values, under similar climatic conditions, these models could be used to forecast the river water temperature in the future over a maximum period equal to the one for which their parameters were fitted. The longer the period used to obtain the model parameters, the more accurate the prediction. However, considering the climate is changing a maximum of half of this period should be used for predicting water temperature with the same values of model parameters.

References

1. G.Garner, I.A. Malcom, J.P. Sadler, D.M. Hannah, J. Hydrol. **553**, 471 (2017)
2. S.J. Dugdale, D.M. Hannah, I.A. Malcolm, Earth-Sci. Rev., **175**, 97 (2017)
3. D. Caissie, Freshwater Biol. **51**(8), 1389 (2006)
4. ***, Global and European temperature – Indicator assessment, Report, 2014, https://www.eea.europa.eu/data-and-maps/indicators/global-and-european-temperature-8/assessment
5. ***, The climate in Romania, https://www.worlddata.info/europe/romania/climate.php
6. A. Dobrinescu, A. Busuioc, M.V. Birsan, A. Dumitrescu, and A. Orzan, Clim. Res. **64**, 213 (2015)
7. A. Dumitrescu, R. Bojariu, M.-V. Birsan, L. Marin, and A. Manea, Theor. Appl. Climatol. **122**, 111 (2015)
8. V. Dima, A. Irimescu, F. Georgescu, D. Mihăilescu, *Heatwaves in Romania* (Printech, 2016)
9. *Romania's 6th National Communication on Climate Change and 1st Biennial Report* (Ministry of Environment and Climate Change, 2013), https://unfccc.int/sites/default/files/6th_nccc_and_1st_br_of_romania%5B1%5D.pdf
10. OpenWeather, https://openweathermap.org/accuracy-and-quality, accessed May 3, 2021
11. G. Chen, X. Fang, H. Fan, J. Hydrol. Eng., **21** (10) , 05016023-1 (2016)
12. S. Zeiger, J. A. Hubbart, S. H. Anderso, M. C. Stambaugh, Hydrol. Process. **30**, 503 (2016)

A method to analyse urban heatwaves. Case study - Bucharest City (Romania)

Daniela-Elena Gogoașe-Nistoran[1,*], *Daniel-Marian* Antohe, *Ioana* Opriș[2] and *Cristina-Sorana* Ionescu[1]

[1]University Politehnica of Bucharest, Faculty of Power Engineering, Dept. of Hydraulics, Hydraulic machines and Environmental engineering
[2]University Politehnica of Bucharest, Faculty of Power Engineering, Dept. of Energy Production and use

Abstract. Long-term variation of hourly air temperature obtained from Open Weather, Romania, was analysed in the center of Bucharest city, over a period of 40 years (1980-2020). A computer program to extract summer heatwaves within the study period was written. Analysing the results an extreme heatwave scenario has been defined within the context of climate change and urban influence, to be used in future air and water temperature models.

1 Introduction

During the last decades, Europe has faced important heatwaves, that present an increasing trend since 1970s, both in magnitude and frequency [1, 2, 3, 4, 5]. Similar trends were as well registered in Romania [6, 7, 8, 9].

The impact of excessive warming on human health and the environment is particularly stronger during summer time within and around city areas. This is due to anthropic activities, that generate the urban heat island effect, giving rise to a substantial air temperature increase throughout summers. Among the most important factors known to trigger the urban island effect are: extended built areas and heavy concentration of pavement and impervious surfaces that absorb and retain heat, intensified road transport, lack of green areas, waste heat through increased use of air conditioning systems, three-dimensional (3D) urban geometry [10, 11, 12, 13] etc.

There are different definitions of heatwaves, which are considered very hot periods for the local climate. The **World Meteorological Organization (WMO)** defines a **heatwave** as a period during which the daily maximum temperature exceeds by 5°C for more than five consecutive days the average multiannual maximum temperature.

Depending on local climate, each country may use different definitions for determining heatwaves. In Romania two threshold values are used to define a heatwave lasting at least five consecutive days, such as : the daily maximum temperatures over 35°C and the night maximum ones over 20°C.

* Corresponding author: daniela.nistoran@upb.ro

Furthermore, in Meteorology and particularily in the urban environments, a *temperature-humidity index* (THI) is used to quantify the human "felt" heatwave. The comfort - associated THI index takes also into account both temperature and the relative humidity. The latter is an important parameter, having a negative influence on the real feel of high temperature perception.

In Bucharest, summers have become increasingly hotter during the last two decades [8]. This has also affected the water temperature of the water bodies within the city, such as the Morii Reservoir which feeds the Dambovița canal and Colentina River. River managers had to cope with the increasing spreading of vegetation bursts occurring during summers along Dambovi'a canal, during the last few years. In august 2017, after a prolonged heatwave local media reported a large quantity of fish died in the Morii Reservoir due to high temperatures. Thus, heatwaves also affect fauna and flora, not only people within the urban environment [14]. However, scientific papers dealing with these late changes on Dambovița River are missing.

The objective of the paper is to analyse the heatwaves in Bucharest city, based on a specially designed computer program in order to define the air and temperature and humidity variation during an extreme heat-wave scenario. The program allows to identify heatwaves characteristics such as their frequency, duration, pattern, maximum temperature, humidity variation, period of occurrence as well as to compute the THI. These data are useful for a water river temperature modeling.

2 Data

Hourly air temperatures were downloaded from OpenWeather.com in the centre of Bucharest city, at Piața Unirii (Unirii Square, 26.086°E and 44.438° N), for the period 1980 - 2020. OpenWeather collects and processes weather data from a vast network of over 40,000 weather stations and from meteorological broadcast services worldwide, such as satellites, radars and even uses global and local weather models (set of algorithms, to improve its quality and accuracy) [15]. The accuracy and quality of OpenWeather temperature data has been evaluated for 371 major cities, including Bucharest (26.10626°E and 44.43225° N) and found to have a mean absolute error (MAE) between two stations being of 0.5°C, a reliability of 90-100% and a mean accuracy of 99%.

By fitting the temperature variation with a linear function, an average increase of 1.5 °C over the 40-years study period was noticed. In Fig. 1 is represented this variation over 30 years, from 1990-2020 only, since yearly heatwaves over 35°C became more frequent. One may observe that after the year 2000, maximum summer temperatures reached and even surpassed 40 °C (in 2000, 2007, 2012 and 2017).

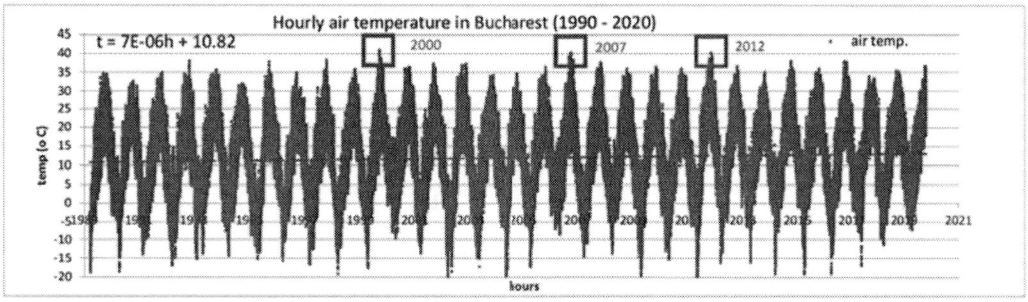

Fig. 1. Hourly air temperatures in Bucharest City centre over 30 years (1990-2020) (satellite data from Openweather.com at 2m from ground)

3 Method

The computer program is based on the Romanian definition of the heatwave. Its flow chart is shown in Fig. 2. The main steps of the conceptual model it relies on are:

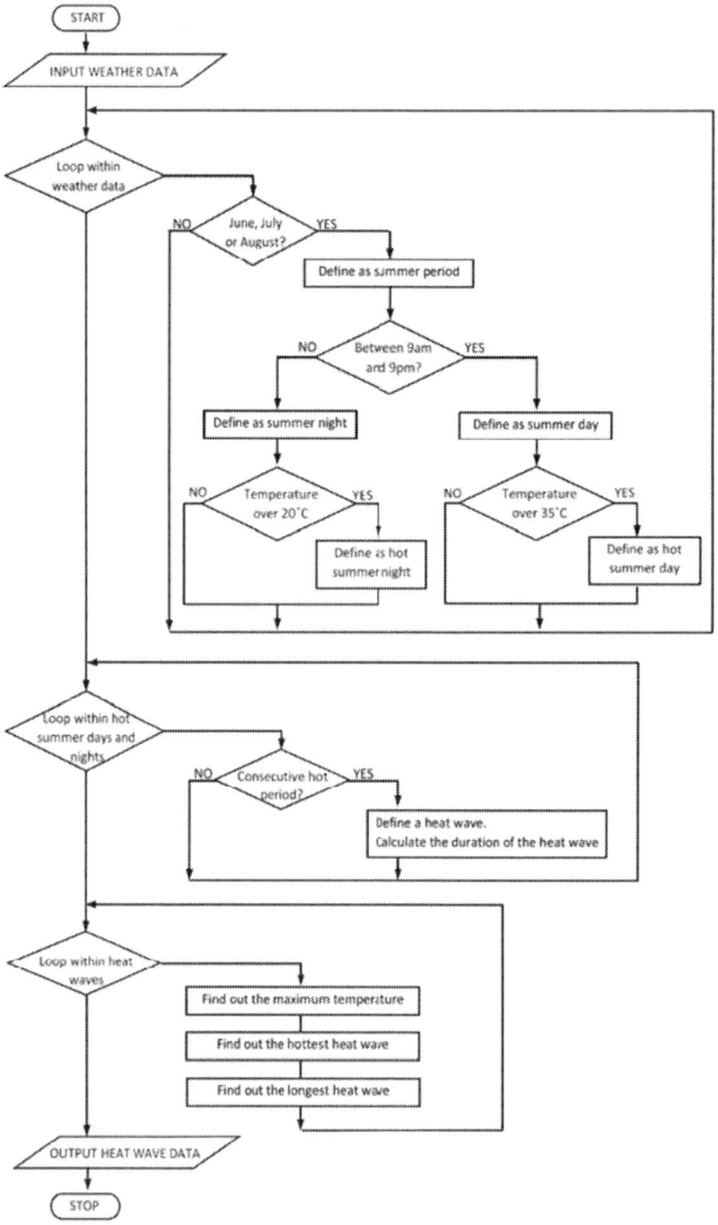

Fig. 2 Flow chart of the computer program

1. Find summer periods from 15 June to 15 September.

2. Create two line vectors with air temperature data during days and nights, defined between 9:01 - 21:00 and 21:01 - 9:00, respectively.
3. Find dates with daily temperatures over 35°C and nights with temperatures over 20°C of the two vectors.
4. Eliminate duplicate data
5. Merge the two-line vectors from step 3 into one with days and nights and create a new line vector with at least 5 consecutive days satisfying both conditions of step 3.
6. Find years with heatwaves.
7. Compute each heatwave duration.
8. Find the longest heatwave.
9. Find maximum temperature for each heatwave.
10. Find most extreme heatwave.
11. Read relative humidity corresponding to each heatwave.
12. Compute temperature-humidity index (*THI*) with the formula given by [16]

$$THI = (1.8 \cdot t + 32) - (0.55 - 0.0055 \cdot RH) \cdot [(1.8 \cdot t + 32) - 58] \qquad (1)$$

where *RH* is the relative humidity in % and *t* is the air temperature in °C. One may observe from this relation that high THI units correspond to high relative humidity values, which makes temperature feel hotter than it is. Values under 65 are in the confort zone, between 65-80 fall into the critical zone, whereas values over the threshold of 80 units are in the disconfort zone responsible for high impact on human health.

4 Results and discussions

During the studied period, the program identified 6 heatwaves, all of them occurring during the months of July and August. In Fig. 3 (a - f) are represented the temperature, relative humidity and THI index for these 6 heatwaves.

One may observe that temperature and relative humidity daily sinusoidal variations are inversely proportional. Also, by fitting these variables with second degree polynomial functions over a heatwave, one may observe that the mean temperature variation is also inverse to the mean relative humidity. The points of intersection of these fitting functions define the start and end of the heatwaves. This is a new way of finding the duration of a heatwave.

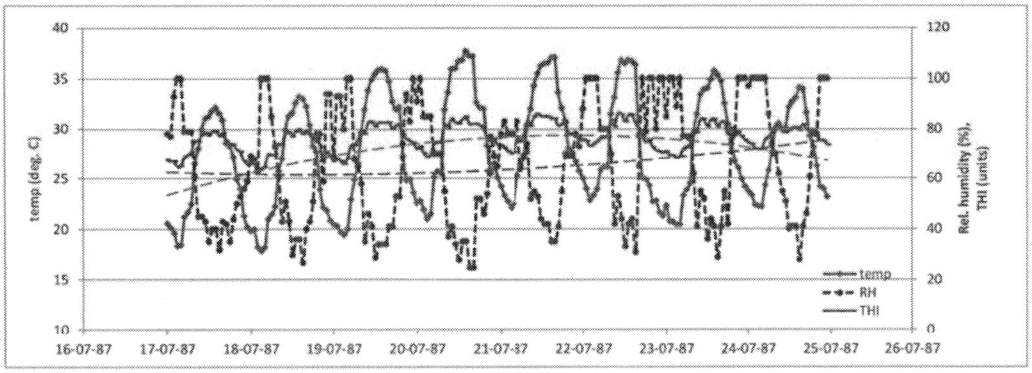

a)

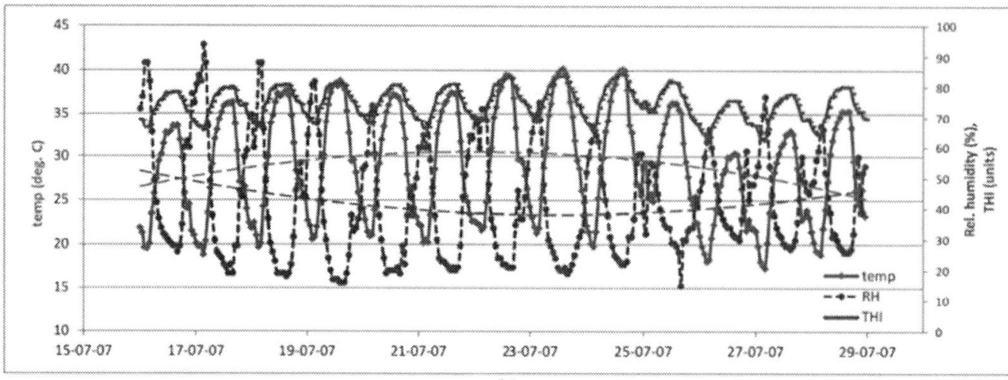
temp (deg. C)
Rel. humidity (%), THI (units)
temp
RH
THI
15-07-07
17-07-07
19-07-07
21-07-07
23-07-07
25-07-07
27-07-07
29-07-07

b)

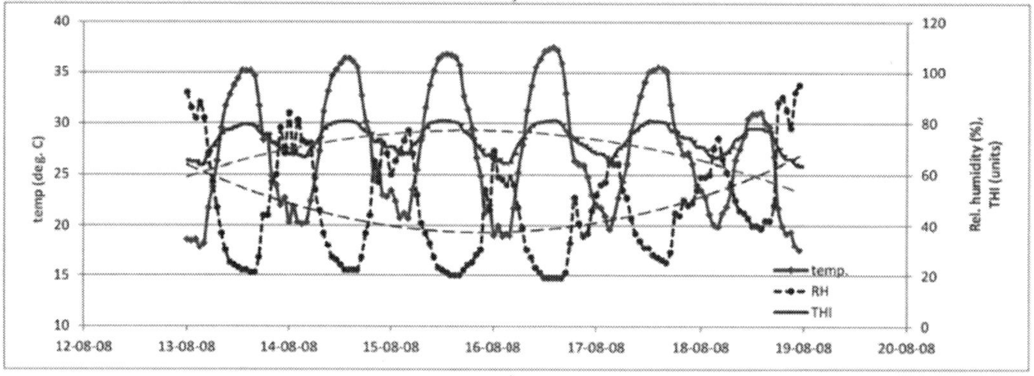
temp (deg. C)
Rel. humidity (%), THI (units)
temp.
RH
THI
12-08-08
13-08-08
14-08-08
15-08-08
16-08-08
17-08-08
18-08-08
19-08-08
20-08-08

c)

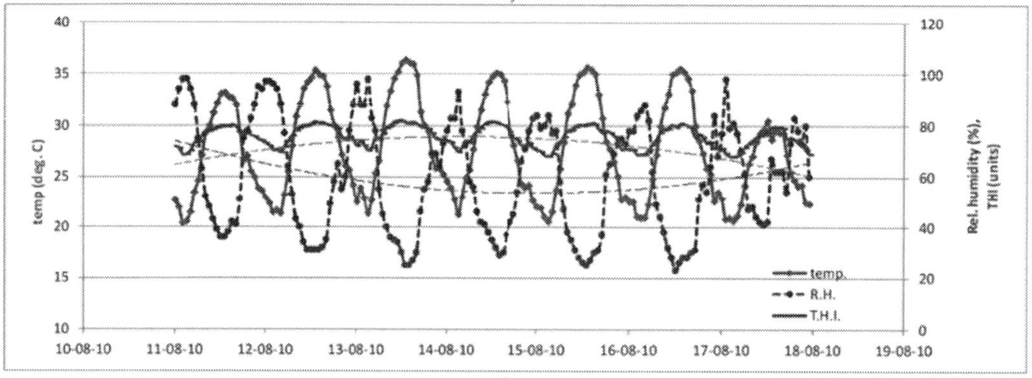
temp (deg. C)
Rel. humidity (%), THI (units)
temp.
R.H.
T.H.I.
10-08-10
11-08-10
12-08-10
13-08-10
14-08-10
15-08-10
16-08-10
17-08-10
18-08-10
19-08-10

d)

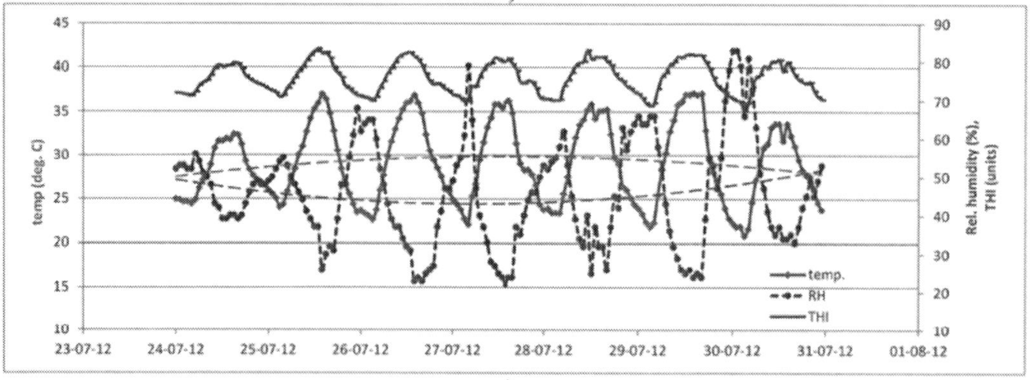
temp (deg. C)
Rel. humidity (%), THI (units)
temp.
RH
THI
23-07-12
24-07-12
25-07-12
26-07-12
27-07-12
28-07-12
29-07-12
30-07-12
31-07-12
01-08-12

e)

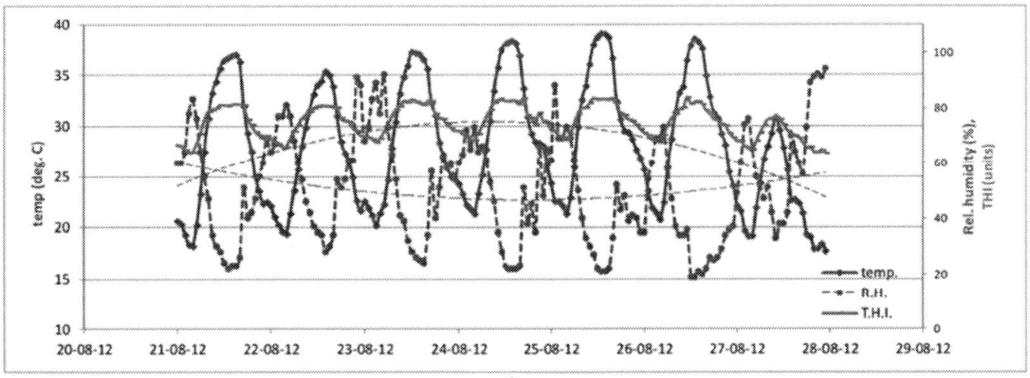

f)

Fig. 3 a) – f) The six heatwaves (temperature, Relative humidity and Temperature – humidity index variation) identified within the 40-years study period in a) 1987, b) 2007, c) 2008, d) 2010, e) July 2012 and f) August 2012,

One may see from Fig. 3 that for all the identified heatwaves the range of calculated THI index overpassed the critical threshold of 80 units during several days.
Table 1 presents the most important parameters of the heatwaves. The temperature-humidity index was calculated with the formula (1).

Tab. 1 The most important characteristics of the six identified heatwaves

Crt. No	Period of time	duration	Max. temp	Relative humidity range	Range of THI
		days	°C	%	units
1	19-24.07.1987	6	37.7	25 - 100	63.4 - 86
2	17-26.07.2007	10	40.2	15 - 94	62.4 – 84.5
3	13-18.08.2008	6	37.5	19 - 95	63.3 - 81
4	12-17.8.2010	6	36.42	23 - 98	68 - 82
5	25-30.07.2012	6	37.12	22 - 83	67 - 83.3
6	21-27.08.2012	7	39.13	19 - 98	61 - 83.7
	Values considered for the extreme scenario	7	**20.4 - 40**	**20 – 94.4**	**68 – 83.4**

The longest and the hottest heatwave is the one from July 2007. The year 2007 was also found by [16] and [17] to be the hottest year since measurements are performed in Romania. However, by analysing the heatwaves of the years 2007, 2012 and 2015, [17] showed that 2012 was an extreme hot summer, particularily for the Bucharest study area, due the extent over nearly two months (during July and August) of the high temperatures coupled with lack of precipitations.

5 Conclusions

The values and ranges of the main features of the six identified heatwaves in Table 1 were used to build a 7-days extreme heatwave, that will be used as a worst-case scenario in future modeling air and / or water temperature time variation. Temperature and relative humidity variation were build as simple sinusoidal functions.

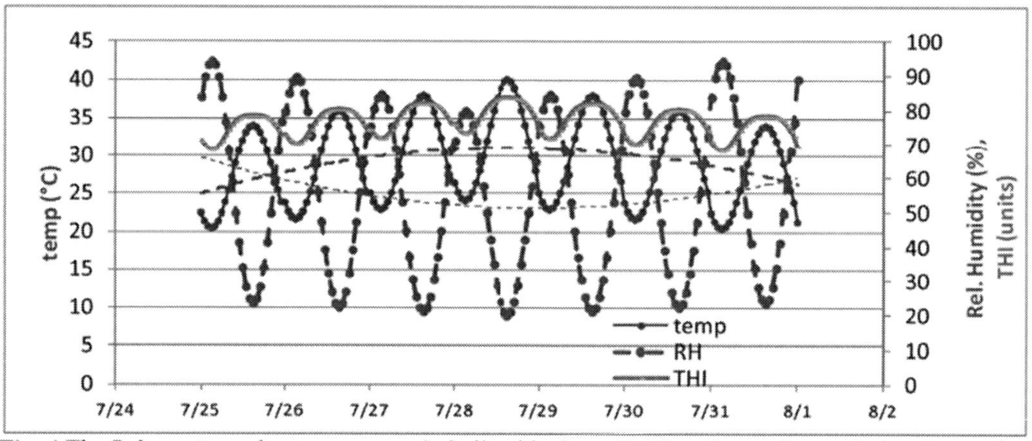

Fig. 4 The 7-day extreme heatwave scenario built with sinusoidal functions, after analysing the results of the heatwaves in Bucharest (extracted over a 40 year-period with the proposed program): air temperature, relative humidity and THI

The necessity of writing present analysis program arose from not finding such a ready-made customizable tool. Therefore, one of the advantages of the computer program built for the present study is that it can be used for any period of time, any location and any definitions of the heatwave to analyse their frequency duration and patterns. The current disadvantage is does not have a user-friendly interface, needing basing programming knowledge.

Also, by presenting the flowchart and the steps of the conceptual model, it may be written in any programming language. However, one drawback is that the results (temperature and humidity over the resulted periods) are easier plotted, processed and formatted by using another specialised graphical software.

The new method of finding the duration of a heatwave by fitting the temperature and relative humidity variations with second order polynomial functions may also be used for other analyses.

References

1. S. Russo, J. Sillmann, E. M. Fischer, Environ. Res. Lett. **10**, 124003 (2015)
2. R. Zhang, C. Sun, J. Zhu, Re. Zhang, Weijing L., npj Clim. Atmos. Sci. **3**, 7 (2020)

3. C. Schär, P. Vidale, D. Luthi, C. Frei, C. Haberli, M. Liniger, C. Appenzeller, Nature **427**, 332 (2004)
4. P. Stott, D. Stone, M. Allen, Nature **432,** 610 (2004)
5. P.M. Della-Marta, J. Luterbacher, H. von Weissenfluh, E. Xoplaki, M. Brunet, H. Wanner, Clim. Dyn. **29**, 251 (2007)
6. A.-E. Croitoru, A. Piticar, A.F. Ciupertea, C.F. Roşca, Glob. Planet. Change **146**, 109 (2016)
7. C.-S. Dragotă, L.-E Havriş, In Proceedings of Air and water—Components of the Environment, Cluj, Romania, 15, 20–22 March 2015.
8. A.-E. Croitoru, R.I. Antonie, A. Rus, In *Proceedings of the 14th International Multidisciplinary Scientific GeoConference,* pp. 375–382 (SGEM 2014)
9. L. Sfică, A. E. Croitoru, I. Iordache, A. F. Ciupertea, Atmosphere **8**(3), 50 (2017)
10. S. Cheval, A. Dumitrescu, Theor Appl Climatol, **121**, 631 (2015)
11. Y. H. Ryu, J. J. Baik, J. Appl. Meteorol. Climatol., **5**, 842 (2012)
12. K. Ward, S. Lauf, B. Kleinschmit, W. Endlicher, Sci. Total Environ., **569–570**, 527 (2016)
13. M. Nastran, M. Kobal, K. Eler, Urban. For. Urban. Green., **37**, 33 (2019)
14. L. Zaharia, G. Ioana-Toroimac, O. Cocoș, F. A. Ghiță, E. Mailat, Urbanization effects on the river systems in the Bucharest City region (Romania), Ecosyst. Health Sustain, **2**, e01247 (2016)
15. OpenWeather, https://openweathermap.org/accuracy-and-quality, accessed May 3, 2021
16. L. Leontie, A. Timofte, D. C. Bostan, S. Bostan, Present Environment and Sustainable development, 2, **23** (2008)
17. V. Dima, A. Irimescu, F., Georgescu, D. Mihăilescu, Valuri de căldură în România, Ed. Printech, Bucureşti, 2016

Theoretical and experimental researches on the implementation of smart location

Mariana Panaitescu[1,*], *Fanel-Viorel* Panaitescut[1] , *Marius-Valentin* Dumitrescu[2], and *Valeriu-Nicolaie* Panaitescu[2]

[1]Electromechanical Faculty, Engineering Sciences in mechanics and Environment Department, 900663 Mircea cel Batran Street, Romania

[2] Doctoral School of Energy Engineering, POLITEHNICA University, Bucharest, Romania

Abstract. The present paper considers the presentation of practical solutions for the realization of smart location, materialized through research on solutions for monitoring parameters and solutions for these locations.In this paper we present: description of the concept of *smart location*, structures and architectures, realization of a pilot installation of *smart location* and data aquisition. For these researches was used to monitor electro-mechanical radiation of its own design. We obtained the following results: realization and design of the elements of the housing components with the 123D Design program, processing with the Cura program, obtaining g-code files for transfer to 3D printing, monitors in real time and detects errors or deficiencies in location. The main objective of this paper is to propose a model of technical solution for an smart location, which can be applicable in the maritime field, respectively to ships in the engine room or other location onboard.

1 Introduction about smart location

The present paper considers the presentation of practical solutions for the realization of smart locations, materialized through research on solutions for monitoring parameters and solutions for these locations.

Now that the construction of "*Intelligent Locations compatible with IoT*" is in full swing, it was considered that a research and understanding of systems and how to operate, followed by a realization at the level of the model would be auspicious [1]. "Intelligent Locations is an Internet of Things (IoT) ..... that provides.... with data insights gained through advanced monitoring of assets, people and processes to solve persistent problems of inefficiency, affordability and accessibility." [2].

By definition: "An intelligent locations represents a fusion of technologies, projectors and energy systems. A smart home has primary and secondary data buses that connect high-level controllers with lower-level controllers, input / output devices, and the user interface

* Corresponding author: panaitescumariana1@gmail.com

(Figure 1) [2]. Primary and secondary data buses can be CAN, BACnet, fiber optic, ethernet, ARCNET, RS-232, RS-485 or wireless ".[1]

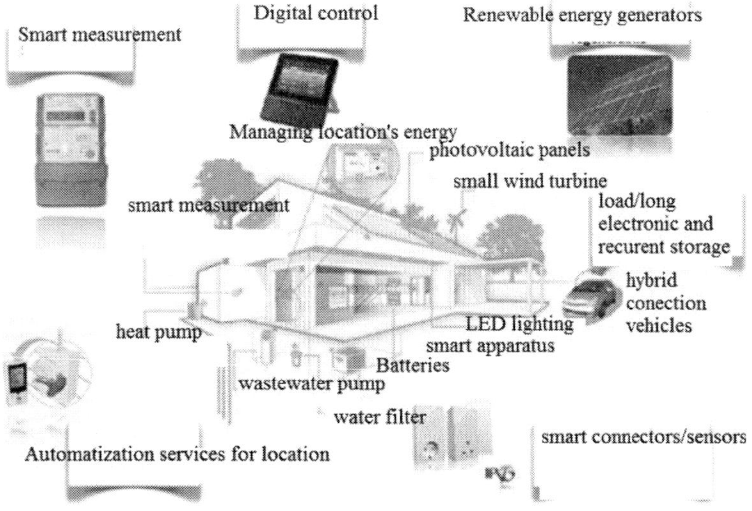

Fig. 1. Intelligent location.

Now we can summarize the concept of "*Smart Location*" as the fusion of technology, design and energy systems (Figure 2). They are well managed, have well-integrated physical and digital infrastructure that provides optimal employment services in a reliable, cost-effective and sustainable way.

The *smart location* helps its users to improve their lifestyle, by improving the technical performance of the location, reducing consumption, implicitly expenses, increasing reliability, optimizing space and last but not least, minimizing the environmental impact of locations. *Smart location* can be :" **Smart Cities** where sensors can offer us services like **Smart Parking** – to find free parking spots in the streets– or managing the intensity of the luminosity in street lights to save energy "[3]. This concept means "*specific sensors and wireless technologies for each application*" [3].

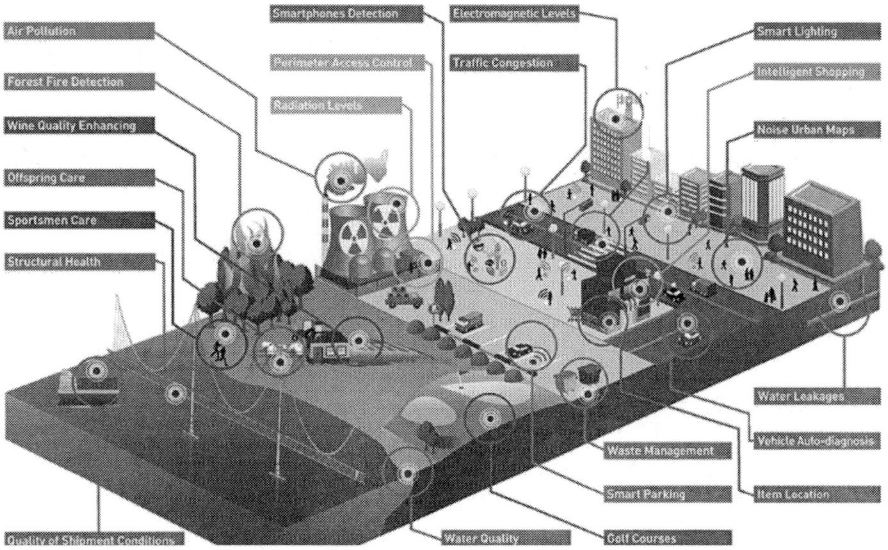

Fig. 2. Smart location.

The main objective of this paper is to propose a model of technical solution for an *smart location*, which can be applicable in the *maritime field,* respectively to ships in the engine room or other location onboard.

A smart location:

- provides data on the performance of building systems;
- monitors in real time and detects errors or deficiencies in construction systems;
- reports in real time and uses operations management, energy and tenant comfort;
- includes tools, technologies, resources and systems to contribute to the conservation of energy and environmental sustainability.

" (IoT) is a "system of systems" in which people, machines, devices, sensors and businesses connect and interact"[4].

In maritime field, onboard, in the engine room or other location, the smart location is workplace, where "as devices continue to connect employees, they are creating much more than an office – they are creating collective intelligence. Applications abound in workplace safety, employee productivity, asset management and building comfort"[4].

A smart location onboard means a part of smart ship which consists of four main elements (Figure 3)[5]:navigation, guidance, physical ship and controls.

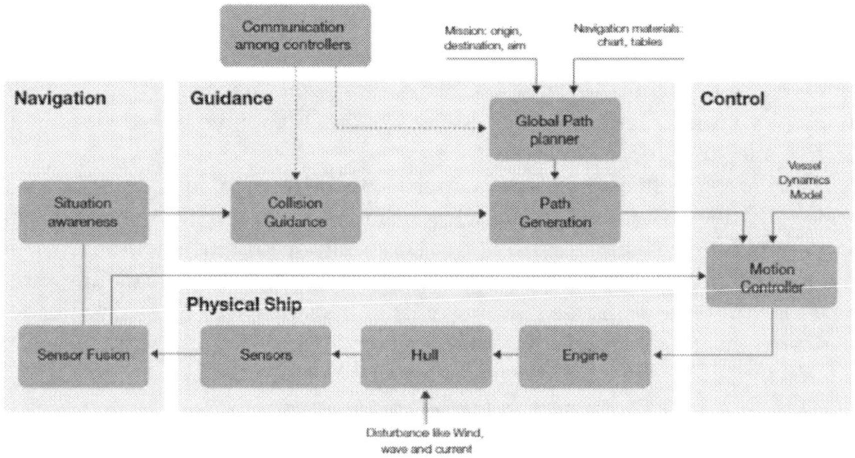

Fig. 3. Smart ship.

In this paper we describe a part of physical ship- engine, a smart pilot installation of this location and data aquisition.

1.1 Architectures and structure

1.1.1 Architectures

The architecture of a sensor network is dependent on the application for which it is used and in which it is integrated. Temperature and humidity sensors, a HID interface (Human Interface Devices)- Display were used and a logic board Esp 32 (Figure 4).

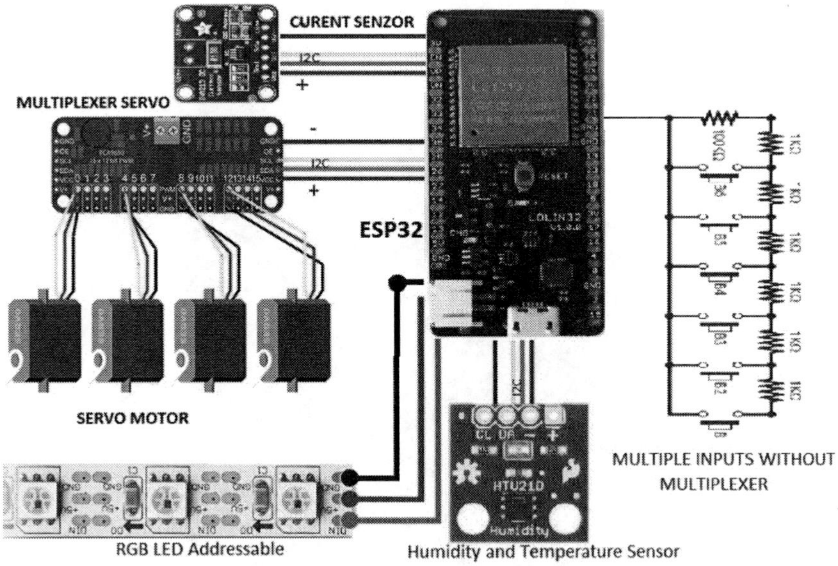

Fig. 4. The architecture of a sensor network.

1.1.2 Structure

The network of automation of an intelligent fall is an interconnection of hardwares and centralized software that monitors and controls the industry, the environment and the environment (Figure 5) [5], [6].

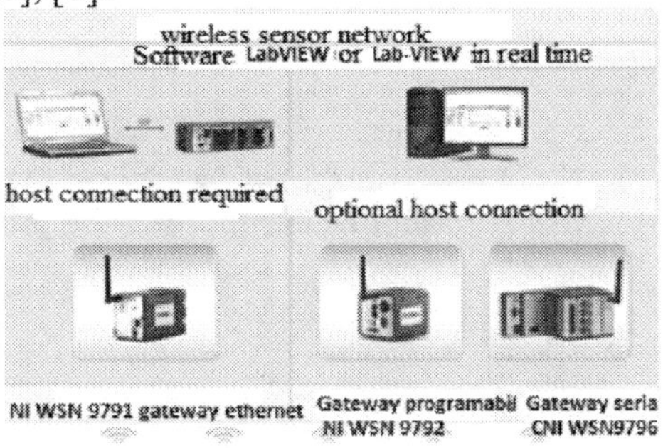

Fig. 5. Wireless sensor network.

1.2 Smart location pilot structure

The location structure: four rooms, a hall and a warehouse, equipped with "gadgets" for transformation into a *Smart Location* [7], [8].
The gadgets that have been used for data acquisition are: various sensors and switches Normally Open [9].
For data main line we will use the I_2C protocol, similar to the CAN and RS232 protocol used in professional systems [10].
For execution we will use addressable LEDs and servomotors, and for interface we will use OLED display [11].
The logic board has an ESP32 processor, compatible with Arduino development boards and the Arduino IDE program [16]. For programming the ESP32 module, the Arduino IDE programming environment was used, helped by the libraries related to sensors and multiplexers.The possibility of internet connection and the existence of the internal memory of the ESP32 module facilitated the creation and implementation of a web page accessible online by accessing the data acquired by the sensors [10]. ESP32 is a series of low-cost, low-power system on a chip microcontrollers with integrated Wi-Fi and dual-mode Bluetooth. It also has the possibility to connect to the internet, in order to transform the compatible IoT project (Figure 6) [11].

Fig. 6. Pilot smart structure.

2 Realization and design of the elements of the housing components

2.1 Design program

The components of the pilot installation were made with the **123D Design Program** (Figure 7) [12]." Autodesk 123D was a suite of hobbyist CAD and 3D modelling tools created by Autodesk."123D Design Program is a simplified program to create 3D models.

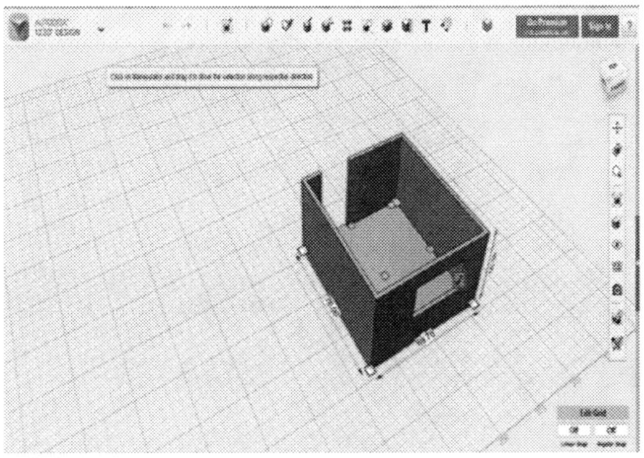

Fig. 7. The hall design.

The created elements were exported in **Stl** files and then processed with the **Cura program** to obtain g-code files." Cura is an open source slicing application for 3D printers"[13]. The latter were transferred to the 3D printer and processed for the final shape. "Ultimaker Cura works by slicing the user's model file into layers and generating a printer-specific g-code. Once finished, the g-code can be sent to the printer for the manufacture of the physical object."[14], [15]

2.2 Engine Room design

The design of the engine rooms was done as follows:
- designing the hall (Figure 7) and the whole smart location (Figure 8);
- design of storage (Figure 9);
- processing with the Cura program, obtaining Gcode files for transfer to 3D printing (Figure 10, Figure 11, Figure 12, Figure 13);
- File transfer and 3D printing (Figure 14).

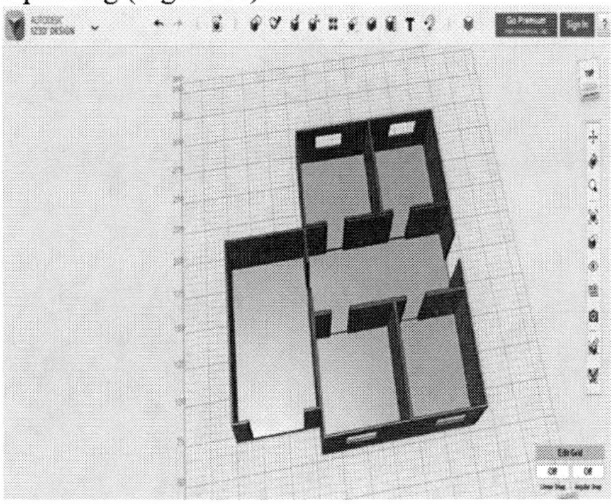

Fig. 8. The whole smart location design.

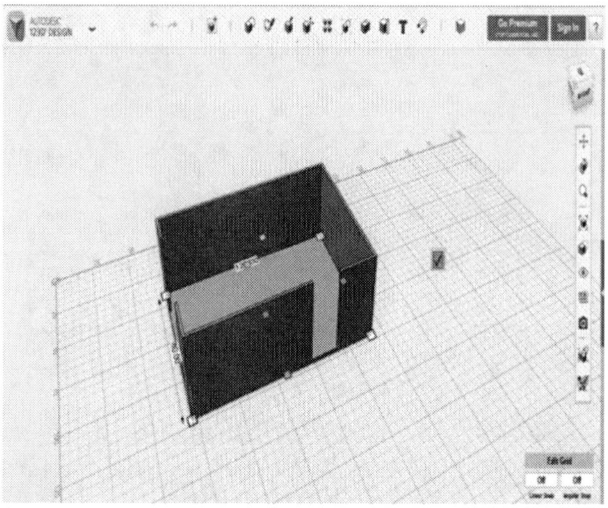

Fig. 9. The store design.

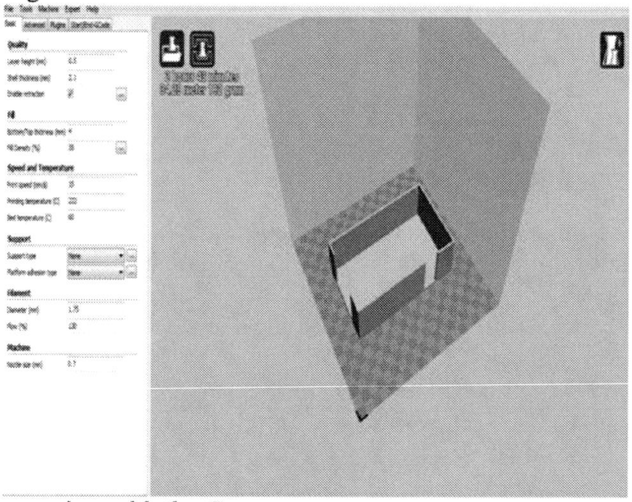

Fig.10. The hall- processing with the Cura program.

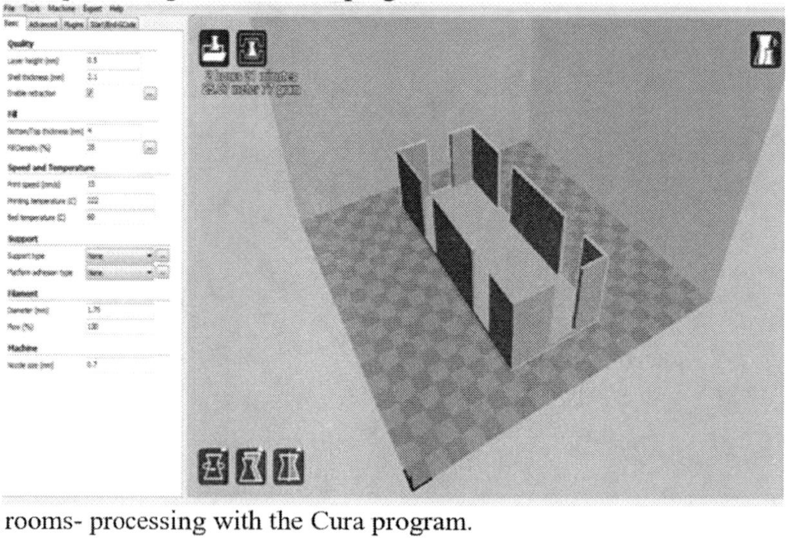

Fig.11. The rooms- processing with the Cura program.

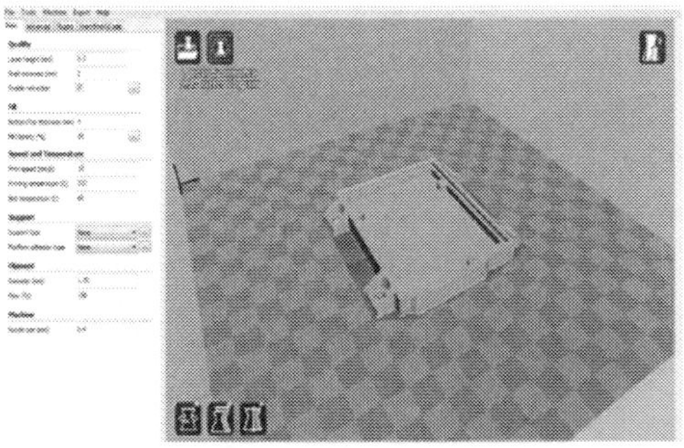

Fig.12. The materboard- processing with the Cura program.

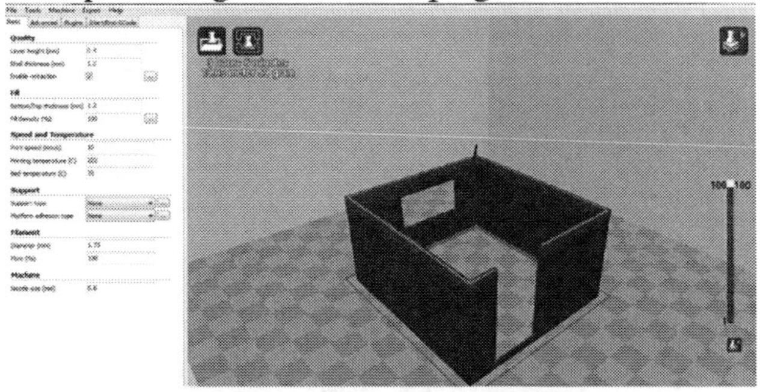

Fig.13. The storage room- processing with the Cura program.

Fig.14. File transfer and 3D printing.

2.3 Assembling the location

Motherboard:

- on a melamine chipboard support.

- Actuation, closing and opening of blinds and storage room doors - with actuators and control microcontacts - actuation and ignition of interior lights with a control such as push on / push off circuit.

Positioning of electronic elements:

The board is acting as a connection panel, interconnecting the respective electronic elements:
Energy sources: photovoltaic panel and network.
Control elements: NO type buttons
Execution elements: Servomotors, through tca9548a i2c multiplexer, LED lighting.
Sensors: Humidity and temperature sensor HTU21D , Current-Sensor INA219
HID (Human Interface Devices): Display
Logic board: Esp 32 or Esp8266.

2.4 Installation of connections and actuators

In order to obtain the final structure of the pilot installation, the connections and actuators are mounted (Figure 15, Figure 16).

Fig.15. Installation of connections.

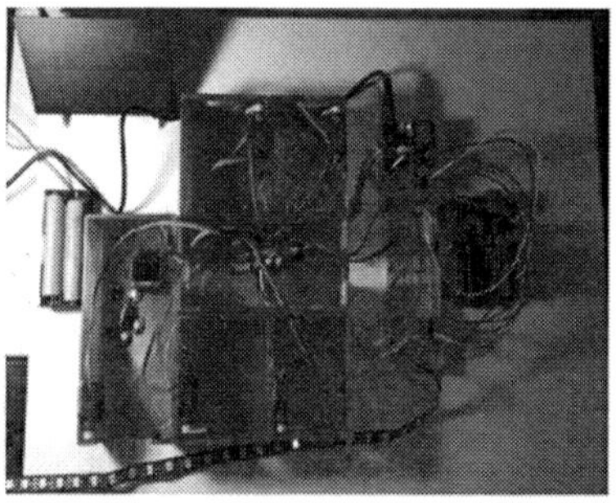

Fig.16. Installation of actuators.

3 Smart location assembly operation

By pressing the NO (round) buttons, an electrical signal is transmitted to the logic board which, through the I2C bus, transmits to the controller for the actuators the opening or closing of the blinds or the garage door.

The actuators can be pre-programmed at which angle to open or close.

By pressing the NO (rectangular) buttons, an electrical signal is transmitted to the logic board, which transmits the on or off LEDs to the addressable LEDs. The color and intensity of the LED light can be preset from the software. They can ignite according to a predefined code for issuing warnings (burglary, fire, etc.).

Also, their software can run in anti-burglary system that mimics the presence of tenants in the house, when they are away from home for a longer period.

The sensors connected to the I2C bus transmit data collected from the home to the logic board, which can be accessed by the owners via the Internet. HTU21D is a digital humidity sensor with temperature output [17]. This sensor provides calibrated, linearized signals in digital, I²C format. HTU21D digital humidity sensors are dedicated humidity and temperature plug. The INA219 Current-Sensor is an I2C interface based zero drift and bi-directional current/power monitoring module [18]. It can sense shunt voltage, current, and power at the same time and submit the data via I2C protocol.

4 Data from the trial period of the project on the mobile phone

Through the internet you can order the start or stop of the different systems that equip the home.

In the realized project we used as elements that simulate actuation, direct current servo motors. Besides the simple on / off functions they can also execute angular movements necessary in partial openings, ventilation systems, positioning systems and others.

Micro Servo Motor SG90 ,used in this project, is a tiny and lightweight server motor with high output power. Servo can rotate approximately 180 degrees ,90 in each direction.

Starting, stopping and setting the home heating, watering systems, animal feed, space ventilation, ambient light, etc. For our smart location we have collected daily and weekly data (Table 1, Figure 17, Figure 18).

Table 1. Daily and weekly variation on temperature and umidity.

Day/hour	Temperature [^{0}C]	Umidity [%]
5/15/2021 9:21:00 PM	22	39.1
5/15/2021 8:21:00 PM	22	39.1
5/15/2021 6:51:00 PM	22.4	38.6
5/15/2021 6:01:00 PM	22.3	39.1
5/15/2021 2:41:00 PM	22	39
5/15/2021 1:21:00 PM	22	39.1
5/15/2021 12:51:00 AM	21.5	40.6
5/15/2021 12:11:00 PM	21.9	39.6

5/15/2021 9:21:00 AM	21.7	40.1
5/15/2021 3:41:00 AM	21.5	40.6

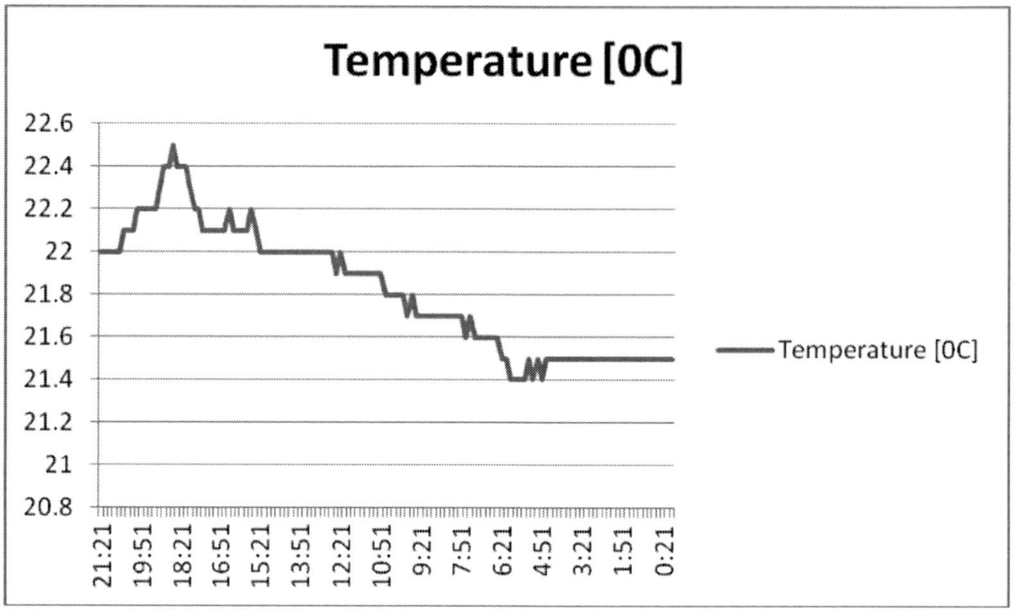

Fig.17. Daily variation on temperature.

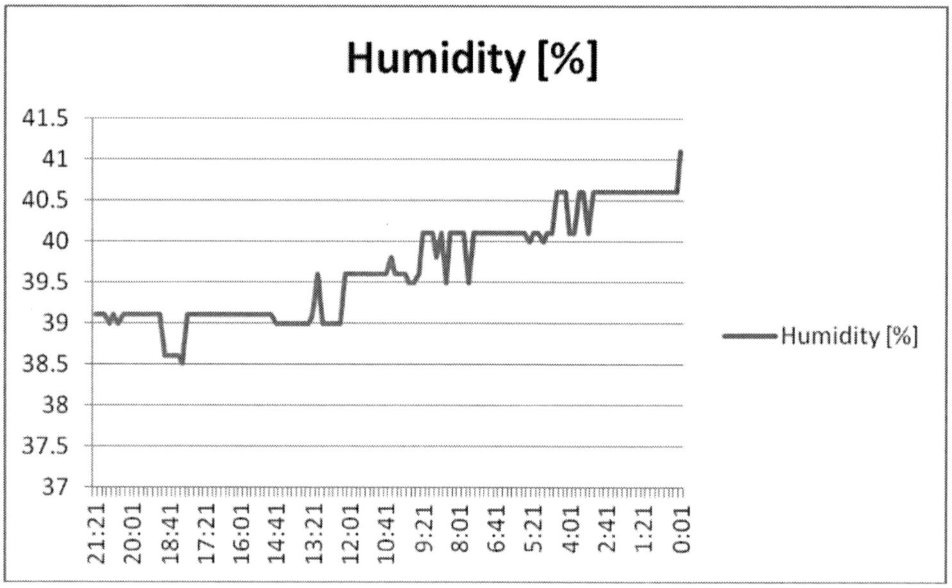

Fig.17. Daily variation on humidity.

The location can be provided with a connection to a photovoltaic system, consisting of a solar panel, buck boost inverter and battery (battery), which provides electricity on sunny days, thus contributing to the application of the concept of environmental sustainability.

Other data we can find on http://5.14.172.250.

5 Conclusions

The purpose of this project, that of finding practical solutions to carry out "smart location" projects without having to invest large sums of money and with logistical resources at the disposal of many entities, has been achieved. Following the experience gained in carrying out the exercise, we can say that the technical solution model created by us can be easily adapted and replicated for other environments. In the naval field, the main objective is to monitor the environment in the engine room or other places of interest on board ships.

The cost reduction was due to the use of 123D Design, Cura and Arduino IDE programs with freeware license. Also, the implementation of the I2C communication protocol, which does not require the payment of a license, is also a cost-reducing solution. This communication protocol allows us to attach a total of 128 acquisition or execution equipment.

The use of cheap, pre-calibrated micro-sensors developed by renowned companies at the urging of the European Community able to offer I2C compatibility was another correct choice in this approach.

The chosen technical solution offered the user access to a wide range of essential data from the monitored environments such as:

- information about energy sources with monitoring of essential parameters, voltage, current.
- information about the parameters of air quality, temperature, humidity, atmospheric pressure,
- the system can issue alarms, both locally and remotely, when the preset values for the monitored elements are exceeded.

The motivation for choosing this type of work for this project was to ultimately serve as a teaching material for our students bringing ease in understanding the operation of IoT systems. We also leave the possibility of their involvement and further development of such a project.

References

1. Daniel Wilson. Smart House-Your So-Called Sci-Fi Life", Popular Mechanics Technology News, Mai (2007)
2. https://www.prnewswire.com/news-releases/intelligent-locations-and-sas-collaborate-to-support-healthcare-organizations-iot-and-contact-tracing-efforts-301114734.html
3. https://www.libelium.com/libeliumworld/top-50-iot-sensor-applications-ranking/
4. https://isg-one.com/articles/iot-and-the-intelligent-workplace
5. Mariana Panaitescu, Marius Valentin Dumitrescu, Fanel-Viorel Panaitescu. Indoor air quality monitoring universal device, Proc.of the 8th International Conference on

Modern Manufacturing Technologies in Industrial Engineer ISSN 2286-4369, Iasi, Romania (June 23-27, 2020).
6. Viegi Giulio. Indoor air pollution and airway disease, Int.J.Tuberc.Lung Disease, 8, 1401 (2004)
7. Dumitrescu Marius-Valentin, Studies on techniques and technologies, 2-86, Bucharest, Romania (2019)
8. Panaitescu Mariana, Panaitescu Fanel-Viorel, Dumitrescu Marius-Valentin, Panaitescu Valeriu Nicolae, The Impact of Quality Air in The Engine Room on the Crew, Int.J.of Marine Navigation and Safety of Sea Transportation **V1,** Isue 3.2, 409-414 (2019)
9. Vasile BAHRIN, Câmpul electromagnetic de radiaţie generat de echipamentele electrice, Electrician J. **06**,6-12 (2014)
10. Espressif System ESP32, Technical Reference Manual **V4.1,** 285-286 (2019)
11. Leszczynski D., Joenv S., Reivinen J, and Kuokka R., Non-thermal activation of the hsp27/p38MAPK stress pathway by mobile phone radiation in human endothelial cells: Molecular mechanism for cancer and blood-brain barrier-related effects, Differentiation J., **70** , 120 – 129 (2002)
12. https://en.wikipedia.org/wiki/Autodesk_123D.
13. "Cura home page". Ultimaker. Ultimaker. Retrieved 31 December 2016
14. https://all3dp.com/cura-2-3-update-new-features-major-improvements-3d-slicer.
15. https://3dprint.com/211217/ultimaker-engineering-materials.
16. https://www.arduino.cc/en/software.
17. https://cdn-shop.adafruit.com/datasheets/1899_HTU21D.pdf.
18. https://www.ti.com/product/INA219.

Acknowledgments

Authors gratefully acknowledge to this material support path received under Project PN-III-P1-1.2-PCCDI-2017-0404/ 31PCCDI / 2018, Holistic on the Impact of Renewable Energy Sources on Environment and Climate-HORESEC.

Dimensional characteristics of grist intermediate products obtained in the first two technological passages in the reduction phase of an industrial milling plant

Gabriel-Alexandru Constantin[1], *Gheorghe* Voicu[1], *Elena-Madalina* Stefan[1,*], *Mariana-Gabriela* Munteanu [1], *Gabriel* Musuroi[1], and *Iulian* Voicea[2]

[1]Politehnica University of Bucharest, Faculty of Biotechnical Systems Engineering, Romania
[2]The National Institute of Research – Development for Machines and Installations Designed for Agriculture and Food Industry – INMA Bucharest

Abstract. After the wheat is coarsely grinded in the breakage technological phase and after a certain percentage of flour and bran has been extracted here, the crushing is continued in the reduction technological phase. The paper presents the flow of grist products at the first two technological passages from the reduction phase of an industrial milling unit. Samples taken from these two technological passages were subjected to a granulometric analysis, and with the experimental data a nonlinear correlation was performed with the Rosin-Rammler law, obtaining correlation coefficients of over 0.954. The paper also discusses the limits of the dimensions between which the particles of each fraction are sorted at the first two technological passages in the reduction phase. The analysis performed in this paper can serve in establishing the fabrics of the sifting frames from the plansifter compartments, respectively when adjusting the roller mills.

1 Introduction

Among the grains, wheat (*Triticum aestivum*) is the second largest production level in the world, most of which being subjected to a process of grinding and turning into flour [1, 2], in the first place being corn, while in third place is rice (for grinding). Wheat is the only cereal with enough gluten content, from whose flour ordinary bread can be obtained without being mixed with other types of flour [3].

The modern roller milling process of wheat can be divided into three main systems: break system, where grist fractions are passed through a series of fluted break rolls and the various resulted milling fractions are separated by being passed through a plansifter compartments; the grading system, where the particles are separated from sizes and endosperm particles are separates from adhering bran; reduction system which reduce pure

* Corresponding author: stefanelenamadalina@gmail.com

endosperm to fine particles of flour by passing through a series of smooth grinding rolls [4 - 6].

The grinding process takes place continuously, by gradually passing the material from machine to machine, in order to obtain flour, semolina and bran. The main machines used in a milling unit are: roller mills (break system and reduction system), plansifter, semolina machine and bran finisher (grading system). Thus, the two principal operations in milling process are grinding and sifting, [6,]. Several technological passages are provided for two phases of grist breakage and reduction, passages that consists in a grinding the same product on one or more rolls pairs followed immediately by sifting in one or more plansifter compartments. The technological diagrams for grinding wheat have in composition, mainly, several breakage technological passages and several reduction technological passages. In the breakage phase are used grinding rolls with flutes, while in the reduction phase the grinding rolls have a smooth surface.

The optimization of the milling processes (reduce energy costs, yield flour, improve quality) requires understanding the behaviour of materials, the material circuit on the technological flow, and machine operations [6]. The factors affecting the milling process and the end-product quality are machines type, the machine operational parameters, and the material properties, as described in papers [7 - 11]. Flour from different grinding passage differ significantly in terms of chemical composition and physical properties [12]. In papers [12, 13] it is shown that at the first two reduction passages was obtained the highest yield in flour with a minimum ash content, but also of the highest quality, the best quality bread was obtained from them. As has been shown in numerous papers [11, 14, 15], the material obtained at the first passage of the reduction phase is composed of endosperm and endosperm with particles of bran adhering. The bran particles are flattened and remain in the coarsest size fractions (>200 μm), which favours the process of separating the endosperm particles from the flour particles.

Modern wheat grinding is an efficient technological process that economically splits wheat seeds to recover high quality flour. The endosperm is the main part of the seed, being composed of about 64-75% starch, 11-16% protein, 1.5% fat, 0.5% minerals, 1.5% dietary fibre, and other components, [16].

In the process of dry grinding, especially in the last breakage and reduction passages, an increase in the content of Alternaria-type toxins was detected, which is mainly concentrated in the peripheral layers [1]. For this reason, first quality flour, which is extracted from the central parts of the core, can be considered safer than inferior quality flour or than milling by-products.

Also, Hajnal et al. [1] shows that fine wheat particles from breakage passages IV, V and VI or those from the last reduction passages lead to dark flour, with lower quality, with a high mineral content (ash). It seems that the by-products have a weight of about 17-24% (bran and short fractions), the dominant share being the big bran (walloping, 14-15%).

In the grinding process, starch granules can be damaged, especially in durum wheat where the starch-protein bonds are stronger, compared to common wheat which bonds are weaker. The aleuronic layer, surrounding the endosperm, having cells overgrown with it, but with strong ties with seed coat, it is difficult to extract in the grinding process, that is why he is largely removed with the brans. If aleuronic layer is included in the flour, the nutritional quality of the flour may increase, but the process requires special care so that the bran does not reach in the flour with the aleuronic particles [1, 4, 17].

The paper analyses the dimensional characteristics of the grist intermediate products from the first two technological passages in the reduction phase of a milling unit of 4.2 t/h.

2 Material and method

Grist samples were taken from S.C. Spicul S.A., Roșiorii de Vede, Romania, for the first two technological passages from the reduction technological phase. Thus, samples were taken from the grist mixture that feeds the two smooth grinding rolls (M1A and M1B), from the mixture that feeds the two plansifter compartments of the first two passages (C1 and C2), but also samples from the fractions sorted inside the two plansifter compartments. The technological flow of the semolina reduction phase is presented in figure 1. The equivalence between the number of the sieve and the dimensions of its apertures, are presented in standards and in paper [18].

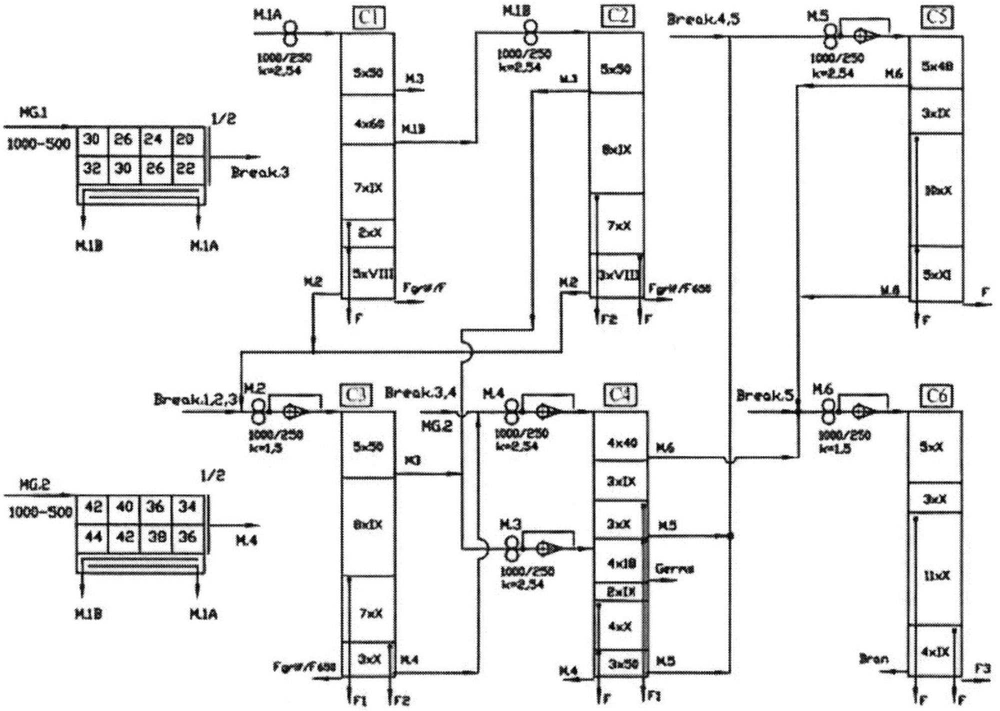

Fig. 1. Technological flow of semolina reduction phase, [19]. *C1–C6 – plansifter compartments; Break 1–5 – breaking rolls; MG1, MG2 – semolina machines; M1A, M1B, M2–M6 – reduction rolls; F – F3 – flour.*

It can be seen from Figure 1 that all grinding rolls of the reduction technological phase have a length of 1000 mm and a diameter of 250 mm and a smooth surface, without flutes. The ratio of the angular velocities of the grinding rolls is, for five pairs of reduction rolls, k=2.54, and for two pairs of reduction rolls k=1.5. This is contrary to the recommendations in the literature, which says that at reduction k = 1.5 or k = 1.

The products to be grinded in the reduction technological phase are products that arrive from the breakage technological phase (Break 1-6) or from semolina machines and bran finishers. Undersized particles from semolina machines, representing semolina that have dimensions below 1.0 mm, are grinded in the first two technological passages (roller mill M1A and plansifter compartment C1, and roller mill M1B and plansifter compartment C2). In diagram, the first fractions (an oversized fractions) sorted in the plansifter compartments of M1A and M1B technological passages are led to the grinding roll M3, which works together with a half of plansifter compartment (half from C4). It should be noted that in grinding rolls that crush smaller endosperm particles (about 0.40 mm), after the grinding rolls are introduced into the technological flow material detachers, due to agglomerations that occur by compressing smaller endosperm particles in the area of action of the grinding rolls.

Semolina fractions grinded at reduction roll M1A are sorted on dimensional fractions inside the C1 plansifter compartment. This compartment consists of five sifting frames packages of metal or plastic fabric.

From Figure 1 it can be seen that the first five sifting frames (with fabric No. 50) are arranged in a package. The oversized particles collected from this package is sent to grinding roll M3, and the undersized particle feed the second package, consisting, in turn, of four sieve frames (No. 60). The oversized particles from this package is sent to grinding roll M1B, and the undersized particle feed the third package. It consists of seven sieve frames with 0.17 mm fabric apertures (flour sieve - no. IX). Fourth package of sifting frames (no. X) has two frames with plastic fabric with 0.15 mm apertures. The undersized particles collected from sifting frames of packages three and four from this compartment are discharged as a flour, and the oversized particles feed the last sifting package of compartment, which has five frames no. VIII (0.18 mm apertures width). The undersized particles from this package is a semolina flour that is discharged as such. The oversized particles are redirected for grinding to reduction roll M2.

After grinding at reduction roll M1B, the semolina is sorted into dimensional fractions in the C2 plansifter compartment, consisting of 4 packs of frames. The first package has five frames with fabric no. 50, at the end of which the first fraction of oversized particles is directed to the grinding roll M3. The undersized particles from this package feeds the second package consisting of eight flour frames no. IX. In turn, the undersized particles of the second package is discharged in the form of flour (F2), having particles smaller than 0.17 mm. The oversized particles from this package is directed to the third package consisting of seven frames of flour sieve no. X which evacuates the undersized particles as flour. The oversized particles of this package reaches the first of the three frames of the last package of the compartment (with fabric no. VIII). The oversized particles of the fourth package is directed for re-grinding to the grinding roll M2, and the undersized particles are extracted in the form of semolina flour.

After sampling on the flow of the milling unit, in the laboratory 100 grams of material was subjected to a particle size analysis with a classifier with sieves Analysette 3 Spartan type, the fineness of the fraction being appreciated by the mean diameter of the grist particles:

$$d_m = \frac{\sum p_i \cdot d_i}{\sum p_i}, mm \quad (1)$$

where: p_i - the percentage of material on sieve i of the classifier; $\Sigma p_i = 100$ – the sum of the percentages of material on the sieves; d_i – the mean particle size of each fraction, considered as the arithmetic mean of the size of the apertures of the sieves enclosing the respective fraction $d_i = (x_i + x_{i+1})/2$.

The grist samples (14 in number) were sorted by sifting on a set of 5 superimposed sieves driven in vibrating motion with an amplitude of 2 mm, the sifting time being 3 minutes. The working methodology is presented in numerous papers in the literature, [20, 21].

The experimental data were subjected to a nonlinear regression analysis, in Microcal Origin. It were thus obtained, variation curves for sifted T (x) and for rejected material, R (x). Nonlinear regression analysis was performed using Rosin-Rammler particle size distribution law, [22, 23], and which is represented by relationships:

$$T(x) = 100 \cdot \left(1 - e^{-\alpha \cdot x^{\beta}}\right) \quad (2)$$

$$R(x) = 100 \cdot e^{-\alpha \cdot x^{\beta}} \quad (3)$$

where *T(x)* – represents the mass percentage weight of the fraction with particles smaller than x (passed through the sieve with size x), and *R(x)* – mass percentage by weight of the fraction with particles larger than x (which do not pass through the sieve with dimension x); *x* – the size of the sieve apertures; α and β – nonlinear regression coefficients.

3 Results and discussions

In table 1 is shown the results of the particle size analysis for the grist that feeds grinding roll M1A, and is then sorted in the plansifter compartment C1 (see figure 1) and the dimensions of the sieve apertures used in the classifier.

Table 1. Weight values p_i (%) of the fractions on the classifier sieves and of the cumulative weights T_i (%) for products collected at the inlet and at the 5 outlets of the plansifter compartment of the first technological passage, but also for the fraction that feeds the grinding roll M1A.

	x	M1A Entr.		x	C1 Entrance		x	C1M3		x	C1M1B	
	(mm)	p(%)	T(%)	(mm)	p(%)	T(%)	(mm)	p (%)	T(%)	(mm)	p(%)	T(%)
0	0.000	0.90	0.00	0.000	34.40	0.00	0.000	0.10	0.00	0.000	0.40	0.00
1	0.180	1.00	0.90	0.180	16.30	34.40	0.100	0.30	0.10	0.100	7.90	0.40
2	0.250	3.70	1.90	0.250	10.30	50.70	0.125	19.40	0.40	0.125	8.70	8.30
3	0.315	33.30	5.60	0.315	15.80	61.00	0.180	50.80	19.80	0.180	30.70	17.00
4	0.500	49.10	38.90	0.500	12.50	76.80	0.250	25.50	70.60	0.250	45.30	47.70
5	0.710	12.00	88.00	0.710	10.70	89.30	0.315	3.90	96.10	0.315	7.00	93.00
d_m	$d_{M1A.E}$ = 0.55			d_{1E}=0.33 mm			d_{1M3}=0.23 mm			d_{1M1B}=0.24 mm		

	x	C1M2		x	C1F		x	C1Fgrif	
	(mm)	p(%)	T(%)	(mm)	p(%)	T(%)	(mm)	p(%)	T(%)
0	0.000	0.20	0.00	0.000	1.30	0.00	0.000	0.60	0.00
1	0.063	4.20	0.20	0.090	35.30	1.30	0.063	9.30	0.60
2	0.090	10.10	4.40	0.125	17.50	36.60	0.090	25.70	9.90
3	0.125	33.50	14.50	0.160	22.90	54.10	0.125	38.60	35.60
4	0.160	40.70	48.00	0.180	15.60	77.00	0.160	20.10	74.20
5	0.180	11.30	88.70	0.200	7.40	92.60	0.180	5.70	94.30
d_m	d_{1M2} = 0.16 mm			d_{1F}=0.15 mm			d_{1Fgrif} = 0.17 mm		

Analysing figure 1 it can be seen that the M1A grinding roll is fed with part of the undersized particles of the two semolina machines (MG1 and MG2) from reduction phase. This mixture of particles that feeds the first grinding roll has most of the particles with dimensions greater than 0.25 mm (98.1 %) and generally represent the categories of semolina (great semolina – 12%, medium semolina – 49.1 % and small semolina – 33.3%) and harsh dunsts– 3.7 %. After grinding, the mixture obtained (fraction C1 Entrance) it will be fed into the plansifter compartment C1 for separation on particle size fractions. After grinding, the percentage of particles larger than 0.25 mm decreases from 98.1% to 49.3% and represents the same categories as above (great semolina – 10.7%, medium semolina – 12.5 %, small semolina – 15.8% and harsh dunsts – 10.3 %). In this mixture there are also particles with dimensions below 0.18 mm (about 34.4%) mainly represented by the C1F and C1Fgrif grist fractions. The resulting fractions, which are sorted at compartment C1, have mean particle sizes corresponding to Table 1. It can also be seen that if the mean diameter of the particle mixture being fed into the M1A grinding roll is 0.55 mm, immediately after grinding the average diameter decreases by 0.22 mm for the fraction that will feed the plansifter compartment C1 (C1 Entrance).

The fraction C1M3 represents an oversized particles fraction with a high coating content. The mean particle size for C1M3 is 0.23 mm, but within this fraction about 80.2%

of the particles have dimensions of more than 0.18 mm. This fraction re-enters the grinding process at the M3 reduction roll.

The fraction C1M1B, also a refuse (of the second frames package), has an mean particle size of 0.24 mm, with a percentage of more than 83% from particles larger than 0.18 mm. The representative category for this mixture of particles is that of dunsts. So, about 50% of these particles are harsh dunsts, and the remaining about 33% are fine dunsts. This fraction is combined with the remaining undersized particles from the semolina machines and, together, re-enters the grinding process at the M1B grinding roll.

The third fraction is also a refusal (of the last package of frames), from the category of dunsts-type products, have 11.8% particles larger than 0.18 mm. Having an average particle size of 0.16 mm, is grinded in the M2 reduction roll of the milling unit.

The two flours extracted in the compartment (the undersized particles from packages 3, 4 and 5) have an mean particle size of 0.15 mm (C1F), respectively 0.17 mm (C1Fgrif). Fraction C1F is extracted as an undersized particles fraction of packages 3 and 4 and has 63% of particles larger than 0,125 mm. Fraction C1Fgrif is a semolina flour and has most of the particles (over 64%) larger than 0.125 mm.

Table 2. Weight values p_i (%) of the fractions on the classifier sieves and of the cumulative weights T_i (%) for products collected at the inlet and at the 5 outlets of the plansifter compartment of the first technological passage, but also for the fraction that feeds the grinding roll M1B.

	x	M1B Entr.		x	C2 Entrance		x	C2M3		x	C2M2	
	(mm)	p(%)	T(%)	(mm)	p(%)	T(%)	(mm)	p (%)	T(%)	(mm)	p(%)	T(%)
0	0.000	0.40	0.00	0.000	7.70	0.00	0.000	1.70	0.00	0.000	3.00	0.00
1	0.130	0.60	0.40	0.125	8.00	7.70	0.090	3.00	1.70	0.045	4.80	3.00
2	0.180	3.90	1.00	0.180	23.80	15.70	0.125	27.00	4.70	0.063	5.10	7.80
3	0.250	22.10	4.90	0.250	35.60	39.50	0.180	21.20	31.70	0.090	20.90	12.90
4	0.315	48.90	27.00	0.315	21.90	75.10	0.200	37.40	52.90	0.125	55.80	33.80
5	0.400	24.10	75.90	0.450	3.00	97.00	0.250	9.70	90.30	0.160	10.40	89.60
d_m	$d_{M1B.E}$ = 0.36			d_{2E}=0.26 mm			d_{2M3}=0.20 mm			d_{2M2}=0.13 mm		

	x	C2Fgrif		x	C2F2		x	C2F	
	(mm)	p(%)	T(%)	(mm)	p(%)	T(%)	(mm)	p(%)	T(%)
0	0.000	3.60	0.00	0.000	39.10	0.00	0.000	10.40	0.00
1	0.045	8.80	3.60	0.090	20.60	39.10	0.045	22.90	10.40
2	0.063	10.60	12.40	0.125	30.30	59.70	0.063	17.60	33.30
3	0.090	18.50	23.00	0.160	5.20	90.00	0.090	29.60	50.90
4	0.125	53.10	41.50	0.180	3.50	95.20	0.125	18.40	80.50
5	0.180	5.40	94.60	0.200	1.30	98.70	0.160	1.10	98.90
d_m	d_{2Fgrif} = 0.13 mm			d_{2F2}=0.10 mm			d_{2F} = 0.09 mm		

In table 2 is shown the results of the granulometric analysis for the fraction that feeds grinding roll M1B, and is then sorted in the plansifter compartment C2 and the dimensions of the sieve apertures used in the classifier.

Grinding roll M1B is, also, fed with the other part of the undersized particles from the two semolina machines (MG1 and MG2) from the reduction phase but also with the second refusal of compartment C1. This mixture of particles feeding the second grinding roll has most 95.1 % from particles larger than 0.25 mm. In generally, this particles represent the categories of dunsts (harsh dunsts- 70% and fine dunsts – 5 %) but small semolina particles are also found – 24.1%. After grinding, the mixture obtained (fraction C2 Entrance) it will be fed into the plansifter compartment C2 for separation on particle size fractions. After grinding, the percentage of particles larger than 0.18 mm decreases from 99% to 84.3% and represents particles in the categories small semolina – 5%, harsh dunsts – about 46 % and fine dunsts - about 33%. In this mixture there are also particles with dimensions below 0.18 mm (about 15.7%) represented mainly, by the fractions C2F, C2F2 and C1Fgrif. It can be seen that if the mean diameter of the particle mixture that is fed into the M1B grinding roll is about 0.36 mm, immediately after grinding the mean diameter decreases by 0.10 mm for the fraction that will feed the plansifter compartment C2 (C2 Entrance).

First fraction extracted from compartment C2 (C2M3) represents a refusal (oversized particles) with a mean particle size of 0.20 mm, 68.3% of the particles from this fraction have dimensions over 0.18 mm, and re-enters the grinding process at the M3 reduction roll.

The fraction C2M2 is also an oversized fraction and have a mean particle size of 0.13 mm, This fraction have more than 10 % of particles larger than 0.18 mm and has a high endosperm content.

The three flours extracted in the compartment have a mean particle size of 0.13 mm (C1Fgrif), 0.10 mm (C2F2) and 0.09 mm (C1F) respectively. 58.5% from C2Fgrif fraction particles (which is a semolina flour) is larger than 0.125 mm, compared to the C2F2 fraction which has 40% in the same range and reaches 19% within the C2F fraction.

In table 3 are presented the values of the coefficients α and β of the Rosin-Rammler distribution function, and the correlation coefficient R^2, values obtained from nonlinear regression analysis.

Table 3. Values of the coefficients α and β of the Rosin-Rammler distribution function, and the correlation coefficient R^2

Fraction / Coefficient	M1A Entrance	C1 Entrance	C1M3	C1M1B	C1M2	C1F	C1Fgrif
	R(x), T(x)	*R(x), T(x)*	*R(x), T(x)*	*R(x), T(x)*	*R(x), T(x)*	*R(x), T(x)*	*R(x), T(x)*
α	0.018	0.088	0.001	0.002	$1\cdot10^{-5}$	$4\cdot10^{-4}$	$6\cdot10^{-6}$
β	-5.702	-1.468	-5.149	-4.124	-5.813	-3.843	-5.773
R^2	0.998	0.997	0.998	0.978	0.981	0.975	0.991
Fraction / Coefficient	M1B Entrance	C2 Entrance	C2M3	C2M2	C2Fgrif	C2F2	C2F
	R(x), T(x)	*R(x), T(x)*	*R(x), T(x)*	*R(x), T(x)*	*R(x), T(x)*	*R(x), T(x)*	*R(x), T(x)*
α	$9\cdot10^{-4}$	0.006	$1\cdot10^{-5}$	$4\cdot10^{-9}$	$1\cdot10^{-7}$	$3\cdot10^{-4}$	0.002
β	-6.301	-3.474	-6.716	-9.277	-7.503	-3.339	-2.367
R^2	0.999	0.986	0.997	0.982	0.954	0.985	0.981

As can be seen from the charts presented in figure 2, the majority of particles for some fractions have dimensions close to the min. value of sieves apertures. Other fractions have particles with mean dimensions close to the max. value of the sieve apertures. The experimental points are arranged to the left or right of the curve. At the same time, it can also be observed that the meeting point for the two particle size curves is rather towards the sieve with larger apertures. We refer here to the sieves used in particle size analysis.

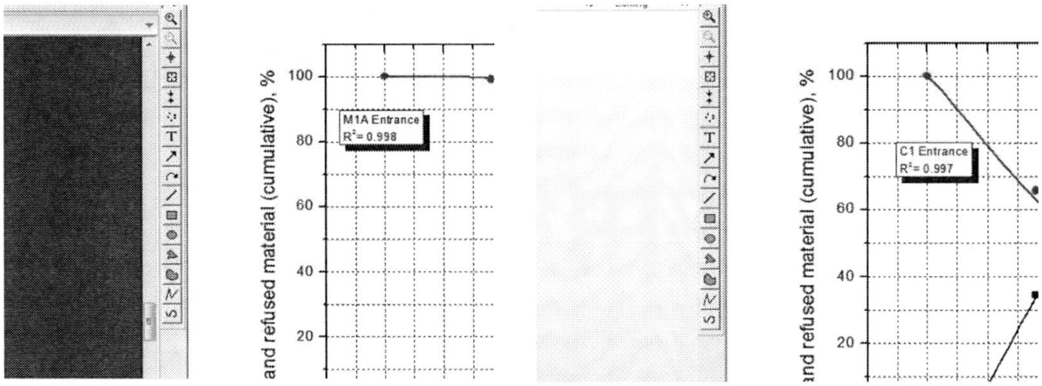

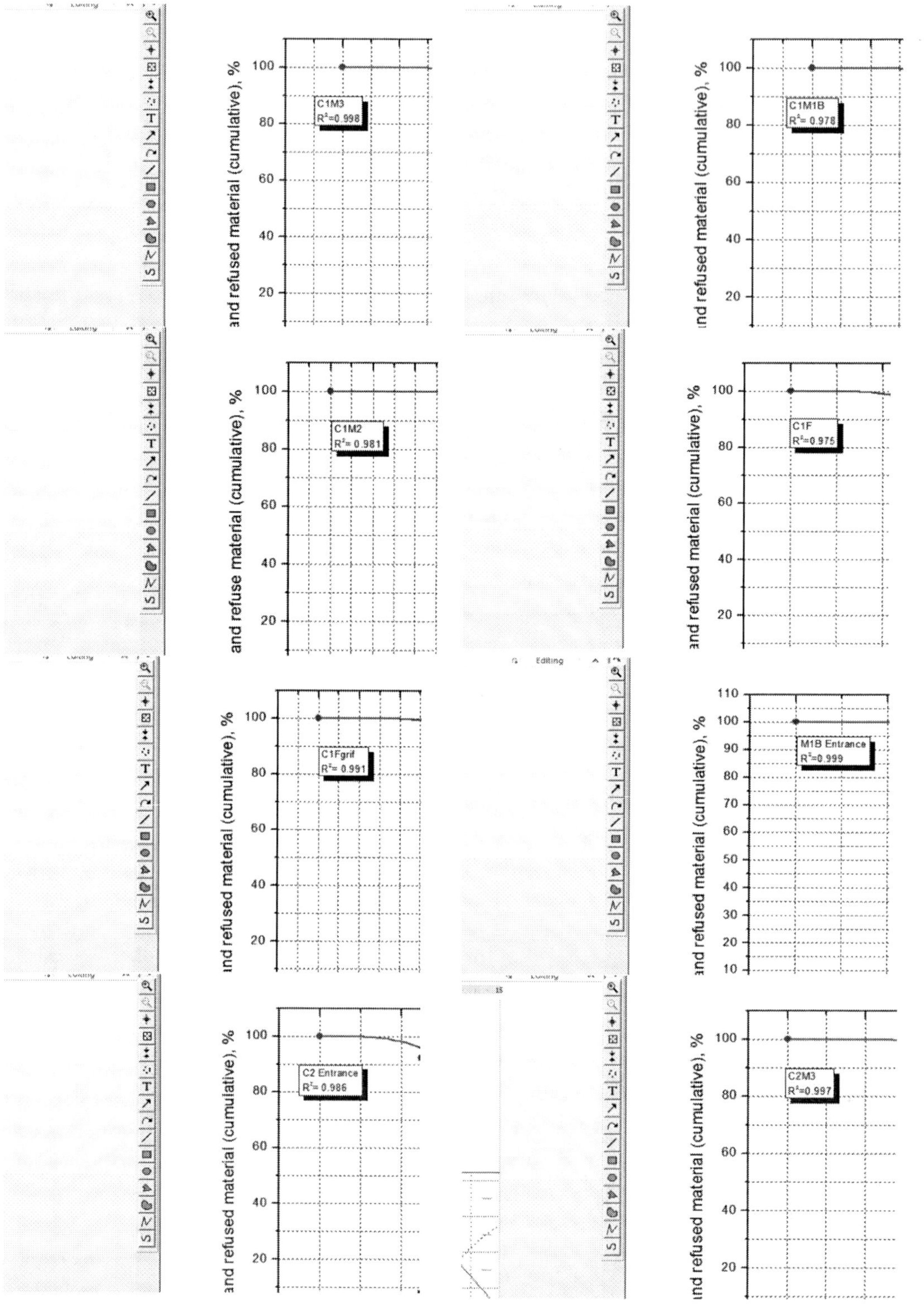
C1M3
R²=0.998
C1M1B
R²= 0.978
C1M2
R²= 0.981
C1F
R²=0.975
C1Fgrif
R²= 0.991
M1B Entrance
R²=0.999
C2 Entrance
R²= 0.986
C2M3
R²=0.997
and refused material (cumulative), %
and refuse material (cumulative), %
Editing

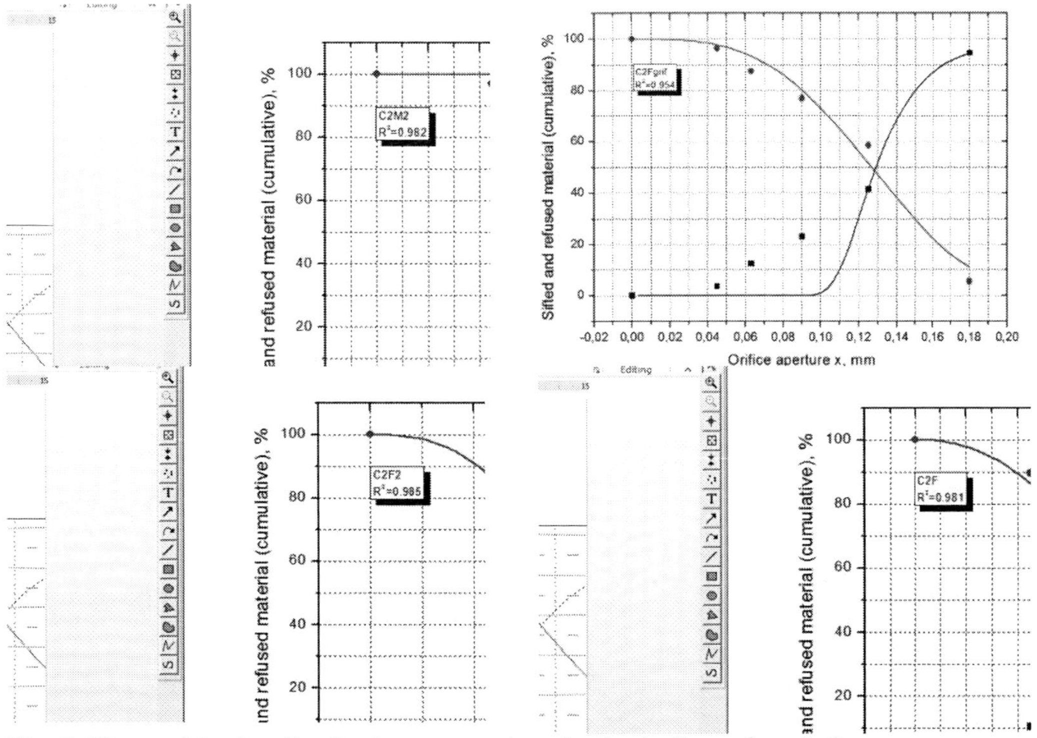

Fig. 2. The particle size distribution curves given by Rosin-Rammler relations;
——— - R(x); ——— - T(x)

The shape of the regression curves is correlated with the data from experiments. This shape depends on the weight of grist collected on the sieves used in the classifier for the particle size analysis.

4 Conclusions

For the two technological passages of the reduction phase analysed in this paper, we can say that the Rosin-Rammler law represents very well the experimental data, the correlation coefficient being of $R^2 \geq 0.954$.

The degree of non-uniformity for the grist particles fractions is indicated by β coefficient from the Rosin-Rammler equations. Form table 3 we can see that its values fall in a rather narrow range. This means that the fractions analysed were uniform, in terms of particle size.

The Rosin-Rammler law used in the paper shows (for the 14 grist fractions analysed) a very good correlation with the experimental. Knowledge of the mean dimensions and size distribution, and the other physical characteristics of the grist fractions are also requirements for adjusting the machines from the flow, and in choosing the fabrics of the sieves from plansifter compartments, from the input to the compartment to the output.

The paper was funded from the project „ *Improving the base of practical applications in solariums, gardens, orchards and vineyards (IMPRASGLIV)" CNFIS-FDI-2021-0021*, from the Ministry of Education through the Executive Agency for Financing Higher Education, Research, Development and Innovation.

References

1. E.J. Hajnal, J. Mastilović, F. Bagi, D. Orčić, D. Budacov, J. Kos, Z. Savic, Toxins, 11, 139, (2019);
2. *** www.statista.com;
3. V. Kanojia, N.L. Kushwaha, M. Reshi, A. Rouf, H. Muzaffar, International Journal of Chemical Studies, 6(4), (2018);
4. G.M. Campbell, C. Fang, I.I. Muhamad, Food Bioprod Process, 85, (2007);
5. G.M. Campbell, *Roller Milling of Wheat, Handbook of Powder Technology*, Vol. 12, (Science Direct, 2007);
6. G. Owens, *Cereals processing technology*, (Woodhead Publishing Ltd., (2001);
7. Q. Fang, E. Haque, C.K. Spilman, P.V. Reddy, J.L. Steel, Transaction of the ASAE, 41(6), (1998);
8. C. Fang, G.M. Campbell, Journal of Cereal Science, 37, (2003);
9. P. Guritno, E. Haque, Transactions of the ASAE, 37(4), (1994);
10. G.M. Campbell, P.J. Bunn, C. Webb and S.C.W. Hook, Powder Technology 115 (2001);
11. M. G. Scanlon, J. E. Dexter, Cereal chemistry, 63, (1986);
12. G. Cacak-Pietrzak, A. Sułek, M. Wyzińska, Agricultural Science, Research for Rural Development, 2, (2019);
13. S. D. Sakhare, A. A. Inamdar, J Food Sci Technol, 51(4), (2014);
14. A.K. Zaalouk, F.I. Zabady, Journal Ag.Eng. 26(1), (2009);
15. A. Z. Fišteš and M. Đ. Vukmirović, APTEFF, 40, (2009);
16. S.P. Galindez Najera, *A compositional breakage equation for first break roller milling of wheat*, University of Manchester – CEAS, PhD thesis, (2014);
17. G.M. Campbell, C. Sharp, K. Wall, F. Mateos-Salvador, S. Gubatz, A. Huttly, P. Shewry, Journal of Cereal Science, 55(3), (2012);
18. G.A. Constantin, *Researches on the sifting and sorting process of grist fractions in a industrial milling plant*, Doctoral thesis, "Politehnica" University of Bucharest, (2014).
19. Gh. Voicu, S.S. Biriş, E.M. Ştefan, G.A. Constantin, N. Ungureanu, G*rinding characteristics of wheat in industrial mills. Chapter 15 in Food Industry Book, Edited by InTech Europe*, (University Campus STeP Ri, 2013);
20. T. Căsăndroiu, Gh. Voicu, G.D. Ţuţuianu, Proceeding of ISBTeh-2002 Conference, (2002);
21. Gh. Voicu, T. Căsăndroiu, G.D. Ţuţuianu. Proceedings International Symposium „Euro-Aliment 2003", (2003);
22. V. Headley, H. Pfost, Transactions of the ASAE, 11 (3), (1968);
23. S. Henderson, Hansen R., Transactions of the ASAE, 11 (3), (1968).

Installation of submarine cables and pipes in shallow waters and connecting them to the shore by the method of horizontal directional drilling

Valeriu-Florian Vasilescu[1,*] and *Dumitru* Dinu[2]

[1]Constanta Maritime University, Doctoral School of Mechanical Engineering and Mechatronic, 104 Mircea cel Batran Street, ZC 900663, Constanta, Romania
[2]Constanta Maritime University, Doctoral School of Mechanical Engineering and Mechatronic, 104 Mircea cel Batran Street, ZC 900663, Constanta, Romania

Abstract: This article presents the horizontal directional drilling (HDD) as one of the most suitable technique used to reduce the impact on the environment during the submarine pipe and cable installation process, especially in the coastal area and shallow waters. HDD is a technique used to drill a subsea tunnel or under other designated area with the goal to pull a pipe or other facility through the drilled underground tunnel. With the development of the offshore natural gas and wind industry, the demand for the construction of transfer pipeline and cables has also increased. The installation of submarine pipelines and cables can have negative effects on the marine environment. This is one of the main reasons that construction and operation of offshore wind energy systems has been and continues to be regarded with scepticism by environmental activists, despite the undeniable benefits of this renewable energy source. The main objectives of this article are to emphasize the importance of using this technique but also to highlight the benefits, in particular by significantly reducing the impact on the marine environment.

1 Short description of horizontal directional drilling (HDD)

HDD is a technique used to drill a subsea tunnel or under other designated area with the goal to pull a pipe or other facility through the drilled underground tunnel. Depending on the desired tunnel's diameter and length, the HDD method is divided into three categories: mini HDD, midi HDD and maxi HDD, their features and applications are shown in Table 1.

Table 1 Classification of HDD method [1]

Type	Diameter range (m)	Crossing length (m)	Pulling force (x 10^3 N)	Machine weight (kg)
Mini	0.05 – 0.3	≤177	>445	≤9000

* Corresponding author: vasilescuvaleriu75@gmail.com

Midi	0.3 – 0.6	≤274	90 - 445	≤18000
Maxi	0.6 – 1.5	≤3058	<90	≤30000

In HDD there are considered three main installation stages (Fig. 1):

- Pilot drilling
- Reaming the initial pilot borehole
- Pulling back the pipeline in the reamed borehole or the cable through the installed pipe

When drilling a wellbore, a pilot hole is drilled mainly for safety reasons. The wellbore is drilled using a drilling fluid, which has the role of cooling and lubricating the drill bit but also to transport the material displaced by the bit to the surface through the annular space outside the drill string. The drilling fluid is pumped into the wellbore through the hydraulic head at a high-pressure rate. Next step is to enlarge the pilot hole using a reamer. The reamer is a down hole tool able to enlarge a borehole gradually, depending on drilling program requirements. At the end of drilling the wellbore, the drill string is pulled out, but during the whole period of drilling and extracting the drill string, the drilling fluid is present in the well.

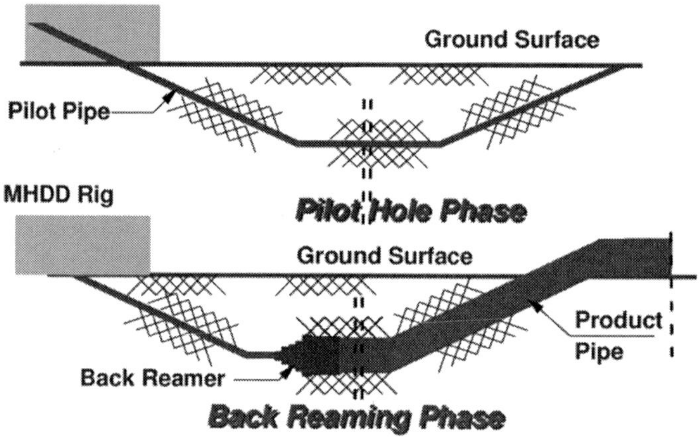

Fig. 1 HDD layout [2]

One of the biggest challenges in an installation process of submarine pipes or cables, by using the HDD technique, is the one due to the lack of qualified personnel. At the same time, insufficient geotechnical information and the lack of drilling profiles can lead to the impossibility of performing a horizontal directional drilling. Due to the technological innovations that HDD has known in recent past years, this technique can meet the latest engineering requirements. By modernizing the pipe handling equipment, it was possible to increase the drilling speed as well as the degree of operational safety.

Before deciding to install submarine cables or pipes using HDD, at least the following aspects must be taken into account:

- the preliminary geological survey report should provide at least the following information: plane map, geological profile, sampling depth, water content, saturation, granularity, standard penetration number, liquid index, plastic index, liquid limit, plastic limit and so on
- the geological conditions that HDD is suitable for crossing include clay, mild clay, well-forming sand, and soft rock
- the design radius of the drilling has to be larger than the minimum allowable bending radius of the cable, pipe and of the drill string

- the curvature radius of the pipeline should be as large as possible, so as to avoid the increase of pullback resistance
- the entrance angle and exit angle should be determined according to the topography, geological conditions and the diameter of the pipeline. Generally, the entrance angle should be controlled at 8 to 18 degrees and the exit angle should be controlled at 4 to 12 degrees. At present, the maximum entrance angle can reach 28 degrees and the exit angle can reach 15 degrees.

The main components of HDD are presented in Fig.2

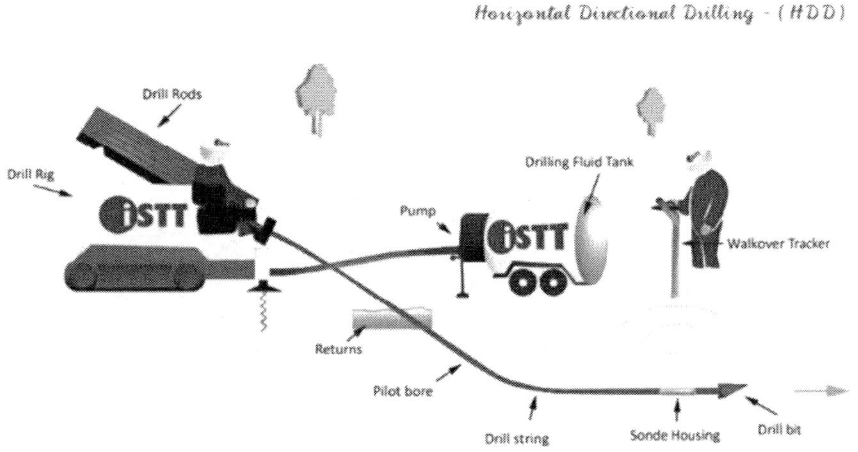

Fig. 2 Components of HDD [3]

2 Design of a drilling radii

The design of a drilling radii plays an important role and the Drilling Contractors Association (DCA) guideline recommends using a design radius, which considers also the drill ability by a constant for the individual geological condition. The drilling radii can be calculated with the following formula [4]:

$$R_{design} = C\sqrt{Dt} \tag{1}$$

Were:
R_{design} = design radius
C = soil characteristic constant (Table 2)
D = outer pipe diameter
t = pipe wall thickness

Table 2 Soil parameters

CONDITION	Cone Penetration Test (CPT)	Standard Penetration Test (SPT)	Elastic modulus Es	Soil characteristic constant C
	qc [MPa]	N_{30} [blows/30 cm]	[MPa]	[-]
Sand (very dense)	› 20	› 50	100-200	8500
Sand (medium density)	10-20	25-50	50-100	9400
Sand (low density)	5-10	10-25	20-50	10200
Clay (dense)	› 2	› 8	10-25	10500
Clay (medium density)	1-2	2-8	5-10	11500
Soft clay (silty)	‹ 1	‹ 2	0-5	12500

The soil conditions are of a paramount importance for the planning and design of a horizontal directional drilling.

In order to better understand this very important aspect take a look over the below simulation of a geotechnical onshore and offshore investigation campaign (Table 3).

Lithologically they all show alternating fine to medium sands underlaid by sandy and silty clays, comparable to the beach borehole.

Following these investigations, both the drilling route and the reaming procedure can be decided.

According with the values from the bellow simulation we can conclude that the procedure of push ream is not applicable due to significant number of SPT values that are between 0 and 8 and only a back reaming procedure will be able to be realized.

Table 3 Soil condition

Distance from entry:	100m	400m	500m	600m	700m	800m	900m	1000m
Lat:	Beach	BH 01	BH 02	BH 03	BH 04	BH 05	BH 06	BH 07
1								
	fine							
0	fine							
	fine							
-1	fine							
	fine							
-2	fine							
	medium							
-3	medium	medium	medium					
	medium	medium	medium	medium				
-4	medium	medium	medium	fine	fine	fine		
-5	silty	medium	medium	fine	fine	fine	fine	

	silty	fine	fine	fine	fine	fine	fine	fine
-6	silty	fine	fine	fine	fine	fine	fine	fine
	silty	fine	fine	fine	fine	fine	fine	fine
-7	silty	fine	fine	fine	fine	fine	fine	fine
	silty	fine	fine	silty	silty	silty	silty	silty
-8	silty	fine	fine	silty	silty	silty	silty	silty
	silty	fine	fine	silty	silty	silty	silty	silty
-9	silty	silty	silty	silty	silty	silty	silty	silty
	silty	silty	silty	silty	silty	silty	silty	silty
-10	silty	silty	silty	silty	silty	silty	silty	silty
	silty	silty	silty	silty	silty	silty	silty	silty
-11	fine	silty	silty	silty	silty	silty	silty	silty
	fine	silty	silty	silty	silty	silty	silty	silty
-12	fine	silty	silty	silty	silty	silty	silty	silty
	fine	silty	silty	silty	silty	silty	silty	silty
-13	fine	silty	silty	silty	silty	silty	silty	silty
	fine	silty	silty	silty	silty	silty	silty	silty
-14	fine	silty	silty	silty	silty	silty	silty	silty
	silty	silty	silty	silty	silty	silty	silty	silty
-15	silty	silty	silty	silty	silty	silty	silty	silty
	silty	silty	silty	silty	silty	silty	silty	silty
-17	silty	silty	silty	silty	silty	silty	silty	silty
	silty	silty	silty	silty	silty	silty	silty	silty
-18	silty	silty	silty	silty	silty	silty	silty	silty
	silty	silty	silty	silty	silty	silty	silty	silty
-19	silty	silty	silty	silty	silty	silty	silty	silty
	silty	silty	silty	silty	silty	silty	silty	silty
-20	clay	silty	silty	silty	silty	silty	silty	silty
	clay	silty	silty	silty	silty	silty	silty	silty
-21	clay	silty	silty	silty	silty	silty	silty	silty
	clay	silty	silty	silty	silty	silty	silty	silty
-22	clay	silty	silty	silty	silty	silty	silty	silty
	clay	silty	silty	silty	silty	silty	silty	silty
-23	clay	silty	silty	silty	silty	silty	silty	silty
	clay	silty	silty	silty	silty	silty	silty	silty
-24	clay	silty	silty	silty	silty	silty	silty	silty
	clay	gravel	gravel	gravel	gravel	gravel	gravel	gravel
-25	clay	gravel	gravel	gravel	gravel	gravel	gravel	gravel
	clay	gravel	gravel	gravel	gravel	gravel	gravel	gravel

3 Finite Element Analysis (FEA)

In this section is analysed, by using the ANSYS Workbench software, the pipe stress under a scenario which respects reality as much as possible. Workbench is an integrated environment software released by ANSYS which keep all the functions of ANSYS classic platform in FEA and integrates powerful geometric modelling functions of computer-aided design (CAD).

In this scenario the size of the pipe is 400 mm × 14 mm (diameter × thickness), and the diameter of the borehole is 650 mm. According to Saint Venant's principle, the soil far away from the object has little influence on the analysis, so the width and height of the soil are set at about 11 times the diameter of the borehole, which is 7.2 m. According to Fig. 3 we can observed be that the object of analysis is symmetrical on the X-axis.

In order to save computing time, only half of the models for soil and pipeline are established on the cross section. In addition, the length of the pipe is taken as 5 m in this analyse. Therefore, the size of the whole model is 3.6 m ×5 m × 7.2 m (width × length × height). The geometric model built with Design Modeler software is shown in Fig.4.

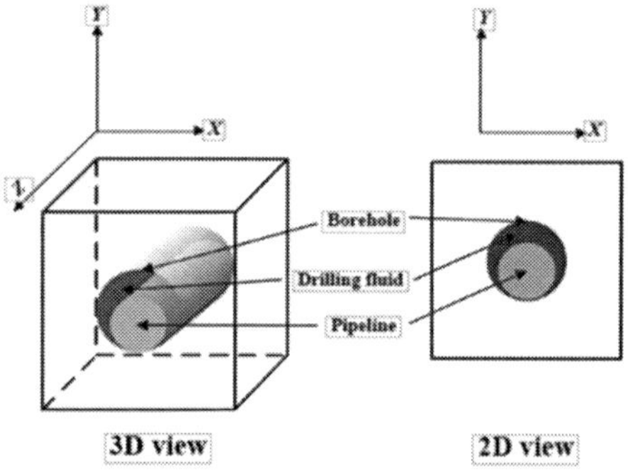

Fig. 3 Physical model of the proposed scenario

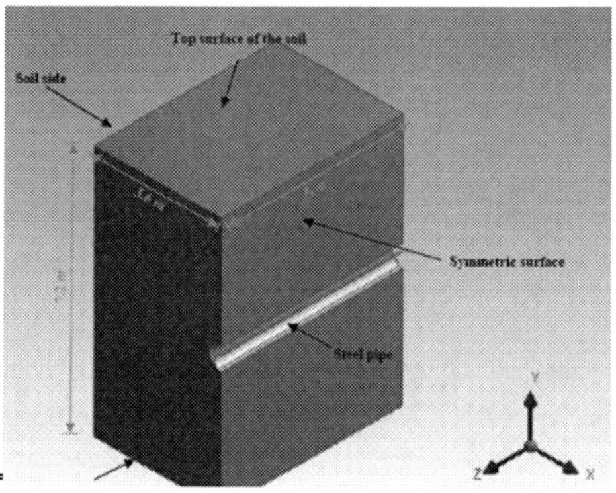

Fig. 4 Pipeline-soil geometric model

For this analyse only mechanical properties of pipe and soil need to be set, as shown in Table 4.

Drucker-Prager model is used for soil, considered closer to the actual scenario.

There is only one contact involved in the model: contact between the inner wall of the borehole and the outer wall of the pipe.

Table 4 : Parameters of the steel pipe and soil

Steel pipe parameter	**Value**	**Soil parameter**	**Value**
Outer diameter (OD)	400 mm	Poisson's ratio	0.2
Wall thickness	14 mm	Elastic modulus	42 MPa
Material	API 5Lx52	Dilatancy angle of soil	4.83°
Elastic modulus	210 GPa (3.046×10^7 psi)	Internal friction angle	9.65°
Bulk modulus	175 GPa (2.538×10^7 psi)	Density	2500 kg/m^3 (156.07 lb/ft^3)
Shear modulus	80.769 GPa (11714553 psi)	Cohesion of soil	12.94 kPa
Poisson's ratio	0.3	Borehole diameter	650 mm
Density	7850 kg/m^3 (490 lb/ft^3)		

 https://doi.org/10.1051/e3sconf/202128604007

Since the bottom of the pipe is in contact with the bottom of the borehole, that is, the normal direction is not separated, the tangential direction may have a small sliding, so that the contact of "frictional" is used, as shown in Fig.5. According to actual engineering data, the coefficient of friction between the pipeline and the soil is 0.24.

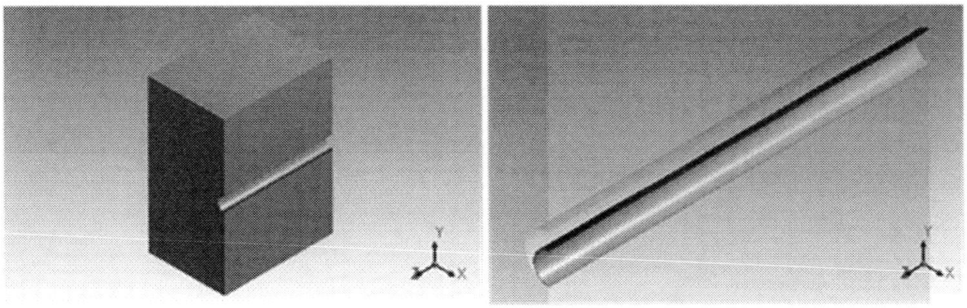

Fig. 5 Contact between the inner wall of the borehole and the outer wall of the pipe

The hydrostatic pressure of a curved surface can be calculated automatically. When the density of drilling fluid is 1200 kg/m3, the calculation results of the hydrostatic pressure acting on the borehole wall and the outer wall of the pipeline are shown in Fig. 6 and Fig. 7.

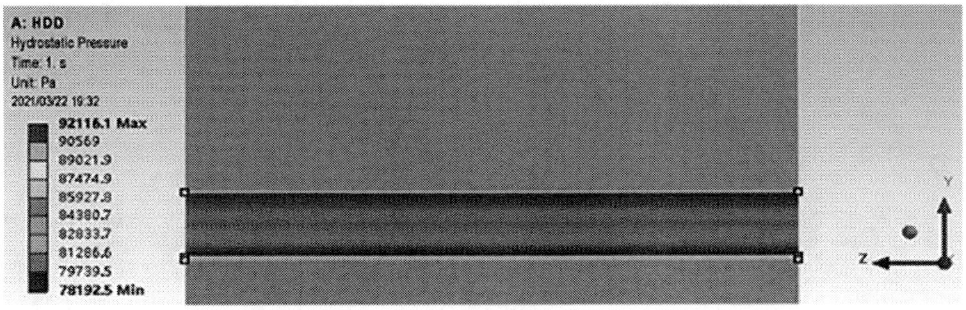

Fig. 6 The hydrostatic pressure acting on the borehole wall

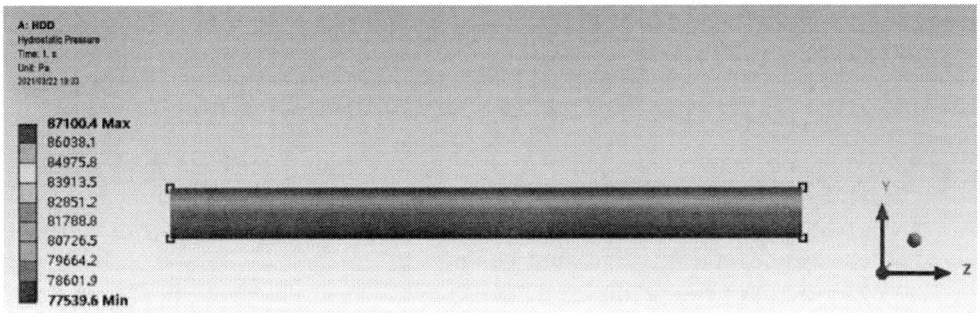

Fig. 7 The hydrostatic pressure acting on the outer wall of the pipe

It shows that the hydrostatic pressure on the upper part of the pipeline is greater than that on the lower part. Based on the above analysis, all loads in this scenario are shown in Fig.8.

 https://doi.org/10.1051/e3sconf/202128604007

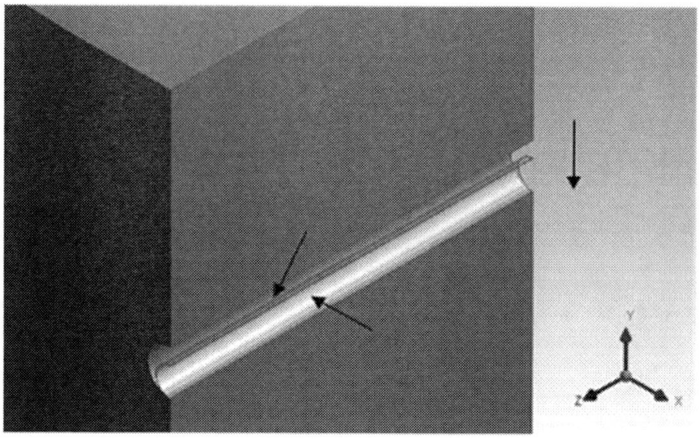

Fig. 8 Loads for this scenario

A free boundary conditions of the pipe-soil system at both ends of the pipeline is presented in the figure 9 (a).

In case of a fixed boundary condition can be observed a stress concentration at both ends of the pipe as is shown in the figure (b).

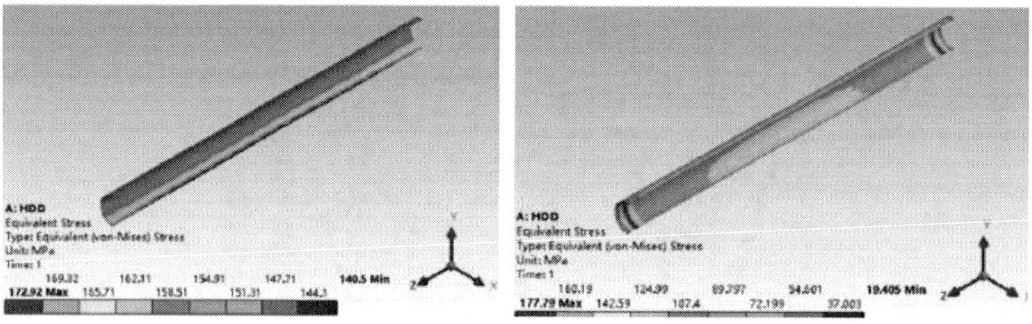

(a) (b)

Fig. 9 Equivalent stress nephogram of pipeline: (a) free boundary conditions; (b) fixed boundary conditions

4 Installation of submarine cables and pipes in shallow water

The installation of submarine cables in deep water is done with specialized ships and state-of-the-art equipment. Unfortunately, the installation and repair of submarine cables in shallow water remains a challenge, and it is necessary to develop new concepts to ensure increased efficiency and reduced environmental impact.

It is observed that in many projects, for the connection to the shore of the submarine cables, the installing's used techniques, often lead to cable damage and increase the costs of these operations, having a negative impact on the environment. This leads to the decision to use various types of pontoons or barges, with various technical capabilities, with or without their own propulsion. Unfortunately, these installation methods involve the use of a large number of tugs and / or other equipment (Fig.10) [5]

Fig. 10 CLV VETAG 8 and crawler excavator into intertidal zone [5]

The Waves and sea currents action can affect submarine cables by the forces developed by the action of water or by redistribution of sediments. The action of the waves is stronger on the shore or in shallow waters and decreases with increasing depth. In general, the action of the waves stops at a depth of about 30 m. Only exceptionally, during strong storms the action of the waves is felt more deeply. Coastal areas, especially sandy ones, are undergoing the most dramatic changes due to the action of waves. In such cases, if the cable or pipes are not enough buried, they can be brought to the surface (Fig. 11) [6]

Fig. 11 Cable unearthed under the action of waves [6]

5 Benefits of using HDD for installation of submarine cables and pipes

Nowadays there is an increase in interest in the use of horizontal directional drilling in the installation of cables and pipes [7]. Bellow I want to show you the most important benefits of using HDD:

- requires a relatively short installation period (the standby time due to weather windows is eliminated)
- relatively low installation costs
- reduces the impact on the environment being a less invasive technique than traditional open cutting
- reduction of carbon emissions due to the low number of tugs and / or other equipment used for installation operations

- minimizes the negative impact on other installations or networks of cables and / or pipes installed in the same area
- does not require interruption of maritime traffic
- reduces the environmental impact of any activities on the sea bottom, above the pipeline or cable (in case of preventive or corrective maintenance's works or even decommissioning's activities)
- increasing the safety of operation (even in the case of anchoring ships in dangerous situations, due to force majeure)
- also makes possible the installation of less resistant pipes to the classic installation operations, respectively liquid hydrogen transport pipes

6 Conclusions

Having in mind that this article is only part of a major research on the installation of submarine cables in the offshore wind industry we like to present here only the method of installing a submarine pipe using HDD. The installation of the cables through these pipes being only a later stage with multiple advantages presented in the previous chapter.

Trusting that this technique will be used more and more often and not only when it is the only technical solution due to constraints, I would like to draw the following conclusions:

The simulation analysis shows that the loads on the pipe during installation and later on during operation do not exceed the design values

Through the influencing factor analysis and stress sensitivity analysis, it is concluded that the sensitivity order (from high to low) of five factors to pipeline stress are as follows: diameter -thickness – pressure - drilling fluid density - buried depth

As it has become the case in our days that suspended networks in large cities are moved underground in the near future will have to do the same with all submarine pipes or cables in coastal and shallow water areas. This technique seems to be the key to success

Even though the horizontal directional drilling technique has become very popular nowadays and most of the submarine cable and pipeline installation operations performed so far have been successful, this technique still cannot guarantee control over the risks from underground. For these reasons, in order to improve this technique with the main purpose of increasing the length of these boreholes horizontally, we consider that encouraging research in this field is of major importance

Performance of a risk analysis can help to control the risks.

7 References

1. T. Iseley, S.B. Gokhale, *Synthesis of highway practice 242: trenchless installation of conduits beneath roadways.* Transportation Research Board, National Academy Press, Washington (1997)
2. A. Atalah, J. Kariuki, *Cost Comparison between Horizontal Directional Drilling and Open-Cut Construction Methods in Nairobi, Kenya,* International Conference on Pipelines and Trenchless Technology (2009)
3. S.J. Jariwala, J. Pitroda, J.J. Bhavsar, *Horizontal Directional Drilling: New Era for Underground Utilities,* Proceedings of National Conference CRDCE13, 20-21 December 2013, SVIT, Vasad
4. R. Wisniowski, P. Lopata, G. Orlowicz, *Numerical Methods for Optimization of the Horizontal Directional Drilling (HDD) Well Path Trajectory*, MDPI July (2020)

5. V.F. Vasilescu, D. Dinu, Installation *of Submarine Cables in The Offshore Wind Industry and Their Impact on The Marine Environment,* Journal of Marine Technology and Environment, **1** (2021)
6. ***https://www.aphotomarine.com/geology_sand_erosion_wave_hub_cable_cornwall.html accessed on 15.09.2020
7. H.J. Bayer, *HDD practice handbook,* Vulkan Verlag, Essen (2005)

Technologies used for food preservation using microwaves

Elena Sorică[1*], *Cristian Marian* Sorică[1], *Mario* Cristea[1], and *Iulia Andreea* Grigore[1]

[1]INMA National Institute of Research - Development for Machines and Installations designed to Agriculture and Food Industry, Bucharest, Romania

Abstract. Food preservation is the process of treating food, with the aim of preserving its qualities for as long as possible. Extending the freshness period for processed foods has been and is a continuing challenge for producers in the food industry. New technologies and conservation methodologies are continuously researched, which will have as little effect as possible on the nutritional value of the products. Microwave food processing is constantly evolving, rapid heating and high energy efficiency are the major advantages of using this technology. The paper presents a study regarding the preservation of food products using microwaves, its acting mechanism and other applications of microwaves for food processing, as well as some installations and equipment that use this technology.

1 Introduction

Use The food industry has experienced a constant development in recent years, and has been identified so far different methods and techniques for food preservation [1].

Even though many food preservation methods have been developed over time, most of the current approaches may impact the quality of foods and/or change the regular form and aspect [2].

The preservation of food products, including those from the vegetable - horticultural sector, is performed according to a series of principles and methods that involve compliance with certain technologies depending on the purpose of the method [3].

Among the newest and most modern methods of preservation we mention the athermic and thermal ones which are different from the classical methods.

The most important athermic preservation methods are:

- preservation using high pressures;
- preservation using the magnetic field;
- preservation using ionizing radiation;
- preservation using the pulsating magnetic field;
- preservation using ultra-short light pulses;

* Corresponding author: postelnicu.elena@yahoo.com

- preservation using ultraviolet radiation.

The most important thermal preservation methods are:

- preservation by microwave heating;
- preservation by ohmic heating;
- conservation by heating with radio frequency waves;
- preservation by indirect heating with Joule effect (actiJoule) [4].

Microwave heating is an emerging alternative direct heating method [5].

Microwaves are high frequency electromagnetic radiation, ranging from 300 MHz to 300 GHz and having wavelengths between 1 m and 1 mm, therefore are ranked among radio waves. Their existence was first mathematically assumed by Maxwell, then experimentally verified in 1885 by Hertz [6].

Fast heating and high energy efficiency are the major advantages of using microwaves in food processing [7].

When used for heating microwaves have significant advantages over conventional methods, due to the volumetric manner in which they generate heat which leads to a more rapid heating compared to conventional methods, particularly when it comes to solid foods [8].

Other notable advantages are: space saving, a better process control, selective heating and a better preservation of the nutritional quality of food. There are however some problems with preserving several properties of the food with this technology and the fact that uneven heating usually occurs [7].

In order to overcome these challenges, especially nonuniform heating, various solutions have been proposed including proper selection of packaging materials, using optimization tools to obtain the best processing parameters as well as holding materials in constant motions corresponding to uniform microwave power absorbed in the cavity, thereby preventing cold/hot regions [9].

Microwave heating can be applied to various food processing operations, namely, drying, baking, roasting, blanching, pasteurization, and sterilization [10].

2 Mechanism and applications of microwaves for food preserving

The microwave heating mechanism differs from other types of heating. The differences may or may not be beneficial depending on the process in which the microwaves are used. For example, the fact that microwave heating time is short is to the advantage of food processing, because nutrient losses are minimal. On the other hand, for bakery products, this aspect becomes negative, because this time interval is not enough for the biochemical reactions to take place in optimal conditions.

To understand the mechanism of microwave heating it is needed to understand the dielectric properties of food. The dielectric properties of food can be assessed according to the value of the dielectric constant (ε ') and the loss factor (ε"). The dielectric constant is the real part, and the loss factor is the theoretical part of the complex permittivity ε_r, which is defined by the following relation:

$$\varepsilon_r = \varepsilon' - j\varepsilon'' \qquad (1)$$

The dielectric constant expresses the property of the material to store microwave energy, while the dielectric loss factor is correlated with the ability of the material to dissipate microwave energy in the form of heat. The dielectric properties of foods depend on their composition, temperature and frequency. There are currently data on dielectric constants for many foods such as fruits and vegetables, meat and ham, starch and glucose solutions, various food proteins, and some types of pasta.

However, there is less data on the dielectric properties of frozen foods, as well as for foods found at temperatures above the boiling point of water. If the frozen foods are important for thawing or tempering processes, the temperatures above the boiling point of water are needed for the use of microwaves for pasteurization and sterilization. Such measurements were made for protein gel from whey, macaroni and cheese mixes at temperatures between 20-121^{0}C and at frequencies of 27, 40, 915 and 1800 MHz. The dielectric properties of food can also be used to control the quality of products. Therefore, microwaves can be used to determine the freshness of the fish, the quality of the frying oil, etc. The dielectric properties give a quick indication of the moisture content of agricultural products. Dielectric measurements can be used to distinguish the bound water from free water.

The density of the thermal power generated by a microwave field is given by the following formula:

$$P_1 = 5.56.10^{-12} f \cdot E^2 \cdot \varepsilon'' \tag{2}$$

where P1 is the density of the thermal power (W / m3), f - the frequency of the microwave field (Hz), (generally 2.45 MHz), E - the intensity of the electric field (V / m) and ε "- the dielectric loss factor of food (the imaginary part of the complex dielectric constant).

The dielectric loss factor of food is a measure that characterize the possibility to transform microwave energy into heat. When electromagnetic waves come in contact with the dielectric (with the food product), a part of the energy will be reflected and the other will enter the product where it will gradually attenuate, turning into heat. The depth of microwave penetration (d) into food depends on both the dielectric constant of the food and the loss factor, according to the following formula:

$$d = \frac{0{,}1125 \cdot \lambda_0}{\sqrt{\sqrt{(\varepsilon')^2 + (\varepsilon'')^2} - \varepsilon'}} \quad \text{or} \quad d = \frac{0{,}159 \cdot \lambda_0 \cdot \sqrt{\varepsilon'}}{\varepsilon''} \ \text{if } \varepsilon'' << \varepsilon' \tag{3}$$

where λ_0 is the wavelength of the microwave in a vacuum (12,2 cm la 2,45 GHz). The two formulas show the importance of the dielectric properties ε "and ε 'on the microwave-food interaction. Some relevant values of these properties, found at the frequency 2.45 GHz are presented in table 1.

The loss factor, ε ", increases significantly with product moisture and salt concentration. As can be seen in Table 1, the ice loss factor is very small, which causes a food loss factor to increase significantly over time. Fats and oils do not absorb microwaves well [7].

Table 1. Dielectric properties of foods [7]

Product (25^{0}C)	**ε'**	**ε"**	**Penetration depth**
Beef	61	17	9 mm
Pork	58	16	9 mm
Potatoes	64	14	11 mm
Carrot	72	15	11 mm
Water	78	12.5	13 mm
Ice (0^0 C)	3,2	0,0029	11,7 m

Even if pure ice is transparent to the microwave, other components show a different behaviour, like the fraction of water left unfrozen, that will absorb the microwave, which is actually why is quite easy to defrost food. At the same time, defrost parts, which are richer in liquid water than parts that are still frozen, will tend to absorb more energy, which will lead to a heterogeneous heating. Therefore, it is important not to defrost food at full power, to allow the transfer through conduction, time to achieve and balance temperatures. Dielectric properties are not the only determining factor for heating rate in a microwave oven. At the air-product interface, like other electromagnetic waves, microwaves are partly reflected by the surface of the product. The metal walls of the oven reflect almost entirely the energy of the microwave. Therefore, virtually all microwave power will eventually be absorbed by the product, even if more reflected wave passes are required.

This happens when the mass of the product is not very small (being less than 100 g). For example, half a liter of oil will heat up faster than half a liter of water. When a product with an appreciable mass is in a microwave oven, its power becomes a limiting factor rather than food loss factor. For the example given above, the specific heat of the oil is twice as low as that of the water, so the oil will heat up twice as fast.

Applications of microwaves in the home use (for reheating, cooking, defrosting) are well known. In the food industry the use of microwaves is limited only by the cost of the needed equipment. Microwave applications in the food industry are numerous: dehydration-drying, heating (blanching, pasteurization, baking, frying, defrosting), detoxification, clarification, structural modification, analytical determinations and others. The most common, however, are defrosting and vacuum drying. In general, microwave installations consist of microwave generators and a cavity in which the product is placed. The microwave generator made of a power supply and transformation into high voltage stabilized current, an electronic tube with speed modulator and a cooling system for the transmitter tube. Depending on the power required, a more powerful generator can be used or several lower power generators can be used. Depending on the size and nature of the products (liquids, pastes or solids), the cavity can be different oven-type or tunnel-type, with variable dimensions depending on the destination. Some examples of the use of microwaves will be presented below.

Defrosting meat and fish in the microwave is completely different from conventional defrosting, because, the heating is no longer done from the surface to the inside by conduction process, but takes place through all the product mass due to dielectric losses. Defrosting takes place very well up to -7 to -5^0C, in the thermal center, becoming difficult between -5^0C and 0^0C. The non-uniformity of temperatures during defrosting is due to several causes:

• non-uniformity of the products, the thinner parts tend to defrost faster than the thick ones;

• non-uniformity of composition, fatter portions absorb less energy than the weakest, and salty parts absorb more energy than unsalted ones;

• non-uniformity of the microwave field.

Even with these observations, microwave defrosting is advantageous because it ensures better product quality (no juice is released and no product contamination occurs due to the speed of the process), the defrost surface is 1/10 of the classic one [7].

Novel defrosting methods, such as dielectric heating [(Microwave (MW) and radio frequency (RF)] have been developed and recently used in domestic and industrial applications due to their volumetric heating capability, which has outstanding properties for food-quality preservation [11].

Applications for Microwave pasteurization and sterilization include both in-pack preservation of packed foods (e.g., ready-to-heat meals), that are heated continuously on a conveyor belt, batch-wise sterilization, as well as tubular in-flow sterilization of semi-fluid pumpable foods [12].

Microwave drying is 20 times faster than conventional drying processes. Other advantages are: there are significant savings in terms of constructed area, savings in electricity and steam, savings in the use of washing agents and maintenance. The quality of the products is improved, the appearance being more attractive and the pathogenic microorganisms (staphylococci and salmon) are eliminated more safely. Such installations are used for drying pasta and obtaining "chips". Microwaves can also be used to dehydrate vegetable products (carrots, onions), the dehydration duration being about 20 minutes, to reduce the humidity from 60% to 20% [7].

Drying is an important operation in the industry processes, usually involving heat applied to separate a liquid from a solid, leading to the evaporation of the liquid phase (water or another solvent). Drying is an energy-intensive activity and dryers are expensive pieces of equipment, therefore drying must be carried out as economically as possible. Utilizing microwaves, that uses a fundamentally different method of energy transfer, the drying process can be more advantageous compared to the conventional drying, due to the selective nature of interaction of the microwaves with the product to be dried [13].

Numerous researches (Table 2) on microwave-assisted air drying (MWAD) have been found in the literature on fruits and vegetables. Nistor et al. (2017) investigated the effect of several drying techniques like free convection (50°C, 60°C, 70°C) and combination of free convection, forced convection (40°C) along with microwave power (315 W) on physical and chemical properties of red beetroot. They reported that combined drying methods were more effective than the convective drying (CD) methods. The increase in drying rate and a decrease in drying time was obtained by increasing the drying temperature followed by microwave heating. The free convection (60°C), followed by forced convection (40°C) combined with microwave power (315 W) was the best drying method for beetroot [14].

Table 2. Numerous Researches Carried Out on Microwave-Assisted Air Drying [14]

Product	Temperature (°C) / Air Velocity (m/s)	Power Rate (W/g) / Power (W)
Apple	50°C, 60°C, 70°C / 2 m/s	0.1–1 W/g
	50°C, 60°C, 70°C / 2 m/s	-
	25°C, 30°C, 40°C, 50°C / 1 m/s	3, 5, 7, 10 W/g
	30°C, 50°C / 2.5 m/s	0.5 W/g
	40°C, 60°C, 80°C / 0.5, 1 m/s	0.5 W/g
Potato	50°C, 60°C, 70°C / 2 m/s	-
	18°C, 65°C / 0.032 m/s	5, 10 W/g
	45°C / 1.5 m/s	0.6 W/g
Carrot	45°C / 1.5 m/s	0.6 W/g
	45°C, 60°C / 1.7 m/s	120, 240 W
Kiwifruit	60°C / 1.29 m/s	210 W
Olive	100°C, 160°C, 225°C / 1 m/s	350, 490, 700 W
Grape	50°C	-
Mushroom	40°C, 60°C, 80°C / 0.5, 1 m/s	0.5 W/g
Orange	60°C / 2 m/s	0.17, 0.36, 0.69, 0.88 W/g
Asparagus	50°C, 60°C, 70°C	2, 4 W/g
Banana	60°C / 1.45 m/s	350, 490, 700 W
Pumpkin	1 m/s	160, 350 W

Vacuum dehydration. Microwaves can also be used for vacuum dehydration of the following products: soups, tomato extracts, vegetable proteins, meat and liver extracts, instant drinks, coffee extracts, concentrated fruit and vegetable juices, malt and hop extracts, beer must, (basic materials for the production of non - alcoholic beer), chocolate mass, vitamin extracts, intermediate materials for vitamins, enzymes and antibiotics, cultures of microorganisms.

Desensitization / disinsection / disinfection of agri-food products and reduction of their number of microorganisms. At a treatment of 9-10 seconds, molds (including spores) such as *Fussarium graminearum*, *Aspergillus flavus*, *Penicillium patulum*, *Penicillium cyclopium*, *Tricothecium roseum*, *Rhizopus negricans* and others can be destroyed. Research has shown that the anti-nutritional factors in some agri-food products, and especially those of a protein nature, can be inactivated by microwaves, as the food is heated through all their mass. Thermostable aflatoxins in peanut mass can be inactivated after 240 seconds of microwave treatment. Microwaves can also be used for lysis of microorganisms along with other processes such as the use of enzymes, chemicals, ultrasound, mechanical shocks, pressure or thermal shocks. The application of microwaves for 150 seconds leads to the thermolysis of various microbial cells, such as *Trichoderma album*, *Geotrihum candium*, *Saccharomyces cerevisiae*, favoring the release of cellular constituents, which is very important in obtaining plasmolysates, autolyses and protein hydrolysates.

Homogenization and stabilization of food emulsions. The analytical determinations where microwaves can be used are: the rapid determination of the dry matter in dairy products, molasses and preserving cans and the determination of the moisture content of granular and powdery products.

Microwave heating, as well as high (radio) frequency heating, is happening due to the transformation into heat by dielectric hysteresis of a part of the energy contained in the electromagnetic wave.

The microwave frequency range is between 3x108-3x1010 Hz and overlaps the telecommunications bands. In order to avoid jamming phenomena, certain frequency bands have been allocated for industrial, scientific and medical applications, also called ISM (Industrial, Medical and Scientific) frequencies. For Europe these frequencies are: 2450 MHz (the most used), 5800 MHz and 22125 MHz (most commonly used in the US). In the microwave field the wavelengths are smaller than the dimensions of the installations (for example, for 2450 MHz, λ=12,2cm) and respects fundamental laws of propagation of electromagnetic waves. In the field of radio frequencies, the wavelengths are larger than the installation dimensions, and the laws of electromagnetism can still be applied in a quasi-stationary regime.

When heating dielectrics at high frequency or in "hyperfrequency", the depth of penetration is defined as the depth for which the power density is reduced for "e" times the value from the surface of the material that is under heating.

In radio frequency dielectric heating, the penetration depth is greater than 1 m and can reach several tens of meters, the heating being considered homogeneous throughout the material. In contrast, in the case of microwaves, the depth of penetration is much lower (for 2450 MHz, Δ=12mm for water with a temperature of 12°C and 1200mm for ice) and that is why it becomes an important factor [7]. Figure 1 presents the penetration depth.

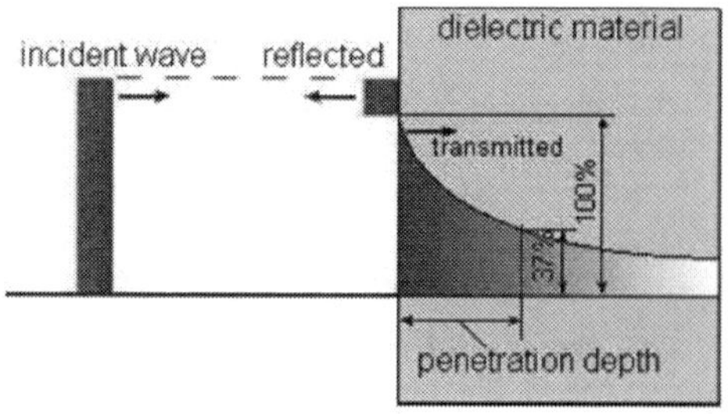

Fig. 1. Depth of penetration [9]

Microwaves are freely radiated into space by an electrode called an antenna and, in order for them to interact as efficiently as possible with the heating material, they must be guided and concentrated. A microwave heater will need the same three main components: the microwave generator, the waveguide, and the applicator [7].

3 Technological equipments using microwaves for food preserving

There are many companies that produce technical equipment for microwave treatment. In this regard, *Mitech S.R.L. (Italy)* is a manufacturer of ovens, equipment and industrial systems using microwave technology.

Microwave technology ensures a drastic decrease in treatment time in the following applications: BAKING - HEATING - DEFROST - DRYING - DEHYDRATION - PASTEURIZATION OF PACKED AND UNPACKED PRODUCTS. The product range includes tunnel-type ovens, with continuous operation and batch furnace systems, with discontinuous operation, used in the industrial field, special ovens for laboratories, vacuum chambers and pressure autoclaves for scientific research.

Continuous tunnel ovens for the food industry are used to dry sweets and vegetable products through microwaves and hot air. Power may vary from 30 to 300 kW.

In this regard, figure 2 presents a system used for drying sweets and vegetable products.

Fig. 2. System used for drying sweets and vegetable products [15]

Pasteurization of packed and unpacked products can be performed using microwaves and cold air. The power varies from 10 to 40 kW. Figure 3 presents the system for pasteurization of packaged and unpackaged products.

Fig. 3. The system for pasteurizing packed and unpacked products [15]

The company also produces detachable microwave chambers, made to facilitate transport and treatment at the place of using. Their power varies from 3 to 12 kW. Figure 4 presents the mobile treatment system.

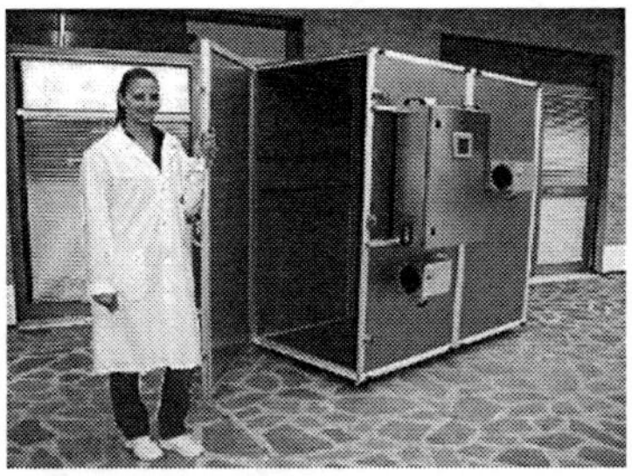

Fig. 4. Mobile treatment system [15]

For laboratory research use, special microwave ovens are sold, having the temperature controlled by an infrared thermometer. The power varies from 1 to 6 kW. Figure 5 presents the laboratory research system.

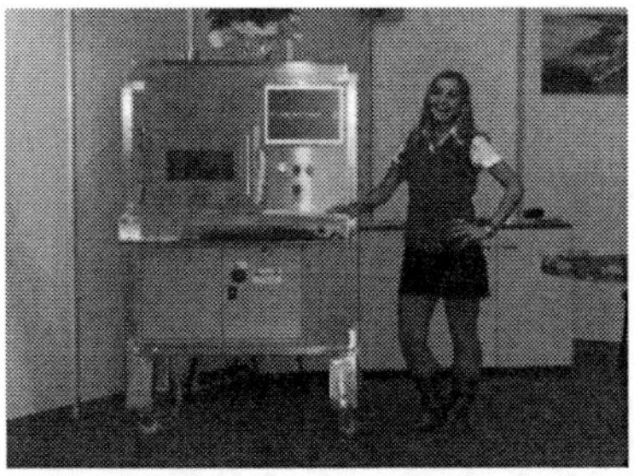

Fig. 5. Laboratory research system [15]

Emitech S.R.L. (Italy), sells patented technology named MISYA Agrofood, for disinfection and / or microwave drying treatments of food such as cereals, legumes, dried fruits.

The installation MISYA AH-36-T (fig. 6) operates continuously, using electromagnetic energy, at specific microwave frequencies, to treat legumes and cereals in the post-harvest phase.

This system is designed especially for continuous disinfestation of sensitive seeds. It consists of a shielded reverberation chamber, containing a slightly inclined tube in which the seeds are irradiated. The rotating tube, which allows the seeds to pass easily from inlet to outlet, also promotes constant mixing by reducing the risk of seed breakage and damage.

The device has a maximum production capacity of 800 kg / hour with a power of 36 kW. All process parameters (temperature, power, speed) are constantly monitored using a PLC-based electronic control and management system.

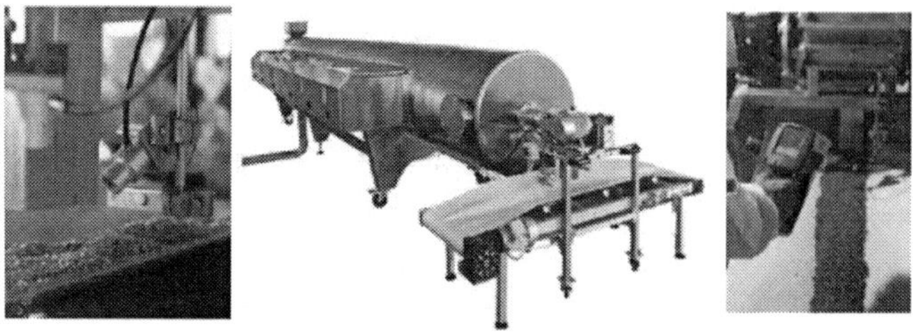

Fig. 6. The system MISYA AH-36-T [16]

Kerone (India) is a company that sells microwave installations, which can be used for blanching, pasteurization-sterilization, defrosting and drying. Figure 7 presents the microwave treatment system for various applications.

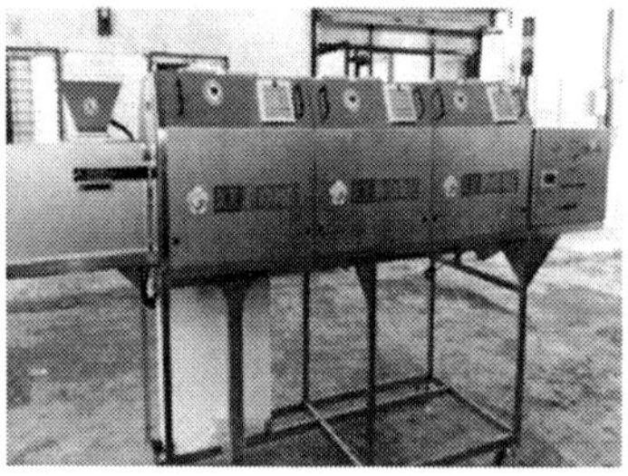

Fig. 7. Microwave treatment system for various applications [17]

4 Conclusions

The application of microwaves energy to enhance conventional or emerging food processing techniques is a viable alternative with high process efficiency and minimal changes to food quality attributes, when compared to traditional methods or usage of these technologies individually. These benefits can open the path for producing highly desirable food products for consumers and for increasing the adoption of microwave technology at commercial scale. The advantages of microwave assisted food processes are numerous.

Fast heating and high energy efficiency are the major advantages of using microwaves in food processing. Other advantages are: space saving, more efficient process control, selective heating and preservation of the nutritional quality of food.

Microwave applications in the food industry are numerous: dehydration-drying, heating (blanching, pasteurization, baking, frying, thawing), detoxification, clarification, structural modification, analytical determinations and others. The most common, however, are defrosting and vacuum drying.

Microwave defrosting is highly advantageous because ensures a better quality of the products (no juice is released and there is no contamination of the product due to the speed of the process), and the defrosting surface represents 1/10 of the normal surface.

Microwave drying is 20 times faster than other conventional drying processes.

This work was supported by a grant of the Ministry of Agriculture and Rural Development on the Sectoral Plan for Research and Development in the field of Agriculture and Rural Development – ADER 2022, contract no. ADER 7.5.1.

References

1. B. Ioan, Preservation of agri-food products, Course support "Gheorghe Asachi" Tehnical University of Iași
2. F. Iordache, I. Gheorghe, V. Lazar, C. Curutiu, L. M. Ditu, A. M. Grumezescu, A., M. Holban, *Food Preservation*, Chapter: 9 Nanostructurated materials for prolonged and safe food preservation, A volume in Nanotechnology in the Agri-Food Industry, Pages 305-335, (2017)
3. E. Roşculete, *Principles and methods for food preservation*, UNIVERSITARIA Craiova Publishing House, Course, (2020)
4. C. Lungu, *General principles of food preservation*, Course, (2002)
5. B. G. Theresa, K. Fabian, O. Jörg - H*inrichs Innovative Food Science & Emerging Technologies*, New experimental set-up for testing microwave technology to continuously heat fouling-sensitive food products like milk concentrates, **65**, (2020)
6. L. Gîtin, *Special food processing technologies* - "Dunarea de Jos" University of Galati, Course, (2008)
7. Project ADER 7.5.1., *Research on the application of emerging technologies based on minimal processing methods in the industrialization of horticultural products*, Sectoral Plan of the Romanian Ministry of Agriculture and Rural Development (2020)
8. T. F. Bedane, G. Quinn, J. G. Lyng - *Innovative Food Processing Technologies*, Chapter: 1.47 Microwave and Radio Frequency Processing of Animal-Related Food Products, Pages 743-766 (2021)
9. Flora-G. C. Ekezie et all, *Trends in Food Science & Technology*, Microwave-assisted food processing technologies for enhancing product quality and process efficiency: A review of recent developments, (2017)
10. G. J. Swamya, K. Muthukumarappanb, *Innovative Food Processing Technologies*, Chapter: 1.46 Microwave and Radiofrequency Processing of Plant-Related Food Products, Pages 731-742, (2021)
11. Y. Lave, N. Sakai, *Food Processing for Increased Quality and Consumption*, Chapter 13 - Dielectric Defrosting of Frozen Foods, Pages 383-422, (2018)
12. B. W. Raaholt, *Food Science*, Microwave Pasteurization and Sterilization, (2020)
13. M. Radoiu, *Chemical Engineering and Processing - Process Intensification*, Microwave drying process scale-up, **155**, (2020)
14. D. K. Verma, N. K. Mahanti, M. Thakur, S. K. Chakraborty and P. P. Srivastav - *Microwave Heating: Alternative Thermal Process Technology for Food Application*, Chapter 2, 2020
15. http://www.mitech-srl.eu/public/it/index.html
16. https://emitech.it/en/misya-agrofood/
17. http://kerone.com/

Aspects regarding the use of ultrasounds in the food industry

Cristian Marian Sorică[1*], *Marian* Vintilă[2], *Laurentiu* Vlădutoiu[1], *Elena* Sorică[1], *Iulia Andreea* Grigore[1] and *Mario Cristea*[1]

[1]National Institute of Research – Development for Machines and Installations Designed to Agriculture and Food Industry – INMA, Ion Ionescu de la Brad no. 6 Blvd., sector 1, Bucharest, Romania
[2]The Institute of Research and Development for Industrialization and Marketing of Horticultural Products - ICDIMPH Horting, Drumul Gilăului no. 5N, sector 4, Bucharest, Romania

Abstract. Ultrasound is a rapidly developping research field that is more and more used in the food domain, both for analysis and processing of food products. This technology has been successfully applied within food industry for various processes like freezing, filtration, drying, separation, emulsion, sterilization and extraction. The paper presents aspects regarding the mechanism and use of ultrasound in food processing, as well as some technical equipment that use ultrasound within a combined technology, for the decontamination of the external surfaces of some horticultural products.

1 Introduction

Minimal processing of foods is defined as a trend to replace classical heat treatment processes with new, athermal processes, including new, milder heat processes.

The food industry is experiencing a strong development and has at its disposal a number of methods and techniques for the food processing [1].

The most important methods of athermal processing are:
- processing using high pressures;
- processing with supercritical fluids;
- processing using ionizing radiations;
- processing using high voltage pulsating electric field;
- processing with ultra-short light pulses;
- processing using UV radiation;
- *ultrasound processing;*
- processing using magnetic field.

The use of athermal processing technologies has increased due to the growing request for high quality foods that still contain the natural taste and flavor, without chemical additives and preservatives. Between the varied athermal processing technologies, ultrasound has a special potential. Whether used alone or combined with other processing

*Corresponding author: cri_sor2002@yahoo.com

methods, ultrasound processing technology obtained considerable positive results on the quality of food products, being considered effective. The influence of ultrasound on the food processes is considered to be related by cavitation phenomenon and mass transfer increasing. This technology has been successfully applied within food industry for various processes like freezing, filtration, drying, separation, emulsion, sterilization and extraction. Different research studies have suggested that using ultrasound increases the efficiency of the process and improves the quality of the food [2].

Nowadays, athermal processing technology, like ultrasound, has begun to be considered as an emerging technology for removing food contaminants [3].

The ultrasound waves have a frequency ranging between 20 kHz and 100 MHz, propagating through any medium, as a mechanical wave. When a high acoustic force is applied, it leads to the formation, development and implosion of cavitation bubbles in a liquid medium, enhancing various food processes [4, 5].The Ultrasound could be considered an essential technology within the effort to achieve the goal of sustainable "green" chemistry and extraction. Within the chemical and food industry domain, the ultrasound is acquainted for considerably affecting the processing rate of various processes used [6, 7]. Ultrasound-Assisted Extraction (UAE) works based on the concept of acoustic cavitation that deteriorates the plant matrix cell walls and therefore supports the process of releasing the bioactive compounds [8].Low energy usage (low power or low intensity) implies the application of ultrasound having frequencies above 100 kHz at intensities smaller than 1 W/cm^{-2}. Applied ultrasound waves, having low intensities, utilizes a low level of power that do not induce chemical or physical modifications within the material it crosses, meaning that it is basically non-destructive.

Ultrasound is also used successfully for non-invasive monitoring of food processing. The most common applications of low frequency ultrasound within the food industry is an analytical technique that provides usefull data regarding the physico-chemical properties of food like composition, structure and physical condition. Ultrasound has the advantage over other classical analytical techniques in that the measurements are fast, non-destructive, accurate, automated and can be performed both, within the laboratory or on the flow line.

Ultrasound waves having high intensities has been used for quite some time to generate emulsions, destroy cells walls and disperse aggregate materials. Other applications have been discovered to have great opportunity for further development, like alteration and conduct of crystallization, liquid food degassing, enzyme inactivation, advanced drying and filtration and induction of oxidation reactions. The benefit of using sound wave energy is achieved through different effects that ultrasound causes within the environment in which it is transmitted.

The physical, chemical and mechanical effects of ultrasound at this level are able to change the properties of the material by generating a huge pressure and a high temperature gradient in the environment in which it propagates. Through the sonication process, the longitudinal waves produced when a sonic wave encounters a liquid environment, create compression regions followed by expansion regions. These pressure regions cause cavities and gas bubbles to appear. During the expansion period, the gas bubbles have a much larger surface area, that enhances the diffusion of the gas, determining the bubbles to expand. It arrives to a point where the supplied ultrasonic energy becomes insufficient to maintain the gaseous phase within the bubble; that is why fast condensation takes place. Condensed molecules collide brutally, causing shock waves, which in turn create regions of raised temperature and pressure, attaining up to 5500 °C and 50 MPa.Cavitations can occure from microemission, being able to increase heat and mass transfer. The capacity of ultrasound to create cavitations is influenced by the properties of the ultrasound wave (frequency, intensity), the characteristics of the product (surface tension, viscosity) and the environment conditions (temperature, pressure). In order to obtain cavitation, the ultrasound intensity needed raises appreciably over 100 kHz.

Currently, more and more uses of ultrasound within various industrial food applications are reported, such as:

✓non-destructive assessment of inner quality and latent defects for whole fruits and vegetables;

✓improving the monitoring of food production processes by accelerating diffusion;

✓measuring the texture, viscosity and concentration of numerous solid or liquid food products;

✓determining the composition of eggs, meat, fruits, dairy products etc.;

✓measuring the consistency, flow and temperature for monitoring and controlling several processes;

✓improving processes such as: surface cleaning, drying and filtration;

✓inactivation of microorganisms and enzymes;

✓degassing of liquids;

✓acceleration of heat transfer and extractive processes.

From these examples, it is obvious that ultrasound technology has several applications within the food industry and their use will intensify in the future.

2 Mechanism and applications of ultrasound in food processing

There are many arguments for using ultrasound in food processing. Nowadays, the food industry is becoming progressively aware of the significance of developing new analytical techniques to investigate complex food materials and to evaluate the characteristics of food during processing; ultrasonic techniques are suitable for these two applications.Ultrasonic instruments can be entirely automated and perform fast and accurate determinations. Ultrasound is not destructive or invasive, it can be adapted for online utilizations and used to analyze optically opaque mediums.

Among various food technologies it can be found many examples of processing where ultrasound can be used. Until lately, ultrasound was used within food technologies mainly for non-invasive analyzes, especially referring to quality evaluation. Applications like these use principles similarly to those developed for medical diagnostics, using low-power and high-frequency ultrasound.The application of these techniques can be exemplified in locating foreign bodies within food, analyzing the size of the drop in emulsions of comestible fats and oils and in determining the degree of crystallization of the drops dispersed within the emulsion.

The relation among the characteristics of food that can be measured ultrasonically (propagation speed, attenuation coefficient and acoustic impedance) and physico-chemical properties (composition, structure and physical condition) is the foundation of ultrasonic analysis. This relation can be determined both, empirically by plotting a calibration curve regarding the characteristic we are interested in measuring ultrasonically, or theoretically using mathematical relations that describe the propagation of ultrasound waves through materials. By measuring the attenuation of an ultrasonic pulse, it has been shown that it can be determined the degree of homogenization for milk fat. The determination of ultrasonic propagation speed in connection with attenuation could be used to appreciate the degree of emulsification in materials alike. Also, it can be determined parameters like the degree of creaminess of a specimen, the movement of solid particles / fats on the surface. Information like that provides details, for instance, on the long-term stability of fruit juices and the stability of emulsions like mayonnaise. The conjunction of measurement of propagation speed and attenuation promises to be a method of assessment of comestible fats and oil and to determine the degree of crystallization and melting of droplets dispersed within emulsions.

Lately, food technicians have focused their attention to the use of ultrasound within food processing. First mention can be found in 1927 when, in an article called "The chemical effects of high frequency sound waves", it was stated that the physical, chemical and mechanical impact of ultrasonic waves at this level are able to change the properties of matter (destroy the physical integrity, speeding special chemical responses) by generating a huge pressure and a temperature gradient in the environment in which it propagates.

The applications of high power ultrasound generally depend on complex effects of inducing vibration within the propagation medium, which produces cavitations in biological fluids or tissues. Besides to cavitations, ultrasound is capable of weakening the physical integrity of the material or environment, providing environmental dimensions that are alike those of the ultrasound waves used. One of the applications of ultrasonic power, of major importance used in industry, is cleaning and has proven to be a highly effective technology. Surface cleaning can be applied to a large number of disciplines and utilizations (sensors, filters, substrates, reactors, catalysts and heat exchangers).

Ultrasound is very efficient for in-situ cleaning in addition with chemical treatment and provides the following advantages:

- reduces the consumption of chemicals;
- reduces the time of direct contact of the worker with dangerous chemicals;
- high cleaning speed;
- the ultrasonic activity takes place at the micro level in nature and attains all the sections for a uniform cleaning;
- automated operations and savings in energy expenditure, labor and work area [9].

The potential uses of ultrasound within the food industry are varied. One of the most recent uses of ultrasound within food processing has been in emulsification. Emulsions obtained by using ultrasound are frequently much more stable than the products obtained conventionally. Research studies determined that using ultrasound as an aid in processing can reduce the production time of yogurt by up to 40%. Moreover, sonication reduces the normal dependence of the process on milk source and also improves the consistency and texture of the product. It has been shown also that fish eggs exposed to ultrasound having a frequency of 1MHz for 35 minutes, three times a day, reduced the incubation time of the Grindel (fish species) from 72 to 60 hours. Also, various reports within the literature imply that treating the seeds before sowing, with ultrasonic, improves crop yield.

One of the incipient uses of ultrasound in biochemistry was to destroy cell walls to release the contents. It was later proven that the ultrasound can activate immobilized enzymes by rising the transport of substrate to enzymes. Regarding enzymes, ultrasound it can also be effective as inhibition method. Chambers revealed that using sonication, the pure pepsin was inactivated, likely as a result of cavitation. The incipient activity of peroxidase, liable for the increasing of brown flavors and pigments, was gradually reduced by 90% asa a result of applying the ultrasound over three hours.

The use of ultrasound notably increases the extraction of organic compounds included within the body or seeds of the plant. The mechanical action of ultrasound ensures a better penetration of the solvent into cellular materials and improves mass transfer. Supplementary benefits outcome from the destruction of cell walls to ease the release of the contents. Ultrasound treatment increases mass transfer as a result of the microstreaming effect, representing a highly efficient method for extracting sugar. Sonication accelerates the diffusion of sugar, obtaining superior levels of dry matter content and sugar content within the juice. In some cases the use of ultrasound has increased the extraction efficiency at low temperatures obtaining a much purer compound in a reduced amount of time.

Due to the use ultrasound, the extraction of tea compounds from the leaves has increased by 20%. The authors observed that most of the compounds were extracted in the first 10 minutes. Zayas reported that a large amount of rennin enzyme was obtained from the calf's stomach by using ultrasound. Moreover, the ultrasound extraction activity has been shown to be slightly increased compared to normal technologies. Ultrasound was also used to filtration. Following the use of ultrasound, the moisture content of the sludge was quickly decreased from 50% to 25%, while usual filtration reached a value of 40% [9]. Acoustic filtration is used effectively to enhance wastewater filtration from industries, process that is usually considered as a gruesome one [10]. While the optimization of ultrasound velocity within filters is extremely important to avoid its damage [7, 11], ultrasound technology used in conjunction with filters increases filter life through prevention of membrane caking and clogging, allowing continuing cavitation around filter surface [12].

As an example of other application of ultrasound, with highly commercial potential, we can present acoustic drying. Advanced ultrasonic drying can be conducted at low temperatures compared to common methodology, decreasing the possibility of oxidation and deterioration of the product. By using ultrasound the heat transfer between a heated solid surface and a liquid is enhanced with about 30-60% [9]. Ultrasound used within the drying process accelerates the drying rates for meat, fish, fruits and vegetables. This way, it reduces the drying time, increases the heat and mass transfer rate and preserves the properties of the product [13]. In vacuum drying [14] and freeze drying [15], ultrasound increases the drying ratio by comparison to the convective drying used within conventional methods.The power of ultrasound has proven effective in freezing food, the benefits being quite large. Also, ultrasound is used within the process of freezing fresh food, leading both to the rising of the freezing rate and to the increasing of the frozen products quality. It is also applied to the ice cream manufacturing process by reducing the size of the crystals, preventing surface fouling. Other applications include improving the processes of flavor extraction, filtration, mixing and homogenization. New applications like oxidation of unsaturated oils, aging of alcoholic beverages, hydration of acetylene, decalcification of bones and hydrolysis of esters are under development. Table 1 presents applications of ultrasound within the food industry.

Table 1. Applications of ultrasound within the food industry[9]

Applications	Common methodes	Ultrasound fundamental	Benefits	Produce
Cooking	Oven; steam bath	Uniform heat transfer	Short processing time; improved sensory qualities and heat transfer	Meat; vegetables;
Freezing / crystallization	Freezer; immersion freezer; by contact	Uniform heat transfer	Shorter processing time; smaller crystals, improved diffusion; rapid drop in temperature;	Meat; vegetables;fruits, dairy products;
Drying	Hot gas stream atomization; spray freezing	Uniform heat transfer	Short processing time; improved sensory qualities and heat transfer	Dehydrated products (fruits, vegetables)
Filtration	Filters	Vibration	Shorter processing time; better filtering	Liquids (juices)
Destruction of foam	Thermal treatment; chemical; electric; mechanic.	Cavitation phenomenon	Shorter processing time; improved hygiene	Carbonated beverages; fermented products
Emulsification	Mechanical treatment	Cavitation phenomenon	Shorter processing time	Emulsions

 https://doi.org/10.1051/e3sconf/202128604009

3Technical equipment using ultrasound for food processing

For the decontamination of the external surfaces of some horticultural products (within the process of washing fruits and vegetables), there are various companies that produce installations based on a combined technology, both ultrasound and ozone.

SonoSteam (Denmark) manufactures installations for decontamination of the outer surfaces of berries (Figure 1). In this way an efficient reduction of microorganisms on the surface of the products is achieved, as well as of pathogens: Campylobacter, Salmonella, E. coli, Listeria and other bacteria. Supplementary, SonoSteam is efficient against yeast and fungi.

SonoSteam for berry decontamination is a continuous treatment system with adjustable capacity to be correlated with most production lines. The installationhas an integrated PLC system for complete process control and monitoring during treatment.

SonoSteam treatment is completely safe and complies with the regulations for the use of organic products, as it does not change the chemical composition and does not leave residues on the products.

Fig. 1. SonoSteam system [16]

Also, the company produces the MultiMini installation, having multiple possibilities of use (Figure 2).

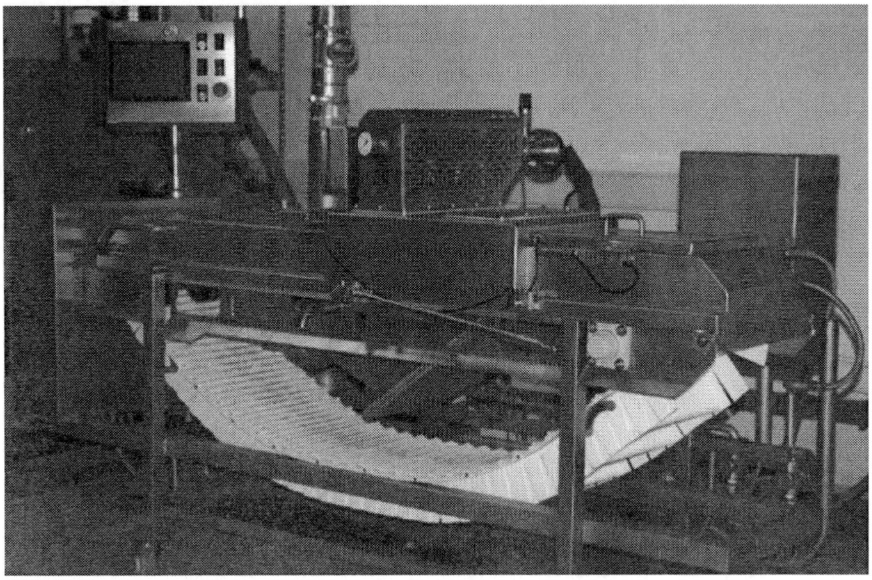

Fig. 2. MultiMinisystem [16]

The treatment zone is insulated to prevent unreasonable heat formation and to prevent steam consumption. Three nozzles are placed above the conveyor belt and one nozzle is placed below. Every nozzle has a flow rate of approx. 6 grams of steam per second. The great majority of food types and non-food products are processed in only one second, which is frequently enough to achieve a high microbial reduction.

The company R K Transonic (India), produces a range of installations (RKT 1000, RKT 1200 FoodSonic Pro, FoodSonic 600) that use both ozone and ultrasound in the process of washing fruits and vegetables (Figure 3, 4 and 5). Technical characteristics of these installations are presented in Table 2, 3 and 4.

Among the main features of these installations are:

- deep, easy and efficient cleaning: the ultrasonic washing machine uses ultrasound to eliminate 98.9% of the existing agricultural chemicals from the surface of fruits and vegetables;
- eliminates odor: the container material is endowed with negative ions, that aid to remove disagreeable odors, chlorine and other chemicals;
- silent operation / facile installation: it works using a quiet vibration, without splashing.

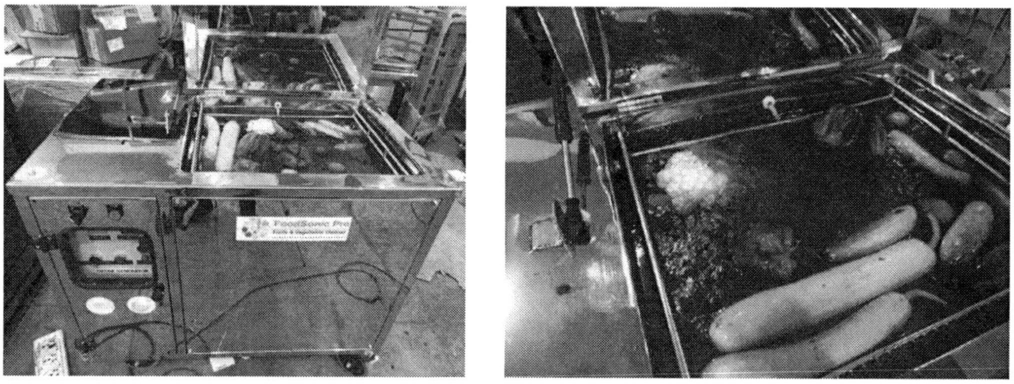

Fig. 3.RKT 1000system[17]

Table 2. Characteristics[17]

Characteristic	Value
Frequency	30 Hz
Model	RKT 1000
Brand	R K Transonic
Container dimensions	24 x 16 x 16 inch
Useful volume	98 liters
Load capacity	15 kg Vegetables/Fruits
Container material	Stainless steel AISI 316
Average power	1000 W
Maximum power	2000 W

Fig. 4.RKT 1200 FoodSonic Pro system [17]

Table 3. Characteristics[17]

Characteristic	Value
Model	RKT 1200 FoodSonic Pro
Container material	Stainless steelAISI 316
Container dimensions	24 x 16 x 16 inch
Useful volume	100 liters
Hourly productivity	100-150 Kgs
Load capacity	10-15 Kg Vegetables/Fruits
Average power	1200 W
Maximum power	2400 W

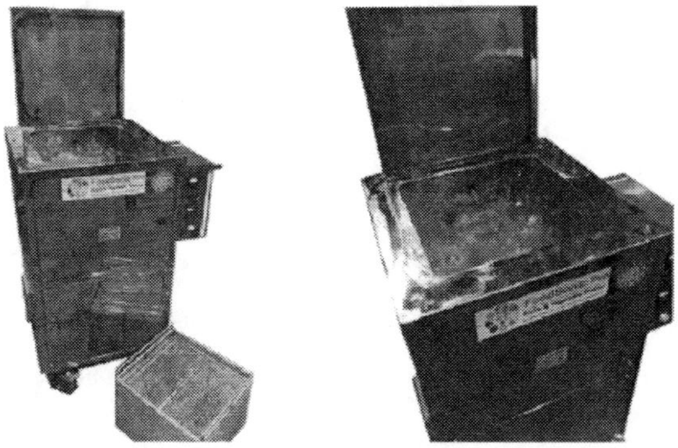

Fig. 5. FoodSonic 600system [17]

Table 4. Characteristics[17]

Characteristic	Value
Model	FoodSonic 600
Use	Vegetable cleaning
Dimensions	15 x 15 x 12 inch
Brand	R K Transonic
Technology	Ozone + Ultrasound
Cleaning capacity	Up to 70 Kg per hour

4 Conclusions

Minimal processing of foods is defined as a trend to replace classical heat treatment processes with new, athermal processes, including new, milder heat processes.Between the varied athermal processing technologies, ultrasound has a special potential.

The potential uses of ultrasound within the food industry are quite varied, like freezing, filtration, drying, separation, emulsion, sterilization and extraction. Certain advantages of using ultrasound technology within these processes could be summarized as follows:

- when applied to freezing process, it leads both to improve the freezing rate and to enhance the quality of frozen products;
- when applied to sludge filtration, the moisture content of the sludge was quickly decreased from 50% to 25%, while usual filtration reached a value of 40%;
- when ultrasound is applied within drying process, the heat transfer between a heated solid surface and a liquid is enhanced with about 30-60%;
- when is used to obtain emulsions, the production time is reduced by 40% and the emulsions are often much more stable than conventionally obtained products;
- when used in combination with ozone, for deep cleaning of fruits and vegetables, it removes almost 99% of the existing agricultural chemicals from the surface of these products;
- when used within extraction process, it has been shown that the extraction activity is slightly increased compared to normal technologies, sometimes with even 20%.

Also, ultrasound is used successfully in analytical determinations having the advantage, compared to other traditional techniques, that the measurements are fast, non-destructive, accurate, automated and can be performed both, within the laboratory or on the flow line.

This work was supported by a grant of the Ministry of Agriculture and Rural Development on the Sectoral Plan for Research and Development in the field of Agriculture and Rural Development – ADER 2022, contract no. ADER 7.5.1.

References

1. I. Băisan, *Preservation of agri-food products*, Gheorghe Asachi Technical University of Iași, Faculty of Mechanics, Course, (2018);
2. M. Singla, N. Sit, *Application of ultrasound in combination with other technologies in food processing: A review*, Ultrasonics Sonochemistry,**73**, (2021);
3. S. Yuana, C. Li, Y. Zhang, H. Yu, Y. X., Y. Guo, W. Yao, *Trends in Food Science & Technology,* Ultrasound as an emerging technology for the elimination of chemical contaminants in food, **109**, pp. 374-385, (2021);
4. Y. Tao, D.W. Sun, *Enhancement of food processes by ultrasound: a review,* Crit. Rev. Food Sci. Nutr., **55 (4)**, pp. 570-594, (2015);
5. Z. Zhu, Z. Chen, Q. Zhou, D.W. Sun, H. Chen, Y. Zhao, H. Pan, *Freezing efficiency and quality attributes as affected by voids in plant tissues during ultrasound-assisted immersion freezing,* Food Bioprocess Technol., **11** (9), pp. 1615-1626, (2018);
6. M. Dastkhoon, M. Ghaedi, A. Asfaram, R. Jannesar, F. Sadeghfar, *Magnetic based nanocomposite sorbent combination with ultrasound assisted for solid-phase microextraction of Azure II in water samples prior to its determination spectrophotometric,* J. Colloid Interface Sci.,**513**, pp. 240-250, (2018);
7. N. Bhargava, R.S. Mor, K. Kumar, V.S. Sharanagat, *Advances in application of ultrasound in food processing: a review*, Ultrasonics Sonochemistry, (2020);

8. B.K. Tiwari, *Ultrasound: A clean, green extraction technology*, TrAC, Trends Anal. Chem., **71**, pp. 100-109, (2015);
9. Project ADER 7.5.1, *Research on the application of emerging technologies based on minimal processing methods in the industrialization of horticultural products*, Sectoral Plan of the Romanian Ministry of Agriculture and Rural Development (2020);
10. H.M. Kyllönen, P. Pirkonen, M. Nyström, *Membrane filtration enhanced by ultrasound: a review*, Desalination, **181** (1–3), pp. 319-335, (2005);
11. X. Wen, P. Sui, X. Huang, *Exerting ultrasound to control the membrane fouling in filtration of anaerobic activated sludge—mechanism and membrane damage*, Water Sci. Technol., **57** (5), pp. 773-779, (2008);
12. M.T. Grossner, J.M. Belovich, D.L. Feke, *Transport analysis and model for the performance of an ultrasonically enhanced filtration process,* Chem. Eng. Sci., **60** (12), pp. 3233-3238, (2005);
13. S.J. Kowalski, D. Mierzwa, M. Stasiak, *Ultrasound-assisted convective drying of apples at different process conditions,* Drying Technol., **35** (8), pp. 939-947, (2017);
14. Z.H. Tekin, M. Başlar, S. Karasu, M. Kilicli, *Dehydration of green beans using ultrasound-assisted vacuum drying as a novel technique: drying kinetics and quality parameters,* J. Food Process. Preserv., **41** (6) (2017), p. e13227, (2017);
15. X.F. Cheng, M. Zhang, B. Adhikari, *Effect of ultrasonically induced nucleation on the drying kinetics and physical properties of freeze-dried strawberry,* Drying Technol., **32** (15), pp. 1857-1864, (2014);
16. https://sonosteam.com/
17. https://www.ultrasonicindia.com/

E3S Web of Conferences **286**, 04010 (2021) https://doi.org/10.1051/e3sconf/202128604010
TE-RE-RD 2021

Pneumatic Actuator Controlled by Proportional Valve. Experimental results

Valentin Nicolae Cococi[1], *Constantin* Călinoiu[1], and *Carmen-Anca* Safta[1*]

[1]University Politehnica of Bucharest, Power Engineering Faculty, 313 Splaiul Independentei, 060042, Bucharest, Romania

Abstract. In nowadays the pneumatic controlled systems are widely used in industrial applications where valves must be operated, where there is a fire ignition risk, or in different automation systems where a positioning action is desired. The paper presents the experimental results of a pneumatic actuator controlled by a proportional control valve. The goal of the paper is to compare the experimental results with the numerical simulation results and to improve the mathematical model associated with the experiment.

1 Introduction

Electropneumatic control systems have the advantage that can be miniaturized and embedded in themselves electronic elements which allow them to be used in complex control systems [1]. So, new structures of pneumatic drives combined with electric components open the possibility for the use of programmable logic controller (PLC). This is an important advantage to use the electropneumatic systems in advanced control strategies which are characteristic to "Industry 4.0" philosophy [2].

This new perspective of electropneumatic devices used in so many industrial applications, medical systems, automotive structures, justify the theoretical and experimental approach of their study.

The aim of this work is to extend the theoretical studies of the pneumatic actuator controlled by proportional valve, in open loop and control loop, with experimental tests. In our previous studies [3-5] we considered the dynamic behaviour of a pneumatic actuator, controlled by a proportional valve, when different types of input signals where applied. Also, the electropneumatic servomechanism was loaded with variable technological force which can appear during the technological process or in an unexpected moment of time [4, 5].

In this paper, we present the results of the experimental study which is made on a pneumatic actuator controlled by a 5/3 ways proportional valve. It was experimented if the actuator can follow the input signal applied with different frequencies and amplitudes to command the proportional valve. A static characteristic of the servo pneumatic system was plotted.

The experimental results were reconstructed by numerical simulation on a numerical model that have constructive parameters corresponding to the devices used in the

* Corresponding author: safta.carmenanca@gmail.com

experimental stand. Are compared the result obtained in the experiments with the ones obtained in numerical simulations, and the conclusions are obtained to improve the experiments program.

The paper is organized in five sections. After a short introduction in which was underlined the aim of the paper, in the second section is presented the numerical simulation model of the pneumatic actuator and proportional valve that are included in the experimental test rig. Numerical simulations results are presented. In the third section is described the experimental bench and the test program. The experimental results and comments of them are the subject of the fourth section. In section five conclusions for further work are mentioned.

2 Numerical simulation model and numerical results

In the description of the numerical simulation model, we used the experience gained in the previous work [3-5] where it was considered a pneumatic actuator with the proportional pneumatic valve (PPV) in closed loop. The technical characteristics of the pneumatic actuator and PPV are the same with those used in the experimental setup.

It was used a proportional controller with proportional coefficient gain K_p =1, and an unloaded pneumatic actuator (the external force is zero).

In Figure 1 is given the numerical simulation scheme.

The pressure of the compressed air supplied in the system is 7 barA. The pneumatic actuator is a double-acting pneumatic cylinder with the length stroke of 450 mm and the piston active area 491,6 mm^2. The mathematical model which describes the actuator behaviour consider the viscous friction coefficient of 75 Ns/m and thermal exchange coefficient of 50 J/m2/K/s with an external temperature of 293.15 K.

The PPV has the area of control orifices of 28,27 mm^2 and the flow control coefficient of 0.72. The valve limit frequency (-3 dB) is 100 Hz at 10 bar maximum pressure value and nominal voltage value at pneumatic mid-position of 5 V DC.

The liniar transducer used in the control scheme has a zero offset and the power supply of 10 V DC with an effective electrical working distance of 457 mm.

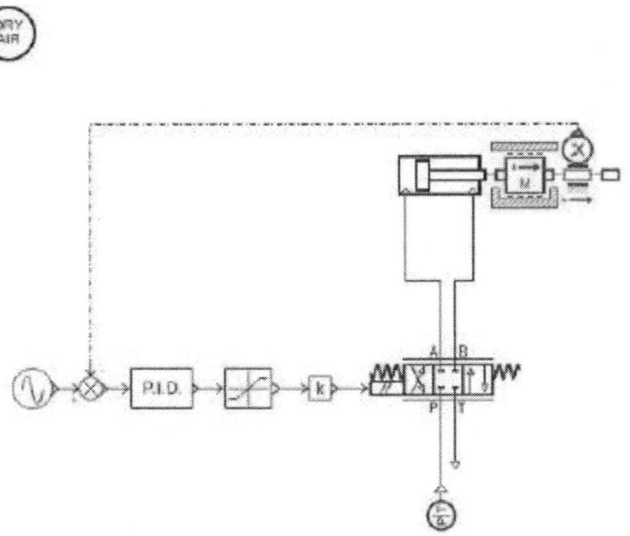

Fig.1. Numerical simulation model

The reference input signal is plotted in Figure 2. The numerical results obtained for the input signal applied to the system are plotted in Figure 3.

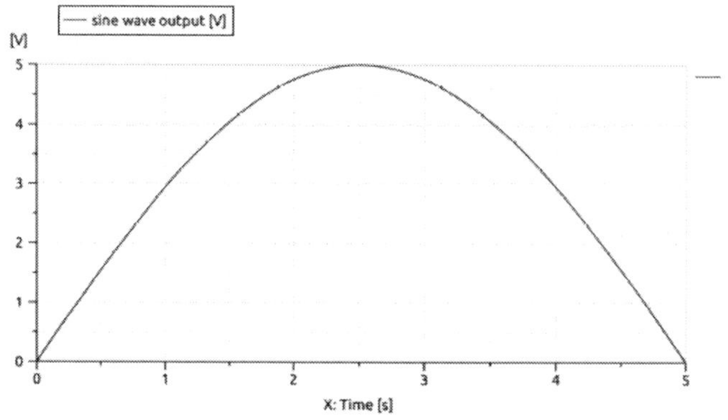

Fig 2. Time variation of the reference (input signal)

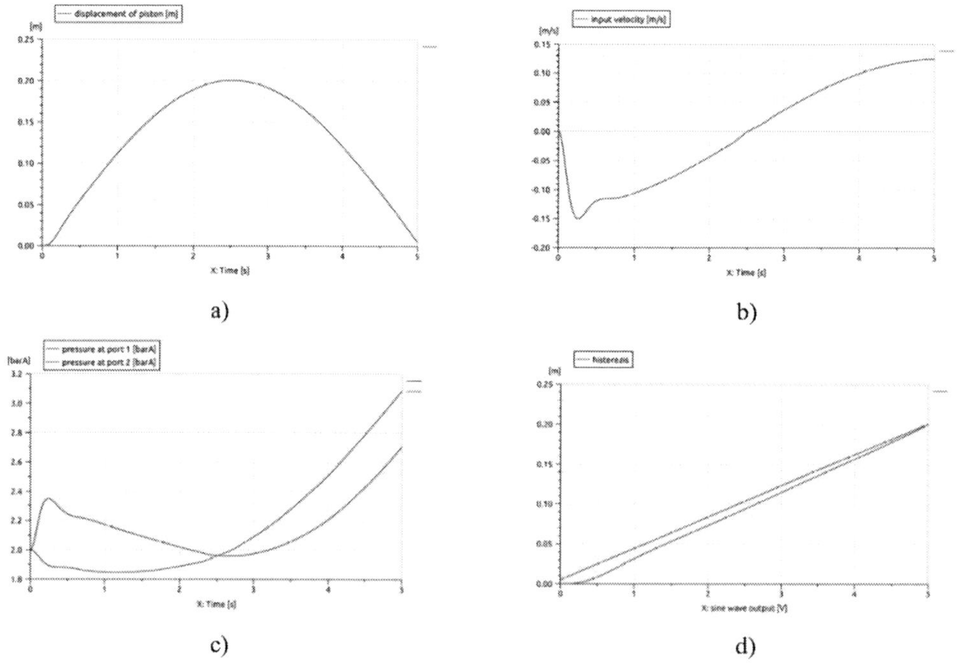

Fig.3. Electropneumatic servomechanism dynamic and static behaviour: a) displacement of the piston; b) velocity of the piston; c) pressures in cylinder; d) static characteristic curve

It is observed that for an amplitude of 5 V of the input signal the maximum stroke obtained by the pneumatic actuator is 200 mm in 2.5 seconds.

In figure 3.b, the velocity pattern follows the displacement evolution and is observed the absence of external force. In the pressure distribution plot, "port 1" of the cylinder corresponds to the left chamber of the cylinder, Figure 1.

The hysteresis curve, Figure 3.d, is obtained by plotting the output signal versus input, the piston displacement. Even hysteresis curve underlines the absence of static friction within the system, caused by no external force.

3 Experimental setup

For experiments we considered a pneumatic servo mechanism having a pneumatic actuator and a proportional pneumatic valve with the characteristics mentioned above. The experimental arrangement is show in Figure 4.a and the bloc diagram of the control system is presented in Figure 4.b.

The pneumatic servomechanism is supplied with air from the air compression unit (noted 1 in Figure 4.a). The proportional valve (noted 4) is controlled by the PID controller (noted 5) electrical supplied by 24 VDC power supply (noted 6). Also, the proportional valve will control the flow to the pneumatic actuator (noted 2) of which displacement is follow up by the displacement transducer (noted 3). The electrical signal from the transducer is compared with the reference input from the controller.

The Arduino board is used as a data acquisition system. The Arduino board is connected to the computer's USB port using a USB A-B cable, available in 1.5 meters or 3 meters. The Software used to read the data from the board is Arduino Integrated Development Environment (IDE) [6, 7].

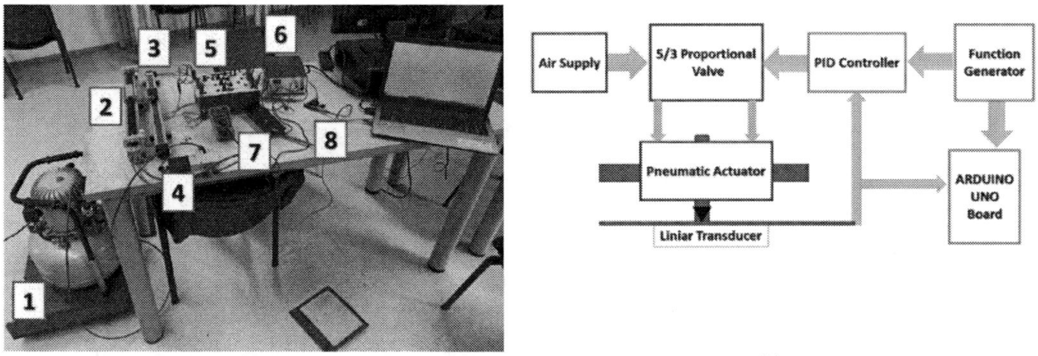

a) b)

Fig. 4. Experimental test stand: a) Position control components: 1-air supply, 2-pneumatic actuator, 3-linear transducer, 4-proportional valve 5/3 ways, 5-PID controller, 6-24V power supply, 7-multimeter, 8-Arduino UNO board; b) Block diagram

4 Experimental results

In the experiments were use input signals with different amplitudes and frequencies, as in Figure 5. The ability of the system to track the input signal was observed. The resolution of the data acquisition system is 10 bits, and the conversion range is 0… 5V.

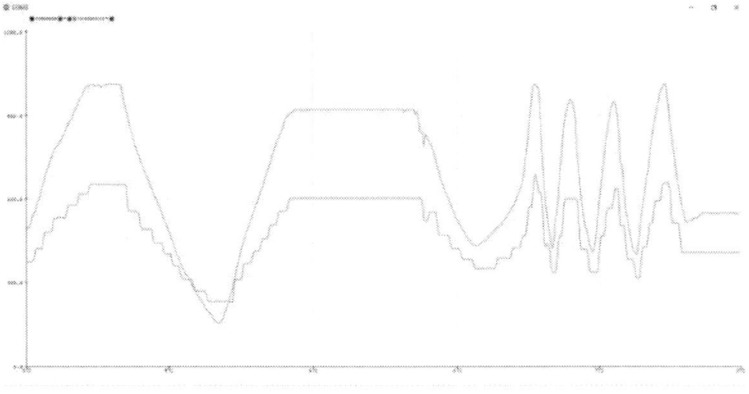

Fig. 5. Signals in bits: green line input signal, violet line-output signal

In Figure 6 is plotted the experimental response of the servomechanism to the input signal. It is observed that the output signal follows the input, but it was found that the area of insensitivity/resolution of the drive system has values in the range of 100 ÷ 250 mV, depending on the position of the piston, with values higher towards the end of the stroke.

The relatively poor resolution (10 mm) of the drive system could be given by friction in the cylinder or by the performance of the servo controller. Another explanation of this result is given by the fact that the signal generator did not allow a control voltage of around 5 V which is the voltage corresponding to the neutral zone of the servomechanism. The experiments were performed with control signals below the value of 5 V and hence the difficulties in the experimental determinations.

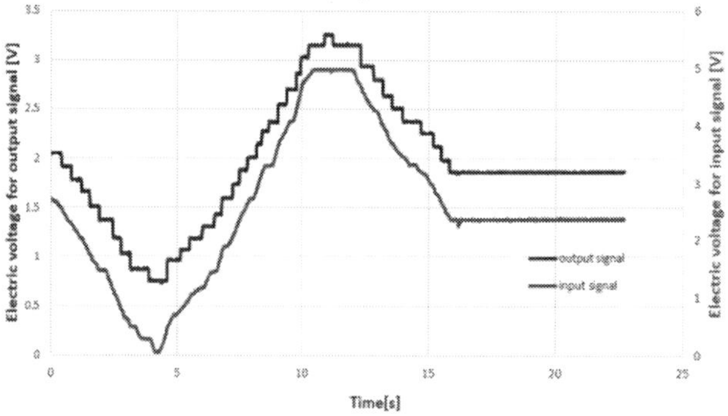

a)

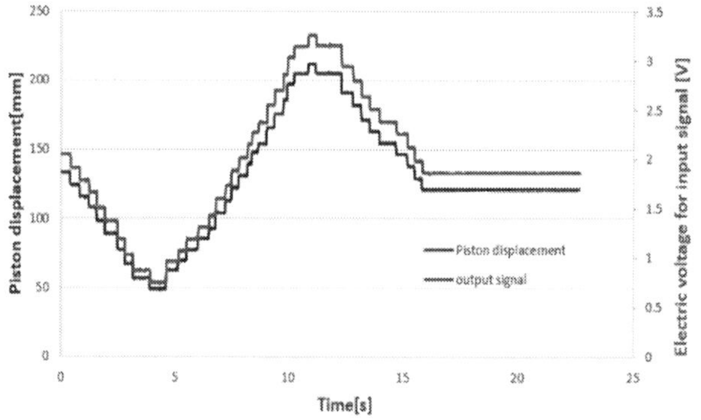

b)

Fig. 6. The variation of input signal and output signal: a) in electric voltage; b) in mm for the output signal

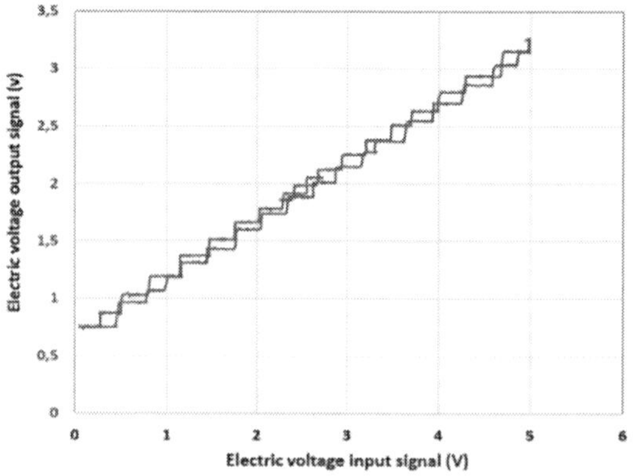

Fig. 7. Hysteresis curve

 https://doi.org/10.1051/e3sconf/202128604010

In Figure 7 is plotted the static characteristic (or hysteresis curve) of the control system. It is observed that he static characteristic of the drive system is quasi-linear.

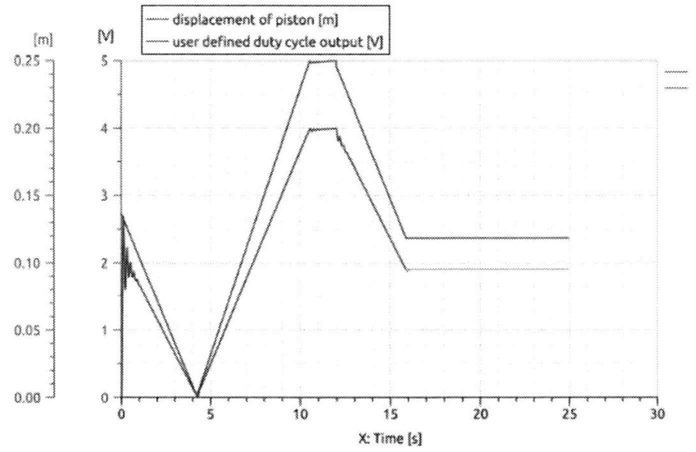

a)

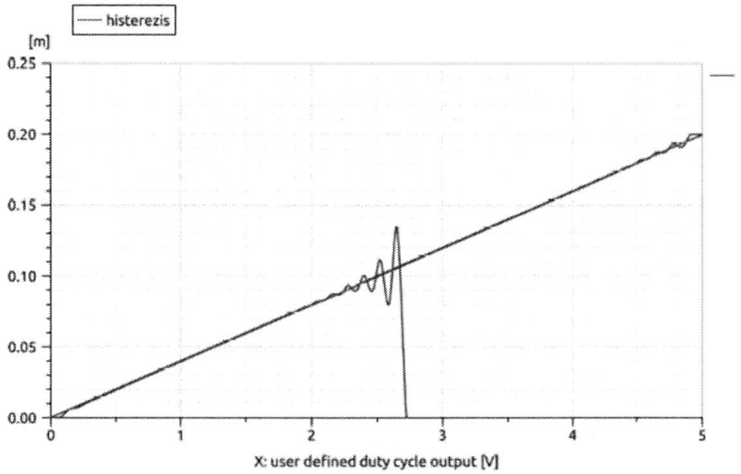

b)

Fig. 8. Simulation results for an input signal similar with the one in the experiments: a) evolution of the piston displacement; b) hysteresis curve

Finally, having the numerical model and the experimental results, we numerically build an input signal with an amplitude of 4.97 V as in Figure 6a which was applied in the numerical model. The response of the system is given in Figure 8. The numerical simulation results are comparable with the experiments (see Figure 6 and Figure 8.a). Even the quasi-linear characteristic of hysteresis curve is approximatively replicated. The drop observed in Figure 8b is given by the sudden openings of the spool valve. How speed is a derivative of the displacement, on short period of time the values are big.

5 Conclusions

Based on the results of previous studies on the behavior of pneumatic positioning servomechanisms using numerical simulation models, experiments have been performed in this paper.

Thus, the positioning system consisting of pneumatic actuator, proportional distributor, displacement transducer and controller was considered. For different input signals, the actuator's ability to track the input signal was observed. The hysteresis characteristic of the system was also found.

The obtained results showed that the positioning servomechanism can follow the input signal, but the resolution of the system is poor. This deficiency of the system may be caused by static friction, or by the characteristics of the controller.

Therefore, a more rigorous testing program is proposed to identify the causes of poor system resolution. It is also necessary to use a more efficient signal generator and to improve the test procedure regarding the dynamic behavior of pneumatic servomechanism.

We would like to thank the company FESTO Didactic Romania for lent us a part of the pneumatic equipment necessary to carry out the experiments.

References

1. W. Kobayashi, N. Kato, S. Dohta, T. Akagi, Position Control of Flexible Pneumatic Cylinder Using Tiny Embedded Controller with Disturbance Observer, Int. J. of Mechanical Engineering and Robotics Research Vol. **6**, No. 4, 318-321 (2017)
2. K Foit, W Banaś, G Ćwikła, IOP Conf. Ser.: Mater. Sci. Eng. **400** 022024 (2018)
3. V.N. Cococi, C.-A. Safta, C. Calinoiu, Dynamic behaviour of pneumatic actuators in open-loop controlled by proportional valves, E3S Web of Conferences, 2020, **180**, 04012
4. V.N. Cococi, C.-A. Safta, C. Calinoiu, Numerical simulation approach of a pneumatic actuator to force perturbation, EPE 2020 - Proceedings of the 2020 11th International Conference and Exposition on Electrical And Power Engineering, 2020, pp. 589–593, (2020) 9305558
5. V.N. Cococi, C.-A. Safta, C. Calinoiu, Parameter tuning process for a closed-loop pneumatic actuator, The 7th Conference of the Sustainable Solutions for Energy and Environment (EENVIRO 2020), (to be published)
6. F. Mohamed, A. A. Dahoud, Integrated Development Environment "IDE" For Arduino (2018) https://www.researchgate.net/publication/328615543_Integrated_Development_Environment_IDE_For_Arduino
7. https://docs.rs-online.com/406c/0900766b8133324b.pdf

Electronic system for control of temperature of exhaust gases and pressure in turbochargers of diesel automobile engines

Seher Kadirova[1], *Stiliyan* Okishelov[1], and *Zhivko* Kolev[2,*]

[1]Department of Electronics, University of Ruse, Ruse 7017, Bulgaria
[2]Department of Heat, Hydraulics and Environmental Engineering, University of Ruse, Ruse 7017, Bulgaria

Abstract. The paper presents design and experimental investigation of an electronic system for control of the temperature of exhaust gases and the turbocharging air pressure in turbochargers of diesel automobile engines. The existing problems are faults in the fuel system of an engine. The indicators are changes in the values of the temperature and pressure in exact areas of the turbocharger. The presented device is a controller that monitors precisely the temperature and pressure, which are so vital for the long operation of the automobile. The control system is based on Arduino microcontroller. OLED Display has been added to visualize the obtained results. A schematic diagram of an electronic module for control of the temperature of exhaust gases and turbocharging air pressure in turbochargers of diesel automobile engines has been synthesized. The system has been investigated in laboratory conditions and practically implemented in a real automobile. As a result of laboratory experimental investigation, results were obtained for the time-monitored parameters temperature of the exhaust gases and turbocharging air pressure in the turbocharger system of a diesel automobile engine.

1 Introduction

There are various studies on the turbocharged diesel internal combustion engines. In [1] a procedure for selection and matching of turbochargers for diesel engines has been developed. Paper [2] presents investigation on the energy distribution in turbocharger system of a diesel engine. Different techniques used in turbocharging processes of diesel engines, in order to improve the engine efficiency and exhaust emissions, have been considered in paper [3]. In study [4] an evaluation method for the transient response performance of turbocharged diesel internal combustion engines has been proposed and analysed. Investigation [5] presents a method for evaluation of the adaptability to different altitudes of turbocharging diesel engine systems. In [6] an experimental study on the performance and economy parameters of a diesel internal combustion engine with variable nozzle turbocharger is accomplished. Paper [7] presents an investigation of the exhaust

* Corresponding author: zkolev@uni-ruse.bg

gases' pressure on the turbocharger performance. Mathematical model for evaluation the coupling of an inverted Brayton cycle – based recovery unit with a turbocharged diesel engine, has been considered in [8]. In [9] a performance analysis of a turbocharged two-stroke diesel engine, operating at various ambient conditions has been presented. The research [10] is focuses on determination of some heat transfer parameters of processes in the intake system of a piston engine with a turbocharger system. Numerical investigation of combustion processes in a turbocharged diesel engine has been considered in [11].

Paper [12] presents a nonlinear controller for the air-path system of a turbocharged diesel engine. One of the controlled parameters is the air pressure in the turbocharger outlet. Measure of pressures and temperatures in selected sections of intake and exhaust circuits of turbocharged diesel engine has been considered in [13]. The pressures have been measured by strain gauge transducers. For measure of the temperatures platinum resistance thermometers and K-type thermocouples have been used. In [14] the heat transfer process in the air duct between turbocharger and intercooler has been studied. One of the measured parameters used for determination of the heat flux, is the pressure of the hot air after the turbocharger. Study [15] presents a procedure of exhaust gas temperature measurements by thermocouples in turbocharged marine internal combustion engines. A linear parameter varying polytopic observer of the exhaust gases' pressure in turbocharged automotive diesel engine has been considered in [16]. In [17] the influence of the air pressure and air temperature after the turbocharger on the pressure in the cylinders of turbocharged a four-stroke diesel engine has been discussed. Investigation of the influence of some operation parameters of a diesel engine on the exhaust gas temperature has been considered in [18]. In [19] an experimental study for calculation the heat quantity by electronic system has been presented. The system reliability also has to be considered. In [20] a case study and some prescriptions have been suggested to increase the operational reliability. The relation between the system of transformation of matter, energy and information, the technical system and the executing bodies of the technical system also has to be considered [21] for having a steady model development.

The paper presents design and experimental investigation of an electronic system for monitoring of the temperature of exhaust gases and the turbocharging air pressure in turbochargers of diesel automobile engines. The presented device is a controller that monitors precisely the temperature and pressure, which are so vital for the long operation of the automobile.

These two factors determine the service life of the turbocharger in every internal combustion engine. With the spread of more stringent measures related to internal combustion engines, the temperature requirements to the components of each internal combustion engine become very high. In this way, when a vehicle is equipped with such a monitoring system, unwanted damage will be avoided and timely alert will be given before it occurs. The monitoring system is an easy to operate and reliable system because it is not equipped with complex components and there is a visual connection with the driver.

2 Methodology

Some of the most important parameters that need to be monitored in the operation of modern diesel engines are the temperature of the exhaust gases in the inlet and the air pressure in the outlet of the turbocharger. The cylinder pressure in the internal combustion engines depends on the fuel-burning rate during the premixed burning phase. The increasing of the cylinder pressure leads to good combustion and increasing of the rate of heat release.

The exhaust gas temperature in the internal combustion engine, measured in characteristic control cross-sections of its thermal and flow system, can be a valuable source of diagnostic information about technical condition of elements limiting particular engine workspaces, including the supercharging system, the fuel supply system, and the working medium exchange system. In standard measuring systems installed on vehicle engines equipped with pulse turbocharging system, the exhaust gas temperature is measured at the outlets of particular cylinders, and at the turbocharger turbine inlet and outlet [15].

The proposed electronic device is implemented on the basis of ATMEGA 2560. It contains all the necessary elements for normal operation, expect of the power supply module and the interface for communication between the microcontroller and the device, through which the program code is set. The easiest implementation of a project containing microcontroller control and data visualization is platform such as Arduino. The platform was used in this project due to the wide range of peripherals and their low price. The problem with this platform is the limited memory of the controller and therefore the optimal option is Arduino MEGA 2560. Blue Oled display SH1106 128X64 - pixels is added as a periphery which is operating on the basis of I2C communication and allowing the use of only 2 pins. The picture is shown in Fig. 1.

Fig. 1. OLED Display - SH1106.

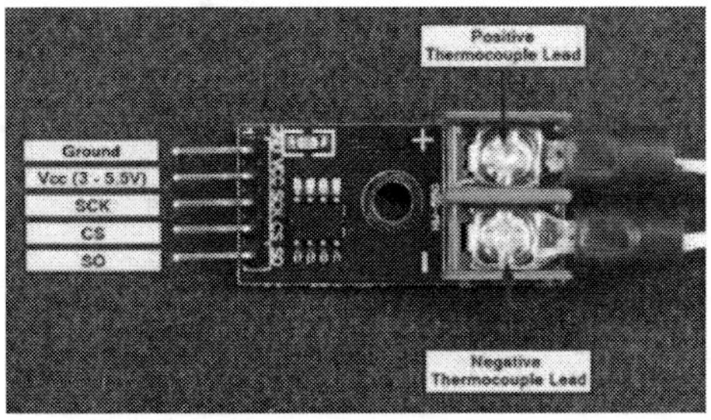

Fig. 2. Thermocouples MAX6675K.

The MAX6675K Thermocouples amplifier each having digital serial data output and supplied along with K-type thermocouple is connected ATM2560. It is selected and complies with the standard used in most of vehicles using BOSCH systems. The graphic is presented in Fig. 2.

The pressure sensor MPX4250AP (Fig. 3), operates with analog signal, in the range -2.5 ~ 2.5 bar. In this case this is a voltage from 0 to 5 V, which is then converted to pressure, by the algorithm in the program.

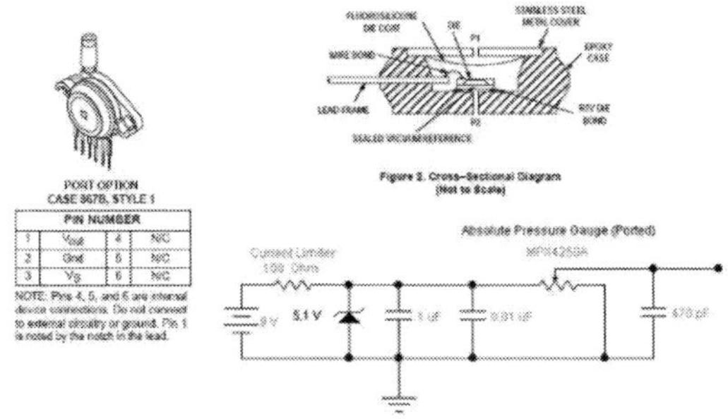

Fig. 3. MPX4250AP Sensor.

Fig. 4. Atmosphere pressure sensor BMP085.

The role of the Adafruit BMP085 Barometric Sensor is to make the necessary correction in the MPX4250AP readings, as the atmospheric pressure changes depending on the altitude, and the weather conditions.

3 Results and discussion

Fig. 5 presents the diagram of a conventional system with the additionally placed sensors. In position 11 is placed the air pressure sensor, and in 13 is the EGT sensor, respectively.

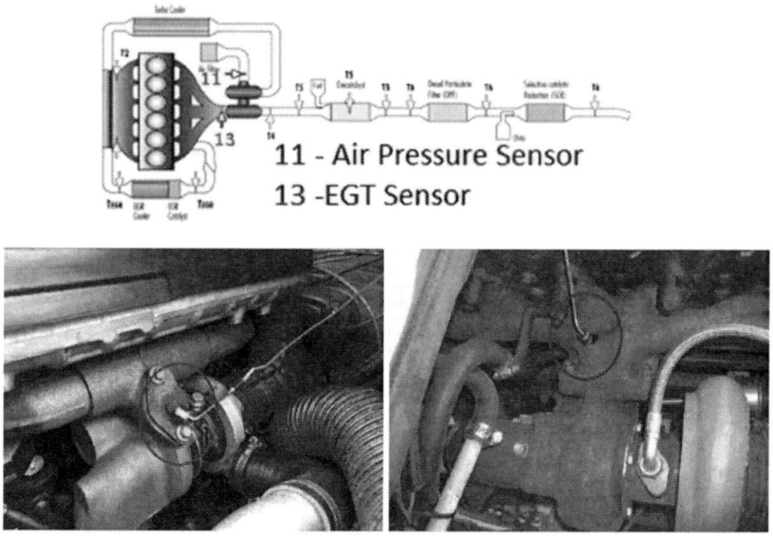

Fig. 5. Diagram of the system with the additionally placed sensors and experimental pictures.

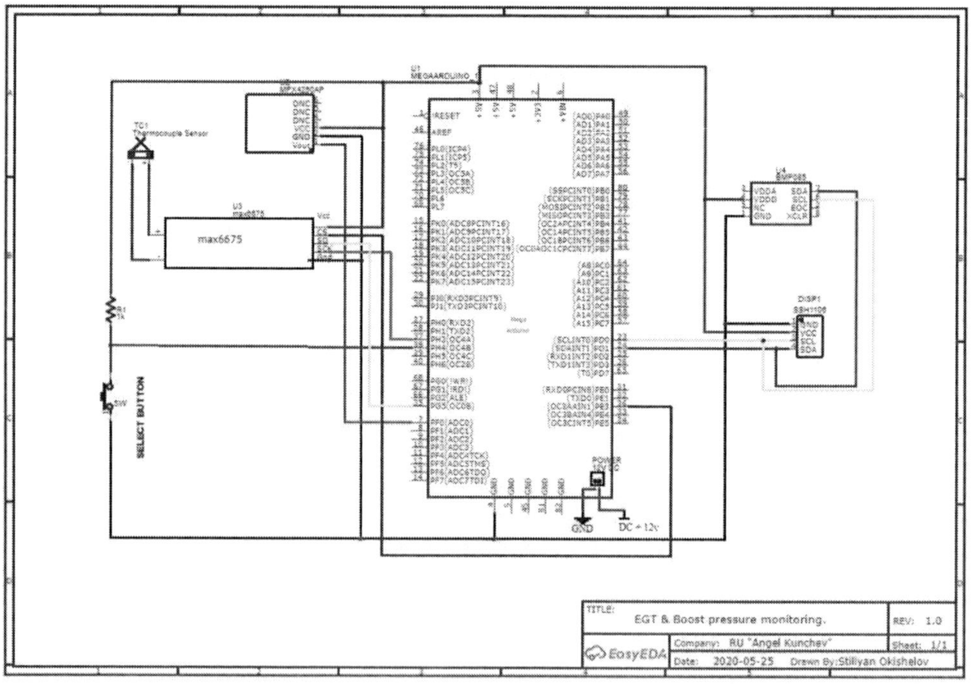

Fig. 6. Electrical schematic diagram of the system.

Fig. 6 presents the electrical schematic diagram of the system. The system has been investigated in laboratory conditions and practically implemented in a real automobile.

The button can be optionally selected, depend on the place where it will be installed to ensure easy access. Its role is to change the measurement mode. When the button is pressed once, it changes the mode of the measuring program. The modes in this case are two "EGT Temperature measurement", and "TCP Turbocharger pressure leaching". This mode switch is circular, by pressing the button again the modes switch between both.

The temperature measured by the sensor, which is in analogic form, is given to the inputs of the digital converter MAX6675. The signal is converted into a digital 12bit SPI signal and transmitted to the microcontroller, which in our case is mounted on the Arduino MEGA2560 platform. The signal is processed by the controller, based on the program compiled in it. After processing the information, the controller sends a signal via the I^2C bus to the display, which displays the read temperature from the thermo-complement pair.

The program has a function to store the minimum and maximum values read by the sensor.

Fig. 7 presents pictures of the experimental test board and results for the temperature and the pressure, achieved during the experiments.

Fig. 7. Test board pictures and results.

The button can be used to switch to another monitored parameter. In our case, this is the pressure of the turbocharger. This pressure is read on the basis of a measured voltage at one of the microcontroller's analogic inputs, converted to a constant "0-255" and then sent to a subprogram in the main program to be calculated for converting the value in bars. In addition to the circuit, the module BMP085 is used, which appears as a correction of the value obtained from the pressure sensor. Then the current value is displayed.

By using an oscilloscope, we examine analogue signals sent to the microprocessor from the real pressure sensor. It is given a voltage signal in the range 0 ~ 5 V. The range of the sensor is (-0.5 ~ 1.5 bar). This means that it is also possible to measure vacuum. In Fig. 8 a) a voltage of 0.6 mV is measures, which after processing by the algorithm in the program, is equal to -0.5 bar. In Fig.8 b) the measured voltage is 1.88 V which corresponds to an absolute 0 bar. This is because as mentioned above the sensor range is -0.5 ~ 1.5 bar and the operating voltage is 0 ~ 5 V. This makes it possible to measure a vacuum. In Fig. 8 c) the measured voltage is 2.88 V which according to the algorithm corresponds to 0.5 bar. In Fig. 8 d) the measured voltage is 3.88 V, corresponding to 1.0 bar. In Fig. 8 e) a voltage of 4.88 V corresponding to the maximum pressure that the sensor can measure (1.5 bar is monitored). Although the sensor operates in the range of 0 ~ 5 V, the entire range is never used to avoid unwanted and unrealistic values, even damage from overloading the components.

Fig. 8 presents analogic signals 0 ~ 5 V for Pressure sensor, taken by an oscilloscope.

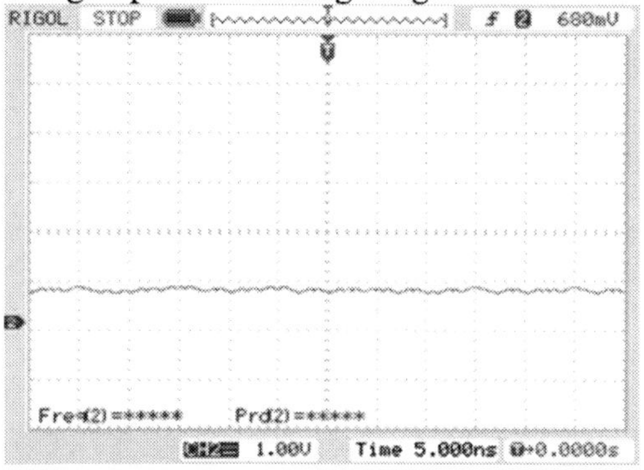

a) Vacuum 0.5 bar

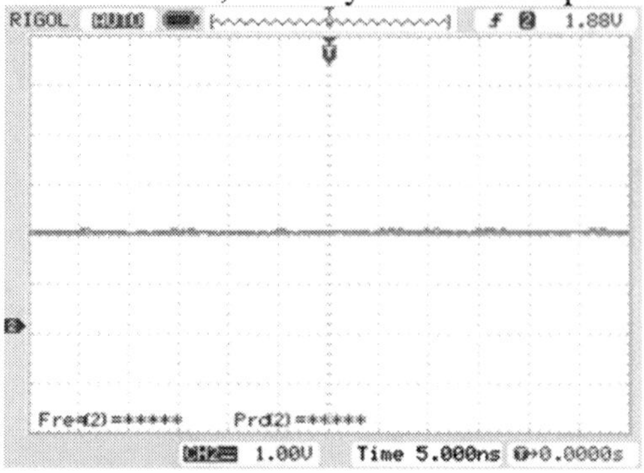

b) Pressure 0.0 bar

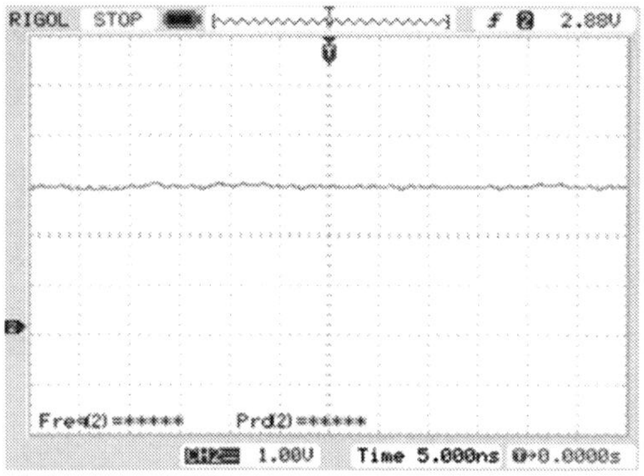

c) Pressure 0.5 bar

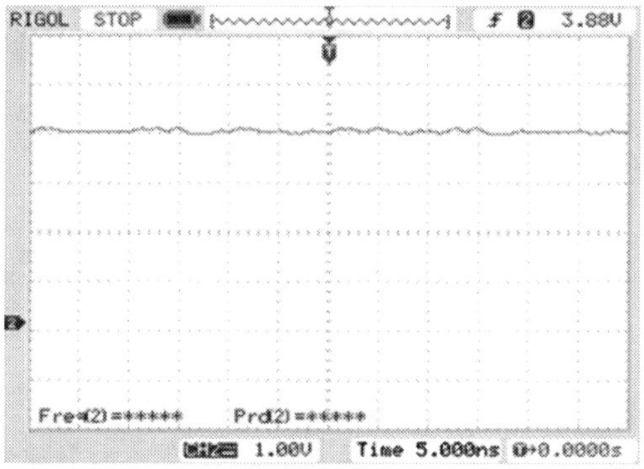

d) Pressure 1.0 bar

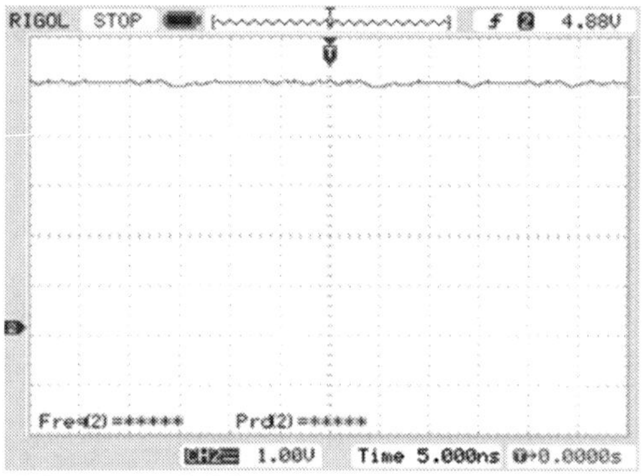

e) Pressure 1.5 bar

Fig. 8. Pressure sensor signals.

Fig. 9 presents results of SPI communication. The waveforms are taken using an oscilloscope. A series of diagrams below visualizing the digital SPI communication, through which the temperature sensor MAX6675 communicates with the microcontroller using the built-in analog-to-digital converter, has been presented. Fig. 9 a) presents the waveforms of a standard SPI communication, as Fig. 9 b), c) and d) the SPI communication signals at 1 V, 5 V and 10 V respectively.

https://doi.org/10.1051/e3sconf/202128604011

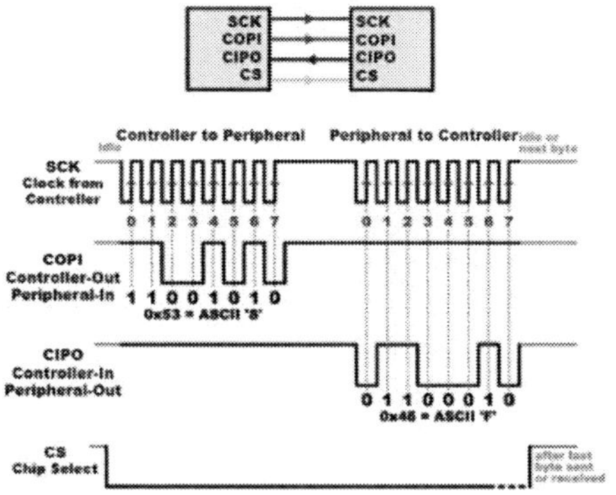

a) SPI Communication waveforms

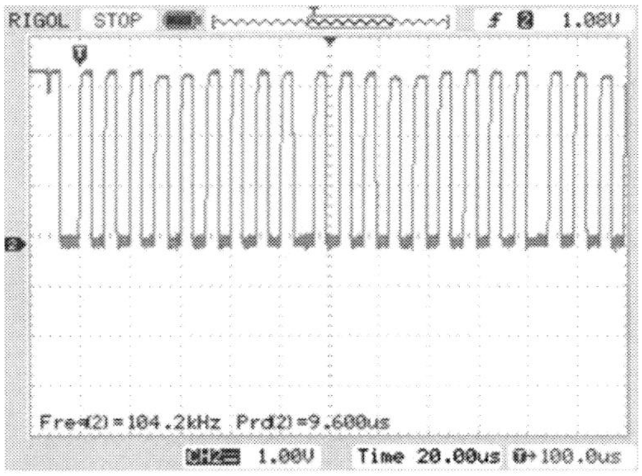

b) 1 V

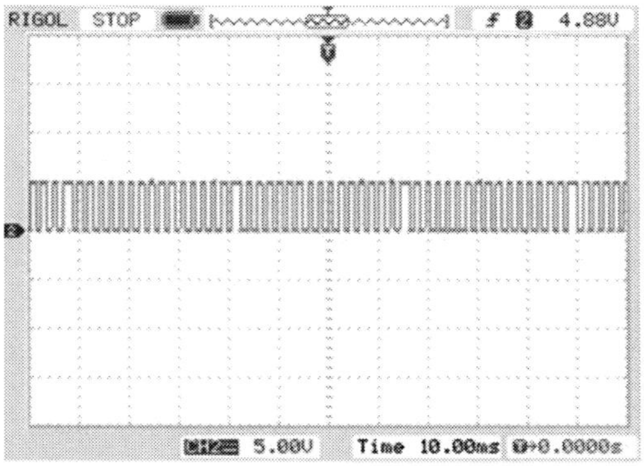

c) 5 V

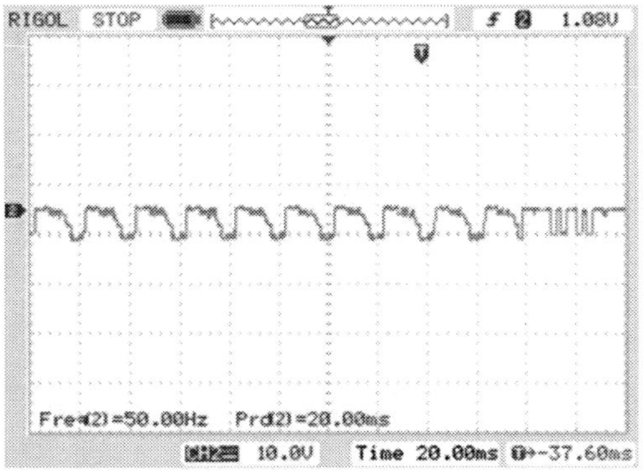

d) 10 V

Fig. 9. SPI signals waveforms.

4 Conclusion

A schematic diagram of an electronic module for temperature of exhaust gases and turbocharging air pressure in turbochargers of diesel automobile engines has been synthesized.

As a result of laboratory experimental investigation, results were obtained for the time-monitored parameters temperature of the exhaust gases and turbocharging air pressure in the turbocharger system of a diesel automobile engine.

Last but not least the developed electronic module, installed in a real automobile continuously monitors the temperature and pressure in the specific zone of the turbocharger. The driver can be informed in time about the malfunction and prevent a fatal result in the turbocharger.

As a future work the project can be have additional options for monitoring the pressure and temperature of the lubricant. It is required due to the imposed environmental requirements that reduce the performance of the components used in modern internal combustion engines. One of the main causes of engine failure is temperature dependence and lack of lubricant or problem with its delivery (the pressure).

References

1. K. Emara, A. Emara, El Sayed Abdel Razek, Turbocharger selection and matching criteria in a heavy duty diesel engine, International Journal of Scientific and Engineering Research **7,** 12 (2016)
2. Jianbing Gaoa, Haibo Chenc, Guohong Tianb, Chaochen Maa, Fei Zhua, An analysis of energy flow in a turbocharged diesel engine of a heavy truck and potentials of improving fuel economy and reducing exhaust emissions, Energy Conversion and Management **184**, (2019)
3. M. Muqeem, M. Ahmad, A. F. Sherwani, Turbocharging of diesel engine for improving performance and exhaust emissions: A review, Journal of Mechanical and Civil Engineering **12**, 4 (2015)
4. Shuo Liu, Yi Cui, Yi Wang, Kangyao Deng, Sheng Liu, An evaluation method for transient response performance of turbocharged diesel engines, Energy **182** (2019)
5. Mingyang Yang, Yuncheng Gu, Kangyao Deng, Zhenhuan Yang, Yangjun Zhang, Analysis on altitude adaptability of turbocharging systems for a heavy-duty diesel engine, Applied Thermal Engineering **128** (2018)
6. Jiaqiang Ea, Xiaohuan Zhaoa, Liangsheng Qiua, Kexiang Weic, Zhiqing Zhanga, Yuanwang Denga, Dandan Hana, Guanlin Liua, Experimental investigation on performance and economy characteristics of a diesel engine with variable nozzle turbocharger and its application in urban bus, Energy Conversion and Management **193** (2019)
7. A. Reihania, J. Hoarda, S. Klinkertc, Chih-Kuang Kuanc, D. Stylesc, G. McConvillec, Experimental response surface study of the effects of low-pressure exhaust gas recirculation mixing on turbocharger compressor performance, Applied Energy **261** (2020)
8. D. Di Battista, R. Carapellucci, R. Cipollone, Integrated evaluation of Inverted Brayton cycle recovery unit bottomed to a turbocharged diesel engine, Applied Thermal Engineering **175** (2020)
9. Ü. Güneş, Y. Üst, A. Sinan Karakurt, Performance analysis of turbocharged 2-stroke diesel engine, *Proceedings of International Conference on Engineering and Natural Science* (2016)
10. L. V. Plotnikov, B. P. Zhilkin, Specific aspects of the thermal and mechanic characteristics of pulsating gas flows in the intake system of a piston engine with a turbocharger system, Applied Thermal Engineering **160** (2019)
11. Z. Petranović, M. Sjerić, I. Taritaš, M. Vujanović, D. Kozarac, Study of advanced engine operating strategies on a turbocharged diesel engine by using coupled numerical approaches, Energy Conversion and Management **171** (2018)
12. M. Hirata, T. Hayashi, M. Takahashi, Y. Yamasaki, S. Kaneko, A nonlinear feedforward controller design taking account of dynamics of turbocharger and manifolds for diesel engine air-path system, *IFAC papers online,* 52-5 (2019)
13. G. Zamboni, A study on combustion parameters in an automotive turbocharged diesel engine, Energies (2018)
14. S. Kaul, T. Joshi, L. Varshney, Analysis of hot air duct between turbocharger and intercooler, *Materials Today: Proceedings* (2020)
15. Z. Korczewski, Exhaust gas temperature measurements in diagnostics of turbocharged marine internal combustion engines Part I Standard measurements, Polish Maritime Research **1(85)**, 22 (2015)

16. D. Dubic, O. Sename, D. Bresh-Pietri, A. Halle, C. Gauthier, Exhaust pressure LVP observer for turbocharged diesel engine on-board diagnosis, *IFAC papers online,* 51-24 (2018)
17. M. S. Oun, S. A. Farhat, M. A. lrabeei, The effect of turbocharger pressure and intercooler temperature on engine performance, Journal of Engineering Research (University of Tripoli, Libya) **23** (2017)
18. B. Wu, Zhi Jia, Zhen guo Li, Guang yi Liu, Xiang lin Zhong, Different exhaust temperature management technologies for heavy-duty diesel engines with regard to thermal efficiency, Applied Thermal Engineering **186** (2021)
19. K. Yordanov, I. Hadzhidimov, P. Zlateva, Electronic equipment measuring device for heat quantity through a flat wall, E3S Web Conf. **207** (2020)
20. J. Garipova, A. Georgiev, T. Papanchev, D. Zlatev, Operational Reliability Assessment of Systems Containing Electronic Elements, *Proceedings of 2nd International Scientific Conference "Intelligent Information Technologies for Industry"* (2017).
21. A. Manukova-Marinova, Design of technical systems by functional-structural analysis, *2020 7th International Conference on Energy Efficiency and Agricultural Engineering, EE and AE 2020 – Proceedings*, (2020)

Developing a comprehensive GIS tool to evaluate representative parameters regarding life comfort

Cristian-Gabriel Alionte[1], *Dana* Rizescu[1,*], and *Ciprian Ion* Rizescu[1]

[1]University POLITEHNICA of Bucharest, Department of Mechatronics and Precision Mechanics, RO-060042, Splaiul Independentei, 313, sector 6, Bucharest, Romania

Abstract. In this paper, we present a tool dedicated to assessing life quality according to the notion of comfort which can link three aspects of noise, light, and air pollution with two aspects of thermal and odour discomfort because the beforementioned aspects are using different calculation methods of the parameters which are distinct and some of them are defined in the national or EU legislation. The authors developed a compressive database to evaluate representative quantities of energy loses. There were considered some performance indicators as: energy class, total specific energy consumption, CO2 equivalent emission index, the overall heat transfer coefficient of the building and characteristic temperatures, the corrected number of degrees days, and the duration of the heating season - at locality level.

1 Introduction

The term quality of life is a complex and multidisciplinary term [1] and, for this reason, it is difficult to identify and establish measurement parameters although there is a great lot of interest in constantly improving the quality of life in the European Union. But the factors affecting the quality of life in the environment are known [2] and are being linked in particular to the health of the person or offering sustainable welfare so they are related to the environment in which they live, and they are transformed into a simple notion of comfort/discomfort.

The software tools used for calculation are based on Geographical Information System (GIS) used for managing, analysing, and storing data and associated attributes which are spatially referenced to a specific Earth coordinates [3] recognized by each nation by legislation. For Romania can be used two different geographic coordinate systems Stereo70 and ETRS89 according to Order of the Minister for Environment and Sustainable Development No. 1830/2007 for the approval of the Guide on the production, analysis, and evaluation of strategic noise maps [4] and can be used also for thermic assessment, light, and air pollution.

The GIS data can be graphical features and attribute data, with relational patterns between them over time and space [5]. GIS data are simplified representations of real

* Corresponding author: dana.rizescu@upb.ro

objects which can be discrete (buildings, fences, roads, point source, line source) or continuous fields (land use, elevation, waterways, etc.).
For the pollution calculation vector images (containing points, lines, or polygons) are used and, for representation support, raster images are used. The vectors are used to model the sources and obstacles and continuous fields are used to model the environment and the surfaces (the terrain or the land use). Also, several internal calculation methods for dispersion, absorption, and path calculation can be applied to individual objects or to summarize calculation or can be used external software for calculation (e.g., R code).
The output can be graphical (raster images) where are presented the level of pollution in colour scales or statistical where are presented the affected persons. In the following chapters we will present some of the specific parameters for acoustic comfort, air quality, and energy performance.

2 Acoustic comfort

Acoustic comfort is defined in terms of annoyance [6] and sleeps disturbance [7]. The L_{den} which is indicator of annoyance and L_{night} used for sleep disturbance are legislated by Environmental Noise Directive (END 2002/49/CE [8]) and are individually assess for four source types through modelling. In the table 1 is presented each standard defined for modelling in Romania according to the Government Decision 321/2005 with amendments and additions [1] which transposes the END 2002/49/CE [7].

Table 1. Setting Word's margins.

Noise source/type	Method
Road	NMPB Routes-96 (SETRA-CERTU-LCPCCSTB), French Standard XPS 31-133 [9]
Railway	Dutch SRM II – 1996 [10]
Aircraft	ECAC.CEAC Doc. 29 [11]
Industrial	ISO 9613-2 - Acoustics — Attenuation of sound during propagation outdoors — Part 2: General method of calculation [12]

For the noise calculation, it can be used some specialized software based on GIS software: CadnaA, IMMI, SPM9613 Community Noise Prediction, CUSTIC, Predictor-LimA Software Suite Type 7810, Acoustic Determinator 7816, MITHRA SIG, SoundPLAN [3].
The formula (1), used for L_{den} is [9]:

$$L_{DEN}[dB] = 10lg\frac{1}{24}(12 \cdot 10^{\frac{L_{day}}{10}} + 4 \cdot 10^{\frac{L_{evening}+5}{10}} + 8 \cdot 10^{\frac{L_{night}+10}{10}}) \quad (1)$$

L_{day}, $L_{evening}$, and L_{night} are statistic factors which mediate all the noise for 12, 4 and 8 hours of the day between 07:00-19:00, 19:00-23:00, and 23:00-07:00.
One of the problems consists in that the noise assessment considers only separate sources not the effect of all sources because when the noise level is exceeded the competent authority can implement a solution for noise abatement. In this way, the whole problem of the noise is not solved.

2 Air quality and odour discomfort

The smell sense is a mechanism which aid human survival because protect the human from noxious gases and must be quick and sensitive. But a correlation between sensitivity of an odour and relevance because information regarding the olfactory sensitivity is limited and depends on person to person. Therefore, another approach must be made and was introduced the notion of air quality. According to the Romanian legislation, the air quality assessment is done using long-time measurements (monitoring) according to Law no. 104/2011 [15]. But using monitoring gives us only punctual values so it can be used calculation air dispersion models to provide information regarding the spatial distribution of the quality of air. For example, it can be used a simple Gaussian dispersion model based on Gaussian plume distribution [27] is described by relation (2) based on the model presented in figure 1:

$$c = \frac{Q}{2\pi\sigma_y\sigma_z u} e^{\frac{-y^2}{2\sigma_y^2}} \left(e^{\frac{-(z-h)^2}{2\sigma_z^2}} + e^{\frac{-(z+h)^2}{2\sigma_z^2}}\right) \tag{2}$$

where:
c - concentration at a given position on field ($\mu g \cdot m^{-3}$),
Q - the source power (grams),
x - the downwind (m/s),
y - the crosswind (m/s),
z - the vertical direction (m),
u - the wind speed (m/s),
h – the height of the release[m].

The σ_y, σ_z deviations describe the vertical mixing of the pollutant which depend on atmospheric stability.

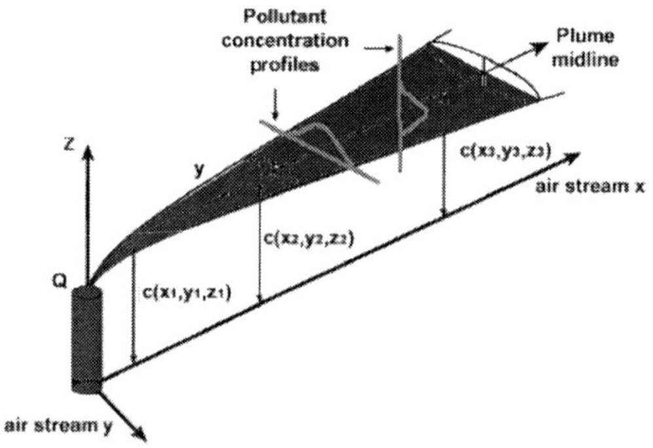

Fig. 1. Gaussian plume distribution model [27]

3 Developing a comprehensive database to evaluate representative quantities regarding energy losses

The building energy performance is starting to consider some performance indicators:

- energy class;
- total specific energy consumption;
- CO2 equivalent emission index;
- the overall heat transfer coefficient of the building;
- characteristic temperatures, the corrected number of degrees days, and the duration of the heating season - at locality level

The application of these parameters in Romania has a great interest because of the great potential [17]. It is known that in South-East Europe (Romania, Bulgaria, Czech, Hungary, Moldova, Poland, Slovakia, etc.) had similar buildings development plans during the last 60 years. An example is the blockhouses developed in the two countries. There were used similar designs, technologies, materials, and plans. Unifying the present rules of building construction and keeping the local apprentice at the same time is one goal. The outcomes of this approach will be a recommendation for local authorities to authorize those technologies which make the best thermal isolation of the buildings with fewer costs and lower fuel consumption per square meter for maintaining indoor air quality. If you want to heat or cool the indoor, energy transfer from indoor to outdoor is a key factor. To uptake the new or improved technologies to reduce or to prevent energy waste, also, reducing the use of hazardous materials in building sector concerning environmental and social ethics conduct the idea to find new assessment tools. For that, the steps are pre-normative research on thermal building expertise, building investigation by thermo-vision, uptake database from local authorities, and develop a comprehensive database to evaluate representative entities regarding energy losses. This method needs gathering data from existing local administration, based on energy consumption levels and architectural or construction structures, considering all types of dwelling. The heterogeneous distributed data that concern different aspects of energy utilizing for heating and cooling of buildings will be considered, organized, and classified in an ontological manner. To build the database is also need:

1. Data collection on building construction and orientation for different types of dwelling units.
2. Collection and evaluation of climatic data.
3. Creation of domain ontology for the semantic description of the data regarding energy loses
4. Analysis of semantic differences for the data from different sources such as local info, thermo-vision expertise, information of the construction structures, and support of interaction under different data schemes for the heterogeneous distributed data
5. Investigate the possibilities of including thermal insulation, double glazed windows to reduce heat transfer indoors outdoor.

The procedure must include the assessment and improvement of the environmental, economic, and institutional impacts of technologies on buildings efficiency and normative proposal elaboration with technical and nontechnical barriers concerning directive 2002/91/EC. Also, must be used models for describing heterogeneous data concerning building construction, orientation, climatic data, building enveloping methods, and thermo-vision.

Thermo graphical and data achievement on a building should respect the existing specifications of EU standards and recommendations as EN ISO/CEI 17025:2005 [19] and ENV ISO 13005:2005[20], that guide the ways of data acquiring, processing, inference, and, finally, decision making according to a certain trusting degree on building efficiency. Based on the beforementioned conditions we made a database that is considered one parameter: the energy class of the building: energy class of the building.

3 Light pollution

Light pollution is unwanted artificial lighting, and it has negative impacts on humans and animals [20]. This type of pollution leads to an increased risk of breast cancer in women since the obtrusive light disturbs the production of melatonin which keeps people from falling into a deep sleep [22]. Also, light pollution can have a significant impact on urban air quality because light pollution breaks down NO3 via photolysis (which is destroyed by sunlight) which serves to cleanse the atmosphere and mitigate the ozone and other atmospheric pollutants. [22].

One method for modelling light pollution is to use fractional logit regression. For this the model assumes that the light sources are points. For the receiving point can be used the equation [23]:

$$X\beta = \ln\left(\frac{D}{1-D}\right) \tag{3}$$

where X is the observation matrix of lighting variables and β is the vector of estimated coefficients.

The disadvantage of using this model is that of lost observations. The least number of observations are lost for those light pollution measures with more moderate threshold values.

4 Experimental results

For the calculation of the pollution parameters, we used QGis [23], Noise Modelling [25], and R code [26]. We build the model using Open Street Map [27]. The source input emission data was measured or estimated. The calculation was performed only for educational purposes. After the calculation, we have the following results.

For the noise map it was built a model for the noise maps road traffic, railroad traffic, and industry, using the definitions and the assessment methods specified in END 2002/49 /EU [8] for Lden and Lnight , figure 2.

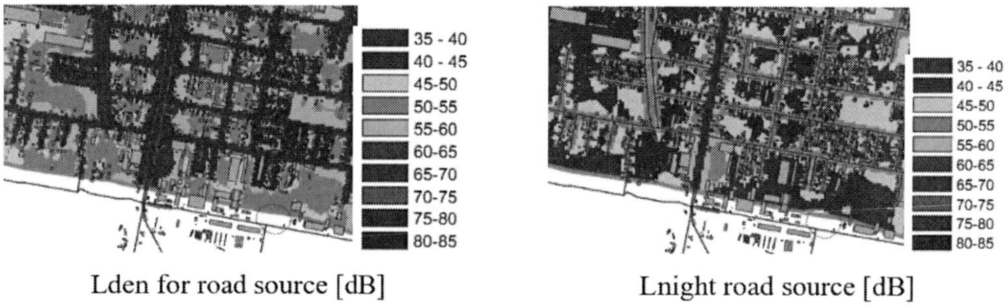

Fig. 2. Road noise pollution

For the odour/air pollution, it was used the Gaussian plume dispersion method. The roads were divided into segments, and after that in points that represents the pollution sources. After that, a grid of points was build which represent the calculated receivers, and for each receiver, it was calculated the distance from each source. The matrix of distances was built and introduced in R code to calculate the dispersion using the equation of Gaussian plume model for each source, figure 3.

Fig. 3. The air pollution map

For each building it was found the center which took the value of the receiver point, so each building is becoming a receiver, figure 4 .

Fig. 4. Air pollution calculated for each building

An important part of solving the problem of light pollution is to have a thorough understanding of its magnitude, and a great way to do that is to measure the brightness of the night sky. It is known that creating and controlling light is a great engineering achievement. Engineers develop artificial sources of light to satisfy specific requirements including good luminance, energy savings, and light fixture design that appropriately directs and reflects light, reduces light waste, and minimizes light invasion. The modern complexity of light installation in buildings and outdoor venues requires the expertise of lighting/illumination engineers to design for specific needs and optimize power consumption.

The same method, but using fractional logit regression was used to calculate the light pollution. The results are presented in figure 5.

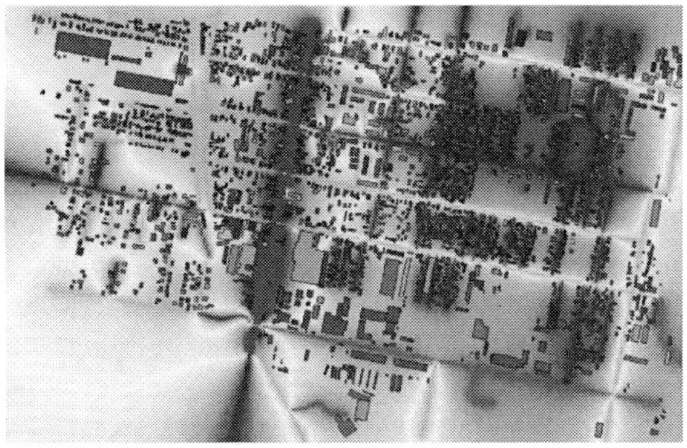

Fig. 5. Light pollution calculation in lux (blue – 75% pollution, yellow – 50% pollution, red – dark zone)

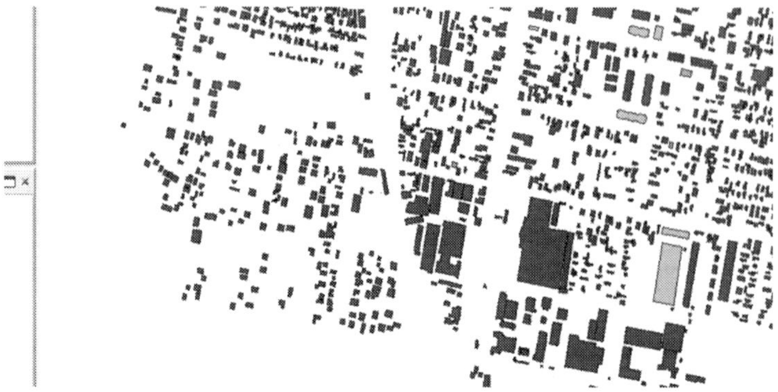

Fig. 6. Thermo vision analyses

Energy performance indices are used around the world to evaluate and monitor residential and commercial building energy performance during design, construction, renovation, and operation. The two most common indices are Asset Ratings and Operational Ratings. Asset Ratings are based on modelled energy use with uniform conditions of climate, schedules, plug loads, occupancy, and energy management.

For building energy performance was introduced the energy class as a building attribute. The calculation method was based on thermo-vision and on building construction material, figure 6.

5 Conclusions

The paper presents a tool dedicated to assessing life quality according to the notion of comfort which can link three aspects of noise, light, and air pollution with two aspects of thermal and odour discomfort because the beforementioned aspects are using different calculation methods of the parameters which are distinct and some of them are defined in the national or EU legislation. Also, the authors developed a compressive database to evaluate representative quantities of energy loses based on the specific parameters for: acoustic comfort, air quality, and energy performance. There were considered some performance indicators as energy class, total specific energy consumption, CO2 equivalent emission index, the overall heat transfer coefficient of the building and characteristic temperatures, the corrected number of degrees days, and the duration of the heating season - at locality level. Countries from Central and East Europe: Romania, Bulgaria, Czech, Hungary, Moldova, Poland, Slovakia etc. have similar buildings developed in the last 60 years. The database, depending on the geographical coverage, will be developed by each country from the project. Each country will consider how many areas are covered, where, and how many sources, how many receivers will be placed. These data can also be provided at the internet address where these results can be consulted. This new database could be developed in the frame of bilateral projects as well as regional projects.

References

1. R. Keles, The Quality of Life and the Environment, Social and Behavioral Sciences 35 23 – 32, Asia Pacific International Conference on Environment-Behavior Studies, Salamis Bay Conti Resort Hotel, Famagusta, North Cyprus, 7-9 December 2011, (2012).

2. N. Görer, and A. Ushowl, An Essay on the Quality of Life and Indicators of the Quality of Urban Life), Localism and Politics: In Honour of Rusen Keles ImagePublishers, Ankara, pp.225-234, (2007).
3. A. Petrovici, C. Tomozei, R. Gey Flores, F. Nedeff, O. Irimia, Noise prediction, calculation and mapping using specialized software, Journal of Engineering Studies and Research – Volume 21, No. 3, (2015).
4. OM MMDD no. 1830/2007
5. S. Rao, and K. Jayasree, Dynamic Geographic Information System for Intelligent Traffic Guidance Systems, Geosciences and Remote Sensing Symposium 2005 , Proceedings. IEEE International, (2005)..
6. J.Y. Jeon, P.J. Lee, J.H. Kim, et al. Subjective evaluation of heavy-weight floor impact sounds in relation to spatial characteristics. J Acoust Soc Am; 125(5): 2987–2994, (2009).
7. G. Belojevic, V. Slepcevic, B. Jakovljevic - Mental performance in noise: The role of introversion, June 2001, Journal of Environmental Psychology 21(2):209-213
8. European Directive 49/2002 (ED 2002/49, 2002)
9. HOTĂRÂRE nr. 321 din 14 aprilie 2005 republicată și actualizată până la data de 28 decembrie 2016 privind evaluarea şi gestionarea zgomotului ambient
10. CERTU, CSTB, LCPC, and SETRA. Bruit des Infrastructures Routières: Méthode de Calcul incluant les Effets Météorologiques, Version Expérimentale NMPB-Routes-96; J. Off. du 10 mai 1995, Artic. 6o French Stand. 'XPS31-133'; Cetur-Centre Etud Transp: Alairac, France, p. 98, (1997).
11. Reken - en Meetvoorschrift Railverkeerslawaai '96, Ministerie Volkshuisvesting, Ruimtelijke Ordening en Milieubeheer, 20 November (1996).
12. https://www.ecac-ceac.org/images/documents/ECAC-ANCAT_Guidelines_Noise_Measurement_Homebuilt_Aircraft_2013.pdf [Accessed 19.05.2021]
13. Law no. 104/2011 - http://legislatie.just.ro/Public/DetaliiDocument/129642 [Accessed 19.05.2021]
14. R. Bhattacharya, Atmospheric Dispersion, https://ansn.iaea.org/Common/Topics/OpenTopic.aspx?ID=13012 [Accessed 19.05.2021]
15. https://ttk.elte.hu/dstore/document/1133/energiafoldrajz%20tankonyv.pdf [Accessed 19.05.2021]
16. S.D. Cîrstea, C. S. Martis, A. Cîrstea, A. Constantinescu-Dobra and M. Timea Fülöp, Current Situation and Future Perspectives of the Romanian Renewable Energy, Energies, 11, (2018).
17. ISO/IEC 17025:2005- General requirements for the competence of testing and calibration laboratories, https://www.iso.org/standard/39883.html [Accessed 19.05.2021]
18. ENV 13005 Guide on Uncertainty in Measurement (GUM) https://boss.cen.eu/media/CEN/ref/cen_13.pdf [Accessed 19.05.2021]
19. C. Rich, and T. Longcore, Ecological Consequences of Artificial Night Lighting. Washington DC: Island Press, (2006).
20. K.J. Navara, and R. J. Nelson. "The Dark Side of Light at Night: Physiological, Epidemiological, and Ecological Consequences." Journal of Pineal Research 43: 215–224, (2007).

https://doi.org/10.1051/e3sconf/202128604012

21. H. Stark, S. S. Brown, W. P. Dube, N. Wagner, T. B. Ryerson, I. B. Pollack, C. D. Elvidge, D. Ziskin, and D. D. Parrish. "Nighttime Photochemistry: Nitrate Radical Destruction by Anthropogenic Light Sources." American Geophysical Union Meeting, (2010).
22. O. Reed, T. Gallaway, D. Mitchell, Modelling US light pollution, Journal of Environmental Planning and Management 57(6), (2014).
23. https://www.qgis.org/en/site/? [Accessed 19.05.2021]
24. https://github.com/Ifsttar/NoiseModelling [Accessed 19.05.2021]
25. https://www.pgm-solutions.com/rcode [Accessed 19.05.2021]
26. https://www.openstreetmap.org/#map=7/45.996/24.981 [Accessed 19.05.2021]
27. https://personalpages.manchester.ac.uk/staff/paul.connolly/teaching/practicals/material/gaussian_plume_modelling/gp_notes.pdf [Accessed 19.05.2021]

Considerations regarding the anti-icing system for the ship propulsion plant with gas turbine

George Iulian Balan[1], *Octavian Narcis* Volintiru[2], *Ionut Cristian* Scurtu[2*], *Florin* Ioniţă[2], *Mirela Letitia* Vasile[1], and *Claudia* Borzea[1]

1. National Research and Development Institute for Gas Turbines COMOTI, Bucureşti, România
2. Mircea cel Batran Naval Academy, Constanta, Romania

Abstract. Vessels that have navigation routes in areas with ambient temperatures that can drop below + 5 [°C], with a relative humidity of over 65%, will have implemented technical solutions for monitoring and combating ice accumulations in the intake routes of gas turbine power plants. Because gas turbines are not designed and built to allow the admission of foreign objects (in this case - ice), it is necessary to avoid the accumulation of ice through anti-icing systems and not to melt ice through defrost systems. Naval anti-icing systems may have as a source of energy flow compressed air, supersaturated steam, exhaust gases, electricity or a combination of those listed. The monitoring and optimization of the operation of the anti-icing system gives the gas turbine power plant an operation as close as possible to the normal regimes stipulated in the ship's construction or retrofit specification.

Anti-icing = to prevent the ice accrual
De-icing = removing the ice from the inlet ducts and/or the inlet flare of gas turbine (highly not recommended during running).

1. Structural details about ST40M gas turbine propulsion system

The system is built around of the (Pratt& Whitney) ST40M gas turbine and it is a retrofitting project named TURBONAV and a main contractor for the research and development was nominated The National Research and Development Institute for Gas Turbines COMOTI.As per the Fig.1., following the gas path there can be identified the ship air intake, air intake duct, ST40M gas turbine, exhaust duct and the ship uptake. Propulsion power is transferred from the power turbine to the torque tube via a high speed clutch and from the torque tube to the primary gearbox.

* Corresponding author: scurtucristian@yahoo.com

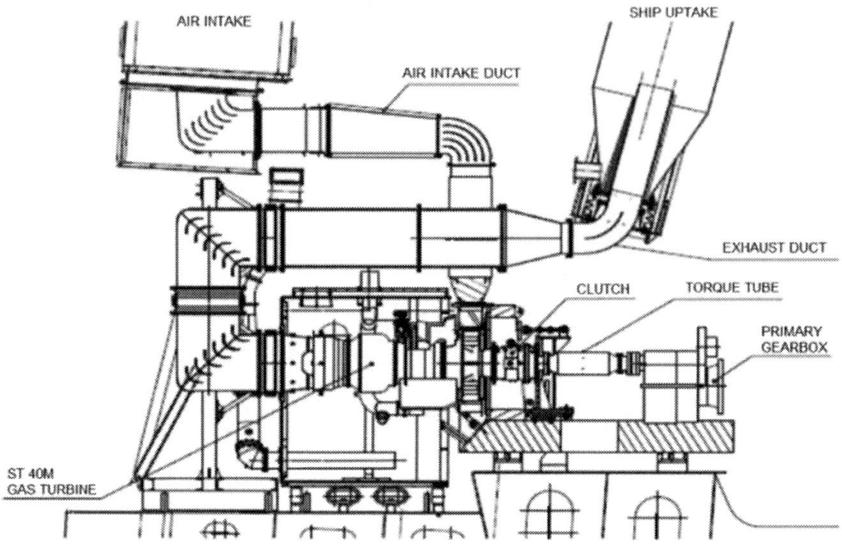

Fig. 1. Energetic system with ST40M gas turbine[19]

2. Air intake parameters

	Wet (1:1 water: fuel)		Dry	
	Power	**Heat rate**	**Power**	**Heat rate**
Peak Rating	7183 shp 5357 kW	7921 Btu/hph 10611Btu/kWh	6576 shp 4904 kW	7595 Btu/hph 10185 Btu/kWh
Base Rating	6089 shp 4547 kW	7916 Btu/hph 10615Btu/kWh	5416 shp 4039 kW	7688 Btu/hph 10310 Btu/kWh

Fig. 2.ST40M Gas turbine functional parameters [20]

- mass air flow: $\dot{m}_a$ [kg /s]
- forecast ambient temperature for icingt_{amb} = (-20°C ÷ +5°C)
- icing conditions:

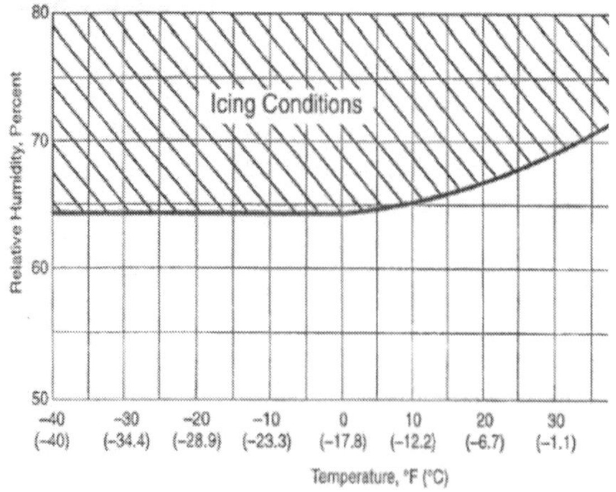

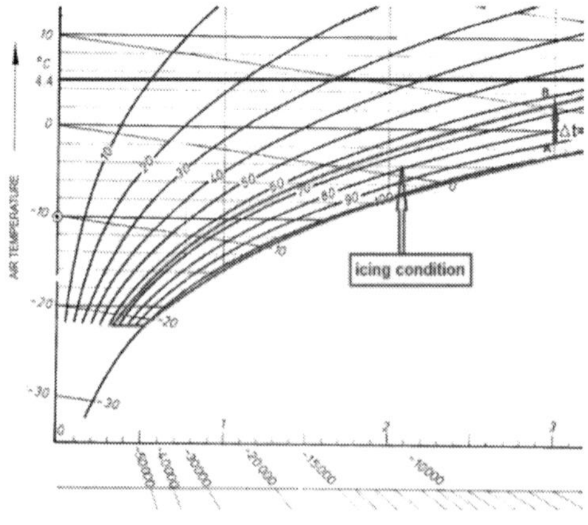

Fig. 3. Icing condition Rolls Royce [8]

Fig. 4. Icing condition in according to the

Mollier diagram [15]

$$t[^0F]=32+1.8t[^0C]$$

$$t[^0C]=(t[^0F]-32)/1.8$$

t^0C	-40	-30	-10	0	5	10	15	30	40
t^0F	-40	-22	14	32	41	50	59	86	104

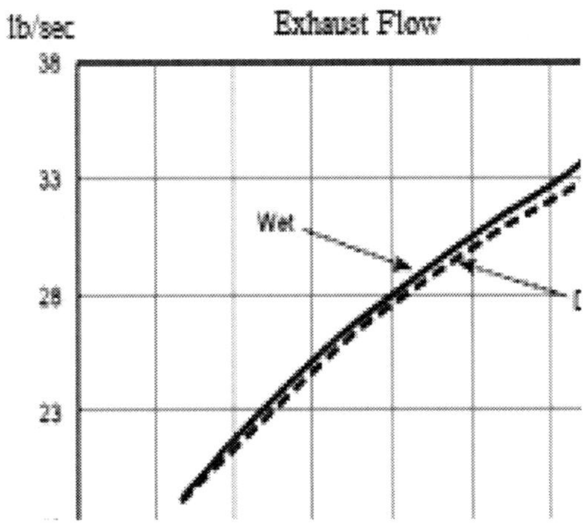

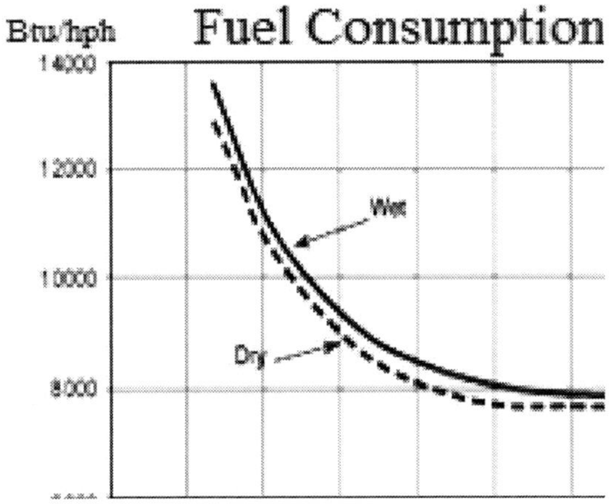

Fig. 5.ST40M Exhaust Flow chart [20] **Fig. 6.**ST40M Fuel Consumption chart [20]

3. Energetically fluxes for the gas turbine air intake

A. Electric energetically flux
B. Hot air energetically flux
  B1. Energy flow exchanger (hot air and intake air)
  B2. Hot air injection in the air intake duct
C. Exhaust gas energetically flux
  C1. Energy flow exchanger (exhaust gas and intake air)
  C2. Exhaust gas injection in the air intake duct
D. Steam energetically flux
  D1. Energy flow exchanger (steam and intake air)
  D2. Steam injection in the air intake duct.
E. Ballast water energetically flux
  E1. Energy flow exchanger (ballast water and intake air)

By using the data from the fuel consumption diagram (Fig. 6.), the sea trial parameters [13] and the following equations we can find the functional parameters for the ST40M gas turbine propulsion system. The calculus is based on F75/F76 NATO fuel, on wet and dry running.

$$1[\mathrm{HP}] = 0.736[\mathrm{KW}] \tag{1}$$

$$10000[\mathrm{BTU}/_{\mathrm{h}}] = 2.93[\mathrm{KW}];\ 10000[\mathrm{BTU}] = 2.93[\mathrm{KWh}] \tag{2}$$

$$10000\left[\frac{\mathrm{BTU}}{\mathrm{HPh}}\right] = \frac{2.93}{0.736}\left[\frac{\mathrm{KWh}}{\mathrm{KWh}}\right] = 3.98\left[\frac{\mathrm{KWh}}{\mathrm{KWh}}\right] \tag{3}$$

$$Qi = 42700\left[\frac{KJ}{Kg}\right] - \text{Lower Heating Value; } C_h\left[{}^{Kg}/_{h}\right] - \text{fuel consumption} \quad (4)$$

$$P_{e_{TGP}} \cong 0.33 \cdot P_{ITG}[KW] \text{- Propulsion effective power} \quad (5)$$

$$\dot{Q}_d = \frac{C_h \cdot Qi}{3600} \quad \text{- available energy flow} \quad (6)$$

$$\dot{Q}_d = RE \cdot P_{ITG} \;\; ; RE - \text{energy ratio} \quad (7)$$

$$C_h = \frac{RE \cdot P_{ITG} \cdot 3600}{Q_I} \text{- fuel consumption} \quad (8)$$

$$\rho_M = 0.84\left[\frac{Kg}{l}\right] \text{- fuel density} \quad (9)$$

$$P_{prop} = C_N \cdot V_N^3[KW] - \text{ship propulsion power} \quad (10)$$

$$C_N = 5.87 - \text{ship resistance coefficient*} \quad (11)$$

* it was determinate after sea trial in 2019

$$1Kn = \frac{1852}{3600}\left[\frac{m}{s}\right] = 0.514\left[\frac{m}{s}\right] \quad (12)$$

$$V_N = \left(\frac{P_{prop}}{C_N}\right)^{1/3}\left[\frac{m}{s}\right] \text{- ship speed} \quad (13)$$

4. Anti-icing parameters

The exhaust gas mass flow rate:

$\dot{m}_{gaze}$=15.7 [kg/s] at ambient temperature 0°C

Specific fuel consumption:

c_e = 1089 [kg/h] = 0.3025 [kg/s]

The air intake mass flow rate:

$\dot{m}_{aer} = \dot{m}_{gaze} - c_e = 15.7 - 0.3025 = 15.3975$ [kg/s]

$\dot{m}_{aer} = 0.3025 \cdot \alpha \cdot 13.8$

From where the air excess coefficient is: α=3.688.

The necessary energetically flux for anti-icing (when the total amount of the air intake is heated):

For $\Delta t = 5$ °C. c_p=1.01[kJ/kg grad]

$\dot{Q}_{nec5} = \dot{m}_{aer} \cdot c_p \cdot \Delta t$

$\dot{Q}_{nec5} = 15.3975 \cdot 1.01 \cdot 5 =$ **77.76 [kW]**

For $\Delta t = 6$°C, c_p=1.01[kJ/kg grad]

$\dot{Q}_{nec6} = \dot{m}_{aer} \cdot c_p \cdot \Delta t$

$\dot{Q}_{nec6} = 15.3975 \cdot 1.01 \cdot 6 =$ **93.3 [kW]**

For $\Delta t = 10$ °C, c_p=1.01[kJ/kg grad]

$\dot{Q}_{nec10} = \dot{m}_{aer} \cdot c_p \cdot \Delta t$

$\dot{Q}_{nec10} = 15.3975 \cdot 1.01 \cdot 10 =$ **155.51[kW]**

For $\Delta t = 20$ °C, c_p=1.01[kJ/kg grad]

$\dot{Q}_{nec20} = \dot{m}_{aer} \cdot c_p \cdot \Delta t$

$\dot{Q}_{nec20} = 15.3975 \cdot 1.01 \cdot 20 =$ **311.02[kW]**

5. Electric anti-icing system proposal, version 1:

Fitting of electric heating coils inside of the transition duct between splitter box and cascade bend of the ST40M gas turbine air intake.

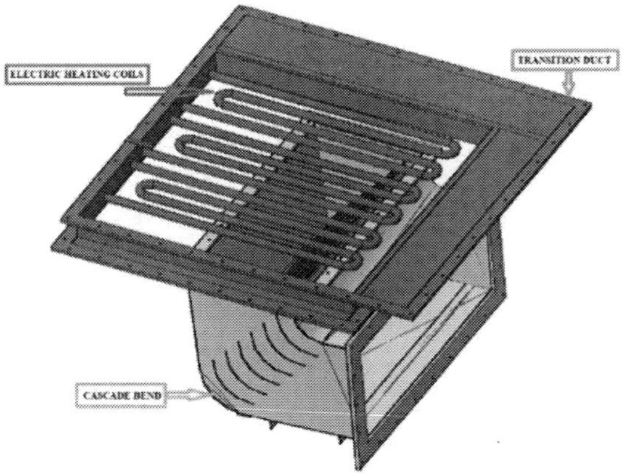

Fig. 7. Anti-icing system version 1

According with the material description sheets from the heating systems producer Chromalox, a single electric heating coil in 3 [m] length, heaving a 29 [mm] diameter it is able to generate an energetic flux of 10 [KW].

For heating-up the total amount of the gas turbine air intake with 6 [°C] there will be necessary to install 9 electric Fintube heating coils. The powering of the anti-icing systems will require 120 [KW], 440 [V] from the ship power grid.

In Fig. 7.there are shown only three electric Fintube heating coils and, therefore, a total energetically flux of 30 [KW] and, for the above example will be necessary to install another two layers of coils (9 heating coils in total) for generating 90 [KW].

6. Electric anti-icing system proposal, version 2

Fitting of electric heating coils outside of the cascade bend.

Using this version of the electric anti-icing system will ensure that the air flow is not perturbed and will be no any additional pressure losses inside of the air intake duct.

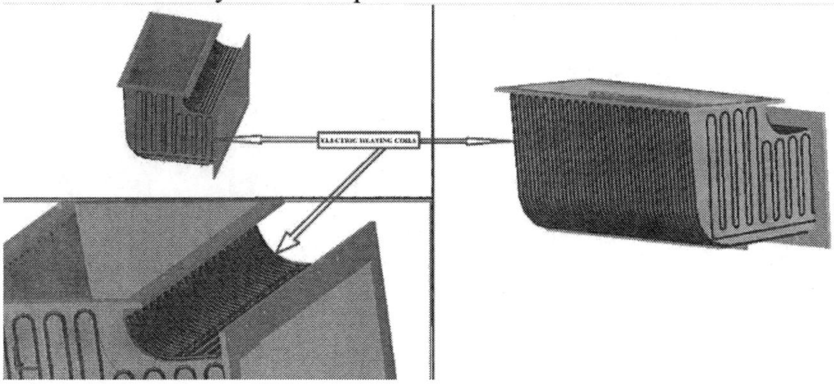

Fig. 8. Anti-icingsystemversion 2

Using this electric heating coils arrangement there will be a 2.5 [m^2] heated area and will act as a heat exchanger. There will be necessary to thermal insulate the cascade bend to minimise the energy flux escaping outside.

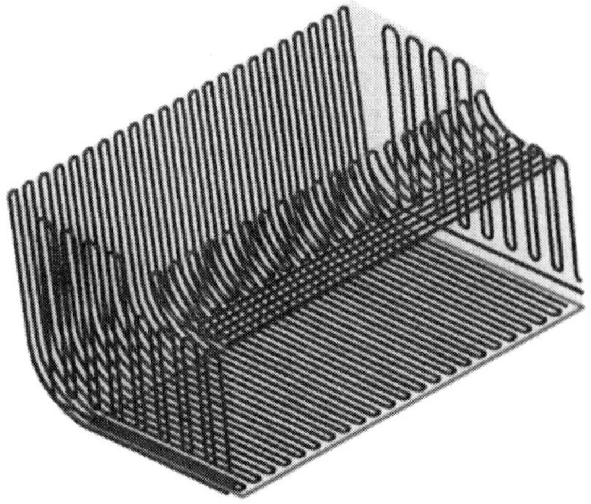

Fig.9. Anti-icing system version 2 - details

7. Thermal anti-icing system - proposal

The most ships propelled by the energetically installations with gas turbine are using the anti-icing systems based on energy fluxes from compressed air, steam or exhaust gas.

Transmission of the energy fluxes from a thermal agent to the intake air can be done by heat exchangers or by direct injection of the thermal agent in the air intake duct.

For the retrofitting project where there is no steam generator already installed, this system will have to be based on compressed air or on exhaust gas.

The exhaust gas can be leaded through a heat exchanger installed inside of the air intake duct. This system will generate structural modifications and will introduce power losses in the air flow.

Exhaust gas direct injection will be a proper solution for the anti-icing due to high temperature thermal agent but it will be necessary to be accepted by the gas turbine manufacturer.

In Fig. 10.is shown an anti-icing system using the energy flow from the direct injected compressed air produced by the gas turbine compressor [3]. This technical solution for the anti-icing system is the most used for the navy ships, is robust, easy in control and easy to be implemented in case of ship retrofitting. By using a certain amount of the gas turbine compressed air it will reduce the delivery power of the entire propulsion system.

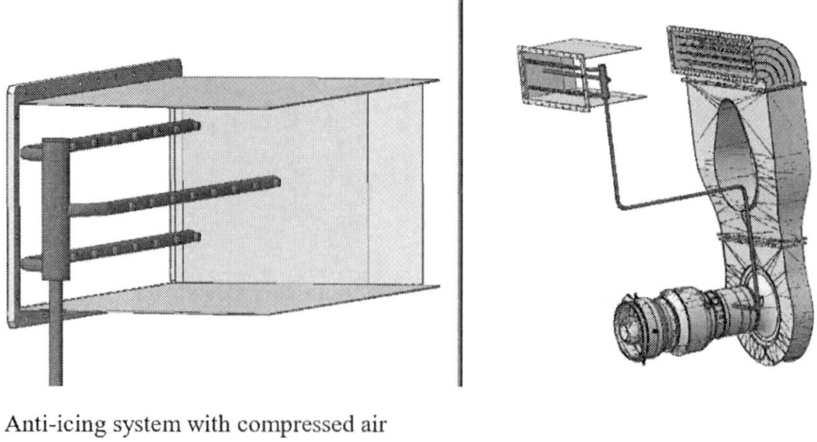

Fig. 10. Anti-icing system with compressed air

Conforming to the ST40M installation manual there is the possibility to extract an amount of maximal 10% of the compressed air mass flow (6% from the LP compressor and 10% from the HP compressor but not more than 10% from the compressed air mass flow in total).

Therefore, that is possible to extract the compressed air from LP system having maxim temperature $\mathbf{t_{LP}}$= +282 [°C]with a mass flow of 6 [%]from total intake air mass flow**(6 [%]·15.3975 [kg/s] = 0.92 [kg/s])**or compressed air from HP system having maxim temperature $\mathbf{t_{HP}}$ = +510 [°C] la with a mass flow of 10 [%]from total intake air mass flow**(10 [%]·15.3975 [kg/s] = 1.54 [kg/s])**.

Calculus of the compressed air mass flow used for the anti-icing system will be done by expressions of the enthalpy balance against of the reference temperature. We will consider the reference temperaturet_{amb} = +15 [°C]and a specific heat of the intake air at constant pressurec_p = 1.01 [kJ/kg grad].

The enthalpy balance equations:

$$\dot{Q}_{intake} = \dot{m}_{intake}\cdot c_p\cdot \Delta T_{intake}$$

$$\dot{Q}_{intake} = \dot{m}_{amb}\cdot c_p\cdot (T_{amb} - 288.15) + \dot{m}_{comp}\cdot c_p\cdot (T_{comp} - 288.15)$$

From where we can find$\dot{m}_{comp}$depending of the temperature difference$\Delta T_{intake}$for the anti-icing system:

$$\dot{m}_{comp} = [\dot{m}_{intake}\cdot \Delta T_{intake} - \dot{m}_{amb}\cdot (T_{amb} - 288.15)]/ (T_{comp} - 288.15)$$

For the case when it is extracted the compressed air from the LP system we will admit ΔT_{intake} = 6 [°K] =Δt [°C] (rising the temperature from -3 [°C] to +3 [°C]) the followings are finding:

$$\dot{m}_{compLP} = [15.3975\cdot 6 - 14.4775 \cdot (270.15 - 288.15)]/ (555.15 - 288.15)$$

$$\mathbf{\dot{m}_{compLP} = 1.32\ [kg/s]} \qquad \mathbf{(14)}$$

For the case when it is extracted the compressed air from the HP system we will admit ΔT_{intake} = 6 [°K] =Δt [°C] (rising the temperature from -3 [°C] to +3 [°C]) the followings are finding:

$$\dot{m}_{compHP} = [15.3975 \cdot 6 - 13.8577 \cdot (270.15 - 288.15)]/ (783.15 - 288.15)$$

$$\mathbf{\dot{m}_{compHP} = 0.69\ [kg/s]} \qquad \mathbf{(15)}$$

From the determination of air mass flow– relations(14) and (15) there can be observed that for rising of the intake air in amount of 6 [°C] (for example: from -3 [°C]to +3 [°C]) is needed a compressed airwith a mass flow equal with **1.32 [kg/s]** from the LP system (but this amount of mass flow is not allowed to be extracted from the gas turbine LP compressor because it represented about **8.57 [%]**from the total intake air mass flow) but, at the other hand the anti-icing can be achieved by using compressed air from the HP systemwith mass flow equal with **0.75 [kg/s]** which is about **4.48[%]**from the total intake air mass flow.

8. Ice accrual sensors for the anti-icing systems

In conformity with the ship classification rules, it is mandatory to use the special designed sensors for monitoring the ice accrual inside of air tank ducts for gas turbine propulsion systems[1] [16].

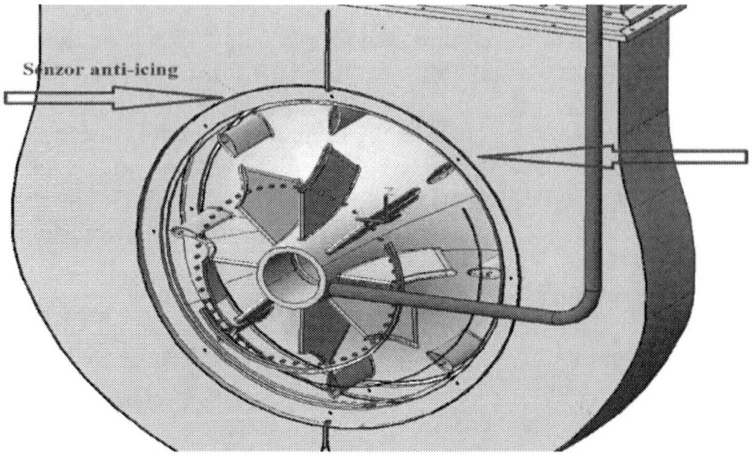

Fig. 11. Anti-icing sensor arrangement

Due to the fact that the air tank struts are the susceptible areas for the ice accrual [8] but also the inlet guide vanes and inter stage stator vanes, it is recommended to install one ice accrual sensor (if the sensors has a failing signal) or two sensors in the other cases.

9. Conclusions

If the electrical balance of the ship is designed to withstand an amount 180 [kW] (90 [kW] for each gas turbine) than, the electric anti-icing systems version 1 and version 2 are the best technical solutions. The main advantages of this two versions are the easiness of the installations, controlling, optimisation and monitoring of the ice accrual without using the energy flow from the propulsion system.

The anti-icing system with air compressed extracted from the gas turbine compressor is very robust and safety to use, but will not allow the propulsion system to run on maximal functional parameters. The effective propulsion power of the gas turbine energetically system will be diminished by few percents (up to 10% in power loss). The timing of the compressed airextraction, the air mass flow and duration of extraction is controlled by using ambient temperature sensors and ice accrual sensors. Due to anti-icing system optimisation the fluctuations of the effective power for propulsion can be maintained at lower level and therefore, the fuel consumption can be properly adjusted.

References

1. Military standard MIL-STD-3045, 2014, USA
2. Davis L.D., Soreng A., Twin Drive Gas Turbines in Single Propulsion Package for Norwegian Fast Patrol Boat, ASME Turbo Expo 2009, ISBN:978-0-7918-4885-2
3. Aye N., McAndtrews G., Mendenhall B., MARINE GAS TURBINE PACKAGE FOR THE KOREAN NAVY PKX PROGRAM, ASME Turbo Expo 2008, Germany, ISBN: 978-0-7918-4317-8
4. Lacey J.J., Turbine Engine Icing and Ice Detection, ASME, 1972, USA
5. Dickson J., Problems Associated with Cold Weather Operation of Gas Turbines, ASME, 1976, USA
6. Swan K.T., A survey of icing condition for marine gas turbines, NAPTC –PE-114, NAVSEC, 1977, USA

7. Sammak M., Anti-Icing in Gas Turbines, 2006, Suedia, ISSN 0282 - 1990
8. Loud R.L., Slaterpryce A.A., Gas Turbine Inlet Air Treatment. GE technical paper GER-3419A, 1991, USA
9. Rashid T., Khawaja H.A., Edvardsen K., Review of marine icing and anti-/de-icing systems, Journal of Marine Engineering & Technology, 15:2, 79-87,

   DOI: 10.1080/20464177.2016.1216734
10. Wilcox M., s.a., GUIDELINE FOR GAS TURBINE INLET AIR FILTRATION SYSTEMS, Gas Machinery Research Council Southwest Research Institute, 2010, USA de Souza J.R.B., s.a., Thermal analysis of anti-icing systems in aeronautical velocity sensors and structures, Brazilia, J Braz. Soc. Mech. Sci. Eng. (2016) 38:1489–1509, DOI 10.1007/s40430-015-0449-7
11. F. Niculescu, C. Borzea, M. L. Vasile, Marine gas turbine for efficient ship propulsion, 2020, ISSN 2668-7003, ISSN-L 2457-5011,

    DOI: 10.37410/EMERG.2020.3.08
12. Niculescu, F., Claudia Borzea, Adrian Savescu, Andrei Mitru, & Mirela Letitia Vasile. (2020). Automation and Electronic Control of Marine Gas Turbine Engine for Ship Revamp. *Technium: Romanian Journal of Applied Sciences and Technology*, *2*(4), 98–108. https://doi.org/10.47577/technium.v2i4.923
13. Purece, C., Vasile Pleşca, & Lilica Corlan. (2020). Technologies for obtaining energy from micro-hydropower resources. *Technium: Romanian Journal of Applied Sciences and Technology*, *2*(4), 124–133. https://doi.org/10.47577/technium.v2i4.837
14. Lupchian, M. (2020). Influence of propulsion installation performance on travel efficiency. *Technium: Romanian Journal of Applied Sciences and Technology*, *2*(7), 50–53. https://doi.org/10.47577/technium.v2i7.1644
15. Crudu, L., Bosoancă, R., & Obreja, D. (2020). A comparative review of the resistance of a 37,000 dwt Chemical Tanker based on experimental tests and calculations. *Technium: Romanian Journal of Applied Sciences and Technology*, *1*, 59–66. https://doi.org/10.47577/technium.v1i.32
16. F.Niculescu,s.a., Automation control for revamping the propulsion system of a navy frigate, U.P.B. Sci. Bull., Series D, Vol. 83, Iss. 1, 2021, ISSN 1454-2358
17. V. Pimsner, C. A. Vasilescu, A. Petcovici, Termodinamica tehnică, 1982, Editura didactica si pedagogică, Bucureşti
18. *** www.lr.org
19. *** www.chromalox.com
20. *** www.heat-trace.com
21. *** www.comoti.ro
22. *** "ST40M-Marine and Industrial Gas turbine"-Fact sheet, Pratt & Whitney, Canada.

Numerical and analytical study of seismic response of structural systems with new formulation using energy and impact methods

Mehdi Babaei[1*], *Somayeh* Mollaei[1], *Amin Moslemi* Petrudi[2], *Maysam* Jalilkhani[3], and *Ionut Cristian* Scurtu[4]

[1]Department of Civil Engineering, University of Bonab, East Azerbaijan, P. O. B. 45139-63391, P. O. B. 55513-95133,Iran.
[2] Department of Mechanical Engineering, IHU University, Tehran, P. O. B. 47155-66195, Iran
[3] Department of Civil Engineering, Engineering Faculty of Khoy, Urmia University,P. O. B. 58159-14853, Iran.
[4] Mircea cel Batran Naval Academy, Constanta, Romania

Abstract. This paper presents two simple and robust technique for response estimating of single-degree-of-freedom (SDOF) structural systems. The impulse method, because it is formulated based on the fundamentals of dynamics; especially, the linear impulse concept, and also the energy method, because the main idea of this method is inspired by energy conservation principles. These methods can strongly cope with linear damped systems for which damping ratio ζ is greater than 0.01.Assessment of SDOF dynamic systems under any arbitrary excitations is easily possible through the proposed methods. There is no error propagation through the solving process. The numerical example reveals the simplicity and robustness of the new technique compared to Duhamel's integral and similar techniques. Finally, a numerical example is investigated to demonstrate the efficiency of the algorithms. The most famous record of El Centro ground motion is applied to the systems. The obtained results show that the new algorithm works exactly enough to compete with a conventional method such as the Duhamel integration method and the Newmark-β method.A comparison between the results of this method with the solution methods used by other researchers is shown to be a good match.

Keywords: Seismic response, SDOF system, Mass-spring oscillator, Impulse, and energy method.

1 Introduction

The response of various dynamic systems, undergoing time-dependent changes, can be described by a second-order ordinary differential equation (ODE). As we know the merely linear differential equation can be analytically solved; and, there are numerous ODEs in applied mechanics, physics, biology, and economics for which analytical solutions are not achievable; especially, those that receive a noisy time-depending signal as input data. For this reason, the application of numerical techniques is indispensable for processing this family of ODEs. So, the numerical field of solving applied ODEs is one of the most active fields in science and engineering. Therefore, solving the governing equations of motion (EOM) for these systems would be of high importance for many applied problems. SDOF odel is the most known simplification of vibrating systems for seismic analysis of buildings and structures. Solving EOM of SDOF oscillators under base acceleration always have been a place of challenge for earthquake engineering researchers from the very early to the recent efforts [1]. Available methods to the solution of linear differential equations, such as EOM of SDOF systems, are mostly classical approaches like Duhamel's integral and frequency-domain method. In the classical solution, the response consists of the sum of the complementary solution and the particular solution. Duhamel's integral is based on representing the applied excitation as a sequence of infinitesimally short impulses. The frequency-domain method takes advantage of the Laplace and Fourier transforms. Besides, another category of numerical methods is known as stepwise methods. They are extended for analyzing nonlinear systems. The methods of excitation interpolation, central difference method, and also Newmark's method is the most commonly used numerical procedures in this family [2,3]. Accuracy of some numerical stepwise procedures used to compute the dynamic structural response of the SDOF system to a time-dependent excitation is evaluated in technical report ITL-97-7 published by the U.S. Army Corps of Engineers. They mainly investigated the Wilson-θ method, linear acceleration method, Newmark-β algorithm, central difference method, the fourth-order Runge-Kutta method, and Duhamel's integral and Piecewise Exact Method [4]. Also, a new exact approach for forced vibration analysis of SDOF systems with arbitrary time-varying mass and stiffness was introduced by Li [5]. To conclude, most of the aforementioned conventional methods of solving SDOF's equation of motion, such as interpolation of excitation, Duhamel integral, central difference method, Wilson-θ, and Newmark-β, have an insufficiently complex algorithm, offer a restricted level of precision, and almost difficult to understand techniques. Sometimes, they need advanced mathematical background and technical expertise. Most of the vibration cases in mechanics and structural engineering can be presented as an ideal single degree of freedom (SDOF) system (Paz & Leigh, 2004). These systems follow an identical differential equation, the so-called equation of motion, which must be solved to identify its kinematic and kinetic parameters. Accordingly, solving differential equations of motion for vibrating systems is of great importance in mechanics and structural engineering. The solution of this equation can be a continuous analytical function or a numerical approximation computed at discrete time instances. Although analytical solutions are more desirable, there are merely some limited cases in applied mechanics that can be analytically solved by indeterminate integration; and, most of the real cases cannot be analytically solved. In these cases, for which indeterminate integration is not possible in its closed form, the implementation of numerical simulation is indispensable. In this regards, numerical solution of the equation of motion for the systems under irregularly changed excitation always has been a place of challenge for engineers from the very early studies [6,7] up to the most recent ones [8-10]. Current approaches to the solution of linear differential equations are classified into two categories of numerical methods in mechanical systems. First, those which use the superposition principle in their formulation such as Duhamel 's integral and Fourier integral methods. These methods necessitate an evaluation of numerous response contributions at each time instance which is considered as a disadvantage with them. Besides, although these methods have theoretically strong formulation, none of them can be applied to nonlinear systems since nonlinear systems do not follow the superposition principle which is the basic assumption in their formulation. In this regard, numerous stepwise techniques have been coming into existence. Some of the most known stepwise methods are the piecewise exact method, finite difference method, integration-based methods. Integration based methods include a wide verity of techniques such as Euler-Gauss integration, Newmark-β method, and Runge-Kutta integration [11-14]. Most of these method employs numerical approximation of integrals at each time step. There is a vast body of contribution in this field [15-20]. A similar issue is followed in this work to formulate new stepwise approaches. Chang tried to thoroughly confirm the performance of Newmark's method in solving nonlinear SDOF systems and its stability with numerical examples [21, 22]. Kazakov studied the dynamic response of the SDOF mass-spring system using Duhamel integral. They treated the problem as a form of Laplace integral transformation [23]. Also, a simple numerical method for computing the response of the SDOF system in terms of Taylor polynomials in the matrix form has been presented by Kurt and Çevik [24]. Li and Wu presented an iterative procedure to solve the nonlinear oscillations of a conservative SDOF system [25, 26]. Also, I introduced a new exact approach for forced vibration analysis of SDOF systems with arbitrary time-varying mass and stiffness [5].Mohsenian investigated the seismic reliability of diagrid structural systems and develop more efficient performance-based design methodologies. Demand and supply response modification factors are calculated for 16, 24, and 32-story buildings with diagrid structural systems using a 65° diagrid angle and designed in compliance with current standards under a set of 12 spectrum compatible earthquakes [27]. Pang investigated to analyze the Magna earthquake sequence and resolve oblique-normal slip on a shallow (30–35°) west-dipping fault at ~9- to 12-km depth. Combined with near-surface geological observations of steep dip (~70°), our results support a curved, or listric, fault shape. High-precision aftershock locations show the activation of multiple, low-angle (<30–35°) structures, indicating the existence of a complicated fault system [28]. Mitseas investigated an approximate stochastic dynamics technique for determining the system response amplitude probability density functions (PDFs) is developed. Firstly, relying on statistical linearization and state-variable formulation the complex eigenvalue problem is addressed through the time-domain [29]. Mitseas investigated a structural system comprising the bilinear hysteretic model serves as a numerical example for demonstrating the reliability of the proposed first-excursion PDF-based stochastic incremental dynamics methodology [30]. The merits of different stepwise methods originated from four main aspects: their convergence, level of precision, stability, and computational expense needed for a certain problem. However, having these advantages, most of the conventional methods have a sort of complicacy with their formulation, need high computational effort, and are sometimes difficult to understand. They need more or less advanced mathematical tools and technical expertise to use. Accordingly, we are to propose significantly easy techniques to deal with this problem. In this article, the simple energy method is introduced. It is motivated by the most known principle of energy conservation. This method introduced here is satisfactorily precise for analyzing linear damped system belonging viscous damping. They facilitate the assessment of SDOF response under any arbitrary irregular excitations and offer very simple platforms which makes it distinctive from its counterparts. Also, the simple method is introduced to solve the differential equation of vibrating SDOF systems. The initial idea of this method was initiated first by these authors working on an approximate solution of the merely mathematical differential equation [31]. However, the new method introduced here is highly precise and facilitates the assessment of SDOF dynamic response under any arbitrary excitations. The introduced procedure, the so-called impulse method, has a straightforward formulation based on the fundamentals of Dynamics, e.g., linear impulse concept. This method mainly targets SDOF linear damped systems undergoing earthquake excitation. The numerical example reveals the simplicity and efficiency of the impulse technique compared with the most known Duhamel's integral and Newmark-β method. All computation is done by MATLAB software.

2 Problem Statement

Shown in Fig. 1 is the most known idealized linear mass-spring system. Commonly speaking, this system is composed of three main components: mass, spring, and damper.
As proved in the literature, if this system was under support movement of magnitude a_g instead of external force F(t), the idealized system could be equalized as shown in Fig. 1. In this system, $-ma_g$is substituted for the external force F(t), where $a_g$denotes the acceleration of the system support and x is the relative displacement of mass (or deformation of spring).

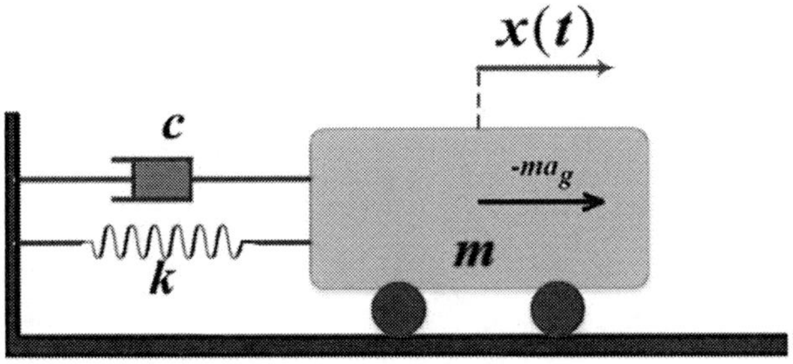

Fig.1 Idealized mass-spring model of vibration under support movement.

As proved in technical literature, denoting the displacement of mass by x and its velocity and acceleration respectively by v and a, and assuming linear behavior of spring, i.e., $F_s = kx$, and idealized viscously damping behavior of damper, i.e., $F_D = cv$, we can write the governing equation of motion [2,14].

$$kx + cv + ma = -ma_g \tag{1}$$

Dividing equation (1) by mass m, the governing equation of motion for vibrating systems with linear components and undergoing ground acceleration is formulated as:

$$\omega_n^2 x + 2\zeta\omega_n v + a = -a_g \tag{2}$$

where$\omega_n = \sqrt{\frac{k}{m}}$, is the natural frequency of the oscillating mass-spring system, $\zeta = \frac{c}{2m\omega_n}$ is the damping ratio, and $a_g(t)$ is the ground (or support) acceleration. This equation indicates that any two systems with the same natural frequency and damping ratio have identicalresponse$x(t)$, regardless of being massive and stiffer relative to each other. This equation is an alternative expression of the vibration equation that we may use in our formulation.

3 Methodology: Proposed Impulse Method

To propose an easily employable technique for dealing with the most known vibration problem, i.e., damping SDOFsystems, we first recall the concept of impulse in mechanics. The product of force and time is defined as the linear impulse of the force:

$$dI = Fdt \tag{3}$$

And the total impulse of the mass m from time instance t_i to t_{i+1} is indicated as:

$$I = \int_{t_i}^{t_{i+1}} Fdt \tag{4}$$

Multiplying Eq. (1) by time differential, dt, we have:

$$kxdt + cvdt + madt = -ma_g dt \tag{5}$$

It means that the differential impulse of the system in dynamic equilibrium is conservative during a time differential dt. Recalling that $v = dx/dt$ and $a = dv/dt$ we have:

$$kxdt + c\frac{dx}{dt}dt + m\frac{dv}{dt}dt = -ma_g dt \tag{6}$$

Integrating Eq. (6) from t_i to t_{i+1}gives:

$$\int_{t_i}^{t_{i+1}} kxdt + \int_{x_i}^{x_{i+1}} cdx + \int_{v_i}^{v_{i+1}} madt = \int_{t_i}^{t_{i+1}} -ma_g dt \tag{7}$$

Assuming linear variation for components x and a_g and employing a trapezoidal integration method, we have:

$$\frac{k\Delta t}{2}(x_{i+1} + x_i) + c(x_{i+1} - x_i) + \frac{m\Delta t}{2}(a_{i+1} + a_i) = -\frac{m\Delta t}{2}(a_{g.i+1} + a_{g.i}) \tag{8}$$

Each term of this equation is the total impulse of a force component of SDOF as well as spring force, friction force, inertia force, and the force from external excitation. It means that the resultant impulse of component forces is conservative during the time step from t_i to t_{i+1}. This formula gives the recursive relationship needed to compute x_{i+1}, v_{i+1} during a time step. For this purpose, we rewrite the equation (8) as:

$$x_{i+1} = x_i + \left[-\frac{m\,\Delta t}{2}\left[(a_{g.i} + a_{g.i+1}) + \omega_n^2(x_i + x_{i+1}) + (a_i + a_{i+1})\right]\right] / c \tag{9}$$

$$v_{i+1} = v_i + \left[-\frac{m\,\Delta t}{2}\left[(a_{g.i} + a_{g.i+1}) + \omega_n^2(x_i + x_{i+1}) + 2\omega_n\zeta(v_i + v_{i+1})\right]\right] /m \tag{10}$$

Where $\omega_n^2 = k/m$ and $2\omega_n\zeta = c/m$. $\omega_n$and ζ are respectively natural frequency and damping coefficient of the system. Simplifing Eq.s (9) and (10) respectively, we obtain:

$$x_{i+1} = x_i - \frac{m\,\Delta t}{2c}[\,(a_{g.i} + a_{g.i+1}) + \omega_n^2(x_i + x_{i+1}) + (a_i + a_{i+1})] \tag{11}$$

$$v_{i+1} = v_i - \frac{\Delta t}{2}[\,(a_{g.i} + a_{g.i+1}) + \omega_n^2(x_i + x_{i+1}) + 2\omega_n\zeta(v_i + v_{i+1})] \tag{12}$$

Finally, we must compute and update the magnitude of the acceleration a_{i+1} while updating x_{i+1} and v_{i+1} Acceleration response of any SDOF systems has very important information on system state and it contains high details of the mass movements in the dynamic system. Therefore, the iterative process isseverely sensitive to its value and if itwas not accurate enough, the iteration process would rapidly diverge. This is a key point while using the impulse technique. At any instance, we compute acceleration by the differential equations of motion (DEM) of the system:

$$a_{i+1} = -(a_{g.i+1} + \omega_n^2 x_{i+1} + 2\omega_n\zeta v_{i+1}) \tag{13}$$

These relations constitute the trunk of the impulse technique algorithm.

Now, the question isthat "what are the values of x_{i+1}, v_{i+1}, while we initialize to calculate x_{i+1}, v_{i+1}themselves? " To answerthis question, we note that the variation of geometricparameters of motion x and v is almost infinitesimal during a time step. So, we can start each instance computation, assuming that x_{i+1}, v_{i+1} are both equal to x_i, v_i, respectively. The step-by-step process of the suggested method is summarized in Table 1.

Table 1 Step-by-step algorithm of the proposed impulse method.

1. For time instance $i = 1$ initialize with *

$$t_1 = 0.\quad x_1 = x(0)\ .\ v_1 = v(0)\ .\ a_1 = -(a_{g.1} + \omega_n^2 x_1 + 2\omega_n\zeta v_1)$$

2. Predict the response at instance $i + 1$

$$x_{i+1} = x_i + \frac{\Delta t}{2}v_i, v_{i+1} = v_i + \frac{\Delta t}{2}a_i$$

3. Update x_{i+1}, v_{i+1}, a_{i+1}:

$$a_{i+1} = -(a_{g.i+1} + \omega_n^2 x_{i+1} + 2\omega_n\zeta v_{i+1})$$

$$x_{i+1} = x_i - \frac{m\,\Delta t}{2c}[(a_{g.i} + a_{g.i+1}) + \omega_n^2(x_i + x_{i+1}) + (a_i + a_{i+1})]$$

$$v_{i+1} = v_i - \frac{\Delta t}{2}[\,(a_{g.i} + a_{g.i+1}) + \omega_n^2(x_i + x_{i+1}) + 2\omega_n\zeta(v_i + v_{i+1})$$

4. Repeat steps 3 and 4 until desirable precision is achieved.
5. Set $i = i + 1$ and repeat steps 2 to 5 for the next time instance.

*For the case of external loads instead, use the followings in steps 1 and 3, respectively
$a_1 = [F_1 - (kx_1 + cv_1)]/m$, $a_{i+1} = [F_{i+1} - (kx_{i+1} + cv_{i+1})]/m$

Despite the extraordinary simplicity of this algorithm, it works accurately for damped linear systems. This is investigated through a numerical example under earthquake records. It should be noted that the current version of the impulse method is merely applicable to linear systems for which the damping ratio ζ is at least greater than 0.01.

There is a clear criterion to test any numerical solution presented for vibration differential equations. If the solution was exact, the unbalanced force diminishes. Denoting the force by the residual of the differential equation we have:

$$Res(t) = kx(t) + cv(t) + ma(t) + ma_g \tag{14}$$

Accordingly, for the numerical solution which is presented at discrete points of time, the value of this function can be the best tool for precision judgment. So, we introduce the absolute value of the error function to check the exactness of oursolutions:

$$ErrFun(t_i) = \left|kx_i + cv_i + ma_i + ma_{gi}\right| \tag{15}$$

If the value of the error function vanishes, the appropriately precise solution is found.

4 Methodology: Proposed Energy Method

The first recall the definition of the work done by a force during the straight pass dx:

$$dU = F\,dx \tag{16}$$

Then, multiplying Eq. (1) by dx we have:

$$kx\,dx + cv\,dx + ma\,dx = -ma_g\,dx \tag{17}$$

Each term of thisequation can be interpreted as the work or energy induced by its corresponding force component. It means, $kx\,dx$ is the differential energy of spring, $cv\,dx$ is the differential energy dissipated by friction force, $ma\,dx$ is the differential kinetic energy of the system, and $-ma_g dx$ can be considered as the differential work done by ground acceleration which is the mere agent of motion. Eq. (17) means that energy income from earthquake force is distributed among the three essential components of the system, i.e., spring, friction, and inertia. Now, to reform this equation some technical modification is conducted. Recalling$v = \frac{dx}{dt}$ and $a_g = \frac{dv_g}{dt}$ and substituting them into Eq. (17),we have:

$$kx\,dx + cvdx + madx = -mvdv_g \tag{18}$$

Dividing the last equation by v and noting that $\frac{dx}{v} = dt$, we can Integrate Eq. (18) from i to $i+1$:

$$\int_{t_i}^{t_{i+1}} kxdt + \int_{x_i}^{x_{i+1}} cdx + \int_{t_i}^{t_{i+1}} madt = \int_{v_{g.i}}^{v_{g.i+1}} -mdv_g \tag{19}$$

Then:

$$\frac{k\,\Delta t}{2}(x_i + x_{i+1}) + c(x_{i+1} - x_i) + \frac{m\,\Delta t}{2}(a_i + a_{i+1}) \\ = -m(v_{g.i+1} - v_{g.i}) \tag{20}$$

Solving the last equation for x_{i+1}:

$$x_{i+1} = -\left[m(v_{g.i+1} - v_{g.i}) + \left(\frac{k\,\Delta t}{2} - c\right)x_i + \frac{m\,\Delta t}{2}(a_i + a_{i+1})\right] / \left(\frac{k\,\Delta t}{2} + c\right) \tag{21}$$

This gives the recursive relationship for x_{i+1}. Similarly, the following can be obtained for velocity:

$$v_{i+1} = v_i - \left[m(v_{g.i+1} - v_{g.i}) + \left(\frac{k\,\Delta t}{2} - c\right)x_i + \left(\frac{k\,\Delta t}{2} + c\right)x_{i+1}\right] / m \tag{22}$$

Eq. (21) and (22)coupledwith the following relation obtained:

$$\ddot{x}_{i+1} = -(a_{g.i+1} + \omega_n^2 x_{i+1} + 2\omega_n \zeta v_{i+1}) \tag{23}$$

constitute the numerical algorithm for computing linear SDOF response. The algorithm of the energy approach is summarized in Table 2. Despite the extreme simplicity with energy methods, it works precisely for linear damped systems.

Table 2 Step-by-step algorithm of the proposede nergy method.

1. For time instance $i = 1$ initialize with *

   $t_1 = 0\ .\ x_1 = x(0)\ .\ v_1 = v(0)\ .\ a_1 = -(a_{g.1} + \omega_n^2 x_1 + 2\omega_n \zeta v_1)$

2. Predict the response at instance $i + 1$

$x_{i+1} = x_i + \Delta t\, v_i, v_{i+1} = v_i + \Delta t\, a_i$

3. Correct the system state using the following

$\ddot{x}_{i+1} = -(a_{g.i+1} + \omega_n^2 x_{i+1} + 2\omega_n \zeta v_{i+1})$

$x_{i+1} = -\left[m(v_{g.i+1} - v_{g.i}) + \left(\frac{k\,\Delta t}{2} - c\right)x_i + \frac{m\,\Delta t}{2}(a_i + a_{i+1})\right] / \left(\frac{k\,\Delta t}{2} + c\right)$

$\dot{x}_{i+1} = v_i - \left[m(v_{g.i+1} - v_{g.i}) + \left(\frac{k\,\Delta t}{2} - c\right)x_i + \left(\frac{k\,\Delta t}{2} + c\right)x_{i+1}\right] / m$

4. Repeat step 3 and 4 until desirable precision

5. Set $i = i + 1$ and repeat steps 2 to 5 for the next time instance.

*For the case of external loads instead, use the followings in steps 1 and 3, respectively

$$\ddot{x}_1 = [F_1 - (kx_1 + cv_1)]/m \quad , \quad \ddot{x}_{i+1} = [F_{i+1} - (kx_{i+1} + cv_{i+1})]/m$$

Important note:The load impulse method does not yield reliable results for undamped systems and the system with$\zeta \leq 0.01$. In these cases, the integration method is advised which will be presented in the next sections. However, most of the applied cases have a damping ratio greaterthan 0.01, i.e., ($\zeta \geq 0.01$).

5 Result and discussion

5.1 The linear damped system under El Centro's earthquake

A general case of the dynamiclinear system underearthquake excitation is investigated. The system to be investigated is a linear damped system:

Mass: $m = 1\ kg$

Natural period:$T_n = \ 0.5\ sec$

Damping ratio:$\zeta = 0.02$

Timestep of the ground motion record:$\Delta t = 0.02\ sec$

Peak ground acceleration of the ground motion:$PGA = 0.31882\ g = 0.31882 \times 386.22 = 123.134\ in/sec\ ^2$

Natural angular frequency:

$$\omega_n \ = 2\pi/T_n = 2 \times 3.1415/0.5 = \ 12.5663\ rad/sec$$

Spring constant:

$$k = m\omega_n^2 = 1 \times \ 12.562^2 = 157.9\ kip/in$$

Damping coefficient:

$$c = 2\ m\ \omega_n \zeta = \ 2 \times 1 \times 12.56\ \times\ 0.02 = 0.5027$$

El Centro's ground acceleration (Fig. 2) is applied to the system which was initially at rest.

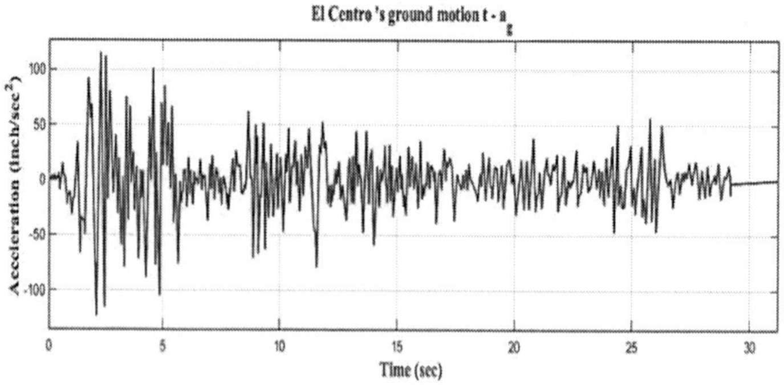

Fig. 2 El Centro's ground motion.

The dynamic response of the system is obtained by the energy method. Then, to have a comparison with conventional methods, the dynamic response of the system is also computed with Duhamel integral and Newmark-βmethod. All responses are presented in Fig. 3. All methods yield approximately close responses.

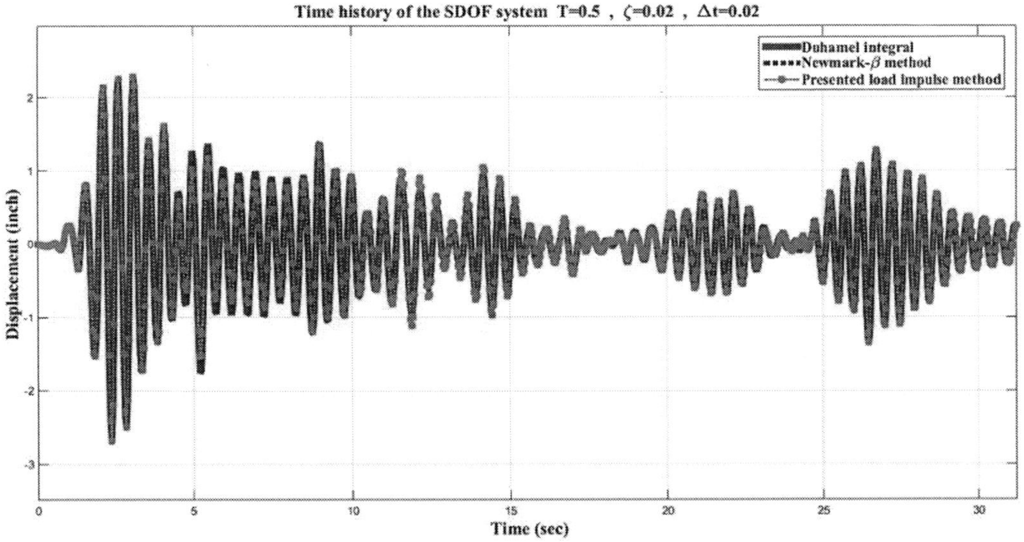

a)

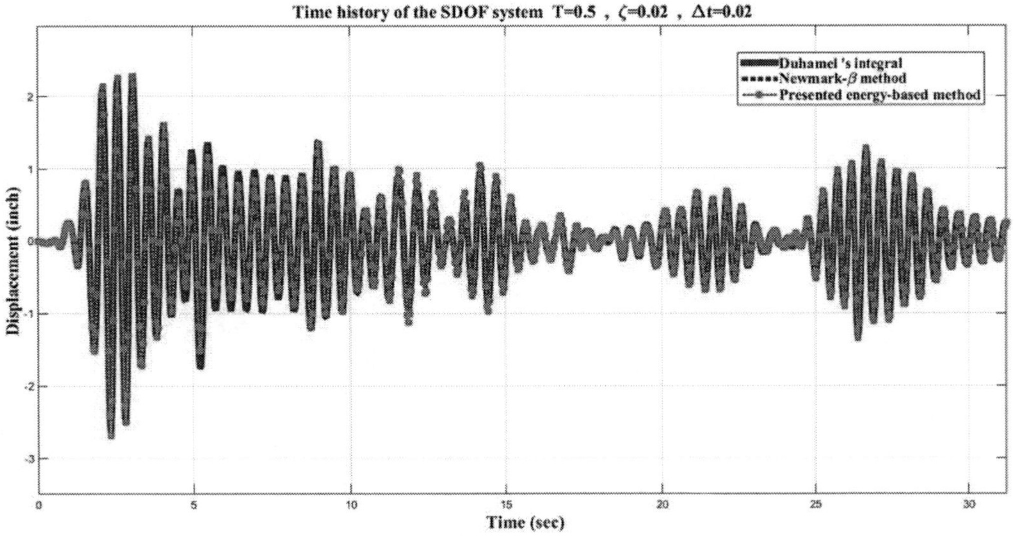

b)

Fig. 3 Response of the linear SDOF system to the El Centro earthquake.

Fig. 3 indicates that the new methods can properly detect the motion even with normal time increment. Moreover, it is observed that any appeared errors at the peak points of responses do not propagate to the proceeding steps; and, errors decay after some limited number of cycles.Numerical results are summarized in Table 2 to transmit a clear insight into solution precision. Duhamel's integral works more precisely than the others; however, it uses computation of order $[O]^2$ which would be very costly for the smaller time step size.

Table 3 Summary results for the response of the linear SDOF system to the El Centro earthquake ($\Delta t = 0.02\ sec$).

Time instance t_i (sec)	Duhamel's integral x_i (inch)	Newmark-β method x_i (inch)	Energy method x_i (inch)	Impulse method x_i (inch)
0	0.0000	0.0000	0.0000	0.0000
0.8	-0.0100	-0.0094	-0.0092	-0.0092
1.6	0.6602	0.6627	0.6650	0.6650
2.4	-2.1659	-2.1811	-2.1975	-2.1975
3.2	0.5313	0.6260	0.7177	0.7175
4	0.9774	0.8689	0.7553	0.7556
4.8	-0.6248	-0.5776	-0.5164	-0.5166
5.6	0.0624	0.1226	0.1619	0.1618
6.4	0.8159	0.7155	0.6275	0.6277
7.2	-0.9830	-0.9046	-0.8240	-0.8241
8	0.7710	0.7535	0.7198	0.7198
8.8	-0.9259	-0.9730	-0.9976	-0.9976
9.6	0.2967	0.3937	0.4739	0.4737
10.4	0.2663	0.1632	0.0639	0.0641
11.2	-0.7026	-0.6424	-0.5687	-0.5689
12	-0.2292	-0.2397	-0.2727	-0.2726
12.8	0.1215	0.1003	0.1018	0.1018
13.6	0.2956	0.3357	0.3594	0.3594
14.4	-0.8320	-0.8687	-0.9000	-0.8999
15.2	0.4913	0.5338	0.5806	0.5805
16	0.0490	0.0026	-0.0548	-0.0547
16.8	0.2190	0.2451	0.2868	0.2867
17.6	0.1503	0.1396	0.1144	0.1144
18.4	0.0520	0.0436	0.0450	0.0450
19.2	-0.1290	-0.1076	-0.0894	-0.0895
20	0.1717	0.1483	0.1215	0.1216
20.8	-0.2613	-0.2425	-0.2142	-0.2143
21.6	0.4351	0.4219	0.3980	0.3981
22.4	-0.5252	-0.5362	-0.5362	-0.5362
23.2	-0.0088	0.0315	0.0643	0.0642
24	-0.0193	-0.0590	-0.0988	-0.0987

24.8	0.1708	0.1959	0.2277	0.2276
25.6	0.0734	0.0516	0.0188	0.0189
26.4	-1.2889	-1.2637	- 1.2265	-1.2266
27.2	1.0664	1.0781	1.0772	1.0772
28	-0.5742	-0.6338	- 0.6836	-0.6835
28.8	0.0001	0.0877	0.1733	0.1731
29.6	0.2521	0.1807	0.1000	0.1002
30.4	-0.2948	-0.2638	- 0.2152	-0.2153
31.2	0.2382	0.2518	0.2457	0.2457

The peak of displacement, velocity, and accelerationisalsopresented in Table 3. This information istakenfrom the responses of time increment$\Delta t = 0.002$ sec. A notable piece of information in this table is the measured run time corresponding to each method. The results show that Duhamel integralis the slowestalgorithmwhichneeds high computationalcost and the Newmark-β methodis the fastest of all.

Table 4 Peak values of responses of various methods with a time increment of $\Delta t = 0.002$ sec.

Item	Duhamel's integral	Newmark-β method	Energy method	Impulse method
Max Disp. (inch)	2.6881	2.6879	2.7033	2.6879
Max Vel. (inch/sec)	NC*	32.2654	32.4484	32.2633
Max Acceleration (inch/sec^2)	NC*	486.421	487.678	486.421
Number of iteration	1	1	5	3
Analyses time (sec)	13.015	0.0019	0.0130	0.0083

**NC: Not computed*

6 Conclusions

In this study, a simple numerical formulation, the Impulse, and energy method were proposed to obtain the dynamic response of linear damped SDOF systems under the seismic loading. Results show that the proposed procedures satisfactorily estimate the seismic response of the linear damped SDOF systems. Another noticeable observation is that the estimation errors from the methods are not accumulated in each analysis step and gradually decreased while analysis proceeds. It can be concluded from the results; the load impulse algorithm can reliably estimate the displacement. There was good agreement between the resultsfrom the impulse method and that one time-history of the SDOF systems from Duhamel integral and Newmark-β methods. Accordingly, the proposed method can be identified as a reliable analysis tool in estimating the seismic demand for linear damped SDOF systems. Further studies in the development of the proposed procedure for

estimating the seismic response of other types of structural systems such as inelastic nonlinear SDOF models are still underway.
The energy conservation principle trigger edits base idea. It is employed to manipulate the original form of the vibration equation in a differential form. Then, the differential terms of the equation are manipulated by kinematic relations of motion to reform the configuration of the equation.
The most important characteristics of the presented methods can be summarized as:

1) Computer programming is very simple for both methods.
2) Current models are simple iterative calculation.
3) Any arbitrary type of excitation such as harmonic, impulse, and even irregular earthquake excitation can be treated.
4) Advanced mathematical tools are avoided and computational operations are composed of very basic operations.
5) Analyzing a linear SDOF system is accessible even with beginners in earthquake engineering.
6) Any appeared errors, regardless of their sources, do not propagate proceeding steps.
7) They have low computation cost in comparison with Duhamel integral.

and limitations with the proposed methods are:

1) It is formulated merely for linear systems; and, they cannot deal with nonlinear systems.
2) it cannot converge for undamped systems and those for which $\zeta<0.01$.
3) Convergence speed is dissatisfying with this algorithm.

The performance of the methods is checked through a general numerous example. Mathematical discussion on the stability of the new algorithmis not included in current research; however, convergence and stability of solution are evident in example diagrams. Further development of energy methods for dealing with nonlinear structural systems is still underway.

References

1. Izadifard, R.A., Mollaei, S. and Omran, M.E.N. "Preparing pressure-impulse diagrams for reinforced concrete columns with constant axial load using single degree of freedom approach", International Journal of Advancements in Technology, 2016.
2. https://doi.org/10.4172/0976-4860.1000173
3. Bathe, K.J."Frontiers in finite element procedures". Klaus-Jurgen Bathe, 2006.https://doi.org/10.4203/csets.35.1
4. Chopra, A.K."Dynamics of structures", (pp. 174-196). Upper Saddle River, NJ: Pearson Education, 2012.
5. Ebeling, Robert M., Russell A. Green, and Samuel E. French. "Accuracy of response of single-degree-of-freedom systems to ground motion",Army Corps of Engineers, 97-7, 1997.
6. https://doi.org/10.21236/ada336674
7. Li, P.S. and Wu, B.S."An iteration approach to nonlinear oscillations of conservative single-degree-of-freedom systems", Acta Mechanica, 170(1-2), pp.69-75, 2004.
8. https://doi.org/10.1007/s00707-004-0112-3

9. Newmark, N.M."A method of computation for structural dynamics", Journal of the engineering mechanics division, 85(3), pp.67-94, 1959.
10. Veletsos, A.S., Newmark, N.M. and Chelapati, C.V. January. "Deformation spectra for elastic and elastoplastic systems subjected to ground shock and earthquake motions", In Proceedings of the 3rd world conference on earthquake engineering (Vol. 2, pp. 663-682), 1965.
11. Wang, Y. and Inman, D.J."Autonomous Gust Alleviation in UAVs", Advanced UAV Aerodynamics, Flight Stability and Control: Novel Concepts, Theory and Applications, p.465, 2017.
12. https://doi.org/10.1002/9781118928691
13. Gawronski, W."Advanced structural dynamics and active control of structures", Springer Science & Business Media, 2004.
14. https://doi.org/10.1007/978-0-387-72133-0
15. Benaroya, H., Nagurka, M. and Han, S."Mechanical vibration: analysis, uncertainties, and control", CRC Press, 2017.
16. https://doi.org/10.1201/9781315118369
17. Paz, M."International handbook of earthquake engineering: codes, programs, and examples", Springer Science & Business Media, 2012.
18. https://doi.org/10.1007/978-1-4615-2069-6
19. Wu, B. S., & Lim, C. W. "Large amplitude non-linear oscillations of a general conservative system", International Journal of Non-Linear Mechanics, 39(5), 859-870, 2004.
20. https://doi.org/10.1016/S0020-7462(03)00071-4
21. Chopra, A. K., Goel, R. K., & Chintanapakdee, C. "Statistics of single-degree-of-freedom estimate of displacement for pushover analysis of buildings", Journal of structural engineering, 129(4), 459-469, 2003.
22. https://doi.org/10.1061/(ASCE)0733-9445(2003)129:4(459)
23. Clough, R.W. and Penzien, J. "Dynamics of Structures (3rd)", Computers & Structures, Inc., California, pp.342-344, 2003.
24. https://doi.org/10.1115/1.3424082
25. Craig, R. R., &Kurdila, A. J. "Fundamentals of structural dynamics", John Wiley & Sons, 2006.
26. Gatti, P. L. "Applied Structural and Mechanical Vibrations: Theory and Methods". CRC Press, 2014.
27. https://doi.org/10.1201/b16697
28. Géradin, M., &Rixen, D. J. "Mechanical vibrations: theory and application to structural dynamics". John Wiley & Sons, 2014.
29. Anderson, J.C. and Naeim, F. "Basic structural dynamics", John Wiley & Sons, 2012.
30. https://doi.org/10.1002/9781118279137
31. Paz, M. and Leigh, W. "Structural Dynamics: Theory and Computation, 5th Ed", Springer Science & Business Media, N.Y, 2004. https://doi.org/10.1007/978-1-4615-0481-8
32. Rao, S. S."Mechanical vibrations in SI units", Pearson Higher Ed, 2017.
33. Chang, S.Y."Studies of Newmark method for solving nonlinear systems:(I) basic analysis", Journal of the Chinese Institute of Engineers, 27(5), pp.651-662, 2004.
34. https://doi.org/10.1080/02533839.2004.9670913

35. Chang, S.Y."Studies of Newmark method for solving nonlinear systems:(II) Verification and guideline", Journal of the Chinese institute of engineers, 27(5), pp.663-675, 2004.
36. https://doi.org/10.1080/02533839.2004.9670914
37. Kazakov, K. S."Dynamic Response of a Single Degree of Freedom (SDOF) System in some Special Load Cases, based on the Duhamel Integral", In International Conference on Engineering Optimization (pp. 01-05), 2008.
38. Kurt, N., &Çevik, M. "Polynomial solution of the single degree of freedom system by Taylor matrix method", Mechanics Research Communications, 35(8), 530-536, 2008.
39. Wu, J. S. "Analytical and numerical methods for vibration analyses", John Wiley & Sons, 2013.
40. Thomson, W. "Theory of vibration with applications", CrC Press, 2018.
41. Mohsenian, V., Padashpour, S. and Hajirasouliha, I. "Seismic reliability analysis and estimation of multilevel response modification factor for steel diagrid structural systems", Journal of Building Engineering, 29, p.101168, 2020.
42. https://doi.org/10.1016/j.jobe.2019.101168
43. Pang, G., Koper, K.D., Mesimeri, M., Pankow, K.L., Baker, B., Farrell, J., Holt, J. "Seismic analysis of the 2020 Magna, Utah, earthquake sequence: Evidence for a listric Wasatch fault", Geophysical Research Letters, 47(18), p. e2020GL089798, 2020. https://doi.org/10.1029/2020GL08979
44. Mitseas, I.P. and Beer, M., 2020. "Fragility analysis of nonproportionally damped inelastic MDOF structural systems exposed to stochastic seismic excitation", Computers & Structures, 226, p.106129.
45. https://doi.org/10.1016/j.compstruc.2019.106129
46. Mitseas, I.P. and Beer, M., 2020. "First-excursion stochastic incremental dynamics methodology for hysteretic structural systems subject to seismic excitation", Computers & Structures, 242, p.106359.
47. https://doi.org/10.1016/j.compstruc.2020.106359
48. Babaei, M."A general approach to approximate solutions of nonlinear differential equations using particle swarm optimization", Applied Soft Computing, 13(7), pp.3354-3365, 2013. https://doi.org/10.1016/j.asoc.2013.02.005
49. [31] Rahmani , M., Scurtu, I. C., & Moslemi Petrudi, A. (2020). Analytical and Dynamic study of Pulled Mass Nonlinear Vibration by Two Cables using Newton's Harmonic Balance Method. Technium: Romanian Journal of Applied Sciences and Technology, 2(2), 79–86. https://doi.org/10.47577/technium.v2i2.343

AUTHOR INDEX

AUTHOR INDEX